CASE STUDIES IN SYSTEM OF SYSTEMS, ENTERPRISE SYSTEMS, AND COMPLEX SYSTEMS ENGINEERING

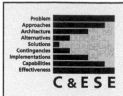

COMPLEX AND ENTERPRISE SYSTEMS ENGINEERING

Series Editors: Paul R. Garvey and Brian E. White

The MITRE Corporation

www.enterprise-systems-engineering.com

Architecture and Principles of Systems Engineering
Charles Dickerson and Dimitri N. Mavris
ISBN: 978-1-4200-7253-2

Case Studies in System of Systems, Enterprise Systems, and Complex Systems Engineering
Edited by Alex Gorod, Brian E. White, Vernon Ireland, S. Jimmy Gandhi, and Brian Sauser
ISBN: 978-1-4665-0239-0

Designing Complex Systems: Foundations of Design in the Functional Domain
Erik W. Aslaksen
ISBN: 978-1-4200-8753-6

Engineering Mega-Systems: The Challenge of Systems Engineering in the Information Age
Renee Stevens
ISBN: 978-1-4200-7666-0

Enterprise Systems Engineering: Advances in the Theory and Practice
George Rebovich, Jr. and Brian E. White
ISBN: 978-1-4200-7329-4

Leadership in Chaordic Organizations
Beverly G. McCarter and Brian E. White
ISBN: 978-1-4200-7417-8

Model-Oriented Systems Engineering Science: A Unifying Framework for Traditional and Complex Systems
Duane W. Hybertson
ISBN: 978-1-4200-7251-8

RELATED BOOK

Enterprise Dynamics Sourcebook
Kenneth C. Hoffman, Christopher G. Glazner, William J. Bunting, Leonard A. Wojcik, and Anne Cady
ISBN: 978-1-4200-8256-2

AUERBACH PUBLICATIONS

www.auerbach-publications.com
To Order Call: 1-800-272-7737 • Fax: 1-800-374-3401
E-mail: orders@crcpress.com

CASE STUDIES IN SYSTEM OF SYSTEMS, ENTERPRISE SYSTEMS, AND COMPLEX SYSTEMS ENGINEERING

Edited by
Alex Gorod • Brian E. White
Vernon Ireland • S. Jimmy Gandhi • Brian Sauser

CRC Press
Taylor & Francis Group
Boca Raton London New York

CRC Press is an imprint of the
Taylor & Francis Group, an **Informa** business
AN AUERBACH BOOK

CRC Press
Taylor & Francis Group
6000 Broken Sound Parkway NW, Suite 300
Boca Raton, FL 33487-2742

© 2015 by Taylor & Francis Group, LLC
CRC Press is an imprint of Taylor & Francis Group, an Informa business

No claim to original U.S. Government works

Printed on acid-free paper
Version Date: 20140418

International Standard Book Number-13: 978-1-4665-0239-0 (Hardback)

Library of Congress Cataloging-in-Publication Data

Case studies in system of systems, enterprise systems, and complex systems engineering / editors, Alex Gorod, Brian E. White, Vernon Ireland, S. Jimmy Gandhi, Brian Sauser.
     pages cm. -- (Complex and enterprise systems engineering)
     Includes bibliographical references and index.
     ISBN 978-1-4665-0239-0 (hardback)
     1. Systems engineering--Case studies. I. Gorod, Alex, editor of compilation.

TA168.C358 2014
620.001'1--dc23
                                                                                                    2014008503

Visit the Taylor & Francis Web site at
http://www.taylorandfrancis.com

and the CRC Press Web site at
http://www.crcpress.com

Contents

SECTION I Preliminaries

SECTION II Commerce

SECTION III   Culture

SECTION IV   Environment

SECTION V   Finance

SECTION VI   Health Care

SECTION VII   Homeland Security

Foreword

For over half a century, the discipline of systems engineering has served to bring order to the tumultuous process of marshaling the industrial commons in service to meeting society's greatest challenges. It was the organizing force behind the systems that put man on the moon, defended our citizenry, and built our transportation, communications, and information infrastructures. It has served as the foundation and framing of the engineering enterprise, which has been the wellspring of our security and our prosperity for generations.

However, as these systems have evolved, they have changed in ways that call into question the very assumptions underlying the traditional approach to systems engineering—that the system is a whole that can be reduced to its independent parts.

Software now delivers more than 90% of the functionality of many complex systems and runs on hardware architectures that have become incredibly complex. Elements of a system, which were once largely uncoupled, are now networked in ways that give rise to unforeseen behavior in the system as a whole. Most significantly, people are themselves an intimately connected network of components, with all of our own messiness factoring into the systems we depend on.

If the whole can no longer be assumed to be the sum of its independent parts, then we need a new paradigm for systems engineering—embracing and channeling complexity, rather than attempting to subdue it with increasing layers of process and concomitantly increasing costs. We need to shift the lens through which we manage systems development to one that puts people at the center and process as a supporting player, rather than the director. The relationships among stakeholders in our community and how they communicate information form the true framework for successfully developing and delivering a complex system.

Since ancient times, when society has needed to communicate complex yet critical concepts, people have turned to the art of storytelling, and this book follows in that tradition. Storytelling through case studies has the potential to explicate the complex systems development process in ways that traditional analyses cannot. I hope you find in these engaging stories insights that will inform and inspire you as, together, we work to ensure our systems engineering community can continue to serve generations to come.

Jeff Wilcox
Vice President, Engineering
Lockheed Martin Corporation
Bethesda, Maryland

Preface

All of us have experience with complex systems as a result of living in a number of personal relationships, which include living in a family, interacting with social and government systems, and relating to an employer, other employees, and clients in a work situation. Many of us also operate in a technical world, whether this embodies information technology, commerce, manufacturing, construction, transportation, defense, health care, education, and/or many other systems or industries. Largely, we have seen these areas as being different from each other via our own personal context, and we have behaved differently in each in order to cope with the complex systems each embodies.

While there is no generally agreed definition of a complex system, the one the authors use in this book is that complex systems are made up of networks of autonomous independent systems, including people, which have a life of their own as they operate. They were not developed to exist in the complex system being considered, and there are no external resources to facilitate their interactions. The only support to any of these systems comes from the complex integrated system of systems (SoS) or enterprise viewed as an external autonomous system.

The study of complex systems engenders new insights into how organizations, both large and small, operate. An important insight is that such study recognizes the integration of technological and human systems underpinning the operation of an organization which is influenced by culture, history, and language, particularly the way words are used in specific contexts.

The separation of focuses in the study of aspects of technology and human systems from an integrated perspective was originated by René Descartes in 1674 when he astutely postulated that the love of knowledge, and its exploration, by all thinkers would lead to limited progress. He thought a better approach was for people to specialize. Currently, we enjoy great depths of knowledge through focused discoveries in physics and chemistry, mathematics, biology, psychology, and many other disciplines, and the engineering application of these for human benefits. This separation, and the related notion of a hierarchy of knowledge, is called reductionism. However, the implication of reductionism is that a system is no more than the sum of its parts. The limitation of this view, when one considers people as the ultimate example of an integrated system, is apparent.

As a result of scientific discovery, great faith was placed on the development and application of science in the mid-twentieth century, which was reinforced by many significant medical discoveries such as penicillin and the subsequent likelihood of reducing or even eliminating infections. Other scientific advances led to the successful space exploration program in the 1960s, which reinforced confidence in scientific discoveries.

However, when attempts were made to apply technology-based solutions, through operations research, to solving human problems, the results were disappointing. Thus, complex systems study attempts to integrate technological and human aspects, drawing upon many social and cultural systems, and even predict behavioral consequences. While there are some aspects of common interpretation and meaning in various homogenous societies, the existence of common meaning in interpreting events does not always exist even at the very local level, when one argues with one's neighbor about the dog barking or the height of the fence. Consequently, it is very unlikely that there will be common meaning and understanding across a broad range of issues among diverse cultural groups. Meaning continues to develop in societies, as evidenced by the role of changing language in cultures. These are influenced by local events and experiences, by history and geography, by political and social contexts, and a range of other influences. It is unlikely that there will be a uniform and common global culture, although improvements in communication are bringing people closer together. Notwithstanding improvements in communication, there is still a need to build societal systems that integrate entities and concepts, including both technological and human aspects, because this integration of systems in a complex environment has been neglected.

Many human systems are still essentially underpinned by a reductionist framework. Leadership was initially thought of as depending on the competencies of the leader. It is only recently that transformational leadership has recognized the relationships between the leader and the followers, including how erstwhile followers can become temporary leaders. Furthermore, specialization is very evident with the development of professions and the narrowing of jobs to specialists. Seldom do we see the existence of dual professionals whose role is to integrate specialist knowledge. The medical engineer and the engineering lawyer are not so common.

While we understand the notion of networks, many of us still think in terms of the entities in a network rather than the relationship among the entities. Once we examine the relationships, we can then attempt to describe their elements, which include communication and the tools to do this, power relationships of one organization or entity over another, and also less tangible aspects such as loyalty, trust, and goodwill. The systemigram tool, used in one case of this book, goes some way to illustrate relationships between organizations and can be applied much more extensively.

A further underlying concept of this case study book is the distinction between complicated and complex systems. In common terminology, these two words are often used interchangeably. However, in terms of this book, a complicated system is one that one can model and predict outcomes in a way that cannot be done with a complex system. As has been mentioned, a complex system is the integration of autonomous, independent systems. A good example illustrating the difference between these terms is that the design and manufacture of jet engines is more at the complicated end of the complexity spectrum while the selling of jet engines is more complex, because customers and competitors are autonomous and independent.

Significant benefit can accrue from recognizing complex systems. Some major world issues have been addressed in a reductionist fashion, resulting in either not gaining the benefits of a systemic approach or hindering the development of a solution. For example, an expert on international peace negotiations commented that, as far as he is aware, the classic approach to solving disputes between warring nations is to identify the top eight or ten issues and then attempt to solve these in isolation; however, the negotiators were surprised that when they tried to put together these separate issues, they did not mesh, or the major entities had changed, or the conflation had different meaning to each of the parties, or the priorities had changed.

The US government recognized the need for an SoS approach when it established the Homeland Security organization, which may be defined as an enterprise, because information and initiatives had slipped between the cracks when the Twin Towers were attacked in 2001. Government departments were then integrated as depicted in Figure P.1.

Here, the federal government is characterized as having relatively large scope but coarse granularity; the local government has small scope but fine granularity, while the state government is somewhere in between. Various segments of society that interact as an SoS are represented by vertical slices in this diagram. The five organizational entities, represented by ovals on the right side of the diagram, are intended to help stitch things together.

In order to defend the nation, the US Department of Defense has a policy of building on existing, largely software-based, defense assets. Each of these assets continues to contribute separately to the original purpose for which they were built, but they also contribute to the SoS. One of the extreme cases is the US Air Force's Air Operations Center (AOC), which includes over 80 autonomous, independent systems, all of them contributing to their original purpose and also contributing to the objectives of the AOC.

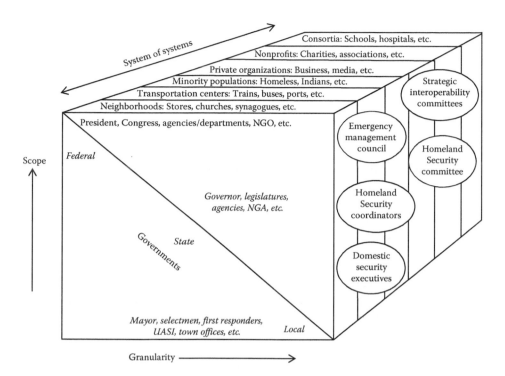

FIGURE P.1   The integrated SoS forming the US Homeland Security Department. *Note:* NGA, National Governors Association; NGO, non-governmental organization; UASI, urban area security initiative. (Redrawn from http://www.emd.wa.gov/grants/images/WAHomelandDefenseHomelandSecurityASystemofSystemsApproach.JPG.)

A "system of systems is a compilation of task oriented or dedicated systems in which their resources and capabilities are bundled together to obtain a newer, more complex system that offers more functionality and greater performance than simply the summation of basic systems" (Chan et al. 2011). An example is provided by wireless networks that allow the sharing of information among a number of independent systems. Linking systems into a joint SoS allows the interoperability and integration of command, control, communications, computers, intelligence, surveillance, and reconnaissance (C4ISR) systems, which is important to the military. The SoS approach provides a new methodology for solving problems, which includes the interaction of technology, organizational policy, resources, and human motivation and behavior. Decision support, among other things, is enhanced.

Health-care systems in many countries are examples of complex systems, which would benefit from better integration of the many independent systems of medical practitioners, hospitals, medical insurance systems, ambulance services, university medical programs, hospital catering suppliers, etc. Other domains have moved toward significant network integration. In the United States, Europe, and Australia, and many other countries, integrated networks of power generation, distribution, and use covering vast territories exist. These networks have great advantages in sharing power reliably over wide areas. However, the detailed operation of the networks leaves quite a lot yet to be understood. Examples of failure include a 2010 power outage in a number of states in the United States and Canada involving behavior that is not fully understood, although the results were very evident. Further study of the behavior of such networks is important.

SoS concepts are also relevant to the operation of federally based governments, in that their states or provinces are independent and autonomous entities with regard to a number of responsibilities, normally including police, education, and to some extent health. An example of the benefits of integrating systems occurs in the Australian continent with large amounts of rain falling in Queensland

and then running through rivers in Queensland, New South Wales, Victoria, and finally into South Australia, in the Murray–Darling system. However, politicians in both Queensland and New South Wales had allocated irrigation rights in excess of the appropriate use of the whole river system, leaving the river in Victoria and South Australia at risk. Finally, the Australian government passed legislation to provide itself with the authority to overrule states and ensure the whole river system operated in an integrated manner. Unfortunately, it neglected to convince the states that had access to the water of the need for the revised allocations. The change emphasized the effectiveness of the system, but the users were not fully convinced.

George W. Bush declared victory in the war in Afghanistan in 2003 from the point of view of the United States, but he was only recognizing the military part of the system and neglecting to recognize 1000 years of cultural conflict between the nominally Christian West and Islam, going back to the Crusades. Subduing the military in Afghanistan was much easier than winning the hearts and minds of the Afghan people. However, the conventional SoS approach does recognize cultural issues, and there is fertile scope to address differences of religion, development, language, education, decision-making, relationships between men and women, and other factors in an SoS model.

Korsten and Seider (2010), in a major study by IBM, make the point that "the world is fraught with inefficiency," and the inefficiency has a total value of $15 trillion, of which $4 trillion could be readily saved through recognizing the SoS nature of many of our actions. For example, their research illustrates the global value of integrated systems shown in Figure P.2.

Korsten and Seider's research indicated that across all of these systems, other systems contribute 47%, on average, of the value of each particular system. They found that the system most dependent on others was the world's transportation and food system, with over 60% dependence on other systems.

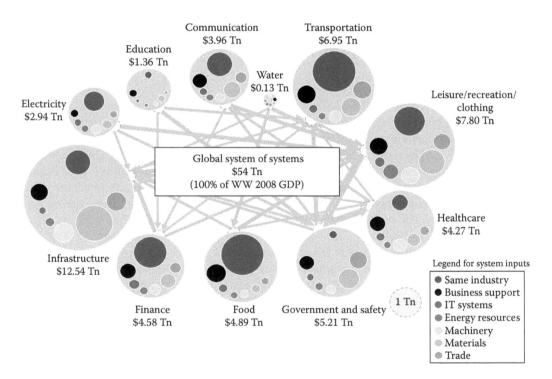

FIGURE P.2 Korsten and Seider's estimate of the total value of the world's interconnected SoS. *Note:* 1. Size of bubbles represents systems' economic values. 2. Arrows represent the strength of systems' interaction. (From Korsten, P. and Seider, C., The world's 4 trillion dollar challenge: Using a system-of-systems approach to build a smarter planet, Ref# GBE03278-USEN, IBM Institute for Business Value Analysis of Organisation for Economic Co-Operation and Development (OECD) data, 2010, http://www935.ibm.com/services/us/gbs/bus/html/ibv-smarter-planet-system-of-systems.html. With permission.)

They make the related point that the world is not particularly efficient, and their research shows that traffic jams or hold-ups in the United States waste 2.3 million barrels of crude oil per year. Their research also shows that the wastage in electric power networks "could power the United States, China and the entire continent of Europe for half a year" ... "Based on in-depth analysis of global GDP and a survey of more than 500 economists worldwide, we estimate that our planet's system of systems carries inefficiencies totalling nearly $15 trillion, or 28 percent of worldwide GDP."

One aspect of the SoS focus that surprises many, and deserves further research, is the concept of *satisficing* (Simon 1956) rather than optimizing. This satisficing of the ideal to the pragmatic is controversial to many, but a quick overview of human behavior shows that many choices are made on the basis of what is satisfactory rather than what is the ultimate best. If occasionally these coincide, that is a bonus.

A study of network systems is important. For example, one characteristic of network systems is that powerful nodes attract more support. We see this in the growth of hub airports. However, we need to recognize that like attracts like, with the implication that there is a tendency toward information systems providing supportive information rather than contrary information. Therefore, using information search engines to seek confirming advice can involve bias. Balancing positive questions with negative questions may correct this tendency.

The authors believe many of the approaches encompassed in this book, if adopted, can facilitate significant progress in human affairs. A number of the case studies focus on addressing real human needs. Diverse approaches such as soft systems skills are shown in application, and other techniques are referenced. Enterprises and individuals have always recognized the existence of autonomous, independent systems. Participants in enterprises have always recognized that customers and competitors were independent and that we need design approaches which address this independence. One of the benefits of SoS is a more formal integration of the various systems, both inside the enterprise and outside of it.

The value of this book, *Case Studies in System of Systems, Enterprise Systems, and Complex Systems*, is the significant contribution of a body of case studies in SoS, enterprises, and complex systems engineering and a guidance of how we as a community of scholars can further a body of cases in these domains (Chapter 3). There is real value in the study of cases because there is limited theory of the management of SoS. There is a strong need to develop useful tools that the practitioners and students of complex systems can adopt and use. Many tools are indigenous to the case studies presented in this book. By seeking case studies from a broad set of domains, the editors have attempted to provide approaches, tools, and techniques for the reader to extend his or her skill set. The study of complex systems is still in its infancy, and it is likely to evolve for decades to come. While this book does not seek to provide all the answers, some useful approaches and techniques are set out for the reader to use and develop.

Alex Gorod
New York, New York

Brian E. White
Sudbury, Massachusetts

Vernon Ireland
Adelaide, South Australia, Australia

S. Jimmy Gandhi
Northridge, California

Brian Sauser
Denton, Texas

MATLAB® is a registered trademark of The MathWorks, Inc. For product information, please contact:

The MathWorks, Inc.
3 Apple Hill Drive
Natick, MA 01760-2098 USA
Tel: 508-647-7000
Fax: 508-647-7001
E-mail: info@mathworks.com
Web: www.mathworks.com

REFERENCES

Chan, P., H. Man, D. Nowicki, and M. Mansouri. 2011. System engineering approach in tactical wireless RF network analysis. System of Systems Symposium, Stevens Institute of Technology, Report on a Summer Conversation, July 21–22, 2004. Potomac Institute, Arlington, VA.

Korsten, P. and C. Seider. 2010. The world's $4 trillion challenge: Using a system of systems approach to build a smarter planet. IBM Institute for Business Value.

Simon, H. A. 1956. Rational choice and the structure of the environment. *Psychological Review* 63(2): 129–138. doi: 10.1037/h0042760.

Acknowledgments

The editors thank all the case study authors who have contributed to the preparation of this book, and, of course, especially those whose case studies appear herein. Taylor & Francis Group's acquiring editor, Rich O'Hanley, is acknowledged and thanked for overseeing the entire process and helping with major decisions of development and production. Amber Donley, the project coordinator, did the excellent work of initial editing and ensuring that the required copyright permissions were received from authors. Lastly, the final editing and layout efforts of Ram Pradap Narendran (of SPi Publisher Services supporting CRC Press) are gratefully acknowledged.

Editors

Alex Gorod is the founder and managing member of SystemicNet, LLC in New York. He is also a visiting fellow at the University of Adelaide and adjunct associate professor at Zicklin School of Business, Baruch College/City University of New York. Alex is a recipient of the Fabrycky–Blanchard Award for Excellence in Systems Engineering Research and Robert Crooks Stanley Doctoral Fellowship in Engineering Management. He holds a PhD in engineering management from Stevens Institute of Technology.

Brian E. White received his PhD and MS in computer sciences from the University of Wisconsin and his SM and SB in electrical engineering from M.I.T. He served in the US Air Force, and for eight years was at M.I.T. Lincoln Laboratory. For five years, Dr. White was a principal engineering manager at Signatron, Inc. In his 28 years at The MITRE Corporation, he held a variety of senior professional staff and project/resource management positions. He was director of MITRE's Systems Engineering Process Office, 2003–2009. Dr. White retired from MITRE in July 2010 and currently offers a consulting service, CAU←SES ("Complexity Are Us" ←Systems Engineering Strategies—website: www.cau-ses.net).

A summary of some of Dr. White's professional activities in recent years follows:

- 2006–present: Coeditor of Complex and Enterprise Systems Engineering Series with Taylor & Francis Group (Boca Raton, FL: CRC Press)
  - Coeditor and Chapter 5, "Enterprise opportunity and risk," author of *Enterprise Systems Engineering: Advances in the Theory and Practice*, Boca Raton, FL: CRC Press, 2011
  - Coauthor of Chapter 3, "Emergence of SoS, socio-cognitive aspects," in *System of Systems Engineering—Principles and Applications*, Boca Raton, FL: CRC Press, 2009
  - Coauthor of *Leadership in Decentralized Organizations*, Boca Raton, FL: CRC Press, 2013
- 2005–present: Authored and presented more than 40 journal articles and conference papers on complex systems engineering and closely related topics.
- 2011–present: Authored and presented "On Principles of Complex Systems Engineering—Complex Systems Made Simple," tutorial at/to
  - *2013 Complex Adaptive Systems Conference,* Baltimore, MD, November 13–15, 2013
  - *23rd Annual INCOSE International Symposium,* Philadelphia, PA, June 24–27, 2013
  - *2013 IEEE International Systems Conference,* Orlando, FL, April 15–18, 2013
  - *Seventh International Conference of INCOSE_IL,* Tel Aviv, Israel, March 5–7, 2013
  - *National Defense Industrial Association (NDIA) Systems Engineering Conference,* San Diego, CA, October 22, 2012
  - Washington [DC] Metropolitan Chapter of INCOSE in McLean, VA, August 27, 2011
  - *INCOSE's Annual International Symposium,* Denver, CO, June 2011

Professor Vernon Ireland, BE, BA, MEngSc, PhD, FieAust, CPEng, EngExec, is director of project management for the University of Adelaide, offering master's degrees in traditional and complex project management.

Previously, he has been corporate development director of Fletcher Challenge Construction, responsible for people and business systems improvement, in the United States, New Zealand, Australia, Pacific, and Asian businesses, in the $2 billion pa company. In academe, he was the foundation dean of design, architecture, and building at the University of Technology, Sydney.

Professor Ireland has been awarded the Engineer's Australia Medal (2008), the Rotary International Gold Medal for contributions to vocational education (2006), and the Magnolia Silver Medal from the Shanghai Government (2000) for contributions to Chinese overseas relations. He has spent almost equal time in academia and industry interspersing these roles.

Professor Ireland was chairman of the Building Services Corporation, the licensing authority in New South Wales, reporting to the Minister for Housing. He supervises nine PhD students, five on linear projects and four on complex projects.

S. Jimmy Gandhi holds a BS in engineering management from Illinois Institute of Technology, an MS from California State University, Northridge, and a PhD from Stevens Institute of Technology in Engineering Management, with a focus on risk management. His specific research interests are in the field of risk management of extended enterprises and also in the field of risk management education in the engineering domain. He is an adjunct faculty with the School of Systems and Enterprises at Stevens Institute of Technology in Hoboken, NJ, and has coauthored and been the editor for several books on entrepreneurship as well as a coeditor for the *Engineering Management Handbook* published by the American Society for Engineering Management. He is also a member of and actively participates in several professional organizations such as the American Society of Engineering Management and the American Society of Engineering Education.

Brian Sauser is an associate professor in the Department of Marketing and Logistics at the University of North Texas (UNT). He currently serves as the director of the Complex Logistics and Behavioral Research Laboratory and as associate director of research for the Center of Logistics Education and Research. Before joining UNT, he held positions as an assistant professor with the School of Systems and Enterprises at Stevens Institute of Technology; project specialist with ASRC Aerospace at NASA Kennedy Space Center; program administrator with the New Jersey—NASA Specialized Center of Research and Training at Rutgers, the State University of New Jersey; and laboratory director with G.B. Tech Engineering at NASA Johnson Space Center.

He teaches or has taught courses in project management of complex systems, designing and managing the development enterprise, logistics and business analytics, theory of logistics systems, systems thinking, and systems engineering and management. In addition, he is a National Aeronautics and Space Administration Faculty Fellow, IEEE Senior Member, associate editor of the *IEEE Systems Journal*, and associate editor of the *Guide to Systems Engineering Knowledge* (a collaboration of the International Council on Systems Engineering [INCOSE], the Institute of Electrical and Electronics Engineers Computer Society [IEEE-CS], and the Systems Engineering Research Center [SERC]).

Dr. Sauser holds a BS in agricultural development with an emphasis on horticulture technology from Texas A&M University; an MS in bioresource engineering from Rutgers, the State University of New Jersey; and a PhD in project management from Stevens Institute of Technology.

Contributors

Sergei Astapov
Tallinn University of Technology
Tallinn, Estonia

Mikhail Belov
IBS
Moscow, Russia

Eusebio Bernabeu
Complutense University of Madrid
Madrid, Spain

Bustamante Brathwaite
Brooklyn College
New York, New York

Barbara Chomicka
EC Harris, Arcadis Group
London, United Kingdom

Claire Cizaire
Optimix
Boston, Massachusetts

Tina Comes
University of Agder
Kristiansand, Norway

Stephen C. Cook
University of South Australia
Adelaide, South Australia, Australia

Hamid R. Darabi
Stevens Institute of Technology
Hoboken, New Jersey

John Q. Dickmann
Sonalysts, Inc.
Waterford, Connecticut

James R. Enos
United States Military Academy
New York, New York

John V. Farr
United States Military Academy
New York, New York

John Findlay
Zing Technologies Private Limited
and
Maverick & Boutique
Sydney, New South Wales, Australia

Eranga Gamage
Brooklyn College
New York, New York

S. Jimmy Gandhi
California State University
Northridge, California

Alex Gorod
SystemicNet, LLC
New York, New York

and

University of Adelaide
Adelaide, South Australia, Australia

Shawn Hall
Emerging Health Information
    Technology
Yonkers, New York

Roelof Hamberg
Embedded Systems Innovation
The Netherlands Organisation for Applied
    Scientific Research
Eindhoven, the Netherlands

Jeffrey Higginson
The MITRE Corporation
Bedford, Massachusetts

Richard J. Hodge
Brooke Institute
Melbourne, Victoria, Australia

Vernon Ireland
Entrepreneurship, Commercialisation and
    Innovation Centre
The University of Adelaide
Adelaide, South Australia, Australia

Maite Irigoyen
Complutense University of Madrid
Madrid, Spain

Khaldoun Khashanah
Stevens Institute of Technology
Hoboken, New Jersey

Michael W. Kometer
The MITRE Corporation
Bedford, Massachusetts

Danny Kopec
Brooklyn College
New York, New York

James Llinas
State University of New York
   at Buffalo
Buffalo, New York

Mo Mansouri
Stevens Institute of Technology
Hoboken, New Jersey

Daniel J. McCarthy
United States Military Academy
New York, New York

Merik Meriste
Tallinn University of Technology
Tallinn, Estonia

Rob Mitchell
Indigo Foundation
Fig Tree, New South Wales, Australia

Leo Motus
Tallinn University of Technology
Tallinn, Estonia

Gerrit Muller
Buskerud University College
Kongsberg, Norway

Raido Pahtma
Tallinn University of Technology
Tallinn, Estonia

Jürgo-Sören Preden
Tallinn University of Technology
and
Defendec
Tallinn, Estonia

Karunya Rajagopalan
City University of New York
New York, New York

Tim Rudolph
United States Air Force Life Cycle
   Management Center
Bedford, Massachusetts

Jon Salwen
The MITRE Corporation
Bedford, Massachusetts

Brian Sauser
University of North Texas
Denton, Texas

Raul Savimaa
Tallinn University of Technology
Tallinn, Estonia

Frank Schätter
Karlsruhe Institute of Technology
Karlsruhe, Germany

Frank Schultmann
Karlsruhe Institute of Technology
Karlsruhe, Germany

Abby Straus
Zing Technologies Private Limited
and
Maverick & Boutique
Sydney, New South Wales, Australia

Jose Luís Tercero
Complutense University of Madrid
Madrid, Spain

Sergio Torres
Information Systems and Global
   Solutions Civil
Lockheed Martin
Rockville, Maryland

Mariusz Tybinski
Ipreo
London, United Kingdom

M. Ulieru
Carleton University
Ottawa, Ontario, Canada

Ricardo Valerdi
The University of Arizona
Tucson, Arizona

John van Trijp
Libertas
Utrecht, the Netherlands

and

VU University Amsterdam
Amsterdam, the Netherlands

Michael J. Vinarcik
University of Detroit Mercy
Detroit, Michigan

Brian E. White
CAU←SES
Sudbury, Massachusetts

Lawrence D. Willey
Northern Power Systems
Barre, Vermont

Section I

Preliminaries

1 Modern History of System of Systems, Enterprises, and Complex Systems

Alex Gorod, S. Jimmy Gandhi, Brian E. White, Vernon Ireland, and Brian Sauser

CONTENTS

INTRODUCTION

This chapter provides a background foundation for the rest of the book. Topics covered include the significance of historical context, a basic introduction to systems, and a discussion of foregoing works in the system of systems (SoS), enterprises, and complex systems (CSs), along with their corresponding engineering efforts.

This chapter emphasizes recent history through briefly introducing and listing the importance of research, which, to the extent readers delve into some of the many references, will further enhance the understanding and heighten the appreciation of the subsequent material of the book. Chapter 2 focuses on the relevant aspects of CSs and includes many additional references from that domain and complexity theory. In Chapter 3, the critical role of case studies in furthering the practice of systems engineering (SE) in all these forms, particularly for engineering leadership, management, and education, is advocated. The remaining book chapters contain the actual case studies.

We author-editors are keenly aware that our contributions, whether original, insightful, interpretive, or even repetitive in nature, are made possible only through our assimilation, consideration, and reformulation of the work of others who went before. Our goal is to build upon, extend, or complement significant historical results in a manner that will be useful to other researchers and practitioners.

Entities that we call systems are central to the entire conversation. A fundamental grasp of what is meant by system is crucial for grasping/comprehending the related nuances associated with the terms *system, SoS, enterprise*, and *CS* used throughout this book.

As will be seen, systems come in many flavors, covering a complexity spectrum that goes from simple to a realm that is much more than merely complicated. A system is often whatever the individual beholder perceives it to be. It is not surprising that two or more people will disagree on the definition of a *system*. Through discussions about any particular system (or systems), everyone's understanding can be enlightened and sharpened by being exposed to different points of view.

For now, at least for the purpose of this book, suffice it to say that a system is the entity that is the focus of a human-made effort to make something better. This entity can include people, not just technological or technical artifacts. As will be seen in later chapters, some type of a system is the particular focus of every case study in the book.

Following a glossary, this chapter is organized to first treat SoS and SoS engineering (SoSE), then enterprises and enterprise systems engineering (ESE), and finally CS and CS engineering (CSE). We end the chapter with a comparative analysis of the level of challenge in engineering of these three types of systems.

Since SoS has taken on a considerable value in recent years, in the literature, in conferences, and in discussion fora, we decided to lead with that topic. Enterprises are talked about less, it seems, but they often, though not always, can be viewed as SoS as well; an enterprise can be more general than a SoS. CS are included last because many of these can be viewed as the most general type of a system.

As a final introductory note, we embrace a multicultural approach to these case studies, particularly since this book is intended for a worldwide audience. Autonomous independent systems of business, law, religion, decision-making, and various other transdisciplinary aspects can be included. We hope you are open to appreciating such a potential scope.

GLOSSARY

ABM	Agent-based model
ANSI	American National Standards Institute
ASBP	Army Software Blocking Policy
BKCASE	Body of Knowledge and Curriculum to Advance Systems Engineering
BoK	Body of knowledge
C4ISR	Command, control, communication, computers, intelligence, surveillance, and reconnaissance
CAS	Complex adaptive system
CASE	Complex adaptive systems engineering
CMMI	Capability maturity model integrated
CS	Complex system
CSE	Complex systems engineering
DAG	Defense Acquisition Guidebook
DAU	Defense Acquisition University
DoD	Department of Defense
DoDAF	DoD Architecture Framework
EAP	Enterprise Architecture Planning
EIA	Electronic Industries Alliance
EOC	Edge of chaos
ESD	Engineering Systems Division
ESE	Enterprise systems engineering

FBI	Federal Bureau of Investigation
FEAF	Federal Enterprise Architecture Framework
GAO	Government Accountability Office
GRCSE	Graduate Reference Curriculum for Systems Engineering
GST	General Systems Theory
IBM	International Business Machines
ICAS	Intelligent Complex Adaptive Systems
IEC	International Electrotechnical Commission
IEEE	Institute of Electrical and Electronics Engineers
ILS	Integrated logistics support
INCOSE	International Council on Systems Engineering
IRS	Internal Revenue Service
ISI	Institute for Scientific Interchange
ISO	International Organization for Standardization
JCIDS	Joint Capabilities Integration and Development Instructions
MAS	Multiagent system
MIT	Massachusetts Institute of Technology
MOD	Ministry of Defence
MODAF	Ministry of Defence Architecture Framework
NASA	National Aeronautics and Space Administration
NCSoSE	National Center for System of Systems Engineering
NECSI	New England Complex Systems Institute
NIST	National Institute of Standards and Technology
NOAA	National Oceanic and Atmospheric Administration
OR	Operations research
POET	Political, operational, economic, and technical
POSIX	Portable Operating System Interface
QA	Quality assurance
R&D	Research and development
SE	Systems engineering
SEAri	Systems Engineering Advancement Research Initiative
SEBoK	Systems Engineering Body of Knowledge
SEI	Software Engineering Institute
SERC	Systems Engineering Research Center
SFI	Santa Fe Institute
SGSR	Society for General System Research
SoS	System of systems
SoSE	SoS engineering
SoSE&I	System of Systems Engineering and Integration Directorate
SoSECE	System of Systems Engineering Center of Excellence
SSRC	Sociotechnical Systems Research Center
TAFIM	Technical Architecture Framework for Information Management
T-AREA-SoS	Trans-Atlantic Research and Education Agenda on System of Systems
TEAF	Treasury Enterprise Architecture Framework
TISAF	Treasury Information System Architecture Framework
TOGAF	The Open Group Architecture Framework
TRM	Technical Reference Model
TRM	Total resource management
TSE	Traditional systems engineering
WWW	World wide web

SYSTEM OF SYSTEMS (SoS) AND SoS ENGINEERING

Figure 1.1 depicts an updated graphical timeline of the major contributions to the system of systems (SoS) and SoS engineering (SoSE) body of knowledge. The original version was published in the *IEEE Systems Journal* (Gorod et al. 2008).

According to research done at Purdue University, the emerging concept of a SoS describes the large-scale integration of many independent, self-contained systems in order to satisfy a global need. When a large number of different systems that affect each other compose a SoS, the task of operating, maintaining, improving, and/or creating/developing the SoS is inherently multidisciplinary. Therefore, combining systems to function well together as a SoS requires a variety of *transdisciplinary* skills, beyond those usually associated with traditional systems engineering (SE). In fact, some of the conventional SE techniques, such as reductionism, do not work well with SoSs (or enterprises or CS) because one cannot necessarily expect overall system improvements by *divide and conquer* methods. SoSs are typically too dynamic and unstable to approach linearly, which will be explained in more detail later in this chapter.

Initial research in the field of SoS goes back to Boulding (Senge and Sterman 1990), Jackson and Keys (1984), Ackoff (2005), and Jacob (1974). Boulding thought of a SoS to create a *spectrum of theories* greater than the sum of the system's parts. Actually, most definitions of even a simple system state the following: The system achieves a whole greater than the sum of its parts. Jackson and Keys conducted research to conclude that *SoS methodologies* should be used to address interrelationships between different systems-based problem-solving methodologies in the field of operations research (OR). This pointed to the notion that SoSs generally require a *system thinking* or *systemic* approach as opposed to systematic analysis or procedures that are more akin to conventional SE and OR. Ackoff considered a SoS as a *unified or integrated set* of systems concepts. The modern term *system of systems*, as we know it now, was first introduced in 1989 to describe an engineered technology system (SE Guide 2013). Please refer to Chapter 3's Appendix, where we document how SoSs were further defined in 2008 as comprising four distinct types.

Eisner et al. defined SoS as follows:

> A set of several independently acquired systems, each under a nominal systems engineering process; these systems are interdependent and form in their combined operations a multi-functional solution to an overall coherent mission. The optimization of each system does not guarantee the optimization of the overall system of systems. (Stevens 2009)

In other words, this questioned the tendency for conventional SE to focus on systematic analysis, while relying on attempts to optimize the entire system by optimizing its components. The systemic approach is more about *understanding* the SoS and focusing on appropriately balancing competing component system interests. Here, a constructionist or holistic synthesizing approach is more effective in *satisfying* (Simon 1956) the overall SoS needs (Gorod et al. 2009; Gorod 2009).

Further research done by Aaron Shenhar related SoS to the term *array* and described it as "... a large widespread collection or network or systems functioning together to achieve a common purpose" (Shenhar 1994; Shenhar and Sauser 2008). Most definitions of SoSs effectively state that each constituent system in the SoS can still achieve its own individual purpose while sharing the overall purpose of SoS.

Manthorpe Jr. (1996) suggested linking command, control, communication and information with intelligence, surveillance, and reconnaissance to form a SoS in order to achieve *dominant battleship awareness*. Another major contribution came from Maier, who proposed the use of the characterization approach to distinguish *monolithic* systems from SoS (Maier 1996). This was updated in 1998 and resulted in an additional attempt to define a SoS: "A SoS is an assemblage of components which

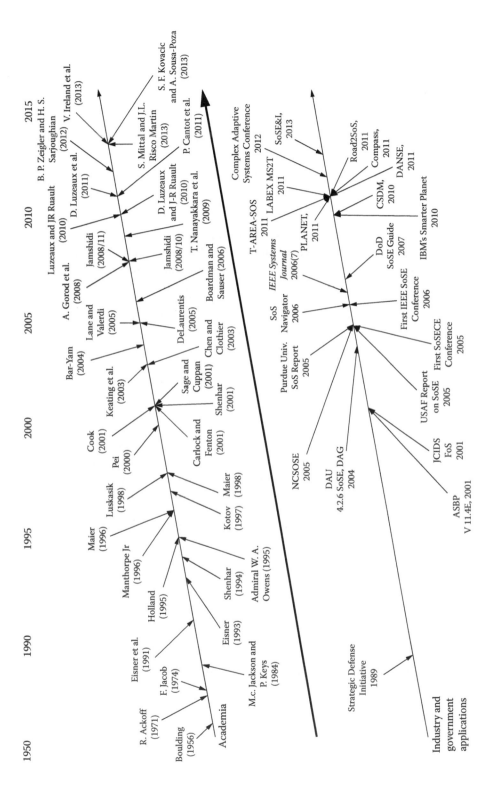

FIGURE 1.1 Modern history of system of systems and SoS engineering.

may be individually considered systems and which possess two additional properties: Operational and Managerial Independence of the Components" (Maier 1998).

Studies in 1995 related to SoS included the one by Holland where he considered a SoS as an artificial complex adaptive system (CAS) that constantly changes through self-organization while taking into consideration local governing rules and the corresponding complexities (Holland 1995; DoD 2007), as well as the study by Admiral W.A. Owens who introduced SoS to the military domain (Owens 1995). Thus, it should not be surprising that a SoS typically evolves as a result of interactions of the component systems with each other and the SoS environment.

The year 1997 marked the contribution of one of the most precise definitions of SoS by Kotov in the application of information technology, modeling, and synthesizing to a SoS (Kotov 1997). Kotov (1997) defined "System of Systems as large scale concurrent and distributed systems that are comprised of CS." Also, in 1998 Luskasik attempted to introduce the SoS concept to the education domain (Luskasik 1998).

In 2000, Pei's work resulted in a new understanding of SoS integration that he described as the ability *to pursue the development, integration, interoperability*, and *optimization of systems* (Pei 2000). The key word here is interoperability, since a SoS cannot (often, if ever) be optimized by optimizing each component system. Conventional SE is usually concerned mainly with the *vertical* integration of a *stovepipe* system. In the realm of SoS, *horizontal* integration is more important for that leads to better interoperability among the systems of the SoS, as well as, potentially, other systems.

The year 2001 marked four major contributions by Sage and Cuppan, Cook, Carlock and Fenton, and Shenhar. This included the work by Sage and Cuppan in which they proposed a new framework by implementing principles of a new federalism (Sage and Cuppan 2001). Cook made a distinction between monolithic systems and a SoS, based on system attributes and acquisition approaches (Cook 2001). Carlock and Fenton developed the concept of "enterprise system of systems engineering (SoSE)," which combined traditional SE activities with enterprise activities of strategic planning and investment analysis (Carlock and Fenton 2001; refer to the section "Enterprises and Enterprise Systems Engineering"). Shenhar further explored the notion of an *array* (Shenhar 2001).

The next significant contribution was in 2003, when Keating et al. did a comparative study of SE and SoSE (Keating et al. 2003). In the same year, Chen and Clothier addressed the need for a SoSE framework which included suggestions that advanced SE practices beyond the traditional project level were needed to focus on the *organizational context* (Chen and Clothier 2003). In 2004, Bar-Yam suggested adding characteristics versus just definitions for a better understanding of SoS (Bar-Yam et al. 2004).

In 2005, the primary contributions came from Jamshidi, Lane and Valerdi, and DeLaurentis. This included a definitional approach to SoS by virtue of collecting different definitions by Jamshidi (2005), a comparative approach to analyzing multiple SoSs in the *cost models* context by Lane and Valerdi (2005), and a description of various SoS traits in the transportation domain of SoS by DeLaurentis (2005).

These contributions were followed by work from Boardman and Sauser (2006), in which they implemented a characterization approach to SoS by identifying patterns and differences in 40 different SoS definitions.

In 2008, the first two books dedicated to SoS were introduced by Jamshidi (2008a,b) and covered a broad spectrum of topics related to SoS.

Among industry contributions, in 2001, the Department of Defense (DoD) defined a SoS as "a collection of systems that share/exchange information which interact synergistically" (Department of Defense [DoD] 2001) and also published a report in which it defined the term *family of systems* in a military context (DoD 2005). The DoD further contributed to the field of SoS in 2004, when they dedicated a section of the Defense Acquisition Guidebook for provisions of guidelines during the acquisition phase of a SoS (DAU 2004).

In 2005, the four major contributions included the establishment of the System of Systems Engineering Center of Excellence (SoSECE) along with the hosting of their first conference.

In addition to that, the SoS Report was created at Purdue University, and the National Center for System of Systems Engineering (NCSoSE) was set up at Old Dominion University. Lastly, in 2005, the US Air Force produced the first report dedicated to SoSE.

In 2006, three significant developments were the publication of the SoS navigator, the launching of a dedicated conference by the IEEE focusing on SoSE and the creation of the IEEE Systems journal.

In 2007, the DoD contributed to the field by publishing their SoSE guide and DeLaurentis et al. (2007) suggested the creation of an International Consortium of System of Systems.

Gorod et al. (2008) did a systematic review of contributions to the SoSE field and discussed a path forward. In 2010, Nanayakkara et al. (2010) published a book titled *Intelligent Control Systems with an Introduction to System of Systems Engineering.*

The year 2010 marked several contributions to the field of SoS. These included the book on SoS by Luzeaux and Ruault (2010), reports related to SoS from IBM titled "The world's 4 trillion dollar challenge—*Using a system of systems approach to build a smarter planet*" (Korsten and Seider 2010), as well as a conference on CS design and management (CSD&M 2013).

The year 2011 was also an illustrious year as far as contributions to the SoS domain were concerned. Luzeaux continued his work in the field of SoS and published another book entitled *CS and System of Systems Engineering* with two other coauthors (Luzeaux et al. 2011). An additional book, *Simulation and Modeling of System of Systems*, was coedited by Cantot and Luzeaux (2011). The year 20011 also marked the initiation of projects for the Trans-Atlantic Research and Education Agenda on System of Systems led by Michael Henshaw (T-AREA-SoS 2011), which includes a series of US–European workshops, as well as the mushrooming of several other entities such as COMPASS, DANSE, PLANET (Road2SoS 2011).

The year 2012 marked the beginning of a Complex Adaptive Systems Conference hosted by the Missouri University of Science and Technology, with a dedicated session on SoS. This year also marked the addition of another book to the SoS bookcase, entitled *Guide to Modeling and Simulation of System of Systems* (Zeigler and Sarjoughian 2013).

In 2013, the US Army started their own SoS engineering and integration directorate (SoSE&I 2013). Furthermore, in the same year, two other books were published in the SoS realm, including one by Mittal and Martin (2013), *Netcentric System of Systems Engineering with DEVS Unified Process*, and another book published that year, *Managing and Engineering in Complex Situations*, by Kovacic and Sousa-Poza (2013).

ENTERPRISES AND ENTERPRISE SYSTEMS ENGINEERING

Figure 1.2 depicts a graphical timeline of the major contributions to the enterprises and enterprise systems engineering (ESE) body of knowledge.

In contrast to other types of systems and systems engineering (SE) discussed in this chapter, this section is devoted to *enterprise* systems and *enterprise* systems engineering (ESE). Although several other definitions exist and will be discussed later, here enterprise is defined as a system (typically more general than a SoS) that has a particular property called homeostasis, which is a form of stability generally attributable to the whole enterprise and not just one or more identifiable components. Body temperature is an example of homeostasis in a human that most would agree is one's personal enterprise. An automobile company, like The Ford Motor Company, is another example of an enterprise. Ford's Job One, the first new car off the assembly line, constitutes or reflects its homeostasis, that is, the essential mission of the company.

Enterprises are not necessarily confined to single organizations, however. Government departments or ministries of defense, for example, usually comprised of many organizations, are enterprises also because they have homeostasis, that is, overall missions to perform, which provides an element of stability everyone involved is working toward.

ESE is the art and science of engineering an enterprise.

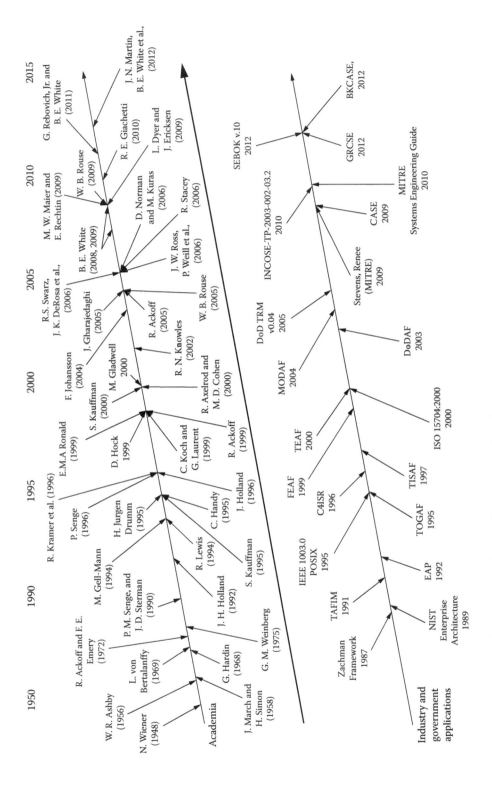

FIGURE 1.2 Modern history of enterprises and enterprise systems engineering.

There are two relatively recent and highly recommended references that are especially relevant to ESE and shall be freely quoted in this section to provide part of the modern history of SE. The first is contained within a Systems Engineering Body of Knowledge (SEBoK) (Various authors 2013), which was created through an intensive effort by dozens of contributors. The second is a book, *Enterprise Systems Engineering—Advances in the Theory and Practice* (Rebovich and White 2011), prominently referenced in the ESE portion of the SEBoK. A synopsis of each of these references is provided in the following text but only with respect to enterprises and ESE. Direct quotes from these sources are indented.

Other important background references by those who can be considered founders of this subject field are also cited. These sources are also deemed helpful in understanding enterprises and ESE.

ON ENTERPRISES AND ESE IN THE SYSTEMS ENGINEERING BODY OF KNOWLEDGE (SEBoK)

This section covers *definitions* of enterprise, the extended enterprise, ESE, and various enterprise and ESE-related *topics*, including non-SEBoK inserts on enterprise architecture and venture capital.

Definitions

An assumption has been made that the SEBoK readership, including many systems engineers whom are members of the International Council on Systems Engineering (INCOSE), will not need definitions of the terms system, engineering or even enterprise, for it has been noted that *INCOSE Systems Engineering Handbook* (INCOSE 2010) has used the terms *enterprise* and *organization* essentially interchangeably. For example,

> [(INCOSE 2010, p. 2)] NOTE: ISO/IEC 15288:2002 used Enterprise Processes that performed the same role as the Organization Project-Enabling Processes of ISO/IEC 15288:2008. For the purposes of this v3.2 Handbook the terms Organization and Organizational are used to be synonymous with Enterprise. (White 2010a)

Refer to the ESE Background section in the following text for further discussion of this point.

Enterprise

Notwithstanding the (preceding and subsequent) discussion of enterprise versus organization, based in part on INCOSE member contributors, for example, refer to Martin et al. (2012), enterprise is further defined in the SEBoK:

> An enterprise consists of a purposeful combination (e.g., a network) of interdependent resources (e.g., people, processes, organizations [Considering the preceding and subsequent discussions above and below concerning the definition of enterprise viz. a viz. organization, the current and future editions of INCOSE's SE Handbook will likely treat a single organization as a simpler or special case of a more general enterprise.], supporting technologies, and funding) that interact with
>
> - Each other to coordinate functions, share information, allocate funding, create workflows, and make decisions, etc.; and
> - Their environment(s) to achieve business and operational goals through a complex web of interactions distributed across geography and time [(Rebovich and White 2011, pp. 4–35)].

The term *enterprise* has also been defined by others as follows:

1. One or more organizations sharing a definite mission, goals, and objectives to offer an output such as a product or service. (Various authors 2000)
2. An organization (or cross organizational entity) supporting a defined business scope and mission that includes interdependent resources (people, organizations and technologies) that must coordinate their functions and share information in support of a common mission (or set of related missions). (FEAF 1999)

3. The term enterprise can be defined in one of two ways. The first is when the entity being considered is tightly bounded and directed by a single executive function. The second is when organizational boundaries are less well defined and where there may be multiple owners in terms of direction of the resources being employed. The common factor is that both entities exist to achieve specified outcomes. (MODAF 2004)

4. A complex (adaptive) socio-technical system that comprises interdependent resources of people, processes, information, and technology that must interact with each other and their environment in support of a common mission. (Giachetti 2010)

Extended Enterprise

As has already been demonstrated, *enterprise* can have varied meanings depending on the context and scope. Nevertheless, many people like to think of relatively limited enterprises, like those of organizations, and then talk about extended enterprises as enterprise-like environments in which those organizations operate.

> Sometimes it is prudent to consider a broader scope than merely the "boundaries" of the organizations involved in an enterprise. In some cases, it is necessary and wise to consider the "extended enterprise" in modeling, assessment, and decision making. This could include upstream suppliers, downstream consumers, and end user organizations, and perhaps even "sidestream" partners and key stakeholders. The extended enterprise can be defined as:

> Wider organization representing all associated entities—customers, employees, suppliers, distributors, etc.,—who directly or indirectly, formally or informally, collaborate in the design, development, production, and delivery of a product (or service) to the end user. (Extended Enterprise 2013)

Enterprise Systems Engineering

A definition of ESE is offered up-front in the portion of the SEBoK discussing enterprises.

> Enterprise systems engineering (ESE) is the application of systems engineering principles, concepts, and methods to the planning, design, improvement, and operation of an enterprise.

> Enterprise systems engineering (ESE), for the purpose of this article, is defined as the application of SE principles, concepts, and methods to the planning, design, improvement, and operation of an enterprise (see Note 3 [in the following]). To enable more efficient and effective enterprise transformation, the enterprise needs to be looked at "as a system," rather than merely as a collection of functions connected solely by information systems and shared facilities (Rouse 2005). While a systems perspective is required for dealing with the enterprise, this is rarely the task or responsibility of people who call themselves systems engineers.

> Note 3. This form of systems engineering (i.e., ESE) includes (1) those traditional principles, concepts, and methods that work well in an enterprise environment, plus (2) an evolving set of newer ideas, precepts, and initiatives derived from complexity theory and the behavior of CS (such as those observed in nature and human languages).

Enterprise and ESE-Related Topics

The enterprise and ESE knowledge area of SEBoK contains the following topics:

- Enterprise systems engineering background
- Enterprise as a system
- Enterprise systems engineering key concepts
- Related business activities
- Enterprise capability management
- Enterprise systems engineering process activities

These topics will be briefly discussed next.

ESE Background

The dichotomy between enterprise and organization (mentioned in the definition of enterprise in the *Definitions* section) is further discussed in SEBoK.

> It is worth noting that an enterprise is not equivalent to an "organization" according to the definition above. This is a frequent misuse of the term enterprise. The [Figure 1.3] below shows that an enterprise includes not only the organizations that participate in it, but also people, knowledge, and other assets such as processes, principles, policies, practices, doctrine, theories, beliefs, facilities, land, intellectual property, and so on.

> Some enterprises are organizations, but not all enterprises are organizations. Likewise, not all organizations are enterprises. Some enterprises have no readily identifiable "organizations" in them. Some enterprises are self-organizing (i.e., not organized by mandate), in that the sentient beings in the enterprise will find for themselves some way in which they can interact to produce greater results than can be done by the individuals alone. Self-organizing enterprises are often more flexible and agile than if they were organized from above. (Dyer and Ericksen 2009; Stacey 2006)

> One type of enterprise architecture that supports agility is a non-hierarchical organization without a single point of control. Individuals function autonomously, constantly interacting with each other to define the vision and aims, maintain a common understanding of requirements and monitor the work that needs to be done. Roles and responsibilities are not predetermined but rather emerge from individuals' self-organizing activities and are constantly in flux. Similarly, projects are generated everywhere in the enterprise, sometimes even from outside affiliates. Key decisions are made collaboratively, on the spot, and on the fly. Because of this, knowledge, power, and intelligence are spread through the enterprise, making it uniquely capable of quickly recovering and adapting to the loss of any key enterprise component. (Business agility 2013)

Enterprise Architecture

Here we include a summary suggesting the rich history of enterprise architecture frameworks. There are many sources of information available online.

Perhaps the most well-known artifact associated with enterprise architecture is the Zachman Framework that first appeared in 1987 (Zachman 1987). This framework is a great attempt at creating a holistic ontology, structure, or schema for the enterprise architecture object—but it is not a methodology for actually creating the specific enterprise architecture one may need.

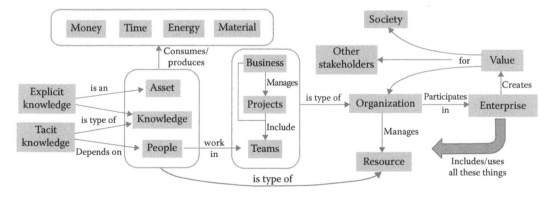

FIGURE 1.3 Organizations manage resources to create enterprise value. *Notes:* 1. All entries shown are decomposable, except people. For example, a business can have subbusinesses, a project can have subprojects, a resource can have sub-resources, an enterprise can have subenterprises. 2. All entries have other names. For example, a program can be a project comprising several subprojects (often called merely projects). Business can be an agency, team can be group, value can be utility, etc. 3. There is no attempt to be prescriptive in the names chosen for this diagram. The main goal of this is to show how this chapter uses these terms and how they are related to each other in a conceptual manner.

Rather this ontology suggests multiple ways of viewing enterprise architecture using a 6 × 6 matrix of classification names (namely, what, how, where, who, when, and why) as columns and audience perspectives (namely, executive, business management, architect, engineer, technician, and enterprise) and model names (namely, scope concepts, business concepts, system logic, technology physics, tool components, and operations instances, respectively) as rows (Zachman 2013). This matrix arguably encompasses all possible aspects one might need to consider in creating enterprise architecture.

Another early implementation of an enterprise architecture framework was the *Technical Architecture Framework for Information Management* (TAFIM 1996). The first draft was completed in 1991 with the TAFIM Technical Reference Model (TAFIM TRM 1996). This technical reference model used open systems and new technologies available in the commercial market to develop a DoD-wide application (Oberndorf and Earl 1998).

The original development of TOGAF (The Open Group Architecture Framework) Version 1 appeared in 1995 (TOGAF 2009). The latest edition, Version 9.1, appeared on December 1, 2011 (TOGAF 2013a). The TOGAF TRM was originally derived from TAFIM, which in turn was derived from the IEEE model 1003.0 (TOGAF 2013b) or POSIX (Portable Operating System Interface) Open System Environment.

A potential benefit to be gained from enterprise architecture is the ability to support decision-making in changing businesses. Because enterprise architecture brings together business models (e.g., process models, organizational charts, etc.) and technical models (e.g., systems architectures, data models, state diagrams, etc.), it is possible to trace the impact of organizational change on the systems, and also the business impact of changes to the systems.

> In spite of this lack of "organization" in some enterprises, SE can still contribute much in the engineering of the enterprise, as described in the articles below. However, SE must be prepared to apply some non-traditional approaches in doing so. Hence the need for embracing the new discipline called [ESE].
>
> Giachetti (2010) distinguishes between enterprise and organization by saying that an organization is a view of the enterprise. The organization[al] view defines the structure and relationships of the organizational units, people, and other actors in an enterprise. Using this definition, we would say that all enterprises have some type of organization, whether formal, informal, hierarchical or self-organizing network.
>
> ESE is an emerging discipline that focuses on frameworks, tools, and problem-solving approaches for dealing with the inherent complexities of the enterprise. Furthermore, ESE addresses more than just solving problems; it also deals with the exploitation of opportunities for better ways to achieve the enterprise goals. A good overall description of ESE is provided by in the book by Rebovich and White (2011; refer to On Enterprises and ESE in *Enterprise Systems Engineering—Advances in the Theory and Practice*).

Enterprise as a System

> An enterprise must strive toward two main objectives: (1) develop things within the enterprise to serve as either external offerings or as internal mechanisms to enable achievement of enterprise operations, and (2) transform the enterprise itself so that it can most effectively and efficiently perform its operations and survive in its competitive and constrained environment.

Enterprise Systems Engineering Key Concepts

Taking the point of view that people, namely, the most relevant stakeholders, are intentionally included within the agreed boundary of the system is perhaps the simplest idea that is crucial to understanding why enterprises are complex. Of course, the system protagonists, that is, those most concerned with ensuring the welfare of the enterprise, must first discuss and decide, for their practical intents and purposes of engineering the enterprise in its environment, where this (often fuzzy) boundary is located. This highly recommended process (The Enneagram developed by Richard N. Knowles (2002) provides an excellent, highly effective template for having discussions of this type that guarantees, if the process is properly followed, that all important aspects, from a systems thinking perspective, are brought to the table) will likely include the identification of the stakeholders to be considered, at least by type, position, and loyalty, and more to the point, by name. Further discussion of the role of stakeholders is included in Chapter 2.

Related Business Activities

The primary purpose of an enterprise is to create value for society, other stakeholders, and for the organizations that participate in that enterprise. This [is] illustrated in [Figure 1.3] that shows all the key elements that contribute to this value creation process.

There are three types of organizations of interest: businesses, projects, and teams (see Note 4 [in the following]). A typical business participates in multiple enterprises through its portfolio of projects. Large SE projects can be enterprises in their own right, with participation by many different businesses, and may be organized as a number of sub-projects.

Note 4. The use of the word *business* is not intended to mean only for-profit commercial ventures. As used here, it also includes government agencies and not-for-profit organizations, as well as commercial ventures. Business is the activity of providing goods and services involving financial, commercial, and industrial aspects.

Venture Capital

As an aside, it is noted that a fair number of businesses start as the result of venture capital investments. It is interesting to compare how money flows in the venture capital arena compared to the military system acquisition venues, for example (Stevens 2009). Typically, if those receiving venture capital do not perform to expectations within a reasonable length of time, the funding sources dry up because the investors are ill disposed to *send good money after bad*. However, in the acquisition world, even when contractors do not perform according to the government-accepted perceived promises, the money continues to flow. Otherwise the program, which usually is still deemed dearly needed by the government sponsor or customer, would stop, that is, fail, and supposedly the public would not be served. Regrettably, some programs are eliminated only after billions of dollars have been expended without result.

A key choice for businesses that conduct SE is to what extent, if at all, they seek to optimize their use of resources (people, knowledge, assets) across teams, projects, and business units.

Optimization of resources is not the goal in itself, but rather a means to achieve the goal of maximizing value for the enterprise and its stakeholders. At one extreme, in a product-oriented organization, projects may be responsible for hiring, training, and firing their own staff, as well as managing all assets required for their delivery of products or services. The term "product-oriente organization" is not meant in the sense of product-oriented SE, but rather in the sense of this being one of the basic constructs available when formulating organizational strategy.

At the other extreme, in a functional organization, the projects delegate almost all their work to functional groups. In between these two extremes is a matrix organization that is used to give functional specialists a *home* between project assignments. A full discussion of organizational approaches and situations, along with their applicability in enabling SE for the organization, is provided in the article called Systems Engineering Organizational Strategy.

The optimization debate can be handled as described in the book called "Enterprise Architecture as Strategy" [(Ross et al. 2006)]. In other words, an enterprise can choose, or not, to unify its operations and can choose, or not, to unify its information base. There are different strategies the enterprise might adopt to achieve and sustain value creation and how ESE helps an enterprise to choose. This is further addressed in the section on Enterprise Architecture Formulation & Assessment in the article called Enterprise Capability Management.

Enterprise Capability Management

In ESE, the emphasis is on adding capabilities incrementally as one is able to develop successful interventions as opposed to focusing on requirements. The latter is more the mantra of traditional or conventional SE. Enterprises are so complex, it is next to impossible to accurately establish requirements and to expect them to hold firmly in place. In essence, due to the human factor, the real world is just too uncertain and changeable. Instead one tries to (1) establish reasonable notions of what is needed; (2) develop what are hoped to be advancements in improving the enterprise; and (3) watch for good opportunities to introduce these enhancements to see what will happen next. Often one must wait a significant amount of time to see results, and even longer to see whether the interventions are really going in the desired direction. If not, one tries to learn from miscues or mistakes and attempts to rectify the problems that arise and try again at a suitable future time.

This process involves the inexact art and science of engineering management. There is a considerable opportunity to develop decision-making heuristics (Maier and Rechtin 2009) to help decision-makers make better decisions. This includes sensing when it is best to wait a little longer before making another intervention, as well as when the time is right to decide to act again.

Enterprise Systems Engineering Process Activities

There are no *cookie-cutter* solutions here. ESE has much to do with trying to understand and define the enterprise and its agreed-to boundary, which is generally very *fuzzy*—as has already been suggested. Arguably two of the most important processes are (1) establishing a leadership/management process for dealing with unexpected situations that undoubtedly will crop up at worst possible moments; and (2) being open to pursuing opportunities at the enterprise level with informed risk taking with the hope of discovering attractive solutions that would otherwise be missed if one is too conservative and risk averse, cf., Chapter 5, "Enterprise Opportunity and Risk," pp. 161–180, of Rebovich and White (2011).

More will be said about a few suggested ESE processes in the section "On Enterprises and ESE in *Enterprise Systems Engineering—Advances in the Theory and Practice*."

Practical Considerations

When it comes to performing SE at the enterprise level, there are several good practices to keep in mind (Rebovich and White 2011):

- Set enterprise fitness as the key measure of system success. Leverage game theory and ecology, along with the practices of satisfying and governing the commons (Hardin 1968).
- Deal with uncertainty and conflict in the enterprise through adaptation: variety, selection, exploration, and experimentation.
- Leverage the practice of layered architectures with loose couplers and the theory of order and chaos in networks.

Enterprise governance involves shaping the political, operational, economic, and technical (POET) landscape. One should not try to control the enterprise like one would in a traditional SE effort at the project level.

The authors' experience is that a good lessons-learned database accessed through keywords is very valuable.

ON ENTERPRISES AND ESE IN ENTERPRISE SYSTEMS ENGINEERING—ADVANCES IN THE THEORY AND PRACTICE

Again, as in the section "On Enterprises and ESE in the Systems Engineering Body of Knowledge (SEBoK)," selected excerpts are freely quoted from Rebovich and White 2011.

In September 2004, … [the Air Force Center of] The MITRE Corporation commissioned a … study [of] Enterprise Systems Engineering (ESE). It was becoming increasingly clear … that the traditional systems engineering processes and tools were inadequate to face the scale and complexity of the systems … government clients needed. … A large amount of money was being spent in the development of these CS, but they seemed to be habitually late-to-market and exceeded the budget. Worse still, many were obsolete before they were deployed or the program was cancelled before delivering anything at all. … Many concerted attempts by the government, by contractors, and by us to better implement systems engineering as we knew it seemed not to improve the situation. Undoubtedly, a new approach was needed. …

Foreword (J. K. DeRosa)

A *partial* list of major cost overruns and/or schedule slips for several US Government acquisition programs circa 2008 is provided in White (2008, 2009, 2010b). Reasons given for this sad state of affairs include mismanagement, a *conspiracy of optimism*, and the *way the world works*. Unfortunately, little has been done to better serve the public interest since; most stakeholders are content with this situation which, after all, stimulates the economy and provides acquisition-contracting jobs.

White (2008, 2009, 2010b) also shows award fee payments despite the lack of mission capability and poor cost and/or schedule performance for several US government acquisition programs. These dismal results serve as motivation for trying an improved SE approach.

Continuing quotations from the Foreword of Rebovich and White (2011):

… something had changed in the way people worked together, both within the general population and the systems engineering discipline. Seldom did isolated groups work on local problems to build stovepiped solutions. Seldom were systems developed or used in a social, political, economic, or technical vacuum. It seemed that everyone and everything were interconnected and interdependent. …

Changes in the sociocultural and technical landscape had repercussions on the practice of systems engineering. Requirements changed faster than they could be met; risks shifted in a never-ending dynamic of actions and reactions within the network of stakeholders; configurations changed with rapidly changing technology cycles; and integrated testing was faced with trying to match the scale and complexity of the operational enterprise. Processes that managed requirements, risks, configurations, and tests, four [aspects] of traditional systems engineering, were confounded by their complex environment. …

The Focus Group's first step was to define terms (White 2007). What is an *enterprise*? What are its *boundaries*? What are the *scales* of the enterprise and the corresponding systems engineering? How do you define *requirements*? These definitions turned out to be more difficult than we had imagined. However one drew the boundary of the enterprise, there were influences coming in from outside that boundary. Systems engineering had to be done coherently at the multiple and seemingly incommensurate levels of the enterprise: the individual systems, groups of systems performing a cooperative mission, and the enterprise as a whole. Requirements changed routinely. We established a baseline taxonomy to begin the work: the enterprise boundary was defined to include all those elements we could either control or influence; we partitioned the system engineering into three engineering tiers—individual systems, system of systems, and enterprise; and the requirements were defined in terms of enterprise outcome spaces without any obvious allocation down to systems or subsystems. There was no established theory to guide our choices.

Next we turned to the practice. After all, these CS had been extensively studied for a number of years, and experienced system engineers were coping with the complexity in their own ways. We consulted our practicing systems engineers who were involved in these CS. This led to the following five methods that seemed to have some success in dealing with complex system and turned them into what we called Enterprise Systems Engineering (ESE) processes [as alluded to in the section "On Enterprises and ESE in the Systems Engineering Body of Knowledge (SEBoK)" under Enterprise Systems Engineering Process Activities].

1. Capabilities-based engineering analysis that focused on the high-level requirements the whole enterprise must satisfy
2. Enterprise architecture that brought a coherent view of the whole enterprise

3. Enterprise analysis and assessment that evaluated the effectiveness of the enterprise rather than the ability of system components to meet specifications
4. Strategic technical planning that laid out a minimum set of standards or patterns that every system in the enterprise would follow
5. Technology planning that looked for cues in the environment that would indicate the emerging technologies that could be implemented next into the enterprise

We fitted these new processes together with the existing traditional systems engineering processes into a new framework. We considered the traditional processes in the usual way from requirements to manufacturing, integrated testing, and deployment. We embedded that as an inner loop within an outer loop of inputs and constraints supplied by the ESE processes. Thus for a single program, the requirements, risks, and so on were fixed through the development cycle, unless and until something in the ESE processes caused them to change. When that happened, the change was made and again the individual system development proceeded as if it were fixed. This seemed to match with reality—our development tools assumed that things were known and fixed, and our experience indicated that things changed abruptly from the outside a number of times throughout the life of a program. In the second year of the Focus Group, we launched several pilot programs to test the effectiveness of those ESE processes. Both the processes and the pilots are reported in the chapters of this book. (Rebovich and White 2011)

There were two main sources of theoretical underpinnings for this new brand of ESE. The first was in the realm of CS. In January 2004, MITRE Technical Report on CS Engineering published by Norman and Kuras reflected the groundbreaking work that Norman had carried out with NECSI. ... It recommended an approach called the system engineering *regimen* based on evolutionary theory and complexity science. More importantly, it opened a floodgate of theoretical knowledge from complexity science, as exemplified by the authors such as Stuart Kauffman, John Holland, Robert Axelrod, Stephen Wolfram, Duncan Watts, and a number of researchers from the Santa Fe Institute. Complexity addressed networks of interrelated nodes just like those of our clients. ... For the first time, we had a mathematical basis for ESE. We had found part of the body of knowledge that was needed to engineer the enterprise.

The second source of theoretical knowledge came from the management literature exemplified by authors such as Russell Ackoff, Jamshid Gharajedaghi, Peter Senge, John Sterman, and Herbert Simon, to name a few. They focused on the complex interactions between people in purposeful endeavors. Their systems were learning and adaptive, and their organizations displayed ordered and explainable social structure. The social and business environments they studied were identical to the socioeconomic environments with which ESE had to cope.

Armed with this new knowledge and the benefit of several ongoing pilots in ESE, those involved in the Focus Group initiative wrote dozens of papers and made numerous presentations at symposia sponsored by professional societies such as the Institute of Electrical and Electronics Engineers, the International Council on Systems Engineering (INCOSE), and the International Symposium on CS. The *Regimen* paper was eventually published by Norman and Kuras as a chapter in a book (Norman and Kuras 2006) edited by Bar Yam and his team from NECSI, and another version was presented by Kuras and White at the INCOSE 2005 Symposium. ... [An updated version called CASE (Complex Adaptive Systems Engineering) (White 2008, 2009, 2010b) was promulgated by Brian E. White.]

Completing this segment of enterprises and ESE with quotes from Rebovich and White (2011):

The roots of ESE theory were the result of deliberate cultivation by some of the pioneers of systems thinking such as Ashby (1956) and Wiener (1948) in the study of Cybernetics, Ludwig Von Bertalanffy (1969), and Weinberg (1975) in the study of General Systems Theory, and Ackoff (Ackoff and Emery 1972) and Simon (March and Simon 1958) in the business literature. In that same era, the field of complexity science was gaining ground from the well springs of biology, chemistry, economics, and evolutionary theory. It came to a nexus in 1984 with the establishment of the Santa Fe Institute (SFI). Gell-Mann (1994), Kauffman (2000), and Holland (1992) are among the many thought leaders associated with SFI.

Chapter 1, Introduction (Joseph K. DeRosa)

TABLE 1.1

Key Elements of the Theory and Practice of ESE

Essential Elements of ESE	Descriptions
Development through adaptation	Deals with uncertainty and conflict in the enterprise through adaptation (adaptive stance): variety, selection, exploration, and experimentation.
Strategic technical planning	Brings a balance of integration and innovation to the enterprise. Leverages the practice of layered architectures with loose couplers and theory of order and chaos in networks.
Enterprise governance	Sets enterprise fitness as a measure of system success. Leverages game theory, ecology, and practices of *satisficing* and governing the commons.
ESE processes	Brings new ways of working horizontally across the enterprise. Leverages business literature and practice.

The elements of the theory and practice of ESE lie somewhere in the milieu of systems thinking, complexity science, and the evolution of the Internet and WWW [World Wide Web]. At this stage in the development of ESE, the practice leads the theory. However, as the two coevolve, each will inform the other in a never-ending virtuous cycle. Although it is not clear at this point in time which concepts and related practices will ultimately prove most useful, I have made a first attempt to cull out those elements of ESE that seem to be gaining favor in the practice and that have some basis in theory. These are given in Table 1.1.

Some important citations, including Gharajedaghi (1999) and (2005), which were also cited in Rebovich and White (2011), make up references (Johansson 2004; Ackoff 2005; Rouse 2009; SE Guide 2013).

COMPLEX SYSTEMS AND COMPLEX SYSTEMS ENGINEERING

Now that we have covered the modern history of SoS and SoS engineering, and enterprises and ESE, we turn finally to CS and CS engineering (CSE). Figure 1.4 depicts a graphical timeline of the major contributions to the CS (CS) and CSE body of knowledge.

COMPLEXITY THEORY DEVELOPMENT OVER THE LAST 50 YEARS

The Greek word *sustema* stood for reunion, conjunction, or assembly. The concept of system surfaced during the seventeenth century, meaning a collection of organized concepts, mainly in a philosophical sense. At the end of the eighteenth century, the philosophical notion of system was firmly established as a constructed set of practices and methods usable to study the real world. Much later, the unavoidable necessity of correlations and mutual interdependence, associated with a complex causality, led naturally to the concept of CS.

The genesis of complex system theory can be traced back to the cybernetics movement, which started during World War II on the east coast of the United States. According to Steve Heims (1991), the cybernetics group in 1942 consisted of eight members from three different fields of science (Abraham 2002):

- Mathematics: Norbert Wiener, John von Neumann, and Walter Pitts
- Engineering: Julian Bigelow and Claude Shannon
- Neurobiology: Rafael Lorente de No, Arturo Rosenblueth, and Warren McCulloch.

Later, the group proliferated to 21 members. They first met in 1942 and were able to sponsor a series of conferences in New York between 1946 and 1953 on the subject of "Circular Causal and Feedback Mechanisms in Biological and Social Systems." Meanwhile, by 1949, three key books were published: Wiener's *Cybernetics: Or Control and Communication in the Animal and*

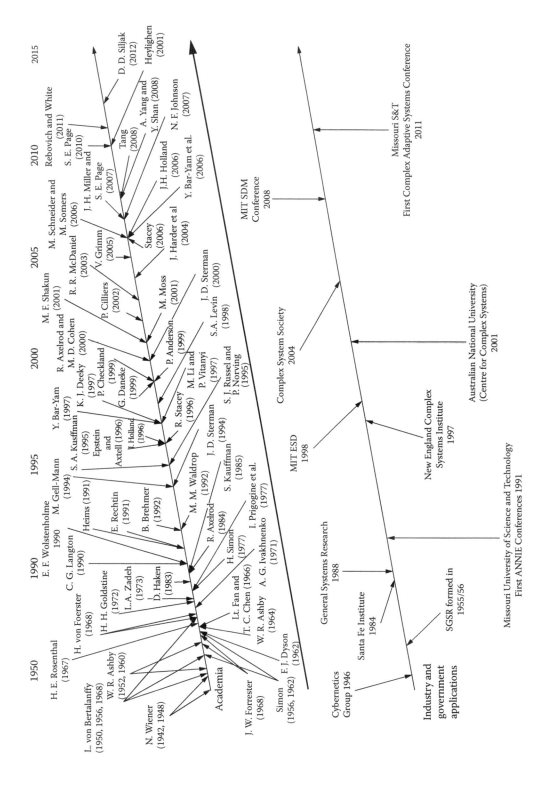

FIGURE 1.4 Modern history of complex systems and complex systems engineering.

the Machine (Wiener 1948), Von Neumann's and Morgenstern's *Theory of Games and Economic Behavior*, and Shannon's and Weaver's *The Mathematical Theory of Communication*. These publications entailed a new science of information and regulation. In 1950, contemporaneous with Von Bertalanffy's emigration to North America, his notions were prelude to the general system theory (GST), but Von Bertalanffy did not compose the theory until 1968, when he punctuated on a new level of description for system properties—one that cannot be solely derived from the behavior of the components (Thelen and Smith 2006). Later, he suggested the following dynamic equations to clarify the principles applying to generalized systems: wholeness or self-organization, openness, equifinality (self-stabilization: in open systems the end state is achievable via various trajectories), and hierarchical organization. Von Bertalanffy's role as a catalyst of the systems view was considered important in two ways: firstly, he clearly mentioned the central concept of systems whereas, on the other hand, he strongly emphasized the *isomorphic laws in science*, deducing the possibility of a new multidisciplinary approach and proposed a *general systems theory* (Davidson 1983). In 1956, Jay Forrester at MIT used digital computers and computer models to simulate the complex dynamic systems (Goldstine 1972). Later, the Society for General System Research (SGSR) was formed in 1955/1956.

Meanwhile, in the first self-organizing conference (Interdisciplinary Conference on Self-Organizing Systems, Chicago, IL, 1959) the British Cybernetician, W. Ross Ashby, suggested what he called *the principles of self-organization*. Based on his notation, dynamic systems, regardless of their type, always tend to evolve toward a state of equilibrium (Ashby 1960). Continuing his work in 1960, he explored the feedback mechanism as a means to understand adaptive systems. In 1959, Heinz Von Foerster denoted the principle of *order from disorder* and explained that the larger the random perturbations (noise) that affect a system, the more quickly it will self-organize (produce *order*) (Ashby 1962). Additionally, in 1961, in his first book called *Industrial Dynamics* (H. Von Foerster 1961) he introduced the concept of *dynamic systems* and proposed a whole new concept to address management problems (Wolstenholme 1990).

A year later, in 1962, Simon defined a complex system as one made up of a large number of parts that have many interactions (Anderson 1999). Furthermore, Simon (1962) also wrote "The Architecture of Complexity" in the *Proceedings of the American Philosophical Society*. In addition, the year 1962 included a contribution from F. J. Dyson, titled "Statistical Theory of the Energy Levels of CS" in the *Journal of Mathematical Physics* (Dyson 1962). In 1964, Ashby formulated the *Law of Requisite Variety*, stating that "The greater the variety of perturbations the system has to cope with, the greater the variety of compensating actions it should be able to perform and the greater the knowledge or intelligence the system will need in order to know which action to perform in which circumstances" (Heylighen et al. 2007). In 1966, Lt. Fan and T. C. Chen published a book entitled *The Continuous Maximum Principle: A Study of CS Optimization* (Fan and Chen 1996). H.E. Rosenthal (1967) developed a graphic method for the determination and presentation of binding parameters in a complex system. Moreover, in 1967, Thompson explained a complex organization in his book *Organizations in Action* as a set of interdependent parts, which together form a whole that is interdependent with its greater environment (Anderson 1999). In 1968, Von Bertalanffy further suggested dynamic equations to clarify the principles applying to generalized systems, which was followed by Forrester's contribution (Forrester 1968), where he described systems as *wholes of elements* cooperating toward a common goal (Schwaninger 2006; Thelen and Smith 2006). Moreover, The Club of Rome was convened by Aurelio Peccei (Abraham 2002). Lastly, in 1968, Forester published his book *Principles of Systems: Text and Workbook*.

Ivakhneko (1971) discussed the polynomial theory of CS in the October issue of the *IEEE Transactions on Systems, Man and Cybernetics*. The following year, in 1972, Goldstine published a book *The Computer from Pascal to Von Neumann* with Princeton University Press. The year 1973 marked the outline of a new approach to the analysis of CS and decision processes as outlined by L.A. Zadeh in the January issue of the *IEEE Transactions on Systems, Man and Cybernetics*. The years 1975 and 1976 noted the contributions by Jong and Alan (1975) through their work,

"An Analysis of the Behavior of a Class of Genetic Adaptive Systems," and a book by Jantsch and Waddington entitled *Evolution and Consciousness: Human Systems in Transition*.

A subsequent contributor to the system theory in 1977 was the Nobel chemist Ilya Prigogine as a result of his noble work on nonequilibrium thermodynamics. "The basis for any natural law describing the evolution of social systems must be the physical laws governing open systems, that is, systems embedded in their environment with which they exchange matter and energy" (Laszlo and Krippner 1998). Later, he introduced the dissipative structures as physical and chemical systems far from thermodynamical equilibrium that tend to self-organize by exporting entropy. The year 1977 also saw the publication of the book *The Organization of CS* by H. Simon.

In 1979, Von Foerster started the *Second-Order Cybernetics* movement with the goal to shift emphasis from objective systems around us to cognitive and social processes by which we construct our subjective models of those systems (Heylighen et al. 2007). In the following year, this concept was reinforced by the concept of autonomy, autopoiesis, and self-organization (Maturana and Varela 1992). The previous endeavors by the second-order cybernetics resulted in statements such as: "...The structure of a system is not given, but developed by the system itself, as a means to survive and adapt to a complex and changing environment" (Heylighen et al. 2007).

In the early 1980s, D. V. Steward (1981) wrote a paper in the *IEEE Transactions on Engineering Management*, related to the design structure system and a method for managing the design of CS. After the advent of computer stimulation, in 1984, the scientists were able to model and simulate systems with various degrees of complexity, which was at the base of the newfangled subject CAS (complex adaptive system) under the study of researchers associating with the Santa Fe Institute in New Mexico, founded in 1984. As part of the research being done in CAS, the biologist Stuart Kauffman studied the development of organisms and tried to discern "how networks of mutually activating or inhibiting genes can give rise to the differentiation of organs and tissues during embryological development" (Kauffman 1985). As a result, he suggested that the self-organization exhibited by Boolean networks is an essential factor in evolution. Moreover, he punctuated on the importance of self-organization in limiting the action of selection in complex networks (Weber 1998). In 1983, German physicist Hermann Haken (2010) suggested the new discipline of *synergetics* to study collective phenomena such as lasers. In 1984, Prigogine published his book entitled *Order Out of Chaos*. This was followed by D. Harel's innovation regarding a visual formalism for CS. Furthermore, Axelrod (1984) stated that agents are egocentric or selfish, which means that they only care about their own goal or fitness while ignoring other agents.

The contribution to follow in 1986 was the book written by Puccia and Levins, *Qualitative Modeling of CS: An Introduction to Loop Analysis and Time Averaging*. According to the International Society of System Sciences (1988), the SGSR was renamed to International Society of Systems Sciences. In addition, Haken (1988) wrote a book entitled *Information and Self Organization: A Macroscopic Approach to CS*, which was followed by Broomhead and Lowe's (1988) work on multivariable functional interpolation and adaptive networks in the *CS* journal.

Langton (1990) called the complexity science *The Edge of Chaos*, which exhibits the nature of a CS (complex system) as characterized by neither order nor disorder, and Wolstenholme authored a book entitled *System Enquiry: A System Dynamics Approach* (Heylighen et al. 2007). The highlighted contributions in 1991 included the publication of *Systems Architecting: Creating and Building CS*, a book by Rechtin, and the First ANNIE conference, which was held at the Missouri University of Science and Technology. Brehmer's contribution in December 1992 was a paper on "Dynamic Decision Making: Human Control of CS." According to Heylighen et al. (2007), in 1992 Waldrop mentioned complexity science, and in 1994 Gell-Mann and Sterman added to the body of knowledge of CS, particularly by Gell Mann's assertion that in complex adaptive systems, the agents follow a set of rules (Heylighen 2007; Anderson 1999). Furthermore, Kauffman

(1995) proposed the *coevolve* characteristics of the agents and described the self-organization as follows: "The system spontaneously arranges its components and their interactions into a sustainable, global structure that tries to maximize overall fitness, without need for an external or internal designer or controller" (Heylighen et al. 2007). Moreover, he described the *bounded instability*, which appears when the number of agents approximates the number of interactions between them.

In the mid-1990s, the agent-based models (ABMs) were recognized as a computational approach to solve problems that were decentralized, changeable, and ill-constructed (Jennings and Wooldridge 1995; Epstein and Axtell 1996). During the same period, the agents were defined as problem-solving entities that can sense and respond to stimuli acquired from their environment or other agents (Tang 2008). According to Tang (2008), pertaining to the work of Russell and Norvig (1995), the intelligent agent (IA) was described as "… agents whose behavior is built on predefined knowledge and their own experiences." Such agents possess the ability to individually generate and adjust their decision-making strategies, in response to a dynamic environment or other agents (Tang 2008). Moreover, this discovery was utilized in future research that was done on intelligent complex adoptive systems (ICAS).

Holland (1996), in his book *Hidden Order*, defined a CAS (currently denoted as Multiagent system [MAS]) as systems composed of several autonomous agents and moreover focused on the relation between the system and environment also known as adoption (Yang and Shan 2008). The same year, Stacey (1996) stated that the agents follow principles in their interactions (schemas) to improve on their behavior. In 1997, Y. Bar-Yam introduced *Dynamics of CS* (Westview Press), and Li and Vitanyi (1997) published the book *An Introduction to Kolmogorov Complexity and Its Applications*. The same year, the New England CS Institute (NECSI) was also founded by Bar-Yam, and according to the organization's website, he held the first conference on CS in the Boston area in September (NECSI 1997). In 1998, Levin (1998) defined emergence as macro-level patterns in a self-organized matter, which are produced as a result of agent–agent or agent–environment interactions (Tang 2008).

In 1998, MIT also established the engineering systems division. They rightly populated the division with professors who maintained their existing memberships in their home departments. This helped ensure the cross-coupling of the many transdisciplines necessary to the advancement of *engineering systems*. The term *systems engineering* was reversed to emphasize the complex nature of the subject, which was not only an expansion of but complementary to existing SE practice. As mentioned in a quote of the section "On Enterprises and ESE in *Enterprise Systems Engineering— Advances in the Theory and Practice*," in March 2004, the engineering systems division hosted an important symposium for engineering leaders to focus on CS engineering.

Gregory Daneke (1999) defined a CAS as "a simulation approach that studies the coevolution of a set of interacting agents that have alternative sets of behavioral rules." Concurrently, Pascale (1999) noted that the interaction between the agents will lead to more *levels* of organization (Daneke 1999; Yang and Shan 2008). By the end of the 1990s, the key elements of CAS models were described as follows: "Agents with schemata, Self-Organizing Networks Sustained by Importing Energy, Coevolution to the edge of chaos, Recombination and System Evolution" (Heylighen et al. 2007).

The twenty-first century started with the Axelrod et al. notation: "The situational decision-making rules (varied schemas) differentiate or generate variety among the agents and the agents are also distinguished based on their geography" (Yang and Shan 2008). Yang and Shan (2008) also stated that the first book published in the twenty-first century, relevant to the CS field, was written by Sterman and titled *Business Dynamics: Systems Thinking and Modeling for a Complex World*. Analogous to the previous notion, Shakun (2001) stated that agents take action to achieve their goals. Moreover, Moss (2001) noted that agents self-organize until they reach more stable patterns of activity. While on the same subject, in 2001 Olson and Eoyang (2001) noted that agents evolve overtime in order to reach fitness. The same year also saw the formation of the Center for CS under the umbrella of the New National Institute of Physical Sciences at the Australian National University, as well as Jennings' creation of an agent-based approach for building complex software systems. Cilliers (2002) made the primary contribution to the field of CS by publishing his book *Complexity and Postmodernism: Understanding CS*. McDaniel et al. (2003) proposed that the interaction between the agents eventually over time leads to self-organization. Additionally, Chiva-Gomez (2003) referred to the work of Stacey and proposed that CASs that are closer to the edge of chaos will experience more self-organization (Yang and Shan 2008).

In 2003 and 2004, some authors believed that the interaction among the agents led to uncertainty. Thus, in 2003 Dent added that the purpose of such interactions is to find fit[ness]. Later, Harder et al. (2004) stated that the varied agents interact to maintain a system as opposed to creating a new one (homeostasis; see "Enterprises and Enterprise Systems Engineering"). Such phrases contend that the agents do not evolve or change, and rather, they enable their CASs to interact, change, and evolve (Yang and Shan 2008). Drawing on previous research, the following statement regarding the core of CASs could be derived: "Agents are the core of CASs, with schemas interacting overtime. The results of those interactions are maximized at the EOC [Edge of Chaos] and are subject to a fit test with the environment" (Yang and Shan 2008).

The year 2004 also experienced the establishment of the Complex System Society, which was launched at the European level on December 7 during the European Conference on CS at the Institute for Scientific Interchange (ISI) in Torino, Italy. In 2005, Grimm et al., published their contributions related to pattern-oriented modeling of agent-based CS in a journal article entitled, "Pattern Oriented Modeling of Agent Based CS: Lessons from Ecology" (Grimm et al. 2005).

The year 2006 was a memorable year from the perspective of advancements made in the CS domain. The contributions included Bar-Yam et al.'s book (2006) on complex engineered systems as well as journal articles by Holland (2006) and Schneider and Somer (2006) on studying complex adaptive systems and on understanding organizations as complex adaptive systems, respectively.

The year 2007 included an additional book on complex adaptive systems by Miller and Page (2007). Furthermore, the year 2007 also produced a book by Neil Johnson (2007) entitled *Simply Complexity: A Clear guide to Complexity Theory.*

MIT hosted the first systems thinking conference in October 2008. Yang and Shan also explained in 2008 that agents are the core of CASs, with schemas interacting overtime.

CURRENT PROSPECTS FOR FURTHER DEVELOPMENT OF COMPLEXITY IDEAS

Four of the more recent and notable contributions to the CS domain include a book entitled *Diversity and Complexity: Primers in Complex Systems*, by Page (2010), the first complex adaptive conference being hosted by Missouri University of Science and Technology in Chicago in November 2011, the 2011 book on ESE (Rebovich and White 2011) (see section "On Enterprises and ESE in *Enterprise Systems Engineering—Advances in the Theory and Practice*"), and D. D. Siljak's published *Decentralized Control of CS*, in 2012.

This book's author-editors have also contributed some ideas toward developing a complex project management body of knowledge (BoK) in 2012 (Ireland et al. 2012).

Referring back to Figure 1.4, there is certainly a high density of major works concerning CS and CSE. Although the cited publications have thinned somewhat in recent years, there is really no reason to believe they will cease to appear. We author-editors believe that seminal works introducing and describing CS will continue to be forthcoming. More importantly, other publications, emphasizing not only potential and experimental improvements in—but also validated methods for—CSE will increase. We believe the present book, containing many actual case studies that illuminate pathways to success in dealing CS, is one such resource.

A further aspect of the operation of CS could be the example of operating in a foreign culture, such as Afghanistan. CS are primarily those that are autonomous and independent, with a number of systems interlinked. The classic example of a well-accepted system of systems is that of Norman and Kuras (2006), which has been talked about in the section "On Enterprises and ESE in the Systems Engineering Body of Knowledge (SEBoK)" of this chapter and describes the operation of the Air Operations Center of the US DoD. In this case, there are 80 autonomous independent systems, all of which were developed for another purpose and all of which have a life in parallel, where they are contributing to other system of systems.

The parallel of a project operating in a foreign culture, such as Afghanistan, is very clear. The autonomous independent systems include language, legal systems, culture, religion, relationship between men and women, decision-making processes at the village level, and a number of others that people from the Western culture would not be aware of. Each of these systems almost exactly mirrors the case of Norman and Kuras. They operate actively outside the system of systems being considered, are autonomous and independent in the sense that they change without relationship to the project, as well as contribute to emergent properties, which again, is a key characteristic of CS.

The authors have included this aspect of CS in order to broaden the concept of a complex system or a SoS so that it can serve as a valuable contribution to the way people think about SoS.

COMPARATIVE ANALYSIS OF SYSTEM, SYSTEM OF SYSTEMS, ENTERPRISE, AND COMPLEX SYSTEM

As elucidated in the introduction to this chapter, a fundamental grasp of the concept of a system is the basis for understanding the more general notions of SoS, enterprise, and CS. More to the point, the case studies of this book have more to do with the SE of these entities. Something more than traditional or conventional SE is needed to deal effectively with SoSs, enterprises, and CS, that is, what is termed SoSE, ESE, and CS engineering (CSE), respectively.

ESE can be viewed as very difficult in engineering management, especially in trying to maintain the homeostasis property. In fact, ESE can be more challenging than trying to engineer a CS in the sense that one might give a CS more free rein to evolve on its own without trying to impose as many constraints or intervene as often as in an enterprise situation. Subsequently, if one wants to pictorially represent the relative intricacy of engineering systems, SoS, CS, and enterprises, respectively, Figure 1.5 is worth contemplating.

On the other hand, some CS are more general than enterprises, for example, if the CS in question has no homeostasis property, that is, truly CS are essentially unstable in that they are continuously evolving, and therefore, if there are relatively stable states, these are quite transient. For example, in contrast to the Ford Motor Company enterprise example given in the section "Enterprises and Enterprise Systems Engineering", the worldwide automobile industry, including new and used cars, the buying public, rules of the road, government emission standards, etc., is a CS that tends to continually evolve while rarely, if ever, reaching a stable state during any period longer than a year. If and when CS stabilize, they cease to evolve and essentially die. Thus, one might prefer the viewpoint of Figure 1.6 that depicts the engineering of a CS as the most difficult.

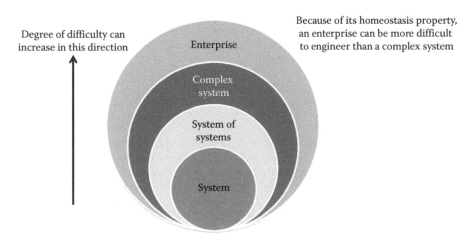

FIGURE 1.5 One view of the relative difficulty of engineering various types of systems. (Refer to Figure 1.6 and Chapter 3 for additional views.)

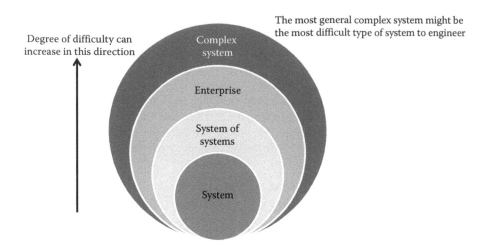

FIGURE 1.6   Alternative view of the relative difficulty of engineering various types of systems.

CONCLUSION

This chapter has introduced systems, SoS, enterprises, and CS and what it means to successfully engineer such systems. The historical perspective has provided a rich background as the basis for more comprehensive understanding of the following chapters of this book.

We encourage you, readers, to glean what you can that fits your own purposes in pursuing SoSE, ESE, and/or CSE, and ask that as you travel these paths, you also endeavor to contribute to the growing bodies of knowledge in this field through your teaching, mentoring, presentations, and/or publications.

REFERENCES

Abraham, R.H. 2002. The genesis of complexity. In Gregory Bateson and Alfonso Montuori, eds., *Mind and Nature: A Necessary Unity*. Part of the series Advances in Systems Theory, Complexity, and the Human Sciences. New York: Hampton Press. pp. 1–17. http://www.goodreads.com/book/show/277145. Mind_and_Nature.

Ackoff, R. 1971. Towards a system of systems concepts. *Management Science* 17(11): 661–672.

Ackoff, R. 2005. *Systems Thinking and Its Radical Implications for Management*. Boston, MA: IMS Lecture.

Ackoff, R.L. and F.E. Emery. 1972. *On Purposeful Systems*. London, U.K.: Tavistock Publications; Aldine.

Anderson, P. May–June 1999. *Perspective: Complexity Theory and Organization Science. Organization Science*, Vol. 10, Issue 3. Special Issue: Application of Complexity Theory to Organization Science, pp. 216–232. http://www.jstor.org/discover/10.2307/2640328?uid=3739696&uid=2&uid=4&uid=3739256 &sid=21102529026117. Accessed July 12, 2013.

Ashby, W.R. 1952. *Design for a Brain—The Origin of Adaptive Behavior*. 1st ed. New York: John Wiley & Sons, Inc; London, U.K.: Chapman & Hall Limited.

Ashby, W.R. 1956. *Introduction to Cybernetics*. London, U.K.: Chapman & Hall.

Ashby, W.R. 1960. *Design for a Brain—The Origin of Adaptive Behavior*, 2nd ed. London, U.K.: Chapman & Hall.

Ashby, W.R. 1962. In H. Von Foerster and G. W. Zopf, Jr., eds., *Principles of Self-Organization: Transactions of the University of Illinois Symposium*. London, U.K.: Pergamon Press, pp. 255–278.

Axelrod, R.M. 1984. *The Evolution of Cooperation*. New York: Basic Books.

Axelrod, R. and M.D. Cohen. 2000. *Harnessing Complexity: Organizational Implications of a Scientific Frontier*. New York: Basic Books.

Bar-Yam, Y. 1997. *Dynamics of Complex Systems (Studies in Nonlinearity)*, 1st ed. Boulder, CO: Westview Press. http://www.westviewpress.com/about.php.

Bar-Yam, Y., M.A. Allison, R. Batdorf, H. Chen, H. Generazio, H. Singh, and S. Tucker. 2004. The characteristics and emerging behaviors system of systems. NECSI: Complex Physical, Biological and Social Systems Project. http://necsi.edu/.

Bar-Yam, Y., D. Braha, A. Minai et al. 2006. In Y. Bar-Yam, D. Braha, and A. Minai, eds., *Complex Engineered Systems: Science Meets Technology (Understanding Complex Systems)*. New York: Springer Publication.

Boardman, J. and B. Sauser. 2006. System of systems—The meaning of OF. *IEEE International Systems of Systems Conference*. Los Angeles, CA.

Boulding, K.E. 1956. General systems theory—The skeleton of science. *Management Science* 2(3): 197.

Brehmer, B. 1992. The dynamic decision making: Human control of complex systems. *Acta Psychologica* 81(3): 211–241.

Broomhead, D.S. and D. Lowe. 1988. Multivariable functional interpolation and adaptive networks. *Complex Systems* 2: 321–355.

Business Agility. 2013. *Business Agility*. Wikipedia, The free encyclopedia. http://en.wikipedia.org/wiki/Business_agility. Accessed April 5, 2013.

Cantot, P. and D. Luzeaux, eds. 2011. *Simulation and Modeling of Systems of Systems*. New York: Wiley. http://www.amazon.com/Simulation-Modeling-Systems-ISTE/dp/1848212348/ref=sr_1_2?s=books&ie=UTF8&qid=1367273440&sr=1-2. Accessed March 16, 2014.

Carlock, P.G. and R.E. Fenton. 2001. System of systems (SoS) enterprise systems engineering for information-intensive organizations. *Systems Engineering* 4–4: 242–261.

Checkland, P. 1999. *Systems Thinking, Systems Practice*. New York: Wiley.

Chen, P. and J. Clothier. 2003. Advancing systems engineering for systems-of-systems challenges. *Systems Engineering* 6(3): 170–181.

Chiva-Gomez, R. 2003. The facilitating factors for organizational learning: Bringing ideas from complex adaptive systems. *Knowledge and Process Management* 10(2): 99–114.

Cilliers, P. 2002. *Complexity and Postmodernism Understanding Complex Systems*, 1st ed. New Fetter Lane, London, U.K.: Routledge.

COMPASS. 2011. http://www.compass-research.eu/approach.html. Accessed March 16, 2014.

2013. *Complex Adaptive Systems Conference*. Baltimore, MD. http://complexsystems.mst.edu/. Accessed July 12, 2013

Cook, S.C. 2001. On the acquisition of systems of systems. *INCOSE Annual Symposium*. Melbourne, Australia.

CSD&M. 2013a. *Complex Adaptive System Conference 2013*, Baltimore, MD. Accessed July 12, 2013.

CSD&M. 2013b. *Complex Adaptive Systems Conference 2013*, Baltimore, MD. Accessed November 13–15, 2013.

CSD&M. 2013. *Complex Systems Design & Management Conference*. Paris, France. http://www.csdm2013.csdm.fr/

Daneke, G.A. 1 October 1999. *Systemic Choices: Nonlinear Dynamics and Practical Management*. Ann Arbor, MI: The University of Michigan Press.

DANSE. 2011. http://www.danse-ip.eu/home/. Accessed March 16, 2014.

Davidson, M. 1983. *Uncommon Sense: The Life and Thought of Ludwig von Bertalanffy (1901–1972), Father of General Systems Theory*. Los Angeles, CA: J.P. Tarcher.

DAU. 2004. *Defense Acquisition Guidebook (DAG)*. Defense Acquisition University. Fort Belvoir, Virginia.

DeLaurentis, D. 2005. Understanding transportation as a system of systems design problem. *43rd AIAA Aerospace Science Meeting*. Reno, NV.

DeLaurentis, D., C. Dickerson, M. DiMario et al. September 2007. A case for an international consortium on system-of-systems engineering. *IEEE Systems Journal* 1–1: 68–73.

Department of Defense. 2001. Army software blocking policy: Version 11.4E. DoD. Arlington, VA.

DoD. 2005. Joint capabilities integration and development instructions. Department of Defense. Arlington, VA.

DoD. 2007. System of systems, systems engineering guide: Considerations for systems engineering in system of systems environment. U. S. Department of Defense. Arlington, VA.

Dyer, L. and J. Ericksen. 2009. Complexity-based agile enterprises: Putting self-organizing emergence to work. In A. Wilkinson et al., eds., *The Sage Handbook of Human Resource Management*. London, U.K.: Sage, pp. 436–457.

Drumm, H.J. December, 1995. The paradigm of a new decentralization: Its implications for organization and HRM. *Employee Relations* 17(8): 29–46.

Dyson, F.J. 1962. Statistical theory of the energy levels of complex systems I. *Journal of Mathematical Physics* Vol. 3, No. 1, pp. 140–156.

Eisner, E. 1993. RCASSE: Rapid computer-aided systems of systems (S2) engineering. In *Proceedings 3rd International Symposium of the National Council on Systems Engineering* [now *International Council on Systems Engineering (INCOSE)*]. pp. 267–273.

Eisner, H., J. Marciniak, and R. McMillan. 1991. Computer-aided system of systems (C2) engineering. In *IEEE International Conference on Systems, Man, and Cybernetics*. Charlottesville, VA.

Epstein, J.M. and I. Axtell. 1996. *Growing Artificial Societies: Social Science from the Bottom Up*. Cambridge, MA: The MIT Press.

Extended Enterprise. 2013. Extendedenterprise. http://www.businessdictionary.com. Accessed July 11, 2013.

Fan, L.T. and T.C. Chen. 1966. *Continuous Maximum Principle: A Study of Complex Systems Optimization.* New York: Wiley.

FEAF. 1999. *Federal Enterprise Architecture Framework (FEAF).* Washington, DC: Chief Information Officer (CIO) Council.

Forrester, J.W. 1968. *Principles of Systems; Text and Workbook.* Cambridge, MA: Wright-Allen Press.

Gell-Mann, M. 1994. *The Quark and the Jaguar: Adventures in the Simple and the Complex.* New York: Holt & Co.

Gharajedaghi, J. 2005. *Systems Thinking, Second Edition: Managing Chaos and Complexity: A Platform for Designing Business Architecture*, 2nd ed. Boston, MA: Butterworth Heinemann. 1999. http://www.butterworth.heinemann.co.uk/. Accessed March 16, 2014.

Giachetti, R.E. 2010. *Design of Enterprise Systems: Theory, Architecture, and Methods.* Boca Raton, FL: CRC Press; Taylor & Francis Group.

Gladwell, M. 2000. *The Tipping Point—How Little Things Can Make a Big Difference.* New York: Little, Brown and Company.

Goldstine, H.H. 1972. *The Computer from Pascal to von Neumann.* Princeton, NJ: Princeton University Press.

Gorod, A. 2009. System of systems engineering management framework: A "satisficing" network approach. PhD Dissertation, Stevens Institute of Technology, Hoboken, NJ.

Gorod, A., M. DiMario, B. Sauser, and J. Boardman. 2009. "Satisficing" system of systems (SoS) using dynamic and static doctrines. *International Journal of SoSE* 1(3): 347–366.

Gorod, A., B. Sauser, and J. Boardman. 2008. System-of-Systems engineering management: A review of modern history and a path forward. *IEEE Systems Journal* 2(4): 484–499. http://ieeexplore.ieee.org/xpl/login.jsp?tp=&arnumber=4682611&url=http%3A%2F%2Fieeexplore.ieee.org%2Fxpls%2Fabs_all.jsp%3Farnumber%3D4682611. Accessed March 16, 2014.

Grimm, V. et al. 2005. Pattern-Oriented Modeling of Agent-Based Complex Systems: Lessons from Ecology. *Science* 310(5750): 987–991. http://www.sciencemag.org/content/310/5750/987.full. Accessed September 21, 2013.

Haken, H. 1988. *Information and Self-Organization: A Macroscopic Approach to Complex Systems*, 1st ed. Berlin, Germany: Springer.

Haken, H. 2010. *Information and Self-Organization: A Macroscopic Approach to Complex Systems.* Berlin, Germany: Springer.

Handy, C. May–June, 1995. Trust and the virtual organization: How do you manage people whom you do not see? *Harvard Business Review* 70: 40–50.

Harder, J., P.J. Robertson, and H. Woodward. 2004. The spirit of the new workplace: Breathing life into organizations. *Organization Development Journal* 22(2): 79–103.

Hardi, G. 1968. The tragedy of the commons. *Science* 162(3859): 1243–1248.

Heims, S.J. 1991. *The Cybernetic Group.* Cambridge, MA: MIT Press.

Heylighen, F. 2001. The science of self-organization and adaptivity, center "Leo Apostel". Brussels, Belgium: Free University of Brussels. http://social3.org/t/the-science-of-self—organization-and-adaptivity-e4980.c. Accessed October 5, 2013.

Heylighen, F., P. Cilliers, and C. Gershenson. 2007. Complexity and philosophy. In J. Bogg and R. Geyer, eds., *Complexity, Science and Society.* Oxford, U.K.: Radcliffe Publishing. http://arxiv.org/ftp/cs/papers/0604/0604072.pdf. http://arxiv.org/abs/cs/0604072. Accessed July 12, 2013.

Hock, D. 1999. *Birth of the Chaordic Age.* San Francisco, CA: Berrett-Koehler Publishers, Inc.

Holland, J.H. 1992. *Adaptation in the Natural and Artificial Systems.* Cambridge, MA: MIT Press.

Holland, J.H. 1996. *Hidden Order: How Adaptation Builds Complexity.* Reading, MA: Addison-Wesley.

Holland, J.H. 2006. Studying complex adaptive systems. *Journal of Systems Science and Complexity* 19(1): 1–8.

INCOSE, Haskins, C., ed. 2010. *Systems Engineering Handbook: A Guide for System Life Cycle Processes and Activities.* INCOSE-TP-2003-002-03.2. *INCOSE Systems Engineering Handbook*, Version 3.2. International Council on Systems Engineering (INCOSE), 7670 Opportunity Road, Suite 220, San Diego, CA, pp. 92111–92222.

International Society for Systems Sciences. 1988. http://isss.org/world/about-the-isss. Accessed October 5, 2013.

Ireland, V., A. Gorod, B.E. White, S.J. Gandhi, and B. Sauser. 2012. A contribution to developing a complex project management BoK. *Project Perspectives.* XXXV: 16–25. The Annual Publication of the International Project Management Association.

Ireland, V., A. Gorod, B. White, J. Gandhi, and B. Sauser. 2013. A contribution to developing a complex project management BOK. *Project Perspectives.* XXXV: 16–25. The Annual Publication of International Project Management Association.

Ivakhneko, A.G. October 1971. Polynomial theory of complex systems. *IEEE Transaction on Systems. Man and Cybernetics* 1(4): 364–378.

Jackson, M.C. and P. Keys. 1984. Towards a system of systems methodologies. *Journal of the Operational Research Society* 35(6): 473–486.

Jacob, F. 1974. *The Logic of Living Systems*. London, U.K.: Allen Lane.

Jamshidi, M. 2005. *System of Systems Engineering—A Definition*. Piscataway, NJ: IEEE SMC.

Jamshidi, M. ed. 2008a. *System of System Engineering—Innovations for the 21st Century*. Hoboken, NJ: Wiley.

Jamshidi, M. ed. 2008b. *System of Systems—Principles and Applications*. Boca Raton, FL: Taylor & Francis Group.

Jamshidi, M. ed. 2008/10. *System of System Engineering—Innovations for the 21st Century*. Hoboken, NJ: Wiley.

Jamshidi, M. ed. 2008/11. *System of Systems—Principles and Applications*. Boca Raton, FL: Taylor & Francis.

Jennings, N.R. and M. Wooldridge. 1995. Applying agent technology. *Applied Artificial Intelligence* 9(4): 357–369.

Johansson, F. 2004. *The Medici Effect*. Boston, MA: Harvard Business School Press.

Johnson, N.F. 2007. *Simply Complexity: A Clear Guide to Complexity Theory*. Oxford, U.K.: Oneworld Publications. http://www.amazon.com/Simply-Complexity-Clear-Guide-Theory/dp/1851686304/ref=sr_1_11?s=books&ie=UTF8&qid=1373835469&sr=1-11&keywords=complex+systems.

Jong, D. and K. Alan. 1975. Analysis of the behavior of a class of genetic adaptive systems. Computer and Communication Science Department. College of Science. The University of Michigan. Ann Arbor, MI.

Kauffman, S. 2000. *Investigations*. New York: Oxford University Press.

Kauffman, S.A. 1985. Self-organization, selective adaptation, and its limits: A new pattern of inference in evolution and development. In D.J. Depew and B.H. Weber, eds., *Evolution at a Crossroads*. Cambridge, MA: MIT Press, pp. 169–207.

Kauffman, S.A. 1995. *At Home in the Universe: The Search for Laws of Self-Organization and Complexity*. Oxford, U.K.: Oxford University Press.

Keating, C., R. Rogers, R. Unal et al. 2003. Systems of systems engineering. *Engineering Management Journal* 15(3): 36–45.

Knowles, R.N. 2002. *Leadership Dance—Pathways to Extraordinary Organizational Effectiveness*, 3rd ed. Niagara Falls, NY: The Center for Self-Organizing Leadership.

Koch, C. and G. Laurent. April 2, 1999. Complexity and the nervous system. *Science* 284: 96–98. http://www.sciencemag.org.

Korsten, P. and C. Seider. March 16, 2010. Building a smarter planet—A smarter planet blog. http://asmarterplanet.com/blog/2010/03/the-internet-of-things.html. http://public.dhe.ibm.com/common/ssi/ecm/en/gbe03278usen/GBE03278USEN.PDF. Accessed March 16, 2014.

Kotov, V. 1997. Systems of systems as communicating structures. Hewlett Packard, HPL-97-124. http://www.hpl.hp.com/techreports/97/HPL-97-124.html.

Kovacic, S.F. and A. Sousa-Poza, eds. 2013. *Managing and Engineering in Complex Situations (Topics in Safety, Risk, Reliability and Quality)*. New York: Springer. http://www.amazon.com/Managing-Engineering-Complex-Situations-Reliability/dp/9400755147/ref=sr_1_12?s=books&ie=UTF8&qid=1367272828&sr=1-12&keywords=system+of+systems+engineering+sose. Accessed March 16, 2014.

Kramer, R., M. Brewer, and B. Hanna. 1996. Collective trust and collective action: The decision to trust as a social decision. Chapter 17. In R.M. Kramer and T.R. Tyler, eds., *Trust in Organizations*. Thousands Oaks, CA: Sage. http://knowledge.sagepub.com/view/trust-in-organizations/n17.xml.

Lane, J.A. and R. Valerdi. 2005. Synthesizing SoS concepts for use in cost estimation. In *IEEE International Conference on Systems, Man, and Cybernetics*. Waikoloa, HI.

Langton, C.G. 1990, Computation at the edge of chaos: Phase transitions and emergent computation. *Physica D* 42(1–3 June): 12–37.

Laszlo, A. and S. Krippner. 1998. Systems theories: Their origins, foundations, and development. In J.S. Jordan, ed., *Systems Theories and A Priori Aspects of Perception*. Amsterdam, the Netherlands: Elsevier. *Science*, Ch. 3, pp. 47–74. http://archive.syntonyquest.org/elcTree/resourcesPDFs/SystemsTheory.pdf. Accessed July 12, 2013.

Levin, S.A. 1998. Ecosystem and the biosphere as complex adaptive systems. *Ecosystem* 1(5): 431–436.

Lewis, R. 1994. From chaos to complexity: Implications for organizations. *Executive Development* 7: 16–17.

Li, M. and P. Vitanyi. 1997. *An Introduction to Kolmogorov Complexity and Its Applications*, 2nd ed. New York: Springer-Verlag.

Luskasik, S.J. 1998. System, systems of systems, and the education of engineers. Artificial Intelligence Engineering Design, Analysis, and Manufacturing 12(1): 55–60.

Luzeaux, D. and J.R. Ruault. 2010. *Systems of Systems*. New York: ISTE Limited and John Wiley & Sons Inc.

Luzeaux, D., J.R. Ruault, and J.L. Wippler. 2011. *Complex System and Systems of Systems Engineering*. ISTE Limited and John Wiley & Sons Inc. http://www.amazon.com/Complex-Systems-Engineering-ISTE/dp/1848212534/ref=sr_1_7?s=books&ie=UTF8&qid=1367273400&sr=1-7&keywords=%22system+of+systems+modeling%22. Accessed July 12, 2013.

McDaniel, R., M.E. Jordan, and B.F. Fleeman. 2003. Surprise, surprise, surprise! A complexity science view of the unexpected. *Health Care Management Review* 28(3): 266–277.

Maier, M. 1996. Architecting principles of systems-of-systems. In *Sixth Annual International Symposium of the International Council on Systems Engineering*. Boston, MA.

Maier, M. 1998. Architecting principles for system-of-systems. *Systems Engineering* 1(4): 267–284.

Maier, M.W. and E. Rechtin. 2009. Appendix A: Heuristics for systems-level architecting. In *The Art of Systems Architecting*, 3rd ed. Boca Raton, FL: CRC Press; Tayor & Francis Group.

Manthorpe, W.H. 1996. The emerging joint system of system: A systems engineering challenge and opportunity for APL. *John Hopkins APL Technical Digest* 17(3): 305–310.

March, J. and H. Simon. 1958. *Organizations*. New York: John Wiley & Sons.

Martin, J.N., B.E. White et al. November 30, 2012. Enterprise systems engineering. http://www.sebokwiki.org/1.0.1/index.php?title = Enterprise_Systems_Engineering. In A. Pyster et al., eds., *Guide to the Systems Engineering Body of Knowledge (SEBoK) v. 1.0.1*. http://www.sebokwiki.org/1.0.1/index.php?title=Main_Page. Accessed April 5, 2013.

Maturana, H.R. and F.J. Varela. 1992. *The Tree of Knowledge: The Biological Roots of Human Understanding*, R. Paolucci, ed. and Trans. Boston, MA: Shambhala. pp. 17–55.

Miller, J.H. and S.E. Page. 2007. *Complex Adaptive Systems: An Introduction to Computational Models of Social Life* (Princeton Studies in Complexity). Princeton, NJ: Princeton University Press.

Mittal, S. and J.L. Risco Martín. 2013. Netcentric system of systems engineering with DEVS unified process. *Netcentric System of Systems Engineering with DEVS Unified Process*. Boca Raton, FL: CRC Press; Taylor & Francis Group. http://www.crcpress.com/product/isbn/9781439827062

MODAF. 2004. *Ministry of Defence Architecture Framework (MODAF)* (Version 2). London, U.K.: Ministry of Defence.

Moss, M. 2001. Sensemaking, complexity and organizational knowledge. *Knowledge and Process Management* 8(4): 217–232.

Nanayakkara, T., M. Jamshidi, and F. Sahin. 2010. *Intelligent Control Systems with an Introduction to System of Systems Engineering*. Boca Raton, FL: CRC Press; Taylor & Francis Group. http://www.amazon.com/Intelligent-Control-Systems-Introduction-Engineering/dp/1420079247/ref=sr_1_7?s=books&ie=UTF8&qid=1367272796&sr=1-7&keywords=system+of+systems+engineering+sose. Accessed July 12, 2013.

NECSI. 1997. New England Complex Systems Institute. http://necsi.edu/html/iccs.html. Accessed August 22, 2013.

Norman, D.O. and M.L. Kuras. 2006. Engineering complex systems. Chapter. 10. In D. Braha, A. Minai, and Y. Bar-Yam, eds., *Complex Engineered Systems—Science Meets Technology*. New England Complex Systems Institute. Cambridge, MA: Springer. pp. 206–245.

Oberndorf, P.A. and A. Earl. 1998. *Department of Veterans Affairs Reference Models*. SEI Carnegie Mellon University. Pittsburgh, PA.

Olson, E.E. and G.H. Eoyang. 2001. *Facilitating Organizational Change: Lessons from Complexity Science*. San Francisco, CA: Jossey-Bass/Pfeiffer.

Owens, W.A. 1995. The emerging U.S. system of systems. In S. Johnson and M. Libicki, eds., *Dominant Battlespace Knowledge*. Washington, DC: NDU Press. pp. ii–viii.

Page, S.E. 2010. *Diversity and Complexity (Primers in Complex Systems)*. Princeton, NJ: Princeton University Press. http://books.google.com/books?hl=en&lr=&id=Mi6zkXss14IC&oi=fnd&pg=PP2&dq=Scott+E.+Page,+2010,+Diversity+and+Complexity+(Primers+in+Complex+Systems),+Princeton+University+Press&ots=vdM1V8otzP&sig=hVB8C7AVJhLXqtanr6y93F_c_YQ#v=onepage&q&f=false. Accessed September 22, 2013.

Pascale, R.T. 1999. Surfing the edge of chaos. *Sloan Management Review* 40(3): 83–94.

Pei, R. 2000. System of systems integration (SoSI)—Smart way of acquiring army C412WS systems. In *Summer Computer Simulation Conference*, Vancouver, Bristish Columbia, Canada. pp. 574–579.

PLANET. 2011. http://www.planet-ict.eu/. Accessed March 16, 2014.

Puccia, C.J. and R. Levins. 1986. *Qualitative Modeling of Complex Systems: An Introduction to Loop Analysis and Time Averaging*. Cambridge, MA: Harvard University Press.

Rebovich, G. Jr. and B.E. White eds. 2011. *Enterprise Systems Engineering—Advances in the Theory and Practice*. Complex and Enterprise Systems Engineering Series. Boca Raton, FL: CRC Press. An Auerbach Book, Taylor & Francis Group.

Rechtin, E. December 1991. *Systems Architecting: Creating and Building Complex Systems*, 1st ed. Englewood Cliffs, NJ: Prentice Hall.

Road2SoS. 2011. http://www.road2sos-project.eu/cms/front_content.php. Accessed March 16, 2014.

Ronald, E.M.A. 1999. Design, observation, surprise!—A test of emergence. *Artificial Life* 5: 225–239.

Rosenthal, H.E. 1967. A graphic method for the determination and presentation of binding parameters in a complex system. *Analytical Biochemistry* 20(3): 525–532.

Ross, J.W., P. Weill, and D. Robertson. 2006. *Enterprise Architecture as Strategy: Creating a Foundation for Business Execution*. Boston, MA: Harvard Business Review Press.

Rouse, W.B. 2005. Enterprise as systems: Essential challenges and enterprise transformation. *Systems Engineering, the Journal of the International Council on Systems Engineering (INCOSE)* 8(2): 138–150.

Rouse, W.B. 2009. Engineering the enterprise as a system. In A.P. Sage and W.B. Rouse, eds., *Handbook of Systems Engineering and Management*, 2nd ed. New York: Wiley & Sons, Inc.

Russel, S.J. and P. Norving. 1995. *Artificial Intelligence: A Modern Approach*. Upper Saddle River, NJ: Prentice Hall.

Sage, A.P. and C.D. Cuppan. 2001. On the systems engineering and management of systems of systems and federations of systems. *Information, Knowledge Systems Management* 2: 325–345.

Schneider, M. and M. Somers. August 2006. Organizations as complex adaptive systems: Implications of complexity theory for leadership research. *The Leadership Quarterly* 17(4): 351–365.

Schwaninger, M. September/October 2006. System dynamics and the evolution of the systems movement. *Systems Research and Behavioral Science*, Vol. 23, Issue 5. John Wiley & Sons, Ltd. Special Issue: ISSS Yearbook: The Potential Impacts of Systemics on Society, pp. 577–711.

SE Guide. 2013. Enterprise engineering. In *Systems Engineering Guide*. The MITRE Corporation Bedford, MA. http://www.mitre.org/work/systems_engineering/guide/enterprise_engineering/. Accessed April 5, 2013.

Senge, P.M.F. 1996. The ecology of leadership. (Copyrighted) Leader to leader, No. 2. Chestnut Hill, MA. http://www.pfdf.org/leaderbooks/121/fall96/senge.htm.

Senge, P.M. and J.D. Sterman. 1990. Systems thinking and organizational learning: Acting locally and thinking globally in the organization of the future. In *Proceedings of the 1990 System Dynamics Conference*.

Shakun, M.F. 2001. Unbounded rationality. *Group Decision and Negotiation* 10(2): 97–118.

Shenhar, A. 1994. A new systems engineering taxonomy. In *Proceedings 4th International Symposium of the National Council on Systems Engineering* [now *International Council on Systems Engineering (INCOSE)*]. pp. 261–276.

Shenhar, A. 2001. One size does not fit all projects: Exploring classical con- tingensy domains. *Management Science* 47(3): 394–414.

Shenhar, A. and B. Sauser. 2008. Systems engineering management: The multidisciplinary discipline. *Handbook of Systems Engineering and Management*, 2nd ed. New York: Wiley.

Siljak, D.D. 2012. *Decentralized Control of Complex Systems*. New York: Dover Publications.

Simon, H.A. 1956. Rational choice and the structure of the environment. *Psychological Review* 63(2): 129–138. doi:10.1037/h0042769.

Simon, H.A. 1962. The architecture of complexity. *Proceedings of the American Philosophical Society* 106(6): 467–482.

Simon, H. 1977. The organization of complex systems. *Boston Studies in the Philosophy of Science* 54: 245–261.

SoSE&I. 2013. System of systems engineering and integration directorate. http://www.bctmod.army.mil/SoSE_/sose_i.html. Accessed July 12, 2013.

Stacey, R. January 15, 1996. *Complexity and Creativity in Organizations,* 1st ed. San Francisco, CA: Berrett-Koehler Publishers Inc.

Stacey, R. 2006. The science of complexity: An alternative perspective for strategic change processes. In R. MacIntosh et al., eds., *Complexity and Organization: Readings and Conversations*. London, U.K.: Routledge, pp. 74–100.

Sterman, J.D. February 2000. *Business Dynamics: Systems Thinking and Modeling for a Complex World with CD-ROM*. New York: McGraw-Hill/Irwin.

Stevens, R. September 24, 2009. *Acquisition Strategies to Address Uncertainty: Acquisition Research Findings*. From MITRE-Sponsored Research, Enterprise Systems Acquisition Using Venture Capital Concepts. The MITRE Corporation. Bedford, MA.

Steward, D.V. 1981. The design structure system: A method for managing the design of complex systems. EEE Transactions on Engineering Management 28(3): 71–74.

Swarz, R.S., J.K. DeRosa, and G. Rebovich. July 9–13, 2006. An enterprise systems engineering model. In *INCOSE International Symposium*, Orlando, FL.

TAFIM. April 1996. Technical architecture framework for information management: 1 and 4. U.S. Department of Defense. TAFIM (1996): Arlington, VA.

TAFIM TRM. April 30, 1996. TAFIM, Vol. 2: Technical Reference Model, Defense Information Systems Agency Center for Standards, Department of Defense Technical Architecture Framework for Information Management (TAFIM), Version 3.0. http://www.everyspec.com/DoD/DOD-General/DISA_TAFIM_VOL2_7540/. Accessed March 16, 2014.

Tang, W. 2008. Simulating adaptive geographic systems: A geographically aware intelligent agent approach. *Cartography and Geographic Information Science* 35(4): 239–263.

T-AREA-SoS. 2011. Trans-Atlantic research and education agenda on system of systems. https://www.tareasos.eu/details.php. Accessed March 16, 2014.

Thelen, E. and L.B. Smith. 2006. Dynamic System Theories. Chapter 6. In W. Damon and R.M. Lerner, eds., *Handbook of Child Psychology*, Vol. 1, *Theoretical Models of Human Development*, 6th ed., pp. 258–312. New York: Wiley. ISBN: 978-0-471-27287-8. http://www.wiley.com/WileyCDA/WileyTitle/productCd-0471272876.html.

TOGAF (TheOpenGroupArchitectureFramework). 2009. Welcome to TOGAF® Version 9.1 ["Enterprise Edition"], an Open Group Standard. http://pubs.opengroup.org/architecture/togaf9-doc/arch/. Accessed July 11, 2013.

TOGAF. 2013a. TOGAF 9.1 White Paper An Introduction to TOGAF, Version 9.1 http://www.opengroup.org/togaf/. Accessed 16 March 2014.

TOGAF. 2013b. TOGAF8.1.1Online > PartIII:EnterpriseContinuum > Foundation Architecture: Technical Reference Model. http://pubs.opengroup.org/architecture/togaf8-doc/arch/chap19.html.

Various authors. 2000. Industrial automation systems—Requirements for enterprise-reference architectures and methodologies. Geneva, Switzerland: International Organization for Standardization (ISO). ISO 15704:2000.

Various authors. 2013. Body of Knowledge and Curriculum to Advance Systems Engineering (BKCASE), Including the Systems Engineering Body of Knowledge (SEBoK) and the "Graduate Reference Curriculum for Systems Engineering (GRCSE) ["Gracie"]. http://www.bkcase.org/. Accessed July 11, 2013.

von Bertalanffy, L. 1969. *General Systems Theory*. New York: George Braziller.

Von Foerster, H. 1961. *Industrial Dynamics*. Cambridge, MA: MIT Press.

Weber, B.H. January 1, 1998. Origins of order in dynamical models. *Biology and Philosophy* 13(1): 133–144.

Weinberg, G.M. 1975. *An Introduction to General Systems Thinking*. New York: Dorset House.

Wendy, L.C. and B. Galliers, eds. Systems thinking. *Rethinking Management Information Systems*, pp. 45–57. Oxford, U.K.: Oxford University Press.

White, B.E. 2007. *Systems Engineering Lexicon*. Taylor & Francis Complex and Enterprise Systems Engineering Book Series web site. http://www.cau-ses.net/html/lexicon.html. Accessed February 28, 2014.

White, B.E. May 12–15, 2008. Complex adaptive systems engineering. MITRE Public Release Case No. 08-1459. In *Eighth Understanding Complex Systems Symposium*. University of Illinois at Urbana-Champaign, Champaign, IL.

White, B.E. March 23–26, 2009. Complex adaptive systems engineering (CASE). In *Third Annual IEEE International Systems Conference*. Vancouver, British Columbia, Canada, pp. 23–26.

White, B.E. July 15, 2010a. INCOSE is Challenged! INCOSE Symposium 2010 Panel. Architecting the Enterprise: Is INCOSE up to the challenge? Organizer and Proposer: James Martin. *INCOSE Symposium*, Chicago, IL.

White, B.E. December 2010b. Complex Adaptive Systems Engineering (CASE). *IEEE Aerospace and Electronic Systems Magazine* 25(12): 16–22. ISSN 0885-8985.

Wiener, N. 1948. *Cybernetics: Or Control and Communication in the Animal and the Machine*. Cambridge, MA: MIT Press.

Wolstenholme, E. F. October 1990. *System Enquiry: A System Dynamics Approach*. Somerset, NJ: John Wiley & Sons, Inc.

Yang, A. and Y. Shan, eds. 2008. *Intelligent Complex Adaptive Systems*. Hershey, PA: IGI Publications. http://trove.nla.gov.au/version/9255774. Accessed July 12, 2013.

Zachman, J.A. 1987. A framework for information systems architecture. *IBM Systems Journal* 26(3). IBM Publication G321-5298. pp. 276–292.

Zachman International Enterprise Architecture (web site). 2013. http://www.zachman.com/about-the-zachman-framework. Accessed July 11, 2013.

Zadeh, L.A. January 1973. Outline of a new approach to the analysis of complex systems and decision processes. *IEEE Transactions on Systems, Man and Cybernetics* 3(1): 28–44.

Zeigler, B.P. and Sarjoughian, H.S. 2012. *Guide to Modeling and Simulation of Systems of Systems*. New York: Springer.

Zeigler, B.P. and H.S. Sarjoughian. 2013. *Guide to Modeling and Simulation of Systems of Systems (Simulation Foundations, Methods and Applications)*. London, U.K.: Springer-Verlag. http://www.amazon.com/Modeling-Simulation-Systems-Foundations-Applications/dp/085729864X/ref=sr_1_154?s=books&ie=UTF8&qid=1367272415&sr=1-154&keywords=%22system+of+systems%22. Accessed 16 March 2014.

2 Relevant Aspects of Complex Systems from Complexity Theory

Vernon Ireland, Brian E. White, S. Jimmy Gandhi, Brian Sauser, and Alex Gorod

CONTENTS

The purpose of this chapter is to address key aspects of complexity from both a theoretical and a practical viewpoint and to identify whether these are relevant to system of systems (SoS) projects. Complex projects will be recognized as part of a hierarchy of simple, complicated, and complex. Complex projects will then be separated into four distinct (but possibly overlapping) types:

COMPLEX PROJECT TYPE

- Type A: The project was essentially integrating existing assets to form a larger system (e.g., the management and control of rivers or road systems by states or food distribution from restaurants to needy users or power networks).
- Type B: A traditional SoS in which one or more existing asset(s), built for another purpose, as described by Dahmann (2009) and Norman and Kuras (2006), usually software based, is(are) included in a project because of the benefit(s) gained.

- Type C: A project operating in a very different culture where previous experience counts for little.
- Type D: Complex projects that require addressing aspects of systems thinking, such as definition of stakeholders, definition of boundaries, and use of Checkland's soft system methodology or system dynamics to identify a solution.

However, before addressing aspects of complexity, we will briefly outline traditional reductionist projects, such as those managed by the Project Management Body of Knowledge (*PMBOK®*) (PMI 2013), Integrating Project Management Standards (IPMA)'s Competence Baseline, and other bodies of knowledge (BOKS) and classical, traditional, or conventional systems engineering (SE) methodologies.

DESCRIPTION OF COMPLEX SYSTEM

Traditional complicated projects such as those managed by *PMBOK*, IPMA's Competence Baseline, and other BOKS and SE as usual are managed under a reductionist framework. Typical reductionist style projects are those in which a top-down command and control structure is effective. Complicated projects, such as space rockets and the man on the moon project, which are primarily governed by the laws or rules of physics, chemistry, engineering, and mathematics, are ideal as complicated projects. The underlying principle here is that knowledge can be partitioned into its elements and these subsets can be addressed separately. We can sometimes explain systems (e.g., the planets around the sun) by reducing the system to simple equations. Project management BOKS are largely based on reductionist principles on the assumption that relationships are fairly linear, the future follows the past, and subsystems have very limited interaction with each other. The notion of work breakdown structure and separately assessing schedule and cost breakdown are examples of this.

Support for the distinction between complicated and complex is provided by Cotsaftis (2007), Glouberman and Zimmerman (2002), and Snowden and Boone (2007).

Furthermore, sponsors of projects attempt to package a new project so that there will be very limited increases in scope, and limited changes to the project, such as cost increases due to cost performance index (CPI), can be handled as change orders. Projects in which there are significant changes to scope, such as the introduction of a common transport ticketing system between rail bus and ferries, in which the organizations had not agreed to either join with each other, or the terms of the joint ticket, are almost certain to end in disaster.

Boardman and Sauser (2008) describe SoS as having the characteristics of

- Autonomy and independence
- Belonging
- Connectivity
- Diversity
- Emergence

From these descriptors, one would think that each system in the SoS is autonomous and independent while belonging to the SoS and is connected to other systems of the SoS. Furthermore, the collection of systems in the SoS is diverse in some sense and can generate unexpected emergent properties. But these concepts require qualification: for example, in a traditional SoS, in which existing assets have been included, these systems were initially independent but are now part of the SoS by engaging a particular version that is not necessarily free to change if that degrades the performance of the SoS. Belonging also needs clarification: does an adult person belong to their parent's family? The answer is both yes and no, thus requiring qualification.

An alternative approach is provided by Jackson (2003, p. 138), who describes complex systems as those that are "interconnected and complicated further by lack of clarity about purposes, conflict, and uncertainty about the environment and social constraints. In tackling wicked problems, problem structuring assumes greater importance and problem solving using conventional techniques often does not work. If problem formulation is ignored or badly handled, managers may end up solving, very thoroughly and precisely, the wrong problem." Mason and Mitroff (1981) added that such ill-structured problem situations are made up of highly interdependent problems.

A further description is provided by Anne-Marie Grisogono (2006), of the Defence Science and Technology Organisation (DSTO) Australia, who described complexity in the following terms:

1. *Causality is complex and networked*: That is, simple cause and effect relationships don't apply—there are many contributing causes and influences to one outcome, and conversely, one action may lead to a multiplicity of consequences and effects.
2. *The number of plausible options is vast*: It is not plausible to optimize.
3. *System behavior is coherent*: There are coping patterns and trends.
4. *The system is not fixed*: The patterns and trends vary, for example, the *rules* seem to be changing—something that worked yesterday may not do so tomorrow.
5. *Predictability is reduced*: For a given action option, it is not possible to accurately predict all its consequences, or for a desired set of outcomes, it is not possible to determine precisely which actions will produce it.

DeRosa et al. (2008) provide a fourth definition of complexity:

1. The situation cannot be unambiguously bounded since there are always significant interactions with elements of the wider context, and some of these may be changing at a rate comparable to that of the situation itself. Moreover, some long latency processes may appear insignificant within the situation but ultimately produce serious consequences.
2. Both the situation and the wider context contain entities (e.g., people, groups, systems) that act in their own interest and react to support or oppose every intervention in the problem, in ways that cannot be precisely predicted (e.g., counterinsurgency warfare, global business operations, web applications).
3. The propagation lengths of disturbances may span the entire situation and its wider context—that is, local changes can have global effects (e.g., the stock market, power grids). As a corollary, the impact of local interventions must be evaluated in the global context (e.g., joint and coalition warfare, network centric operations, food networks, team sports).
4. Most seriously, the number of possible *solutions* grows at least exponentially with the number of entities in the situation creating a huge possibility space that cannot be prestated or analyzed in any compact way (e.g., asset-to-task problem, software assurance, system design).

DeRosa et al. (2008) pick up the issues of a difference between complicated and complex, pointing out that the root of the word complicated means to fold whereas the root of the word complex means to weave. Snowden and Boone (2007) echo this distinction.

DeRosa et al. (2008) concluded that complex systems require "*self-organisation* that is structures, patterns of behaviour through dynamic system processes in the absence of any external controller or direct design" and then add that self-organization is ubiquitous in complex systems (2008, p. 4). They further add that *adaptation* covers "all the various processes that result in complex adaptive systems changing their behaviour, structure and function in ways that improve their success in their environment." Following these changes, the effects are evaluated and the changes are accepted or eliminated.

They comment that interdependence implies that reduction by decomposition cannot work, because the behaviors of each component depend on the behaviors of the others. Writing from a military background, they add four further elements to the list of aspects of complexity:

Complexity builds on and enriches systems theory by articulating additional characteristics of complex systems and by emphasising their inter-relationship and interdependence.

Mitleton-Kelly (2003, p. 3)

Complexity is not a methodology or a set of tools (although it does provide both). It certainly is not a 'management fad'. The theories of complexity provide a conceptual framework, a way of thinking, and a way of seeing the world.

Mitleton-Kelly (2003, p. 4)

I take an example of Al-Qaeda to demonstrate complexity. It is *non-linear*, in that the whole is certainly greater than its parts, with different rationalities combining for 'success'. It has *'strange attractors'*, building on group identity, the need to join the cause. It exhibits *self-organized criticality*, searching for supremacy, and co-evolving. It understands *amplification and initial conditions*, with the attack on the World Trade Centre being a classic example of a huge effect – still not fully played out. It has highly effective *information systems* across all the cells and networks. And it operates at the *edge of chaos*, searching for emergence, a new world order.

Davies (2003, p. 2)

FURTHER EXAMPLES OF COMPLEX SYSTEMS

SUPPLY CHAINS

Supply chains are a classic example of a complex system with autonomous software existing in the original supplier of the product or service, in the carrier to a distribution center, in the distribution center itself, in the second carrier from the distribution center, and in the customer's establishment. While manual interfacing and transfer of information between systems can occur, such rekeying of information is essentially very inefficient, and ideally, the whole supply chain should be integrated in terms of information, such as product identity, dispatch date and time, and number of items of the product, and other relevant information should be readily transferred through the whole system without manual intervention.

OPERATION OF GOVERNMENTS

Unless governments actively integrate systems in each of the separate government departments, there is a major issue of complexity. A simple example is governments being aware of their overall financial position, in terms of profit or loss. However, integration includes many more issues such as the number of people served by various government departments, measures of operational performance, and many other aspects.

OPERATION OF AN ENTERPRISE

The separately developed systems of the various departments in an enterprise need to be integrated as they were normally developed for an independent purpose. This issue becomes even more acute when there is a takeover of one enterprise by another. One of the most extreme examples of the need to integrate is the booking and flight management system in two airlines that are being integrated through an amalgamation.

DISASTER MANAGEMENT

In many countries, various government departments, which are independent of each other, have been given responsibility for disaster management. An example in one state in Australia is over

150 organizations having such statutory responsibility. Coordination of each of these is required in order to deal with a disaster such as one caused by weather. Obviously, integration is required prior to the event in order to ensure an efficient response.

POWER SYSTEMS

Integrated networks of power generators, distributors, and customers have evolved in many countries, and the total number of people being served by the integrated network can be in the hundreds of millions. While the inputs and outputs of power are very carefully balanced, the full operation of the system has not been fully documented or understood. This has led to failures that created an outage affecting over 100 million people (Wikipedia 2014) (Genscape 2014). Each of the generators in the network and each of the distributors have aspects of autonomy and independence that need to be managed.

WORLD FINANCIAL SYSTEM

The catastrophic events of late 2008 and early 2009 illustrate the autonomy of aspects of the world financial system and the need to integrate elements such that the failure of independent and autonomous banks and the reduction of liquidity do not recur. Aspects such as loan incentives based purely on the size of the loan, and with no reference to the value of the property, or the ability of the recipient to repay the loan are examples of extreme autonomy and independence. Rating integrated products of such loans at high levels also illustrates both the autonomy of the system and the need to integrate a number of elements.

AUSTRALIAN CHARITY BUILDING SCHOOLS IN AFGHANISTAN

The Australian charity indigo foundation is currently financing the construction of its fourth school in Afghanistan. The autonomous and independent systems include cultural practices, the legal system, possible conflicts between tribal groups, the language spoken in the supply chain, and attempts to provide the normal controls that would exist in a construction project in which the sponsor was seeking to get value for money.

DISPUTES BETWEEN WARRING NATIONS

This is one of the most complex of issues to deal with as there are examples of the people of warring nations having quite different views of the significance of previous events, of the actions of the other party, and certainly developing an agenda to ensure peace. When there is little common meaning between participants in a process, we have an extreme example of autonomy and independent systems.

ADDRESSING CLIMATE CHANGE

Given the fact that all nations contribute to the increase in greenhouse gases, it is important that there be some degree of coordination in addressing the problem. However, achieving coordinated action between more than 100 nations is not a trivial task and illustrates the attempt to integrate autonomous independent systems.

HIERARCHY OR NETWORKED

Since complex projects include integration of autonomous and independent systems, they are essentially arranged in a network rather than a hierarchy.

Arrangement in a network means that suppliers and subcontractors are not in the same top-down relationship with the project manager as occurs on traditional linear style projects.

Systems have structures as follows. Traditional projects are hierarchical; complex systems are networked.

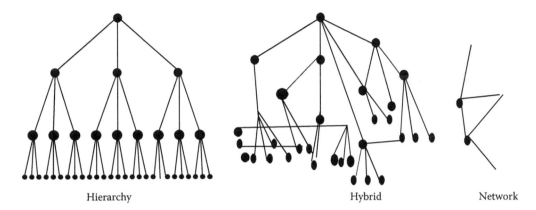

Hierarchy   Hybrid   Network

When possible, there is benefit in realizing the advantages of hierarchy, that is, being able to break a task into subtasks, perform work on those subtasks, and reassemble them to good effect.

Holarchy* recognizes that systems can be both hierarchical and networked. In reality, most SoS are hybrid.

PROJECT ENVIRONMENT

Both the external environment and the internal context of the project play a very important part in complex systems. Bar-Yam comments: "the most basic issue for organisational success is directly matching [a] system's complexity to its environment" (2004, p. 91). Part of this analysis is identifying external forces, as well as the internal forces, acting on a system (2004, p. 118). The external forces acting on the system can be identified by the use of Boardman's and Sauser's systemigram (2008, pp. 87–89). Bar-Yam sees the key aspect of teaching as being the need to match the complexity of the external environment to the needs of the child (2004, p. 165). Bar-Yam also comments that systems can become dependent on connections to their environment (2004, p. 208).

UNCLEAR STAKEHOLDERS

Mason and Mitroff (1981) see a problem with conventional planning and problem solving in that they fail to recognize the value that can be obtained from entertaining different world views. They believe that most organizations fail to deal properly with wicked problems because they find it difficult to challenge accepted ways of doing things, and approaches that diverge from current practice are not given serious consideration (Jackson 2003, p. 141).

They developed strategic assumption surfacing and testing (SAST), which attempts to surface conflicts and to direct them productively as the only way of eventually achieving a productive synthesis of perceptions. Their methodology is based on four key principles: participative,

---

* Holarchy is a word coined by Arthur Koestler. It is a combination between the Greek word *holos* meaning whole and the word *hierarchy*. It is a hierarchically organized structure of units or entities that are called *Holons*. Each Holon could be regarded as either a whole or as a part depending on how one looks at it. A Holon will look as a whole to those parts beneath it in the hierarchy, but it will look as a part to the wholes above it. So, a Holarchy is then a whole that is also a structure of parts that are in themselves wholes.

adversarial, integrative, and managerial mind supporting. The debate can be guided by asking the following questions:

- How are the assumptions of the groups different?
- Which stakeholders feature most strongly in giving rise to the significant assumptions being made by each group?
- Do groups rate assumptions differently (e.g., as to their importance for the success of the strategy)?
- What assumptions of the other groups do each group find the most troubling with respect to its own proposals?

A technique is to judge each assumption on the basis of most certain to least certain and most important to least important and obviously only deal with the certain and important assumptions (Jackson 2003, pp. 143–146).

Jackson criticizes the SAST process in that Mason and Mitroff appear to assume that formulating a problem is synonymous with dealing with it (Jackson 2003, pp. 143–152). However, in the authors' view, the SAST process supports soft system methodology and is consequently of benefit.

UNCLEAR BOUNDARIES

A key difference between SE and SoS engineering is that complex systems boundaries are unclear and can change. One way that sponsors reduce the effects of this issue on projects is that they solve most of the external issues before putting the project to tender. In other words, they package out difficult issues so that contractors can identify the scope and hence an appropriate price.

Ulrich, a PhD student of Churchman, developed critical systems heuristics because he felt that, in trying to grasp the whole system, we inevitably fall short and produce limited accounts and suboptimal decisions based on participative presuppositions. To correct this, we need to unearth the partial presuppositions that underpin the *whole system* judgments we make.

Ulrich's view is that if we are to improve social reality, we must add an additional dimension of purposefulness. In a purposeful system, the ability to determine purposes must be spread throughout the system. Knowledge should be produced relevant to purposes and debate about purposes should be encouraged (Jackson 2003, p. 217).

Ulrich focuses on the nature of boundary judgments that must inevitably enter into social systems design, which need to be significantly tested, because it is only by making them explicit does it become possible to reflect critically on the presuppositions conditioning a social system design (Jackson 2003, p. 218).

Ulrich's 12 boundary questions are as follows:

1. Who ought to be the client (beneficiary) of the system?
2. What ought to be the purpose of the system?
3. What ought to be the system's measure of success?
4. Who ought to be the decision maker (i.e., who has power to change the system's measures of success)?
5. What components (resources and constraints) of the system ought to be controlled by the decision taker?
6. What resources and conditions ought to be part of the system's environment (i.e., *not* to be controlled by the system's decision taker)?
7. Who ought to be involved as designer of the system?
8. What kind of expertise ought to flow into the design of the system?
9. Who ought to be the guarantor of the system (i.e., where ought the designer seek the guarantee that the design will be implemented and will prove successful, as judged by the system's measure of success)?

10. Who ought to belong to the witness representing the concerns of the citizens that will or might be affected by the design of the system (i.e., who among those affected should be involved)?
11. To what degree and in what way ought the affected be given the chance of emancipation from the premises and promises of the involved?
12. On what world view ought either the involved or the affected system's design be based?

Jackson comments that particular attention needs to be paid to ensuring representation of those affected by the proposed approach but not involved in their formulation (Jackson 2003, p. 226). He also comments that Ulrich should develop the appropriate social theory to genuinely assist emancipation; otherwise, his approach could be seen as utopian (Jackson 2003, p. 227).

AUTONOMOUS AND INDEPENDENT SYSTEMS

Firesmith (2010, p. 12) comments: a SoS is "any system that is a relatively large and complex, dynamically evolving, and physically distributed system of pre-existing, heterogeneous, autonomous, and independently governed systems, whereby the system of systems exhibits significant amounts of unexpected emergent behaviour and characteristics." Firesmith comments that these large systems are a result of a department of defense (DoD) policy of building on existing assets.

Examples of independent and autonomous systems, which may be included in complex projects, include the global positioning system, access to defense systems such as the army accessing air force or navy systems, or use of software for air traffic control, bank monitoring, health department monitoring, and a range of other possibilities. Operating under a different legal system (business in China) or a different moral code is a further example.

Norman and Kuras (2006, p. 209) provide an example by the Air (and Space) Operations Center (AOC) of the United States, which provides tools to plan, task, and monitor military (particularly airborne) operations in theaters of war such as Afghanistan and (formerly) Iraq.

AOC is composed of 80 elements of infrastructure including communication balance, application, servers, and databases. The systems

- Don't share a common conceptual basis
- Aren't built for the same purpose or used within specific AOC workflows
- Share an acquisition environment that pushes them to be stand-alone
- Have no common control or management
- Don't share a common funding that can be directed to problems as required
- Have many customers in which the AOC is not the only one
- Evolve at different rates subject to different pressures and needs

Norman and Kuras make a number of observations:

1. The AOC system of systems is an opportunistic aggregation, not a design.
2. Integration enabling technologies (glueware) are drafted onto these elements and integration developments are undertaken after delivery of the component systems to their foreign customers.
3. Funds for the integration are limited.
4. Plans and planning as a primary SoS strategy has problems:
   - The focus [is] on the future but is based [on] the past
   - (Plans tends to mature early [and are] likely incorrect and incomplete)
   - [What a]ctivities tend to is the reality (subject to unplanned change) [in contrast] to the plan (static, based on past beliefs)
   - Expectations, and dependencies, on partially interested participants
   - Design implied in the plan is based on today's understandings. As things change in the world all the elements to be composed are subject to different pressures on decision which will likely not align

PROJECT CONTEXT

An unfamiliar project context can create complexity. An example is creating a project in China when all you have been used to are projects in your own country. The complexity could be due to different legal systems, different governance arrangements, and even lack of familiarity with suppliers, that is, lack of a project network.

EMERGENCE

Emergence occurs as system characteristics and behaviors emerge from simple rules of interaction: individual components interact and some kind of property emerges, something you couldn't have predicted from what you know of the component parts; surprises that cannot even be explained after the fact are of greatest interest (White 2007; McCarter and White 2009). Emergent behavior then feeds back to influence the behaviors of the individuals that produced it (Langton, cited in Urry 2003 and referenced in Ramalingam et al. 2008).

The interactions of individual units driven by local rules cause emergence, and this is not globally coordinated (Marion 1999). In distinguishing complex systems from complicated systems, emergent properties can be used. These are the unexpected behaviors that stem from basic rules that govern the interaction between the elements of a system (Ramalingam et al. 2008, p. 20).

The emergent properties or patterns and properties of a complex system that emerge can be difficult to predict or understand by separately analyzing various *causes* and *effects* or by looking just at the behavior of the system's component parts (Ramalingam et al. 2008, p. 20). Emergent properties can be seen as the result of human action and not of human design.

Examples of emergent properties are structure, processes, functions, memory, measurement, creativity, novelty, and meaning. While the nature of the entities, interactions, and environment of a system are key contributors to emergence, there is no simple relationship between them. Emergence has been used to describe features such as social structure, human personalities, the Internet, consciousness, and even life itself, as one lucid account has it (Newell 2002):

> …metaphors [are] useful. The music created by an orchestra may be envisaged as an emergent phenomenon that is a result of the dynamic, temporal interactions of many musicians at a point in time… the process that is the music may alter a listener's actions or behaviours in the real world… this viewpoint supports the proposition that complex systems, such as people, have multiple emergent levels, each level generating phenomena that is more than the sum of the parts and is not reducible to the parts….

Newell (2003) cited in Ramalingam et al. (2008, p. 21)

Emergence, then, describes how a system emerges from the interactions of individual units driven by local rules (Marion 1999). The dynamic feedback between the parts crucially shapes and changes the whole (Haynes 2003). Models of flocking birds can be created by using three basic rules: separation of flight path from that of local birds, alignment of steering with the average direction of the local flock, and cohesion to move towards the average position of the local flock (Ramalingam et al. 2008, p. 21).

The principles of emergence mean that overcontrolling or top-down approaches will not work well within complex systems. In order to maximize system adaptiveness, there must be space for innovation and novelty to occur. While this may be obvious, this is often the reverse of what happens in the real world, because of a tendency to overdefine and overcontrol rather than simply focus on the critical rules that need to be specified and followed (Morgan 1986 cited in Ramalingam et al. 2008, p. 21).

NONLINEARITY

Complexity science generally states that human systems do not work in a simple linear fashion because feedback processes among interconnected elements and dimensions lead to relationships that exhibit dynamic, nonlinear, and unpredictable change (Stacey 1996). Nonlinearity is a direct result of the interdependence between elements of complex systems. Clear causal relations cannot be traced because of multiple influences (Ramalingam et al. 2008, p. 24):

> Linearity describes the proportionality assumed in idealised situations where responses are proportional to forces and causes are proportional to effects.
>
> **Strogatz (2003)**

Linear problems can be broken down in a reductionist fashion, with each element analyzed separately, and the separate aspects can be recombined to give the right answer to the original problem. In a linear system, the whole is exactly equivalent to the sum of the parts:

> However, linearity is often an approximation of a more complicated reality – most systems only behave linearly if they are close to equilibrium and are not pushed too hard. When a system starts to behave in a nonlinear fashion, all bets are off.
>
> **Strogatz (2003)**

Ramalingam comments that the linear idea of cause and effect (*if activity A is done, output B will result, leading to outcome C and impact D*) is profoundly at odds with the implications of complexity science and, indeed, the experiences of many development practitioners (2008, p. 26). He points out that "the nonlinearity concept means that linear assumptions of how social phenomena play out should be questioned. ... Nonlinearity poses challenges to analysis precisely because such relationships cannot be taken apart—they have to be examined all at once, as a coherent entity."

There are three important aspects that need to be recognized in dealing with nonlinearity:

1. The concept of an independent variable is wrong: this is an approach taken in many research studies and underpinning the way many researchers think; the notion of holding all other variables constant in a system of interconnected and interrelated parts, with feedback loops and adaptive agents in emergent properties, is almost impossible as everything cannot be held constant and there is no independent variable.
2. The assumption that changes in the system output are proportional to changes in input does not work in complex systems; for example, providing more aid to a foreign country, or more social security within a country, does not necessarily create an improved situation as major feedback loops, adaptive behaviors, and emergent dynamics within the system may mean that the relationship between input and output is not linear.
3. The assumption of linearity is that the system output that follows from the sum of two different inputs is equal to the sum of the outputs arising from the individual inputs (Ramalingam et al. 2008, p. 25). To say the least, nonlinear systems do *not* behave in this fashion.

ARCHITECTURE

While traditional SE projects are focused on requirements, in developing a new product or service, the architecture needs to be emphasized (Figure 2.1).

Wilber (2009) illustrates the architecture on the Boeing 787 (cf., Chapter 23 of the present book) that includes communication with the airport, the industry, the airline, private and public networks, and the general manufacturing process. It includes major elements of the passenger environment, cabin and airline services, maintenance, open networking, avionics data, and the flight deck.

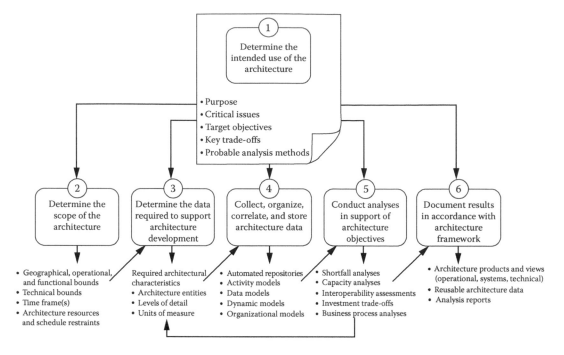

FIGURE 2.1 Architectural development process. (From DODAF V2.0, DoD Architectural Framework Version 1.0, the Deskbook and Volumes I & II, Department of Defense Architecture Working Group. Superseded by DoDAF V 1.5, 2003. With permission.)

Other elements include support of maintenance performance, aeroplane health management, airline flight operations, and materials management through the supply chain.

Various descriptions can be used to describe architecture for complex systems:

DESCRIPTION A

A possible design of the architecture of a SoS is illustrated in Figure 2.1 (DODAF 2003).

DESCRIPTION B

While both requirements and architecture are important, architecture needs to be recognized as having a dominant role in the development of a SoS. Architecture is particularly important for the following reasons given by Maier and Rechtin (2009):

1. Architecting deals largely with the unmeasurable using nonquantitative tools, whereas engineering uses the hard sciences.
2. The word architecting refers only to the process; architecting is an invented word to describe how architectures are created, much as engineering describes how engines and other artifacts are created.
3. The history of classical architecting suggests that the process of creating architectures began in Egypt more than 4000 years ago with the pyramids.
4. However, designing a modern computer program also requires architecting.
5. This architecting is a response to complexity in the system and explains why a single optimum solution never exists: there are just too many variables (2009, p. 6).

6. However, this complexity also requires a systems approach specifically linking value judgments and design decisions.
7. Critical details in the architect's design will be the system's connections and interfaces and the system's components; this combination produces unique system level functions; this is because the system specialists are likely to concentrate most on the core and least on the periphery of their subsystems.
8. Modeling is the creation of abstractions or representations of the system in order to predict and analyze performance, costs, schedules, and risks and to provide guidelines for systems research, development, design, manufacture, and management (2009, p. 12).
9. Architecting is the embodiment of project strategy.

DESCRIPTION C

Crawley et al. (2004) make the following points:

1. Architecture is an abstract description of the entities of the system and interconnections and interaction.
2. It is the link between form and function.
3. It provides rules to follow when creating a system.
4. It is important because it provides guidance and structure to later phases of system design.
5. It provides handles for addressing:
   - Alternative form
   - Substructure and modules
   - Complexity
   - Flexibility
   - Other *illities*
6. It invites abstract thinking about unifying themes based on structure and arrangements.
7. It is used to
   - Define and manage interfaces
   - Standardize components
   - Standardize interfaces
   - Systematize integration, verification, and validation of the lower levels

An important first step in developing architecture is deciding whether the complex system is more of Type A, B, C, or D.

DESCRIPTION D

Dahmann (2009), addressing Type B projects, addresses the following areas:

1. Translating SoS capability objectives into high-level requirements over time
2. Understanding the components of the SoS and their relationships over time
3. Assessing the extent to which the SoS meets capability objectives of the time; developing, evolving, and maintaining a design for the SoS
4. Monitoring and assessing potential impacts of changes on SoS performance
5. Addressing the requirements on SoS and solution options
6. Orchestrating upgrades to the SoS

Description E

Gharajedaghi (2006, pp. 152–184) proposes the following approach to designing business architecture:

1. Identifying the system's boundaries and business environment
2. Developing the system's purpose:
   - This is not always obvious because "emerging consequences contradict expectations because the operating principles are rooted in assumptions that belong to different paradigms"
   - The basic business model is relevant here as this defines how the business generates value, creates a deliverable package, and exchanges it for money or other forms of reward
3. Identifying functions:
   - Selecting a product/market niche is important here and competitive advantage has to be considered
4. Identifying structure:
   - Traditional organizations are based on structurally defined tasks and segmentation and hierarchal coordination of functions
   - SoS requires more distributed leadership
5. Outputs dimensions:
   - "The output dimensions or platform consists of a series of general purpose, semiautonomous, and ideally self-sufficient units charged with all the activities responsible for achieving the organisation's mission and outputs"
6. Market dimensions:
   - This is the interface with customers
7. Organizational or business processes:
   - Processes are required to create integration, alignment, and synergy among the organization's parts
8. Planning, learning, and control system
9. Measurement system

Description F

Dagli and Kilicay-Ergin (2009) see SoS architecting as including the following:

Architecting Properties
- Abstract, metalevel
- Fuzzy uncertain requirements
- Network centric
- Software intensive
- People intensive
- Intensive communication infrastructure
- Networks of various stakeholders
- Collaborative emergent development
- Dynamic architecture

Architecting Constraints
- Emphasis on interface architecting to foster collaborative functions among independent systems
- Concentration on choosing the right collection of systems to satisfy the requirements
- Scalability
- Interoperability
- Trustworthiness
- Hidden cascading failures
- Confusing life cycle context

Legacy Systems

- Abstraction level determines the integration of legacy systems to other systems.
- Balance of heuristics, analytical techniques, and integration modeling.

DESCRIPTION G

Revolution presents challenges due to changes in the system context. Dynamic changing require-ments increase uncertainty, and systems need to be designed for fuzzy attributes. Key questions arise including the following:

- How can we assure trustworthiness?
- How can we assure interoperability?
- How can we reassure large-scale design along with distributed testing?
- How can we assure evolutionary growth?
- How can we deal with hidden interdependencies?
- How can we guard against cascading failures?

The plug and play concept of assembling and organizing coalitions from different systems provides flexibility to respond to a changing operational and environmental situation.

The information architecture is the backbone of the SoS architecture.

OTHER QUESTIONS

There are some important questions that need to be addressed about the role of architecture in com-plex projects. The following will attempt to gather data to address relevant issues:

1. What are the differences among the architectures required for project Types A, B, C, and D?
2. Was the vision for the project clearly expressed in the architecture?
3. Were values addressed in the architecture?
4. Were requirements addressed in the architecture?
5. Were the core and peripheral elements in the system design and structure recognized?
6. Was a design structure matrix (DSM) created in developing the architecture?
7. If a DSM was developed, were design rules also developed?
8. To what extent was there an attempt to modularize the software in the system architecture:
   - Not much
   - A little
   - A lot
   - It was a dominant approach

Building these issues into the architecture of system of systems engineering (SoSE) initially appears to be the most secure approach. "Architecture dictates the form and function by which design deci-sions can be made." Azani (2008) endorses the view that "there seems to be a consensus that the architecture is the structure not only of the systems, but of its functions, the environment in which it will move, and the processes by which it will be built and operated."

Khoo (2009) supports this approach in commenting that the use of closed requirements to spec-ify systems must be amended to become open and flexible; however, he comments that there lacks the express power to rigorously capture complexity. Khoo also comments that government contract-ing and procurement processes are too restrictive to assist this process.

Two of Maier's and Rechtin's comments need to be remembered: "Architecture is the embodiment of strategy" and that one clarifies the problem as one develops the solution (Maier and Rechtin 2009).

However, developing architectures of SoS and SoSE has been fraught with difficulty. Meilich comments that "an adaptive architecture will be required to support operations that may change on the battlefields of the future" (Meilich 2006). He sees a problem in developing a template for architecture of SoS because they have to operate in changing contexts and adapt to changing mission. He finally concludes that developing architecture of SoSE is too challenging a task at this stage.

Bjelkemyr et al. (2007) endorse the need for architecture of SoS because this will help define a clear purpose and apparent boundaries. They recognize that the "internal properties of SoS resemble those of an organism." However, while contributing concepts on architecture, they conclude that significantly more research is required. DeLaurentis and Crossley (2005) provide a hierarchical framework for describing SoS and a taxonomy for guiding design methods. However, they conclude that "a comprehensive set of methods for design in a SoS context do not yet exist and more development effort is needed."

Given the previous discussion that there is no agreed architectural template for SoSE, it is concluded that the contributions to SoSE from complex systems research need to be recognized as processes, rather than embedded in architecture at this stage, as this as far as we can go at present.

REQUIREMENT COMPLEXITY

Sarah Sheard, in her PhD on complex systems (Sheard 2013), found that there were three aspects of requirements that contributed to complexity at significant levels. These were the following:

REQUIREMENT DIFFICULTY

Difficult requirements are considered difficult to implement or engineer, are hard to trace to the sources, and have higher degrees of overlap with other requirements. If there are a number of difficult requirements, it may be questionable that the system can be built at all. Higher numbers of difficult requirements typified projects that had more changes in needs and more problematic stakeholder interactions.

COGNITIVE FOG

Cognitive fog is primarily when the project finds itself in a fog of conflicting data and cognitive overload. Projects without cognitive fog were more likely to deliver a product that had significantly

- Less cost overruns, scheduled delays, and replanting
- Fewer easy and nominal requirements
- More architectural precedence
- Less conflict among technical requirements and between technical requirements and cost and schedule constraints
- Fewer changes in limbo at any given time, less change in stakeholder needs during the project, and more stakeholder concurrence

Cognitive fog relates to the prior and later variables and is exacerbated by

- A larger number of decision makers
- Larger amounts of stakeholder conflict and changes in stakeholder needs
- A high degree of requirements and changes in limbo
- Political arguments
- Instability and conflicting data

Cognitive fog is illustrated in Figure 2.2.

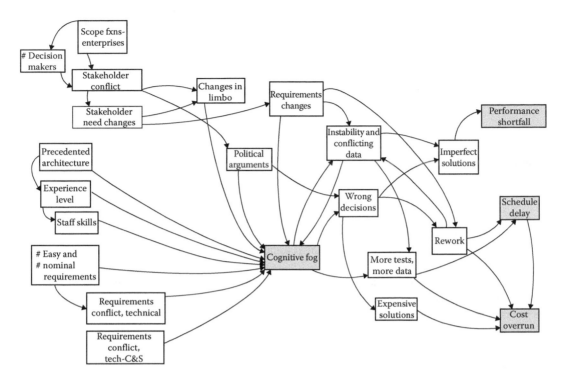

FIGURE 2.2   Relationship of cognitive fog to other variables. (From Sheard, S., Assessing the impact of complexity attributes on system development project outcomes, PhD dissertation, Stevens Institute of Technology, Hoboken, NJ, 2013. With permission.)

Stable Stakeholder Relationships

Projects with stable stakeholder relationships had

- Less replanning;
- Less stakeholder conflict;
- More concurrence among stakeholders;
- Lower cost overruns, scheduled delays, and performance shortfalls;
- Higher subjective success;
- Less conflict among technical requirements;
- Less cognitive fog;
- Less change in stakeholder needs;

SENSITIVITY TO INITIAL CONDITIONS

The initial conditions of complex systems determine where they currently are, and, consequently, two complex systems that initially had their various elements and dimensions very close together can end up in distinctly different places due to their nonlinearity of relationships. Changes are not proportional and small changes in any one of the elements can result in large changes in the current position (Ramalingam et al. 2008, p. 27).

Phase space addresses the evolution of systems by considering the evolution process as a sequence of states in time (Rosen 1991). A state is the position of the system in its phase space at a given time. At any time, the system's state can be seen as the initial conditions for whatever processes that follow. The position of a system in its phase space at any time will have an influence on its future evolution. All interactions are contingent on what has previously occurred (Ramalingam et al. 2008, p. 27).

"This is closely related to the notion of 'path dependence', which is the idea that many alternatives are possible at some stages of a system's development, but once one of these alternatives gains the upper hand, it becomes 'locked in' and it is not possible to go to any of the previous available alternatives." For example,

> ... many cities developed where and how they did not because of the "natural advantages" we are so quick to detect after the fact, but because their establishment set off self-reinforcing expectations and behaviours.

(Cronon, cited in Jervis 1997 and then Ramalingam et al. 2008, p. 28)

Complexity, nonlinearity of behavior, and sensitivity to initial conditions make it extremely difficult to separate the contributions to overall behavior that individual factors provide. Therefore, any notion of *good practice* requires a detailed local knowledge to understand why the practice in question is good.

It is necessary to incorporate an acceptance of the inherent levels of uncertainty into planning. Both simple and intricate processes carry uncertainty of prediction. A realistic understanding of this uncertainty is required, and there is a need to build a level of flexibility and adaptability into projects to provide greater resilience. This is because certain levels of uncertainty are unavoidable when looking into the future (Ramalingam et al. 2008, p. 30).

DEGREE OF CONTROL

At one extreme, your project has a stable environment and you have a reasonable degree of control. Given this, when the project context is fairly familiar, you can reasonably confidently plan aspects of scope and hence predict time and cost of activities. Then, you can control the project by regular review of the scope, time, and cost and make relatively minor changes to the management to achieve the planned targets.

At the other extreme, your project appears to move through continually new states, with change as a constant in a kind of unending turbulence. It would move around one loop, spiraling out from the center, and when it got close to a particular point, it would spiral around again in a chaotic dynamic and develop through a series of sudden jumps.

SELF-ORGANIZATION

Self-organization is a form of emergent property and supports the notion that complex systems cannot be understood in terms of the sum of its parts, since they may not be understood from the properties of individual agents and how they may behave when interacting in large numbers.

Mitleton-Kelly (2003, pp. 19–20) points out that self-organization, emergence, and the creation of new order are three of the key characteristics of complex systems. She reminds us that Kauffman, in the *Origins of Order: Self-organization and Selection in Evolution* (1993), focuses on Darwinian natural selection as a *single singular force* and argues that "It is this single-force view which I believe to be inadequate, for it fails to notice, fails to stress, fails to incorporate the possibility that simple and complex systems exhibit order spontaneously" (Kauffman 1993, p. xiii). However, Pinker (1997) seems to espouse the inclusion of that possibility. Spontaneous order is self-organization.

Mitleton-Kelly (2003, pp. 19–20) adds that "emergent properties, qualities, patterns, or structures, arise from the interaction of individual elements; they are greater than the sum of the parts and may be difficult to predict by studying the individual elements." Emergence is the *process* that creates new order together with self-organization. Mitleton-Kelly (2003, pp. 19–20) reminds us that Checkland defines emergent properties as those exhibited by a human activity system "as a whole entity, which derives from its component activities and their structure, but cannot be reduced to them" (Checkland 1981, p. 314). The emphasis is on the *interacting whole* and the *nonreduction* of those properties to individual parts.

Self-organization may be described in an arrangement when a group spontaneously comes together to perform a task.

The concept of self-organization echoes emergent properties and the fact that a complex system cannot be understood as the sum of its parts, since it may not be discernible from the properties of the individual agents and how they may behave when interacting in large numbers. For example, studies have shown how highly segregated neighborhoods can arise from only low levels of racism in individuals (Schelling 1978). The market is probably the exemplary self-organizing system. As the Nobel Laureate Ilya Prigogine has put it (in Waldrop 1994): "the economy is a self-organising system, in which market structures are spontaneously organised by such things as the demand for labour and demand for goods and services."

Westley et al. (2006) argue that "Bottom-up behaviour [leading to self-organization] seems illogical to Western minds ... we have a hierarchical bias against self-organisation

> ... [which is displayed in our common understanding of how human change happens, especially in organisations]. Our popular management magazines are filled with stories of the omniscient CEO or leader who can see the opportunities or threats in the environment and leads the people into the light. *However*, self-organisation is critical to achieving *change*."

Ramalingam et al. (2008) point out that self-organization describes how the adaptive strategies of individual agents in particular settings are able to give rise to a whole range of emergent phenomena, including the emergence of resilience (2008, pp. 49–50). They further note that self-organization need not necessarily be about change as it can be about resilience in the face of change. They see resilience being about continuous and often simultaneous stages of release, reorganization, exploitation, and conservation, including the possibility of destruction of some existing organizational structures. This frees up essential resources and enables growth in new areas. Cycles of destruction in economies release innovation and creativity. Reorganization is where there is competition for available resources that are then exploited by the dominant species or winning proposal.

EXAMPLE OF SELF-ORGANIZATION FROM MITLETON-KELLY (2003, pp. 10–11)

"Another key concept in complexity is dissipative structures, which are ways in which open systems exchange energy, matter, or information with their environment and which when pushed 'far-from-equilibrium' create new structures and order. The Bénard cell is an example of a physico-chemical dissipative structure. It is made up of two parallel plates and a horizontal liquid layer, such as water. The dimensions of the plates are much larger than the width of the layer of water. When the temperature of the liquid is the same as that of the environment, the cell is at equilibrium and the fluid will tend to a homogeneous state in which all its parts are identical (Nicolis and Prigogine 1989; Prigogine and Stengers 1984). If heat is applied to the bottom plate, and the temperature of the water is greater at the bottom than at the upper surface, at a threshold temperature, the fluid becomes unstable. 'By applying an *external constraint* we do not permit the system to remain at equilibrium' (Nicolis and Prigogine 1989, p. 11). If we remove the

(continued)

EXAMPLE OF SELF-ORGANIZATION FROM MITLETON-KELLY (2003, pp. 10–11) (continued)

system farther and farther from equilibrium by increasing the temperature differential, suddenly at a critical temperature, the liquid performs a bulk movement that is far from random: the fluid is structured in a series of small convection *cells* known as Bénard cells.

Several things have happened in this process: (a) the water molecules have spontaneously organized themselves into right-handed and left-handed cells. This kind of spontaneous movement is called *self-organization* and is one of the key characteristics of complex systems; (b) from molecular chaos, the system has emerged as a higher-level system with *order* and *structure*; (c) the system was pushed far from equilibrium by an *external constraint or perturbation*; (d) although we know that the cells will appear, 'the direction of rotation of the cells is unpredictable and uncontrollable. Only chance in the form of the particular perturbation that may have prevailed at the moment of the experiment, will decide whether a given cell is right- or left-handed' (Nicolis and Prigogine 1989, p. 14); (e) when a constraint is sufficiently strong, the system can adjust to its environment in several different ways, that is, *several solutions* are possible for the *same parameter values*; (f) the fact that only one among many possibilities occurred gives the system 'a *historical dimension*, some sort of *memory* of a past event that took place at a critical moment and which will affect its further evolution' (Nicolis and Prigogine 1989, p. 14); (g) the homogeneity of the molecules at equilibrium was disturbed and their *symmetry was broken*; and (h) the particles behaved in a *coherent* manner, despite the random thermal motion of each of them. This coherence at a macro level characterizes *emergent* behavior, which arises from microlevel interactions of individual elements.

In the Bénard cell, heat transfer has *created new order*. It is this property of complex systems to create new order and coherence that is their distinctive feature. The Bénard cell process in thermal convection is the basis of several important phenomena, such as the circulation of the atmosphere and oceans that determines weather changes (Nicolis and Prigogine 1989, p. 8).

Ilya Prigogine was awarded the 1977 Nobel Prize for chemistry for his work on dissipative structures and his contributions to nonequilibrium thermodynamics. Prigogine has reinterpreted the second law of thermodynamics. Dissolution into entropy is not an absolute condition, but 'under certain conditions, entropy itself becomes the progenitor of order.' To be more specific, '… under non-equilibrium conditions, at least, entropy may produce, rather than degrade, order (and) organisation… If this is so, then entropy, too, loses its either/or character. While certain systems run down, other systems simultaneously evolve and grow more coherent' (Prigogine and Stengers 1984, p. xxi)."

"In dissipative structures, the tendency to split into alternative solutions is called *bifurcation*, but the term is misleading in that it means a separation into *two* paths, when there may be several possible solutions" (Mitleton-Kelly 2003, p. 11).

"Non-equilibrium may allow a system to avoid thermal disorder and to transform part of the energy communicated from the environment into an ordered behavior of a new type, a new *dissipative structure* that is characterized by symmetry breaking and multiple choices. In chemistry, *autocatalysis* (the presence of a substance may increase the rate of its own production) shows similar behaviors, and the Belousov–Zhabotinski (BZ) reaction, under certain nonequilibrium conditions, shows symmetry breaking, self-organization, multiple possible solutions, and hysteresis (the specific path of states that can be followed depends on the system's past *history*) (Nicolis and Prigogine 1989; Kauffman 1995). Furthermore, *self-reproduction*, a fundamental property of biological life, is 'the result of an autocatalytic cycle in which the genetic material is replicated by the intervention of specific proteins, themselves synthesized through the instructions contained in the genetic material' (Nicolis and Prigogine 1989, p. 18). In one sense, complexity is concerned with systems in which evolution—and hence history—plays or has played an important role, whether biological, physical, or chemical systems.

Similarly in a social context, when an organization moves away from equilibrium (i.e., from established patterns of work and behavior), new ways of working are created and new forms of organization may emerge" (Mitleton-Kelly 2003, p. 13).

"In addition, humans can also provide help and support for a new order to be established. If the new order is 'designed' in detail, then the support needed will be greater, because those involved have their self-organising abilities curtailed, and may thus become dependent on the designers to provide a new framework to facilitate and support new relationships and connectivities. Although the intention of change management interventions is to create new ways of working, they may block or constrain emergent patterns of behaviour if they attempt to excessively design and control outcomes. However, if organisation re-design were to concentrate on the provision of *enabling infrastructures* (the socio-cultural and technical conditions that facilitate the emergence of new ways of organising), allowing the new patterns of relationships and ways of working to emerge, new forms of organisation may arise that would be unique and perhaps not susceptible to copying. These new organisational forms may be more robust and sustainable in competitive environments" (Mitleton-Kelly 2003, p. 14).

LEADERSHIP

The role of leadership is important in terms of self-organization of complex systems. Marion and Uhl-Bien (2001) suggest that in complex systems, there is a need for different leadership qualities. Ramalingam et al. (2008) point out that the traditional perspective of leadership (*in Western society?*) is based on a view of organizations as mechanical systems (Capra 1996; Stacey 1995). This is made up of highly prescriptive rule sets, formalized control, and hierarchical authority structures. Leaders are expected to contribute to stabilization through directive actions, based on planning for the future and controlling the organizational response. However, in complex systems, the future cannot be predicted, and change cannot be directed.

Leaders of self-organized adaptive agents are characterized by their ability to (1) disrupt existing patterns by creating and highlighting conflicts, (2) encourage novelty, and (3) use *sensemaking*. Each of these is worth exploring in a little more detail (Plowman et al. 2007a,b cited by Ramalingam et al. 2008, p. 51). Leaders disrupt existing patterns by acknowledging and embracing uncertainty, refusing to back away from uncomfortable truths, talking openly about the most serious issues, and challenging institutional *taboos*, which encourage more open thinking about the new ideas and patterns to emerge. An effective method for communicating in this way is provided in Knowles (2002).

Mary Uhl-Bien et al. (2007) developed the concept of complexity leadership that is quite different to traditional command and control leadership and is suitable to fast-paced volatile contracts in the knowledge era. Complexity leadership uses the concept of complex adaptive systems (CASs) and "recognises that leadership should not be seen only as position and authority but also as an emergent, interactive dynamic—a complex interplay from which a collective impetus for action and change emerges when heterogeneous agents interact in networks in ways that produce new behaviour or new modes of operating." The question for the purposes of this chapter is the extent to which complexity leadership applies in SoS projects.

Uhl-Bien et al. (2007) propose a leadership framework that they call complexity leadership theory (CLT). Within CLT, they recognize three broad types of leadership: (1) leadership grounded in traditional bureaucratic notions of hierarchy, alignment, and control, which they call administrative leadership; (2) leadership that structures and enables conditions such as CAS and is able to optimally address creative problem solving, adaptability, and learning; and (3) leadership as a generative dynamic that underlies emergent change activities, which they call adaptive leadership.

Complexity leadership is embedded in context, which is the "ambience that spawns a given systems dynamic personae." This complex systems persona refers to the nature of interactions and interdependencies among agents (people and ideas), hierarchical divisions, organizations, and environments. CAS and leadership are socially constructed in and from this context. The theory distinguishes between leadership and leaders in which leadership is an emergent, interactive dynamic that is produced out of adaptive outcomes. This theory recognizes that leadership theory has largely focused on leaders and their actions as individuals and seeing leadership as a dynamic systems concept and processes that comprise leadership. Complexity leadership occurs in the face of adaptive challenges that are typical of the knowledge era rather than technical problems or characteristic of the industrial age.

McCarter and White (2012) contain additional reference material relevant to complex leadership.

A fresh look at leadership recognizing it is embedded in context and goes beyond the notion of individual leaders to recognize leadership processes. Complexity leadership recognizes the difference between complicated and complex organizations, in which a jumbo jet is complicated but a rainforest is complex (Cilliers 1998). Complexity leadership is intensely adaptive and innovative (Cilliers 1998; Marion 1999), which is assisted by being loosely coupled. Coupling imposes restrictions on adaptation (Kauffman 1993; Marion 1999) and bottom-up behavior has more opportunity to succeed. Informal emergence can occur if internal controls do not hinder demands imposed by environmental exigencies. Such constraints are valuable for allocating resources, controlling costs, and coordinating action.

CASs are unique and desirable in their ability to adapt rapidly and creatively to environmental changes. Complex systems enhance their capacity for adaptive response to environmental problems or internal demand by diversifying their behaviors or strategies. Diversification, from the perspective of complexity science, is defined as increasing internal complexity (number and level of interdependent relationships, heterogeneity of skills and outlooks within the CAS, and attention to the point of, or exceeding, that of competitors or the environment). McKelvey and Boise (2003) identify requisite complexity referring to Ashby's requisite variety. "This law simply states that the system must possess complexity equal to that of the environment in order to function effectively" (Uhl-Bien et al. 2007, p. 301). To do so, it must release the capacity of a neural network of agents in pursuit of such optimization—that is, the system must learn and adapt. By contrast, traditional organizations have done the opposite in simplifying adaptation.

LIMITATIONS ON CURRENT LEADERSHIP THEORY

Current leadership theory largely focuses on how leaders can motivate others to achieve organizational goals. At best, they do so by using transformational leadership that is effective.

COMPLEXITY IN THE LEADERSHIP THEORY

This framework includes three leadership functions that are called adaptive, administrative, and enabling:

CLT provides an overarching framework that describes adaptive leadership, administrative leadership, and enabling leadership; it also provides for entanglement among the three leadership roles and in particular between CAS and bureaucracy.

1. *Adaptive leadership* is adaptive, creative, and learning action that emerged from the interactions of the CAS as they strive to adjust to tensions (constraints or perturbations).

Adaptive leadership is an emergent, interactive dynamic that produces adaptive outcomes in a social system. It is a collective change movement that emerges nonlinearly from interactive exchanges or, more specifically, from the spaces between agents. That is, it originates in struggles among agents and groups over conflicting needs, ideas, or preferences. Adaptive leadership is an emergent, interactive dynamic that is the primary source by which adaptive outcomes are produced in a firm.

2. *Administrative leadership* refers to the actions of individuals and groups in formal managerial roles to plan and coordinate activities to accomplish organizationally prescribed outcomes. Administrative leadership structures tasks and organizes planning, builds visions, and allocates resources to achieve goals and manage crises. Administrative leadership is a top-down function based on authority and position; it possesses the power to make decisions for the organization. However, administrative leadership sensibly exercises its authority with consideration of the firms' need for creativity, learning, and adaptability for its actions have significant impact on these dynamics. Administrative leadership is the actions of individuals and groups in formal managerial roles who plan and coordinate organizational activities (bureaucratic functions). If an interaction is largely one sided and authority based, then the leadership event can be labeled as top-down. Its authority asymmetry is less one sided and more preference oriented, and the leadership event is more likely based on interactive dynamics driven by differences in preferences.

3. *Enabling leadership* serves to help manage the entanglement between administrative and adaptive leadership (by fostering enabling conditions and managing the innovation to organization interface). These roles are entangled within and across people and actions. Enabling leadership works to capitalize on conditions in which adaptive leadership can thrive and manage the entanglement between bureaucratic (administrative leadership) and emergent (adaptive leadership) functions of the organization. Managing entanglement involves creating appropriate organizational conditions (or enabling conditions) to foster effective adaptive leadership in places where innovation and adaptability are needed and also facilitating the flow of knowledge and creativity from adaptive structures into administrative structures.

In CLT, these three leadership functions are intertwined in a manner that we refer to as entanglement. *Entanglement* describes a dynamic relationship between the formal top-down, administrative forces (bureaucracy) and the informal, complexity adaptive emergent forces (i.e., CAS) of social systems. Enabling leadership can work to augment the strategic needs of administrative leadership, it can rebel against it, or it can act independently of administrative leadership. To accomplish this, it must tailor behaviors of administrative and adaptive leadership so that they function in tandem with one another.

The role of enabling leadership at the strategic level then is to manage the coordination of rhythms, or oscillations, between the relative importance of top-down, hierarchical dynamics and emergent CASs.

In *adaptive leadership*, change is produced by the clash of the existing but seemingly incompatible ideas, knowledge, and technologies. It takes the form of new knowledge and creative ideas or adaptation. Impact can be independent of significance because impact is influenced by the authority and reputation of the agents who generated the idea, the degree to which an idea captures the imagination or to which its implications are understood, or whether the idea can generate enough support to exert an impact. Adaptive leadership is not an act of an individual but rather a dynamic of interdependent agents (i.e., CAS). Adaptive leadership must be embedded in an appropriately restructured network of CAS and agents and exhibit significance and impact that generates change in the social system.

Enabling Conditions That Capitalize Adaptive Leadership

One function of enabling leadership is to catalyze CAS dynamics that promote adaptive leadership by creating conditions that are interactive, moderately interdependent, and infused with tension. This includes the following:

Fostering Interaction

- Fostering interdependency
- Injecting adaptive tension

This will occur through

- Open architecture of workplaces
- Self-selected workgroups
- Electronic workgroups
- Management-induced scheduling or rule structuring

Fostering interaction with the environment enables importation of fresh information into the creative dynamic and it broadens the organization's capacity to adapt to environmental changes beyond the adaptive capacity of strategic leadership acting alone. Adaptability of terrorist networks provides a very good example (see Boal and Schultz 2007). Individuals can enlarge their personal networks to increase the amount of access and network resources they can bring to the table. They can keep themselves informed and knowledgeable on issues important to the firm and their field. They can monitor their environment (political, economic, social, national, international, technological) to understand the nature of the forces that are influencing adaptive dynamics.

Interdependency

Interdependency's potency derives from naturally occurring emergent networks of conflicting constraints. Conflicting constraints are manifested in the well-being of one agent and inversely dependent on the well-being of another or when information broadcast by one agent isn't incompatible with that broadcast by another agent. Such pressures induce agents to adjust their actions and to elaborate their information. One useful tool for promoting interdependency is to allow measured autonomy for informal behavior. Autonomy permits conflicting constraints to emerge and enable agents to work through those constraints without interference from formal authorities.

Frequent coordination of agents who are operating interdependently and in small groups will assist. This applies to programmers who periodically run their code against the code of other programmers.

Tension

Finally, tension creates an imperative to act and to elaborate strategy, information, and adaptability; enabling leadership also works to foster tension. Homogeneity pressures agents to adapt to their differences. At the upper level echelon and organizational levels, enabling leadership promotes heterogeneity by building an atmosphere in which such diversity is respected and by structuring workgroups to enable interaction of diverse ideas. Enabling an atmosphere that tolerates dissent and diverging perspectives on problems, one in which personnel are charged with resolving their differences and finding solutions to their problems, is critical. Upper and middle level leaders inject tension with managerial pressures or challenges by distributing resources in a manner that supports creative movements and by creating domains for results. Enabling leaders can create perturbations by fostering learning and creativity including introducing seeds that include ideas, information, judiciously placed resources, new people, and the capacity to access unspecified resources to gateways that permit exploration.

Managing the Entanglement between Adaptive and Administrative Structures

A key role of enabling leadership is to manage the entanglement between CAS dynamics and formal administrative systems and structures. This requires using authority where applicable and access to resources and influence to keep the formal and informal organizational systems working in tandem rather than counter to one another. It is not clear whether creativity is enabled or hampered by administrative planning; however, unrestrained adaptive behavior could be expensive to support and could compromise rather than enhance the organization's strategic mission. Organizational plans should impose limits that assure creative emergence is consistent with the core competencies or themes of the system. Planners should separate the creative process from the structure in which it occurs, which can occur by the creative process itself: that is, adaptive behaviors should not be unduly managed or constrained by administrative planning and coordination. Increasing the availability of information resources and dependence on flows of information resources, such as access to electronic databases, is effective.

Enabling leadership managers create conditions consistent with the strategy and mission of the organization by articulating the mission to project teams. Strategy and mission consistency are fostered by discouraging nonuseful adaptation. Enabling leaders deal with crises that threaten to derail adaptive functions.

Managing the Innovation to Organization Interface

This is the second role of enabling leaders and includes

- Overcoming social and political pressures imposed and converting these to their advantage
- Demonstrating personal commitment and promoting ideas through informal networks
- Willingly risking their position and reputation to ensure success
- Maintaining contact with top management to keep them informed and enthusiastic about the project
- Exerting social and political effort to galvanize support for concepts

Formal organizational systems are often not structured to foster internal dissemination of innovation—rather, they tend to inhibit it because formal structures present obstacles for innovation to organization transfer, and power is needed to facilitate, orchestrate, and ensure innovation ideas and outcomes throughout the organization. Unless product innovation has an explicit, organizational-wide powerbase, there is no generative force, no energy, for developing new products continuously and leaving them functioning. A proinnovation approach is required by moving beyond reliance on networks of personal power to focus on individuals and towards an organization system base of power.

In summary, the implications for complex leaders from considering leadership of complex projects include the following:

- Focus on how to foster and speed up the emergence of distributed intelligence instead of viewing leadership just as interpersonal influence leadership.
- Influence networks, creating atmospheres for formation of aggregates and meta-aggregates (e.g., the emergent structure concepts of complexity theory to be discussed in the following), in ways that permit innovation and dissemination of innovations.
- Focus on interactions among ensembles or interfaces.
- Involve exerting interpersonal influence (e.g., relationship-oriented behavior), but part of it may not be this (hence, the broader definition of leadership).
- Move away from *providing answers* or providing too much direction (e.g., initiating structure) to creating the conditions in which followers' behaviors can produce structure and innovation.
- Cultivate largely undirected interactions; focus on global interactions rather than controlling local events.
- Foster interaction to enable correlation; enable people/workgroups to work through conflicting constraints that inhibit their need preferences.

- Develop skills that enable productive surprises—these may involve role changes.
- Encourage emergence of structure and behavior in the project team.
- Reduce command and control behavior as much as possible (minimal leadership fiats [rules, agendas, powerful leadership vision, etc.]).
- Encourage networked intelligence of its constituent units, including encouraging creation of catalysts (or tags), which encourage development.
- Discourage strong coupling patterns within units of the organization, that is, encourage loose structures.
- Foster network building.
- Drop seeds of emergence.
- Encourage systems thinking.
- Encourage empowerment (Uhl-Bien et al. 2007).

COEVOLUTION

"When adaptable autonomous agents or organisms interact intimately in an environment, such as in predator–prey and parasite–host relationships, they influence each other's evolution. This effect is called coevolution, and it is the key to understanding how all large-scale CASs behave over the long term. Each adaptive agent in a complex system has other agents of the same and different kinds as part of its environment. As the agent adapts to its surroundings, various elements of its surroundings are adapting to it and each other. One important result of the interconnectedness of adaptive bodies is the concept of coevolution. This means that "the evolution of one domain or entity is partially dependent on the evolution of other related domains or entities" (Kauffman 1995). A commonly cited example, elephants thrive on acacia trees, but the latter can only develop in the absence of the former. After a while, the elephants destroy the trees, drastically changing the wildlife that the area can sustain and even affecting the physical shape of the land. In the process, they render the area uncongenial to themselves, and they either die or move on.

Work in biology on *fitness landscapes* is an interesting illustration of competitive coevolution (Kauffman 1995). A fitness landscape is based on the idea that the fitness of an organism is not dependent only on its intrinsic characteristics but also on its interaction with its environment. The term *landscape* comes from visualizing a geographical landscape of fitness *peaks*, where each peak represents an adaptive solution to a problem of optimizing certain kinds of benefits to the species. The *fitness landscape* is most appropriately used where there is a clear single measure of the *fitness* of an entity, so it may not always be useful in social sciences" (Ramalingam et al. 2008, pp. 53–54).

Also refer to Dawkins (2006) for a profound discussion of coevolution.

MODULARITY

Modularity is typically defined as a continuum describing the degree to which a system's components may be separated and recombined. It refers to both the tightness of coupling between components and the degree to which the *rules* of the system architecture enable (or prohibit) the mixing and matching of components. Its use, however, can vary somewhat by context: it refers to the idea that a system is composed of independent, closed, domain-specific processing modules.

Modularity makes the complexity of the system manageable by providing divisions within the system, thus enabling parallel work (Baldwin and Clark 2004, p. 6).

When faced with a risky design process, which has a wide range of potential options, it is worthwhile experimenting with the modular breakdown. Options interact with modularity in a powerful way. By definition, a modular architecture allows modular designs to be changed and improved over time without undercutting the functionality of the system as a whole. A layered architecture, which in addition insists that the interfaces between any two adjacent layers remain relatively simple, is generally even better than a modular architecture.

A series of design rules can be created by experimenting with modularity.

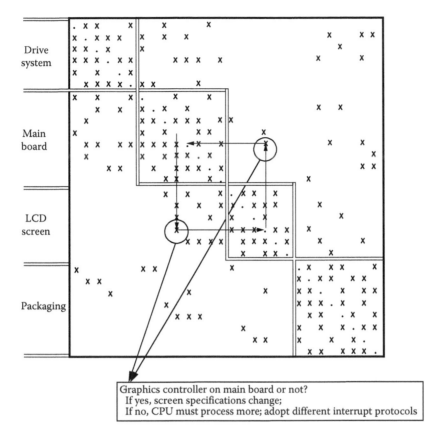

FIGURE 2.3 DSM map of a laptop computer. (From Bartolomei et al. Analysis and application of design matrix, domain mapping metrics and engineering system metrics frameworks, MIT. With permission.)

The design of complex systems can be mapped through the use of a DSM (Bartolomei 2007) that integrates the interactions between various components of the system. Thus, changes can be made to the membership of components that will support and enhance modularity.

Key research questions include as follows: Was a DSM created for the project? Were there adjustments made to the DSM? Were design rules created as part of the DSM experimentation? Was there a special design rules the team created? While this will probably have been done for defense projects, it may not have been done for health or transport projects.

It is worth noting that the complex system design proceeds from being interdependent to be modular through this process. The architect of the system must first identify the dependencies between distinct components and address them through design rules. Thus, close coupling is reduced. An example is shown in Figures 2.3 and 2.4.

COUPLING

Normally, core units of architecture are coupled, whereas noncore elements are not coupled.

VIEWS

Designing a complex project is a challenge that is made easier by considering the project from a number of viewpoints. Viewpoints can include the value of attaining the project goals, the business processes, the system architecture viewpoint, the viewpoint of multiple different users (White 2007; McCarter and White 2009), and viewpoints of the task domain.

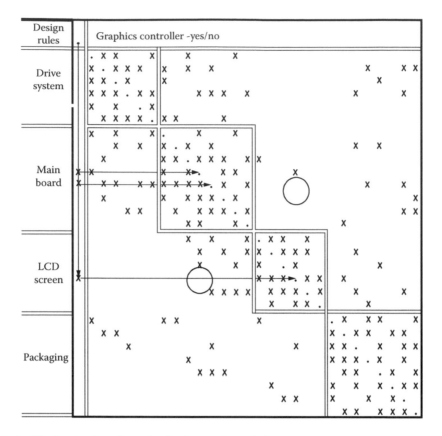

FIGURE 2.4 Eliminating interdependencies by creating a design rule.

A view is a representation of the system from the perspective of related concerns or issues (ANSI/ IEEE 1471–2000). White (2007) and McCarter and White (2009) define {view} as composed of any combination of the four-tuple {scope, granularity, mind-set, and time frame}.

A viewpoint is a template, pattern, or specification for constructing a view (ANSI/IEEE: 1471–2000).

SCALE, COMPLEXITY, AND STRUCTURE

Scale refers to the number of parts of a system that act together in a strictly coordinated way. An army has a larger scale than a battalion, and an aircraft carrier, surrounded by destroyers and other craft, has a larger scale than a single destroyer.

However, complexity interacts with scale: for example, an aircraft carrier in the open ocean, surrounded by its support ships, is in a less complex environment than in a bay in which speedy shore-based craft could threaten it. An army with the objective of overpowering and demolishing the opposition, with its support tanks and missiles, has large scale. However, a commando unit, with the task of attempting to distinguish between villagers who are friendly and those who are potential suicide bombers, is small scale but a complex task.

Some key points are as follows:

1. A calculation of aggregate force that can be applied by a system is a characterization of the largest scale of potential action.
2. When forces are organized hierarchically, the number of possible actions at a small scale increases as the number of small units (e.g., fire teams) increases.

3. The number of possible actions at a large scale increases as the number of larger units (e.g., battalions) increases.
4. Thus, the complexity profile roughly corresponds to the number of units at each level of command (individual, fire team, squad, company, or battalion).

However, it also depends on the following:

5. How independent the individuals are within teams?
6. How independent teams are within squads?
7. How independent squads are within companies and how independent companies are within battalions?
8. When the units at a particular level of organization are more independent, the complexity is larger at that scale. However, the possibility (complexity) of larger-scale action is smaller.

We consider *all of the units at each level*:

1. When considering the capabilities of forces in information age warfare, military technology should not be evaluated separately from force organization.
2. In a well-designed force, technology and force organization are inseparable.
3. Indeed, the command, control, communications, computers, intelligence, surveillance, and reconnaissance targeting (C4ISRT) system should be designed in conjunction with military organization.
4. The role of information and information processing is tightly linked to functional capabilities since the specific information needed (and not too much more) must be present in the right place at the right time to enable effective system functioning.
5. Thus, in complex systems, the distinction between physical and informational aspects of the system is blurred.

Thus, a relationship exists among scale, complexity, and structure (shown in the following):

A command and control structure is effective for large-scale and less complex tasks, whereas a small-scale operation, which is highly complex, requires more independence of the operatives.

Scale	Complexity	Structure
High	Low	Traditional top-down command and control: • Analyze and respond • Create panels of experts • Listen to conflicting ideas
High	High	Leaders with abstract reasoning, business acumen, and a good understanding of the strategic issues: • Probe, sense, and respond • Increase levels of interaction and communication • Use methods that can help generate ideas • Open up discussion through large group methods • Create environments and experiments that allow patterns to emerge • Encourage dissent and diversity and manage starting conditions • Monitor for emergence
Low	Low	Traditional top-down command and control
Low	High	• High-capability people or units • Diverse backgrounds • Comfort with ambiguity • Comfort with being challenged • Encouraging challenges • Emotional intelligence

The Helmsman (2009) study of all 32 complex defense projects found examined the skills required of project participants on SoS projects and found that the following skill sets were deeply important, as well as comfort with ambiguity and emotional intelligence.

RECOGNITION THAT THERE ARE UNKNOWN UNKNOWNS

Traditional risk management identifies a number of risks to the project and develops a risk response plan. While this is very useful, it can provide the project with a degree of false security in that project team members believe risks are adequately dealt with. In developing a risk management plan, people are largely responding to the known unknowns, whereas the unknown unknowns are neglected. The *flip side* aspect of risk mitigation that is also often neglected, especially in complex military environments, is a pursuit of opportunities (White 2006) (White 2007).

Lock et al. (2007) recommend:

Do Multiple Parallel Trials: Selectionism

> In the face of uncertainty, one launches several solution attempts, or sub-projects, each with a different solution strategy to the problem in hand. If the solution strategies are sufficiently different, one would hope that one of them will succeed and lead to a useful outcome. Success depends on generating enough variations so that, ex post, we obtain desirable results.

Loch et al. (2007, p. 124)

Loch et al. (2007, p. 131) refer to a study of 56 new business development projects in which the one key difference between firms that are able to adapt to a changing environment and those that fail to do so lies in their ability to apply selectionism, that is, creating a variety of solution approaches. As the degree of environmental change increases, that is, as the number of unknown unknowns increases, selectionism increases in importance and produces better solutions than continuous improvement.

WHAT MAKES SELECTIONISM WORK?

Loch et al. identify the following reasons for selectionism to be successful (2006, p. 133), which are categorized as questions:

1. In what space are we going to form alternatives? What is the set space of feasible and practical solutions?
2. How many options, sets, or experiments can one afford to carry out simultaneously?
3. When do we stop options or projects?
4. How does one ensure that the selection indeed happens, and how does one create a commitment to the selected outcome?
5. Key to the success of selectionism is the ability to integrate learning across the projects. How does one leverage the learning or other benefits from the nonselected experiment?

Toyota is very careful to determine the set space. Functional departments within Toyota are required to develop their systems simultaneously with defining feasible regions from their perspective. In parallel, they put the primary design constraints on the system based on their experience, analysis, experimentation, and testing, as well as outside information. These design constraints are translated into engineering checklists, which are used throughout the project to filter possible trials or sets.

HOW MANY TRIALS IN PARALLEL?

Loch et al. (2007, p. 134) comment that the answer to the question depends on four drivers:

1. What is the cost of one trial or experiment?
2. How soon can one decide to stop a trial?

3. What is the degree of uncertainty, or unk unks, and complexity one needs to overcome?
4. How strong is the managerial capability of the organization to live with ambiguity of the existence of parallel projects?

Clearly, the more complex the problem, or subproblem, the more trials are required. This is strongly backed by research literature.

Clearly, the organization needs a manager who can manage multiple projects, which means juggling multiple balls at once. Good system architecture is required.

LEVERAGING THE BENEFITS OF NONSELECTED OUTCOMES

Even though a trial may not have led to a result that will be used on the current project, there will be benefits that can be enjoyed. These benefits are usually embedded in people, and getting these benefits means careful career management of these people.

SELECTIONISM AND LEARNING IN PROJECTS

In order to gain the best combination of selection and learning, Loch et al. (2007, p. 145) outline four different approaches. "The Darwinian approach is pure selectionism with projects running in parallel and allowed to compete, the unk unks are revealed, and the best project is chosen after this."

VALUE OF DARWINIAN AND SEQUENTIAL LEARNING

For the Darwinian selection process operated by offering multiple models of a manufactured product, the benefit of the information is significant but the cost of developing multiple solutions and projects is quite high. Clearly, Darwinian selection is favored when the cost of running multiple trials is relatively cheap and/or the cost of delays is quite high.

GOVERNANCE

Morris, Place, and Smith comment

> Systems of systems introduce complications for information technology (IT) governance because their individual system components exhibit considerable autonomy. This technical note examines the ways in which six key characteristics of good IT governance are affected by the autonomy of individual systems in a system of systems. The characteristics discussed are (1) collaboration and authority, (2) motivation and accountability, (3) multiple models, (4) expectation of evolution, (5) highly fluid processes, and (6) minimal centrality. This report examines each characteristic in detail and, where possible, provides guidance for the practitioner.
>
> **CMU/SEI-2006-TN-036 (2006)**

SoS governance processes must take into account the governance policies of many, primarily autonomous, organizations. This allowance will require the adoption of more democratic governing processes. Adopting those processes will not be easy, because individual systems are often components of multiple SoS. In these cases, an organization may be a party to negotiations for multiple SoS governance policies.

According to the IT Governance Institute, the "overall objective of IT governance … is to understand the issues and the strategic importance of IT, so that the enterprise can sustain its operations and implement the strategies required to extend its activities into the future. IT governance aims at ensuring that expectations for IT are met and IT risks are mitigated" (ITGI 2003).

"Collaborative system-of-systems governance involves abandoning the notion of rigid top-down governance of IT processes, standards, and procedures and adopting peer-to-peer approaches.

Such collaborative system-of-systems governance is clearly at odds with the natural tendency of business and military organizations, because it means that the 'chain of command' must evolve to a 'web of shared interest'." Collaborative SoS governance requires cooperation between separate authorities, even when there is no formal agreement. Carney and associates observed in a case study on infrastructure replacement that distrust between the two government organizations led to initial difficulties in the relationship between contractors (Carney 2005). In addressing and amending the characteristics of collaborative governance among public and private sector entities, Freeman (1997) provides a model of SoS governance that we have adapted here:

A problem-solving orientation. This viewpoint brings relevance and focus to the SoS governance activities.

Participation by interested and affected parties in all stages of decision-making processes. This democratic process facilitates effective problem solving and buy-in.

Provisional solutions. Policies are recognized as being subject to revision, which requires willingness to move forward under conditions of uncertainty and to reconsider goals and solutions.

Accountability. Traditional top-down oversight may be supplemented or replaced by self-disclosure and monitoring through community and independent (third-party) organizations.

A flexibly engaged agency. It works in many roles, as appropriate—including convener and facilitator of negotiation processes, provider of incentives for participation and sharing, technical resource provider, and funding source (Freeman 1997).

While SoS governance policies will certainly differ by role, there might be additional variations. For instance, there could be a shift in focus from governance primarily at design time (e.g., *use standard X, document design by standard Y*) to governance for deployment and use of capabilities. The US Army software blocking policy (SWB) (JITC 2001) offers one instance of this aspect. It follows that there is also need for a model of governance at runtime, providing policies on how capabilities can, or should, be used.

SYSTEMIGRAM

Boardman and Sauser's systemigram (Boardman and Sauser 2008) is a simple concept in which all of the entities that have an effect on the project are recognized. The software then arranges these and shows arrows between an organization and the organization on which it has an effect. On each arrow is a verb describing the effect. Hence, a project manager can receive a full picture of where their organization sits on the hierarchy (Figures 2.5 and 2.6).

EPILOGUE

During a later stage in the preparation of this book, the first author of this chapter prepared a survey on relevant aspects that he distributed to those who the coeditors felt were credible experts in the field of complex systems. *The following contributors* (listed in alphabetical order by last name) *to this survey are thanked for their participation*:

Braha, Dan
Clemens, Walter
Cotsaftis, Michel
Dagli, Cihan
DeRosa, Joe
Dove, Rick
Ergin, Nil
Glouberman, Sholom

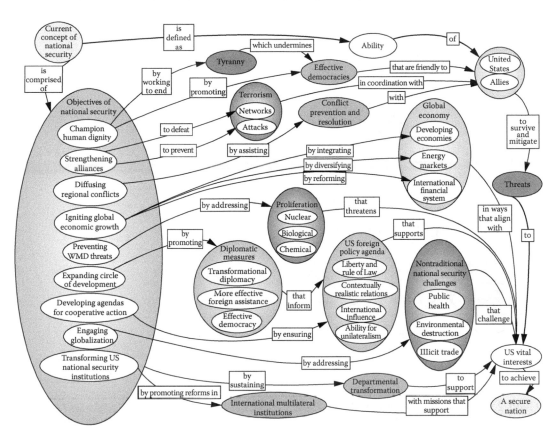

FIGURE 2.5 Example of a systemigram untransformed. (From Boardman, J. and Sauser, B., *Systems Thinking: Coping with 21st Century Problems*, CRC Press, Boca Raton, FL, www.Boardmansauser.com., 2008. With permission.)

Grisogono, Anne-Marie
Hayenga, Craig
Hitchins, Derek (declined)
Krob, Daniel
Marion, Russ
Nelson, Amy
Sheard, Sarah
Sousa Poza, Andres
Uhl-Bien, Mary

Sixteen of the 17 contributors (indicated by scrambled index numbers anonymously representing the contributor list earlier) completed the survey with results that are documented in the following. Only one contributor added a few alternative definitions of some of the terms so they and his associated ratings are omitted here.

Originally, the coeditors intended to take the results of the survey, extract those relevant aspects that seemed to be of greatest importance in the opinion of the experts surveyed, and ask authors of this book's case studies to address at least those relevant aspects in their case studies. This turned out to be too ambitious a task, considering the urgency in finishing the book without unduly stretching its preparation schedule. Thus, the authors of this chapter decided to present, analyze, and summarize the survey and let readers of the book judge for themselves how these relevant aspects affected the various case studies.

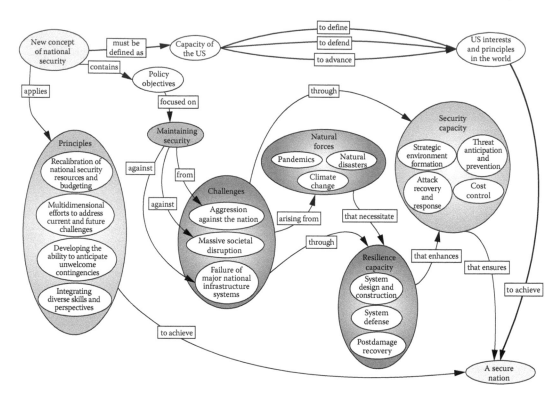

FIGURE 2.6   Example of a transformed systemigram. (From Boardman, J. and Sauser, B., *Systems Thinking: Coping with 21st Century Problems*, CRC Press, Boca Raton, FL, www.Boardmansauser.com., 2008. With permission.)

SURVEY AND RESULTS

The actual verbatim survey request except for the publication date in the *Introduction* below and the responses (covering five pages of this epilogue and organized by an anonymous key index code that shows individual responses via the same integer index) are shown in the following. The survey's rather long length and the time required to complete the survey were among the concerns of the coeditors. This attests well of the dedication of the experts who responded.

CONSOLIDATED

Relevant aspects of complex systems from complexity theory

We mean relevant to managing a complex system

Dear complex systems expert,

Introduction: The results from the survey in the succeeding text will form part of a book of case studies on complex systems that will be published late in 2014 by Taylor and Francis. If you complete the survey in the following, your response will be kept anonymous, but you will be named in the book as one of the recognized authorities on complex systems.

The editors are as follows: Dr. Alex Gorod, Dr. Brian White, Dr. Jimmy Gandhi, Dr. Brian Sauser, and myself (Professor Vernon Ireland).

Directions: Please indicate the degree to which the concepts in the following table are relevant to an understanding of complex systems.

Please mark up the following table with an "X" in the appropriate column for each concept. The columns are as follows:

0 = Not relevant
1 = Useful concept
2 = Somewhat relevant
3 = Quite relevant
4 = Very relevant
5 = Essential concept

Please add any concepts that you feel should be added and rate these, as well. Feel free to suggest a better definition, if you wish, at the bottom of the table.

Then, send the marked-up table to vernon.ireland@adelaide.edu.au.

Analysis

The second author of this chapter created an Excel spreadsheet for replicating and processing the response data. A score representing the expert responses received along the Likert-like scale (0, 1, 2, 3, 4, or 5) of the six columns for each concept was calculated. This row score is defined as the sum of the six column factors of a given row, where each factor was the product of the scale number (0, ..., 5) times the number of respondents that selected that number. Naturally, the average score for a concept is the row score divided by the number of respondents for that row. Similarly, the variance is defined as the sum of the squares of the difference between the scale number and the average score, all divided by the number of respondents for that row. The standard deviation is the square root of the variance.

A bar chart showing the average scores and corresponding standard deviations for many of the concepts is shown in Figure 2.7 (following the survey results of Table 2.1). The number of the concept from the Table 2.1 is shown vertically, and the scale number is shown horizontally. Because there are so many concepts, it was decided to only include those concepts that received an average score of 3 or better, that is, only those concepts that the respondents felt at least met the *3 = Quite relevant* criteria (on the average). Even so, 34 concepts qualified as a *3 = Quite relevant*, *4 = Very relevant*, or *5 = Essential* concept.

It is noted in passing that 3.1–3.13 and 3.15–3.19 (various architecture pronouncements), 5 (autonomous and independent systems), 6 (causal loop diagrams), 7 (cognitive fog), 8 (chaos), 12–16 (complex systems A, B, C, D, and E), 17 (complicated projects), 18 (context), and 26 (governance) did not make the list of at least *3 = Quite relevant*.

The reader is invited to peruse the bar chart and the tables to ascertain which concepts received the higher (or lower) scores and the lower (or higher) standard deviations. The authors of this chapter note that

1. (In decreasing order of average scores) Concepts 1 (adaptation), 31 (nonlinearity), 22 (emergence), 4 (attractor), 28 (interdependence), 9 (coevolution), 18 (context), and 34 (resilience) received average scores of *4 = Very relevant* or more
2. (In increasing order of standard deviations) Concepts 1 (adaptation), 9 (coevolution), 19 (coupling), 18 (context), 4 (attractor), 31 (nonlinearity), 34 (resilience), 28 (interdependence), and 36 (sensitivity to initial conditions) had the smallest standard deviations
3. Concepts 37–43 exhibited significantly higher standard deviations, indicating that the experts had wider disagreement on the relative importance of those concepts, that is, several felt that these concepts were merely *0 = Not relevant*, *1 = Useful concept*, or *2 = Somewhat relevant*

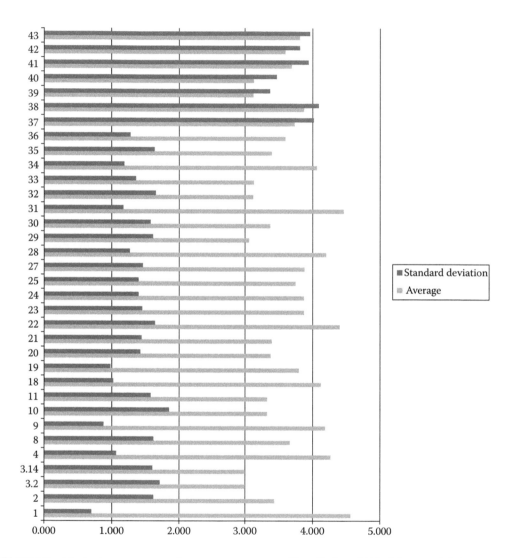

FIGURE 2.7   Most relevant concepts from survey responses, *Note*: Refer to Table 2.1 to match index numbers to the associated complex systems concept.

TABLE 2.1

Results of Survey About Relevant Aspects of Complex Systems

Complex Systems Concept	Your Rating of Concept					
	0	1	2	3	4	5
1 Adaptation — System responding to changes in its external environment.				1, 6	3, 13, 15	2, 4, 5, 7, 8, 10, 11, 12, 14, 16, 17
2 Agents — A collection of properties, strategies, and capabilities for interacting with artifacts and other agents (Wikipedia).	2	8, 10	4	6, 14, 15	3, 5, 13	1, 7, 11, 12, 16, 17
3.1 Architecture — Architecture is the embodiment of strategy.	1, 7, 11, 16	8, 12, 15, 17	3, 4, 13	6, 14		5, 10
3.2 Architecture is an abstract description of the entities of the system and interconnections and interaction.	11, 16	8, 12		6, 13, 14, 15, 17	3, 7	1, 4, 5, 10
3.3 Architecture is the link between form and function.	11	1, 8, 12, 17	15	13, 14,	7, 16	3, 5, 10
3.4 Architecting allows the problem and solution to be developed in parallel.	7, 11	13, 1, 16, 17		1, 3, 4, 6, 10, 12, 14	5	
3.5 Architecture addresses values.	1, 7, 11, 17	6, 8, 12, 13, 15	3, 4, 8, 16	14		5, 10
3.6 Architecture provides rules to follow when creating a system.	11	3, 6, 8, 16	4, 7, 12	1, 14, 15, 17		5, 10
3.7 Architecture is important because it provides guidance and structure to later phases of system design.	11	3, 8	6, 12, 16	1, 4, 15, 17	7, 13	5, 10, 14
3.8 Both requirements and architecture are important; however, architecture needs to be recognized as having a dominant role in the development of a complex system.	7, 11, 15	8, 13, 14	3, 12, 16	4, 6, 17		1, 5, 10
3.9 Architecting deals largely with the unmeasurable using nonquantitative tools, whereas engineering uses the hard sciences.	1, 7, 8, 11, 17	12, 13, 14	10, 15, 16		3, 4, 6	5
3.10 The word architecting refers only to the process; architecting is an invented word to describe how architectures are created, much as engineering describes how engines and other artifacts are created.	7, 8, 11, 17	12, 14	3, 15, 16	4	5, 13	1, 10

(continued)

TABLE 2.1 (continued)
Results of Survey About Relevant Aspects of Complex Systems

Complex Systems Concept		Your Rating of Concept					
		0	1	2	3	4	5
3.11	Architecting is a response to complexity in the system and explains why a single optimum solution never exists: there are just too many variables.	7, 11, 17	3, 8, 13	16	1, 4, 12, 15	6, 14	10
3.12	Critical details in the architect's design will be the system's connections and interfaces and the system's components; this combination produces unique system level functions; this is because the system specialists are likely to concentrate most on the core and least on the periphery of their subsystems.	11	6, 7, 16	8	3, 12, 13, 15, 17	4, 14	1, 5, 10
3.13	Architecting provides tools to define interfaces.	11	3, 6, 7, 8, 12, 16	4, 13, 15, 17		14	1, 10
3.14	Architecting invites abstract thinking about unifying themes based on structure and arrangements.	11	3, 8, 12	16	15	4, 6, 7, 13, 14, 17	1, 10
3.15	Architecture translates SoS capability objectives into high-level requirements over time.	3, 11	7, 8, 12	6, 16	1, 4	14, 17	10
3.16	Architecting identifies the system's boundaries and business environment.	11	3, 6, 12, 13, 16	8	1, 7, 15, 17	4, 14	10
3.17	Architecting develops the system's purpose.	3, 7, 11	12	1, 8, 13, 16, 17	15	6	10
3.18	Architecting develops the organizational or business processes.	11	3, 12, 14	1, 4, 7, 8, 15, 16, 17		13	
3.19	Architecting addresses evolutionary change of requirements.	3, 7, 11	12, 17	8, 15, 16	1, 4	13, 14	6, 10
4 Attractor	Islands of stability in a sea of chaos; attractor basins can be seen as stable states or where the system ends up; appearance of radical novelty; emergence cannot be associated with the sum of the parts of a system.			7, 15	16	8, 13, 14	1, 3, 4, 5, 6, 10, 11, 12, 17
5 Autonomous and independent systems	Systems within a SoS that are built into the system but were designed for another purpose and retain degrees of independence.	6	2, 8, 13, 17	7, 11, 16	1, 4, 12, 15	3, 14	10

#	Term	Description						
6	Causal loop diagrams	Expressions of systems operating that may not be obvious, which aid in visualizing how interrelated variables affect one another; the relationships between these variables, represented by arrows, can be labeled as positive or negative.	2, 6, 13, 16		1, 7, 8	11, 15, 17	3, 4, 10, 14	5, 12
7	Cognitive fog	Primarily when the project finds itself in a fog of conflicting data and cognitive overload; projects without cognitive fog were more likely to deliver a product.	6, 7, 16	2, 8, 14	1, 4, 11, 12, 13		15, 17	5, 10
8	Chaos	A state of disorder: there are both recursive and nonlinear behavior.	3, 16		7		4, 5, 8, 10, 11, 13, 14	1, 2, 6, 12, 17
9	Coevolution	The influence of one system's evolution on another's evolution: examples are adaptable autonomous agents or organisms interacting intimately in an environment, such as in predator–prey and parasite–host relationships.				3, 13, 14, 15, 16	4, 6, 8	1, 2, 5, 7, 10, 11, 12, 17
10	Complexity	An intricate arrangement of the system in which the elements can change in response to each other within the system and the external environment.	6, 10	15	3, 7	4, 8	2	1, 5, 11, 12, 13, 14, 17
11	Complex environment	An environment that is difficult to understand and predict how it will operate; frequently changing environment.	6	7	3, 4, 8	10, 15	2, 13, 14	1, 5, 11, 12, 17
12	Complex systems A	The project was essentially integrating existing assets to form a larger system (e.g., the management and control of rivers or road systems by states or food distribution from restaurants to needy users or power networks).	3, 5, 6, 16	12, 17	4, 10, 14	13	8, 11, 15	1
13	Complex systems B	A traditional SoS in which an existing asset, built for another purpose, usually software based, is included in your project because of the power gained.	3, 5, 6, 7, 16	8, 12	1, 4, 10, 11	13, 14, 15	17	
14	Complex systems C	A system operating in a foreign environment where the law and culture are different.	3, 5, 6, 7, 16	1, 8, 12	4, 10, 15, 17	11, 13, 14,		
15	Complex systems D	A system with lack of unity of command.	6, 7, 16	8	3, 4, 10, 11, 15, 17	5, 12, 13, 14	1	

(continued)

TABLE 2.1 (continued)
Results of Survey About Relevant Aspects of Complex Systems

Complex Systems Concept		Your Rating of Concept					
		0	1	2	3	4	5
16	Complex systems E	A system that requires use of aspects of systems thinking, such as definition of stakeholders, definition of boundaries, and use of Checkland's soft system methodology or system dynamics to identify a solution.					
		6, 7, 16		4, 5, 8, 10	3, 11, 12, 13	14, 15	1, 17
17	Complicated projects	Traditional projects such as those managed by PMBOK®, IPMA's Competence Baseline, and other BOKS and SE as usual are managed under a reductionist framework; typical reductionist style projects are those in which a top-down command and control structure is effective; complicated projects such as space rockets, the man on the moon project, and others that are primarily governed by the rules of physics, chemistry, and engineering are ideal as complicated projects; examples of complicated projects include jet aircraft and vehicles.					
		5, 7, 15, 16	14	3, 17	4, 6, 10, 13	8, 11	1, 12
18	Context	The environment of a system.					
				8, 16	3	4, 6, 10, 13, 15, 17	1, 2, 5, 7, 11, 12, 14
19	Coupling	The degree of knowledge or relationship of one component with another in a system.					
				7, 8	2, 15, 16	3, 4, 11, 13 14, 17	1, 5, 10, 12
20	Degree of control	The extent to which system leaders can control outcomes.					
			1, 8	15	14, 16, 17	3, 4, 11, 13	6, 10, 12
21	Developing the problem and the solution in parallel	An important approach recognizing the benefits of simultaneously integrating problem understanding and solution development, which is in contrast to SE and traditional project management that responds to requirements or a brief.					
		12		15	2, 3, 4, 11, 14	1, 7, 10, 13, 16, 17	5, 6

#	Aspect	Definition						
22	Emergence	Emergence occurs as system characteristics and behaviors result from simple rules of interaction: individual components interact and some kind of property appears, something you couldn't have predicted from what you know of the component parts; surprises that cannot even be explained after the fact are of greatest interest; emergent behavior then feeds back to influence the behaviors of the individuals that produced it.			7	3, 15	4, 17	1, 5, 6, 8, 10, 11, 12, 13, 14, 16
23	Emergent properties	These can appear when a number of simple entities (agents) operate in an environment, forming more complex behaviors as a collective.	16		10	7, 15, 17	3, 4, 6, 8	1, 2, 5, 11, 12, 13, 14
24	Environment	The external forces acting on a system.	15		7, 16	10, 17	3, 4, 6, 8, 13, 14	1, 2, 5, 11, 12, 13, 14
25	Fitness	The ability to cope with complexity: to survive challenges and make the most of opportunity, a fit organism can process information about and deal with many variables.	16		3, 7	4, 10, 15	6, 8, 13, 17	1, 2, 5, 11, 12, 14
26	Governance	The act of governing: relates to decisions that define expectations, grant power, or verify performance. It consists of either a separate process or part of management or leadership processes (WIKI).	7	8, 11, 14	1, 3, 15	4, 6, 12, 13, 16		2, 5, 10, 17
27	Hierarchy or networked	Description of the arrangement of systems in a multisystem; hierarchical is top-down; networked in equal membership; hybrid is a combination of networked and hierarchical.	6, 7		8, 16	11	3, 15, 17	1, 2, 4, 5, 10, 11, 12, 13, 14
28	Interdependence	A relationship in which each member is mutually dependent on the others.	8			6, 7, 15	3, 13, 16	1, 2, 4, 10, 11, 12, 14, 17
29	Leadership style	Definition of whether the system leadership is primarily top-down command and control or distributed among a number of project participants.	7	8, 14	6, 15, 17	3, 16	1, 2, 4, 10, 13,	5, 11, 12

(continued)

TABLE 2.1 (continued)
Results of Survey About Relevant Aspects of Complex Systems

Complex Systems Concept	Your Rating of Concept					
	0	1	2	3	4	5
30 Modularity — Typically defined as a continuum describing the degree to which a system's components maybe separated and recombined; it refers to both the tightness of coupling between components and the degree to which the *rules* of the system architecture enable (or prohibit) the mixing and matching of components.		2, 6, 8	7, 14	3, 15, 16	12, 13	1, 4, 5, 10, 11, 17
31 Nonlinearity — The incidence of not operating in a simple linear fashion because feedback processes among interconnected elements and dimensions lead to relationships that exhibit dynamic, nonlinear, and unpredictable change.			7	10, 15		1, 2, 3, 4, 11, 12, 13, 14, 16, 17
32 Reductionism — An approach to understanding the nature of complex things by reducing them to the behaviors of their parts or to simpler or more fundamental things.	3	8, 17	1, 7, 10, 15	16	2, 6, 13	4, 5, 11, 12, 14
33 Requirement difficulty — Requirements that are considered difficult to implement or engineer are hard to trace to the sources and have higher degrees of overlap with other requirements; if there are a number of difficult requirements, it may be questionable that the system can be built at all; higher numbers of difficult requirements typified projects that had more changes in needs and more problematic stakeholder interactions.		7, 8	2, 6, 11, 17	3, 13	6, 12, 14, 15	1, 5, 10
34 Resilience — The ability of a complex system to cope and survive with stress and diversity.		7		3, 15, 17	4, 6, 8, 14	1, 2, 5, 10, 11, 12, 13
35 Scale — A measure of size including number of units, such as people, lines of code, tanks, or other aspects.		7, 8, 15, 17		6, 11, 14	3, 4	1, 2, 5, 10, 12, 13
36 Sensitivity to initial conditions — The initial conditions of complex systems determine where they currently are, and, consequently, two complex systems that initially had their various elements and dimensions very close together can end up in distinctly different places due to their nonlinearity of relationships.		7	3, 10, 15	11	2, 4, 6, 8, 14, 17	1, 5, 12, 13

#	Term	Description						
37	Self-organization	When a group of entities spontaneously comes together to perform a task.		7	15	10	2, 3, 4, 6, 8, 14, 17	1, 5, 11, 12, 13
38	System	A set of interacting or interdependent components forming an integrated whole or a set of elements (often called *components*) and relationships that are different from relationships of the set or its elements to other elements or sets (Wikipedia).		8	3, 6, 7	15, 17	2, 16	1, 4, 5, 10, 11, 12, 13, 14
39	SoS	Boardman and Sauser (2008) describe SoS as having the characteristics of autonomy and independence, belonging, connectivity, and diversity; when the extent of something is not clear.	6	3	4, 5, 8, 16	7, 15	1, 2, 10, 11, 13, 14, 17	12
40	Unclear boundaries	This can occur with regard to the sponsor, stakeholders, the systems in use, and other aspects.	7	1, 15	4	2, 8, 13, 14	3, 6, 11, 17	5, 10, 12
41	Unknown unknowns	Risks that one does not anticipate as being relevant.		7	4	13, 14	2, 3, 6, 11, 17	1, 5, 10, 12
42	Views	Multiple separate or concurrent ways of examining a system: viewpoints can include the value of attaining the project goals, the business processes, the system architecture viewpoint, the viewpoint of multiple different users, and viewpoints of the task domain.			7, 13	5, 8, 15, 17	2, 3, 6, 11, 14	1, 4, 10, 12
43	Wicked problems	A phrase originally used in social planning to describe a problem that is difficult or impossible to solve because of incomplete, contradictory, and changing requirements that are often difficult to recognize. The term *wicked* is used, not in the sense of evil but rather its resistance to resolution: moreover, because of complex interdependencies, the effort to solve one aspect of a wicked problem may reveal or create other problems.		13	7	11, 16	2, 3, 4, 6, 8, 12, 14, 15	1, 5, 10, 17

REFERENCES

ANSI/IEEE 1471-(2000). Recommended practice for architecture description of software-intensive systems; Institute of Electrical and Electronic Engineers.

Azani, C.H. (2008). System of systems architecting via natural development principles. *International Conference on System of Systems Engineering*, Singapore, 2–4 June 2008.

Baldwin, C. and Clark, K.B. (2004). Modularity in the design of complex engineering systems. www.mendeley.com/.../modularity-in-the-design-of-co

Bartolomei, J., Cokus, M., Dahlgren, J., de Neufville, R., Maldonado, D., and Wilds, J. (2007). Analysis and application of design matrix, domain mapping metrics and engineering system metrics frameworks, Engineering systems division. MIT.

Bartolomei, J.E. (2007). Qualitative knowledge construction for engineering systems: Extending the design structure matrix methodology in scope and procedure. Doctoral dissertation, M.I.T., Cambridge, MA. http://www.dtic.mil/cgi-bin/GetTRDoc?Location = U2&doc = GetTRDoc.pdf&AD = ADA470222. Accessed July 23, 2007.

Bar-Yam, Y. (2004). *Making Things Work—Solving Complex Problems in a Complex World.* Boston, MA: NECSI Knowledge Press.

Bjelkemyr, M., Semere, D., and Lindberg, B. (2007). An engineering systems perspective on system of systems methodology. *1st Annual IEEE Systems Conference,* Honolulu, HI, April 2007.

Boal, K.B., and Schultz, P.L. (2007). Storytelling, time, and evolution: The role of strategic leadership in complex adaptive systems. *The Leadership Quarterly* 18: 411–428.

Boardman, J. and Sauser, B. (2008). *Systems Thinking: Coping with 21st Century Problems.* Boca Raton, FL: CRC Press. www.Boardmansauser.com. Accessed 3 March 2014.

Braha, D., Minai, A.A., and Bar-Yam, Y. (2006). *Complex Engineered Systems.* Cambridge, MA: Springer.

Byrne, D. (1998). *Complexity Theory and the Social Sciences: An Introduction.* London, U.K.: Routledge. Accessed February 25, 2014

Capra, F. (1996). *The Web of Life.* London, U.K.: Flamingo/Harper Collins.

Carney, M. (May 2005). Corporate governance and competitive advantage in family controlled firms. *Entrepreneurship Theory and Practice* 29(3): 249–265.

Checkland, P.B. (1981). *Systems Thinking, Systems Practice.* Chichester, U.K.: John Wiley.

Cilliers, P. (1998). Complexity and postmodernism: Understanding complex systems. New York: Routledge.

CMU/SEI-(2006)-TN-036, *An Examination of a Structural Modeling Risk Probe Technique.* Pittsburgh, PA: Carnegie Mellon University.

Cotsaftis, M. (2007). What makes a system complex? An approach to self-organisation and emergence. mcot@ece.fr. http://arxiv.org/ftp/arxiv/papers/0706/0706.0440.pdf. Accessed 3 March 2014.

Crawley, E., de Weck, O., Eppinger, S., Magee, C., Moses, J., Seering, W., Schindall, J., Wallace, D., and Whitney, D. (2004). *System Architecture and Complexity.* Cambridge, MA: MIT.

Dagli, C.G. and Kilicay-Ergin, N. (2009). System of systems architecting. In Jamshidi, M. (ed.), *System of Systems: Innovations for the 21st Century*, Chapter 4, pp. 77–100. Hoboken, NJ.: Wiley & Sons Inc.

Dahmann, J. (2009). *Systems Engineering for Department of Defence System of Systems* (Jamshidi 2009), Chapter 9, pp. 218–231. Hoboken, NJ.:Wiley & Sons Inc.

Davies, L. (2003). *Conflict and Chaos: War and Education.* London, U.K.: Routledge Falmer. Preparing for Peace, The website of the Westmorland General Meeting 'Preparing for Peace' initiative. http://www.preparingforpeace.org/davies.htm. Accessed: 3 March 2014.

Dawkins, R. (2006). *The Selfish Gene: 30th Anniversary Edition—With a New Introduction by the Author.* New York: Oxford.

DeLaurentis, D. and Crossley, W. (2005). A taxonomy-based perspective for system of systems design methods. *IEEE International Conference on System, Man, and Cybernetics Conference*, Waikoloa, HI, October 10–12, 2005. Paper 0-7803-9298-1/05.

DeRosa, J.K., Grisogono, A.-M., Ryan, A.J., and D. Norman, D. 2008. A research agenda for the engineering of complex systems. SysCon 2008. *IEEE International Systems Conference (SysCon 2008).* Montreal, Quebec, Canada. April 7–10, 2008.

DoDAF V2.0. (2003). DoD Architectural Framework Version 1.0, the Deskbook and Volumes I & II, Department of Defence Architecture Working Group. Superseded by DoDAF V 1.5. US Department of Defense; Deputy Chief Information Officer, McLean Virginia.

Firesmith, D. (2010). Profiling systems using the defining characteristics of system of systems. Technical Note CMU/SEI 2010-TN-001. Pittsburgh, PA: Software Engineering Institute, Carnegie Mellon University.

Freeman, J. (1997). Collaborative governance in the administrative state. *UCLA Law Review* 45: 1–98.

Genscape. (2014). What caused the power blackout to spread so widely and so fast? Genscape's unique data will help answer that question. http://www.prnewswire.com/news-releases/what-caused-the-power-blackout-to-spread-so-widely-and-so-fast-genscapes-unique-data-will-help-answer-that-question-70952022.html. Accessed March 2, 2014.

Gharajedaghi, J. (2006). *Systems Thinking—Managing Chaos and Complexity: A Platform for Designing Business Architecture*, 2nd edn., Burlington, MA: Butterworth-Heinemann.

Glouberman, S. and Zimmerman, B. (2002). Complicated and complex systems: What would successful reform of medicare look like? Discussion Paper 8, Ottawa, Ontario, Canada: Commission on Future Health.

Grisogono, A-M. (2006). The implications of complex adaptive systems theory for C2. *CCRTS 2006: State of the Art and the State of the Practice*. San Diego, CA: Department of Defence Command & Control Research Program.

Haynes, P. (2003). *Managing Complexity in the Public Services*. Berkshire, England: Open University Press.

Helmsman Institute. (2009). A comparison of project complexity between defence and other sectors. Sydney, New South Wales, Australia: Helmsman Institute.

Holland, J. (1995). *Hidden Order: How Adaptation Builds Complexity*. New York: Helix Books.

IPMA. (2006). ICB—IPMA competence baseline for project management. Version 3.0. Nijkerk, the Netherlands: International Project Management Association.

ITGI. (2003). IT Governance Global status report. London, U.K.: PricewaterhouseCoopers.

Jackson, M.C. (2003). *Systems Thinking—Creative Holism for Managers*. Chichester, U.K.: John Wiley.

Jamshidi, M. (2009). *System of Systems Engineering—Innovations for the 21st Century*. Hoboken, NJ: John Wiley.

Jervis, R. (1997). *System Effects: Complexity in Political and Social Life*. Princeton, NJ: Princeton University Press (Mason and Mitoff 1981).

JITC. (2001). Department of Defense Dictionary of Military and Associated Terms. jitc.fhu.disa.mil/jitc_dri/pdfs/jp1_02.pdf. as amended through to 2009. Accessed February 26, 2014.

Kauffman S. A. (1993). The Origins of Order: *Self-organization and Selection in Evolution*. New York: Oxford University Press.

Kauffman, S. (1995). *At Home in the Universe: The Search for the Laws of Self-organisation and Complexity*. London, U.K.: Penguin Books.

Khoo, T. (2009). Domain engineering methodology, *IEEE SysCon 2009, 3rd Annual IEEE International Systems Conference*, Vancouver, British Columbia, Canada.

Knowles, R. N. (2002). *The Leadership Dance: Pathways to Extraordinary Organizational Effectiveness*. 3rd edn. Niagra Falls, NY: The Center for Self-Organizing Leadership.

Loch, C.H., De Meyer, T.A., and Pich, M.T. (2007). *Managing the Unknown: A New Approach to Managing High Uncertainty and Risk in Projects.* Hoboken, NJ: Wiley.

Maier, M.W. and Rechtin, E.T. (2009). *The Art of Systems Architecting*. Boca Raton, FL: CRC Press/Taylor & Francis Group.

Marion, R. and Uhl-Bien, M. (2001). Leadership in complex organizations. *Leadership Quarterly* 12(4): 389–418.

Mason, R.O. and Mitroff, I.I. (1981). *Challenging Strategic Planning Assumptions*. Chichester, U.K.: John Wiley & Sons, Inc.

Marion, R. (1999). *The Edge of Organization: Chaos and Complexity Theories of Formal Social Systems*. Thousand Oaks, CA: Sage.

McCarter, B.G. and White, B E. (2009). Chapter 3: Emergence of SoS, sociocognitive aspects. In Jamshidi, M. (ed.) *Systems of Systems Engineering: Principles and Applications*, Boca Raton, FL: CRC Press.

McCarter, B.G. and White, B.E. (2012). *Leadership in Chaordic Organizations*. Boca Raton, FL: CRC Press.

Meilich, A. (2006). System of systems (SoS) engineering & architecture challenges in a net centric environment. *IEEE/SMC International Conference on System of Systems Engineering*, Los Angeles, CA, April.

Mitleton-Kelly, E. (2003). Ten complex systems and evolutionary perspectives on organizations. In *Complex Systems and Evolutionary Perspectives on Organisations The Application of Complexity Theory to Organisations*! Oxford, UK: Elsevier Science Ltd. ISBN: 0-08-043957-8.

Newell, D. (2003). Concepts in the study of complexity and their possible relation to chiropractic healthcare. *Clinical Chiropractic* 6: 15–33.

Nicolis, G. and Prigogine, I. (1989). *Exploring Complexity*. New York: W.H. Freeman.

Norman, D. and Kuras, M. (2006). Engineering complex systems, Chapter 10 in Complex Engineered Systems. In Braha, D., Minai, A., and Bar-Yam, Y. (eds.) Science + Business Media. Berlin, Heidelberg, New York Springer 206–245. ISBN 3-540-32831-9. http://necsi.edu/publications/engcontents.html. Accessed: 16 March 2014.

Pinker, S. (1997). *How the Mind Works*. New York: W. W. Norton.

Plowman, D.A., Baker, L.T., Beck, T.E., Kulkarni, M., Solansky, S.T., and Travis, D.V. (2007b). Radical change accidentally: The emergence and amplification of small change. *Academy of Management Journal* 50(3): 513–541.

Plowman, D.A., Solansky, S.T, Beck, T.E., Baker, L., Kulkarni, M., and Travis, D.V. (2007a). The role of leadership in emergent, self-organization. *The Leadership Quarterly* 18: 341–356.

PMI (2013). *A Guide to the Project Management Body of Knowledge PMBOK® Guide*, 5th edn. Harrisonburg, VA: Project Management Institute.

Prigogine, I. and Stengers, I. (1984). *Order Out of Chaos: Man's New Dialogue with Nature.* London, U.K.: Heinemann.

Ramalingam, B., Jones, H., Reba, T., and Young, J. (2008). *Exploring the Science of Complexity Ideas and Implications for Development and Humanitarian Efforts.* London, U.K.: ODI, Working Paper 285.

Schelling, T. (1978). *Micromotives and Macrobehavior.* New York: Norton.

Sheard, S. (2013). Assessing the impact of complexity attributes on system development project outcomes. PhD dissertation, Stevens Institute of Technology, Hoboken, NJ.

Snowden, D.J. and Boone, M.E. (2007). The leaders framework for decision making. *Harvard Business Review* 85: 69–76.

Stacey, R. (1995). The science of complexity: An alternative perspective for strategic change processes. *Strategic Management Journal* 16: 477–495.

Stacey, R. (1996). *Complexity and Creativity in Organisations.* San Francisco, CA: Berrett-Koehler Publishers.

Stanford Encyclopaedia of Philosophy. (2006). Emergent properties. http://plato.stanford.edu/entries/properties-emergent/.

Strogatz, S. (2003). *Sync: The Emerging Science of Spontaneous Order.* London, U.K.: Penguin Books.

Uhl-Bien, M., Marion, R., and McKelvey, B. (2007). Complexity leadership theory: Shifting leadership from the industrial age to the knowledge era. *The Leadership Quarterly* 18(4): 298–318.

Urry, J. (2003). *Global Complexity.* Cambridge, U.K.: Blackwell Publishing Ltd.

Waldrop, M. (1994). *Complexity: The Emerging Science at the Edge of Order and Chaos.* London, U.K.: Penguin Books.

White, B.E. (2006). Enterprise opportunity and risk, *INCOSE 2006, 16th International Symposium*, Orlando, FL, July 9–13, 2006.

White, B.E. (2007). On interpreting scale (or view) and emergence in complex systems engineering. *1st Annual IEEE Systems Conference*, Honolulu, HI, April 9–12, 2007.

White, B.E. (2011). Co-Editor (with George Rebovich) and Chapter 5, 'Enterprise Opportunity and Risk,' author of book, *Enterprise Systems Engineering: Advances in Theory and Practice*, Boca Raton, FL: CRC Press.

Wikipedia. (2014). List of major power outages. Wikipedia, the free encyclopedia. http://en.wikipedia.org/wiki/List_of_major_power_outages. Accessed February 28, 2014.

Wilber, G.F. (2009). Boeing's SoSE approach to e-enabling commercial airlines. In Jamshidi M. (ed.), (pp. 232–256).

3 Application of Case Studies to Engineering Leadership/ Management and Systems Engineering Education*

*Brian E. White, Brian Sauser, Alex Gorod,[†]
S. Jimmy Gandhi, and Vernon Ireland*

CONTENTS

---

* Two earlier conference papers (White et al. 2012, 2013) were based on this chapter.

[†] Alex Gorod deserves major credit for (1) generating (in late 2010) the original idea, proposal, and impetus for writing this book, (2) assembling the coeditor team, (3) most of the list of possible case study authors, and (4) the initial outline suggested for use by case study authors. Brian White expanded this outline to make it more comprehensive. The proposed case study outline, which is the subject of the principal contribution of this chapter, was refined in late 2012 by the coeditor team.

SYNOPSIS

Case studies provide a means for highlighting and extracting practical principles and methods for shaping and accelerating progress in solving pressing real-world problems. Case studies inform burgeoning theories such as those associated with complex systems engineering where people are considered part of the system to be conceived, developed, fielded, and operated or extant systems targeted for improvement upgrades. This chapter advocates for the importance and value of case studies. Motivations and ideas for increasing the use of case studies in the world's most difficult present and future systems engineering environments are suggested. In addition, the importance of case studies to engineering management and systems engineering education and frameworks for their implementation are described. A case study template that can be used as a guideline in engineering education is presented.

INTRODUCTION

Engineering is about solving human-made problems. In this endeavor, scientific understanding; the laws of physics, chemistry, biology, etc.; and the formulations of mathematics are applied to effect appropriately as possible. Systems engineering (SE) is multidisciplinary, even transdisciplinary (Sage 2013), and like many other disciplines continually evolves to become even more relevant and effective as a practical approach to achieving worthwhile objectives. As of this writing, the case study technique has not yet been widely applied to or broadly used in engineering, engineering management, or SE education, at least not with respect to traditional or conventional SE. The premise is that case studies are even more important in complex, system of systems (SoS), and enterprise SE (ESE). Before proceeding further into the main chapter text, the reader may wish to consult Appendix 3.A to consider how various terms are defined.

Historically in education, the underlying theory and the reasoning behind it are taught first. Then students are asked to apply the theory in practice, initially in solving homework problems and conducting laboratory exercises, for example, and later in their real-world jobs. Another approach, as famously espoused by business schools (Harvard 2013, Paper Chase 2013), for instance, is to teach through case studies, the philosophy being that established theories do not work so well in arenas where animate objects like humans exercise free will in unpredictable ways. So case studies promise to open a new realm of trying to address *wicked* problems (Conklin 2005) at the *messy* frontier (Stevens 2008), in megasystem (Stevens 2011) SE environments.

Useful Notions to Keep in Mind

A key notion in addressing complex projects is not being held strictly to the paradigm established by the Project Management Body of Knowledge (Project Management 2008) in managing their nine knowledge areas of integration, scope, time, cost, risk, quality, human resources, communications, and procurement. Again, it is important to recognize that the Project Management Institute's reductionist techniques that were applied with notable success in going to the moon and subsequent space exploration projects, for example, have limitations when applied to many current earth-bound problems. Complexity is more profound in human-made crises such as religious–extremist terrorism, the enduring Middle Eastern disputes, world financial meltdowns, global-warming-induced climate changes, unbounded material growth, and overpopulation, to name a few.

One continually questions whether these vast problems can be usefully addressed. Worriers about these crises are clearly crying out for solutions. But it is abundantly clear that such problems cannot be solved by the traditional methods applied to normal projects. Traditionally, project sponsors and customers have attempted to package projects into controllable, confining spaces to obtain clear scopes, well-defined requirements, and bounded costs and schedules, all of which reduces the opportunities for evolutionary change and truly effective solutions. When engineers attempt to solve such problems with reductionist techniques, they and their customers, sponsors, and especially the system users and/or operators are usually disappointed.

Instead, concepts of systems thinking are central to successfully addressing such complex projects as those noted earlier. This also has strong implications for the future direction of engineering education. The following notes briefly describe a number of systems thinking techniques (Jackson 2003) that contribute to the proper definition of project stakeholders, their objectives, underlying assumptions, and possible methods for solving these wicked (Conklin 2005) problems.

Strategic assumptions surface testing recognizes the benefits of various stances of a range of participative, adversarial, integrative, and managerial-minded stakeholders and locates them on a certainty/importance scale.

Soft systems methodology is a most powerful technique for solving wicked problems especially using rich pictures developed from many conceptual models of the real world and enhancing these by using additional perspectives (or modalities) (Bergvall-Kåreborn 2002) including faith, love, justice, social intercourse, feeling, and sensory perception. These are especially relevant in cross-cultural and/or international conflicts, for example.

Critical systems heuristics relates to the partial presuppositions that underpin system judgments. This methodology provides Ulrich Werner's twelve boundary questions that affect project scope and focuses on who is marginalized and suggests techniques that allow these groups to be heard. Emotive forces in groups are recognized.

Postmodern systems thinking recognizes conflict among groups and critically questions (1) power relationships, (2) the role of language, (3) the extent to which people are self-determining, and (4) the roles of signs and images and provides a technique for first- and second-phase deliberation, debate, and decision.

Total systems intervention asserts that the traditional approach has been focused on the functional aspects, whereas Linstone's approach (Linstone 1999) focuses on technical, organizational, and procedural aspects that act as filters through which systems are viewed.

Such techniques as those named earlier will significantly broaden the education of engineers and make them much better grounded and equipped to help solve complex problems.

Abstract theory is fine and a critical part of everyone's ongoing lifelong education about why the world works the way it does. Indeed, that's the main purpose of science, the illumination and dispelling of mysteries and the deepening of our understanding of natural and human-made phenomena. But don't ignore storytelling as a powerful activity that creates awareness and motivation! There is no substitute for practical examples showing what really works well—or what doesn't. This is why good case studies in SE are advocated, particularly in the less familiar categories of complex systems, SoS, and enterprises. Case studies have been used effectively for many years in educational fields of not only business but also social studies, psychology, medicine, and health care.

In a growing era of complex systems, where solutions to problems are more systemic and less linear—and often quite nonlinear—case studies can and will have a vital, pervasive, and stable home in engineering education.

Examples of How Case Studies Can Benefit Engineering Management Education

As indicated earlier, the use of case studies has not yet permeated the practice of conventional SE or SE education. However, there are plenty of signs that leading academic institutions are bent on rectifying this situation.

A simple Google search yields sources to numerous examples of case studies already utilized in engineering education. For example, MIT's Engineering Systems Division is on the forefront of educating graduate students in complex systems engineering (CSE) and enterprise systems engineering (ESE). Many of their case study efforts are publically available (MIT 15 February 2013). In addition, MIT's System Design and Management (SDM) master's degree program produces case studies (MIT 23 April 2013). Similarly, doing a Google search for case studies in engineering management yields another publically available resource (Engineering 2014). The Stevens Institute of Technology hosts a plethora of case studies and related material (Stevens 2013). There is more on engineering case studies at the University of Vermont (Vermont 2013), the University of Virginia (Virginia 2013), and the University of Texas (Texas 2013). The University of Southern California's (USC's) Viterbi School of Engineering offers courses in case studies (USC 2013). There is less available on case studies (it seems) at the University of Illinois (U of I 2013).

It is left to the reader to explore any of these sources for case studies of interest or for additional references on how to conduct a case study. A classic book (Yin 2003) on preparing case studies is also recommended as a reference. A fundamental paper on case studies in more conventional SE and systems management is also recalled (Friedman and Sage 2004). Another case study effort underway, by the IEEE SoS Technical Committee for the Trans-Atlantic Research and Education Agenda in Systems of Systems (T-AREA-SoS) project, is also worth noting (IEEE 2003).

Changing Mindsets

Engineering has long been regarded as a set of technical processes (based upon the application of practical methods and scientific knowledge) that is used by some people (e.g., engineers) in attempting to solve a broad class of problems intended to improve the lives of human beings. As technology, our capabilities, and human-made constructions, that is, systems, have evolved, further advances have become increasingly more complex (not just more complicated) to where many of our intentions have far outrun the availability of effective methodologies and artifacts for accomplishing the goals. Likewise, the complexity of engineered systems has increased over the years precipitating numerous complex systems project failures primarily due to the lack of an appropriate systemic perspective (White 2008). The word *complex* or *complexity* is often used to describe this state of affairs (as in the earlier text and many peoples' vernacular). Nonetheless, a fundamental (paradigm) shift in our mindset* is required to grasp the true nature of the issue, namely, that people need to be considered part of any engineered system. The world is far beyond the machine (or even information) age; hence, why are tools still talked about instead of *chaordic* (Hock 1999) *agents* or some other more appropriate word or phrase? Global concerns and sociotechnical systems are now much stronger influences. Accordingly, the technical influences of political, operational, economic as well as technological aspects must be embraced. Beyond these factors, psychology, sociology, organizational change theory/management, and even philosophy (Boardman and Sauser 2008) must be considered if there is to be much hope in achieving accelerated progress in the engineering of systemic solutions.

---

* Although mindset is being used, *mindsight* is a better word, perhaps, since that term emphasizes perspective and suggests more flexibility [due to Daniel J. Siegel, MD, in (Hawn 2011: xiv)].

However, beyond emphasizing the crucial direct influences of people in engineering and leadership/management, endemic effects that may be only indirectly related to or even independent of human behavior must be kept in mind. Complex systems do not necessarily respond to reductionist approaches or follow predictable paths such as that occur in many engineering and science projects, especially when dealing with systems over which there is little or no control, for example, the *war on terror* since 9/11.

EDUCATING YOUNG PEOPLE

As has been demonstrated throughout history, younger generations have opened the doors to many cultural and technological paradigm shifts. In recent years, this demographic has been much more attuned to some of the key traits of complex human systems, for example, sharing information, collaborating, and self-organizing. Whether this perspective will be sustained into adulthood and be more conducive in solving humanity's major problems compared to previous generations remains to be seen. There is an obligation to educate and train current and future technical leaders to sustain these traits not only to better address traditional technological problems but also to recognize and contribute to solving the world's more complex problems. A concerted effort must be made to create appropriate learning conditions and facilitate further educational development along these lines.

Unfortunately, most schooling methods, particularly in the United States, are more linear in their philosophies and approaches where students come to believe that every question has a known (or perhaps unknown) answer. Methods must be found to enhance their learning by allowing students to ask and contemplate questions for which there may not be a single answer or when potential solutions are unconstrained by political, dogmatic (or other types of) biases. A profound understanding of complexity should be nurtured and instilled into our children's minds, so eventually they can better deal with improving the quality of human life and the sustainability of the planet. The United States, for one, has seen a decline in engineering and technical talent, so more must be done to influence young people toward entering the engineering profession.

EDUCATING PRACTITIONERS

The solution to the challenge of engineering and leading/managing complex systems is itself a complex system. As part of the solution, educating and training our current and future technical leaders and providing suggested changes in their mindsights are imperative. Case studies in education can be a platform through which the analysis, knowledge application, and drawing of conclusions can occur to facilitate coping with the most complex systems. Case study learning has proven successful in the training of business leaders (Case Studies 2012) with real-life examples of the strategies and tactics used by leading businesses to succeed globally. A valuable characteristic of case studies is that they can support a holistic understanding and interpretation of the systems of action or interrelated activities engaged in by the participants. This book purports to help educate technical leaders/managers in how to engineer and lead/manage complex systems. This chapter lays a foundation and suggests a framework for case study development. Later, a case study template is offered that not only has been used by many authors of the case studies presented in this book but also could be used by others as a guideline for developing case studies for engineering education.

WHY DO CASE STUDIES?

Systems engineers can have the best ideas for promising ways forward in engineering difficult systems. They view and present their suggestions as abstract, general, theoretical, logical, reasonable, practical, rational, and emotionally compelling ideas. Frequently, they do not even think of case studies as being needed. Instead, they simply draw upon their experience in performing SE.

This learning is usually not acquired in an academic environment* but is accumulated on the job, hopefully while being mentored by a good systems engineer. And good program managers listen to their systems engineers and often take their advice—but not always!

Prove It!

Program managers, project leaders, bosses, and other key stakeholders, especially those that have already achieved a measure of career success in their current organizational positions by operating differently, want proof these SE ideas will work, especially those associated with CSE that may seem more radical. Before committing to, supporting, and promoting any suggested course of action, they ask the systems engineer "Prove to me this will work!"

In his or her defense, the systems engineer might say "This is not a mathematical theorem. People are not predictable and their behaviors are difficult, if not impossible to model accurately. Therefore, you need to have some faith and take an informed risk in pursuing promising opportunities in trying these complex system interventions." To help alleviate these stakeholders' reluctance or fears, the systems engineer can promote modeling and simulation as a (perhaps with a lower negative consequence outcome if it does not work) means of exploring the ideas to discover what might happen in practice.

Modeling and Simulation

Agent-based modeling (ABM) (Bonabeau 2002) is an established method that may show skeptics or naysayers whether the systems engineer's bright ideas are truly promising. This is done by giving several autonomous actors or agents that represent system constituents, including stakeholders, a few simple rules of interaction. Then the model is exercised, with the agents collectively interacting with each other, usually over a large number of iterations. One advantage of this is that unexpected events are likely to occur that could not have been predicted even when each step of the simulation is completely deterministic, for example, as in the appearance of a *glider* in game of life experiments involving simple 2D automata (Abbott 2005).

Case studies can be used as initial conditions for modeling and simulation. What about connecting case studies to ABM? The results of case studies can be used to guide ABM through tailoring the rules of engagement among the actors. Then the simulation can be rerun to see how well the model mimics the behaviors of the subject system of the case study. Of course, this means time and effort must be invested in defining, selecting, and conducting suitable case studies unless one is lucky enough to already have a good and relevant case study on hand.

Advantages

What better way is there than to thoughtfully and mindfully detail real life in case studies to show what works and does not work? If this is done properly, the case study results will inform theory and cause it to be shaped or expanded to better validate good practice. In other words, theory follows practice in case studies. Paradoxically, many people tend to believe that applications must follow theory!

Various advantages of case studies come to mind. One can

- Build upon or enhance a body of knowledge
- Suggest things to try
- Highlight errors to avoid
- Learn[†] from others' mistakes
- Compare specific aspects across case studies

---

* However, more and more, SE is being taught at the graduate level in college and sometimes even at the undergraduate level.
† One only learns from mistakes, one's own or mistakes of others (Ackoff 2004). If one knows what to do already, there is no learning.

- Show the importance of including people in the system that needs to be engineered
- Dig into specific situations and extract ideas from participants that can be generalized into principles for others to apply

STORYTELLING

Some extant case studies tout success focusing on the wonderful technology that is involved in the system. Unfortunately, what made the system successful in human terms is largely ignored. The latter aspects are what need to be especially emphasized for the benefit of other practitioners.

In storytelling, one must first gain the attention of the listeners, usually by scaring them about bad things that might happen (Denning 2011). Once they are paying close attention, to the extent the story is compelling, the listeners will naturally translate the story into something more meaningful to them personally while they are listening. That's how the mind works.

HOW TO DO CASE STUDIES

The authors' previous papers on this topic (White et al. 2012, 2013) emphasized case studies in engineering education and management. Here, the focus is more on case studies in complex, SoS, and enterprise engineering.

PLANNING

The first step in preparing a case study is the anticipation of doing one. So when starting an important project involving a complex system, prepare an initial plan for gathering and recording data that can be used later. Recognize that some of the project participants will likely come and go, so it is critical to preserve the project record and not rely entirely on people's memory. This record must be kept up to date continually, preserving the issues, decisions, actions, and results of each step or stage of the development or upgrade process. Even the plan should be updated and modified as required.

DOCUMENTATION

Routine aspects associated with dates of occurrence can be documented in any convenient format. The more interesting developments should be described in concise essay form. Any nuggets that can be immediately gleaned from unexpected events, in particular, will provide valuable case study fodder later on. Every so often, this chronological record should be reviewed retrospectively to gain additional insights that may be useful in going forward. The ultimate objective is to end up with sufficient raw material to shape into a case study that would be of considerable interest to other practitioners.

GUIDELINES

Case studies must be written honestly, highlighting the realities of the challenges and the struggles in seeking solutions. Sometimes, case study descriptions must be watered down to gain publishing approval of sponsors or customers. Because most organizations are risk averse and do not want to jeopardize their proprietary *crown jewels*,* be criticized by outsiders, or lose their funding to competitors, obtaining public release properly can be a challenge.

Objectivity is important. But one must realize that emotions are also relevant. Humans cannot claim to be entirely rational, especially in decision making (White 2012). Descartes was wrong! (Damasio 2005).

Next, a suggested framework for case studies and an associated outline to follow are discussed.

---

* Nevertheless, it is quite possible to share much information to one's own advantage and still protect the core business (Tapscott and Williams 2006).

FRAMEWORK FOR CASE STUDY DEVELOPMENT

Referring to Figure 3.1, the suggested main case study sections (each indicated by an index integer) within the same shaded and letter-coded node are considered relatively tightly coupled; any two sections that are part of different nodes are considered loosely coupled. The following remarks simply list these sections, which will be explained in more detail later.

The node labelled **B** contains the basic facts associated with the case study.

1. Case study elements
2. Background
3. Purpose
4. Constituents

The node labelled **R** contains further detail about the subject system, emphasizing the level(s) of difficulty.

5. (Complex) system/SoS/enterprise
6. Challenges

The node labelled **K** contains the heart of the case study, focusing on the elements of

7. Development
8. Results

The node labelled **V** contains elements about what it all means.

9. Analysis
10. Summary
11. Conclusions

Finally, the node labelled **G** contains suggestions as to where to go next.

12. Suggested future work
13. References

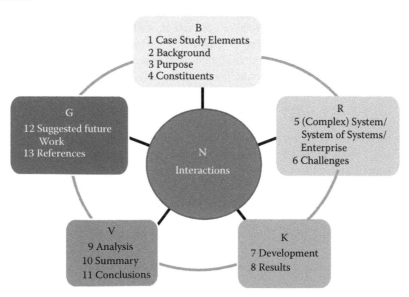

FIGURE 3.1   Relationships among case study sections.

TABLE 3.1

Principal Objectives of Case Study Sections

Index	Main Section of Case Study	This Section's Principal Case Study Objective	Letter-Label[a]
1	**Case Study Elements**	Provide enough concise information to enable a researcher/ practitioner or general reader to decide whether this case study is of particular interest.	**B**
2	**Background**	Further define the case study, especially for those that are not yet sure if it is relevant to their interests.	**B**
3	**Purpose**	Capture the impetus behind and specific reasons for the case study, and show some of the passion that drove or is driving this transformation.	**B**
4	**Constituents**	Characterize people and institutions interacting in the case study, and illuminate their motivations, e.g., what incentives drove or drive them?	**B**
5	**(Complex) System/System of Systems/Enterprise**	Provide a clear and complete but focused description of the subject complex system, SoS, or enterprise.	**R**
6	**Challenges**	Highlight the principal aspirations and difficulties.	**R**
7	**Development**	Show just how transformational change can occur.	**K**
8	**Results**	Answer the "So what?" questions.	**K**
9	**Analysis**	Provide suggestions to others in how to interpret results of their system transformation and what it all means.	**V**
10	**Summary**	Complement mainly the Case Study Elements, Background, and Purpose sections, especially for those that only want to skim the case study and not delve into the details.	**V**
11	**Conclusions**	Whet a reader's appetite to revisit the case study body for more detail.	**V**
12	**Suggestions for Future Work**	Motivate additional effort to further understand complex systems and advance CSE.	**G**
13	**References**	Lend credibility to the case study, and highlight relevant literature from related bodies of knowledge.	**G**

[a] Refer to the code labels of the nodes of Figure 3.1.

Figure 3.1 represents the nonlinear process nature of complex systems with feedback and interactions (depicted by the node labelled **N**) among the various sections of the case study. As one progresses in trying things and seeing what happens, often with considerable delay in the observable feedback, there may be changes of direction, emphasis, expectation, and strategy. Although the outline and the *wheel* of Figure 3.1 (read in a logical clockwise and numerical order fashion) suggests a linear progression, that is not the intent because a linear mindsight is rarely effective with complex systems. Refer to Chapter 1 for a *background* and *history* of complexity, complex systems, SoS, enterprises, and ESE; Chapter 2 for *relevant aspects* of complex systems and CSE; and Appendix 3.A for some related *definitions*.

The principal objective of each case study section of Figure 3.1 is presented in Table 3.1. Again, it does not matter much in which order these objectives are obtained. Their collective impact is what is important. However, for the purposes of easing the formidable problem of comparing multiple case studies, it helps if each case study of a given set followed roughly the same outline.

CASE STUDY OUTLINE

As just indicated to the extent an outline template is followed in preparing a collection of case studies, the comparing and contrasting of problems (and the methods and techniques of seeking solutions) across different case studies is possible.

A suggested outline to follow in preparing a case study is provided in Appendix 3.B. This version of the outline is quite similar but updated from the outlines presented in White et al. (2012) and White et al. (2013). Each main section and principal subsections (indicated in **bold**-faced type) of this outline is explained in detail as follows. Relatively minor sections and additional optional aspects (that may be pursued by the author(s) are indicated in brackets [...]) are <u>not</u> **bold**ed.

Case Study Elements

This first section is intended to be quite brief so readers can rapidly assess their potential level of interest in reading the case study. A *bulletized* executive summary is suggested that can be (1) used for sorting among all case studies and (2) scanned quickly to understand the nature of the case study.

Under **Fundamental Essence** and **Topical Relevance**, respectively, in this portion of the outline, the author(s) should briefly indicate (3) what the case study is about and (4) why it matters. Tell what **Domain(s)** is(are) represented by the case study or to what Domain(s) does it apply. Sample Domain(s) are Academia, Commerce, Government, Industry, and Other, the latter requiring specification. The **Country of Focus** and the **Interested Constituency**,* respectively, that is, the country (or countries) most involved and, for example, stakeholders who cared or care the most, need to be indicated. If appropriate, distinguish which ones are Passive/Active and Primary/Secondary. The **Primary Insights** or main *takeaways* should be summarized.

Keywords, Abstract, and Glossary

These features serve to help the reader determine whether to go further into the case study. Consistent with current conference paper practice, we suggest **Keywords** (or short phrases) be listed alphabetically and separated by commas. The **Abstract** should be informative but concise, perhaps no more than about 200 words. The **Glossary** is used to define acronyms or abbreviations; these terms should also be alphabetized.

Background

This section is intended to provide additional information beyond that of the **Case Study Elements** and **Keywords**, **Abstract**, and **Glossary** sections. An executive summary (textual as opposed to bulletized) style should still be utilized.

Explain the **Context**, that is, how this case study arose or arises and why. What are the **Relevant Definitions**? Terms likely to be unfamiliar to most readers should be defined. What theoretical knowledge or **Pertaining Theories** were applied. What **Research Nuggets** (both past and present) emerged? Any important research results employed should be at least mentioned explicitly. In addition, an overview of the supporting literature may be provided (as an option). Another useful topic would be any notable **Existing Practices**, including extant methods, available artifacts, and/ or proven processes, to be recommended. Describe the **Guiding Principles**, that is, the main principles, precepts, and/or tenets of the case study. A couple of suggested aids for **Characterizations** (first, the case study environment and, second, the SE activities employed) are embodied in Figures 3.A.3 and 3.A.4 of Appendix 3.A. Even if these specific profiler formats are not used, the case study should be **characterize**d in terms of its Type of System, its System Maturity (as to legacy, upgrade, or new system), the system Environment (before and after), and the Systems Engineering Activities (before and after) involved. The author(s) should compare and contrast the before and after nature of the System Description. **"As Is" System Description** and **"To Be" System Description** should include High-Level Diagrams (and possibly representative Performance Graphs as options); these descriptions could also include before and after versions of Figures 3.A.3 and 3.A.4.

---

* Relevant definitions of constituency: (1) parts of a whole, (2) a component part of something, and (3) people or things that combine to form a whole, such as an overall complex system.

Most of this **Background** section is primarily intended to provide information that will be useful in better understanding the later descriptions of the complex system, SoS, or enterprise studied. By documenting the starting and end points (of the before and after depictions mentioned earlier), the reader will be given an overview of what changed.

This section should cover a short answer to the *what* question.

PURPOSE

This section should cover a short answer to the *why* question.

Describe what is behind the appearance of or change in the system documented in this case study, that is, what was the **Purpose**? Provide a brief **History** of the system and what prompted the attention to its creation or improvement and describe how it evolved. This would include the starting point, the **Then Current Situation** and an initial view of the system's **Known Problem(s)**, and its intended or ultimate **Mission and Desired or Expected Capabilities**. The latter would cover Objectives and might include, as options, the associated Vision, Goals, and specific Targets that surfaced along the way. Finally, the basic reason for the system's **Transformation Needed and Why** should be explained.

(COMPLEX) SYSTEM/SYSTEM OF SYSTEMS/ENTERPRISE

This is the first section of the outline that really introduces the case study's complex system, SoS, or enterprise in sufficient detail.

As a relatively compact and summary introduction, a specially tailored context diagram is highly encouraged. The generic figure to be tailored is provided in Figure 3.2. Readers that relate better to

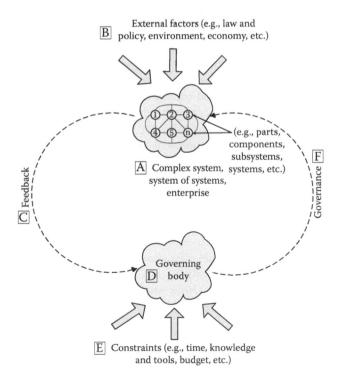

FIGURE 3.2 Context of complex system/SoS/enterprise. (Adapted from Gorod, A., System of systems engineering management framework: A 'satisficing' network approach, PhD dissertation, Stevens Institute of Technology, Hoboken, NJ, 2009; Gorod, A. et al., System of systems engineering management: A Review of Modern History and a Path Forward. *IEEE Syst. J.*, 2(4), 484, 2008.)

pictures than text should be able to ascertain the principal elements of the case study just from this. Ideally, this figure will capture the most important points of what has already been written.

The following aspect labels, A, B, …, F, corresponding to the capital letters in Figure 3.2, should be elaborated upon:

A—A short description about the complex system, SoS, or enterprise. Please explain the logic for the selection and also explain the parts, components, subsystems, systems, etc., that make up the high-level choice. So if this is an SoS, please define and explain the systems that constitute the SoS.

B—Please provide a short description of the external factors that could affect the parts, components, subsystems or systems, etc., in the complex systems, SoS, or enterprise. Examples of external factors include law and policy, environment, and economy. Explain briefly the effects and/or results on (especially the interconnectedness of) the parts under consideration.

C—Please include details of the feedback provided by the complex system, SoS, or enterprise to the governing body.

D—Describe the governing body of the complex system, SoS, or enterprise.

E—Describe the constraints that could affect the governing body. Examples of constraints can be time, knowledge and tools, and budget.

F—Based on the constraints affecting the governing body, there will be governance provided by the governing body. Please describe this outcome in the case study as well.

Describe each of the following aspects of the subject case study in sufficient detail. What was or is the system's **Environment**, that is, in what was or is the system embedded, what were or are its external factors, and what (e.g., funding, changing *requirements*, new capabilities) flowed or flows back and forth between it and its environment? How big was or is the system's **Scope** (purview, span of control, or influence) noting the influence on **Boundaries** in the following? What was or is the system's architectural **Structure** (subsystems and their interrelationships, perhaps characterized by the functions of network nodes and links)? How were or are the system **Boundaries** defined, noting the implications on **Scope** mentioned previously? Amplify the system's most significant **Internal Relationships**, noting the dependence on the aforementioned **Structure**. What were or are the most influential **External Factors** coming from the **Environment** mentioned earlier? Surely, there were or are **Constraints** from the **Environment** that limited the system's **Scope, Structure, Boundaries, Internal Relationships**, and potential capabilities and/or operations. What were or are some of them? Add any **Other Descriptors** that may not be well covered by the earlier aspects. Who was or is involved in the system transformation, that is, the **Constituents**, namely, the Sponsor(s), Customer(s), Governing Body (describe the Governing Body's relevance, e.g., Interactions With, Decision-Making Process, and Methods/Artifacts/Practices/Principles Used), and optionally, Teams/ Groups/Employers, Operators/Users, Countries/Organizations? Identify the relevant organizations, principal players, as appropriate, and their objectives and roles. Here is a good place for telling *war stories*, suitably sanitized, of course, that is, without divulging personal, private, proprietary, or sensitive/classified information. Also, cover any **Other Stakeholders**.

CHALLENGES

What were or are the greatest hopes or worries people, particularly the main proponents and stakeholders, had or have about the entity's transformation, that is, what kept or keeps people awake at night? Make a distinction between the **Challenges** that were **Anticipated** (particularly those that did not materialize) from others that were **Actual**, with special emphasis on those that were unexpected.

This section of the outline offers a place to make cogent points about how difficult the case study not only seemed to be but also how hard it was to develop and bring to fruition.

DEVELOPMENT

How was or is the system's transformation accomplished? Conventional SE methods may apply here to some extent, but presumably several nontraditional processes were invented, tried, or adopted, as well. With an eye toward SE innovation from which others may benefit, emphasize the degree to which nonconventional approaches were successful and why.

This is an area that should be expanded upon greatly. For example, there are various generally relevant traditional categories, for example, **Program Management** (primarily Planning) (narrow-sense, traditional, or conventional **Systems Engineering**, as opposed to CSE) and **Change Management**.

Under Planning, there are many possible nuances, and those that are the most relevant in the case study should be given some prominence. For example, one must plan for Contingencies and how Information Management, especially regarding Sharing and Security, was handled. How much attention was or is allocated to an overall guiding Strategy and the expenditure of Resources, to include classically; Staffing, particularly contributing Roles; and the Budget that often is revised based upon continually incremental funding. Compared to conventional SE, in CSE, one should strive to move more toward rewarding results as opposed to paying up front for perceived promises (White 2010a). To what extent were desired results rewarded and publicized? Of course, every program needs a planned Schedule, although unanticipated events can and usually do disrupt planned schedules, necessitating continual managerial flexibility in creating updated schedules. Paraphrasing General Dwight D. Eisenhower's famous statement, all plans are (essentially) worthless in combat but (continuous) planning is invaluable. It is also good practice to institute User/Operator Involvement up front and throughout the development process, primarily to gain their inputs and insights and to minimize surprises when testing and fielding the system being developed. Clearly, both developers and users/operators can benefit from these interactions. An optional feature for the case study is the listing and explanation of any significant Processes that were Instantiated.

Good traditional **Systems Engineering** includes the construction of a guiding system Architecture that does not change very much or often compared to the system under development. If the architecture is too unstable in this sense, it cannot be a good one! Describe the architectural approach used in the system's development. Alternatives Analysis is a critical aspect of SE. This involves several System Approaches with their Descriptions, the Technologies, including the levels of Technical Readiness (optionally) and Technologies Selected for each approach. One can view the Alternatives Analysis process as paying attention to Opportunities as well as Risks. Once one or more Alternatives are discarded, more attention is naturally focused on mitigating Risks, and even that can sometimes lead to new Opportunities. It is important, especially in CSE, to carry at least two Alternatives well into the development process to mitigate risk and protect against unexpected events. Describe the Alternative Analysis process employed in the case study.

A balanced Opportunity and Risk Management process is paramount in CSE because one is likely to go offtrack frequently due to unforeseen events or influences from the system environment, necessitating adaptation in the form of revising requirements, rescoping the job, pursuing other Opportunities, etc. Arguably, the biggest risk in complex, SoS, and enterprise engineering is not pursuing opportunities (White 2011b), at least until it becomes clear that such an opportunity leads to a significant new risk, in which case something else should be tried. This is made more difficult to the extent that there are delays in observing the effects of system interventions through (often heuristic) decision making.

Once the Selected Approach(es) to pursue is(are) decided, the development continues in earnest with detailed Design(s). The more favored designs are also implemented in the Implementation(s) phase and at least partially into the Integration phase to the extent possible. The subsystems of these implementations should be integrated *vertically* as part of the same system. In addition, at least the *hooks* for integrating these subsystems *horizontally* to enable potential interfaces and interactions with other systems with which the subject system will be interoperating should be included. Describe the Design, Implementation, and Integration processes used in the case study.

Testing should begin as soon as possible on both types of integration so that the inevitable errors in design or unanticipated consequences can be fixed or mitigated with lesser impacts on cost, schedule, and performance, compared to what would happen if these flaws were left undiscovered and unattended until later in the development. In parallel, plans for Fielding the system or upgrade should proceed with the hope of minimizing *glitches* or further delays in achieving smooth operation once the system is made available to users. A formal mechanism for gathering and acting on feedback from the field operations is critical for ensuring user acceptance. Optional aspects would include a description of system Sustainment and its eventual Retirement. Describe the Testing and Fielding of the system.

The **Development** is likely to be more successful if **Change Management** is taken seriously, that is, how was it instituted (implemented and integrated) into the **Development**? Greater complexities increase the likelihood of unforeseen events perturbing **Development**. Thus, whether a formal contingency procedure was established in advance should be of great interest in the case study. Optionally, Philosophy, Policies, and/or Organizations of Change Management could be discussed. The system's environmental effects of Politics, Operations, Economics, and Technologies on Change Management should be considered. Describe the Change Management aspect of the case study.

RESULTS

So what happened from all this effort? Describe what emerged (particularly the unexpected results) from the aforementioned **Development** including the major improvements, added capabilities, user or operator satisfactions, setbacks, shortfalls, and unintended consequences.

Describe the system **Transformation Accomplished** in terms of the system's Functions, Services, and Other Assets or Capabilities. The **Final System Description** should include a High-Level Diagram and possibly (as an option) Performance Graphs.

ANALYSIS

This section should contain a summary of the technical assessments performed as part of the **transformation** effort. More importantly, the "Why?" questions should be answered, for example, what were the root causes of the results from an analytical point of view? If it was or is not possible to readily determine the causes precisely, do a credible job of explaining the primary set of conditions responsible.

Analytical Findings include principal Activities (i.e., the key tasks and their interactions), Time Frame/Line aspects including the Sequence of Events (an option), Significant Delays Incurred and Why, and Methods Employed (and their efficacies). **Lessons Learned**, a very important topic in case studies, should include answers to the questions How Were Biggest Challenges Met? What Worked and Why? What Did <u>Not</u> Work and Why? What Should Have Been Done Differently? and To What Extent Were Lessons Applied to Subsequent Programs/Projects? Also, include what changes in policies or procedures were or are being implemented so that these lessons are <u>really</u> learned and not forgotten? **Best Practices** recommended for the benefit of others are especially critical. Addressing how practical this case study might become for **Replication Prospects**, including Necessary Conditions and Proposed Action Steps, should be assessed.

SUMMARY

Provide a concise overview of the problem, its proposed solution, the **transformation**al approach and the results, all with the benefit of hindsight. This **Summary** might include (as an option) an **Epilogue** describing any significant events that might have happened after the designated end point of the case study.

CONCLUSIONS

Detail or list the most important ideas that the reader should take away from this case study. This could be modeled after an *elevator speech* where one must convey the essence of the topic in a very short time.

SUGGESTED FUTURE WORK

Undoubtedly, there are several unanswered questions that arose or arise during the system's **transformation**. These questions can be shaped into suggestions for **Future Work** in the form of research, experimental practices or processes, postulated precepts or principles, **lessons** to be **learned** and exercised, etc. Endeavor to stimulate further progress in this or **transformation**s by preparing some compelling reasons to continue improvements.

This should include **Questions for Discussion** (the really only <u>mandatory</u> item of this outline) and suggestions for **Additional Research**. These are intended primarily for educational and academic purposes. Such **Questions** could enliven SE classroom discussions, for instance. Those interested in adding to the SE body of knowledge might be stimulated by future research topics.

Endnotes <u>are optional</u>. Footnotes can be used in the case study, but if there are many of these side comments, collecting them at the end of the case study in the form of **Endnotes** might be more convenient for the readers.

REFERENCES

It is always useful to at least some readers to provide outstanding **References** (and/or a bibliography) that completely cite and fully document previous work upon which the present work depended or depends upon. These references should help justify the present transformation by explaining or supporting assumptions or statements made in the case study. They should also be rich in offering additional background detail.

The case study citations might include both primary and secondary **References**. The standard format used by the publisher (in this case Taylor & Francis) should be utilized.

APPENDICES AND INDEX

These features are optional.

CONCLUSION

The authors have advocated that case studies are important and a valuable means for understanding what works and doesn't work in addressing the most difficult SE problems. They urge the SE community to focus more on improving the practice of SE by creating and learning from case studies. In their view, the better theories of CSE will follow the successful applications of its case studies.

They strongly believe in the power of case studies for furthering the good and more effective practices of SE within the ever-increasing complexity of the problem spaces facing humanity and the global community. For example, if the practice of SE in helping to conserve our Earth's resources is not advanced, the unsustainability of unlimited material growth and overpopulation will eventually drag us all down to much lower standards of living (Meadows 2008). We must greatly advance our understanding of complex systems and our application of CSE principles to the world's problems. Clearly, pursuit of this goal will be accelerated if there is concentration on educating and motivating emerging technical leaders and those already recently in or entering the engineering profession about modern SE, that is, CSE. This education can also extend into transdisciplinary fields of philosophy, psychology, sociology, economics, politics, organizational change management, and chaordic (Hock 1999) leadership, for example.

APPENDIX 3.A

It is assumed that certain terms, like system, engineering, and enterprise, do not really need to be defined for the general engineering audience. An extant collection of definitions of these and many other related terms exists (White 2007) for those interested in delving further. However, the goal here is to simply explain how other terms like *complex* and *complexity*, which are not so widely agreed to in their nuanced vernacular, are used in this book.

DEFINITIONS

So what do we mean by system, SoS, enterprises, and complex systems? Further, what do we mean by SE, SoS engineering, enterprise engineering, and CSE? Finally, what's the difference between systematic and systemic? Let's take these questions in order. Readers should understand that these definitions are not meant to be sacrosanct; however, they were constructed with three basic goals in mind: (1) relative brevity, (2) essential essence, and (3) nonviolation of somewhat different and nuanced definitions from other respected sources.

System: An interacting mix of elements forming an intended whole that is greater than the sum of its parts. *Features*: These elements may include people, cultures, organizations, policies, services, techniques, technologies, information/data, facilities, products, procedures, processes, and other human-made or natural entities. The whole is sufficiently cohesive to have an identity distinct from its environment. *Note*: In general, a system does not necessarily have to be fully understood, have a defined goal or objective, or have to be designed or orchestrated to perform an activity. However, in the present definition, *intended* means an understood/defined goal/objective and designed/orchestrated to perform a useful activity.

SoS: A collection of systems that functions to achieve a purpose not generally achievable by the individual systems acting independently. *Features*: Each system can operate independently and is managed primarily to accomplish its own separate purpose. An SoS can be geographically distributed and can exhibit evolutionary development or emergent behavior. *Note*: There's more about different types of SoS in the following.

Enterprise: A complex system in a shared human endeavor that can exhibit homeostasis, that is, relatively stable equilibria or behaviors among many interdependent component systems. *Features*: An enterprise may be embedded in or overlap with a more inclusive complex system. External dependencies affecting an enterprise may impose environmental, political, legal, operational, economic, legacy, technical, and other constraints. *Notes*: An enterprise usually includes an agreed-to or defined scope/mission or set of goals or objectives.

Complex and **Complexity**: Complex is more than complicated; the latter term represents a notion that is on the lowest rung of a discrete or even continuous scale of increasing complexity. Many people, including specialty engineers and systems engineers, use complex and complicated interchangeably or, worse, use complex when they mean only complicated. Here, complex refers to a range of difficulty that is more, and often much more, than merely complicated.

Complex System: An open system with continually cooperating and competing elements, in which there are autonomous and independent systems. *Features*: This type of system continually evolves and changes its behavior, often in unexpected ways, according to its own condition and external environment. Changes between states of order and chaotic flux are possible. Relationships among its elements are imperfectly known and are difficult to describe, understand, predict, manage, control, design, or change. *Notes*: This suggests examining the role of a system's boundary in differentiating between open and closed. A closed system is merely a system that has been defined with respect to a boundary that contains the totality of its

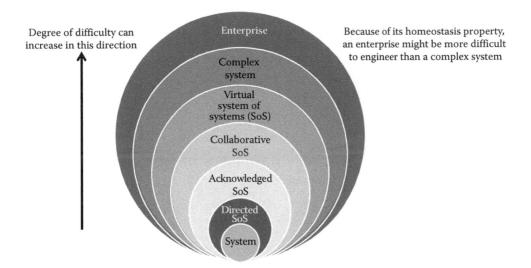

Degree of difficulty can increase in this direction

Because of its homeostasis property, an enterprise might be more difficult to engineer than a complex system

FIGURE 3.A.1 One perceived Venn diagram relationship among systems. (Adapted from White, B.E., On principles of complex systems engineering-complex systems made simple tutorial, *INCOSE Symposium*, Denver, CO, June 20–23, 2011a.)

interactions. Inside that boundary, that same system might look *open*. On the other hand, even an open system has a boundary, or else, there would be no *external* to define or identify the system. Defining the bounds of a system is a critical early step in any good SE process. It is sometimes possible to make the system open or closed by appropriately defining the boundary. A complex system is not merely complicated (as noted earlier under **Complex** and **Complexity**). It is nonlinear, and chaotic behavior can be an intrinsic property of the system that connotes the sensitivity of the system to perturbations of the initial conditions. Also, when such a system has given inputs from an aggregation of random processes, the result appears complex, for example, unpredictable; a system that is predominately linear <u>can</u> be predictable, even if the inputs are random. A complex system (e.g., that is not homeostatic) is not necessarily an enterprise. A complex system is often distinguished from simple, uncomplicated, or merely technologically complicated systems by intentionally including people, for example, key stakeholders, as part of the system. It is much more difficult to successfully conceive, develop, improve, field, and operate such systems. This is fundamentally because this type of complex system evolves on its own in a self-organized fashion in response to its (usually many) internal and environmental interactions.

It may be helpful to think of a system, a SoS, an enterprise, and a complex system in terms of the Venn diagrams of Figures 3.A.1 and 3.A.2. Here, SoS is subdivided into four types defined in Table 3.A.1.

CONTINUING WITH OTHER DEFINITIONS

SE: An iterative and interdisciplinary management and development process that defines and transforms requirements or desired capabilities into an operational system. *Features*: Typically, this process involves environmental, economic, political, and social aspects. Activities include conceiving, researching, architecting, utilizing, designing, developing, fabricating, producing, integrating, testing, deploying, operating, sustaining, and retiring system elements. *Notes*: The customer for or user of the system usually states the initial version of the requirements. The SE process is used to help better define and refine these requirements. Further, the requirements often change as new decisions

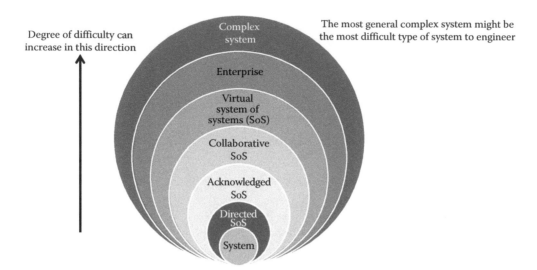

FIGURE 3.A.2   Another perceived Venn diagram relationship among systems. (Adapted from White, B.E., On principles of complex systems engineering-complex systems made simple tutorial, *INCOSE Symposium*, Denver, CO, June 20–23, 2011a.)

TABLE 3.A.1
SoS Type Definitions

Virtual: Virtual SoS lack a central management authority and a centrally agreed-upon purpose for the SoS. Large-scale behavior emerges—and may be desirable—but this type of SoS must rely upon relatively invisible mechanisms to maintain it.

Collaborative: In collaborative SoS, the component systems interact more or less voluntarily to fulfill agreed-upon central purposes. The Internet is a collaborative system. The Internet Engineering Task Force works on standards but has no power to enforce them. The central players collectively decide how to provide or deny service, thereby providing some means of enforcing and maintaining standards.

Acknowledged: Acknowledged SoS have recognized objectives, a designated manager, and resources for the SoS; however, the constituent systems retain their independent ownership, objectives, funding, and development and sustainment approaches. Changes in the systems are based on collaboration between the SoS and each member system.

Directed: Directed SoS are those in which the integrated SoS is built and managed to fulfill specific purposes. It is centrally managed during long-term operation to continue to fulfill those purposes as well as any new ones the system owners might wish to address. The component systems maintain an ability to operate independently, but their normal operational mode is subordinated to the central managed purpose.

Dahmann, J.S. et al., *CrossTalk J. Defense Softw. Eng.*, 21(11), 4, November 2008, http://www.stsc.hill.af.mil/crosstalk/2008/11/index.html.

are made as a result of SE. This definition does not imply that a successful system is always realized. The word *integrated* is not included in this definition because SE efforts are not always that well integrated. Systems engineers whom work with complex systems are called complex systems engineers; they are not really in control of very much because of the nature of complex systems. They often can only influence the system and decide to intervene again at some point if the system moves in undesirable directions.

ESE: A regimen for engineering *successful* enterprises. *Features*: ESE is SE that emphasizes a body of knowledge, tenets, principles, and precepts concerning the analysis, design, implementation, operation, performance, etc., of an enterprise. Rather than focusing on parts of the enterprise, the enterprise systems engineer concentrates on the enterprise as a whole and how its design, as applied, interacts with its environment. An ESE approach focuses on how these parts interact within the system and with its outside environment. *Notes*: Here, *regimen* means a prescribed course of engineering for the promotion of enterprise success.

CSE: ESE that includes conscious attempts to further open an enterprise in order to create a less stable equilibrium among its interdependent component systems. *Features*: In CSE, critical attention is paid to emergence of phenomena or conditions in the system that could be desirable or not, especially because of the tendency for complex systems to become more open. Thus, it is important to attempt to manage as quickly (beware of unexpected consequences that may emerge after some time delay) and deliberately as possible, the natural processes that shape the development of complex systems. *Notes*: For some, a better working definition might be SE that attempts to exploit complexity and control or manage the results of complexity. Neither *open* nor *closed* is a determinant of complex or emergent behavior.

A proposed methodology for accomplishing CSE exists and has been advocated (White 2008). A primary purpose of performing additional case studies is to further validate such methodologies.

COMPLETING THIS SET OF DEFINITIONS

Systemic has much to do with systems thinking (Gharajedaghi 2006), a holistic, fundamental approach to considering all the important aspects of a complex issue, before formulating a solution that may be too simplistic or limited to solve the essential problem. This applies primarily in domains where one is not in control of much but can only hope to influence, because of the people (especially stakeholders) involved. Here, one cannot hope to prespecify accurately or predict outcomes, even if events are deterministic because conditions change too fast; the (complex) system evolves on its own whether one intervenes or not. There are unknown unknowns as well as known unknowns. One must pursue opportunities (White 2011b), try to intervene constructively, and watch what transpires, as opposed to a single rigid approach and focusing just on managing risk and trying to stay on some preconceived track no matter what.

Systematic is more about turning the crank on something that is well understood like the classical, traditional, or conventional SE process of defining the problem, establishing the requirements, constructing an architecture, analyzing alternatives, developing a solution, testing and integrating the subsystems, and testing, fielding, and eventually retiring the system. Systematic is more about doing things right, rather than doing the right things, as in systemic.

One's complex system environment can be characterized using the ESE profiler (Stevens 2008) of Figure 3.A.3. This is done by selecting one of the three rings in each of the eight sectors by placing a dot. Then the eight dots are connected as shown in this specific example to create a *spider diagram*. The more this pattern extends outward, the greater the difficulty of the environment of the associated complex system. If the SE actions taken (these may be characterized by another spider diagram) are not well matched to the nature of the first diagram, then things must change or it's likely that less progress will be made. Another way to characterize SE actions is embodied in the next profiler.

The systems activity profiler (SEA) (White 2010b) of Figure 3.A.4 is a way of holistically characterizing the SE approach being taken in any systems acquisition effort at any point in time. This is a self-characterization method for helping to assess where one is on the SE continuum from conventional to complex. The farther the sliders (the black rectangles in the figure) are to the right, the more one is involved in CSE. Here, the *sliders* are set in hypothetical positions for each of the nine SE activities associated with a typical SE process.

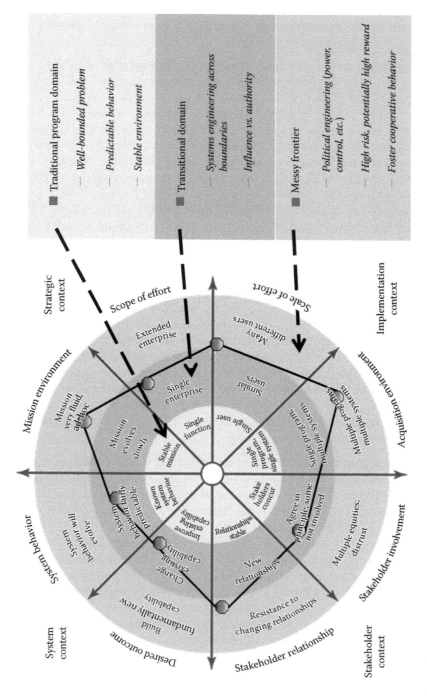

FIGURE 3.A.3   ESE profiler. (Courtesy of The MITRE Corporation, Bedford, MA.)

Version 4 – 4 Jan 09

Typical systems engineering activity	Left end of Slider	Left Intermediate interval	Center intermediate Interval	Right intermediate interval	Right end of slider
Define the system problem	Establish system requirements	Adapt to changing requirements; ReScope	Revise and restate objectives	Try to predict future enterprise needs	Discover needed mission capabilities
Analyze alternatives	Conduct systems trade-offs	Model/simulate system functionalities	Perform systematic cost-benefit analyses	Include social and psychological factors	Emphasize enterprise aspects
Utilize a guiding architecture	Apply an existing framework	Develop architectural perspectives (views)	Really define (not Just views of) architecture	Adapt architecture to accommodate change	Embrace an evolutionary architecture
Consider technical approaches	Employ available techniques	Research, track, and plan for new technologies	Research and evaluate new technical ideas	Proactively plan for promising techniques	Explore new techniques and innovate
Pursue solutions	Advocate one system approach	Consider alternative solution approaches	Investigate departures from planned track	Iterate and shape solution space	Keep options open while evolving answer
Manage contingencies	Emphasize and manage system risks	Mitigate system risks and watch opportunities	Sort, balance, and manage all uncertainties	Pursue enterprise opportunities	Prepare for unknown unknowns
Develop implementations	Hatch system improvements OffLine	Prepare enhancements for fielding	Experiment in operational exercises	Develop in realistic environments	Innovate with users safely
Integrate operational capabilities	Test and incorporate functionalities	Work towards better interoperability	Advance horizontal integration as feasible	Advocate for needed policy changes	Consolidate mission successes
Learn by evaluating effectiveness	Analyze and fix operational problems	Propose operational effectiveness measures	Collect value metrics and learn lessons	Adjust enterprise approach	Promulgate enterprise learning

Convenient labels (Only; interpret them): Traditional systems engineering (TSE) — Complex systems engineering (CSE)

Aggregate assessment of above slider positions

FIGURE 3.A.4 SEA profiler. (Courtesy of The MITRE Corporation, Bedford, MA.)

Once familiar with the underlying descriptions of nine basic areas of SE (rows) shown in the first column, rapid profiling is possible. For each area (row), a *slider* is placed at the appropriate point (one of seven possible discrete locations in the third through fifth columns) to represent the degree that conventional (or traditional) versus complex (or enterprise) SE is performed.

The patterns that result can stimulate conversations among practitioners in comparing and contrasting SE approaches across time within the same program or across separate programs. Again, if certain aspects of a pattern are not well matched to the SE environment (e.g., characterized by the ESE profiler of Figure 3.A.3), changes in the SE approach is warranted.

APPENDIX 3.B

(Complex) System/SoS/Enterprise Case Study Outline [optional items are indicated in brackets]
Case Study Elements (bulletized, for sorting purposes)
Fundamental Essence (briefly, what's this about?)
Topical Relevance (briefly, why does this matter?)
Domain(s) (choose one)
Academia
Commerce
Government
Industry
Other (specify)
Country of Focus (country most involved)
Interested Constituents (e.g., Stakeholders)
Passive/Active

Primary/Secondary
Primary Insights (takeaways)
Key Words (alphabetized, separated by commas)
Abstract (no more than 200 words)
Glossary (abbreviations and acronyms, alphabetized)
Background
Context (how did this arise, and why?)
Relevant Definitions (define unfamiliar terms)
Pertaining Theories (theoretical knowledge applied)
[Literature Overview]
Research Nuggets (past and present)
Existing Practices (extant methods, available artifacts, and/or proven processes)
Guiding Principles (applicable principles, precepts, and/or tenets)
Characterizations
Type of System
System Maturity (legacy, upgrade, or new)
Environment (before and after)
Systems Engineering Activities (before and after)
"As Is" System Description (before)
High-Level Diagram
[Performance Graphs]
"To Be" System Description (after)
High-Level Diagram
[Performance Graphs]
Purpose
History (describe previous situation and evolution)
Then Current Situation
Known Problem(s)
Mission and Desired or Expected Capabilities
[Vision
Goals]
Objectives
[Targets]
Transformation Needed and Why
(Complex) System/System of Systems/Enterprise (describe each in sufficient detail) Complete a specific instantiation of Figure 3.2.
Environment
Scope
Structure
Boundaries
Internal Relationships
External Factors
Constraints
Other Descriptors
Constituents
Sponsor(s)
Customer(s)
Governing Body
Interactions With
Decision-Making Process
Methods/Tools/Practices/Principles Used

[Teams/Groups/Employers]
[Operators/Users]
[Countries/Organizations]
Other Stakeholders
Challenges (what kept people awake at night?)
Anticipated
Actual
Development (emphasize <u>non</u>conventional aspects)
Project/Program Management
Planning
Contingencies
Information Management
Sharing
Security
Strategy
Resources
Staffing
Roles
Budget
Schedule
User/Operator Involvement
[Processes Instantiated]
Systems Engineering (in narrow sense)
Architecture
Alternatives Analysis
System Approaches
Description
Technology
[Technology Readiness]
Technologies Selected
Opportunity and Risk Management
Selected Approach(es)
Design(s)
Implementation(s)
Integration
Testing
Fielding
[Sustainment]
[Retirement]
Change Management (how implemented and integrated?)
[Philosophy]
[Policy]
Politics
[Organization]
Operations
Economics
Technologies
Results
Transformation Accomplished
Functions
Services

Other Assets or Capabilities
Final System Description
High-Level Diagram
[Performance Graphs]

ANALYSIS

ANALYTICAL FINDINGS

Activities (key tasks and their interactions)
Time Frame/Line
[Sequence of Events]
Significant Delays Incurred and Why
Methods Employed (and their efficacies)

LESSONS LEARNED

How Were Biggest Challenges Met?
What Worked and Why?
What Did Not Work and Why?
What Should Have Been Done Differently?
To What Extent Were Lessons Applied to Subsequent Programs/Projects?
Best Practices (what would be recommended to others?)
Replication Prospects (how practical might this case study become?)
Necessary Conditions
Proposed Action Steps
Summary (provide concise overview of what happened after the fact)
[Epilogue] [what significant events have occurred since?]
Conclusions (construct an *elevator speech*)
Suggested Future Work
Questions for Discussion*
Additional Research
[Endnotes]
References (primary and secondary)
[Appendices]
[Index]

REFERENCES

Abbott, R. J. 2005. Emergence explained: Getting epiphenomena to do real work. Department of Computer Science, California State University, Los Angeles, CA, 12 October.

Ackoff, R. L. 2004. Transforming the systems movement. *The Systems Thinker*, 15(8): 2–5. Pegasus Communications (October). www.pegasuscom.com.

Bergvall-Kåreborn, B. 2002. Enriching the model-building phase of soft systems methodology. *Systems Research and Behavioral Science*, 19(1): 27–48, January/February. Article first published online: 12 December 2001, doi: 10.1002/sres.416. Wiley On-Line Library, http://onlinelibrary.wiley.com/doi/10.1002/sres.v19:1/issuetoc.

Boardman, J. and B. Sauser. 2008. *Systems Thinking: Coping with 21st Century Problems*. Boca Raton, FL: CRC Press.

Bonabeau, E. 2002. Agent-based modeling: methods and techniques for simulating human systems. *Proceedings of the National Academy [PNAS] of Sciences of the United States of America*, 99(Suppl. 3): 720–727, May 14. doi:10.1073/pnas.082080899.

* Three to five case-based questions that could be used for a discussion in a classroom environment are required.

Case Studies. 2012. *Harvard Business Review*, http://hbr.org/search/Case%252520Studies/; MIT Sloan http://mitsloan.mit.edu/search.php?cx=005593932814951658129%3An5_vflfbe-k&cof=FORID%3A11&q=case+studies. Accessed February 26, 2014.

Conklin, J. 2005. Chapter 1: Wicked problems and social complexity. In *Dialogue Mapping: Building Shared Understanding of Wicked Problem*. CogNexus Institute, http://www.cs.uml.edu/radical-design/uploads/Main/WickedProblems.pdf; © 2001-2005 CogNexus Institute http://cognexus.org R4b Pages 1–25 (accessed 26 February 2014).

Dahmann, J. S., G. Rebovich, Jr., and J. A. Lane. 2008. Systems engineering for capabilities. *CrossTalk: The Journal of Defense Software Engineering*, 21(11): 4, November. http://www.stsc.hill.af.mil/crosstalk/2008/11/index.html.

Damasio, A. 2005. *Descartes' Error: Emotion, Reason, and the Human Brain*. New York: Penguin Books.

Denning, S. 2011. *The Leader's Guide to Storytelling: Mastering the Art and Discipline of Business Narrative*. San Francisco, CA: Wiley.

Engineering. Engineering management case studies. http://www.bing.com/search?q=case+studies+in+engineering+management&qs=n&form=QBRE&pq=case+studies+in+engineering+management&sc=0-26&sp=-1&sk=. Accessed February 26, 2016.

Friedman, G. and A. P. Sage. 2004. Case studies of systems engineering and management in systems acquisition. *Systems Engineering*, 7–1: 84–97.

Gharajedaghi, J. 2006. *Systems Thinking: Managing Chaos and Complexity: A Platform for Designing Business Architecture*, 2nd edn. Burlington, VT: Elsevier, Inc.

Gorod, A. 2009. System of systems engineering management framework: A 'satisficing' network approach. PhD dissertation, Stevens Institute of Technology, Hoboken, NJ.

Gorod, A., B. Sauser, and J. Boardman. 2008. System of systems engineering management: A review of modern history and a path forward. *IEEE Systems Journal*, 2(4): 484–499.

Harvard. HBS case method (MBA). Boston, MA: The Harvard Business School. http://www.hbs.edu/mba/academic-experience/Pages/the-hbs-case-method.aspx. Accessed February 15, 2013.

Hawn, G. (with Wendy Holden). 2011. *10 Mindful Minutes*. New York: Penguin Books.

Hock, D. 1999. *Birth of the Chaordic Age*. San Francisco, CA: Berrett-Koehler Publishers, Inc.

IEEE. IEEE SoS Technical Committee (Roadmapping Exercise), T-AREA-SoS project, point of contact: S. A. Henson@lboro.ac.uk. Accessed April 2003.

Jackson, M. C. 2003. *Systems Thinking: Creative Holism for Managers*. New York: Wiley, November. ISBN: 978-0-470-84522-6.

Linstone, H. A. 1999. *Decision Making for Technology Executives: Using Multiple Perspectives to Improve Performance*. Boston, MA: Artech House Publishers.

Meadows, D. H. 2008. *Thinking in Systems: A Primer*, Diana Wright (ed.). White River Junction, VT: Chelsea Green Publishing.

MIT. The Massachusetts Institute of Technology (http://esd.mit.edu/); also, refer to SEAri (http://seari.mit.edu/) and LAI (http://seari.mit.edu/): http://search.mit.edu/search?q=case+studies&btnG=go&site=mit&client=mit&proxystylesheet=mit&output=xml_no_dtd&as_dt=i&as_sitesearch=esd.mit.edu&ie=UTF-8&ip=127.0.0.1&access=p&sort=date:D:L:d1&entqr=3&entsp=0&oe=UTF-8&ud=1&is_secure=. Accessed February 15, 2013.

MIT. The Massachusetts Institute of Technology (http://sdm.mit.edu/): http://search.mit.edu/search?site=sdm&client=mit&proxystylesheet=mit&output=xml_no_dtd&as_dt=i&q=case+studies. Accessed April 23, 2013.

Paper Chase. 1973. "Paper Chase," (movie). YouTube. http://www.youtube.com/watch?v = qx22TyCge7w. Accessed February 15, 2013.

Project Management. 2008. *A Guide to the Project Management Body of Knowledge: (Pmbok Guide)*, 4th edn. Newton Square, PA: Project Management Institute, 31 December. http://www.pmi.org/PMBOK-Guide-and-Standards/Standards-Library-of-PMI-Global-Standards.aspx.

Sage, A. P. (ed.). Wiley Series in Systems Engineering and Management, authoritative treatments of all foundational areas central to this *transdisciplinary* subject, New York: Wiley, http://www.wiley.com/WileyCDA/Section/id-397384.html?sort=DATE&sortDirection=DESC&page=3. Accessed February 15, 2013.

Stevens, R. 2008. Profiling complex systems. *Proceedings of the IEEE International Systems Conference*, Montreal, Quebec, Canada, 7–10 April.

Stevens, R. 2011. *Engineering Mega-Systems: The Challenge of Systems Engineering in the Information Age*. Complex and Enterprise Systems Engineering Book Series. Boca Raton, FL: Auerbach Publications/CRC Press/Taylor & Francis Group.

Tapscott, D. and A. D. Williams. 2006. *Wikinomics: How Mass Collaboration Changes Everything.* New York: Penguin.

Texas. The University of Texas. http://www.engr.utexas.edu/search?option=com_googlesearch_cse&n=30 &cx=010165778158772724462%3A83i3acrhdki&cof=FORID%3A11&ie=ISO-8859-1&q=engineerin g+case+studies&hl=en&sa=. Accessed April 23, 2013.

The Stevens. The Stevens Institute of Technology (http://www.stevens.edu/): http://www.stevens.edu/sit/search.cfm? cx=001246728825567149980%3Amlcoplzz4w4&cof=FORID%3A11&q=case+studies&sa.x=14&sa.y=13. Accessed April 23, 2013.

U of I. The University of Illinois: http://engineering.illinois.edu/search/node/engineering%20case%20studies. Accessed April 23, 2013.

USC. The University of Southern California: http://search.usc.edu/?cx=017196764489587948961%3A0uzwq g1rcr4&ie=utf8&oe=utf8&q=engineering+case+studies. Accessed April 23, 2013.

Vermont. The University of Vermont: http://www.uvm.edu/search/?q=case%20studies. Accessed April 23, 2013.

Virginia. The University of Virginia: http://www.google.com/cse?it+is+a+tool+under+custom+searches=this+ custom+google+search+engine++was+set+up+by+Rich+Gregory&cx=001051378019901782744%3A aryse_cxxlw&ie=UTF-8&q=engineering+case+studies&submit1.x=28&submit1.y=5#gsc.tab=0&gsc. q=engineering%20case%20studies&gsc.page=10. Accessed April 23, 2013.

White, B. E. 2007. *Engineering Lexicon. Complex and Enterprise Systems Engineering Book Series.* Boca Raton, FL: Taylor & Francis Group, http://www.enterprise-systems-engineering.com/lexicon.htm.

White, B. E. 2008. Complex adaptive systems engineering. *8th Understanding Complex Systems Symposium,* University of Illinois at Urbana-Champaign, Champaign, IL, 12–15 May. http://www.howhy.com/ ucs2008/schedule.html.

White, B. E. 2009. Complex adaptive systems engineering, (CASE). *3rd Annual IEEE International Systems Conference,* Vancouver, British Columbia, Canada, 23–26 March.

White, B. E. 2010a. Complex adaptive systems engineering (CASE). *IEEE Aerospace and Electronic Systems Magazine,* 25(12): 16–22 December. ISSN 0885-8985

White, B. E. 2010b. Systems engineering activity (SEA) profiler. *8th Conference on Systems Engineering Research (CSER),* Hoboken, NJ, 17–19 March.

White, B. E. 2011a. On principles of complex systems engineering-complex systems made simple tutorial. *INCOSE Symposium,* Denver, CO, 20–23 June.

White, B. E. 2011b. Chapter 5: Enterprise opportunity and risk. In Rebovich, G. Jr. and B. E. White (eds.), *Enterprise Systems Engineering: Advances in the Theory and Practice.* Complex and Enterprise Systems Engineering Book Series, pp. 161–180. Boca Raton, FL: Auerbach Publications/CRC Press/Taylor & Francis Group.

White, B. E. 2012. Systems engineering decision making may be more emotional than rational! *22nd Annual INCOSE International Symposium,* Rome, Italy, 9–12 July.

White, B. E., A. Gorod, S. J. Gandhi, V. Ireland, and B. Sauser. 2012. Application of case studies to engineering management and systems engineering education. *American Society for Engineering Education (ASEE) Annual Conference,* San Antonio, TX, 10–13 June. Copyright © 2012 ASEE Annual Conference & Exposition. ASEE grants permission to reuse portions of this cited paper in this book.

White, B. E., A. Gorod, S. J. Gandhi, V. Ireland, and B. Sauser. 2013. On the importance and value of case studies—Introduction to the special session on case studies in system of systems, enterprises, and complex systems engineering. *2013 IEEE International Systems Conference (SysCon) 2013,* Orlando, FL, 15–18 April. Copyright © 2013 IEEE International Systems Conference. IEEE grants permission to reuse portions of this cited paper in this book.

Wicked. 2013. http://www.cs.uml.edu/radical-design/uploads/Main/WickedProblems.pdf; © 2001–2005 CogNexus Institute http://cognexus.org R4b Pages 1–25; you could also change. Accessed February 15, 2013 to accessed February 26, 2014.

Yin, R. K. 2003. *Case Study Research: Design and Methods,* 3rd edn., Vol. 5. Applied Social Research Methods Series. Thousand Oaks, CA: Sage Publications.

Section II

Commerce

4 MediaCityUK
Corporate Framework for Complex Construction

Barbara Chomicka

CONTENTS

CASE STUDY ELEMENTS

Fundamental essence

- The strength that comes from human collaboration is the central truth behind the success of complex construction projects.
- Concepts of systems thinking are central to successfully delivering complex construction projects, because the large gap in time from the realization of the need for a development to the final handover of the finished product means that many complex projects are effectively untested prototypes that are nonetheless expected to conform to standards prevailing at some future date.
- Complex construction projects are made of flesh, not concrete—they are works in progress that never stop, adjusting to the changing needs, technology, and aspirations of their end users.
- An insight into successful complex project delivery in the construction industry is offered by an analysis of Europe's first purpose-built media development (completed December 2011) in Salford, United Kingdom. The BBC's new *face in the North*—MediaCityUK— allowed Salford, in the Metropolitan Borough of Greater Manchester, to join Singapore, Seoul, and Abu Dhabi on the world's media map.
- Keeping such a large, complex project viable during a period of construction cost inflation and subsequent general economic turmoil required continual postcontract scope changes to be accommodated.
- The up-ending, in the case study presented, of the defined, linear, contractual, and adversarial approach prevailing in the construction industry in favor of a truly complex, cooperative framework created competitive advantage and resulted in a project completed on time, under planned cost, and conforming to standards that did not even exist when construction commenced.

Topical relevance

- While there are various project management tools used in the construction industry to manage complex projects, most of such tools rely on ex ante identification and evaluation of all factors for which preventive methods and response strategies are incorporated into contractual agreements. Phase 1 of the MediaCityUK project presented an unusual combination of circumstances for which existing project management tools and processes proved inadequate from the start: (1) unforeseen market circumstances, that is, construction price inflation preceding the global economic crisis and the subsequent market downturn; (2) the management and technical complexity of the project (this involved a concurrent construction of offices, high-definition (HD) studios, including the largest HD studio in Europe, supermarket, shops and leisure facilities including bars, restaurants, healthcare services, public park and events space, new apartments, hotel, car park, a new bridge, a new tram link to Manchester city centre, and a highway junction—most of which were located on the constrained [by the surrounding canals] site); (3) the nature of the product to be delivered (*media cities* are recognized to be a *geographically complex phenomenon* (Krätke, 2003)); and (4) the end-user requirement that the end product conforms to standards unknown at the time, which would prevail at some future date.
- This combination of circumstances required a mandate for the team delivering this complex project to operate with extremely high flexibility, responding to constantly changing needs, scope, program, budget, and strategy.

Domains

- Construction
- Management in the construction industry

Country of focus

- United Kingdom

Interested constituency

- The construction industry landscape has changed dramatically in recent years, with financial institutions placing very tight lending constraints on the industry's products and companies. Construction industry clients have to drive significant operational and asset performance savings to generate revenue and protect shareholder value. In such a context, the construction industry's clients are demanding more from their delivery teams. To gain competitive advantage in this new landscape, or even just to survive, everyone involved in complex construction projects delivery needs to think and act very differently. To meet this challenge, service providers in every sector of the construction industry have to integrate their knowledge, expertise, technology, resources, and market insight to create new solutions.

Primary insights

- The successful delivery of large-scale, complex construction projects, where service processes are highly interdependent, uncertain, and time-constrained, requires all members of the team to adopt a certain state of mind and to hold a particular expectation about other team members' behavior that they will perform in a mutually acceptable manner, providing excellent customer service.
- Firms involved in virtual corporations' (VCs') delivery vehicles employ people with strong leadership abilities, which allow them to interact and make strategic decisions within their area of expertise, to deal confidently with paradox, to operate from an outcome-oriented perspective, and to act boldly and wisely.
- The case presented demonstrates that complex project delivery and effective cooperation are facilitated by difficult circumstances, whether this involves time pressure, a position of mutual dependence, or exposure to liability.

KEYWORDS

Complex projects, Cooperation, MediaCityUK

ABSTRACT

Large-scale, complex construction projects, where service processes are highly interdependent, uncertain, and time-constrained, present an unusual combination of circumstances. Existing project management tools and processes, relying on ex ante identification and evaluation of all factors for which preventive methods and response strategies are incorporated into contractual agreements, have proved inadequate in these circumstances. The delivery of Phase 1 of MediaCityUK presented in this case study was set against an unusual combination of unforeseen market circumstances, such as construction price inflation preceding the global economic crisis and the subsequent market downturn and the significant management and technical complexity of the project. This meant that the team delivering this complex project was required to operate with an extremely high level of flexibility, responding to constantly changing needs, scope, program, budget, and strategy. The case presented demonstrates that complex project delivery and effective cooperation are facilitated by difficult circumstances, whether this involves time pressure, a position of mutual dependence, or exposure to liability, and that to gain competitive advantage in this new landscape, or even just survive, everyone involved in the delivery of complex construction projects must integrate their knowledge, expertise, technology, resources, and market insight to create new solutions.

GLOSSARY

(abbreviations and acronyms, alphabetized)

APM   Association for Project Management
BBC   British Broadcasting Corporation
BIM   Building information modeling
BLL   Bovis Lend Lease
CAD   Computer-aided design
CPE   Construction project extranet
CT   Chapman Taylor
HD   High definition
MC   Management contracting
NAO   National Audit Office
RACI   Responsibility Assignment Configuration Item (Matrix)
RIBA   Royal Institute of British Architects
TMO   Temporary multiple organization
TSI   Total systems intervention
UK   United Kingdom
VC   Virtual corporations

BACKGROUND

Context

The British Broadcasting Corporation (BBC), as part of its plans for the renewal of its Royal Charter, announced in December 2004 its strategic decision to decentralize its operations and to move whole channels or strands of production from London to Manchester. In response to this opportunity, several VCs emerged, presenting potential development opportunities in the north of England. During 2005, the BBC narrowed down an initial list of 18 sites to a shortlist of four and, late in 2006, chose the VC led by Peel Group (the developer) as the *preferred bidder*. The vision for the Peel Group, Salford City Council, the Central Salford Urban Regeneration Company, and the Northwest Regional Development Agency was to create, on the 200-acre (81 ha) Salford Quays site, at the eastern end of the Manchester Ship Canal on the site of the former Manchester Docks, a significant new media city capable of competing on a global scale with developments in Copenhagen and Singapore.

The BBC's move northwards was conditional on a satisfactory license fee settlement from the UK government, which required the developer to take significant risks well ahead of the formal contractual agreement sign-off to secure timely delivery. This involved assembling the team, preparation of a planning application to develop the site, and, as soon as the detailed planning consent was granted (in May 2007), commencement of the construction of the development in June 2007—almost 6 months before the formal contractual agreement with the BBC was ready to be signed.

The BBC allowed only three years to design and build its new center of operations, giving the developer a date of December 2010 for the completion of Phase 1. In addition, at the eleventh hour, the corporation notified the developer that it did not like the master plan or the concept design of the buildings. Following the change of design team, development of the new concept, and long contractual sign-off process, the short timescale remained the greatest challenge to the successful delivery throughout the whole design and construction phase. However, the additional challenge associated with the nature of the product to be delivered cannot be underestimated. Media cities are widely recognized as a *geographically complex phenomenon* (Krätke, 2003). They are centers of cultural production and media industry with geographical links at both the local and global level. Media, as an industry, revolves around the unstoppable and accelerating march of technology, while cities and buildings embody decisions taken some time earlier, in distant circumstances, often for obsolete reasons. The location, size, internal layout, and materials involved in a construction project generally reflect legislation and standards prevailing when that project's brief was realized, which means that at completion, it is already outdated. During the construction phase, building codes have often evolved, sustainability standards have often changed, and some building materials have since been determined to be a health hazard (Brand, 1984). To deliver a true media city, the team adjusted the scope of the project as it progressed to grasp new opportunities and improve project performance. This resulted in BBC buildings being delivered ahead of contractual dates and, according to National Audit Office (NAO), under the planned cost (NAO, 2011).

The scheme proved a success. After the first trial shows (from November 2010) and the first programs filmed at MediaCityUK, from February 2011, the BBC commenced transfer of its employees to the development in May 2011, starting a 36-week staff migration process. Following the successful BBC start, Channel 4 has expressed interest in moving some of its activities to MediaCityUK and ITV commenced delivery of its production center across the bridge from the BBC offices. The Phase 1 developer is currently preparing plans for development of Phase 2.

PERTAINING THEORIES

There are four concepts of systems thinking that have been used in this case study presentation to assist in explaining the successful delivery of MediaCityUK:

- Wicked problem
- Game theory
- Scenario planning
- Total Systems Intervention (TSI)

The identification of the complex construction project as exhibiting elements of *wickedness*—or as a *wicked* problem (Conklin, 2005)—is driven by the difficult nature of some of the problems encountered by such projects, which are seemingly impossible to solve. The challenge for the successful delivery of a complex project stems from incomplete, contradictory, and changing requirements and complex interdependencies, which result in the effort to solve one aspect revealing or creating other problems.

Game theory is a branch of applied mathematics that has been widely recognized as an important tool in many fields: social sciences, economics, engineering, political science, international relations, and so forth. Game theory attempts to mathematically capture behavior in strategic situations, in which a particular party's success in making choices depends on the choices of others (Aumann, 1987). By applying game theory as a theoretical framework to explain the possible mechanism driving the team members' performance, it is possible to demonstrate that an anticipation of future behavior facilitates the emergence of interfirm alliance structures, which are necessary to successfully deliver a complex construction project.

Scenario planning (also known as scenario thinking or scenario analysis) is a strategic planning method that in the delivery of complex construction projects permits the preservation of uncertainty: making flexible long-term plans and entertaining multiple futures rather than trying to predict the future. The aim of scenario planning is to bind the future but in a flexible way that permits learning and adjustment as the future unfolds.

TSI thinking's application to the delivery of complex construction projects is linked to the TSI cyclical mode of inquiry. In the traditional, linear mode of project delivery, as set out in established project delivery plans, for example, the Royal Institute of British Architects (RIBA)'s Plan of Work or the Association for Project Management (APM)'s Project Life Cycle, the delivery of a project revolves around linear progress, fully aligned with the assumed strategy and methodology, through established stages, with each stage being concluded with some form of sign-off and closure. A TSI method assumes a certain amount of movement back and forth among the phases and milestones and between methodologies as the project progresses.

GUIDING PRINCIPLES

Timely delivery was the main principle driving the setup and delivery of the MediaCityUK project. As a result of the criticality of the time factor, the construction commenced five months before the tripartite agreement between the client (BBC), the developer (Peel Group), and the contractor (Bovis Lend Lease [BLL]) was ready to be signed. Based on experience in earlier projects, the contractor was instructed by the developer to proceed with construction despite the lack of a formal agreement with the BBC. Both parties, making certain assumptions as to what would be required, signed a one-page, two-paragraph letter setting out their intention to proceed together and the contractor was granted access to the site. The construction proceeded on the basis that all the VCs' partners continued to follow an agreed path *unless instructed otherwise* by the developer.

CHARACTERIZATIONS

There are two key characteristics associated with complex construction projects that are illustrated by this case study. The first is linked to the type of organization delivering it, and the second is linked to the management complexity of such projects. The third characteristic of complex construction projects, technical complexity, which in the case study presented was linked to the delivery of the required level of power resilience, high-performance spaces such as the largest HD studio in Europe and radio studios with glass walls located on the exterior elevations (to address the BBC's requirement for *transparency* of operations, allowing members of the public to see radio programs in the making from the nearby public square), has been omitted due to the size constraints of the case study overview.

The first characteristic (the type of organization) results from the fact that in the United Kingdom, most complex construction projects are delivered through VCs—temporary and project-specific organizations of clients, consultants, and construction firms that have neither a central office nor an organizational chart. These highly adaptable and opportunistic structures, composed of independent companies, often have no hierarchy and no vertical integration (Byrne, 1993). Their business meetings are usually held in portable steel container site offices, participation entailing the playing of various roles in completing tasks or deliverables for a project or business process described by a Responsibility Assignment Configuration Item (RACI) (Matrix) (Chomicka, 2011b). Each partner contributes its *core competency* to those of other companies, resulting in a virtual syndicate of all the skills necessary to complete the project. The main objective of such a syndicate is to configure a *best of everything* corporation that takes advantage of economies of scale and economies of scope, as no single company has all the in-house expertise and spare capacity required to quickly launch a large, complex construction project. Instead, in response to a specific opportunity, a group of collaborators quickly organizes itself to exploit it. Once the opportunity has been exploited, the venture disbands, unless another opportunity arises for the team, for example, a move to a follow-up project. The evolution of construction organizations toward the VC model is driven by fast-moving markets, innovation, the shortening of product life cycles, globalization (Kupke and Latteman, 2006), competition, the increasing need for specialization, and the increasing size and complexity of construction projects.

The second characteristic (management complexity) stems from the fact that complex projects in the construction industry are usually characterized by complexity linked to the business context and organizational aspects of the project rather than the technical aspects (which constitute technical complexity). Their complexity also arises from the dynamic nature of the business context and the dynamic nature of VCs that deliver such projects. In the context of complex project delivery, changes to the business context and organization are often not *unplanned disturbances* to project performance, as described by many researchers, but actively pursued business opportunities (Chomicka, 2011b). The instability of the requirements of such projects results in a lack of predictability in principle, which is often not a consequence of incomplete knowledge, such as incomplete scope definition or any other controllable cause, but a natural response to the timescales involved in the delivery of such projects and opportunities emerging midproject. Also, the leadership within VCs is very different to many linear projects—there is no clear line between the *leaders* and the *led*, a situation described by Joseph Raelin (2003) as leadership leaderful practice. Organizations involved in complex project delivery employ people with leadership abilities to interact and make strategic decisions within their area of expertise: these are people who deal confidently with paradox and ambiguity and operate from a creative, outcome-oriented perspective (Raelin, 2003).

Type of system: Complex, large-scale mixed-use construction scheme
System maturity: New build
Environment: Private construction development
Systems engineering activities: Constantly evolving scope, schedule (program), and cost

As Is System Description

Traditionally, construction project sponsors, delivery teams, and customers attempt to package projects into well-defined, controllable requirements, scopes, costs, and schedules, all of which reduce the ability to pursue opportunities emerging midproject and the possibility of evolutionary change taking place. The large gaps in time between the realization of need and procurement and final handover of the finished product make change to the original scope and assumptions unavoidable. Such change is traditionally perceived as a time-consuming and expensive disruption to the project, as opposed to an opportunity that drives better solutions and cost savings.

To Be System Description

The delivery of MediaCityUK proceeded on the basis that the team continued to follow an agreed path *unless instructed otherwise* by the developer. Numerous postcontract changes to the scope of the project, the program, and the location and shape of buildings required intense interaction between the developer and its partners. The reliance on the prior experience of team members of working together in virtual delivery vehicles informed many decisions, minimized risks, and allowed progress even in the absence of static scope, requirements, and performance information. The team continuously revisited the proposed solutions, some of which were already executed on site, to see whether they still brought the best value.

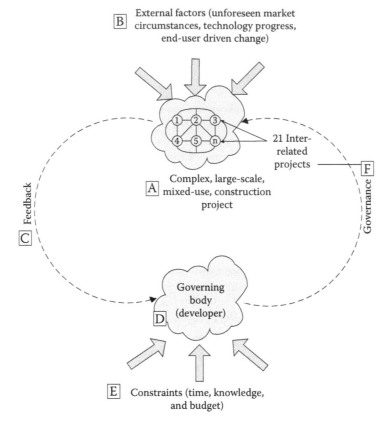

FIGURE 4.1   Overview characterization of the MediaCityUK large-scale construction project.

(Complex) System/System of Systems/Enterprise

This generic Figure 4.1 summarizes the complexity of this MediaCityUK case study.

A—This complex system was broken down into 21 interrelated subprojects with carefully managed interfaces, completion, and handover dates.

B—External factors are characterized by largely unforeseen markets burgeoning technology and changes driven by the vagaries of end-user desires and objectives.

C—The MediaCityUK developer provided direction and redefined the project—its scope, time, and cost—depending on emerging opportunities, encountered problems, and external factors affecting the scheme as the works progressed.

D—The governing body was the MediaCityUK developer.

E—The main constraints were time, knowledge, and budget.

F—Due to complex interdependencies, the effort to solve one aspect revealed or created other problems providing feedback to the developer.

CONSTITUENTS

In the case study researched, the main driver behind the project was the developer of MediaCityUK, the Peel Group. The Peel Group is a diverse and dynamic enterprise and is the leading infrastructure, transport, and real estate company in the United Kingdom, with assets of almost £6bn. What makes the Peel Group successful is its approach to doing business. Peel is a loyal client that invests for the long term, with strong authority and established methods of cooperation. The Peel Group's business model revolves around the assumption that payoffs in the future will be greater than today (The Peel Corporate Brochure, 2009), which allowed the company to build up a successful business and deliver, for example, a £1.4bn shopping and leisure destination, the Trafford Centre, in Manchester, which has over 30 million visitors per year, before venturing into delivery of the £650m MediaCityUK. Following the completion of Phase 1 of MediaCityUK, and possible market recovery, Peel's efforts will be likely to focus on Phase 2 and its new vision—a £50bn regeneration initiative, Ocean Gateway.

Similar to the majority of complex construction projects in the United Kingdom, MediaCityUK was delivered through a VC—a project-based, temporary organization of clients, consultants (engineers, architects, planners, and so forth), and construction firms. The concept of the VC originated in the notion of the *temporary multiple organization* (TMO), coined in relation to construction management in the early 1980s. The steady evolution of construction organizations toward temporary delivery vehicles is driven by the increasing complexity of construction projects and an increasing need for specialization (Vrijhoef, 2004). Current organizational and procurement developments are increasingly focused on more effective management of the VC through better integration of the firms involved in the construction effort through the acquisition and formation of strategic partnerships, team selection strategies based on appointment of one-stop, multidisciplinary design firms and various forms of flexible procurement strategies. This is happening in other venues as well (Senge et al., 2008).

In its selection of VC partners, the developer followed its usual strategy. Despite time being the main critical factor, rather than appointing one or two one-stop, all-service firms, the developer relied on a large number of local, highly specialized firms that had provided excellent customer service on earlier projects undertaken by the developer. This approach can be explained thus: diversity helps to achieve the best possible outcome, because it adds perspectives that would otherwise be absent and because it takes away, or at least weakens, some of the destructive characteristics of group decision making. According to Surowiecki (2004), "Homogeneous groups become cohesive more easily than diverse groups, and as they become more cohesive they also become more dependent on the group, more insulated from outside opinions, and therefore more convinced that the group's judgment on important issues must be right." (Surowiecki, 2004, p. 36) The selection of partners was a complex process and involved finding partners with compatible goals and values who would provide the temporary organization with the right breadth and depth of skills. The developer involved the contractor right from the start, on the strength of their longstanding relationship, forged through working on 26 earlier projects together. This specific team selection strategy stemmed directly from the developer's experience and knowledge acquired during earlier projects. The necessary flexibility and openness of mind were facilitated by the developer's ongoing reviews of what had worked before, in similar or different circumstances, which allowed it to mitigate the risks associated with continual postcontract scope changes. Working with this developer required an ability to take a phone call at midnight before construction was due to commence the next morning, communicating the news that a building is to be of a different type from that originally envisaged, as was the case with one of the residential towers developed as a hotel, a few stories lower (e.g., the office tower) or higher (a multistorey car park), or built on a different plot from that originally planned (the Energy Centre).

An early foundation of trust among MediaCityUK's VC partners and the decrease in opportunistic behavior was facilitated by a significant upfront investment from each team member required to secure a deal with the BBC. In line with the way in which a deal is secured in construction, the contract award requires a significant upfront investment from each member (preparation of bid at risk). As uncooperative behavior by a member of such a social unit would risk damage not only to that member's reputation but to that of all the delivery vehicle members, the partners were willing to cooperate well. The shared risk lessened opportunistic behavior and built an early foundation of trust among all partners.

In addition, the attributes of the VC's employees associated with the dynamic nature of their workplace, such as their flexibility, openness of mind, and leadership abilities, contributed to the adoption of systems thinking on the MediaCityUK project and a higher fluctuation of knowledge than is the case in more traditional organizations (Pribilla, 2000). VCs' partners' specializations enable them to concentrate on their core competencies, increasing efficiency and comparative advantage over traditional companies in their particular field or area of expertise (Werther, 1999).

SCOPE

The first £650m phase of MediaCityUK consists of 700,000 sq ft of office space, including office accommodation for the BBC and University of Salford; 250,000 sq ft of HD studios, including the largest HD studio in Europe and one dedicated to the BBC Philharmonic; 60,000 sq ft of shops and leisure facilities, including bars, restaurants, and healthcare services; a public park and an events space twice the size of Trafalgar Square; 378 new apartments; a 218-bedroom Holiday Inn; retail units, including a supermarket; a 2200-space car park; a new bridge; a new tram link to Manchester city centre; and a highway junction.

STRUCTURE

The design and build procurement strategy, which is commonplace in complex construction projects, was abandoned in favor of one of the most challenging procurement methods in construction—management contracting (MC).

INTERNAL RELATIONSHIPS

The client appointed the design team and a managing contractor (who managed the construction stage and entered into contracts with suppliers and works package contractors).

EXTERNAL FACTORS

Period of construction cost inflation and subsequent market downturn.

CONSTRAINTS

Time

GOVERNING BODY

MediaCityUK was delivered through a VC—a project-based, temporary organization of clients, consultants (engineers, architects, planners, and so forth), and construction firms. Its focal partner, the developer (the Peel Group), had both implicit and explicit knowledge, and this knowledge was dispersed into the VC, resulting in a key partner possessing the entire knowledge about the project. This situation avoided the knowledge loss described by Kupke and Latteman (2006), which stems from the frequent reconfiguration of the VC's constituent partners and personnel mobility (caused by the long duration of the project and the economic downturn).

The MediaCityUK delivery team VC's time-limited nature resulted in a life cycle, which can be divided in four phases: (1) preparation, (2) establishment of links among partners, (3) active work on a project, and (4) the dissolution of the VC.

The MediaCityUK delivery team operated as a social unit, taking what could be termed an *adaptive* approach that permits flexible response to unplanned circumstances. The relationship between the delivery team and the BBC followed a similar path. Rather than using the clauses in the agreement for lease (the contract signed between Peel and the BBC) in a legal sense, the parties tended to refer to them only when they needed opportunities for some give and take.

CHALLENGES

Three project characteristics—a very large number of collaborating partners, a procurement strategy that is very rarely adopted nowadays, and the extensive amount of data that were required to be produced on a project of this scale—created an unprecedented amount of information to be exchanged. To resolve the strategic problems arising from these circumstances, there was an extensive amount of interorganizational learning taking place (Senge et al., 2008). The necessary interorganizational learning followed the main characteristics of learning on construction projects. In the construction industry, learning is based on a movement away from known problems rather than toward imagined goals—it is based on experienced unfitness (defects) rather than planned fitness for purpose (Brand, 1984).

The issues associated with the large number of partners and the MC procurement strategy have been described earlier. The coordination of all the activities carried out by various independent parties is one of the most important problems encountered in the delivery of large, complex projects by temporary organizations. The heavy dependence on the exchange of large and complex data and information is one of the main recognized risks, and the amount of information to be exchanged increases rapidly with the number of organizations involved in the project.

Nowadays, there can be more than 100 companies involved in the delivery of a large-scale scheme. It is apparent that interorganizational coordination, especially accuracy, effectiveness and timing of communication, and the exchange of information and data among members of the project team—having everyone working to the same up-to-date information—will be of decisive importance to the success of complex projects.

To facilitate successful collaborative working, every VC utilizes some form of business integration tool, which includes all the functions of project management for the enterprise and its partners. The MediaCityUK team relied on the construction project extranet (CPE)—a web-based project collaboration tool that usually provides web-based access, user interface, document storage, configurable process workflow (a ready-made tool to manage the flow and control of documentation), and audit trail reporting (the facility to track status and progress across activities and the entire workspace). Extranets are powerful tools and allow sophisticated manipulation of vast volumes of information.

There are many application service providers, including BuildOnline, Causeway, Cadweb, Sarcophagus, CONJECT, 4Projects, Business Collaborator, and Bidcom, which are currently combining to form an industry group (Suchocki, 2006). The MediaCityUK team used CONJECT.

The amount of data uploaded on to the extranet that had to be managed, referenced, and reviewed grew from 4,277 documents uploaded in the first year (2007) through to 19,459 documents uploaded in the second year (2008), to reach 25,884 in the third year (2009), declining following several sectional completions in the fourth year (2010) to 10,833 document uploads (Chomicka, 2011a). Apart from documents, there was also a significant amount (105,000) of forms and processes that had to be reviewed, controlled, and monitored—3,106 in year one and 29,353 in year two. It was not possible to establish statistics for the third and fourth year as every attempt to generate a list caused the extranet system to encounter a fault that prevented the request being completed, probably due to the amount of data exceeding the system's capacity. The majority of case studies available in the literature are concerned with extranets operating on a database of up to 15,000 documents, and the teams involved in them provided a positive feedback, save for occasional issues with internet speed and the need to download a large amount of data at the end of a project. Considering the difference in the quantity of documents and processes (15,000 versus 165,000), it is possible that the problem of uncontrollability of data did not occur in the case studies reviewed to date. At the end of the second year, the cumulative sum of all documents and processes exceeded 56,000, prompting an ineffective, and later abandoned, attempt to restructure the data management strategy. This raised concern about how the optimum or equilibrium cluster size can be determined within the project and whether there is an algorithm that governs this interaction. Of course, some of the drawings have been superseded and some processes have been deactivated; nevertheless, the sheer amount of information to be processed at any one time posed a management problem and caused a significant amount of rework on site.

Experience on the MediaCityUK project suggests that certain problems associated with the exchange of information, such as uncontrollability and the reduced interoperability of data, pose one of the greatest challenges to successful project delivery. The wide variation in specialisms, expertise, educational backgrounds, professional skills, computer familiarity, and working environments among the project participants impedes the information management and communication of the project team, leading to mistakes that are very expensive and difficult to rectify, such as the construction of high-performance partition walls in the wrong location or the accidental omission of drainage in posttensioned slabs.

The majority of studies of such web-based project collaboration tools conclude that one of the main advantages of using extranets is that it ensures that all members of the project team have access to the most up-to-date versions of the various project documents, so that traditional mistakes generated from someone working from an old document or drawing are in theory removed or, at the very least, reduced. However, the experience recorded on the case study presented raised concerns about the translation of this potential benefit into actual use.

The combination of the extended fragmentation of the construction industry, the huge amount and wide dissimilarity of the information that is involved in the construction process, and the wide variation in specialisms, expertise, educational backgrounds, professional skills, computer familiarity, and working environments among the project participants impedes the information management, and communication on extranets leads to mistakes that are very expensive and difficult to rectify. Also, the large scale of projects, team selection, and procurement strategies, resulting in further increase in project data, contributes to the problem of data uncontrollability.

In addition, as observed on smaller projects reviewed by extranet provider Cadweb (2010) during the performance evaluation of their product, there was a significant amount of spamming created by extranet users. Cadweb noticed that on projects run with their product, every file uploaded is distributed to 11 people rather than 5. The amount of spamming on the MediaCityUK project confirms this observation. The significant amount of data—100 documents a day being issued on each working day in the third year of site operations—resulted in a problem faced by many environments, both real and virtual; the system reached a point on the scale where growth turns cancerous: when any association becomes too large, beyond a certain point, growth without deterioration is not possible. It is generally recognized that coordination becomes impossible when there are too many people trying to participate in it. Oversized populations generally require additional mechanisms, such as pattern-matching tools and feedback (Johnson, 2002), both mechanisms yet to be contemplated by extranet providers.

The two challenges that were also described earlier are time and the combination of the construction price inflation (at the start of the project) and economic downturn (halfway through the delivery process).

DEVELOPMENT

- The perpetual changes required flexibility while maintaining the high level of professional standards and competence necessary to keep the scheme viable at a time of construction cost inflation.
- The construction proceeded on the basis that the team continued to follow an agreed path unless instructed *otherwise* by the developer.

RESULTS

The unique approach to complex project delivery adopted on the Phase 1 of the MediaCityUK project allowed not only the delivery of the BBC buildings ahead of contractual dates, under the planned cost (Plunkett, 2010), but also to achieve standards that did not even exist when construction commenced—MediaCityUK has achieved the first world-recognized BREEAM Sustainable Community Award (BREEAM, 2010).

ANALYSIS

The interpretation of the results will follow the four concepts of systems thinking, which were identified as suitable for this case study presentation: (1) wicked problem, (2) zero-sum game, (3) scenario planning, and (4) TSI.

WICKED PROBLEM

The suggestion that the MediaCityUK project delivery can be compared to the solution to a wicked problem is based on the difficult nature of the problems that were addressed by this project. The challenge to this complex project stemmed from incomplete, contradictory, and changing requirements and complex interdependencies, which resulted in the effort to solve one aspect revealing or creating other problems. Examples of MediaCityUK's *wickedness* range from the business strategy pursued to aspects of design decision making and the knowledge management issues encountered by its delivery team. According to Rittel and Webber's (1973) suggestion, the formulation of these wicked problems can be explained by the existence of 10 characteristics: (1) There was no definitive formulation of these wicked problems (the definition of these wicked problems is itself a wicked problem); (2) the problems had no stopping rule; (3) the solutions to these problems were not true or false but better or worse; (4) there was no immediate and no ultimate test of a solution to these problems; (5) every solution was a *one-shot operation* because there was no opportunity to learn by trial and error, and every attempt counted significantly, especially in financial terms; (6) the potential solutions to some problems could not be fully priced beforehand as enumerable (or fully scoped); (7) the problems were unique to this development; (8) every problem was a symptom of another problem; (9) the nature of the problem's resolution was determined by the choice of explanation; and (10) the team was liable for the consequences of the actions they generated.

For example, in one instance, the BBC needed more time to provide the specific requirements in relation to the soft floor finishes, especially the specification and the preferred installation stage of carpets, but wanted the team to keep its final delivery date. This created a problem relating to the timely supply of the required quantity of the product, as it exceeded the capacity of the largest suppliers in the country. It also created a problem of acoustic testing, as certain acoustic criteria can be met only once the soft floor finishes have been installed. The team, in collaboration with the soft flooring works package contractor, secured the production slot on the basis that the specification of the carpet to be produced would be confirmed just before production commenced and purchased only 100 m^2 of carpet to allow the testing of all areas that had to achieve specific acoustic criteria, on the basis that this 100 m^2 would be moved around all the rooms in three buildings, as required, to enable acoustic testing.

GAME THEORY

There are two central theories that can be used to understand the behavior of exchange parties on complex construction projects. One revolves around the notion of trust as a driving force in economic relations, and the second focuses on contracts as governance structures for managing relationships between commercial parties (the transaction cost theory of contracting). The role of trust in facilitating efficient exchange relations has been considered by numerous researchers, including Macaulay (1963), Lyons and Mehta (1997), and Lorenz (1999). This traditional approach can be described using Sako's (1992) study of interfirm relations in Britain and Japan. Sako described this behavior in economic relationships using three major types of trust: contractual trust, competence trust, and goodwill trust.

Contractual trust is based on a legal agreement between parties who enter into a contract, creating a universal ethical standard and an expectation of promised delivery.

Competence trust revolves around the belief in the other party's ability to do the job, conform to professional standards, and complete the task.

Goodwill trust refers to an expectation of open commitment between parties.

It has been demonstrated that efficient exchange relations are facilitated when the parties trust each other; the presence of trust transforms an exchange relationship by reducing the costs of specification, monitoring and guarding against opportunistic behavior, encouraging better investment decisions, and ensuring rapid and flexible responses to unforeseen events (Lyons and Mehta, 1997).

The transaction cost theory of contracting, on the other hand, developed by Williamson (1985), assumes that interfirm transactions are governed by contracts. This theory predicts that any disturbances that could upset the relationship would be contemplated beforehand, allowing the development of mechanisms that would facilitate joint adaptation with the aim of protecting the relationship. These mechanisms would be incorporated into the contract and include, for example, administrative procedures aimed at dispute prevention or the resolution and distribution of costs and benefits (Williamson, 1985). The theory assumes a competitive economic environment and makes equilibrium forecasts, predicting, for example, that firms that misquote for their services (*misdesign* contracts) will perform poorly in the market. Over time, poor performers will be forced to exit the market, so that in equilibrium, only those who make a profit on their service provision (*alert agents and well-designed contracts*) will survive (Mayer and Argyres, 2004).

Using game theory as a theoretical framework to explain the possible mechanism driving the team members' performance, it is possible to demonstrate that an anticipation of future behavior facilitates the emergence of an interfirm alliance structure associated with commitment to provision of the excellent customer service.

In game theory, as well as in economic theory, zero sum denotes a situation where one party's gains or loss is exactly balanced by the gains or losses of other parties, so the total gains and losses of all participants add up to zero. In contrast, non-zero-sum refers to a situation in which the interacting parties' aggregate gains and losses are either less or more than zero.

In non-zero-sum games, a gain by one player does not necessarily correspond with a loss by another, and the success of all players depends on their ability to come together and cooperate. If the game is played only once and players do not have to fear retaliation from their opponents, they often play differently from the way they would if they played the game repeatedly. In the game's theoretical context, the pursuit of self-interest by an individual or an organization leads to a poor outcome for all (Axelrod, 1984). The possibility that players may meet again (the *shadow of the future*) gives rise to cooperation.

The importance of cooperation is well established in the organizational literature (Kogut and Zander, 1996). The evolution of interfirm relations toward cooperation is being driven by competitive forces and the intensification of competition in the product markets, which has accompanied increasing demands by buyers for greater customization and flexibility in the production and delivery of goods and services (Arighetti et al., 1997). However, groups cooperate well under certain circumstances and less well under others. It is important, therefore, to establish what increases the likelihood of successful cooperation and makes parties go that *extra mile* in delivering a project to the highest standards.

There are two types of cooperation (Welling and Kamann, 2001): perfunctory and consummate.

Perfunctory cooperation is an extent of cooperation that can be imposed through either legal agreement or the threat of sanctions.

Consummate cooperation occurs when parties work together to a mutual end, sharing skills and information and responding flexibly. Since consummate cooperation cannot be enforced, the circumstances in which it can be achieved are important.

In cooperation, according to Axelrod (1984), individuals do not have to be rational, nor do the players have to exchange messages or commitments, as their deeds speak for them; there is no need to assume trust between the players, as the use of reciprocity can be enough to make defection unproductive; altruism is not needed, as successful strategies can bring out cooperation even from an egoist, and no central authority is required, as cooperation based on reciprocity can be self-policing. Therefore, what makes it possible for cooperation to emerge is the fact that players might meet again; the choices made today not only determine the outcome of this particular interaction but can also influence future opportunities for the players. The future *casts a shadow* on the present, affecting the current strategic situation of the parties.

According to Welling and Kamann (2001), parties in construction supply chains seem to have been entangled in a kind of an iterated non-zero-sum game for many years already. If there is no future to influence (the game displays a known, finite number of interactions), the results are likely to include adversarial behavior, low profit margins for construction firms, and a short-term, project-oriented vision, recognized, for example, in the Latham Report (Latham, 1994).

The findings of the Latham Report—a review of the construction industry's procurement and contractual arrangements—still largely inform the popular view of the industry. Entitled *Constructing the Team*, this report describes industry practices as *ineffective, adversarial, fragmented, incapable of delivering value for its customers*, and *lacking respect for its employees*. Since its publication nearly 20 years ago, the Latham Report is still widely cited in press reports. Also, many contemporary government-funded reports rely on the Latham Report's findings when rendering advice to the industry (e.g., the Government Construction Strategy report published in May 2011 that tells the sector *to work more collaboratively*).

The traditional applications of game theory attempt to find equilibrium in a particular game. In equilibrium, each player of the game has adopted a strategy that they are unlikely to change. In a construction context, this can be compared, for example, with the strategy adopted by the executive architect on MediaCityUK scheme—Chapman Taylor (CT). According to CT's director, Tim Partington, the company is committed to providing excellent customer service and project completion, irrespective of the transaction cost, as despite best efforts to reach an optimal level of fee income in contractual relationships, the occasional misquotation is inevitable. To enable the provision of excellent customer service, the company strategy of drawing fee installments is based on an assumption that there is always more work to do at the end of the project. This *extended foresight* cash flow approach adopted by CT helps with cash flow and facilitates the provision of high-quality service even if the company has misquoted for its services or provided more service than originally envisaged to assist the client (and for strategic purposes is not seeking fees renegotiation).

According to Axelrod, the future is less important than the present, for two reasons: "The first is that players tend to value payoffs less as the time of their obtainment recedes into the future. The second is that there is always some chance that the players will not meet again. An ongoing relationship may end when one or other player moves away, changes jobs, dies, or goes bankrupt. For these reasons, the payoff of the next move always counts less than the payoff of the current move" (Axelrod, 1984, p. 12). However, for some organizations, especially those with long-term orientation, the future is more important than the present. For example, the Peel Group business model revolves around the assumption that payoffs in the future will be greater than today (Peel Group, 2009), which allowed the company to build up a successful business and deliver MediaCityUK.

In line with Axelrod's (1984) suggestion, one of the conditions that facilitate cooperation is the shared feeling of a *fair deal* between both partners in the arrangement. The perceived fairness of the transactions between the Peel Group and their delivery teams is not guaranteed by the threat of a legal process but rather by the anticipation of mutually rewarding transactions in the future. However, whatever the anticipation among parties is based on, there has to be an opportunity of future contracts in the first place. In other words, one of the key aspects of game theory relevance for the construction industry is a very specific shadow of the future: the real possibility that individuals might meet again.

Apart from interfirm cooperation at a strategic level, Geerink (1998) also found that in construction practice, the allocation of individuals seems to play an important role in establishing cooperative behavior. Repeated personal contact across organizational boundaries supports a minimum level of courtesy and consideration between the parties. An interesting aspect of this fact can be observed when construction and professional services firms, following appointment for the provision of services on large-scale construction projects, seek to employ individuals who have previous experience of working with a particular business partner. For example, on the MediaCityUK project, the executive architect approached one of the managing contractor's design managers, who, upon agreement with his firm, joined the executive architect's firm to facilitate interfirm dealings.

The future is therefore important for the establishment of the conditions for cooperation; however, the past is important for monitoring of actual behavior. Although service providers and construction firms monitor their own behavior as part of their ISO 9001 quality assurance system, it is the most recent actual customer service experience that determines whether the client will use the particular service provider again.

SCENARIO PLANNING

Scenario planning (scenario thinking or scenario analysis) is a strategic planning method that in the complex construction projects delivery permits preservation of uncertainty: the making of flexible long-term plans and the entertaining of multiple futures, rather than trying to predict the future. The aim of scenario planning in complex projects delivery is to bind the future but in a flexible way that permits learning and adjustment as the future unfolds.

Scenario planning involves aspects of systems thinking, specifically the recognition that many factors may combine in complex ways to create sometimes surprising results, due to nonlinear feedback loops, characteristic to systems thinking.

This chief value of scenario planning is that it allows decision maker(s) to make and learn from mistakes without risking project failures in real life: an opportunity to *rehearse the future*, despite the fact that every action and decision counts (and costs money). Although in the case study researched, the delivery team could not make these mistakes in a safe, unthreatening, gamelike environment while responding to a wide variety of real situations faced by it, scenario planning helped the team understand how the various strands of a complex tapestry move if one or more threads are pulled. The team's simplistic guesses were surprisingly good most of the time but in some cases failed to consider the additional risks and changes that could affect the project.

In the case study researched, scenario planning started by dividing the team's knowledge into two broad domains: (1) things the team believed they knew something about and (2) elements the team considered uncertain.

The first component (often referred in academic writing as *trends*) consisted of must-haves, for example, the floor areas of buildings and program and performance specifications specified by the BBC; the second component (referred to in academic literature as *true uncertainties*) involved mapping out possible ways of achieving these set objectives to ensure they meet the required performance criteria.

Scenario planning involved blending the known and the unknown into a limited number of internally consistent views of the future that spanned a very wide range of identified possibilities. Scenario planning was applied to a wide range of issues, from simple, tactical decisions to the complex process of strategic planning and vision building.

The team used two techniques to arrive at two or three alternative scenarios, addressing each of the problems investigated by brainstorming and detailed assessments based on quantitative and qualitative analysis. Brainstorming sessions were held in the site office and involved six to ten team members, and quantitative and qualitative assessments were carried out by respective team members within their narrow areas of expertise and combined together by the developer's project management team. Each of the two used a method that usually resulted in two or three scenarios: a good and a bad, an optimistic and a pessimistic, and one in between the two extremes.

A good example of scenario planning was the provision of power—as the construction works got underway, it became apparent that the most cost-effective way of providing power to the development would be to abandon the strategy originally pursued and, instead, to drill a tunnel under the Manchester Ship Canal, the largest navigation canal in the world, to pull the cables. This operation normally requires months of preparation and advanced booking of equipment, as there are only two rigs of sufficient size available in the country. As one of these rigs was potentially available within weeks of the realization of the need for these works, it was immediately booked and concurrently, with the company operating the rig, planning a possible route that would allow the transportation of this oversized plant to the MediaCityUK site, while the team was designing the best possible drilling route and the incorporation of the revised power supply strategy into the overall scheme. The first drilling attempt was unsuccessful, as the drill hit a century-old decommissioned culvert that was not identified on maps. Within days, a new route was identified and the rig operator, working nonstop for a couple of days to enable completion of the task without delaying the next assignment, finished the required works.

TOTAL SYSTEMS INTERVENTION

TSI application in the delivery of complex construction projects is linked to its cyclical mode of inquiry. In the traditional, linear mode of project delivery, as set out in project delivery plans established by organizations that set the standards in the construction industry such as the RIBA's Plan of Work or the APM's Project Life Cycle, the delivery of a project revolves around linear progress through established stages, with each stage concluded with some form of a sign-off and closure. The TSI method assumes a certain amount of movement back and forth among the phases and between methodologies, as the project progresses.

TSI assumes three phases, (1) creativity, (2) choice, and (3) implementation and changes between methodologies, as the project progresses. In Phase 1, creativity, MediaCityUK's delivery team and other stakeholders considered creatively improvements to particular aspects of the project, particularly in problem situations and other areas of concern. In Phase 2, choice, using the form of a grid developed around the nature of problem being considered, together with the assumptions underlying a range of problem-solving methodologies, the team selected the most appropriate course of action. Phase 3, implementation, was a set of change proposals along with the set of instructions issued by the contract administrator to orchestrate the change in accordance with agreed objectives, methodology, and solution.

A good example of the application of TSI thinking was the introduction of a trigeneration scheme (the simultaneous production of heat, cooling, and power using a centralized energy center). MediaCityUK achieved a significant saving through the use of a trigeneration system that is capable of delivering heating and cooling energy through the built network from future energy sources, whatever they might be. The MediaCityUK site had been designed with an infrastructure that did not initially incorporate the concept of trigeneration. The idea of the utilization of this new technology came about very late, and the integration of the heating and cooling network within the infrastructure and buildings that was already in progress, on a site where the buildings were built at the same time as the pipework was laid, required a substantial amount of redesign and construction rework as well as major reprogramming of site operations.

SUMMARY

The unique approach to complex project delivery adopted in Phase 1 of the MediaCityUK project allowed the BBC buildings to be delivered ahead of contractual dates and to achieve standards that did not even exist when their construction commenced. The main driver of an excellent team and project performance was the time criticality of the delivery of the building. The end user of the key buildings forming part of the Phase 1 of MediaCityUK, the BBC, not only allowed 3 years to design and build its new center of operations, but at the eleventh hour, it notified the developer that it did not like the master plan or the concept design of buildings. The additional challenges—the nature of the product to be delivered and the global economic crisis that emerged halfway—also cannot be underestimated.

The successful delivery was made possible by the adoption of a flexible strategy that allowed the evolution and adaptation of both the organization (the VC delivering the project) and the project itself.

The VC model allowed the delivery team to face new challenges in a complex environment. This delivery vehicle brought diverse advantages but also displays several perceived disadvantages and limitations, especially its temporary, project-oriented nature and the frequent reconfiguration of its partners. These perceived disadvantages led to better, more dynamic people choosing to be part of the MediaCityUK project.

To facilitate successful collaborative working, the MediaCityUK VC used a business integration tool, the CPE, to coordinate all the activities carried out by various independent parties and the exchange of large and complex data and information.

The experience of using this solution suggests that certain problems associated with reliance on extranets, such as uncontrollability and reduced interoperability of data, contribute to the range of defects and lead to mistakes that are very expensive and difficult to rectify. This means that complex construction projects require a solution that goes beyond the vector-based computer-aided design (CAD) software and sharing of secure files: a combination of 3D shared models, pattern-matching tools, and feedback.

CONCLUSIONS

MediaCityUK was a successful complex systems project in the construction industry that succeeded due to an appropriate systemic perspective. This case study contributes to existing complexity theories by showing that successful delivery and effective cooperation among team members delivering complex projects are facilitated by difficult circumstances, whether these relate to time pressure, a position of mutual dependence, or exposure to liability. The findings also suggest that, in contexts characterized by uncertainty, firms' team members might enter into appointments with levels of commitment that are more bounded than approaches based on trust or transaction cost theory assumptions, providing that the interfirm relationship is durable. This implies that in the context of complex construction projects, ongoing commitment management, including frequent evaluations of future opportunities, might be critical in moving parties toward efficient cooperation.

The durability of the interfirm relations and their future strategic position (the shadow of the future) has a far more important and beneficial impact on delivery of complex projects than is currently recognized. Each party's desire to maintain an ongoing relationship within the MediaCityUK VC and their willingness to make sustained efforts to maintain this connection in the face of numerous disturbances associated with the project's complexity contributed to the success of this enterprise. The findings from this MediaCityUK study suggest that the success of teams delivering complex projects is based on intensive, adaptive learning processes.

The feedback from professionals involved in the MediaCityUK VC suggests that working as part of the VC reduces psychological distance among project members, providing organizations and individuals alike with the opportunity to learn from each other's experiences. Similarly, findings from the MediaCityUK case study confirm the proposition of Borgatti and Cross (2003) that the perceived proximity provides people with the opportunity to learn who knows what, so that members know where to search for relevant knowledge and information across interorganizational boundaries. Such perceived proximity thus creates a higher velocity of knowledge exchange, which is disseminated, recorded, stored, and available in real time via CPE to all parties involved in the project. When the project comes to an end and the VC disbands, CPEs contain a record of accumulated experience that can be accessed and reused on subsequent projects.

It is possible that some of the problems associated with the delivery of complex construction projects such as those described in this case study, including issues associated with the controllability of data, can be resolved by the utilization of building information modeling (BIM). Following the completion of the MediaCityUK scheme, the lead design team commenced work on the £450m redevelopment of Leeds city centre (Leeds Trinity) utilizing BIM as a structured process for generation, sharing, use, and reuse of information. BIM involves the creation and management of digital representations of physical and functional characteristics of a building, such as thermal performance parameters, maintainability, or safety. BIM allows each project team member not only to add to and reference back to all the information they acquire during their period of contribution to the BIM model but also to test proposed design solutions and assess buildability. Although at the time of writing this case study the project is still in the early stage on site, the feedback suggests that certain data uncontrollability and reduced data interoperability issues encountered on the MediaCityUK project may have been avoided by the deployment of BIM. BIM methodology also has the potential to facilitate many of the systems-thinking methodologies described in this case study, such as scenario planning.

The case study also outlines the gap between current working practices in construction and the behaviors assumed by architects of information exchange systems on complex projects. One of the document controllers summarized the feature of extranets, which is most frequently mentioned in the academic literature—accessible real-time data across the globe—in the following manner: "Sure, it's nice to send documents half way across the world in seconds, or to be actively commenting on a drawing through a super-trendy interactive whiteboard but, at the end of the day, all these systems exist to stop someone being sued" (Wilkinson, 2011). What is needed are information exchange systems that would allow the wicked problems of complex construction schemes to be dealt with in an interactive manner that allows the accessing of a larger number of project participants—to go beyond the closed four walls of a room where a brainstorming session involving *an optimum number* of six team members is taking place.

SUGGESTED FUTURE WORK

The MediaCityUK case study contributes to existing theory by showing that complex projects of this scale require a solution that goes beyond the linear processes and procedures defined by professional bodies such as RIBA or APM and that there might be a need for redefinition of these established ways of doing things.

The MediaCityUK case study contributes to existing theory by showing that projects of this scale require a solution that goes beyond the vector-based CAD software and the sharing of secure files. It is possible that on complex projects, considerable risks associated with the uncontrollability of data can be resolved by the combination of 3D shared models, pattern-matching tools, and feedback.

An important area for future research, therefore, is to gain a better understanding of the adaptive delivery processes and how current information exchange platforms can evolve to facilitate the delivery team's interorganizational learning processes and project delivery.

QUESTIONS FOR DISCUSSION

1. What is required, from a project management perspective, to successfully deliver large-scale, complex construction projects?
2. What are VCs and why have they emerged in the construction industry?
3. What types of information exchange problems are encountered in complex construction projects?

REFERENCES

Anumba, C. J. and K. Ruikar. 2002. Electronic commerce in construction—Trends and prospects. *Automation in Construction* 11: 265–275.

Argote, L., B. McEvily, and R. Reagans. 2003. Managing knowledge in organisations: An integrative framework and review of emerging themes. *Management Science* 49(4): 571–582.

Arighetti, A., R. Bachmann, and S. Deakin. 1997. Contract law, social norms and inter-firm cooperation. *Cambridge Journal of Economics* 21(2): 171–195.

Aumann, R. 1987. Game theory. In *The New Palgrave: A Dictionary of Economics*, 2nd edn., eds. J. Eatwell and M. Millgate, pp. 460–482. New York: Palgrave Macmillan.

Axelrod, R. 1984. *The Evolution of Cooperation*. New York: Basic Books.

BBC Trust. 2009. Annual report and accounts 2007/2008. [Online] Available at: http://www.bbc.co.uk/annualreport/ (accessed 20 May 2010).

Bjork, B. 1992. A unified approach for modelling construction information. *Building and Environment* 27(2): 173–194.

Bjork, B. 1998. Information technology in construction: Domain definition and research issues. *International Journal of Computer Integrated Design and Construction* 1(1): 1–16.

Borgatti, S. P. and R. Cross. 2003. A relational view of information seeking and learning in social networks. *Management Science* 49(4): 432–445.

Brand, S. 1984. *How Buildings Learn: What Happens after They Are Built*. New York: Penguin Books.

BREEAM. 2010. Available at: http://www.breeam.org/page.jsp?id=346 (accessed 15 April 2014).

Burati, J. L., J. J. Farrington, and W. B. Ledbetter. 1992. Causes of quality deviations in design and construction. *ASCE Journal of Construction Engineering and Management* 118: 34–49.

Byrne, J. A. 1993. The virtual corporation, *Business Week*, pp. 37–41.

Cadweb. 2010. http://www.cadweb.co.uk/ (accessed 20 March 2013).

Chassiakos, A. P. and S. P. Sakellaropoulos. 2008. A web-based system for managing construction information. *Advances in Engineering Software* 39: 865–876.

Chien, H. J. and S. Barthrope. 2010. The current construction project extranet practices in the construction industry. *Global Journal of Researches in Engineering* 10(3): 23–33.

Child, J., D. Faulkner, and S. Tallman. 2005. *Cooperative Strategy: Managing Alliances, Networks and Joint Ventures*. Oxford, U.K.: Oxford University Press.

Chomicka, B. 2011a. Intractability of data on complex construction projects caused by the reliance on Construction Project Extranets: A case study. Salford, U.K.: RICS COBRA.

Chomicka, B. 2011b. Learning in evolving corporate models in the construction industry: A case study. Phoenix, AZ: IACCM EMEA.

Chomicka, B. 2011c. Shared commitment to excellence. Inter-firm alliance structures in complex construction projects: A case study. Lille, France: ICCPM Research and Innovation Seminar.

Chomicka, B. 2012. *Constructing the Team*. London, U.K.: IACCM EMEA.

Conklin, J. 2005. Chapter 1: Wicked problems and social complexity. *Dialogue Mapping: Building Shared Understanding of Wicked Problems*. Chichester, U.K.: CogNexus Institute. http://cognexus.org/wpf/wickedproblems.pdf (accessed 15 February 2013).

Geerink, S. 1998. Projectonafhankelijke samenwerking—De oplossing voor de knelpunten in het bouwproces? MSc thesis. Enschede, the Netherlands: The University of Twente.

Johnson, S. 2002. *Emergence: The Connected Lives of Ants, Brains, Cities, and Software*. New York: Scribner.

Kogut, B. and U. Zander. 1996. What firms do? Coordination, identity, and learning. *Organization Science* 7(5): 502–518.

Krätke, S. 2003. Global media cities in a worldwide urban network. *European Planning Studies* 11(6): 605–628. [Online] Available at: http://www.lboro.ac.uk/gawc/rb/rb80.html (accessed 03 April 2010).

Kupke, S. and C. Latteman. 2006. The strategic virtual corporation—A new approach to bridge the experience gap. *IADIS International Conference on Web Based Communities*, San Sebastián, Spain, after http://www.iadis.net/dl/final_uploads/200602L006.pdf (accessed on 20 July 2011).

Latham, J. 1994. Constructing the team: Final report of the Government/industry Review of Procurement and Contractual Arrangements in the UK Construction Industry. London, U.K.: Department of the Environment.

Lorenz, E. 1999. Trust, contract and economic cooperation. *Cambridge Journal of Economics* 23: 301–315.

Lyons, B. and J. Mehta. 1997. Contracts, opportunism and trust: Self-interest and social orientation. *Cambridge Journal of Economics* 21(2): 239–257.

Macaulay, S. 1963. Non-contractual relations in business: A preliminary study. *American Sociological Review* 28: 55–67.

Mayer, K. and N. Argyres. 2004. Learning to contract: Evidence from the personal computer industry. *Organisation Science* 15(4): 394–410.

National Audit Office. 2011. National Audit Office Annual Report 2011. http://www.nao.org.uk/report/national-audit-office-annual-report-2011/ (accessed 20 March 2013).

Peel Group. 2009. The peel corporate brochure. [Online] Available at: http://www.peel.co.uk/media/peelbrochure_0909.pdf (accessed 03 April 2010).

Plunkett, J. 2010. NAO findings on BBC development schemes—Project by project—The National Audit Office report's verdict on Broadcasting House, Salford Quays and Pacific Quay in Glasgow. *The Guardian* [internet] 25 February. Available at: http://www.guardian.co.uk/media/2010/feb/25/bbcspent-100m-more-broadcasting-house (accessed 03 April 2010).

Pribilla, P. 2000. Führung in virtuellen unternehmen. *Zeitschrift für Betriebswirtschaft* 2: 1–12.

Raelin, J. A. 2003. *Creating Leaderful Organisations: How to Bring out Leadership in Everyone.* San Francisco, CA: Berrett-Koehler.

Rittel, H. W. J. and Webber, M.M. 1973. Dilemmas in a General Theory of Planning. *Policy Sciences* 4: 155–169.

Sako, M. 1992. *Prices, Quality & Trust: Inter-Firm Relations in Britain and Japan.* Cambridge, U.K.: Cambridge University Press.

Senge, P., B. Smith, N. Kruschwitz, J. Laur, and S. Schley. 2008. *The Necessary Revolution: How Individuals and Organizations Are Working Together to Create a Sustainable World.* New York: Doubleday.

Suchocki, M. 2006. Construction industry collaboration challenges. Atkins, Epsom, U.K.

Surowiecki, J. 2004. *The Wisdom of Crowds: Why the Many Are Smarter Than the Few and How Collective Wisdom Shapes Business, Economies, Societies and Nations.* New York: Anchor Books.

Vrijhoef, R. 2004. Understanding construction as a complex and dynamic system: an adaptive network approach. *12th Annual Conference on Lean Construction,* Copenhagen, Denmark.

Welling, D. and D. Kamann. 2001. Vertical cooperation in the construction industry: size does matter? *Journal of Supply Chain Management* 37(4): 28–33.

Werther, Jr., W. 1999. Structure-driven strategy and virtual organization design. *Business Horizons* 42: 13–18.

Wilkinson, P. 2011. blog www.extranetevolution.com (viewed: 20 February 2011).

Williamson, O. 1985. *Economic Institutions of Capitalism.* New York: The Free Press.

5 IBS Group, Eastern European IT Services
Capability-Based Development for Business Transformation

Mikhail Belov

CONTENTS

CASE STUDY ELEMENTS

Fundamental essence

The case study is about drastic transformation of the strategy and business model of an enterprise that incorporates autonomous but interrelated business units. New capabilities creation was the transformation objective; additional transformation outcome was the *extended enterprise*, which has been formed in the business environment of the enterprise.

Topical relevance

Such an enterprise (consisting of autonomous units) exemplifies a system of systems (SoS). All SoS aspects (complexity, emergency, adaptability, self-organization, network-rather-than-hierarchy, etc.) are natural and typical for the enterprise. The autonomy of the business units engenders deeper involvement of business leaders and strengthens the management team. On the other hand, autonomy impedes centralized control and practically blocks directive control approaches. All stages of the transformation (concept development, planning, and implementation) constitute a set of exemplary SoS engineering activities, as well as exemplary enterprise system (ES) activities.

Domain

SoS and ES in the information technology industry.

Country of focus

Russia.

Interested constituency

Those concerned with engineering and transforming corporations and enterprises.

Stakeholders

Active stakeholders (shareholders, managers, and employees) really operate and directly affect the enterprise during transformation both by real actions and concerns. Passive ones (vendors, contractors, clients, auditors, and regulators) influence the enterprise only by their concerns that are considered by the governing body.

Primary insights

The approaches described might be applied to engineer or to transform the corporations and enterprises. The approaches to enterprise transformation include the following:

- Capability-based development approach and capability-based architecting (organizational structure, concept of the interfaces between business units, and concept of the control by enterprise management).
- Agile-stylized program of project management. The combination of the soft and creative style of development with very strong execution of schedule monitoring and control is a key element of the approach. The approach might be used to solve very different fuzzy and ambiguous tasks of different scales in all areas of systems engineering (SE), enterprise systems engineering (ESE), and SoS engineering (SoSE). The specific details and practical issues of the approach are represented in the "Program Management and Management of Uncertainty" section and the "Lessons Learned" section.
- Definition of key enterprise areas that are critical to form new enterprise capabilities and development of systems in those areas to support new capabilities. Three of them (a system of business unit growth, a management accounting system, and a motivation system for the employees of all units) are critically important for companies all over the world; a fourth area and respective project implementation system are also critical for many companies. Thus, this case study experience might be interesting for many of the companies.

KEYWORDS

Agile program management, Capability-based engineering, Enterprise system, Enterprise system architecture, Extended enterprise, System of systems

ABSTRACT

The case study is about the drastic transformation of the strategy and business model of the Russian information technological enterprise that incorporates autonomous but interrelated business units. The creation of new capabilities was the transformation objective; in 2001, top management of the IBS company made the substantial decision to change the company's business model and strategy and to switch to the information technology (IT) services and consulting area. A capability-based development approach and architecting, as well as an agile-stylized program of project management approach, was successfully applied to achieve the transformation goals described earlier. Also, an extended enterprise was formed in the business environment of the enterprise. An ES/SoS nuclear model has been introduced to describe the transformation. The model represents ES/SoS as the hard business nucleus or kernel (major business agents, who really make the business), a soft business shell (immaterial environment: governing media, competences, experience, knowledge, etc.—intangible assets), and the cloud of extended enterprise (clients, partners, contractors, etc., interlinked by common across business processes). The great majority of transformation activities and outcomes lie in the ES/SoS's soft shell, which includes new capabilities, governing relations, competences, etc.

GLOSSARY

BI	Business intelligence
BU	Business unit
BoU	Back-office unit
CEO	Chief executive officer

CGC   Corporate governing center
CRM   Customer relationship management
ERP   Enterprise resources planning
ES/SoS Enterprise system/system of systems
ESE   Enterprise systems engineering
GAAP   Generally accepted accounting principles
HR   Human resources
IPO   Initial public offering
IT   Information technology
SE   Systems engineering
SoSE   System of systems engineering
SU   Sales unit
TU   Technological unit

BACKGROUND

CONTEXT

In 2001, top management of the IBS company made the substantial decision to change the company's business model and strategy. The decision was mainly based on external factors analysis: the current status of the Russian economy and trends of its further development. It was, in fact, a proactive reaction to detected and expected business environment changes.

At that time, the company was one of the key players in the Russian IT systems integration field, with the headcount of about 950 employees and annual revenues of around $80 million.

IT infrastructure projects (complex computing systems, multiservice networks, etc.) took the huge share of the business at that time; IBS distributed hardware and software (as a local partner of worldwide vendors) as well as executed engineering projects to deliver integrated solutions; and *box-moving* business (software and hardware reselling) took a considerable share in the company revenue.

In 2000, a new separate company Luxoft (a member of IBS group) was spun off, based on the software development department of IBS. It was decided that Luxsoft would focus on off-shore software development, while the IBS company on IT systems integration.

Two main factors drove the Russian economy at that time:

1. The early 2000s were the time of fast economy growth—the Russian economy was recovering after national 1998 financial crisis, when the default of the sovereign debt occurred. Default caused dramatic contraction of the economy and nation currency devaluation (five to six times to the USD). Devaluation of national currency considerably increased Russian economy competitiveness and forced economic growth, especially in oil, gas, and natural resources industries.
2. The biggest Russian corporations started their overseas expansion and started active borrowing from international financial markets. These activities required improvement of corporate efficiency as well as implementation of internationally accepted corporate governance and finance management approaches.

Based on these factors, IBS management predicted considerable growth of IT services and the consulting market. The following logical sequence supported the prediction:

1. Implementation of enterprise development, efficient growth, and internationally accepted approaches and standards would sophisticate corporate processes and procedures.
2. Business processes sophistication would cause increased complexity of the enterprise organization and IT systems. New software systems would be implemented, first of all ERP systems. Legacy systems would be upgraded as well. The changes in software systems would force development and sophistication of hardware.

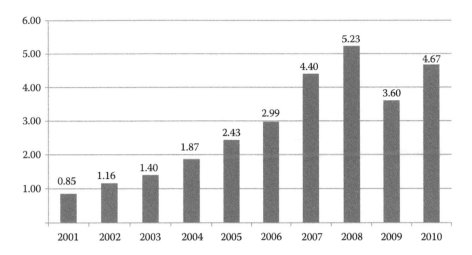

FIGURE 5.1 Russian IT services and consulting market ($ billion) in 2001–2010.

3. The growth in complexity of business processes, software, and hardware systems should require more consulting and IT services in the areas of management consulting, business application implementation, and sophisticated computing and networking engineering.

This forecast turned out to be correct; Figure 5.1 represents the diagram of Russian IT services and consulting market growth in 2001–2010 (based on the report prepared by marketing department of the IBS company in 2010).

Based on the forecast described, IBS management created a new company strategy aimed at the switch to IT services and consulting business, namely, the share of IT services and consulting should be increased from 25% to 50% or more over one year. Further growth of this area of business is planned in a long-term development analysis.

The consulting and IT services business is considerably more marginal in comparison to box-moving. However, it is very complicated technologically and organizationally and requires different and sophisticated employee skills.

To achieve this strategy goal, the company's management defined new capabilities required to learn how to sell and to execute big and complicated multidisciplinary consulting and services projects. Sales and execution processes should be treated as absolutely standardized, regular and routine procedures.

It was evident that such dramatic changes of corporate capabilities might not be done in evolutionary fashion. Thus, a fundamental transformation was necessary, and such a transformation was executed during 2002.

RELEVANT DEFINITIONS

During the transformation period (as well as in 2013), IBS was represented as a set of quite autonomous business units (BUs), the constituent systems. Any of the major BUs is a virtual, independent business or firm:

- Profit and Loss (P&L) reporting is required for each BU according to management accounting procedures.
- BU management establishes and independently conducts human resources (HR), technology, and product policy.
- Centralized back-office units (BoUs) were organized to provide back-office functions for each BU. Thus, BUs do not have back-office; they rely on and *pay* (in terms of management accounting) a corporate governing center (CGC) for these services.

Such an organizational model as well as IBS being an enterprise lets us consider the company as an exemplar SoS and ES at the same time without intermingling the models and approaches. A combined abbreviation ES/SoS will be used further to stress the dual nature of the company and the proximity of the ESE and SoSE approaches in the case study.

The transformation of the ES/SoS to form new capabilities is the key topic in the case study.

"Capabilities" is used in terms of Accelare: "A business capability is what the company needs to execute its business strategy. These capabilities are operational in nature. Another way to think about capabilities is as a collection/container of people, processes, and technologies that is addressable for a specific purpose."

It will be demonstrated in the case study that both parts of the definition are applicable:

- Company's mission was decomposed by capabilities.
- The processes and technologies (and supporting tools) were developed and implemented; and the employees who carry out the capabilities were trained.

Pertaining Theories

A thorough ES/SoS transformation was executed as a set of activities: mission analysis and capabilities decomposition, (business) architecting, planning of the project program and implementation of the new business model. Many of those activities and processes were later described in such papers as SoS_DoD and ESE_processes, as well as in the standards as [15288], despite the transformation being conducted long before the papers and standards were issued. That is very natural because each object requires an approach that is appropriate to its nature, irrespective as to whether or not the approach had yet been described and systemized.

Thus, the successful transformation of IBS might not have been completed without appropriate SoS and ES approaches even though the approaches themselves were described and systemized later, they have even been systemized by now.

Business architecting is a very important stage of the transformation. The (business) architecture and the architecture description are considered in terms of 42010:

Architecture (system) – fundamental concepts or properties of a system in its environment embodied in its elements, relationships, and in the principles of its design and evolution.

Architecture description – work product used to express architecture.

The architecture description was the set of views (each view consisted of one or several models, described in the "Development" section).

Business process modeling and finance modeling, as well as business processes reengineering and improvement, were used as business tools during architecting.

The implemented activities are very appropriate to the ones described in the documents:

I. *SoS_DoD*:
  - Translating SoS capability objectives (mentioned earlier) into high-level SoS requirements (to BU) over time
  - Understanding the constituent systems (BUs) and their relationships over time
  - Developing, evolving, and maintaining an architecture for the SoS (business architecture of the enterprise)
  - Addressing SoS requirements and solution options
II. *ESE_processes*:
  - Technology planning
  - Capabilities-based engineering
  - Enterprise architecture

III. *15288*
- Stakeholder requirements definition process
- Requirements analysis process
- Architectural design process

The transformation itself was executed as program of project, so management activities, which are appropriate to Organizational Project-Enabling Processes, Project Processes, and Technical Processes of ISO15288 15288 standard, were executed:

IV. *SoS_DoD*:
- Orchestrating upgrades to SoS
- Assessing extent to which SoS performance meets capability objectives over time

V. *ESE_processes*:
- Enterprise analysis and assessment

VI. *15288*:
- Project planning process
- Project assessment and control process
- Decision management process
- Risk management process
- Information management process
- Measurement process

Organizational changes and organizational transformation approaches were used as the transformation tools as well.

GUIDING PRINCIPLES

IBS management was based on the following principles [White_1] [White_2] during the whole transformation:

- *Follow holism*—IBS was considered as an ensemble of BUs, as the whole SoS. The holistic idea was employed, for example, to develop *end-to-end* or *crossing* business processes and integrated motivation policy and procedures for BU managers and all employees.
- *Achieve balance*—Transformation team did not try to reach *the optimum*; understanding that cannot be achieved in the organizational and business area. For example, a quite reasonable economic model was developed, and no attempts were done to find *the best solution*.
- *Utilize transdisciplines, embrace POET*—ES/SoS transformation employs different disciplines: political, operational, economic, sociological, and philosophical, as people are the key *manufacturing agents* of consulting or IT company.
- *Nurture discussions, foster trust, create interactive environment*—Transformation team tried to involve the employees of all levels into transformation very widely by using democratic type of leadership in the company. First of all, such involvement lowered the natural resistance to the changes. Secondly, it enabled the testing of the transformation options by discussions and *brain storming*. By the end, it dramatically enlarged *creative engine* of the transformation.
- *Stimulate self-organization*—Transformation team encouraged *horizontal* communications among people and BUs to employ self-organization capabilities. BUs were forced to find compromise solutions when the corporate center ordered their development; however, the means and techniques were not defined for them.

CHARACTERIZATIONS

Before and after transformation, IBS was an exemplar directed SoS (in terms of SoS_DoD): it is governed by the corporate center, and the resources are controlled by this center. The BUs are autonomous, but their operations are supervised by CGC. The same time IBS also has considerable features of an acknowledged SoS (in terms of SoS_DoD): the constituent systems (BUs) retain their independent development and sustainment approaches, and changes in the company are based on collaboration between the SoS (corporate center) and each constituent (business unit). It is important that even operations of BUs are not controlled but only supervised or governed by the corporate center through *soft* recommendations and coordination.

IBS was a quite mature ES/SoS before the transformation, and it was thoroughly upgraded to form new capabilities of the whole system as well as of the constituents.

Enterprise systems engineering profiler, (Figure 5.2).

Strategic context. IBS is considered a virtual extended enterprise that had been formed after transformation ("Boundaries, *Extended Enterprise*" section). The corporate mission was defined before the transformation, but ambiguity or uncertainty of the transformation generated the risk of mission evolution or even mission fluidity.

Implementation context. The transformation process required considerable efforts: a lot of different users (employees of BUs) were involved, and all of them needed to be integrated in the process.

The acquisition environment was represented by the only project program (transformation program) implemented in the set of constituents (BUs).

Stakeholder context was comparatively simple: Although major stakeholders (owners, corporate management, BUs' managers, employees) have some contention, all of them are partners and contribute to company success in general; also they were deeply involved in transformation.

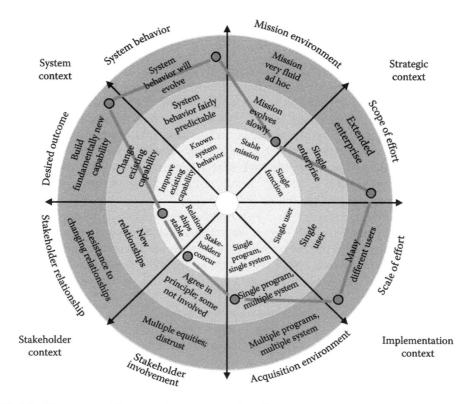

FIGURE 5.2 The case study's enterprise systems engineering profile.

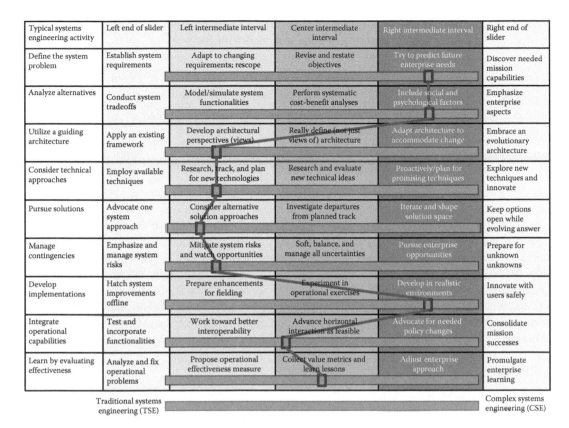

Typical systems engineering activity	Left end of slider	Left intermediate interval	Center intermediate interval	Right intermediate interval	Right end of slider
Define the system problem	Establish system requirements	Adapt to changing requirements; rescope	Revise and restate objectives	Try to predict future enterprise needs	Discover needed mission capabilities
Analyze alternatives	Conduct system tradeoffs	Model/simulate system functionalities	Perform systematic cost-benefit analyses	Include social and psychological factors	Emphasize enterprise aspects
Utilize a guiding architecture	Apply an existing framework	Develop architectural perspectives (views)	Really define (not just views of) architecture	Adapt architecture to accommodate change	Embrace an evolutionary architecture
Consider technical approaches	Employ available techniques	Research, track, and plan for new technologies	Research and evaluate new technical ideas	Proactively/plan for promising techniques	Explore new techniques and innovate
Pursue solutions	Advocate one system approach	Consider alternative solution approaches	Investigate departures from planned track	Iterate and shape solution space	Keep options open while evolving answer
Manage contingencies	Emphasize and manage system risks	Mitigate system risks and watch opportunities	Soft, balance, and manage all uncertainties	Pursue enterprise opportunities	Prepare for unknown unknowns
Develop implementations	Hatch system improvements offline	Prepare enhancements for fielding	Experiment in operational exercises	Develop in realistic environments	Innovate with users safely
Integrate operational capabilities	Test and incorporate functionalities	Work toward better interoperability	Advance horizontal interaction as feasible	Advocate for needed policy changes	Consolidate mission successes
Learn by evaluating effectiveness	Analyze and fix operational problems	Propose operational effectiveness measure	Collect value metrics and learn lessons	Adjust enterprise approach	Promulgate enterprise learning

Traditional systems engineering (TSE) ———————————————————— Complex systems engineering (CSE)

FIGURE 5.3 SEA Profiler in the case study.

The chief executive officer (CEO) of the company played a key role in all stages of transformation: the concept development, planning, and implementation.

Systems context. New capabilities were being systematically formed during transformation. Neither IBS nor any other Russian company had such capabilities before (talking about *projects proceeded on the conveyor* or mass execution of large, complex, multidisciplinary consulting projects). Before the transformation, management team supposed that system behavior would be predictable, but some *nonengineered* examples of BUs' interactions happened during and after transformation, which demonstrates the emergence of SoS.

SEA profiler, (Figure 5.3).

Define the systems problem. The transformation was caused by defining future/required needs and the definition of new capabilities.

Analyze alternatives. The transformation embraced social and psychological factors because fundamental changes in the employees' competences and company's capabilities were needed.

Utilize a guiding architecture. Architectural description in the form of the set of views was one of the first transformation stages.

Consider technical approaches. New business technology to sell and execute complex projects (as a set of processes specially tailored to company needs) was researched, developed, and implemented during transformation.

Pursue solutions. Nearly always only one solution was chosen after considering the alternatives at any decision-making point. Transformation was executed with the working, living company, and therefore more than one option could not have been implemented or tested.

Manage contingencies. Resource limitations forced transformation team to establish very strong risk management system combined with careful watching for opportunities.

Develop implementation. The living and working ES/SoS was transformed so all changes were implemented in realistic environment.

Integrate operational capabilities. Considerable attention was paid to the integrated operations of major business agents, as well as advanced horizontal interactions.

Learn by evaluating effectiveness. A lot of different value metrics have been developed (financial outcome of the project and different utilization metrics). The metrics were used in transformation activities and in operations later to monitor the performance and to improve the efficiency.

So, the majority of SE activities executed belong to Left Intermediate and Center Intermediate Intervals, which is natural for directed and acknowledged SoS (SoS_DoD), and which is appropriate to the definition made in the section earlier. But three *eruptions* to the right intermediate intervals demonstrate the serious complexity of transformation process, which is appropriate to a collaborative SoS. It is explainable: human and social factors play key roles in the ES/SoS analyzed. It will be shown in the "Analysis" section that the transformation dealt mainly with the *soft shell* (competences, governing media, etc.) of the company, which is less directed and is closer to a collaborative SoS.

PURPOSE

KNOWN PROBLEMS

The starting point of transformation focused on problem definition and analysis. Consulting businesses existed in the company before the transformation but they were very small, and consulting projects were quite rare and singular. Those very few projects served as research field to detect problems and to find the way of further development.

Company management understood that a little consulting experience was not enough to bring to light all problems, so they supposed to monitor and analyze company performance during and after transformation to detect problems and tailor approaches to problems found.

Initially detected problems appeared as expenditures exceeding resources, slow delivery of the projects and reworking.

Later, as it was expected, new problems appeared, for example, disinterest of BUs' managers in developing new technologies or raising qualified employees' motivation.

All those problems were solved during transformation and during further ES/SoS development.

MISSION AND DESIRED CAPABILITIES

The first step of the transformation included strategic analysis and mission-to-capabilities decomposition. The decomposition scheme is shown in Figure 5.4.

Five major capability groups to be focused on were defined:

1. Deliver consulting and services. New business lines development was planned in the areas of new core technologies, and capabilities were defined (examples):
   - Deliver management and financial consulting
   - Implement ERP systems of different scales—for large enterprise, medium and small business
   - Implement CRM systems
   - Implement BI systems
   - Deliver services in other technological areas

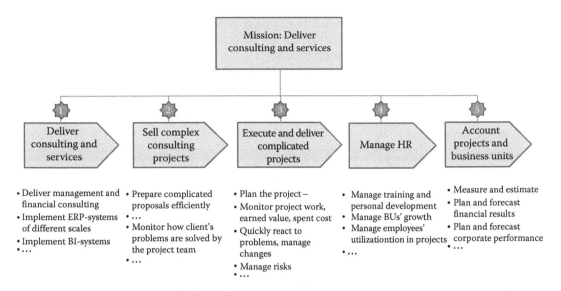

FIGURE 5.4 Mission and capabilities desired.

2. Sell complex consulting projects. New sales capabilities were defined (examples):
    • Prepare complicated proposals efficiently
    • Consider all client's requirements carefully
    • Make adequate pricing
    • Monitor how client's problems are solved by the project team
    • Monitor *day-by-day* customer relations and client's satisfaction
    • Control other marketing and customer relations issues
3. Execute and deliver complicated multidiscipline consulting projects. Project execution and delivery should be very effective and highly standardized process, so the following capabilities should be formed (examples):
    • Plan the project—prepare work breakdown structure, schedule, stuffing, contractors, etc., also make re-planning in case of changes
    • Monitor project work, earned value, spent cost
    • Quickly react to problems, manage changes
    • Manage risks
    • Utilize resources in the project in optimal way
    • Execute logistic support and other project processes effectively and transparently
4. Manage human resources effectively. Highly skilled and experienced employees are the key performing engine of the consulting business, so it is critically important to manage such resources rationally through the following corporate capabilities (examples):
    • Manage training and personal development of project managers and consultants through corporate career development system
    • Manage BUs' growth—hiring and adaptation of new stuff
    • Manage employees utilization in projects
    • Stimulate employees to develop their work skills and expertise
    • Manage other HR issues effectively
5. Measure and account projects, BUs' and employees' performance. Target business model is very dynamic, so online measurement and forecasting of key performance indicators are critically important, realizing by appropriate capabilities (examples):
    • Measure and estimate work-in-progress, value earned, spent cost; do this *online*
    • Plan and forecast financial result of the projects based on initial plans and progress measurements

- Plan and forecast corporate financial and other performance indicators based on the forecasting of financial results of the projects and other indicators measured *in progress*
- Monitor and measure employees' utilization
- Execute finance reporting and analysis procedures effectively as well as finance management function in general

The process for achieving the objectives needed—capabilities forming—is described in the "Development" section.

ENTERPRISE SYSTEM/SYSTEM OF SYSTEMS

EXTERNAL ENVIRONMENT, GENERIC FACTORS

The IBS company, as an ES/SoS, operates in a very sophisticated external environment that combines policy and law, economy, society and culture, technology, and even nature.

The environment and its elements influence the ES/SoS through different factors. This section is devoted to the generic factors, which are typical for the great majority (or even all) of the companies (nearly) all over the world; the factors that are appropriate to Russian IT market of 2000 are described in the next section.

Figure 5.5 demonstrates the environment and main influencing factors examples of each group. All factors might not be listed because the complex environment (which is in turn an SoS) shows its character differently each time, and this manifests itself in SoS emergence.

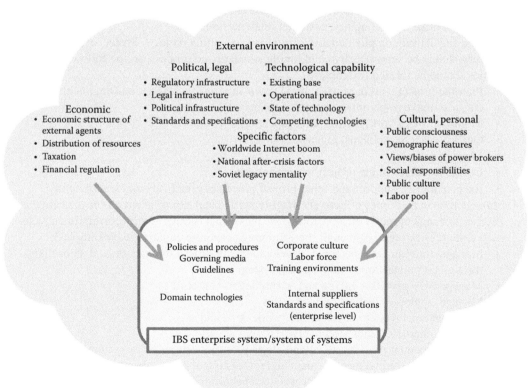

FIGURE 5.5 IBS company and its environment.

Nature influences ES/SoS through physical factors, but the influence of physical factors on IBS is absolutely equivalent to the influence on all other companies and on the rest of environment; therefore, we will not analyze physical factors separately but will consider their influence through other groups of factors.

ES/SoS's elements and environment interact on several different levels: IBS as the whole; business units; projects and employees. And not only does the environment affect ES/SoS but also ES/SoS influences the environment.

Policy and law considerably affect consulting and IT companies all over the world. IT and consulting are among a few wealthy industries in Russia, so the companies attract the attention of politicos and regulators as well as the press and public opinion. All this creates substantial constraints on activities of company management, which is especially sensitive to such attention.

Economics affects through economic structure and economic *wealth* of customers and partners, and resources distribution in national economics. Economics influences employee wealth, financial results of projects, business units, and P&L of the company as the whole. Vice versa, the company affects quite a large set of individuals and firms creating jobs, paying taxes, and purchasing the goods.

Technology dramatically influences high-tech companies like IBS. The technology domain changes very rapidly and forces companies to react proactively on the changes. The transformation goal is to change the technological focus of the company and force company growth based on technology changes. Also, the company affects the technological domain through development of its own technological proposition.

Society and culture affect the business of consulting and IT companies very heavily because the business objective is closely linked with people and society. Due to this, the whole IT domain frequently faces very different challenges, must adapt to them, and respond adequately. For example, current demographic shortfalls cause HR market shortages in general, especially in highly qualified personnel.

EXTERNAL ENVIRONMENT, SPECIFIC FACTORS

Besides generic factors described earlier specific national and industrial factors affected the transformation:

The *worldwide Internet boom* created very high business expectations of IT sector growth and attracted investors; also, new technologies appeared very rapidly, which expanded the IT market very fast. These factors forced IT companies (IBS as well) to follow new technological trends and open new departments to try to capture a niche in the growing market. For example, IBS management at that time established several BUs (CRM systems, Internet technologies, etc.), and some of them were reformed or closed later.

After crisis factor (after national crisis of 1998)—Fast economic recovery, devaluated currency, and international expansion of Russian firms, on the one hand, initiated and pushed the transformation and, on the other hand, were embarrassing due to the fast-growing labor cost. Labor cost is the main component in the cost of a consulting and IT firm, so the fast growth of an average salary created an economic problem during the transformation.

Post-Soviet legacy played a considerable role in society (even now). The transformation was conducted only 10 years after the USSR's dissolution; destruction of entrepreneurship during the Soviet period dramatically lowered business activities for the majority of employees, which complicated necessary changes.

STRUCTURE AND CONSTITUENTS

The IBS structure is shown in Figure 5.6. It consists of CGC and autonomous BUs of one of three types: sales units (SUs), technological unit (TUs), back-office-units (BoUs).

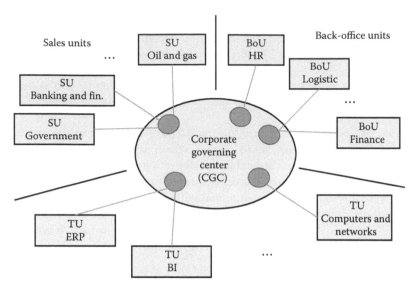

FIGURE 5.6 IBS structure in 2001–2012.

SUs executes the whole sales function, they are responsible for

- Marketing monitoring and analysis (new customer definition, customer needs analysis, etc.)
- Customer relations development and management
- Sales execution and support (agreement preparation, deal monitoring and closure, etc.)

SUs are organized by industries: oil and gas, machinery, energy, utilities, government, retail, banking, etc. Each SU moves to the market all product and services of IBS.

TUs delivers the product and services through project execution, they do

- Marketing monitoring and analysis from product and technology point of view
- Product/technological sales support
- Project (product/services) execution and delivery

TUs are organized by technologies and products, for example, complicated computing platforms and networks, ERP systems, CRM systems, datacenter, etc.

BoUs provide back-office support for SUs and TUs: finance and accounting, HR, employee training, logistic support, administrative support, etc.

BoUs realize outsourcing approach being quite autonomous and providing back-office functions to SUs and TUs.

The *CGC* consists of the CEO and his deputies, who run the company, supervising and coordinating BUs' activities rather than controlling them by directives. The CGC is not the top root of control hierarchy; it is more center of the *star* of autonomous BUs.

Practically any BU may play at the open market independently. For example, sometimes the employees of TUs might *be sold* under *time-and-materials* terms; also, corporate internal training center (one of BoUs) occasionally provides services to external customers.

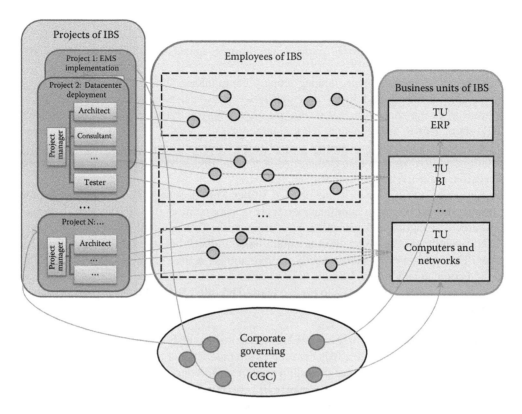

FIGURE 5.7 Key business agents and the relationships among them.

But the IBS business structure is more than *BUs + CGC*: projects (project teams) and employees also play considerable roles. Projects (project teams) and employees as well as BUs are key business agents of the company (Figure 5.7).

BUs serve in reality as the resources pool that form project teams.

The same employee may play different roles in different projects; for example, the head of TU may be assigned as director of one project and executes architect role in another one (being at the same time TU manager), and this is not a rare occasion.

Thus, project management and administrative management approaches are utilized in parallel in the company.

The organizational structure of IBS did not change considerably after transformation. Several new TUs responsible for some new technologies were organized. Also, BoUs were reformed to cut back-office cost and special BoUs (Corporate Technologies BoU) were established to the development and implementation of new business processes supporting new capabilities.

A fundamental organizational principle (SU organized according to industries, TU—to technologies) was employed. BoUs played the role of corporate service center; CGC performed in the *soft* but not directive supervising and governing manner.

These basic principles demonstrated their effectiveness and are being used now.

Also, project teams have been forming ad hoc from TU resource pools.

Company structure finally formed as a hybrid, combining hierarchical elements (in administrative sense) with a network (in project management sense); and *soft* horizontal relations among employees, projects, and BUs dominated.

INTERNAL RELATIONSHIPS

Different types of relationships existed among major business agents of ES/SoS (Figures 5.6 and 5.7):

- *Administrative management*—among employees, managers, and BU heads. These relations control hiring and firing of employees, vacations, employee position changes, employee assignment to project teams, etc. Actually, this relations control the pools of HR—pool growth and reduction, HR assignments.
- *Operational management*—among project team members, inside project teams. These relations realize project management: project planning, scheduling, resource planning, execution control, change management, etc.
- *Supervising*—of BUs by CGC members (CEO and his deputies). These relations realize strategic management: to coordinate strategic development of BUs, to monitor corporate resource distribution and utilization, to coordinate the full set of all project activities and BUs development. These relations are quite *weak* and *soft* because they are not directive ones. They are very important: exactly these relations *accumulate* BUs into the whole and united company and forms ES/SoS.
- *Service providing*—from BoUs to project teams and other BUs. This is *horizontal* type of relations, supporting relations.

There are not any direct relations between any BUs (constituents), they are linked via their employees' participation in joint project teams, BUs themselves are independent to each other.

BoUs link to other BUs by service providing relations—like independent providers and customers. BoUs serves also project teams (e.g., finance BoU executes management accounting for project teams), and logistic BoU serves only project teams, but not BUs.

In general, our ES/SoS might be characterized as *gently* related system, with weak hierarchies.

Being a directive SoS, our ES/SoS acquired considerable features of a collaborative SoS during transformation. Soft and nonhierarchical relations make ES/SoS more adaptive and stimulate each constituent system to grow, and reinforce the business motivation of each BU leader. But such relations also complicate centralized control, make impossible directive control: CGC cannot issue directives, but needs to recommend or *sell the idea*. This style of governing was also employed in transformation.

Such governance approach is not *one way* process; the feedback from the governed constituents—major business agents of the ES/SoS—is natural and appreciable element of the governance. Also, CGC considers the feedback as an important source of improvement of the transformation and company operation in general. So CGC regularly analyzed the feedback information very carefully during the transformation (and have been analyzing till now) and made correctives in the activities. Such analysis has been executing in different ways. The most demonstrative example is the annual planning and budgeting process, when CGC initially issues strategic thesis or guideline for the coming year; BU's management propose plans and budgets and final BU's plans and budgets are formed iteratively by the negotiation between CGC and BUs.

The great majority of the project before transformation was not complex—each of them was executed by the project team formed from the employees of only one BU. In such cases, governing relations being the network in general had some hierarchical elements: projects encapsulated *inside one BU*, which caused hierarchies *BU-project-employee*.

Considerable sharing of complex projects that appeared after transformation weakened these hierarchical senses.

The relations in general are very complex, like network (Figure 5.7), which connects BUs, projects, and employees by different links (administrative or operational, project management, supervising, etc.); the weakening of the hierarchies during transformation increased the complexity of the relations and the ES/SoS.

Boundaries, Extended Enterprise

On the one hand, ES/SoS boundaries might be described very clearly: all BUs and all employees (including part time) are included in ES/SoS; also all executing projects should be considered (Figure 5.7). The ES/SoS is a firm, and so its boundaries are defined by national regulatory documents (law).

On the other hand, the company has a lot of partners (vendors and contractors) and customers.

Very frequently (nearly always) joint project teams are organized to execute complex consulting projects. Such teams usually include employees of the consulting company, its partners, and customers. Thus, the company involves employees of other firms in projects, *extends boundaries* and forms an *extended enterprise*; major agents and relations among them are represented in Figure 5.8.

Different types of relations are established inside an *extended enterprise*, which are equivalent ones described in the "Internal Relationships" section. In practice, such relationships are realized by means of integrated processes of project management and project accounting, which cover joint project teams formed of employees of different firms.

Development and implementation of these processes is described in the "Development" section.

The *extending of the enterprise* derives from joint project teams forming. Such teams are temporal objects (team lifetime equals project duration); so the boundaries of *extended enterprise* are variable and temporal.

Extended enterprise was the complementary result of transformation: before that the share of consulting and services projects was too small, and ES/SoS matched Figure 5.7; after transformation that sharing increased dramatically, and ES/SoS *extended* to Figure 5.8.

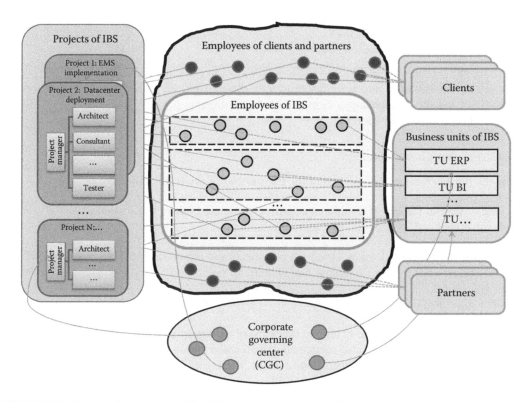

FIGURE 5.8 Extended enterprise of the IBS company and their major business agents.

SCOPE

The scope of transformation covers ES/SoS—the IBS company—as it was in 2001–2003; also, *extended enterprise* was formed around IBS as described in the "Boundaries, *Extended Enterprise*" section; the appropriate data are represented in the following table. The estimates in the "Extended Enterprise" column have been formed based on the following assumptions:

- External companies and their employees are involved in the *extended enterprise* through joint teams of complex consulting projects; external companies play role of HR pools as BUs in such sense.
- The average number of such projects at any moment is estimated as 20 ... 30.
- Usually 2 ... 3 employees (up to 15, but seldom) from customer side participate in each project team. So (20 ... 30) * 4 = 80 ... 120 *virtual* employees and 20 ... 30 *virtual* constituent systems are added by customers.
- Partners (vendors and contractors) are involved quite seldom—in around 20% teams. Usually, the number of employees from the partners side is around 30%, so paso ly 30%uaners side is usual s side is usual estimated "rtners supplement around 6% of total project teams (approximately 66% of TU headcount) and 4 ... 6 constituent systems.

		IBS, Y2001	IBS, Y2003	Extended Enterprise, Y2003
The whole ES/SoS	Total headcount	917	927	>1047 ... 1087
	Number of BUs	30	30	>52 ... 63
	Headcount of CGC	13	15	15
Constituents	Number of TUs	12	15	>37 ... 48
	Headcount of TUs	637	674	>794 ... 834
	Number of SUs	6	6	6
	Headcount of SUs	73	84	84
	Number of BoUs	12	9	9
	Headcount of BoUs	194	154	154

In conclusion, the *extended enterprise* insignificantly differs from the IBS company by headcount (around 20%), but the average supplement of *virtual* constituents is really huge (Number of BUs, Number of TUs around twice or more).

SPONSORSHIP AND GOVERNANCE

Among the most critical success factors of the transformation were the following:

- One or several leaders who initiated activities, stimulated others to take action
- Involvement and sponsorship by major stakeholders

The transformation was realized as a program of projects and organizational structure of the transformation corresponds to company's structure (Figure 5.7):

- Special project teams were organized to develop and implement procedures, software systems, etc..
- The CEO played a key role being the executive of the transformation. He was one of ideologists and initiators and also participated in all major transformation activities.
- All members of CGC were deeply involved in heading comities and working groups, in supervising processes implementation in BUs.
- BUs' heads, managers, and employees took part in project teams, developed and implemented new processes, policies, software systems.

Shareholders supported the transformation initiative; the board of directors talked over the transformation concept, solutions and progress, quarterly.

STAKEHOLDERS AND STAKEHOLDERS' CONCERNS

Shareholders

The transformation as well as company development in general was sponsored by shareholders. In 2001–2003, IBS was a private company owned by

- Founders, who also run the company as top-managers and owned the main share
- Several foreign funds, invested in high-tech sector and emergent markets, that had some considerable share but voted separately
- Employees (managers), who had a very small share

Also, in 2001–2002 management prepared the initial public offering (IPO) to use the synergy of *Internet boom.*

The board of directors consisted of the founders and representatives from funds.

Shareholders considered the transformation as one of the stages of company development. They wanted this stage to be executed as fast as possible, in parallel with ongoing business, without any interruptions or delays. Also, the shareholders wanted to minimize any involvement of external contractors in the transformation to save money and to retain privacy and confidentiality. Shareholders are active stakeholders; their concerns might be summarized as follows:

- The company must be transformed to consulting/services firm.
- The transformation process should not interrupt or break ongoing business and should be completed as fast as possible.
- Involvement of external consultants (or any other contractors) should be minimized.

Employees

The transformation was executed for the benefit of shareholders but also employees who are really interested in stability and efficiency of the company-employer.

Employees are one of three business agents, so they considerably influence company business. They are a group of key and active stakeholders. Employees' concerns are

- The company should not cut jobs, salary, or social benefits.
- The company should be more efficient and steady.
- The company should be more attractive as an employer (providing interesting jobs, friendly social media, etc.).

Vendors

IBS had a very wide partnership network. For a long time, the company has been a *golden partner* of *preferred partner* of such vendors as IBM, Microsoft, Cisco, HP, SAP AG, Oracle, etc.. Such vendors, being global companies, have very restrictive requirements to local partners; all such requirements are very specific and diversified. Vendors are not active stakeholders but affect the company quite appreciably.

Contractors

IBS business involves a lot of contractors due to the following reasons: the full set of IT expertise might not be developed in one company (especially in case of rare or niche competences); outsourcing approach is very valuable and convenient especially in case of routine work. For example, the main share of logistical support was outsourced to a partner company: initially only transportation and custom clearance were outsourced—later also full warehousing functions. IBS staff executes only documentation functions. Considerable attention was paid to establishing an integrated *end-to-end* workflow and document flow through outsourcer, logistic support BoU and TUs. Contractors' functions are very important to IBS, so contractors concerns (how to organize integrated *crossing* processes) was considered during transformation.

Clients

A lot of Russian companies from any industry and of any scale were clients of IBS, starting from small shops and gas stations up to giants like Gazprom or Sberbank (included in worldwide FORBES 1000) as well as different federal and municipal offices. All customers have their own requirements to the contractors (financial and legal, documentary, disclosure, etc.). They have different payment schedules that affect IBS income. All such requirements as well as requirements for joint project activities were considered as their concerns during transformation.

Auditors and Regulators

Regulatory documents and regulators' and auditors' requirements were of concern during transformation. Foreign shareholders and IPO preparation processes influenced IBS: for example, the GAAP requirement was considered as well as Russian accounting regulations.

To summarize vendors', partners', auditors', and regulators' concerns, let us list them:

- Full set of commercial, legal, financial, and logistic features of vendors' and contractors' agreements
- Features of joint workflow and document flow through logistic outsourcer and BUs
- Full set of commercial, legal, financial, and logistic features of clients' agreements
- Different clients' payments schedules
- National and foreign regulators' and auditors' requirements

All stakeholders' concerns were considered during transformation.

Stakeholders differ from each other considerably; also, stakeholders might be classified into two groups: active and passive. Active stakeholders really operate and directly affect the company during transformation (like shareholders and employees) both by real actions and concerns. Passive ones influence the company only by their concerns that are considered by governing body.

CONSTRAINTS

During transformation, IBS management faced a lot of constraints, the most serious were

- At that moment, management did not have any experience of such large-scale transformation
- Management did not have experience to run consulting and services companies
- The lack of professional consultants available at the labor market who are appropriate to new business model

Time and resource limitations also existed: as with requirements to move as soon as possible, do not interrupt ongoing business and avoid external expenses. But time/resources restrictions are absolutely traditional so they are not analyzed specifically.

The lack of experience and knowledge and shortage of professionals were very serious and caused considerable risks that needed to be managed. The approach to the management of these risks is described in the "Development" section.

CONTEXT OF ENTERPRISE SYSTEM/SYSTEM OF SYSTEMS

The description of the entire ES/SoS context of the transformation (Figure 5.9) summarizes the "Enterprise System/System of Systems." section.

The ES/SoS approach makes it possible to review the whole context, to consider all problems and aspects, and employ maximum information about the system and the task itself and its surroundings:

- *External Factors*, generic ones (political, economic, and other), as well as specific industrial and national ones (worldwide Internet boom, after national crisis of 1998 factor, post-Soviet legacy), considerably affected the ES/SoS; all these factors are specified in the "External Environment, Genetic Factors" and "External Environment, Specific Factors" sections.
- *ES/SoS.* Business units, projects (project teams), and employees represent core business agents that constitute the ES/SoS. Internal relationships inside ES/SoS are very diverse and complicated, the ES/SoS structure in general is not hierarchical but network one. *Extended enterprise* formed during transformation expanded the boundaries of ES/SoS. The structure and constituents, internal relationships, and boundaries of the ES/SoS of interest are described in the "Structure and Constituents" and "Boundaries, Extended Enterprise" sections.
- *Governing Body* of the ES/SoS is represented by CGC of the IBS company composes of CEO and his deputies; it is described in the "Structure and Constituents" section.
- *Governing*, which was employed to run the ES/SoS, is not command or strong control one; the constituents are supervised but not ordered by governing body. The governing is described in the "Relevant Definitions" and "Internal Relationships" sections.

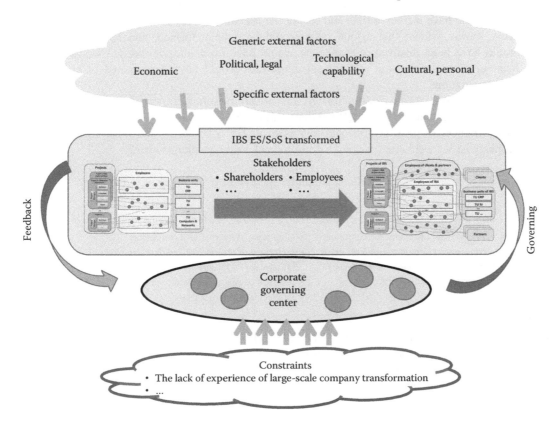

FIGURE 5.9 **(See color insert.)** Context of ES/SoS.

- *Constraints* influencing CGC (the lack of experience and knowledge and shortage of professionals) are depicted in the "Constraints" section.
- *The feedback* from the governed constituents—business units, projects (project teams), and employees—is natural and appreciable complement to the governance, is described in the "Internal Relationships" section.

CHALLENGES

The lack of experience in ES/SoS transformation and capability-based ESE (even the lack of any textbooks or guides in those areas) were the major challenges that IBS management faced. The task to be solved did not come to organizational changes (well-developed and described area), but belonged to ES/SoS engineering. In spite of the lack of the experience, it was decided to prepare and execute the transformation based on the companies' employees without external consultants' involvement. The following arguments supported the decision:

- The task to be solved was not typical, so there were no widely used and well-tested algorithms or methods and there were not a lot of consultants being experienced in exactly such things. So only consultants with general experience (strategy consulting, organizational management) might be hired.
- Russian consulting industry in 2001–2002 was not well developed (see Figure 5.1), so Russian consultants with appropriate experience might not be hired at all. Only foreign professionals were available. But foreign consultants would have needed first of all to study Russian economic specifics (systems environment analysis demonstrated earlier the considerable importance of the external influence on the transformation). Such study, naturally, would have increased time and cost of the transformation.
- Also, it was evident that a joint team would have been formed, and IBS employees would have been involved: management would have been interviewed and would have been involved in decision making; all employees would have participated in changes implementations.
- External consultants are *external people*; they are not stakeholders; so their level of interest in success might not be very high and their output also might not be outstanding.
- Unwillingness to open professional secrets to direct competitors and other intellectual property issues were other factors that limited external consultants hiring.

So the option to hire external consultants might be specified by following pros and cons:

Pros	Cons
• Involvement of generalized consulting experience (focused ones did not exist) • Insignificant reduction of management diversion from the business	• High price of international consultants • Increasing of project cost and duration due to initial study of the specific • Not high output—consultants are not stakeholders • Intellectual properties issue

The final decision was made based on the comparison of pros and cons: to execute the transformation without involvement of external consulting resources. Special BoU responsible for business processes development was established and *agile-stylized* program management approach was applied to take the challenges and to mitigate risks.

Another challenge dealt with the transformation objective: a very high complexity IBS as ES/SoS. Management recognized that the company was very complex, with a lot different agents—constituents and a lot of relationships and that ES/SoS might become even more complex after transformation. This complexification happened due to the company becoming an *extended enterprise*, the governing hierarchies weakened and relations became more sophisticated, which is demonstrated in Figures 5.7 and 5.8.

Another challenge in business area was the risk of mistaken forecast of economic IT market development: expected growth of the consulting and services market might have not happened and in this case the transformation would have been senseless. This challenge generated additional emotional stress for management.

DEVELOPMENT

ENTERPRISE SYSTEMS/SYSTEM OF SYSTEMS ENGINEERING PROCESS

The SE task in the case study was established in the form: to develop required capabilities ("Mission and Desired Capabilities" section, Figure 5.4) for the ES/SoS – the IBS company (whose structure is described in the "Structure and Constituents" section). The SE process of the transformation might be represented by the following specific *V model* (Figure 5.10).

1. *Analysis and capabilities decomposition*: In the initial analysis, the mission was translated to capabilities (Figure 5.4); and *understanding the constituent systems (BUs) and their relationships* was executed. Transformation team found that capabilities might not be directly translated to any business agent. Neither BUs (they serve as resource pools), nor projects (being temporal elements), nor employees (each of them has very limited set of skills, experience, responsibilities, etc.) might realize necessary capabilities.

2. *Key areas definition*: Realizing this, the transformation team defined several key areas of company's operations or activities that were supposed to be changed to form new capabilities. It was planned in each of the areas to engineer and to implement tools which support new capabilities (operational rules, technologies, processes, tutorials, guidebooks, software systems). The areas were defined accordingly to the groups of capabilities (Figure 5.4), but it appeared that one project implementation area covers two groups—*Sell complex consulting projects*, and *Execute and deliver complicated projects*—because both groups deal with the projects at different stages of their life cycle. And vice versa, two areas are appropriate to the group *Manage HR*: one area, employees development and motivating, considerably differs from another one, BUs growing and resource pool management.

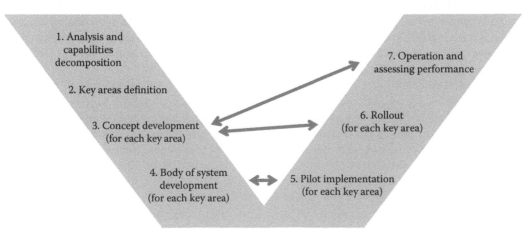

FIGURE 5.10   *V model* of systems engineering process of the transformation.

Thus, five key areas of operations or activities were defined:

1. Core technologies area—consulting and services area
2. Project implementation area
3. Business unit growth area (hiring and newcomer integration)
4. Motivation of the employees area
5. Management accounting area

For each of the areas, appropriate systems to support new capabilities were developed and implemented (the systems described in the "Systems Developed to Support New Capabilities" section); these new capabilities support systems formed exactly the corporate infrastructure of new business model.

3. *Concept development*: For each key area, a set of design documents were developed, describing approaches, polices, system. The whole set of conceptual documents of all key areas actually formed the business architecture description of the company, although this term was not used at that moment. The architectural description presented the manner in which the mission and capabilities should be projected (or reflected) to the elements of the company, ES/SoS. Such a description consisted of the following views (in terms of 42010):

View	Its Content
Organizational	Organization structure. General governing policy. The concept of business units/human resources fast growing.
Technological	The list and description of the technologies and products of each TU.
Marketing or clients	The list of industries and key customers for each SU
Functional	The set of key business processes: project implementation processes and others
Economical	Management accounting concept: indicators' definitions, relations among indicators and other accounting issues.
Motivation	Motivation concept: polices and rules.

The architecture description links or connects all key areas together and defines target operation model of the company after transformation. The architecture description represents multiple views at the company, which reflects ES/SoS nature of the IBS company.

4. *Body of the new capabilities support systems development (for each key area)*: After architecting the capabilities, support systems were designed: the processes *to be* were developed, guides and other documents were issued, and software systems were prepared to be implemented. Processes and rules were developed as *end-to-end* and *crossing* ones to integrate all BUs, project teams, and employees. In reality, all systems were developed in parallel, according to *small V model*, so the whole transformation might be represented as *multiple V model* (Figure 5.11).

5. *Pilot implementation (for each key area)*: The new capabilities support systems (procedures, guides, documents, software systems) were implemented initially in pilot zones to save the time for the testing and for *reimplementing* after changes. Quite frequently transformation team piloted not the whole procedure or system but only the most critical and new fragments (supposing that well-known fragments should not create real problems to be used in production).

6. *Roll-out (for each key area)*: After piloting the new capabilities, support systems were rolled out to the full *extended enterprise*. All systems were focused on new capabilities and were initially developed as an integrated set of systems using common data and common processes. For example, time sheets of employees are used by the motivation system, the project implementation system, and the management accounting system. The special integration stage is not shown separately in the *V model* because integration was considered an

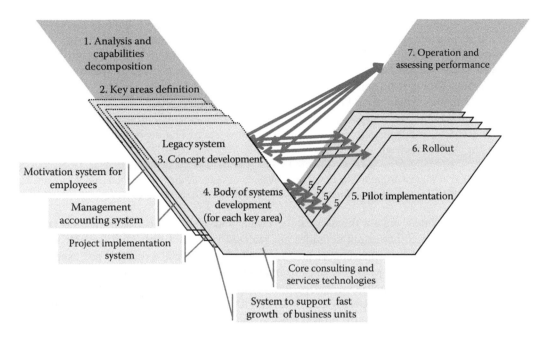

FIGURE 5.11 *Multiple V model* of the transformation.

internal, natural aspect of development, although the development of different processes and software systems was executed by different project teams but inside one company. Such geographical and organizational closeness as well as initial integration requirements make it possible to explicitly omit the integration stage. Definition of a wide set different indicators and measures was a very important engineering job at these stages. The indicators and measures used to meter projects', employees', and BUs' performance were implemented to monitor transformation progress.

7. *Operation and assessing performance*: Fragmental testing during piloting was accepted in consideration of the idea of continuous improvements during operation. Transformation team supposed that company's performance should be monitored based on indicators' measuring and improvements should be developed and implemented (arrows from Stage 7 to Stage 3 in Figure 5.10). Such iterations have been executed in practice not only during transformation but also later, when procedures, guides, and the whole systems were updated.

In addition, it is necessary to mention the existence of legacy systems, processes, guides, and procedures in all areas. They were not appropriate to required capabilities; but such a backlog was considered during development stages for any of key areas systems.

SYSTEMS DEVELOPED TO SUPPORT NEW CAPABILITIES

New capabilities support systems were developed for the five key areas of company's operations or activities:

1. *Core consulting and services area*: First of all, vendors and products were selected for each of consulting and services business lines (ERP implementation, BI, CRM, management consulting, etc.). COTS products and platforms were analyzed, considering maturity, prevalence, vendor's presence, and other criteria. Product/services implementation approaches

(guides, handbooks, documents' templates, and other tools) were developed for each business line. In some cases, vendors' approaches were tailored; in others, original ones were developed. Also, a lot of efforts were focused on employees training. Such groundwork served as the foundation of mass project implementation in new consulting and services areas, that is, supported transformation goal.

2. *Project implementation area*: Project implementation support system was developed as an integrated set of processes (process model and guiding documents); project data and documents (reports, templates, etc.); supporting software system. The *process* is considered in terms of (MBSE review):

> *A Process is a logical sequence of tasks performed to achieve a particular objective. A process defines "WHAT" is to be done, without specifying "HOW" each task is performed.*

Project teams' members are the parts of the system and users of the system just the same time. Practically all important activities are realized in project paradigm, several types of projects were defined to support this approach: "commercial projects" are executed for clients; "internal projects" are used to improve company's organization and business processes; "investment projects" are focused on new products/services development, etc. Project implementation support system covers all lifecycle stages of projects starting from very early presale activities, till project closure, including scheduling, resource planning, reporting, risk management, change management, and other project processes.

3. *Business unit growth area (hiring and newcomer integration)*: Management of the pool of professional resources is one of the key (if not the most) success factors in consulting business. *Consulting resources scissors* effect is well known: you need to demonstrate and to have skilled and trained consulting team to sell consulting project. But excessive (not utilized) stuff causes expenses (salary) without income. So the expertise to regulate the pool of professionals is critically important. Special business unit growth system was implemented to solve this problem; the system includes rules and procedures of hiring and integrating of new employees, involving professionals on *time and material* basis, as well as expanding project teams by employees of clients and partners (extended enterprise).

4. *Motivation of the employees area*: Motivation system was implemented to enforce all stuff to meet the goals of the company. Motivation system consists of the rules, procedures, documents, and software system to manage employees' bonuses, employees' targets, to monitor employees' performance. Project implementation support system, management accounting system combined with motivation system translate the company's strategy goals into employees' project targets; annual and quarterly targets. This scheme involves maximum stuff into company's strategy implementation.

5. *Management accounting area*: Management accounting system provides planning, forecasting, monitoring, and reporting of economic indicators of projects, BUs, and employees, also project pricing. The system (consisting of rules, polices, procedures, documents, and software system) serves to financial measurement and control all business agents online. So key requirements were (a) to consider all sophisticated relations between business agents, all interactions among projects, BUs, employees, and company as the whole; and (b) to execute management and control at any time—without closing of financial period.

These systems support the following decision-making processes, which are critical for new capabilities forming, that address the issues:

- Planning—how to utilize the resources in the best way to execute projects, and which source to select
- Monitoring—how to avoid run out of the budget, ungrounded expenditures, how to keep the reserves at proper sufficient, but not excessive level

- Estimating—what are real project expenses, what is working efficiency, and how efficient is management
- Pricing—what is the minimal price to stay in a profitable zone, where is breakeven point, what is cost–volume–profit ratios for BUs and the whole company
- Motivation—how efficient is each employee, and how the output of each employee related to the whole company results

Although company's governing is decentralized, the set of performance indicators is needed to govern the whole company based on managing projects, BUs, and employees; such set of the indicators is based on correlations between half-way outcomes of the business agents and the whole company's bottom line. The systems in key area described earlier utilize exactly the correlations mentioned and therefore are very appropriate to support new capabilities.

PROGRAM MANAGEMENT AND MANAGEMENT OF UNCERTAINTY

The transformation was in reality quite a compact program of projects. The project teams were formed of company's employees, the program was governed by the members of CGC according to single program plan. The task did not look complicated from a project and program management point of view. Organizational change implementation was quite a standard activity with a well-known personnel resistance problem and tested approaches to overcome this resistance. Deep involvement of top management headed by CEO guaranteed effective project management and execution as well as organizational management implementation.

The set of activities in all previously described key areas had to be executed to the transformation. All those activities were conducted in parallel to save time and resources. Integration and interoperability of the new capabilities support systems (developed in key areas) required a thorough integration of parallel development jobs. So joint workgroups were formed of the employees at the level of low officers; and CGC played the role of integrated workgroup at the management level. Actually, a multilevel integrated workgroup was formed.

The major complexity and risks derived from the challenges are described in the "Challenges" section: the lack of experience, the shortage of professionals, and the risk of mistaken market prediction.

The quick and effective creation of solutions and their practical testing is a very natural and rational approach to manage such risks. Initial conditions and the approach mentioned lead to plenty of changes in implementation process, so it is necessary to manage them fast and effectively.

Transformation team developed and used an approach that is very similar to the Agile Development approach (SE_hbk) to address those risks. The following principles were used to manage the portfolio of projects in case of lack of experience and ready-to-use algorithms and methods:

- Form solution as fast as possible (but not due to pure quality) to test it in practice faster.
- The failures are unavoidable, perceive them easily, and react rationally.
- In case of failure, analyze the situation and find new solution, generate changes, update plan.
- Work in parallel, verify and coordinate intermediate results.
- The schedule might be corrected and updated but should not fail due to improper execution.
- Make and test the most critical and most questionable solutions at first.
- Start from pilot area and then rollout to the whole scope.
- Use high-qualified monitoring and *manual control mode* for piloting to avoid waste of the resources and to test developed solution but not additional aspects.

Following those principles, a quite strong executing discipline, a high level of the sponsorship, and the involvement of all employees enabled the transformation to be completed in time and without hiring consultants while keeping and developing ongoing business.

RESULTS

The accomplishment of the mission (to form needed capabilities in full) was the major result of transformation of ES/SoS. The CEO reported at the board of directors meeting devoted to the results of Y2002: "We have become a real service company. We increased the share of all kinds of services. Large complex projects for Russian treasure, Russian railways, Lukoil and other top companies are being executed. Sales personnel have learned to sell consulting and services. The company's mentality is changing."

Shareholders and management recognized that new capabilities had been formed, that company can deliver consulting and services, sell and execute complex projects, manage consulting resources effectively, measure its performance, plan and forecast financial results.

Created capabilities are emergent in some sense because they are not directly related to concrete constituents (BUs), but are realized by means of integrated end-to-end processes and functions, which are executed in the projects by employees. Comparison of ES/SoS status before and after transformation is represented in Figure 5.12.

It was discussed in previous sections that the transformed ES/SoS (the IBS company) has a very complex internal organization:

- *BU* are the constituents (they can operate as separate systems-companies), they form an SoS, but they do not play a key role in the IBS business.
- *Employees* (they might incidentally be also considered as constituents because they may operate independently) are really *the main engines* of the business; they are bearers of expertise, knowledge, competences, etc.; but each of the employees does not do the business separately.

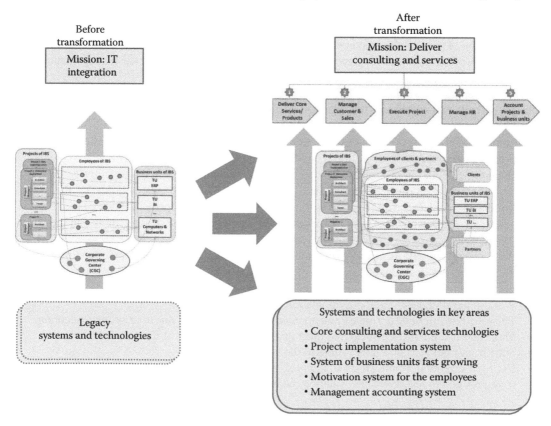

FIGURE 5.12   Results of transformation.

- *Projects (project teams)* definitely are the main business entities, but they are temporal ones.
- *Capabilities* are not directly related to any concrete business agent but at the same time are related to all of them.

The systems organization did not dramatically change during transformation; *visible structure* was not practically changed: no new types of business agents appeared, existing types did not change much. Those factors did not create new capabilities. Target capabilities were formed as the result of development and implementation of, it would seem, auxiliary and supporting tools—new capabilities support systems, described in the "Systems Developed to Support New Capabilities" section.

It is important to note that development of these systems was the only material change in ES/SoS in new capability creation. All other changes were done in the intangible areas of governing relations, personnel skills and expertise, training, etc.

Summarizing the results of ES/SoS transformation might be listed as follows:

- The mission was accomplished: New capabilities of BUs and the company as the whole were formed in areas of sales and delivery of complex projects.
- Business focus was switched to consulting and services instead of *box-moving*; the business model and company strategy dramatically changed.
- A considerable number of complex projects (with integrated joint team) appeared, and those projects started to play a key role in the business.
- Specially developed auxiliary and supporting systems serve as the tools to support new capabilities.
- *Extended enterprise* was formed around the company through involvement employees of clients and partners into project teams.

ANALYSIS

Different elements of ES/SoS and their evolution in transformation were considered in previous sections. It appears that new capabilities were formed mainly by the changes in the intangible areas of governing media, corporate culture, relations, and personnel competences, as well as by the creation of new capabilities support systems, without considerable changes in main company's business agents. Such a logical construct leads to the following *ES/SoS nuclear model* (Figure 5.13):

- *Hard business-nucleus* (business units, projects, employees)—The central or core part of the ES/SoS, including main business agents. It is the matter that exactly does the business. This is what is described by legal and financial documents, what is controlled by officials and regulators.
- *Soft business-shell*—Intangible surroundings of main business agents, which do not execute business activities, it is something without which the business might not run. This represents intangible substance: governing media, corporate culture, relations, and personnel competences. This is what is not accounted in the accounting books.
- *Cloud* of clients, partners, contractor, vendors—the *extended enterprise. The cloud* around ES/SoS of interest that covers other ES/SoS; the indicator of involvement into *the cloud* is the participation in joint integrated *crossing* processes. For example, suppliers (or clients) who sell (or buy) standard goods or services at *open market* without joint project teams organizing are not included in *the cloud*.

In terms of this model, the transformation represents the example of effecting *the soft shell* and creation *the cloud*; all material changes were carried out with tools: software systems, documents, guides, and processes.

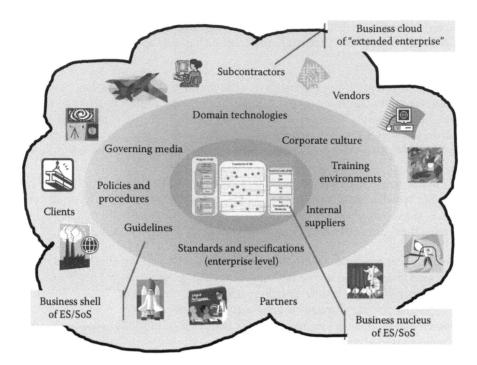

FIGURE 5.13   (See color insert.) ES/SoS nuclear model.

ANALYTICAL FINDINGS

Activities

The following activities were executed in the transformation (Figure 5.14):

1. *Mission to capabilities decomposition* represented the mission as the hierarchical set of capabilities and subsequent formulation as the processes in key areas.
2. *Architecting* was not executed as separate stage; all architecting jobs were done in "Key areas definition" and further coordinated in "Concept development for each key area." Architecting is a very important task because exactly appropriate architecture enabled further parallel and integrated development of the systems and technologies in key areas.
3. *Concurrent development and implementation of the new capabilities support systems in key areas.* Systems and technologies development itself was executed as *body of system development, pilot implementation, roll-out,* and *operation and assessing performance*; all these steps are described in the "Development" section.

Methods Employed

There were found at early stages of the transformation that the main *transformation focus* is situated in *soft business shell* but not in *hard business nucleus*, and that the main efforts should be concentrated not on business agents, but on intangible entities. But in spite of intangible matter of *soft business shell*, the influence on it should be quite concrete because the changes in the shell must be fixed to be utilized in productive process.

Appropriate processes (with supporting guides, documents, templates, and software systems) were used as such fixing means. The *soft business shell* was affected according to the scheme: Mission to capabilities decomposition (Mission = > Capabilities) = > Architecting (Capabilities = > Key areas = > Integrated processes) = > Development and implementation of the new capabilities support systems (Integrated processes = > Tools – guides, templates, software platforms, etc.).

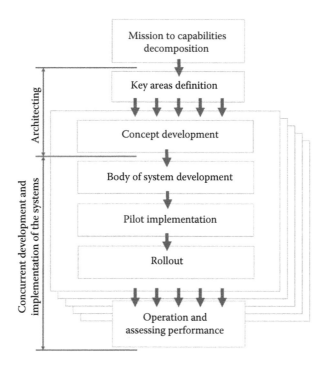

FIGURE 5.14   Main activities of the transformation.

Time Frame/Line

The whole transformation starting from the initial decision till the reporting at the board of directors was completed in less than one calendar year. Transformation team did not establish very strong general time schedule due to described specific of the task and utilized approaches like agile-stylized program management method. After very short stages "Analysis and capabilities decomposition" and "Key areas definition" (took a total of about one month), all key area activities were executed in parallel.

LESSONS LEARNED

How Were the Biggest Challenges Met?

The main challenges that management faced (the lack of experience and the ambiguity of market growth forecast) made the uncertainty factor the critical one in the transformation.

Uncertainty was aggravated by another aspect: the transformation was mainly focused on the *soft business shell*, which was quite fluid. Management used an *agile-stylized program management* approach as the most appropriate to the task: *soft integrated approach to soft fuzzy task*. *Soft*, an adaptive approach with wide horizontal and vertical integration of project teams, with free information exchange, with fast reactions on the changes, with very strong time schedule control, was used in the conditions of very soft, fluid, fuzzy, uncertain task with the lack of proven roadmap and with the necessity to implement the solutions immediately in ongoing business.

What Worked and Why?

Agile-stylized program management in general demonstrated its efficiency and applicability to *soft and uncertain* tasks; the main aspects of the approach are

- Senior and credible sponsors
- Multilevel integrated project team(s)
- Open information exchange

- Partnership and collaboration
- Proactive and motivated parties and constituents
- Creative and innovative way of development
- Prioritizing and focusing on the most ambiguous elements of systems design
- Piloting and subsequent roll-out in realistic environments
- Strong project scope control
- Strong project execution control—time schedule and resources control

What Did Not Work and Why?

Perhaps, corporate knowledge base development was the only more or less serious task that was not solved in transformation. The company's management understood the usefulness of knowledge accumulation and further alienation from the carriers in utilizing their business knowledge, so the goal of developing their own knowledge base was established.

Special database and software system were developed with appropriate guides, reports, and data collection forms. However, formal regulation to fill in engineering knowledge accumulation templates did not work. The gap of interests of the employees who fill in the form and the employee who further utilizes the knowledge was not overcome.

Knowledge is a very personal instance, and formalities did not work in this sphere: the templates were filled in nominally but the collected data might not be utilized in practice. In reality each employee has (at least in his head) a personal knowledge base—his/her experience and knowledge, but such bases are not open and are not available to colleagues. The attempts to force employees to share knowledge with their colleagues, to transfer the practice of knowledge accumulation from personal level to corporate level failed. The formal and *hard* approach did not work in this *soft* situation: the attempt to dictate knowledge sharing rules and format failed.

However, later this issue progressed quite naturally and simply: common folders were established to store project data in free formats. Such folders served to accumulate knowledge but in flat, unstructured form. Certainly, this is not a full-function knowledge base but knowledge is accumulated and might be utilized. Should proper indexing and a search engine be developed, this storage will serve as real and convenient knowledge base.

BEST PRACTICES AND REPLICATION PROSPECTS

The following methods and approaches were proven as efficient and convenient in transformation:

1. *Capability-based development approach and capability-based architecting* might be recommended to be utilized in creation and transformation of ES/SoS. The approaches focus all efforts on the required capabilities and involve in systems engineering process very important relations from enterprise mission and value to capabilities and to functions.
2. *Agile-stylized program management* might be used to solve wide range of fuzzy and ambiguous problems of different scale in the areas of SE, ESE, and SoSE with the huge uncertainty and the lack of expertise and the lack of proven methods and algorithms to solve it. The combination of soft and very creative design with strong planning and progress control is the key element of the approach.
3. *Key area definition and development appropriate new capabilities support systems* (core consulting and services technologies, project implementation system, system of business unit growth, management accounting system, motivation system). Definition exactly these areas and development integrated system in these areas might be considered as quite common for wide group of ES. The integrated bundle of three systems (motivation system, management accounting system, business unit growth system) is definitely appropriate to development or updating of practically any ES from any industry. The fourth system (project implementation system) might be used in considerable share of enterprise, so it is also common.

SUMMARY AND EPILOGUE

New capabilities were developed in ES/SoS transformation, an *extended enterprise* was formed around the IBS company. Active stakeholders (shareholders, management, and employees) demonstrated real interest in the transformation and contributed many efforts. The following activities were executed:

1. Analysis and capabilities decomposition
2. Key areas definition
3. Concept development and body of the new capabilities support systems development for each key area
4. Pilot system implementation and roll-out for each key area
5. Operation and assessing performance

Capability-based development and capability-based architecting, as well as agile-stylized program management approaches, were successfully utilized to accomplish the transformation.

Core activities and main results of transformation belong to *soft business shell* of ES/SoS.

The wisdom of the decision to transform the company has been proven by practical aspects of evolution of Russian economics and the IBS company development in Y2002–2012.

The IT services and consulting market has grown dramatically, and IBS has been leading it having the biggest market share for the years 2009–2012.

The core architectural decisions made in 2002 for the transformation demonstrated its efficiency and functionality; these decisions are still being utilized; updated new capabilities support systems in key areas are still being based on them. These technological basics have supported company development in recent years, including overcoming of the crisis Y2008, several mergers, and acquisitions.

The contemporary business environment is very dynamic, and many changes that happened since 2002 create new challenges for company management. A new transformation might be needed to respond these challenges, but that would be a topic of another case study.

CONCLUSIONS AND FUTURE WORK

The following areas seem to be interesting for further research and development:
A *ES/SoS nucleus model* might be developed to analyze the set of ES/SoS that interact to each other (Figure 5.15).

The companies, which are involved in the *extended enterprise*, are also ES/SoSs and also form their own *nucleuses, shells, clouds* with their own specifics. *The cloud* of one ES/SoS (e.g., IBS) covers several hundred (sometimes thousand) other ES/SoS (companies).

The following hypothesis seems to be very interesting to study:

Regularities, which are appropriate to directed and acknowledged SoS, rules in *the nucleus*, a collaborative SoS—in *the shell* and a virtual SoS—in *the cloud*. This looks like different zones, with the strongest relations and influences in the center, and the weakest in the periphery.

If such zones are common, regular, and systematic, it makes sense in ES/SoS engineering activities to define the zone in which we are working and to apply appropriate approaches.

The example, which is described in the "Characterizations" section, demonstrates how the SEA Profiler was applied to analyze ESE activities. The majority of activities were attributed as Directive SoS and Acknowledged SoS. Three *spikes* into the *Right Intermediate Interval* were detected, which is appropriate to a Collaborative SoS: *Include Social and Psychological Factors* when considering *Analyze Alternatives*; *Try to Predict Future Enterprise Needs* when addressing *Define the System Problem*; *Develop Implementation in Realistic Environments* when doing *Develop Implementation*.

It also seems that the forming of the *extended enterprise* also weakened governing hierarchies more and *shifted* SoS more to the collaborative domain from directive one (SoS_DoD).

Agile-stylized management approach. Its applicability was demonstrated to solve nonstandard task, which are typical for the ESE, SoSE, and CSE domains. So the approach might be further developed

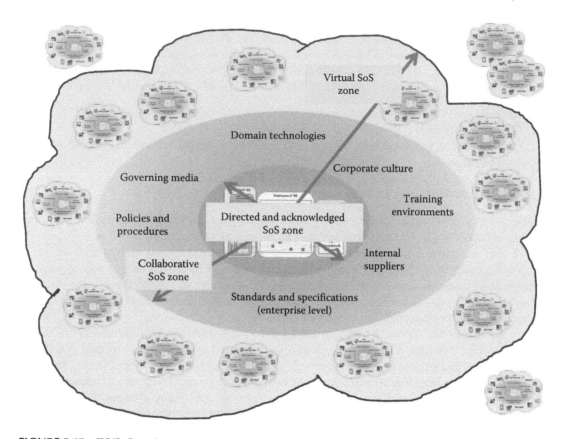

FIGURE 5.15 ES/SoS nucleus model and business society.

to create an agile SE framework consisting of the application conditions, main principles and tools, practical use recommendations, etc.

The development of *integrated auxiliary new capabilities support systems* is another research and development area. Such systems (which support key areas) are very critical for a great majority of companies. It is reasonable to create the framework for integrated key areas—management accounting, motivation of employees, growing of business unit, and project implementation. Such a framework will include integrated methods and approaches, as well as supporting software platforms with embedded standard functions in the areas mentioned.

Issues to discuss:

1. Which methods or approaches or takeaways might work in other industries? For example, in high-tech systems (e.g., in aerospace)? Or in construction?
2. What stage or step or activity is the most critical in the transformation described? Why?
3. Why and how is the *V model* normally suitable for SE also applicable to SoSE?
4. Are the methods or approaches or takeaways robust in response to any contingency?
5. If so, what exactly makes the transformation process robust?

ACKNOWLEDGMENTS

I would like to thank Leonid Zabezhinsky and Sergey Matsotsky, my friends and colleagues, who inspired me to improve and to transform the company.

Thanks also to Anna Belova, my friend and wife, who inspires me to improve and to transform myself.

REFERENCES

Accelare Capability-Based Management. The next revolution in productivity. Randolph, MA: Accelare, Inc., 2009 http://www.accelare.com/SitePages/dynamic_business_architecture_defined.aspx. Accessed 20 March 2013.

SoS_DoD Office of the Deputy Under Secretary of Defense for Acquisition and Technology, Systems and Software Engineering. Systems engineering guide for systems of systems, Version 1.0. Washington, DC: ODUSD (A&T)SSE, 2008.

ESE_processes R. S. Swarz, J. K. DeRosa. A framework for enterprise systems engineering processes, *ICSSEA*, 2006, http://mitre.org/work/tech_papers/tech_papers_06/06_1163/06_1163.pdf. Accessed 20 March 2013.

15288 ISO/IEC 15288:2008(E) IEEE Std 15288™-2008 (Revision of IEEE Std 15288-2004).

42010 ISO/IEC FDIS 42010, Systems and software engineering—Architecture description.

MBSE review INCOSE, Survey of model based systems engineering (MBSE) methodologies. INCOSE-TD-2007-003-02. Passadena, CA: International Council on Systems Engineering (INCOSE), May 23, 2008.

SE_hbk INCOSE, Systems engineering handbook v. 3.2. INCOSE-TP-2003-002-03.2. San Diego, CA: International Council on Systems Engineering (INCOSE), January 2010.

Prof R. Stevens, Profiling complex systems. *IEEE Systems Conference*, Montreal, Quebec, Canada, April 7–10, 2008.

White_1 B. E. White, A personal history in system of systems. Special session on System of Systems (SoS). *International Congress on Ultra Modern Telecommunications and Control Systems (ICUMT-2010)*, Moscow, Russia, October 18–20, 2010.

White_2 B. E. White, On principles of complex systems engineering—Complex systems made simple tutorial. *The 7th International Conference for Systems Engineering of the Israeli Society for Systems Engineering*, Herzlia, Israel, March 4–5, 2013.

White_3 B. E. White, Systems engineering activity (SEA) profiler. MITRE Public Release Case No. 09-2555. *8th Conference on Systems Engineering Research (CSER)*, Hoboken, NJ, March 17–19, 2010.

6 Global Distribution Centers
Multilevel Supply Chain Modeling to Govern Complexity

Gerrit Muller and Roelof Hamberg*

CONTENTS

* This work has been carried out as part of the FALCON project under the responsibility of the Embedded Systems Institute with Vanderlande Industries as the carrying industrial partner. This project is partially supported by the Netherlands Ministry of Economic Affairs under the Embedded Systems Institute (BSIK03021) program.

CASE STUDY ELEMENTS

- *Fundamental essence*: The application of multilevel models to guide the development of enterprise-level systems of systems (SoSs).
- *Topical relevance*: The approach is critically relevant in the sense that emergent behavior appearing at higher levels in systems is often unexpected and not as intended.
- *Domain*: Industry, logistics domain.
- *Country*: The Netherlands.
- *Stakeholders*: Enterprise architects for industrial SoSs.
- *Primary insights*: A federation of connected explorative models at different levels of abstraction is necessary to cope with the complexity at the SoS levels of logistical supply chains, in general, and the associated distribution centers, in particular. This is especially relevant with respect to the aspects of dynamics and uncertainties.

KEYWORDS

Goods flow simulation, model-based engineering, multiple levels of abstraction, system-of-systems

ABSTRACT

Distribution centers comprise many layers of decomposition in systems, subsystems, and components, while they are part of larger enterprises. The relative independence of the development of all these layers results in emergent behavior that confronts systems engineers with the challenge to keep a comprehensive overview. A short introduction of the logistic domain leads to a summary of the essential characteristics of distribution center systems and their relation to the emergence of behavior. A series of modeling examples is offered to illustrate how explorative models can be applied to govern the complexity of developing these systems. In order to bridge the multiple levels of system scope, these models are connected through essential parameters that carry the de facto abstractions of subsystems to their representation at a higher level.

GLOSSARY

ACP	Automated Case Picking
AIP	Automatic Item Picking
B	Total buffer space in the system
CAD	Computer-aided design
DC	Distribution center
DOF	Degrees of freedom
ED	Enterprise dynamics
FIFO	First-In-First-Out
SoS	System of systems

BACKGROUND

Context

This case study is based on the Falcon research project on warehouses of the future (see Hamberg 2012, Falcon 2011). Vanderlande Industries, the Embedded Systems Institute, and several partners cooperated in this research project with the objective to "find and develop efficient means to be able to analyze, design, and implement layered systems which shall comply with stringent requirements on performance, reliability, cost, development time…." The overall challenge was to combine top-down and bottom-up reasoning of business developments, system-level solutions, subsystem feasibility estimations, and the development of promising new technologies. One of the areas of research was the application of modeling to achieve the goal, dealing with this challenge. As one could have anticipated, the achieved project results are applicable not only for the future but also for current developments that take place in the domain of warehousing.

The structure of this case study is as follows: First, the domain of distribution centers (DCs), the logistics supply chain, and the role of DC suppliers are described in a global way. This provides the context for understanding the systems engineering challenges, which are addressed in a later section. Customer-level scenarios facilitate a first exploration of these challenges by making the problems and possible solutions more tangible. A number of modeling examples are discussed next to amplify the insights at different levels of the systems of system (SoSs), systems, and subsystems. After a reflection on the utility of these models, the case study is concluded with some guidelines for systems engineers entering the era of SoSs.

RELEVANT DEFINITIONS

In logistic chains, DCs serve as decoupling and buffering points when transporting goods from suppliers (production centers, such as factories) to outlets (such as retail shops).

An *item* denotes the smallest defined unit in the logistic chain. We use the term *case* for the immediate packaging around a set of goods; it contains one or multiple items. A case can be a carton box or plastic foil. For example, a carton box containing one or multiple identical items, or a number of bottles shrink-wrapped with plastic foil. Cases are stacked on *pallets*, which can be packed in foil, too. Items can also be kept together more loosely in so-called *totes*. Totes contain either multiple identical items or different items depending on their role in the logistic process, that is, efficient, temporary storage of individual items, or a way to collect items that belong to a single customer order. Totes have a uniform shape and size in order to optimize their handling by logistic equipment. They are stacked on pallets or *dollies* for efficiency during transport outside the DC.

PERTAINING THEORIES

A term that is frequently used in the SoS context is *emergence*. The aggregation of systems may show behavior or properties that emerge, that is, that *cannot be localized to a single independently acting constituent or to a small constant number of constituents. Emergent properties arise from the cumulative effects of the local actions and neighbor interactions of many autonomous entities* (quoted from Fisher 2006). Emergent properties can be desired or undesired and expected or unexpected by creators and integrators (Kopetz 2011).

Models with *higher levels of abstraction* (that better match the higher levels of complexity of SoSs, Heylighen 2001) are a means to cope with increased complexity. A higher level of abstraction means to represent essential properties in a fundamental way without losing their characteristic dependencies, and to leave out nonessential properties altogether. System-level properties that are known to relate to emergent behavior require extra caution in this abstraction process, as the emergent properties may derive from lower level actions and interactions.

Systems engineers must find proper levels and ways of abstraction. They need to relate models at various abstraction levels to reason about specification, design, and engineering challenges. We hope that discussion of several examples will illustrate the potential utility of modeling for SoSs.

We advocate that reasoning about SoSs needs to be sufficiently specific and quantitative, even more than in the case of systems due to the increased level of parallel autonomous developments. In this case study, we illustrate the use of scenarios for that purpose. Quantification and scenarios can help to manage the distance between model and reality by calibrating and validating the numbers.

EXISTING PRACTICES: DISTRIBUTION CENTERS AND THE LOGISTICS WORLD

In this section, we introduce the domain of DCs and their role in logistics of goods. DCs facilitate aggregation and repackaging of goods for later transportation.

Figure 6.1 shows a simplified logistics chain with its goods and information flows. A wide variety of cargo transportation means, ranging from trucks to planes, are used to transport goods from suppliers to DCs. Similarly, these means of transportation are used to transport the goods to their destinations. The type of transportation used is adapted to specific circumstances; for example, in inner cities, small vans will be used. A complex network of information systems controls and monitors the goods flow. Sales and purchase departments place orders for products based on actual sales and planning forecasts. Administrative and financial departments manage the related money flows.

We depict Figure 6.1 as a single chain while, in reality, the whole process might consist of multiple producer-consumer steps that are co-mingled within intertwined logistics chains. The focus of this

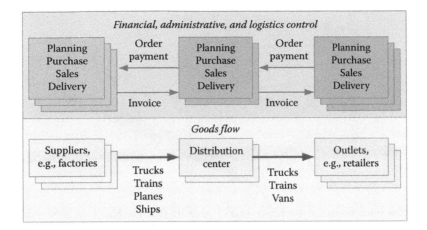

FIGURE 6.1 The logistics chain consists of goods and information flows. These flows are strongly coupled. Return flows of goods are not shown for the sake of simplicity.

case study is on the DC, a complex SoS. We need an understanding of the key drivers in the logistics chain to fully understand the trade-offs of the specification and design of DCs. The simplification of Figure 6.1 allows us to discuss most logistics key drivers without too much loss of realism.

CHARACTERIZATIONS: PROJECT AND PROCESS OF DISTRIBUTION CENTER BUSINESS

The DC design and realization business is a typical project business. The DC suppliers and their customers (logistic service providers, who operate DCs) discuss specifications and conditions during a tender phase. The customer selects a DC system supplier based on the offers from multiple suppliers. Once the customer places an order, the supplier starts a delivery project. After installation and acceptance, the customer pays for the main part of the contract. DCs have an operational lifetime of many years to even decades. During the lifetime, upgrades can occur via a similar tendering and delivery process. Figure 6.2 shows a typical project life cycle.

Business models are evolving in this business. In the past, projects usually finished after delivery. An increasing amount of business is generated through services during DC operation, where service-level agreements define performance and payment levels.

The tendering phase is for the suppliers an initial investment where senior engineers start the systems engineering activity. In this business, suppliers use visualizations and simulations to communicate with customers about the operations in the DC and to estimate performance in terms of throughput and meeting shipment deadlines. These simulations, together with the high-level design of the quoted system, are the starting points for engineering once customer and supplier sign the contract.

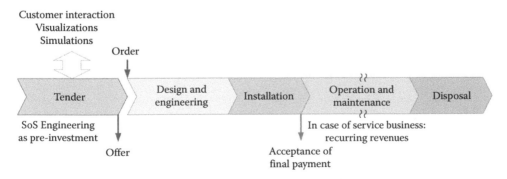

FIGURE 6.2 Project life cycle for DCs.

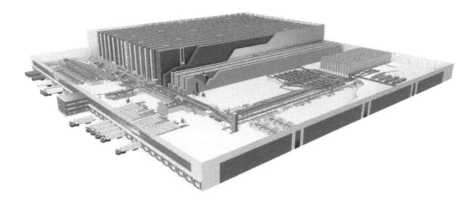

FIGURE 6.3   **(See color insert.)** Artist impression of an entire DC. The blue area is the main pallet store, transport, and pick and place; the green area is a tote store for pick and place; the orange area is storage of oddly sized goods that are handled differently. (Courtesy of Vanderlande Industries B.V., Veghel, the Netherlands.)

GUIDING PRINCIPLES

Central to the case study is the tender phase as sketched in Figure 6.2. The pre-investments that are done by systems engineering (SoS engineering) before actually selling the DC system play a critical role in the system's business success. Insight in the system-level characteristics is crucial for a sound balance between trustworthy system concepts, on the one hand, that are not over-dimensioned, on the other hand. In our opinion, the manageable and communicable connection between (model-based) understanding at different levels of SoS aggregation is a well-suited means to achieve this goal.

DESIGN OF DISTRIBUTION CENTERS

Various types of DCs support a wide range of different supply chains with different business processes. Business process variations include distribution of order sizes and their frequencies, and the range of different goods to handle such as retail, frozen food, fresh food, clothing, jewelry, and spare parts.

To design efficient and effective DC shop-floor processes, that is, the internal DC's daily operations, DC designers have a large range of different design patterns at their disposal. These design patterns are used to compose a complete system from a large range of standard material handling components and engineered specials where necessary. A flexible control system finally integrates all components and controls the shop-floor processes.

In addition to designing the material handling system, designers also have to decide how human operators (performing material handling functions) do fit as part of the system, that is, designing the user interfaces of the system that facilitate performing the manual tasks.

Figure 6.3 gives an artist impression of an entire DC. The inward- and outward-bound shipping docks are at the left-hand side, where the trucks are sketched. The large area with high scaffolds is the bulk storage area. The low areas at the right-hand side are the pick-and-place workstations.

Although DCs have many different types and different shop-floor processes, their global functional view is quite stable. In Figure 6.4, we sketch such a functional view that already hints toward different implementations for different flows.

PURPOSE

HISTORY

Conventional DC suppliers have their origin in other types of businesses, which are mostly related to one of the constituent systems of the current DC systems. This means that the responsibilities for the operations at DC system level are relatively new to most DC suppliers; in earlier days, these were

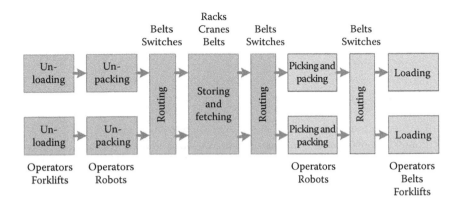

FIGURE 6.4 Generic functional block diagram of a DC.

residing with the owner of the warehouse. As the need for more effective and efficient operations became stronger over time, several companies seized the opportunity and expanded their business from being single system supplier toward integrating SoSs supplier with a servicing business on top of it. This is where systems engineering of the larger scoped SoS came into play.

CURRENT SITUATION

Nowadays, DC suppliers offer complete DC systems to their customers. The role of system engineers is to generate and validate system concepts that fulfill the customer needs within the constraints of a valid business case for both parties. Generation and validation is done by experience and performing fairly detailed simulation experiments with a limited set of options. DC suppliers use these simulations to show the customer how the proposed system will actually operate. Different user scenarios are preferably based on real user data and might represent contractual value, that is, play a role in the customer acceptance test of the delivered system.

KNOWN PROBLEMS

Goods Flow Characteristics at Different Levels

The whole logistics supply chain is highly dynamic. Sources and destinations may change and varying quantities and types of goods are flowing, which are seasonal or fashion-dependent, for example. As a result, logistic service providers continuously extend and adapt their DCs. These companies will also make trade-offs between suppliers and DC locations (distance to outlets) and order sizes and frequencies. The size of the packaging units in its turn will depend on order size and frequency as well.

Logistics Supply Chain and the Dynamic Role of DCs

Many stakeholders with varying concerns and interests are present in the chain. Incentives for making changes to streamline the logistics flow are not always in place. For example, packaging has a huge impact on the goods handling during transportation and in the DC. However, the stakeholder responsible for cost of packaging might have no interest in simplified product handling in the DC.

MISSION AND DESIRED OR EXPECTED CAPABILITIES

The placement of products on a pallet or in a tote is a process with a wide variety of stakeholder needs. The recipients of the products want undamaged products, for example, heavy products stacked on top should not crush fragile products. Transporters need a stable stack on a pallet or tote that will not deform or shift during handling, and they need minimum use of space, that is, pallets and totes packed as full as possible, because transporting *air* is very expensive. Retailers and drivers may prefer a stacking order of products that fits the sequence of delivery. Human operators place products with knowledge of these stakeholder needs.

SYSTEM-OF-SYSTEMS

The context of a distribution center as a SoSs is shown in Figure 6.5. The DC itself (A) is contained in the larger context of enterprises. In this chapter, we have mainly focused on the conception phase of the DC, in which case the governance F, feedback C, constraints E, and external factors B in Figure 6.5 guide the what-if scenarios that have to be considered by the system engineers. In this phase, the governance body D is mainly the system engineer. In an operational setting, a clear governance body is often absent or spread over diverse parties.

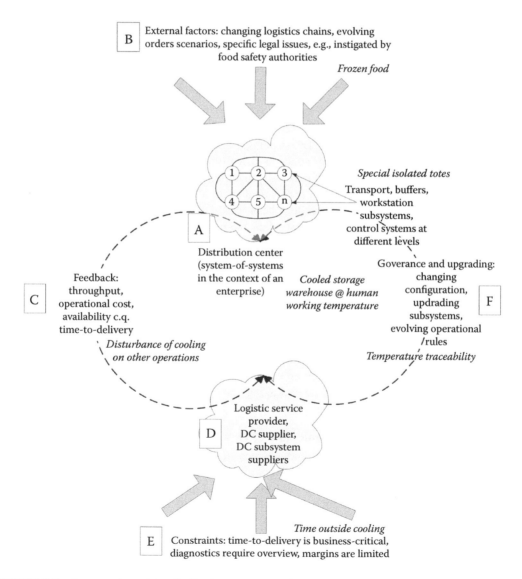

FIGURE 6.5   An overview of the distribution center as an SoS. In italics a specific example is shown of the impact of adding the handling frozen food to an existing warehouse.

As example, we can look at the impact of adding the handling of frozen food, an external need (B), on an existing distribution center (A). This requires the addition of a cooled storage area for the frozen food. The supplier packs the frozen food in special totes that isolate the food from the warmer external environment. There are regulations (F) for the traceability of the food temperature. If the food in the isolated totes is too long in warmer environments, then the food temperature may rise too much. The suppliers (D) will all ensure that the totes are not too long exposed to higher temperatures. The special totes and cooled storage area imposes constraints on the operation. Suppliers will communicate (C) impact of the addition of frozen food handling to the distribution center (A).

ENVIRONMENT

Different logistic service providers have different strategies in choosing locations for their DCs: de-centralized, centralized, or a hybrid approach. The location strategy depends on the types of goods and business, and involves a trade-off between warehousing and transport in the logistics chain. The end customer ultimately pays the cost of transport and warehousing for products. Pricing strategies and business models determine how cost and margin are distributed over the chain. This distribution may change over time.

SCOPE

The scope of the system for this case is the DC itself. Considerations about the logistics chain and enterprises around the DC are inputs to the specification and design.

STRUCTURE

The main functions of a DC are to receive, store, and redistribute goods, the last step probably in other combinations of goods than the first step. In order to fulfill this functionality, it is relevant to look into the basic packaging concepts. The left-hand side of Figure 6.6 shows how an operator composed an order consisting of multiple cases onto a single pallet. A truck transports the pallet to its destination. The right-hand side of Figure 6.6 shows an alternate approach to packaging. The DC designer introduced a standard box, a so-called *tote*. All DC functions can use totes as standardized means to transport, handle, and store items.

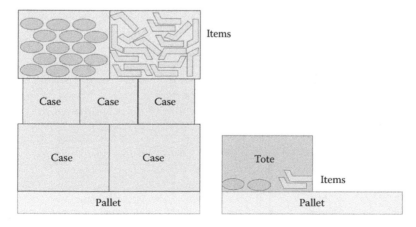

FIGURE 6.6 Packaging options of outgoing orders. Left-hand side is case based (cases delivered by suppliers), right-hand side is tote based (tote composed at the DC).

BOUNDARIES

The packaging concepts of Figure 6.6 play an important role at the boundaries of the DC. Next to the clear use of pallets outside the DC, the use of totes goes beyond DC boundaries as well; they may be transported to and from retailers and in a few cases are used by suppliers. This decreases the amount of superfluous goods handlings as well as increases the quality of transport.

Next to the physical part of the boundaries, there are information interfaces for order input, tracking, goods traceability, etc. Typically, this is a transactional interface at the ERP level. For DC suppliers this is mainly limited to an interface to the ERP system that is owned by the logistic service provider.

INTERNAL RELATIONSHIPS: CROSS-DOCKING, CASE AND ITEM PICKING TO MANAGE GOODS FLOW

Retailers place orders at the DC with a well-defined number per product. In the DC, operators compose the order in the so-called picking process. The rudimentary form is cross-docking: orders consist of complete pallets with goods. In cross-docking, the DC is used as temporary, for example, 1 day, storage of pallets that are retrieved on the basis of a customer order. The next form is case picking: orders consist of cases with cases containing multiple items. Picking is easier when it just involves cases, since cases are easier to handle. However, case picking constrains order flexibility. Item picking is the third approach that can fulfill orders exactly, because the picking is done on the smallest defined unit.

In the last two approaches, operators fetch the proper number of cases or items from the stock and place it in an order tote or on an order pallet. A transportation and routing system brings cases or totes with one type of item from the storage to a pick-and-place workstation (see Figure 6.7) and returns remaining stock to storage if needed.

Figure 6.8 shows an actual pick-and-place workstation. The six cases left and right each contain items for one order; the tote in the middle contains the stock. This increases operator productivity: for fast-moving products, the stock in one product tote can fulfill the demand for multiple orders in the workstation. The central monitor shows the total quantity to pick from the product tote, and six small displays on the sides show the quantities to place in the order totes.

EXTERNAL FACTORS: SOME BASIC CONCEPTS OF GOODS FLOWS

Suppliers package quantities of homogeneous goods (items) to deliver them to DCs. Appropriate packaging (i.e., efficient, secure, and safe) facilitates transportation and handling. The destinations of the goods, for example, retail shops, need a mix of various goods, probably from different suppliers.

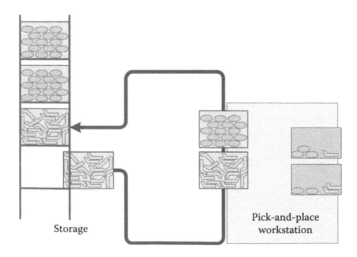

Storage   Pick-and-place workstation

FIGURE 6.7 A typical pick-and-place flow of goods.

FIGURE 6.8 An item pick-and-place workstation. Totes at the left and the right with the small displays on top contain orders; the tote in the middle with the large display contains the stock items. (From PICK@EASE.4 Vanderlande Industries B.V., see http://www.vanderlande.com/. With permission.)

Dependent on the quantities required by the outlet channel (the retailer), different packaging units can be delivered. The information systems that control the DC send the required goods to a workstation, where the operator picks the right number and composes them into an order.

CONSTRAINTS

Items may have all kinds of shapes, might be fragile, slippery, or the opposite: rough. Cases provide some protection for fragile items, and provide some size and shape standardization, but are not always optimized with respect to the logistics chain's requirements. Suppliers typically determine the case size and material, probably based on requirements from the retailers and cost considerations.

SPONSOR OF GOVERNING BODY IN CONCEPT PHASE

In most cases for this type of SoSs, the governing body is ill defined in its operational phase: it is spread over a multitude of parties, including the DC supplier that also defines the SoS. This situation is taken into account by the system engineers that create the SoS concept and validate it against different (change) scenarios. These system engineers form the de facto governing body in the SoS's conceptual phase. This role and the related activities that we describe in this chapter are sponsored by the DC supplier organization at senior management level due to the visibility the DC supplier has in the resulting SoS once it has become operational.

CHALLENGES

Starting from the domain characteristics as sketched in the background section, we will highlight a number of challenges that DC suppliers and the involved systems engineers face. These challenges are further aggravated by the need to include technological developments to address changing customer needs.

DC Suppliers: Challenges and New Technology Development

DC customers span a wide variety of needs, in size, capacity, capital cost, operational cost, types of products, rate of change, timing needs, etc. The result is a lot of diversity in composition of DCs. A core challenge is to determine whether a specific system configuration works and is economical. Currently, systems engineers of DC suppliers mainly address this challenge by experience. A complication is a continuous shift of boundaries of the system, for example, suppliers of systems as one-time deliverers of components, such as conveyors, up to service providers using service level agreements. Many suppliers of DCs have a history in systems like conveyors and evolve into system providers, and later into solution providers.

The key drivers of DC operation are the following:

- Throughput
- Time to delivery
- Operational cost
- Capital expenditure

These key drivers form a field of tensions. For example, many possible measures to increase throughput lengthen the time to delivery, especially in the presence of (physical) errors. Similarly, decrease of operational cost often requires an increase of capital. Personnel cost dominates operational cost in the current state of DCs in Europe and North America. The pick-and-place process is most costly, followed by unpacking. Lack of personnel availability, certainly in Western countries, threatens throughput and time to delivery.

DC providers are therefore looking into increased automation. Advances in robotics technology in combination with its decreasing cost may facilitate further automation. Main challenges for the economic use of automation are coping with:

- A variety of products, for example, in size, shape, rigidity, fragility, slipperiness, etc.
- Continuous changes in products and product mix
- Continuous changes in logistics process, product handling, and transportation

The SoS perspective on this transition is that in the current situation human operators fulfill a crucial role in making processes work with existing systems. Operators recognize fragile products and know to handle such products carefully. These human operators adapt their workflow to cope with such constraints. For example, an operator might park fragile products at the side temporarily so that they can be stacked on top later. What will happen when systems get more tightly integrated? What will happen when no or less human cognition and intelligence is available to cope with unforeseen issues? For certain, the systems will have to become much more intrinsically robust, at each individual level as well as across different levels of operation.

To provide an example for this in the area of the picking functions, Figure 6.9 shows that replacement of humans by robots trigger paradigm shifts for many subfunctions:

- Intelligence and adaptability of the human brain are replaced by analysis and control functions in software.
- Eyes and human senses are replaced by cameras and sensors.
- Arms and hands with many degrees of freedom (DOF) and fine control are replaced by robot arms (or other mechanical topologies, such as x, y, z oriented actuators).
- Legs for transportation may be replaced by increased range of the robot, or by any kind of transportation system; pick and place can be done at completely different places in the DC.

Most changes create new challenges mostly related to intelligence, adaptability, and flexibility. However, other benefits may offset the challenges, for example, indefatigable, strong, fast, large

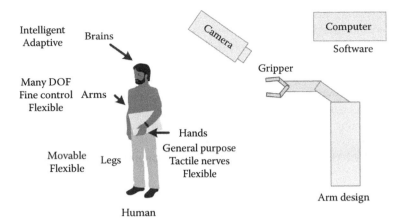

FIGURE 6.9   Replacing intelligent and flexible humans by robots is a paradigm change for many subfunctions.

range, and robust against noise, temperature (e.g., handling frozen goods), reduced lighting and heating to accommodate human operators, avoiding limitation of labor regulations, and many other environmental aspects. DCs can potentially operate continuously with lower operational cost by replacing humans by robots.

SYSTEMS ENGINEERING CHALLENGES

SoSs comprising autonomously developed systems have a number of particular challenges for systems engineers. The independence of individual systems complicates systems integration. Will the aggregation of individual systems behave and perform as intended? The high level of complexity of an SoS limits the overview that designers have of the system and may threaten their ability to reason about individual systems and the SoS as a whole.

Systems engineering of a DC suffers from the well-recognized systems engineering problem of many components with various degrees of freedom and levels of uncertainty. The mapping of business processes to actual components proceeds along various levels of abstraction. An example of these levels in the case of DCs is given in Figure 6.10, where the enterprise level is connected through the warehouse SoS to specific systems, such as the pick-and-place station to be automated,

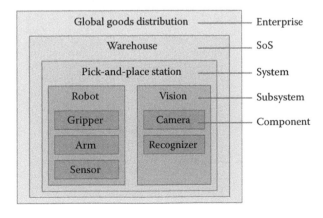

FIGURE 6.10   Summary of system levels in this case from components such as grippers and cameras to an enterprise for global goods distribution.

which consists of several subsystems and their constituent components. Characteristic for the SoS situation is that more abstraction levels are present and that part of the levels and systems are outside the scope of control and may even be undefined.

Learning by extensively testing system candidates is and will remain too expensive. How are we to guide the systems engineering process quantitatively in a cost-effective way? How can we get input data, and how can we validate concept choices? We present two related methods here. The first, exploration by scenarios, is well suited to scope the problem at a sufficiently high level of the system, while at the same time it indicates how the challenges translate to lower levels. The second method, which encompasses modeling, simulation, and analysis, is illustrated by a number of inter-related examples, which connect the scenario-based analyses at the SoS level to the levels below in a more quantitative way.

DEVELOPMENT

SYSTEMS ENGINEERING: EXPLORATION BY SCENARIOS

Systems engineering promotes the use of stories and scenarios to problem and solution spaces. We will discuss a limited number of scenarios with increasing degrees of difficulty to explore the challenges formulated before, and to see how different concepts can contribute to fulfillment of the key drivers.

Scenario 1: High-Volume Drugstore Chain

Scenario 1 comprises a chain of drugstores where a high volume of products is flowing through the DC. Individual drugstores receive cases with products, so individual item picking is not needed. Figure 6.11 shows the scenario at the top. This particular case can be served by a robot with x, y, and z as movement directions, and a gripper with only one degree of freedom. Placing cases with a limited range of sizes and weights on pallets is still challenging to do in a completely automated way. The main success factor for this scenario is effective planning and control across the logistics chain.

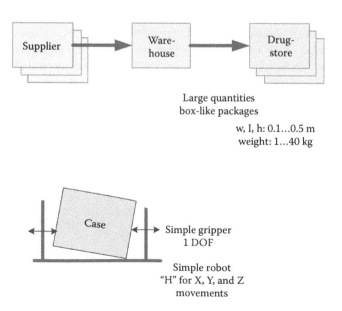

FIGURE 6.11   Scenario 1: high-volume drugstore. The DC distributes medium-size cases with a limited weight range in high quantities.

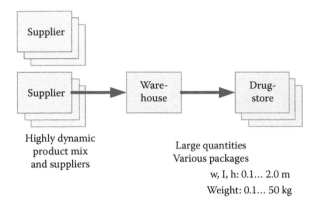

Highly dynamic
product mix
and suppliers

Large quantities
Various packages
w, I, h: 0.1... 2.0 m
Weight: 0.1... 50 kg

Multiple grippers needed?

There is no time to teach (program)
the robots how to handle package variety

FIGURE 6.12   Scenario 2: a chain of drugstores that has extended its business causing much more dynamics in the product mix and in product sizes.

Scenario 2: High Dynamics Extended Drugstore

Some drugstore chains have extended their business (Scenario 2 of Figure 6.12) by selling all kinds of additional products. These products are often one-time specials, season (e.g., Christmas), event (e.g., Olympic Games), or fashion related (e.g., sunglasses). This extension has several consequences:

- There is more variation and dynamics in suppliers; suppliers may appear and disappear in a short period.
- There is more variation in the size and weight of items and cases.

Humans can still handle sizes and weights. However, the basic mechanical system of the high-volume drugstore is inadequate. The sizes vary so much that probably multiple grippers are needed to cope with size differences. Shapes might be less uniform as well, complicating grippers even more.

A main characteristic of such a DC is its continuous operation, where there is no time or competence to adapt parts/systems to cope with some new variety. Solutions where a robot or gripper has to be programmed or taught how to handle packages do not fit. Similarly, introduction or change of suppliers should happen seamlessly and instantaneously.

Scenario 3: Mail Order Company

Another extreme scenario is a mail order company, where orders typically are one or only a few items. An additional complication is that customers frequently return orders that have to be processed and if possible restored in the DC. The variation and dynamics in products is similar to Scenario 2, the dynamic extended drugstore. The consequence is that the DC now needs item picking as functionality, which increases the requirements for the gripper further. We cannot longer assume minimal case sizes or shapes, and we may have to cope with fragile items. Surface texture of items might be an issue; what can the gripper grasp? What can the vision system reliably recognize?

The limitations of grippers and vision systems could probably be mitigated by packaging of individual items, that is, by other systems in the chain. However, packaging serves other purposes too, such as branding, and additional packaging means additional cost and processing steps. How can we achieve a globally improved chain, with a distributed ownership of the systems within an SoS?

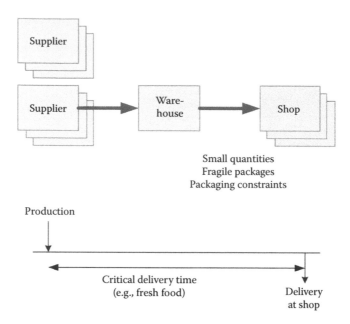

FIGURE 6.13 Distribution of fresh food, where the food may be fragile, packaging has to meet special constraints, and where delivery time is critical.

Scenario 4: Fresh Food Distribution

Distribution of fresh food (see Figure 6.13) introduces new needs, such as the capability to cope with fragile products (eggs, tomatoes), packaging constraints (such as transparency, waterproof, airtight), strict ordering rules for expiration dates, and fast integral delivery time. Fragile products and packaging constraints again affect concept choices for automated pick and place. For instance, the vision system becomes more complicated; can it work sufficiently reliable and fast?

The integral delivery time depends on all functions in the chain from harvesting or production to final positioning for display. Concepts that may improve DC capacity, such as large batch sizes and long belts serving many pick-and-place stations, may increase delivery time.

Management of Uncertainty

Systems engineers use scenarios to make uncertainties explicit and tangible. Within any given scenario further uncertainties are present: how will the business develop in the future, both in terms of customers and the orders that they will issue, how strongly will the assortment change, what are expected properties of goods that have to be handled, etc. Not all of these uncertainties are considered independently: in order to manage them, they are limited by considering certain combinations. On the one hand, this leads to a classical risk management approach (what are chances that certain combinations will occur, and what would be their impact), while, on the other hand, historical DC data are mined to obtain a realistic view of variations that will occur over time.

Systems Engineering: Modeling, Simulation, and Analysis

In this case study, we argue that systems engineers should not use what seems like unmanageable system complexity as an excuse to set their analytical capabilities aside. We propose the use of multilevel models to reason about individual systems, critical design choices in systems, and the expected performance and behavior of aggregated SoSs.

The design of DCs can be characterized by a large degree of design options, for a wide variety of customers. The challenge for the DC supplier is to design sufficiently trustworthy in the tender phase (see Figure 6.2) to be able to make a feasible offer (performance, cost, time) that still meets overall business goals. Suppliers typically build simulation models to analyze design options and to communicate with customers. Simulators thus serve a dual purpose:

- Facilitating design and mitigating project risks by validating the predicted performance characteristics of the system that is being proposed
- Communicating with customers and convincing them of the value proposition in the offer by visualizing the system and showing them the behavior and performance of the system

The consequence of this dual purpose is that suppliers tend to make simulators at a moderate level of detail. The underlying models are dynamic and executable, with inputs mimicking the real world. Nice visual front-ends ensure engagement of customers.

What Tool or Model at What Level of Detail?

Any system, such as a pick-and-place workstation can be viewed at various levels of detail. Figure 6.14 (text and figures in this section are based on [Muller 2011, Sections 2.4 and 4.1]) visualizes this as a pyramid with the vertical axis an exponential scale for the number of details. The most abstract way to look at a system is to see it as, for instance, *a pick-and-place workstation* (at position 10^0). The system can be specified in its ~10 key performance parameters and functions (position 10^1). Elaborating these key performance parameters further results in a system specification (somewhere between 10^2 and 10^3). The bottom of the pyramid represents all details produced by engineering that define the system: mechanical properties in CAD-M files, electrical data in CAD-E and software source code. The bottom of the pyramid depends on the type of system; many systems nowadays have over 10 million details. These details are typically defined, stored, and maintained per discipline. The detailed designs tend to be less extensive (position 10^6). From system specification to detailed design, a multidisciplinary design takes place: partitioning in subsystems, components etc., functional design, function allocation, interface definitions, etc.

The DC supplier delivers many systems; hence, the pyramid reflecting all delivered systems will be significantly larger and reach the order of magnitude of 1 billion details. However, due to its SoS nature, there is a lot of detail outside the delivered system that is relevant for the specification and design of a DC. Figure 6.15 visualizes this by adding an upside-down pyramid with a similar scale for the number of details. As systems engineers we simplify the outside world to its main stakeholders and interfacing systems. However, we can model the context in more detail

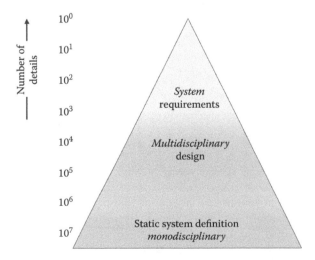

FIGURE 6.14   Various perspectives at a single system, from highly abstracted (the pick-and-place workstation) to all static engineering details as stored in engineering databases (see Muller 2011).

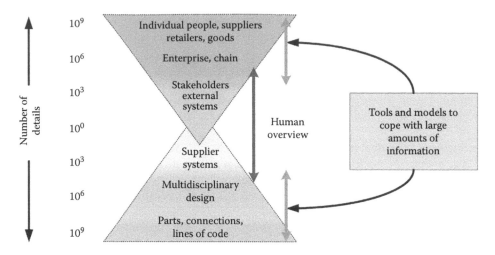

FIGURE 6.15   More perspectives for the systems delivered by the supplier at the bottom, and the context of the supplied systems at the top (see Muller 2011).

using business process models and chain and enterprise models. However, the real context is full of unique, individual people, suppliers, retailers, and goods, all with their own characteristics. The context can easily exceed one billion details as well.

Figure 6.15 illustrates that human reasoning and communication typically is based on abstractions. System developers use computer assistance to manage all engineering details. Human reasoning is based on a rich pallet of representations and models, for example, physical and functional, graphs, and formulas.

In Figure 6.16, we have positioned the simulators that are typically used in the tender phase and afterward for DC projects. These simulators contain a moderate level of detail such that customers and designers perceive the simulator outcome as realistic. The level of detail of these simulators also has some disadvantages:

- Effort of creating, verifying, using, and maintaining is roughly proportional to the level of detail.
- The cycle time to explore concepts is also roughly proportional to the level of detail.

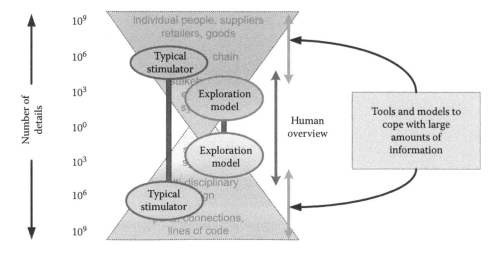

FIGURE 6.16   Positioning of simulators and exploration models.

Architects of the models strive for partitioning of the models and standardization of information and interfaces to minimize the effort to use and adapt models. Well-architected models suffer less from these disadvantages then ad hoc models. However, even well-architected models with lots of detail can slow down exploration. The effort increases especially when the exploration goes outside the foreseen exploration space, for example, for new concepts.

RESULTS

EXAMPLES OF MODELS

In this section, we will discuss multiple models at various levels of abstraction. The purpose is to show the relation between the objectives of an individual model and its level of abstraction. Most of the models are used in combination with other models. For example, modelers can integrate a crane performance model into a DC performance model, or they can use the crane performance model to parameterize the DC-level model properly. Similarly, models of production and order flows can feed a DC performance model.

For most models, we provide an estimate of size and effort to make this particular model. Often, the creation of the model reuses code from similar activities. Simple models may have a size of tens or hundreds lines of code. Larger simulations may build on tens of thousands reused lines of code. Reused code facilitates fast creation of models. However, it may slow down developments when the reused code does not fit the problem. Most modeling effort relies on huge amounts of *black-box* tools and toolboxes, for example, MATLAB®, Eclipse, or other simulation environments.

One of the concepts proposed for a customer similar to Scenario 1 is to simplify the packing by delivering all cases in a fixed predefined sequence. For this customer, the stacking of the cases is critical. Figure 6.17 shows how the picking function is realized by fetching from the storage, the sequencing function by inserting cases in the right place in the transport stream and the packing is performed at the end by an in-sequence palletizer.

ACP (Automated Case Picking) Model to Analyze Effect of Fixed Sequencing on Performance

At several places in the goods flow, we need buffers to decouple operations; for example, between fetching goods from scaffolds and inserting goods in the transport stream. Buffers are typically first-in-first-out (FIFO) buffers, since these are much simpler and lower cost than random access

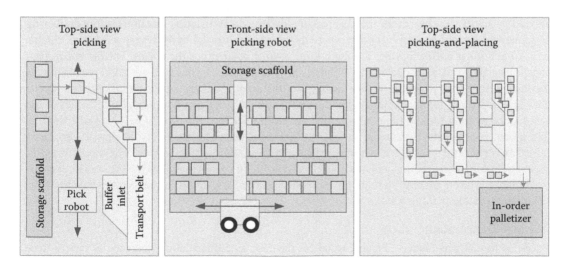

FIGURE 6.17   Model to analyze the effect of fixed sequencing in case picking on performance.

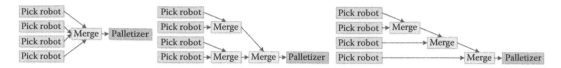

FIGURE 6.18 Examples of transport topologies with FIFO buffers at every merge step. These configurations served as inputs to the performance model sketched in Figure 6.17.

buffers, especially for buffers dealing with physical goods. Random access buffers would have provided more freedom in scheduling the operations in the goods flow.

The Falcon researchers made a model to see the impact of the in-sequence requirement of the palletizer at the end on the performance for different transport configurations of the SoS without random access buffers. This model is a highly abstracted dynamic model. It only takes the picking, merging, and insertion into account to study the impact of various topologies. Figure 6.18 shows a number of example transport topologies to indicate the type of variations that were studied with the performance model. The model simplifications facilitate a comparison of concepts without achieving absolute performance numbers. That is, the absolute performance figures will exhibit large accuracy spreads due to many neglected influences, but the relative performances for different concepts are significant due to sufficient precision in the model analyses.

In retrospect, one simplification, expressing intermediate buffer sizes (how many cases can be temporarily stored before being processed) in a number of pallet tasks instead of cases, turned out to be counterproductive in terms of communicating about the model and its results.

The model made very clear how large the effect of intermediate buffering is on throughput. A static model of the penalty of small buffers was derived showing that the penalty is roughly proportional to $1/\sqrt{B}$, where B is the total buffer space in the system, evenly spread over the merging subsystems. However, cost, space, and probably cycle time are proportional with B, showing trade-offs to be made.

This model is described in more detail in Chapter 6 of [Hamberg 2012] and in [Roode 2011]. The model contains less than 200 lines of code. Construction of the model and performing analyses involved 3 days of work. Such models have influenced the choice for configurations of Automated Case Picking (ACP) systems.

Crane Performance Model

This model zooms in on a picking robot as shown in the middle of Figure 6.17. The picking robot is realized by an off-the-shelf crane that can pick multiple cases at different locations before delivering the sequence of cases at a deposit point where the cases are inserted in the transport stream. Such a series of actions is called a cycle. The high-level control determines the cycles. Falcon researchers made a model to see what cycles should be given to the cranes in order to get good performance. This model forecasts the performance penalty for different number of picking tasks in one cycle. In a second variant, two-sided picking is considered as well, quantifying the adverse effect of the in-sequence picking requirement on the crane's performance as a function of case size distribution. This model is a quantitative model in a script language, in which we apply stochastic variations and many repetitions to arrive at average values and their standard deviations. The model is at 10^2 level of detail. Its construction and analysis effort amounted to less than 2 days. The resulting crane performance is used as input in higher level models of the SoS, and it influenced the design of the buffer layout of the picking robot.

Case Size Distribution

A concise model was made to be able to reflect the case size distribution in customer orders for a limited number of concrete prospect customers, whose order data were available for a given period. The model serves to generate payload input to the crane performance model as well as to the higher level models of the SoS. It consists of a discrete probability distribution from which samples can be drawn. The model is at 10^1 level of detail, while the time spent on it was less than a day.

ACP with *Lumped Control*

Automated case picking is proposed for a new market so at the start one does not have the experience of previously built systems to rely on. Falcon researchers made more detailed models for an SoS similar to Figure 6.17. The more detailed models aimed to facilitate the choice between different SoS concepts covering the functions *pick case from storage*, *transport case*, and *place case on pallet*. These concepts have to be compared on different aspects. One aspect that is difficult to estimate, but highly relevant is throughput. Static models that describe performance and behavior of such SoS only exist in few situations, as dynamic interactions caused by finite buffers and phenomena such as dieback (waiting caused by full buffers) hamper such stationary approximations. Dynamic simulation models facilitate exploration of such effects.

This *lumped control* model is less abstract and more realistic than the model used to explore the impact of fixed sequencing. It works at the level of cases that are moving individually through the system. The size of cases is modeled and size distribution (dependent on the specific customer) is also taken into account. On the other hand, the modelers abstracted high-level controls to a global, constant work-in-progress control (keep the same amount of work in the system at all times)—this fits the regular operation of DCs in general, provided replenishment does not interfere, as the modelers assumed. This assumption is a simplification that might not hold for many concept choices; nevertheless, the results were considered useful because the effect of different layout concepts and order characteristics could be numerically studied. The model had a size of about 650 lines of code, which puts it below the 10^3 level of detail. Construction and analysis together covered about 12 days of work. With the model, some actual proposals for customers were evaluated in a very early phase of tendering for different choices of buffer sizes and transport network configurations.

ACP with Detailed Control Mechanism on Planning Layers and Crane Breakdowns

One of the major factors impacting DC throughput and cycle time is the exception handling. For example, failure of one or even few of the systems should have a small impact on operations. High-level controls together with the topology and concept choices determine the robustness for local failures and exceptions. The perception was that the conventional approach with a strong centralized controller lacked the capability to adjust to problems. The Falcon researchers worked on a model to support a redesign including model-driven configurability based on the hardware layout of the system. In order to do this, distributed variants of the control rules are considered, including replenishment back orders to keep the case picking system operational and responsive (by having sufficient stock). The model of the system included the control layers to validate SoS behavior, as well as behavior under breakdown and repair of specific components in the system. We conducted a broad set of analyses in order to build up knowledge about the effects of changing values of design parameters (e.g., component performance ratios, work-in-progress level) and their interactions. Designers typically study the effects on spreads in pallet completion times as well as the utilization levels by these models in order to gain the insight in the SoS. The implicit assumption is that spread in pallet completion time affects the integral delivery time and variation of it, and that utilization levels correlate with investment and operational costs.

The model contained over 1400 lines of code and was developed in stages. The gross development time, including tests and analyses, covered about 6 months. The complete analyses were collected in a technical report, which was used by system engineers to understand the characteristics of the distributed control rules and their compositions. The fact that this was a new approach for the SoS-level controls, requiring building up new insights, hardly justifies the effort put in the model and its analyses. More regular reflection on challenges at different levels was required in retrospect and would have given rise to additional focal points.

ACP in a Logistics Simulation Tool Using Enterprise Dynamics (ED) with Detailed Layout

Normally, a DC supplier is used to build models on a modeling infrastructure that facilitates detailed layout, dynamic behavior based on the planned high level, and animation capabilities for a realistic visualization of the resulting DC SoS. The modelers mimic the structure of the control software in the model. The functions of this model are

- To support the tendering process in the analysis of specification and design, for example, concept selection, dimensioning, costing
- To communicate with customers with a realistic and trustworthy model

The model is highly detailed to make the visualization as realistic as possible. For example, libraries of product cases are available with properties and appearance so that the animation can show real cases. Traces from real good flows are used to simulate inputs and order streams, again to be as close to reality as possible. The consequence of this approach is that this model requires more effort to make or change, depending on the level of reuse that can be applied. The effort to construct and to analyze such a model is a limitation for fast exploration, but its increased accuracy (few percent level) as well as level of realism make it indispensable for the start of engineering and important customer quotations. The size of the model is estimated to be six thousand lines of code, while its (total) development time was about 10 months. The development of the model went hand in hand with the development of the system control design.

AIP (Automatic Item Picking) Workstation Models

Heling describes several models used to support design of an automatic item picking workstation (Heling 2011).

1. Models to study the order size distribution—samples from a Poisson probability distribution and different (typical customer) scenarios including a worst-case scenario (one item per order). This is a static, level 10^0 details, model providing quantitative insights, while it is also used in a more detailed model of the workstation.
2. A local model of the out-of-sequence arrival at the workstation that characterizes arrivals as an adapted Gamma probability density function. Domain experts recognized and verified the model results. This is also a static, level 10^0 details, model, providing quantitative insights, also used in a more detailed model of the workstation.
3. A kinematic model with only few parameters that computes travel times for the robot. It is based on formulas relating distance to maximum speeds and accelerations. The model can be plugged in other models.
4. Models to facilitate analysis of different layouts and topologies, such as lane orientation and lane flow direction in the workstation. The models help to compare cost and performance, for instance, for many short lanes or few long lanes. These models do not give absolute quantitative results, but rather support qualitative reasoning. Each of the models has only a few parameters or options, while the gross time spent on them is typically in the order of a few days.
5. A quantitative model of the cycle time for the complete pick-and-place action to explore multilevel or single-level workstation concepts, for example, using the vertical dimension. This is an example of a model that uses the aforementioned robot travel time model. In its turn, this model is used in the more detailed model of the AIP workstation. The model contained less than 10 parameters.
6. A more detailed model of the AIP workstation that given a configuration of the workstation facilitates analysis of the control by the information systems. This simulation can be used to compare different strategies, for example, different customer scenarios (models 1 and 2)

are compared to each other. Typically, the DC designer will study many dependent variables or workstation properties in order to learn how control parameters influence the overall performance. As a result, the designers can tune control strategies according to specific customer cases. The size of the model is about 750 lines of code, while the gross time spent to develop the model as well as to perform analyses was about 4 months. For exploration and making sensible trade-offs between several system-level qualities of the workstation such as throughput and cost, this model is vital. In the scope of the overall development time of the workstation, the balance between effort and impact of the model is all right.

OBJECTIVES ACCOMPLISHED AND NOT ACCOMPLISHED

The DC supplier has several means, for example, models, to explore solution alternatives in a limited amount of time. These means help to deal with complexity and especially emergence. Models for exploration decrease the number of unexpected and undesired emergent properties. The models help to increase the desired emergent properties.

Next challenge is the embedding in the organization. Modeling for exploration is not a cookbook-based activity. It requires a lot of domain knowledge, both logistics and technical, and skills in balancing the need for details and accuracy versus exploration speed. From methods point of view, we have to realize that SoSs are moving targets where all stakeholders and systems have their own dynamics. Hence, embedding is not a static challenge, but a continuous effort.

ANALYSIS

ANALYTICAL FINDINGS

Reflection on Models for Exploration

The conventional approach to DC design is to make a simulator with detailed layout and system information. The benefit of such modeling is that the closeness to reality helps customers in trusting the model. Disadvantage is that configuring and adapting the simulator is time and effort intensive, and that the simulator structure limits the exploration of alternatives to the preprogrammed concepts. The increasing SoS nature of the logistics chain induces more variation in the interaction between systems, with more uncertainties and an increase in unforeseen emergent behavioral properties. Falcon researchers were complementing the traditional approach with explorative models. An explorative model typically focuses on one or a few aspects (e.g., effect of fixed sequence, crane performance, or *lumped control*). This focus on limited aspects makes it possible to simplify the model. This simplification helps to reduce the effort of making and adapting the model. However, maybe even more important, it helps designers to reason about this aspect. One of the risks of an SoS approach is that designers lose the ability to reason about different aspects, or worse, that they hide behind the complexity and unpredictability argument to not do the analysis. Figure 6.19 shows the models that have been discussed on the scale with number of details. This figure shows clearly that most explorative models are typically at the 10^3 level of detail or below.

The simplified explorative models worked well in making trade-offs and exploring concepts in the sense that the effort spent is judged to be in line with the impact of their analyses. At the same time, designers and other stakeholders should be aware of the simplifications that went into the explorative models. These simplifications limit the validity and accuracy of the models. This nuance complicates communication with customers.

LESSONS LEARNED

The explorative models that we have shown sometimes simplify to static models. For example, the model relating throughput and buffer sizes is ultimately a mathematical formula that entirely

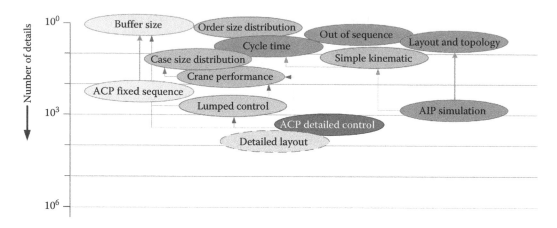

FIGURE 6.19 The models discussed in the results section positioned on a scale with the number of details. The dependencies between models are always qualitatively indicated.

ignores the dynamics of the goods flow. Some of the models are dynamic and do take variations in goods flow or systems into account. Dynamic models are inherently more detailed. The inclusion of the dynamics allows analysis of effects such as peak loads, resonances, and disturbances. Dynamic models tend to be closer to reality. These properties being advantageous, they also invite or provoke to put more details into them than required for a specific analysis. This is a trap of using dynamic models, encountered by ourselves as well in a single case. The remedy to this is a regular reflection on all levels of the SoS, that is, explicitly zooming in and out. Altogether, it is very well possible to apply dynamic models in explorative phases, provided the goals of the models are sufficiently crisp, having a limited scope.

Similarly, we have shown qualitative models; models that give a relative answer (e.g., better, faster, more). Quantitative models give absolute answers (e.g., cases per hour), but when the systematic errors are known to be large, they still can be used to produce relative answers. Qualitative models facilitate reasoning, even when the system properties are still unknown. Qualitative models can help to understand concepts and trade-offs, before lots of effort is spent in quantitative understanding. In later phases, quantitative models are required to facilitate concept and technology choices.

Best Practices

Multilevel Modeling

The models that we have discussed are mostly multidisciplinary models. Allocation of functionality to specific disciplines, such as mechanical, electrical, or software, is not yet relevant. Most models belong to a particular system level (e.g., DC, pick-and-place, or crane or robot). A model at one level can *feed* models at a next level. For example, the crane performance model can provide input to pick and place models. Such model connections are often not literal; the results at one level can often be simplified or abstracted at a next level manually. The virtual world of exploratory models can be viewed as a micro-cosmos of SoSs.

Multiview Modeling

Most models need multiple views to be explained and created. We especially see the use of physical, functional, and quantified views as generic views. Most stakeholders associate with physical models. Physical models are concrete and tangible. Functional models show how systems work. Functional models often show dynamics by some flow (e.g., of goods, energy, or information). The dynamics of functional models complements the partitioning oriented view of physical models.

Quantification facilitates analysis and reasoning, and makes discussions more concrete. These three generic views are applicable throughout all levels of systems from enterprise down to component. These generic views often need additional complementary views for further communication and discussion.

Abstraction and Simplification

A higher level of abstraction means to represent essential properties in a fundamental way without losing their characteristic dependencies and to leave out nonessential properties altogether. A challenge for systems engineers is to find proper levels and ways of abstraction. They need to relate models at various abstraction levels to reason about specification, design, and engineering challenges. We hope that discussion of several examples will illustrate the potential utility of modeling for SoS.

REPLICATION PROSPECTS

The approach using models to explore is one of the main methods in architecting SoSs. The T-AREA-SoS project identified multilevel modeling and conceptual modeling (*insight through simple models*, T-AREA-SoS 2013) as key research topics. Architects have used models for exploration at system and enterprise level in the past. The transition to SoS level amplifies the need for applying the best practices: multilevel modeling, multiview modeling, and abstraction and simplification.

SUMMARY AND CONCLUSION

DCs are typical SoSs, embedded in a logistics enterprise. Stakeholder key performance parameters can be analyzed at all layers, and conversely, technology and concept properties can be studied at all layers. Designers need sufficient insight in the relation between key performance parameters and technology and concept properties to come to a robust, affordable, and well-performing DC. We assert that explorative models, for example, models that can be made with limited effort and time, are necessary to cope with the complexity, dynamics, and uncertainties of the SoS world. These models focus at different levels of the system hierarchy, have limited and clear goals, and regularly use each other's results. Frequent and explicit reflection on the modeling activity is one way to avoid the trap of putting more details into the models than they require for achieving their goals.

We need to see more of their use to accelerate the evolution of the logistics enterprise, which can be enabled by deepened insights across several system levels, starting from traditional logistic components such as conveyors, which are well known to most DC suppliers, all the way up to expected trends in the world of global goods distribution.

SUGGESTED FUTURE WORK

FURTHER QUESTIONS FOR DISCUSSION

Imagine a distribution center for a supermarket chain serving the city of New York. For this distribution center, make the following estimates:

- The number of different products that flow through such center
- The order of magnitude of the amount of goods flowing through the center (in quantity, volume, weight, or value)
- What goods require special treatment, for example, cooling, handling fragile goods, etc.
- The number of suppliers of goods and the number of super markets served
- Show for one specific product what happens to it from supplier to consumer

ADDITIONAL RESEARCH

We have observed that experienced systems architects and engineers make explorative models by nature. However, many *fresh* systems engineers tend to dive and disappear into detailed models; they struggle with fundamentals and get overwhelmed by complexity. What guidelines and methods can we provide to assist less experienced systems engineers to make explorative models?

REFERENCES

Falcon 2011 Falcon Project, 'System-of-systems' performance and reliability in logistics. http://www.esi.nl/falcon/, accessed October 2012.

Fisher 2006 D. A. Fisher, An emergent perspective on interoperation in systems of systems, technical report CMU/SEI-2006-TR-003, 2006.

Hamberg 2012 R. Hamberg, J. Verriet (eds.), *Automation in Warehouse Development*, Springer, London, U.K., 2012.

Heling 2011 J. W. E. Design of an automated item picking workstation, master thesis at Eindhoven University of Technology, June 2011.

Heylighen 2001: The law of requisite variety, in: F. Heylighen, C. Joslyn and V. Turchin (editors): Principia Cybernetica Web (Principia Cybernetica, Brussels), URL: http://cleamc11.vub.ac.be/REQVAR.html Accessed February 22, 2014.

Kopetz 2011 H. Kopetz, *Real-Time Systems: Design Principles for Distributed Embedded Applications*, Springer, New York, 2011.

Muller 2011 G. Muller, *Systems Architecting: A Business Perspective*, CRC Press, Boca Raton, FL, 2011.

Roode 2011 V. Roode, Exception handling in automated case picking, SAI technical report, Eindhoven University of Technology, Eindhoven, the Netherlands, 2011.

T-AREA-SoS 2013 T-AREA-SoS project, Agenda for expert workshop, January 2013. https://www.tareasos.eu/, accessed February 2013.

7 Humanitarian Relief Logistics
Multicriteria Decision Analysis and Simulation for Supply Chain Risk Management

Tina Comes, Frank Schätter, and Frank Schultmann

CONTENTS

CASE STUDY ELEMENTS

- *Fundamental essence*: Humanitarian relief logistics are prone to complexity and uncertainty. We provide a case study that focuses on the problem of locating warehouses to illustrate the impact of uncertainty on the results of the models and show how scenarios can be used for robust planning and decision support under fundamental uncertainty.
- *Topical relevance*: The overwhelming uncertainty and complexity are the reasons why computational models from operational research or business supply chain management (SCM) are often rejected in the context of disaster management. Computational tools and models can, however, be very helpful when large amounts of information need to be analyzed and processed. This chapter shows how both paradigms can be reconciled.
- *Domain*: Academia.
- *Country of focus*: Haiti.
- *Interested constituency*: Emergency management authorities and nongovernmental organizations (NGOs) and, in a larger context, all organizations prone to large-scale crises.
- *Primary insights*: Scenarios and heuristics from operational research provide the basis for timely, robust, and flexible decisions respecting multiple objectives that enable decision-makers to manage complex and uncertain problems.

KEYWORDS

Complexity, Emergency management, Flexible and robust decision support, Humanitarian relief logistics, Multicriteria decision analysis (MCDA), Scenarios, Supply chain risk management

ABSTRACT

The twenty-first century has been termed the "century of disasters": the *Financial Times* included *disaster management* in the top 10 challenges facing science in June 2011.* Population growth, urbanization, and concentration have exposed more people and assets to natural disasters. Climate change causes more and more extreme weather events, such as heat waves, droughts, wildfires, and floods. Additionally, the modern world's interconnectedness causes system failures to propagate widely. Therefore, the need for supporting emergency managers who need to make decisions while facing uncertainty and complexity arises. Particularly in the early phases, decision-makers need to respond quickly although information is sparse while acknowledging that their current decision will impact all future decisions.

This chapter presents a case study that focuses on the humanitarian relief logistics after the 2010 Haiti earthquake. We analyze the key challenges, provide an overview about the status quo, and present a systematic way to support the design of robust disaster relief supply chains. A robust supply chain performs well under a variety of scenarios and thus enables coping with fundamentally different situation developments. To determine the quality of the network, techniques from multicriteria decision analysis (MCDA) and simulation models from operations research are combined.

GLOSSARY

Attribute tree: Hierarchical structure showing how abstract, higher level goals can be broken down into less abstract subgoals until the level of measurable attributes is reached. Attribute trees are typically used in multiattribute decision making to rank and prioritize decision alternatives.

CI: **Critical infrastructure**

---

* See http://www.ft.com/cms/s/2/bedd6da8-9d37-11e0-997d-00144feabdc0.html#axzz1RyBGYDzb.

Decision alternative:	Option for action. Decision-makers choose one of these options for implementation.
FAO:	**Food and Agriculture Organization**
HC:	**Humanitarian country**
ICC:	**Intercluster coordination**
ICT:	**Information and communication technology**
MADM:	**Multiattribute decision making**
MATLAB®:	**Matrix laboratory** (a numerical computing environment and fourth-generation programming language)
MAVT:	**Multiattribute value theory**
MCDA:	**Multicriteria decision analysis**
NFI:	**Nonfood item**
NGOs:	**Nongovernmental organizations.** Relief organizations can be divided into three categories: organizations operating under the United Nations (UN) (e.g., the World Food Programme), international organizations (e.g., the International Federation of Red Cross and Red Crescent Societies operating as a federation with country offices that are auxiliary to country governments), and other NGOs (e.g., World Vision International and Welthungerhilfe).
NP:	Nondeterministic polynomial time
OCHA:	United Nations **Office for the Coordination of Humanitarian Affairs**
OR:	**Operations research**
PROMESS:	**Programme de Médicaments Essentiels**
SBR:	**Scenario-based reasoning**
Scenario:	Description of a system or part of it at a given time and how it unfolds into the future given specific assumptions about the external environment or decisions made.
SoS:	**System of systems**
Supply network:	Set of three or more organizations or individuals directly involved in the upstream and downstream flows of products, services, finances, or information from a source to a customer. While in the past, supply chains were assumed to be largely linear, today's complex economic structures can rather be characterized as networks, including multiple dependencies and feedback loops.
Supply chain management (SCM):	The systemic, strategic coordination of functions and processes as well as the tactics across these functions within a particular organization and across organizations within the supply chain, for the purposes of improving the long-term performance of the individual organizations and the supply chain as a whole. Note that this definition of *good* SCM will depend on the choice of performance measures. In **humanitarian relief SCM**, the objectives, performance measures, and preferences differ from business SCM. In the humanitarian field, the aim of SCM is to provide humanitarian assistance in the forms of food, water, medicine, shelter, and supplies to areas affected by large-scale emergencies.
UN:	**United Nations**
UNDP:	**United Nations Development Programme**

UNHCR: **United Nations High Commissioner for Refugees**
UNICEF: **United Nations Children's Fund**
UWLP: **Uncapacitated warehouse location problem**
WFP: **World Food Programme**
WHO: **World Health Organization**

BACKGROUND

Characterization of Decision Situations in Emergency Management

Emergency situations, which can be man-made or natural, require a coherent and effective management. Decision-makers in these situations have to consider many conflicting objectives and must set priorities while the various perspectives of many stakeholder groups must be brought into some form of consensus [29,42]. Along with the rise of modern information and communication technology (ICT) systems, information that stems from various sources is available and can provide help and guidance to emergency management authorities [100,120]. The information available is, however, very heterogeneous in terms of format, quality, and uncertainty—or even completely lacking [28,119]. Additionally, the emergency itself as well as the information about it may evolve dynamically. Nevertheless, actors are under pressure to make their decision quickly, which may cause cognitive overload and biases to occur [2,30,66].

Trends such as globalization, urbanization, and migration, as well as the rapid development of ICT systems, lead to a high degree of integration and coupling of systems. This is particularly true for critical infrastructure (CI) networks whose continuous functioning is required to provide basic necessities such as food, water, and medicine during an emergency [75]. The possible collapse of the physical infrastructure networks and social service infrastructure that are provided and operated by diverse organizations and authorities raises fundamental questions of responsibility and collaboration: That is, which agency is responsible for which type of process or decision? How can the plans of different organizations be aligned while taking into account their respective goals to achieve compliance?

A difficult aspect of such large, highly integrated systems is that the effects of single events can have dramatic consequences that propagate rapidly and widely through the (globalized) society [80]. Decisions to control, manage, and steer these networked systems are difficult. In the domain of policy assessment and socioeconomic systems, the term *wicked problem* has been coined to refer to problems that are characterized by their fundamental uncertainty about the nature, scope, and dynamic behavior of the underlying systems [106,115]. Wicked problems are difficult to solve because the consequences of actions at individual, organizational, and societal level are unclear; causal relations are numerous, hard to identify, and even harder to quantify.

Supply Chain Risk Management

SCM aims at efficiently aligning demand and supply by organizing the flows of goods, capital, and information to provide a high level of product or service availability to the customer while keeping costs low. Global supply chains and lean and just-in-time production are a source of competitive advantages in the sense that they enable delivering what is needed and where it is needed at a minimum cost or as quickly as possible.

Coupled with these benefits are, however, the increasing complexity and uncertainty associated with the multitude of parallel physical and information flows to be managed. It has been argued that along with the growing maturity of SCM, the vulnerability and consequently the risks that managers face increase [112]. Although supply chains are usually designed to be robust to small perturbations (variations of several standard deviations) [48], disruptions may be caused by extreme events (e.g., earthquakes, floods), whose frequency and severity have increased considerably in recent years [58]. Coupling effects are prominent: while natural hazards may increase the need of supplies rapidly, supply chains are affected in terms of destroyed infrastructures (warehouses, production sites, transportation networks) or due to indirect impacts that result from the impairment of material and information flows [69].

As foreseen by Perrow [79], the modern world's interconnectedness increases the worldwide vulnerability to many risks and global system failures, for example, pandemics, terrorism, and financial crises. Due to the growing vulnerability of societies and the increasing frequency of disasters [15,34], the importance of risk in SCM has increased recently [3,31,56]. There is, however, no standard definition of best practices yet [111]. Rather, there is a plethora of conflicting and ambiguous definitions of the underlying concepts, such as risk or vulnerability. Likewise, there is no consensus about the ways to manage and reduce risks to or stemming from modern supply networks. Often, techniques from (financial) risk management are adopted [49,56], and classical decision-analytical techniques are applied, both of which do not account for the particular characteristics of supply chain risk: risks are always negative, a hedging and limitation of risk is often not possible, and, as supply chains grow more and more interlinked, there is an increasing need for collaboration. At the same time, particularly for disaster risks, uncertainties are severe (i.e., not quantifiable) [9,28]. Probabilistic techniques that are often used to solve decision problems cannot be applied in these settings [29,118]. As the uncertainty is most obvious and unquestionable in humanitarian relief supply chains, we believe that business supply chain risk management can benefit from the considerations presented in this chapter to be better prepared for the increase in natural hazards and supply chain disruptions.

NEED FOR DECISION SUPPORT IN HUMANITARIAN RELIEF SUPPLY CHAINS

The objective of logistics and SCM in humanitarian relief supply chains is to provide assistance in the forms of food, water, medicine, shelter, and supplies to areas affected by large-scale emergencies. Particularly in the early phases (usually the first 72 h), relief goods and services need to be supplied as quickly as possible in order to minimize human suffering and death [5]. Still, inefficient procedures lead to a waste of resources that could have helped further people if they had been used appropriately. Hence, the need for humanitarian relief SCM that is both efficient *and* effective arises [5]. Figure 7.1 shows the situation with respect to medical care, the aspect on which this paper focuses. As most infrastructures are destroyed, tent hospitals need to be set up in adequate places, and transportation of equipment and pharmaceuticals to these places needs to be organized. Moreover, specific requirements such as cooling of some pharmaceuticals or the fragility of instruments need to be respected.

Problems in humanitarian relief supply chains relate to uncertainty and complexity and challenges like short lead times, unpredictable demands, and disrupted or destroyed logistic infrastructures, and lacking or ambiguous information need to be overcome. Furthermore, various stakeholders with different norms, goals, and value judgments are involved in the planning and execution of alternate supply chains [37].

Unlike most business supply chains, humanitarian relief supply chains emerge ad hoc and are evolving dynamically. Natural hazards typically turn into disasters because they are unexpected shocks [97]. As immediate response is required, supply chains need to be designed and deployed at once even though the knowledge of the situation is very limited. Hence, the focus during disaster response lies less on the optimization of relief supply chains and more on their formation, establishment, and stabilization [37].

Diverse actors need to combine their capacities and capability ad hoc to respond to the disasters, although they typically have conflicting interests and do not usually share plans, procedures, or communication protocols [20,99]. Actors comprise governmental and public authorities, military, the civil society, and various NGOs. Often, individual organizations do little to coordinate their efforts while simultaneously allocating their own resources inefficiently due to insufficient information [60]. In summary, the supply chain can be disrupted due to faulty assessments of the

FIGURE 7.1   Grace International tent hospitals in Grace Village, Haiti. (From Grace International, Progress in Haiti, http://graceintl.files.wordpress.com/2010/04/tents-from-above-hospital.jpg. Accessed 20 September 2013.)

population's needs or infrastructure capacities, but it may also be unstable for a lack of coordination or failures in procurement [59].

In both business and humanitarian relief SCM, aims are multifold. However, the importance of goals such as quality, specificity, and cost is less relevant in humanitarian relief SCM that focuses on time, place, and quantity. After the immediate response (ramp-up), most organizations focus on implementing their programs, and cost and efficiencies gain importance [99]. Despite the target modification in humanitarian relief SCM, costs are not negligible. By keeping costs low, organizations and governments can supply aid to a higher number of people. As around 80% of emergency management funding is spent on logistics, any decrease in cost leads to the ability to provide aid for more people assuming donations are stable [1].

Supply chains need to be adaptive and agile to manage humanitarian relief logistics: they must be designed to enable quick response and delivery. Requirements for responding quickly and adequately are valid logistic information, stable and reliable ICT systems, and the security of transportation infrastructures. Furthermore, an effective and efficient coordination between different organizations and countries is essential. This need is located along the supply chains.

HUMANITARIAN RELIEF LOGISTICS FROM A COMPLEX SYSTEMS PERSPECTIVE

TYPE OF SYSTEM

The number and diversity of actors and organizations need to be coordinated in humanitarian relief logistics. Even groups that are seemingly similar, such as NGOs, vary to a great extent: they comprise organizations with different focus (e.g., housing and shelter, water and sanitation, pharmaceuticals) and donors and often compete for public attention [20]. Better coordination of resources and activity flow will increase the accountability, effectiveness, and impact of aid [72]. Due to the sheer number of organizations and the range of resources and activities they offer, the difficulties in coordination have been viewed as a systemic problem for the humanitarian aid [59]. Additionally, the socioeconomic that is affected by the disaster is in itself complex, and the short- and long-term impact of the disaster or the impact of relief operations is hard to predict or assess, even in hindsight [97]. More precisely, complex humanitarian relief supply chains share the following characteristics [41,47,51,54,94]:

- A *large number of interrelated components*, each of which can be understood as an individual system (really, system of systems [SoS]); the behavior of each individual system (characterized, e.g., by the behavior of various actors and groups and their response to the developments) is needed to model and understand the overall SoS; each (sub-)system is in itself characterized by individuality and diversity.
- *Information, knowledge,* and *interactions* are *localized* (individual and specific), and relations between and within the systems can be nonlinear.
- *Emergence*: Local controlling processes create and maintain self-organization (spontaneous coordination and synchronization); in this sense, complexity creates (emergent) order and disorder.
- *Impossibility of accurate prognoses*: Although the system has a history and is retrospectively coherent, hindsight does not necessarily lead to foresight, and equal initial conditions will not necessarily result in equal end states.

For these reasons, the underlying system can be described as a complex system. Additionally, due to the lack of a central managing authority, the system's behavior emerges; mechanisms to steer or even predict the behavior are hard to implement and even harder to control.

System Maturity

When a disaster occurs, humanitarian relief supply chains need to be constituted ad hoc. Coordination is difficult due to the diversity of interests of the different organizations who need to participate. In few cases, it has even been reported that some NGOs failed to report their presence and activities to governmental authorities [72]. Additionally, context and needs, infrastructure disruptions, availability of resources, cultural aspects, and constraints vary from disaster to disaster.

Due to the diversity of disaster situations, the problems occurring, and organizations involved, there is a lack of understanding, trust, and cross-organizational cooperation. Although some attempts to achieve better coordination by common disaster response planning have been made, there is still room for substantial improvement [20,99]. The lack of preparedness and interorganizational training and the ad hoc setup of networks and operations from scratch for each disaster (i.e., the organizational novelty) contribute to a large extent to losses that could be avoided if plans were aligned and common systems and infrastructures and continuous collaboration between organizations were established [99]. The flaws in supply chain coordination have been attributed not only to conflicting interests but also to the aid sector's regard for logistics as a necessary expense (rather than an important strategic component of their work), the lack of depth in operational knowledge (due to high employee turnover), and the lack of investment in technology and communication [6,7].

Environment

Disasters are characterized by their sudden onset and the lack of information about the systems' current state, their future developments, and the underlying cause–effect chains [48]. Humanitarian relief supply chains therefore operate in an environment that tends to be complex or chaotic [41].

In such an unstable and dynamically evolving environment, forecasts are hard or impossible to make. Firstly, it is often unclear if the disaster's cause (e.g., a natural hazard) will be unique or if it will cause further natural or technological hazards (such as aftershocks or release of hazardous matter). Secondly, the behavior of the socioeconomic systems that are affected is hard to predict. Even for highly developed and organized countries such as Japan, the uncertainty about individual and organizational behavior in the aftermath of the Tohoku earthquake was overwhelming [16]. Thirdly, due to the growing dependency of modern societies on CI networks (such as electricity supply or ICT networks), the indirect consequences of disasters have increased [58,69]. While in the past, the impact of a disaster used to be contained locally, in today's globalized world, the impacts propagate through highly interconnected networks [83]. Due to the inherent complexity of these networks [61], the consequences on specific infrastructures or regions are hard to predict [97].

Despite the uncertainty that makes it hard to assess the impact of any decision in this context, the high stakes and time pressure require that decisions are made near real time, forestalling the possibility of further investigations or waiting until more or better information is available.

To analyze the status quo of the most frequently applied decision support systems in this context from a systems engineering perspective, we need to distinguish the formal models applied in operations research (OR) from the techniques used in humanitarian relief. OR methods aim at predicting the system's future states and to optimize the strategy with respect to an objective function. In other words, OR approaches refer mostly to existing knowledge, technologies, and algorithms to solve the decision problem. NGOs and public authorities refer often to best practices and guidelines. Additionally, they try to integrate the specific context and cultural background. In this sense, they tend to emphasize the uniqueness of each disaster and aim at developing strategies that are tailored to the respective situation. The latter can lead to inefficient and even ineffective planning. Contingencies and uncertainties are mostly modeled in terms of probability distributions in OR [7,85]. In humanitarian relief operations, uncertainty is typically viewed as an inherent qualitative aspect of the systems that is hardly predictable [118].

In summary, OR tools and methods help decision-makers in well-defined environments to plan optimal (i.e., efficient and effective) strategies. They usually do not take into account the sociocultural context, the goals and preferences of all actors and stakeholders involved, or the fact that severe uncertainties or residual risks that cannot be quantified may have an important influence. Contrarily, humanitarian relief operations are usually exploratory in the sense that ad hoc networks are built that are—if possible—successively adapted (e.g., when new information is available). The lack of planning and preparedness may, however, lead to inefficient and ineffective solutions implying that resources that could be used to save a greater share of the population or to provide help quicker are not used appropriately [109].

The system we present in this chapter aims at combining methods from both realms to acknowledge for the facts that complexity, severe uncertainties, and the dynamics change what is perceived as the *optimal* disaster relief strategy from disaster to disaster as well as over time within each single disaster. Yet our approach aims at providing humanitarian aid in an efficient and effective manner by enabling decision-makers to choose *robust* strategies that perform well under a variety of *scenarios* (i.e., plausible future developments). Additionally, the strategies should be flexible in the sense that they can be (quickly) adapted to the problem at hand.

CONTEXT OF THE COMPLEX SYSTEM

In the following, the background of our case study is summarized with reference to Figure 7.2:

A: *Complex system*—The use case focuses on humanitarian relief supply chains in a complex (or even chaotic) environment that is characterized by numerous, diverse, interrelated, and individual components—SoS. These systems involve, for instance, multiple and diverse actors that need to be coordinated and affected CI networks. To demonstrate our approach, we consider the problem of warehouse locations after the Haiti earthquake 2010 as they play a significant role in designing supply chain networks. Our use case particularly addresses

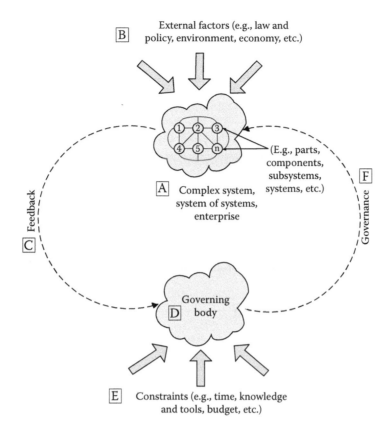

FIGURE 7.2 Context of the humanitarian relief supply chain complex system.

the lack of information about supply and demand and uncertain and dynamically changing information about the disaster environment, infrastructure network, logistic strategies, available resources, and population needs due to the disaster.

B: *External factors and constraints* refer to the nature, scope, and dynamic behavior of the underlying systems. In our use case, the initiating even was an external shock that disturbed the (complex) system (A). Moreover, the development of the systems themselves depends heavily on policies and regulations and interventions of governments or NGOs that need to be coordinated. Additionally, aftershocks or epidemics can aggravate, and in the case of Haiti have aggravated, the situation. Constraints are set by the reaction of donors and organizations providing disaster relief or recovery aid. While in the first—most spectacular—days after the earthquake there was massive media coverage and an unprecedented willingness to help and volunteer, the situation in Haiti is still far from resolved. The nature of the problem has changed from a sudden onset event to a creeping disaster.

C: *Feedback*—We aim at providing decision support to actors in humanitarian relief logistics such as assistance in the forms of food, water, medicine, shelter, and supplies to areas affected by large-scale emergencies. Therefore, it is crucial to monitor the situation and to measure the actual consequences of the decisions made. If they deviate too much from desired or required goals, corrective actions need to be implemented. The formal decision-analytical model that operationalizes vague goals and enables the definition of performance indicators facilitates this. The use of scenarios favors flexible solutions that can be adapted to changes in the environment or the information about it.

D: *Governing body*—Multiple actors such as governmental and public authorities, military, the civil society, and various NGOs operate in humanitarian relief logistics and need to combine their capacities and capability ad hoc to respond to the disaster. The involved actors vary in their focus, donors, and public attention. Moreover, they are characterized by multiple norms, goals, and value judgments. Hence, it is essential to coordinate resources and activities between and within those groups to increase the overall impact of aid.

E: *Constraints for the governing body*—The involved actors need to make decisions near real time under time pressure without the possibility of further investigations or waiting until more or better information is available. Conflicting interests, the lack of depth in operational knowledge, and the lack of investment in technology and communication flaw supply chain coordination.

F: *Governance*—Most actors have headquarters that advocate humanitarian action, raise funds, and ensure adherence to standards. Cooperation and collaboration with those actors are essential for decision support to gain detailed information about the disaster environment as well as about their goals and preferences. Based on the strategic decisions provided in the feedback loop, the actors are reasonable to implement the warehouse locations.

USE CASE: THE HAITI EARTHQUAKE

NEED FOR HUMANITARIAN RELIEF SCM

On January 12, 2010, at 16:53 local time, a severe earthquake of magnitude 7.0 hit Haiti. The earthquake's epicenter was about 17 km west of the country's capital, Port-au-Prince [116], where about 25% of the Haitian population had lived before the earthquake [121]. This external shock (B) caused an extremely severe human toll due to its location. Additionally, the earthquake was followed by about 70 severe aftershocks, two of them with magnitude of 6.0 or higher and sixteen with magnitude of 5.0 [105].

The impact on the underlying complex system (A) was unprecedented: the shocks caused further significant losses in terms of human life (between 200,000 and 300,000 or about 15% of the population of the Port-au-Prince agglomeration [14]), the displacement of about a million inhabitants [36], and severe damage to the country's economy [82]. Assessments of the direct economic consequences vary from US $8 mio to 14 mio [22]. The damage to CIs such as communication, electricity supply, transportation, food and water supply, and health-care services led to severe indirect impacts [108].

Most importantly, from a logistic point of view, Haiti's transportation infrastructure, which had already been in poor condition before the earthquake, was seriously damaged. The main routes connecting Port-au-Prince with the Dominican Republic were destroyed [53]. Minor roads are usually unsurfaced and mountainous [101], such that on the whole the aid goods could not be distributed or transported by land in large parts of the country. Beyond the difficulties of distributing aid goods in the country, also, the supply of goods to Haiti was hindered as Port-au-Prince's airport control tower and harbor were severely damaged [4].

Governance (F) was constrained by lacking information and capacity constraints (E). In the days following the earthquake, access to food and water became increasingly difficult, particularly in the areas directly affected. There was an immediate sharp rise in staple food prices, and by the end of January, the price of wheat flour had risen by nearly 70%, local maize and black beans by 30%–35%, and imported rice by 20%–30% [25]. Due to insufficient supplies, more than 660,000 people left the areas most destroyed by the earthquake and fled to more rural areas, increasing the burden on those regions that were already food insecure [67].

Together, these developments resulted in security problems. However, the military succeeded in restoring airport operations, and joint police forces were able to keep looting and violence largely under control [82]. As an important part of the initial emergency management, effort was dedicated to security reinforcement. The distribution of aid goods like food, water, and health care was slowed down further [101].

As time passed, the efforts of donors led to large volumes of goods that arrived in the main harbors and airports. Due to the disrupted infrastructures, a lack of coordination, and mistakes in planning, however, most food supplies could not be used for months. Most prominently, by the end of March, the government called to cease general food distributions due to their perceived negative impact on Haiti's agricultural economy [25]. At the same time, due to a lack of storage capacity, it became impossible to import further critical goods such as pharmaceuticals [82].

Moreover, there was not one unique governing body (D). In general, disaster response requires close coordination across the various agencies and institutions [18]. Yet, in the case of Haiti, coordination was even more complex than usual due to the sheer number of actors involved and their diversity, the role of the media, and competition between several countries and organizations. Even before the earthquake, 8000–9000 NGOs with different aims, budgets, resources, and sponsors had worked in Haiti [121]. More than 1 week was needed in order to solve the question who should lead and coordinate disaster relief operations as a basis to start an effective distribution of aid [101].

In summary, the direct and indirect consequences of the 2010 earthquake were disastrous: the death toll was high, and destroyed infrastructure as well as a lack of coordination and planning hampered the efficient distribution of disaster relief goods and services. In the following, we will investigate in more depths how the logistics system was planned and managed, before proposing our approach for robust humanitarian relief logistics.

HUMANITARIAN RELIEF LOGISTICS, SCM TECHNIQUES, AND KNOWN PROBLEM(S)

The United Nations (UN) managed humanitarian relief logistics after the Haiti earthquake 2010 and in this sense complemented the local governance (D). To facilitate coordination and make responsibilities and competences transparent, different processes and tasks were assigned to specific *clusters* ranging from camp coordination management, early recovery, education, food, health, logistics, nutrition, to shelter and nonfood items (NFIs) plus water, sanitation, and hygiene [77]. Ten UN operating agencies like the United Nations Children's Fund (UNICEF) or the World Health Organization (WHO) were involved to humanitarian relief logistics after the earthquake. Furthermore, 200 national organizations, 195 international NGOs, and the Red Cross movement took part in humanitarian management tasks [78]. The intercluster coordination (ICC) was established under the leadership of the Office for the Coordination of Humanitarian Affairs (OCHA) to align organizations and manage the cooperation with national and local authorities. The ICC was supported by the Humanitarian Country (HC) on response and preparedness activities [103]. The unprecedented amount of donations and, consequently, the enormous influx of international NGOs—some of them very small and each with their own agendas—complicated the coordination of humanitarian aid [76,104].

The UN World Food Programme (WFP) managed food supply after the earthquake. To organize distribution of food, the WFP relied partly on existing warehouses and established some further centers in the form of jumbo tents or mobile storage units. Those warehouses were located (ordered by their capacity) in Port-au-Prince, Gonaives, Cap-Haïtien, Jacmel, Leogane, and Les Cayes. They supplied the population with cereals, pulses, vegetable oils, and salt [117]. Due to the severely damaged transportation infrastructure (e.g., destroyed roads or bridges), it was impossible to supply all regions, and additional warehouses needed to be established in Port-de-Paix, Les Cayes, and Jacmel. Despite the immediate need, operations of these warehouses could only start by the end of July 2010 [117]. Hence, the earthquake and associated problems in humanitarian relief logistics led to a hunger crisis in large parts of Haiti.

Another challenge was the supply of pharmaceuticals and medical equipment. In Haiti, the supply of this equipment to all public health facilities and NGOs has been managed by the system Programme de Médicaments Essentiels (PROMESS) with a central warehouse located approximately 8 km away from the airport of Port-au-Prince [45]. The earthquake destroyed roads around the warehouse. Additionally, communication between the warehouse, health facilities, and NGOs was disrupted [45]. The main concern, however, remained the lacking preparedness to accommodate the drastic increase in demand that led to logistical problems [45].

CONSTITUENTS

Typically, three types of organizations provide professional humanitarian relief after a natural disaster: the UN manage the relief measures, the UN's assistance agencies are responsible for different fields of humanitarian aid (see below), and international organizations and NGOs participate in assistance activities [7,98].

UN OCHA leads the strategic management. Their tasks encompass establishing security agreements among different agencies, division of responsibilities, and policy development and coordination. Furthermore, there are several UN operating agencies whose tasks are related to specific aspects of relief assistance; examples are the United Nations Development Programme (UNDP), UNICEF, the United Nations High Commissioner for Refugees (UNHCR), and the Food and Agriculture Organization (FAO) [102].

NGOs are humanitarian assistance organizations that act independently of governmental control [7]. They are active within both relief and development efforts. Relief activities have to be handled in large-scale emergencies and are usually short term like the supply of goods [7]. Development activities are longer term, for example, the formation of new permanent and reliable transportation infrastructures [7]. NGOs can be classified into two types: nonprofit or for-profit organizations. Whereas for-profit organizations may use the revenues they earned from customers paying for goods to their own benefit, nonprofit organizations are affected by governmental limitations that dictate how revenues may be used [7]. For-profit organizations try to increase their revenue, whereas nonprofit organizations pursue the goal of reaching a financially balanced budget in order to be successful in their missions [7].

Additionally, there is a plethora of stakeholders with diverse interests. Local or national governments of the disaster-affected country are typically responsible for relief operations. As they may have no experience and knowledge to manage measures effectively [4], they frequently turn to military. Both national and international military are involved in disaster relief measures as they have high competences in supply chain deployment [4], including logistical and material resources like fuel or medicine [46]. Furthermore, they are experienced in the setting up of communication networks and the providing of logistics assets and information on infrastructure and security. Although military focuses on security reinforcement and thus differs from other relief organization's aims, their cooperation is essential [4]. Besides government and military, commercial companies and economy in general are involved into disaster relief operations. As natural disasters go along with a high demand of goods, humanitarian relief is a big profit market. Hence, commercial companies operate most commonly as supplier and transportation providers [4].

The interaction of all stakeholders is characterized by a high complexity. In essence, there are three groups of relief chains: international relief actors, local relief actors and institutions, and the private sector [4]. The coordination of those three groups, including their entities—humanitarian organizations, NGOs, governments, military, and companies—is the basic challenge of successful humanitarian relief logistics.

(COMPLEX) SYSTEM

Our aim is to develop systems and approaches that provide decision support in complex and highly uncertain situation. To this end, it is necessary that the solutions account for the uncertainties present and facilitate consensus building and coordination among different organizations. While the first goal is achieved by a scenario-based approach that enables fundamentally different developments to be explored in a systematic manner, we use techniques from multicriteria decision analysis (MCDA) for the latter.

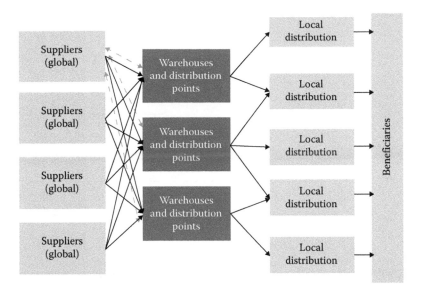

FIGURE 7.3 Schematic representation of a disaster relief supply network. (Adapted from Balcik, B., *Int. J. Prod. Econ.*, 126(1), 22, 2010.)

To demonstrate our approach, this case study focuses on the problem of warehouse location, which sets the framework for all logistics operations and can therefore be considered as the basis for efficient distribution of resources and relief goods. Decisions of warehouse locations, or more general facility locations, play a significant role in designing supply chain networks. Beyond the geographical location, decisions typically include the number and capacity of warehouses. Therefore, warehouse location decisions are of strategic nature, that is, they have a long-term effect and determine the material flow within the supply network to a great extent (e.g., lead times, capacities, accessibility) [34].

Figure 7.3 highlights the prominent role of the warehouse location in disaster relief logistics. Warehouses act as the intermediate buffer between incoming goods (and should be easily accessible and provide sufficient capacity) and the local distribution (that should be fast and correspond to the needs of the affected population). Additionally, Figure 7.3 illustrates the coordination problem. Various organizations and actors will deliver and distribute supplies without necessarily being aware of what is delivered or required by further organizations or the beneficiaries. While the flow of goods is directed (highlighted by black edges in Figure 7.3), the flows of information are bidirectional (highlighted by gray dotted arcs, only shown for one supplier for clarity's sake). Depending on the information systems used and the degree of coordination between the parties, warehouses may be the focal point for exchange of information: here, the supplies available and the demand from the local centers need to be matched. If there are no common IT systems, short lead times are even more important than usual, as they imply—in that case—also better flows of information and are a requirement for quick and adequate response to newly discovered needs.

The efficient and effective supply of relief goods in Haiti was particularly difficult due to the lack of information about (available) supply and demand (required goods), combined with uncertain information about the infrastructure network.

This case study focuses on the aim of satisficing the demand completely. Changing infrastructure status, needs of the population, available resources, and corresponding logistic strategies are the main challenges we address. To make the decision where to place the warehouses, information is required about the available resources (in terms of supplies and transportation capacities), the infrastructure capacities, the development of demand over time, and the desired number of distribution warehouses.

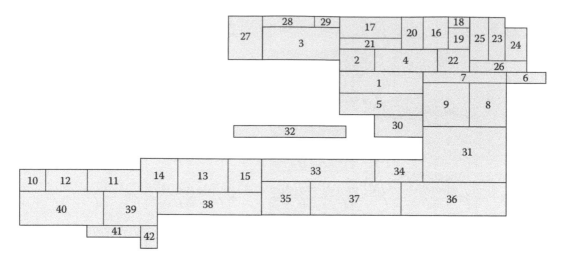

FIGURE 7.4 Schematic representation of Haiti and the regions. Details about the estimated needs in each region can be found in Table 7.1.

Haiti is divided into 10 départements with 42 arrondissements (called *sections* in the following). Figure 7.4 shows an abstract structure of Haiti. To deliver supplies from one region to another, one can refer to road or air transportation. Since there is only little and heterogeneous information available about alternative transportation modes like water or railways, the latter are not considered here, but can, in principle, be added to the model analogously to the framework for road and air.

On the basis of the inland infrastructure network (roads and airports), the expected durations for intact infrastructure between Haiti's sections are calculated with the help of navigation and GPS data. The underlying assumptions are that the infrastructure network is largely intact and that the most important areas within each section are the cities. Therefore, we use transportation time between cities as a proxy for the overall time to transportation goods from one section to another. Specific infrastructures within a section are not considered separately.

This case study focuses on health care. To provide a background, the general medical situation in Haiti before the earthquake hit in 2010 is presented: in 1998, 0.25 physicians per 1000 people were registered officially [44], whereas on average (2000–2009), there are 2.5 physicians per 1000 people for the Americas [44]. Compared to Europe with an average of 3.3 physicians per 1000 inhabitants [44], about 30,000 additional physicists would be needed to reach the European health-care standards. We use this baseline demand, which is a challenge for the development of the country. This baseline lack in health-care services is a particularly important challenge after a disaster, when the demand for health care increases significantly while at the same time the health care system itself is affected and only partially functioning.

To accommodate the need for additional health-care services, a decision has to be made how many additional health-care facilities should be set up and where they should be located such that a better health-care service level can be achieved in an efficient manner (i.e., with minimum waiting and traveling times). To assess the additional health-care demand, we use the difference to the European health-care level (note that more care may be required in for disaster relief and recovery) and population data.

After the earthquake, there were important changes in the geographical distribution of the population: there were considerable migration flows from the Port-au-Prince region to other, more rural départements. For instance, more than 180,000 people moved from Port-au-Prince to seek refuge in the Artibonite and Central regions, typically referring to friends or family, and thereby increasing stress on the small, overcrowded communities that mostly lack access to latrines and clean water [113].

To account for the additional needs for health-care and medical services in these regions, we refer to a study that used the movements of two million mobile phone SIM cards to assess the population distribution after the earthquake [11]. The result shows that approximately 570,000 people left the Port-au-Prince region, and more than 100,000 moved to the départements Sud and Ouest, respectively, whereas the fewest movement was to the département Nord-Est. Table 7.1 shows the additional health-care demands for both situations, before the earthquake 2009 and after the earthquake 2010, and summarizes Haiti's demand for additional health care (measured in service units) per department in 2009 ($demand_1$) and after the earthquake 2010 ($demand_2$).

Note that the demand will also evolve over time, particularly with the outbreaks of epidemics such as cholera due to the poor hygienic conditions in both camps for displaced people and rural regions [113]. To represent this dynamic evolution of the situation, all sections have a special health-care demand in each time step (immediate response, relief, recovery).

For the purpose of this chapter focusing on disaster relief and recovery, we restrict the considerations to the disaster-related temporal health-care demands (e.g., tent hospitals). This case study does not address further development assistance such as the construction and equipment of permanent hospitals.

The most influential external factors are uncertainties in the disaster's impact on the population (demand of health-care services and regions where this demand is realized), the infrastructure (e.g., available goods and health-care services, disruptions of main routes), and further environmental developments such as potential aftershocks. Cascading effects can occur increasing the impact on sectors and regions that were not initially affected by the earthquake itself. An example is the migration of the population to regions that are very poor, increasing dangers of malnutrition or the spread of diseases due to poor hygienic conditions. A particularly relevant problem is the disruption of transportation infrastructures, which obstructs the efficient exchange of goods and services between and within regions. For budget and efficiency reasons, the number of warehouses is limited to a number significantly below the number of regions. Therefore, isolated or separated sections can cause severe difficulties in satisfying the population's demand.

The following use case investigates the location of warehouses to distribute goods and services in an efficient and effective manner. To account for the supply of disaster relief goods that typically will be delivered to Haiti by ship, warehouses are preferably located in sections that provide direct access to the coast. To model the distribution of disaster relief aids within Haiti, our considerations of the road network are limited to larger tarred roads like highways or interstates, as those are the only roads traversable by trucks.

CHALLENGES

The main problems in humanitarian aid logistics are uncertainty and complexity. Our aim is to support emergency management authorities facing these situations. To this end, it is necessary that the solutions account for the uncertainties present and facilitate consensus building and coordination among different organizations. While the first goal is achieved by a scenario-based approach that enables fundamentally different developments to be explored in a systematic manner, we use techniques from MCDA for the latter. To demonstrate our methods and approaches applied and analyze their impact, this case study is divided into three steps: (1) problem structuring and definition, (2) application for Haiti, and (3) assessment of results.

Coordination in humanitarian aid is one of the most important problems due to lack of trust or even competition among various organizations and stakeholders [4]. Particularly, NGOs have no formal arrangements to promote coordination, both within and across NGOs [20]. On the strategic level, most NGOs have headquarters that advocate humanitarian action, raise funds, and ensure adherence to standards. To make the advantages of cooperation and collaboration transparent—a basic requirement for willingness to cooperate—decision support systems that combine information and preferences from various sources and actors can be helpful [39,68,90].

TABLE 7.1

Population Distribution and Estimated Health Care Needs in Haiti in 2010 [56]

Département		Arrondissement	Population 2009	Demand₁ (Service Units)	Population January 1, 2010	Demand₂ (Service Units)
Artibonite	1	Dessalines	375,499	1145	395,576	1207
	2	Gonaïves	411,692	1256	433,705	1323
	3	Gros-Morne	209,471	639	220,671	673
	4	Marmelade	171,485	523	180,654	551
	5	Saint-Marc	402,873	1229	424,414	1294
Center	6	Cerca-la-Source	108,906	332	115,165	351
	7	Hinche	240,939	735	254,786	777
	8	Lascahobas	153,401	468	162,217	495
	9	Mirebalais	175,380	535	185,459	566
Grand Anse	10	Anse-d'Hainault	89,597	273	100,116	305
	11	Corail	119,643	365	133,690	408
	12	Jérémie	216,638	661	242,072	738
Nippes	13	Anse-à-Veau	139,721	426	163,045	497
	14	Baradères	42,797	131	49,941	152
	15	Miragoâne	128,979	393	150,510	459
Nord	16	Acul-du-Nord	117,456	358	122,242	373
	17	Borgne	106,220	324	110,548	337
	18	Cap-Haïtien	324,572	990	337,796	1030
	19	Grande-Rivière-du-Nord	58,759	179	61,153	187
	20	Limbé	96,580	295	100,163	305
	21	Plaisance	112,429	343	117,010	357
	22	Saint-Raphaël	154,479	471	160,773	490
Nord-Est	23	Fort-Liberté	55,139	168	56,524	172
	24	Ouanaminthe	133,214	406	136,560	417
	25	Trou-du-Nord	104,582	319	107,209	327
	26	Vallières	65,342	199	66,983	204
Nord-Ouest	27	Môle-Saint-Nicolas	223,340	681	232,438	709
	28	Port-de-Paix	306,149	934	318,621	972
	29	Saint-Louis-du-Nord	133,288	407	138,718	423
Ouest	30	Arcahaie	180,564	551	196,358	599
	31	Croix-des-Bouquets	431,789	1317	469,558	1432
	32	Gonâve	79,188	242	86,115	263
	33	Léogâne	463,140	1413	503,651	1536
	34	Port-au-Prince	2,509,939	7655	1,940,939	5920
Sud-Est	35	Bainet	123,491	377	149,679	457
	36	Belle-Anse	143,760	438	174,247	531
	37	Jacmel	308,042	940	373,367	1139
Sud	38	Aquin	198,091	604	211,864	646
	39	Les Cayes	314,903	960	336,797	1027
	40	Chardonnières	71,305	217	76,263	233
	41	Côteaux	53,306	163	57,012	174
	42	Port-Salut	67,155	205	71,824	219

In the first step, a complex problem is structured and further characterized. The model pursues the goal to set up a given number of distribution warehouses in the affected region in order to satisfy the overall health-care demand efficiently. Efficiency here comprises lead and transportation times (particularly in the early phases) as well as cost. The first aspects ensure that quick relief can be achieved and it becomes possible to flexibly react to changes in demand (e.g., due aftershocks or an epidemic).

The model has been implemented in a matrix laboratory (MATLAB) as an uncapacitated warehouse location problem (UWLP) that minimizes the sum of durations (time until demand can be satisfied) and costs related to the warehouse locations (transportation and fixed cost) [32]. In fact, three variables, (1) time to satisfy the demand, (2) warehouse locations of tent hospitals, and (3) information regarding the transport infrastructure, are defined. The first variable characterizes the total time (h) that is needed to supply all people with health care by using various warehouse location possibilities. In the context of our case study, warehouses correspond to centers where health-care units are located, for example, tent hospitals. After a disaster, the demand of health-care services may be increasing, and thus, it is essential to set up tent hospitals within regions that ensure the reachability of all people. For the planning of best warehouse locations, information respective of the transport infrastructure status such as disrupted or damaged roads is crucial. Warehouses may be considered as focal point for the exchange of information and need to take into account dynamical changes of the disaster situation.

In general, any location on Haiti can be chosen as a location. Here, we require that the locations are chosen such that the demand can be fulfilled for all eventualities (dominating criterion, see section *Robustness*, *Flexibility*, and *Risk*). The standard UWLP assumes that the demand is stable and known, warehouses have infinite capacity, and information about transportation infrastructures is perfect. Furthermore, the distribution of one homogeneous good or service is analyzed. That means all warehouses are of the same type, and potential exchange between the warehouses is neglected.

These abstractions may seem unrealistic, yet they are important from a computational perspective: even this form of UWLP is Non-deterministic Polynomial-time hard (NP-hard), and it requires considerable effort to solve it numerically [35]. The advantage of using heuristics instead of exact solvers is the gain in flexibility in respecting uncertainty and integrating new information.

One distinguishes construction from improvement heuristics [35]. Within the first heuristic type, a solution is calculated from scratch. Examples are the ADD (greedy) heuristic or the DROP (stingy) heuristic [35]. The second type requires an initial solution as input that is improved successively by the algorithm. Examples for improvement heuristics are the ADD and DROP (interchange) heuristic and metaheuristics like neighborhood search or genetic algorithms [70].

In order to identify a suitable method, the trade-off between the required precision in the algorithm and the required flexibility in performing numerous calculations to cope with the uncertainties has to be made. As in our example, uncertainties in demand, infrastructure, and available resources and goods are overwhelming, such that precision and accurate predictions are unlikely to be achieved even with exact algorithms; we focus on the construction of scenarios to *explore possibilities* rather than predicting one exact optimum. Here, it is necessary to calculate large numbers of scenarios. Therefore, we chose one of the quickest options and chose the ADD heuristic.

We suppose that due to budget and resource restrictions, the number of warehouses is known. To calculate best locations for the warehouses given a description of the infrastructure status and demand, an impact matrix is required that describes the duration and the transportation costs of one section supplying any other location. The impact matrix is defined through the combination of durations between two sections by using one of the transportation modes, air and road, and the fuel costs for these modes. The results of one simulation encompass the total costs, the best warehouse locations, and the allocation of subregions to the warehouses.

The case study considers Haiti under changing environmental conditions within each of the disaster phases: early response, relief, and recovery. That means a series of scenarios has to be simulated in a short time. One of the major challenges is the geographical representation of infrastructure systems, resources, and needs, which is the basis for the warehouse location.

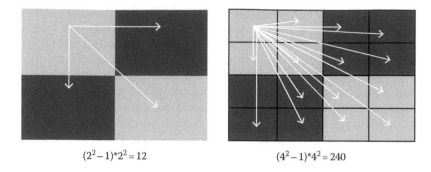

$$(2^2 - 1)^*2^2 = 12 \qquad\qquad (4^2 - 1)^*4^2 = 240$$

FIGURE 7.5   Exponentially growing computational power required.

Figure 7.5 illustrates the trade-off between more precise representation of a specific region and the growing effort of computation. While in the left representation the grid chosen is rather coarse, there are only three edges per subregion whose cost and duration need to be determined and are respected as potential links in the network by the computation. The right part of Figure 7.5 shows a more fine-grained resolution: each subregion was split into four. The result is that instead of performing calculations for 12 edges, now, 240 edges need to be considered. A more detailed grid requires also the edges within the regions to be more precisely known; for instance, information about the roads and airways need to be determined by analyzing navigation and GPS data. Due to heterogeneous and incomplete information—especially in road infrastructure—this requires important efforts, and, due to the dynamically changing environment, also, the effort for the continuous updating increases. Therefore, the trade-off between computational effort (and time) and precision needs to be made carefully: the abstraction level of the regional structure is essential in order to keep the problem solvable within an acceptable time. At the same time, the regions considered have to be small enough to provide meaningful decision support.

To ensure that calculations can be performed (relatively) quickly, we focus on the level of sections (cf., Figure 7.4) and identify the most appropriate sections to place the warehouses. The exact location within each section can subsequently be addressed with the help of local sources of information about the infrastructures (access to roads, airports, and harbors) and needs (highest concentration of beneficiaries, safety, and security).

After defining the grid as shown in Figure 7.4, demands (in terms of health-care service units required) are allocated to each section. As one of Haiti's sections is an island, the problem solution is enlarged by the need for handling separated sections.

Although the problem has been simplified, it is impossible to simulate all potential scenarios. Scenarios, ideally, cover all potential situation developments [26]. As the information is, particularly in the early phases, very uncertain or lacking completely, the number of scenarios is overwhelming. Additionally, due to the problem's complexity, it is impossible to model and simulate all interdependencies. In summary, decision-makers need to focus on a set of scenarios—ideally those that are the most relevant for the problem at hand [29]. The purpose of using scenarios is the identification of robust alternatives, that is, alternatives that have a good performance for a variety of scenarios.

DEVELOPMENT

INTEGRATING CONTEXT AND PREFERENCES: MULTICRITERIA DECISION SUPPORT

In situations with multiple objectives, MCDA has often been chosen as the basis for decision support for its coherent support in complex situations [40]. MCDA's popularity is mainly due to the intuitive and transparent evaluation of alternatives by refining evaluation goals in terms of a number of relatively precise but generally conflicting criteria, on which preference relations can be expressed [95].

A view shared by many MCDA practitioners is that one of the principal benefits from the use of this well-structured approach is the *learning* about the problem itself and the value judgments and priorities of all involved parties contributing to an increased respective understanding [8]. First, experts and decision-makers may want or need to explain the rationale for their decision to the public and thus to promote acceptance and trust. Second, stakeholders can be actively involved in the decision-making process. With the rise of the new communication technologies like mobile phones and the internet, the public has not only to comply with the decisions made, but it is as well a source of information [93]. Therefore, communication with the public becomes more and more important in emergency management.

Multiattribute decision making (MADM) forms the basis of our approach as it allows trade-offs to be explicitly represented. Multiattribute decision support systems provide aids to choose one alternative out of a (finite) list of feasible options respecting multiple criteria. To this end, the alternatives are ranked based on information about the situation and preferential information. Due to its success in strategic decision making [13,23], we use multiattribute value theory (MAVT) for the evaluation of alternatives. In MAVT, the decision process starts by structuring the problem taking it from an initial intuitive understanding to a description that facilitates quantitative analysis [110]. The problem structuring phase results in an attribute tree that hierarchically orders the decision-makers' aims. The attributes, at the lowest level of the tree, are means to quantitatively estimate the consequences arising from the implementation of any alternative [95]. The question how the consequences are determined is left to the decision-makers, and often, specific computational decision support aids are developed [50,52,74].

Although uncertainties are present during the whole process—in the data, parameters, computational models, and the preferences—standard MAVT assumes that all attribute scores are well defined and known with certainty [38]. An approach that addresses the uncertainty in data and preferences comprises sensitivity analysis [12,84], Bayesian approaches recurring to probabilistic models [38,114], or fuzzy logic to capture ambiguities and imprecision of natural language [65,88]. While Bayesian and fuzzy approaches represent the uncertainty of data and parameters explicitly, sensitivity analyses are ex post by varying one or several input parameters and compare the respective results (e.g., via simulation) [89]. All aforementioned methods are targeted at testing the stability of results to perturbations of the data and model parameters rather than exploring fundamentally different developments. Moreover, the validity of the models used is not questioned. Therefore, the applicability of the analyses in situations of fundamental uncertainty is limited.

Due to its transparency and ease of application, we use MAVT to evaluate the alternatives, which are, in our case, different warehouse locations. The key challenge is, however, the integration of multiple fundamentally different situation developments: scenarios. The next section gives a precise definition of the term scenario and describes scenario-based reasoning (SBR) processes.

SCENARIO-BASED REASONING

The growing importance of scenarios to support decision-makers in very uncertain situations is reflected by a vast amount of terminologies that has evolved around the concept: from scenario thinking and planning to scenario analysis, construction, and generation, a number of techniques with various purposes did arise [24,91]. Originally, scenarios were developed in response to the difficulty of creating accurate longer-term forecasts [57]. Today, in the simplest case, a scenario refers to the expected continuation of the status quo. Statements often encountered are "under the current scenario, we anticipate that…" [92,107]. Another common use refers to a set of values for exogenous factors that enter a (simulation) model and provide the background for the actual calculations [17,43]. In our understanding, one scenario is a well-structured account of one possible future. In essence, scenarios are dynamic stories that capture key ingredients of our uncertainty about the future of a system. To distinguish this way of using scenarios from alternative approaches, we use the term SBR.

SBR uses plausible scenarios to help decision-makers think about the main uncertainties they face and the drivers for change *before* they actually occur; in this sense, scenarios help decision-makers to reveal the implications of current paths and facilitate devising strategies and plans [71]. Multiple scenarios offer the possibility to take into account several situation developments, which are considered regardless of their likelihood. Therefore, SBR is particularly suitable for reasoning under severe uncertainty (i.e., in situations when the likelihood of a series of events cannot be quantified) and when creative solutions (*out of the box thinking*) are required [19].

ROBUSTNESS, FLEXIBILITY, AND RISK

Along with the increasing interest in scenarios and severe uncertainty, the concept of robustness has attracted more and more interest [9,21,27,29,33,64,81,86]. The key principle is that decision-makers do not try to find the optimal alternative for one (best guess) scenario. By acknowledging that uncertainties abound, they aim at selecting an alternative that performs (sufficiently) well for a variety of scenarios [10,65,109].

Standard optimization methods assume that the result of a decision can be determined with certainty (deterministic problem) or described by a probability distribution. The difficulty of deterministic problems consists mainly in the infinite (or very large) number of alternatives and results, which need to be compared and ranked [87]. An extension of this approach is provided by techniques for decision making under (probabilistic) risk. These approaches mostly assume that a best-estimate description of the problem exists. Uncertainties are modeled as sets of probability distributions over the model's input parameters [63] and prioritize the alternatives according to their expected utility (i.e., overall performance or *score*) [73]. Robustness has been described as a counterpart to optimization [87]. It can be understood in two ways referring to the stability or quality of results. When robustness is used to refer to the stability of results, it is a means to address the question how flawed or defective the models and data can be without jeopardizing the analyses' quality [9]. In the second sense, the focus is on the quality of an alternative: it is required that a robust alternative reaches a minimum required performance under all eventualities [109]. We follow an approach that combines both aspects: a robust decision aims at identifying an alternative that performs relatively well when compared to further alternatives across a wide range of scenarios.

Supply chain flexibility is, in general, intended to determine a supply chain's adaptive capacity [96]. Flexibility comprises agility (response to short-term changes in demand or supplies) and adaptability, which enables the accommodation of longer-term changes [62]. In summary, flexibility aims at designing supply chains that achieve a set of goals in an environment of continuous and hardly predictable changes. In the framework considered here, the aims are supplying health care to the population affected (fulfilling the demand) in minimum time and at minimum cost. Therefore, flexibility can be measured according to the achievement of the following three criteria:

1. *Time*: How long does it take until the demand is fulfilled for varying scenarios?
2. *Cost*: Which cost is implied by a disruption of the transportation infrastructure?
3. *Volume*: To what extent can the loss of planned supply quantities be compensated for?

These issues will be covered in the scenarios calculated. In this manner, robust and flexible humanitarian relief SCs can be created. The third criterion is of particular importance in humanitarian relief logistics and dominates time and cost considerations. Basically, it is crucial that supplies can be delivered to everybody, that is, that full compensation is possible for any (plausible) scenarios. If there are alternatives (warehouse locations) a_i that fail with respect to (3) and there are further alternatives a_j that enable supplying all regions, then all alternatives a_j are preferred over any a_i.

IMPLEMENTATION

The simulations for this case study have been implemented in MATLAB. Warehouse locations are calculated for various scenarios (S_i), each composed of an alternative (chosen warehouse locations), a development of the demand in each section, and one out of three environments E_i relative to the transportation infrastructure status. Note that the scenarios are actually determined in a dynamic manner: the optimal locations are determined by considering a first set of extreme and best guess scenarios and then tested for robustness and flexibility. To enable efficient scenario generation (targeted at identifying key weaknesses [30]), we adapt the environments such that the path immediately linking the areas where the warehouses are located to the surrounding sections or the most critical paths given the current warehouse location are disrupted.

Following the disaster management cycle, there are four time steps t_1–t_4 covering the situation at the outset, response, relief, and recovery phase [48]. The health-care demand is modeled for each of the phases. The underlying rationale is the following: the original health-care demand in t_1 is specified in Table 7.1. As we assume that there is no uncertainty about the initial situation, these are equal in each scenario. In t_2 (first response), the regions around Port-au-Prince that were most affected by the earthquake have the highest demand: in the very early phases, migration is yet to start, which implies a sudden increase of the additional health-care demand due to injured people. Moreover, transportation from or to these sections is difficult because of damages and disruptions to the infrastructure. In the relief phase (t_3), first emergency management measures are introduced. This implies that camps for displaced people are established, and the migration of the population to more rural regions becomes more prominent. Hence, the additional health-care demands shift to further regions. In time step t_4 recovery, longer-term impact, such as potential outbreaks of epidemics such as cholera, needs to be considered. As the migration patterns and the impact of the disaster on the population or additional health-care demand due to evacuation and escape cannot be predicted with certainty (in fact, they cannot even be modeled by probabilities), we construct a set of demand scenarios that will be the basis for the warehouse location.

The three investigated environments (E_1, E_2, E_3) correspond to the status of transportation infrastructure. Environment E_1 characterizes Haiti's existing transportation infrastructure conditions before the earthquake by means of the duration of transporting a good from one region to another by air or road. In order to represent the earthquake's impact, all durations for transporting goods between the calculated warehouse locations in E_1 and their neighbor sections are doubled in environment E_2. Furthermore, the most vulnerable path within Haiti's transportation infrastructure is identified for the locations chosen under E_1. The most vulnerable path is the series of (connected) road or air links whose failure has the highest impact on the goals in terms of duration and cost. In the third environment E_3, the most vulnerable path per location is supposed to fail.

In summary, we create a set of scenarios that consist of alternatives a_j (warehouse locations), environments E_k, and changes in demand $d_i(t)$ for each phase t, that is, $S_i = S_i(a_j, E_k, d_i(t), t)$.

RESULTS

For this case study, we simulated 30 demand scenarios. The four time steps and according demands $d_i(t)$ are analyzed for different environmental conditions characterizing the transportation infrastructure (environments). The demands $d_i(t)$ describe the overall demand increase and thus the disaster intensity within the affected areas (see Figure 7.6). In demand scenario S_1, the earthquake leads to a relatively low overall demand increase, whereas the disaster-affected areas face high demands $d_{2-4}(t)$ in S_{30}.

The goal of the case study is to identify the best locations for four temporary health-care facilities (warehouses). Within each demand scenario for environment E_1, the most vulnerable paths are identified in order to construct the most relevant scenarios for risk and robustness assessment, which is especially relevant for the road sector.

The generated results contain information about warehouse locations, their respective costs, and the allocation of other sections to the warehouses (i.e., the sections that receive their supplies from each warehouse ideally are identified). The natural disaster is assumed to be an earthquake with its epicenter, as was the case in the Haiti earthquake in 2010, in the Port-au-Prince section. For each scenario,

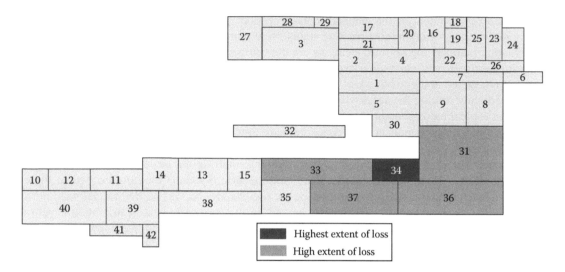

FIGURE 7.6   Schematic representation of Haiti's regions including disaster impact assumptions.

the best warehouse locations are calculated under the conditions as in E_1 (best available information on the transportation infrastructure at the onset of the disaster), for the deteriorated conditions in E_2, and for the critical path disruptions in E_3 relative to transportation infrastructure disruptions.

The most prominent goal of humanitarian relief is to supply aid to all the people in need. Therefore, situations that do not allow supplies to be delivered to all regions are considered as prohibitive: a solution that enables delivering aids to all regions will be preferred over a warehouse location that is more efficient in terms of cost but cannot guarantee reaching all regions under the set of scenarios chosen.

In this use case, we consider two modes of transportation: air and road. Transportation via road is usually the preferred option, as it enables large volumes of goods to be transported at rather low cost. Transportation via air is less dependent on infrastructure networks and very fast. Therefore, it enables responding quick reactions to changing information about the infrastructure (road blockages) or the demand (flexibility). Due to the high cost and need of resources (planes or helicopters) usually, this mode of transportation is only chosen if alternative transportation modes fail. To account for the different requirements and options per mode, we first investigate both options separately (as *extreme scenarios*), investigate the implications for the alternative modes of transport, and discuss how the results can be combined.

Three trends increasing the health-care demand are considered here:

1. The total demand increases in the early phases due to the injuries caused by the earthquakes and the aftershocks and in the later phases also due to the poor hygienic conditions in the evacuation camps or the rural regions.
2. The increased demands in the disaster-affected sections are reduced in the later phases due to population movements toward less harmed sections.
3. At the same time, these migration patterns increase demand in the not directly affected sections.

To take these patterns into account, different demand scenarios for each section are constructed.

As some warehouse locations proved optimal for different demand scenarios, we classify the results according to a set of *demand levels* that depend on the absolute demand in the region and the relative demand compared to the region with the maximum as well as the total demand. Figure 7.7 illustrates the concept by showing that the scenarios can then be classified and clustered according to the demand levels. In this manner, it becomes transparent which demand levels d_1 for the

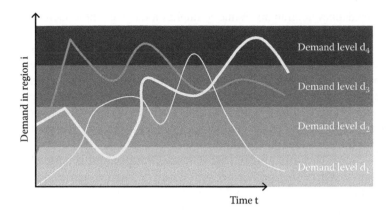

FIGURE 7.7 Concept of demand levels to classify scenarios.

respective regions each scenario covers. This enables more efficient scenario construction, as it is sufficient to consider scenarios that impact different demand levels for decision support (in terms of choosing the optimal warehouse locations).

ANALYSIS

TRANSPORTATION OF GOODS VIA AIR

The analysis of the 30 scenarios for the transportation mode air shows modifications of the locations within each environment relative to different demand levels in the disaster-affected sections. In essence, the demand levels have been assessed. Table 7.2 summarizes the warehouse locations for the three environments E_1, E_2, and E_3 and each demand level. Furthermore, most critical links or paths are given for E_1. The selected warehouse locations for the road transportation mode are highlighted in bold in Table 7.2; Road Transportation is described in the Section "Transportation of Goods via Road".

Four demand levels d_{1-4} are identified, which lead to modifications of the warehouse locations due to increased (relative) additional health-care demands. Demand level d_1 represents the original additional health-care demand levels before the earthquake occurred (in total, 30,267 additional health-care service units are required). Doubling this demand (d_2, in total 60,524 health-care service units) implies primarily an increased additional health-care demand in the disaster-affected sections. Due to migration, demand increases also further sections. For the demand triplication for d_3 (in total 90,801 health-care service units), optimal warehouse locations change again. After setting Haiti's overall demand to its fivefold value (d_4), the optimal warehouse locations reach their final status.

As explained, the 30 scenarios are specified by different, in general, increasing total additional health-care demands that characterize various earthquake intensities and further situation developments

TABLE 7.2
Optimal Warehouse Locations for Varying Demand Levels and Environments (Air)

Demand Level	Warehouse Locations Environment 1	Critical Paths Environment 1	Warehouse Locations Environment 2	Warehouse Locations Environment 3
d_1	1, **3**, **34**, **39**	31–34	31, 32, 33, **34**	**3**, 30, **34**, **39**
d_2	**3**, 33, **34**, **39**	36–34–33–31–34	2, 31, **34**, 41	**3**, 30, 31, **39**
d_3	**3**, **32**, **34**, 37	11–32–34	31, **32**, 33, **34**	31, **32**, 33, **34**
d_4	**32**, **34**, 36, 37	2832–34	**32**, 33, **34**, 36	31, **32**, 33, **34**

Note: Locations that are optimal for more than one environment are highlighted (bold).

(e.g., aftershocks, epidemics, migration). While S_1 implies a relatively little increase in demand, scenario S_{30} describes an extreme scenario that requires important efforts in additional health care.

Modifications of warehouse locations can be investigated best for the sections that are hit most severely by the earthquake (31, 33, 34, 36, and 37; shaded in gray in Figure 7.5). Results for E_1 imply that the higher the (relative) demand in the sections directly affected, the more warehouse locations are set up within those sections. While only one warehouse location in the critical area is needed in d_1 (in the Port-au-Prince section 34), ideally, additional warehouses are set up for d_2 and d_3 in sections 33 and 37, respectively. For demand level d_4, three of the four warehouses ideally would have to be placed in the crisis area (sections 34, 36, 37).

As warehouses cannot be shifted in the course of the emergency, it is basically now the task to determine a location that performs sufficiently well for all demand scenarios and emergency management phases. This is easiest for section 34, which is among the chosen locations in almost all cases (see Table 7.2). Further choices are made according to the similarity of regions. Here, similarity refers to the question how easy it is to get from one section to the other in terms of cost, duration, and, if possible, likelihood of a disruption of the link. In this respect, 33 and 37 can be considered sufficiently similar. As section 33 occurs twice, it is chosen as a further location. Similarly, one proceeds for 3 and 1 and chooses 3 as a location.

To make a decision about the remaining sections 39 and 32, basically the assumptions about reachability via further transportation modes become crucial. Section 39 can be reached via land, as it is part of the main island, whereas section 32 can only be reached via air or water. Due to the missing alternative links, section 32 is chosen.

On the basis of the E_1 locations and conditions, the environments E_2 and E_3 are determined. For E_2, double durations for transportation to the neighboring areas of the sections, in which the warehouses are located, are assumed. Hence, more time is needed to satisfy the overall demand and the optimal warehouse location change. While in the initial phase the critical section warehouse is still located in section 34, all other warehouses should be shifted to the sections 31, 32, and 33, that is, closer to the crisis area. In comparison, fewer modifications would need to be managed for d_4. Only the warehouse in section 37 is shifted to section 33. This implies that in terms of warehouse locations, infrastructure disruptions have a strong impact for smaller demand increases. Note that only section 34 remains in the set of ideal locations in environment E_2.

Furthermore, most critical links and paths are identified in E_1 in order to determine E_3. Although critical paths are less significant for air transportation, as it is relatively easy to switch to another route, some changes of the optimal warehouse locations can be observed. An example is discussed for d_1 with its most critical path (31–34). If this path is disrupted, it is no longer possible to supply Haiti's northern sections via the section 34 warehouse. Hence, one of the E_1 warehouses in the northern sections has to be shifted closer to section 34 in the south. In the example, one in section 30 replaces the warehouse in section 1.

Considering the overall impact across all scenarios and using the same rationale as explained earlier (similarity assessment to resolve conflicts), one replaces section 39 by a warehouse in 33 to achieve the best possible coverage considering risk and uncertainty.

As previously mentioned, link disruptions are not the most significant for the transportation mode air as sufficient alternative routes exist. This effect is confirmed by regarding the costs. Warehouse locations of each environment are applicable for the remaining two environments. Hence, each solution is valid under all investigated environmental conditions, and the goal of the overall satisfaction of all demands can be met in each scenario.

TRANSPORTATION OF GOODS VIA ROAD

Table 7.1 shows the calculated warehouse locations for the transportation mode road. Three demand levels (d_1, d_2, d_3) have identified that lead to modifications of the ideal warehouse locations. Demand level d_1 implies equal warehouse locations for all scenarios in t_1. From scenario 2

TABLE 7.3

Optimal Warehouse Locations for Varying Demand Levels and Environments (Road)

Demand Level	Warehouse Locations Environment 1	Critical Path Environment 1	Warehouse Locations Environment 2	Warehouse Locations Environment 3
d_1	**3**, **32**, **34**, 39	34–31–34–33–15	2, **32**, **34**, 39	**3**, **32**, 36, 39
d_2	**3**, **32**, **34**, 37	34–31–9–8–9	2, **32**, 33, 34	8, 31, **32**, **34**
d_3	**32**, **34**, 36, 37	34–31–9–8–9	**32**, **34**, 36, 37	8, 31, **32**, **34**

Note: Locations that are optimal for more than one environment are highlighted (bold).

to scenario 26, all phases partly have to handle d_2 defined by tripling the original health-care demands. The quicker the demand increases (i.e., the earlier the scenarios enter the highest demand levels, cf., Figure 7.6), the earlier their warehouse locations switch. To make the differences with respect to transportation via air transparent, Table 7.3 highlights the chosen locations for the air transportation mode in bold.

For environment E_1, the ideal warehouse locations for each demand level are shown in Figure 7.7. All demand levels require warehouses to be placed in sections 32 and 34. This choice is also supported by the results for the air case. As section 32 is an island without any roads connecting it to further parts of Haiti, a warehouse is mandatory. This warehouse is dedicated solely to supplying section 32 via road and provides goods to further sections by transportation via air. Hence, this warehouse is relatively expensive. For d_1, the remaining two warehouses are ideally located in section 3 to supply Haiti's northern parts and 39 for the southern parts. The warehouse in 39 should ideally be shifted to the crisis area in d_2 (see Figure 7.8). By a further demand increase (d_3), a third warehouse in the crisis area would be ideal; the warehouse in section 3 should be shifted to section 37. Although this implies higher costs for supplying Haiti's northern sections, the overall costs decrease to the relatively high demand in the crisis region.

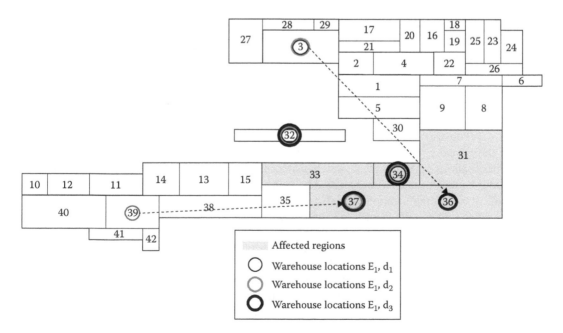

FIGURE 7.8 Impact on varying scenarios on optimal warehouse locations for different initial scenarios.

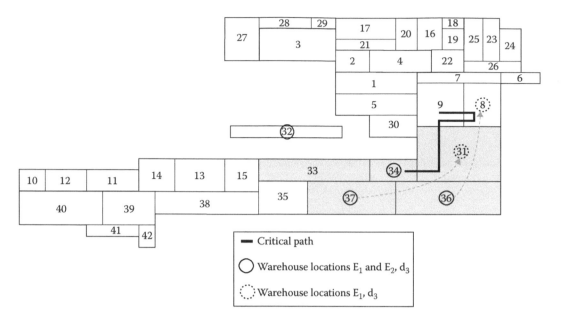

FIGURE 7.9   Warehouse locations, example 2.

As previously mentioned, warehouses are located for the final demand level d_3 within sections 32, 34, 36, and 37. In the case of E_2, durations to pass a road between the E_1 solution and their neighboring sections are doubled in order to create deteriorated conditions. The recalculated warehouse locations show only small modifications. For d_1, the warehouse in the northern section 3 is replaced through the neighboring section 2, due to the new transportation infrastructure conditions between sections 2 and 3. In the second demand level, warehouse 37 is additionally replaced by warehouse 33. For d_3, the recalculated warehouse solution shows no change. Although more costs are required in order to satisfy Haiti's overall demand, the warehouses stay in the same sections as in E_1.

E_3 is defined through the disruption of the most critical path (34–31–9–8–9). In the recalculated solution, a modification of two locations can be observed. Warehouses in sections 37 and 36 have to be replaced through 31 and 8. The shift from 37 to 8 has its reason in the supply of the northern sections. As the south–north axe is disrupted because of the outage of the link (8–9–8), a new warehouse is needed in order to supply this part of Haiti. The reason for the switch of section 37 to 31 is cost based.

Figure 7.9 shows the solution's modifications for d_3 if environmental conditions change.

The E_3 warehouse location is admissible in all E_1 and E_2 scenarios. However, high costs arise. In summary, the E_1 and E_2 warehouse locations are not adaptive to E_3 when the demand in the crisis areas increases above d_3. While those warehouse locations can be used in E_3 in scenario 1, Haiti's overall health-care demand can't be satisfied by using E_1 and E_2 warehouse locations.

To establish the overall locations, similar considerations as for the air case are made: first, constraints are taken into account. Sections 32 and 34 are incontestable. Sections 2 and 3 are considered similar, and due to the dominance of 3 (both in terms of road and air scenarios), 3 remains in the set of warehouse locations—and enables the northern part of the island to receive supplies even in E_3. For the fourth location, section 8 cannot be reached via the sea; it is not well positioned to accommodate freight from overseas and deleted from the set of admissible options. As indicated earlier, section 37 is the gateway to the southern part of Haiti. Therefore, it is preferred over 33, 31, and 36. Due to its closer proximity to the crisis area, it is also preferred over 39 (located in the southern part) and chosen as the final location.

SUMMARY

We presented an approach to evaluate alternatives in humanitarian relief logistics. In general, emergency management is characterized by fundamental uncertainty about the situation and complexity that forestalls understanding and modeling all relevant cause–effect chains. To support decision-makers in these very difficult situations, which are additionally characterized by stress and time pressure, we captured the most prominent uncertainties in scenarios that are targeted at assessing the consequences of different warehouse locations with respect to multiple criteria.

The presented use case illustrated our approach in the field of humanitarian relief logistics. One of the most prominent decisions in this context is the location of facilities and warehouses. This strategic decision sets the framework for the tactical planning and the execution of all supply chain operations, such as scheduling or routing. We focused on identification of robust locations for a set of health-care centers in the aftermath of the Haiti earthquake in January 2010. The results were derived on the basis of real data available from the UN or derived via detailed literature studies.

To address the overwhelming uncertainties, the precision of the model is less important than the ability to perform large numbers of calculations to explore varying future developments. Therefore, we chose a UWLP to determine the consequences of different locations for establishing health-care centers. In this context, uncertainties comprised the disaster's impact on the population (development of health-care demand in all regions over time), infrastructure (transportation mode, resources available, damages, and disruptions), and environmental conditions. Robust locations were identified by assessing the performance for varying scenarios in terms of fulfillment of needs (dominating criterion), duration, and cost by using techniques from MCDA.

CONCLUSIONS

By investigating the case of the humanitarian relief logistics after the earthquake that hit Haiti in January 2010, this chapter has developed an approach to robust and flexible decision making in humanitarian relief logistics. To this end, we adapted and combined approaches from scenario thinking, modeling and simulations, and decision-analytical techniques. Quick heuristics enable simulations and evaluations of fundamentally different situation developments. Combinations of these results into scenarios facilitate communication and transparency. Additionally, the scenarios enable illustrating what could go wrong—or particularly well—and what implications of a specific decision under varying scenarios might be. In that sense, they challenge current mind-sets and enable creative *out of the box thinking*, which is particularly relevant in complex situations.

Our aim is to enable decision-makers and emergency management authorities to make the advantages and drawbacks of potential emergency response plans transparent in order to achieve a consensus. This is particularly relevant given the large number of organizations with different aims, procedures, and best practices that need to work together to achieve efficient and effective disaster relief. By using techniques from MCDA, the trade-offs between different goals and the preferences of diverse actors can be made transparent. Additionally, sensitivities (such as the impact of changes in goals and preferences) can be investigated and become part of the debate. In summary, this approach provides a structured way for *probing* as recommended by Snowden and Boone [94] for solving complex problems by systematically testing and evaluating solutions that may not be optimal for all eventualities, but perform well under varying conditions.

SUGGESTED FUTURE WORK

There are, of course, some very interesting questions that will be part of our future research. First, the systematic combination of results for various modes of transportation will need to be investigated. In the same manner, it will be analyzed if what implications of constraints with respect to available resources (e.g., in terms of trucks, planes) will be. While we followed the actual number of four major warehouses that were used in the Haiti response, the question of the adequate number of warehouses should be addressed. To this end, further scenarios assuming that more or less warehouses are established and the implication with respect to cost and service levels should be researched.

Second, the question of a decision's reversibility should be addressed. While we assumed that the warehouses remain stable throughout the disaster—from its onset until recovery—it is crucial to know which changes in the environmental conditions, demand, infrastructure, or the decision-makers' goals and preferences are so important that the location must be changed.

Lastly, the question how to best communicate the results of this approach requires further research. Given that, especially in the early phases, decision-makers are under time pressure and stress, a very clear communication is mandatory. Still, this communication should not obscure relevant aspects such as uncertainties and potential implications of scenarios for different goals. Here, we work on the development of various options for visualizations and the generation of complementary natural language reports.

REFERENCES

1. H. Abidi, S. Zinnert, and M. Klumpp. Humanitäre Logistik—Status quo und wissenschaftliche Systematisierung, ild Schriftenreihe Logistikforschung, vol. Band 18, pp. 1–40, ild Institut für Logistik- & Dienstleistungsmanagemen, Essen, Germany, 2011.
2. D. Ariely and D. Zakay. A timely account of the role of duration in decision-making. *Acta Psychologica* 108(2): 187–207, 2001.
3. B.E. Asbjornslett. Assessing the vulnerability of supply chains. In: G.A. Zsidisin and B. Ritchie, Eds., *International Series in Operations Research and Management Science*, vol. 124, pp. 15–33. Boston, MA: Springer, 2009.
4. B. Balcik, B.M. Beamon, C. Krejci et al. Coordination in humanitarian relief chains: Practices, challenges and opportunities. *International Journal of Production Economics* 126(1): 22–34, 2010.
5. B. Balcik and B. Beamon. Facility location in humanitarian relief. *International Journal of Logistics: Research and Applications* 11(2): 101–121, 2008.
6. B. Beamon and S. Kotleba. Inventory modelling for complex emergencies in humanitarian relief operations. *International Journal of Logistics* 9(1): 1–18, 2006.
7. B.M. Beamon and B. Balcik. Performance measurement in humanitarian relief chains. *International Journal of Public Sector Management* 21(1): 4–25, 2008.
8. V. Belton and J. Hodgkin. Facilitators, decision-makers, D.I.Y. users: Is intelligent multicriteria decision support for all feasible or desirable? *European Journal of Operational Research* 113(2): 247–260, 1999.
9. Y. Ben-Haim. Robust rationality and decisions under severe uncertainty. *Journal of the Franklin Institute* 337(2–3): 171–199, 2000.
10. Y. Ben-Haim. Uncertainty, probability and information-gaps. *Reliability Engineering and System Safety* 85(1–3): 249–266, 2004.
11. L. Bengtsson. Improved response to disasters and outbreaks by tracking population movements with mobile phone network data: A post-earthquake geospatial study in Haiti. *PLoS Medicine* 8(8): e1001083, 2011.
12. V. Bertsch, J. Geldermann, and O. Rentz. Preference sensitivity analyses for multi-attribute decision support. In: *Operations Research Proceedings 2006*, pp. 411–416. Heidelberg, Germany: Springer, 2007.
13. V. Bertsch, J. Geldermann, O. Rentz et al. Multi-criteria decision support and stakeholder involvement in emergency management. In: Waldmann, Karl-Heinz and Stocker, Ulrike M. Eds., *International Journal of Emergency Management* 3(2–3): 114–130, 2006.
14. R. Bilham. Lessons from the Haiti earthquake. *Nature* 463(7283): 878–879, 2010.
15. J. Birkmann. Risk and vulnerability indicators at different scales: Applicability, usefulness and policy implications. *Environmental Hazards* 7(1): 20–31, 2007.
16. T. Bonapace, S.K. Srivastava, and S. Mohanty. Reducing vulnerability and exposure to disasters: The Asia-Pacific disaster report 2012. Bangkok, Thailand: ESCAP and UNISDR AP, 2012.
17. P.G.M. Boonekamp. Price elasticities, policy measures and actual developments in household energy consumption—A bottom up analysis for the Netherlands. *Energy Economics* 29(2): 133–157, 2007.

18. E. Brattberg and B. Sundelius. Mobilizing for international disaster relief: Comparing U.S. and EU approaches to the 2010 Haiti earthquake. *Journal of Homeland Security and Emergency Management* 8(1): 1–22, 2011.

19. B.P. Bryant and R.J. Lempert. Thinking inside the box: A participatory, computer-assisted approach to scenario discovery. *Technological Forecasting and Social Change* 77(1): 34–49, 2010.

20. D. Byman, I. Esser, B. Pirnie et al. *Strengthening the Partnership: Improving Military Coordination with Relief Agencies and Allies in Humanitarian Operations.* Santa Monica, CA: RAND Corporation, 2000.

21. J.M. Carlson and J. Doyle. Complexity and robustness. *Proceedings of the National Academy of Sciences of the United States of America* 99(Suppl 1): 2538–2545, 2002.

22. E. Cavallo, A. Powell, and O. Becerra. Estimating the direct economic damages of the earthquake in Haiti. *The Economic Journal* 120(546): F298–F312, 2010.

23. Y.H. Chang and C.H. Yeh. Evaluating airline competitiveness using multiattribute decision-making. *Omega* 29(5): 405–415, 2001.

24. T.J. Chermack, S. Lynham, and W. Ruona. A review of scenario planning literature. *Futures Research Quarterly* 17(2): 7–31, 2001.

25. CNSA. Rapid post-earthquake emergency food security assessment. Port-au-Prince, Haiti: Coordination Nationale de la Sécurité Alimentaire (CNSA), 2010.

26. T. Comes. Decision maps for distributed scenario-based multi criteria decision support. PhD thesis, Karlsruhe Institute of Technology, Karlsruhe, Germany, 2011.

27. T. Comes, M. Hiete, N. Wijngaards et al. Enhancing robustness in multi-criteria decision-making: A scenario-based approach. *Proceedings of the Second International Conference on Intelligent Networking and Collaborative Systems*, Thessaloniki, Greece, 2010.

28. T. Comes, M. Hiete, N. Wijngaards et al. Decision maps: A framework for multi-criteria decision support under severe uncertainty. *Decision Support Systems* 52(1): 108–118, 2011.

29. T. Comes, N. Wijngaards, M. Hiete et al. A distributed scenario-based decision support system for robust decision-making in complex situations. *International Journal of Information Systems for Crisis Response and Management* 3(4): 17–35, 2011.

30. T. Comes, N. Wijngaards, and F. Schultmann. Efficient scenario updating in emergency management. *Proceedings of the Ninth International Conference on Information Systems for Crisis Response and Management.* Vancouver, British Columbia, Canada, 2012.

31. C.W. Craighead, J. Blackhurst, M.J. Rungtusanatham et al. The severity of supply chain disruptions: Design characteristics and mitigation capabilities. *Decision Sciences* 38(1): 131–156, 2007.

32. T. Cura. A parallel local search approach to solving the uncapacitated warehouse location problem. *Computers and Industrial Engineering* 59: 1000–1009, 2010.

33. L.C. Dias. *A Note on the Role of Robustness Analysis in Decision-Aiding Processes,* vol. 7, pp. 53–70. Paris, France: Universite Paris Dauphine, 2007.

34. M. Dilley, R.S. Chen, U. Deichmann et al. *Natural Disaster Hotspots: A Global Risk Analysis.* Washington, DC: World Bank, 2005.

35. W. Domschke and A. Drexl. *Logistik: Standorte: Oldenbourgs Lehr- und Handbücher der Wirtschafts- und Sozialwissenschaften.* München, Germany: Oldenbourg R. Verlag GmbH, 1996.

36. D. Echevin, F. Lamanna, and A.-M. Oviedo. Who benefit from cash and food-for-work programs in post-earthquake Haiti? MPRA Paper 35661. Munich, Germany: University Library of Munich, 2011.

37. M. Eßig and S. Tandler. Disaster Response Supply Chain Management (SCM): Integration of Humanitarian and Defence Logistics by Means of SCM. In: W. Feichtinger, M. Gauster, and F. Tanner, Eds., *Economic Impacts of Crisis Response Operations – An Underestimated Factor in External Engagement,* pp. 283–310. Vienna, Austria: Landesverteidigungsakademie (LVAk) / Institut für Friedenssicherung und Konfliktmanagement (IFK). 2011.

38. N. Fenton and M. Neil. Making decisions: Using Bayesian nets and MCDA. *Knowledge-Based Systems* 14(7): 307–325, 2001.

39. N.V. Findler and U.K. Sengupta. Multi-agent collaboration in time-constrained domains. *Artificial Intelligence in Engineering* 9(1): 39–52, 1994.

40. S. French. Multi-attribute decision support in the event of a nuclear accident. *Journal of Multi-Criteria Decision Analysis* 5(1): 39–57, 1996.

41. S. French. Cynefin, statistics and decision analysis. *Journal of the Operational Research Society* 64, 547–561, 2013.

42. J. Geldermann, V. Bertsch, M. Treitz et al. Multi-criteria decision support and evaluation of strategies for nuclear remediation management. *Omega* 37(1): 238–251, 2009.

43. B. Girod, A. Wiek, H. Mieg et al. The evolution of the IPCC's emissions scenarios. *Environmental Science and Policy* 12(2): 103–118, 2009.

44. Global Health Observatory Data Repository. Health workforce, infrastructure, essential medicines: Health workforce by WHO regions. http://apps.who.int/gho/data/#. 2012. WHO. 5–11–2012. Accessed August 27, 2012.

45. Harvard University and NATO. The Haiti case study. Working paper of the collaborative NATO-Harvard project: Towards a comprehensive response to health system strengthening H in crisis-affected fragile states. NATO, 2012.

46. G. Heaslip, A.M. Sharif, and A. Althonayan. Employing a systems-based perspective to the identification of inter-relationships within humanitarian logistics. *International Journal of Production Economics* 139: 377–392, 2012.

47. D. Helbing. Managing complexity in socio-economic systems. *European Review* 17: 423–438, 2009.

48. D. Helbing, H. Ammoser, and C. Kühnert. Disasters as extreme events and the importance of network interactions for disaster response management. In: S. Albeverio, V. Jentsch, and H. Kantz, Eds., *Extreme Events in Nature and Society (The Frontiers Collection)*, pp. 319–348. Berlin, Germany: Springer, 2006.

49. K.B. Hendricks and V.R. Singhal. The effect of supply chain disruptions on shareholder value. *Total Quality Management and Business Excellence* 19(7/8): 777–791, 2008.

50. M. Hiete, V. Bertsch, T. Comes et al.. Evaluation strategies for nuclear and radiological emergency and post-accident management. *Radioprotection* 45(5): 133–148, 2010.

51. D. Hitchins. Autonomous systems behaviour. In: O. Hammami, D. Krob, and J.L. Voirin, Eds., *Proceedings of the Second International Conference on Complex Systems Design and Management (CSDM)*, pp. 41–63. New York: Springer, 2011.

52. J. Hodgkin, V. Belton, and A. Koulouri. Supporting the intelligent MCDA user: A case study in multi-person multi-criteria decision support. *European Journal of Operational Research* 160(1): 172–189, 2005.

53. J. Holguín-Veras, M. Jaller, and T. Wachtendorf. Comparative performance of alternative humanitarian logistic structures after the Port-au-Prince earthquake: ACEs, PIEs, and CANs. *Transportation Research: Part A* 46(10): 1623–1640, 2012.

54. C.S. Holling. Understanding the complexity of economic, ecological, and social systems. *Ecosystems* 4(5): 390–405, 2001.

55. Institut Haitien de Statistique et d'Informatique. Population totale, population de 18 ans et plus menages et densites estimes en 2009. Port au Prince: Institut Haitien de Statistique et d'Informatique, 2009.

56. U. Jüttner. Supply chain risk management: Understanding the business requirements from a practitioner perspective. *The International Journal of Logistics Management* 16: 120–141, 2005.

57. H. Kahn and A.J. Wiener. The next thirty-three years: A framework for speculation. *Daedalus* 96(3): 705–732, 1967.

58. P.R. Kleindorfer and G.H. Saad. Managing disruption risks in supply chains. *Production and Operations Management* 14(1): 53–68, 2005.

59. G. Kovács and K.M. Spens. Humanitarian logistics in disaster relief operations. *International Journal of Physical Distribution and Logistics Management* 37: 99–114, 2007.

60. G. Kovács and P. Tatham. Responding to disruptions in the supply network—From dormant to action. *Journal of Business Logistics* 30(2): 215–229, 2009.

61. W. Kroeger. Critical infrastructures at risk: A need for a new conceptual approach and extended analytical tools. *Reliability Engineering and System Safety* 93(12): 1781–1787, 2008.

62. H.L. Lee. The triple-A supply chain. *Harvard Business Review* 82(10): 102–113, 2004.

63. R.J. Lempert, D.G. Groves, S.W. Popper et al. A general, analytic method for generating robust strategies and narrative scenarios. *Management Science* 52(4): 514–528, 2006.

64. G.S. Liang and J.F. Ding. Fuzzy MCDM based on the concept of alpha-cut. *Journal of Multi-Criteria Decision Analysis* 12(6): 299–310, 2003.

65. M.A. Matos. Decision under risk as a multicriteria problem. *European Journal of Operational Research* 181(3): 1516–1529, 2007.

66. A.J. Maule, G.R. Hockey, and L. Bdzola. Effects of time-pressure on decision-making under uncertainty: Changes in affective state and information processing strategy. *Acta Psychologica* 104(3): 283–301, 2000.

67. D. Maxwell and J. Parker. Coordination in food security crises: A stakeholder analysis of the challenges facing the global food security cluster. *Food Security* 4(1): 25–40, 2012.

68. D. Mendonca, T. Jefferson, and J. Harrald. Collaborative adhocracies and mix-and-match technologies in emergency management. *Communications of the ACM* 50: 44–49, 2007.

69. M. Merz, M. Hiete, and F. Schultmann. An indicator framework for the assessment of the indirect disaster vulnerability of industrial production systems. *Proceedings of the International Conference on Disaster Risk Reduction (IDRC)*, Davos, Switzerland, 2010.

70. L. Michel and P. Van Hentenryck. A simple tabu search for warehouse location. *European Journal of Operational Research* 157: 576–591, 2004.

71. G. Montibeller, H. Gummer, and D. Tumidei. Combining scenario planning and multi-criteria decision analysis in practice. *Journal of Multi-Criteria Decision Analysis* 14(1–3): 5–20, 2006.

72. S. Moore. International NGOs and the role of network centrality in humanitarian aid operations. *Disasters* 27(4): 305–318, 2003.

73. M.G. Morgan and M. Henrion. *Uncertainty: A Guide to Dealing with Uncertainty in Quantitative Risk and Policy Analysis*, Cambridge, U.K.: Cambridge University Press, 1990.

74. J. Mustajoki, R.P. Hamalainen, and K. Sinkko. Interactive computer support in decision conferencing: Two cases on off-site nuclear emergency management. *Decision Support Systems* 42(4): 2247–2260, 2007.

75. G.A. Nick, E. Savoia, L. Elqura et al. Emergency preparedness for vulnerable populations: People with special health-care needs. *Public Health Reports* 124(2): 338–343, 1974.

76. OCHA. Haiti earthquake situation report #21. Geneva, Switzerland: United Nations Office for the Coordination of Humanitarian Affairs, 2010.

77. OCHA. Haiti: One year later. http://www.unocha.org/issues-in-depth/haiti-one-year-later. New York: United Nations Office for the Coordination of Humanitarian Affairs, 2011. Accessed November 7, 2012.

78. OCHA. OCHA in 2012&2013 plan and budget: Haiti. http://www.unocha.org/ocha2012-13/haiti. 2012. United Nations Office for the Coordination of Humanitarian Affairs. Accessed November 6, 2012.

79. C. Perrow. *Normal Accidents: Living with High Risk Technologies*. New York: Basic Books, 1984.

80. J. Rasmussen. Risk management in a dynamic society: A modelling problem. *Safety Science* 27(2–3): 183–213, 1997.

81. H.M. Regan, Y. Ben-Haim, B. Langford et al. Robust decision-making under severe uncertainty for conservation management. *Ecological Applications* 15(4): 1471–1477, 2005.

82. N. Rencoret, A. Stoddard, K. Haver et al. Haiti earthquake response context analysis, London, U.K.: ALNAP, 2010.

83. S.M. Rinaldi, J.P. Peerenboom, and T.K. Kelly. Identifying, understanding, and analyzing critical infrastructure interdependencies. *IEEE Control Systems Magazine* 21(6): 11–25, 2001.

84. D. Ríos-Insua and S. French. A framework for sensitivity analysis in discrete multi-objective decision-making. *European Journal of Operational Research* 54(2): 176–190, 1991.

85. H. Rogers, K. Pawar, and C. Braziotis. Supply chain disturbances: Contextualising the cost of risk and uncertainty in outsourcing. In: H.K. Chan, F. Lettice, and O.A. Durowoju, Eds., *Decision-Making for Supply Chain Integration*, vol. 1, pp. 145–164. London, U.K.: Springer, 2012.

86. J. Rosenhead. Planning under uncertainty: II. A methodology for robustness analysis. *The Journal of the Operational Research Society* 31(4): 331–341, 1980.

87. J. Rosenhead, M. Elton, and S.K. Gupta. Robustness and optimality as criteria for strategic decisions. *Journal of the Operational Research Society* 23(4): 413–431, 1972.

88. G.F. Royes and R. Bastos. Political analysis using fuzzy MCDM. *Journal of Intelligent and Fuzzy Systems* 11(1/2): 53, 2001.

89. A. Saltelli, M. Ratto, T. Andres et al., *Global Sensitivity Analysis: The Primer*. Chichester, U.K.: Wiley-Interscience, 2008.

90. C. Sapateiro, P. Antunes, G. Zurita et al. Supporting unstructured activities in crisis management: A collaboration model and prototype to improve situation awareness. *Lecture Notes in Computer Science*, pp. 101–111. Berlin, Germany: Springer, 2009.

91. P.J.H. Schoemaker. Multiple scenario development: Its conceptual and behavioral foundation. *Strategic Management Journal* 14(3): 193–213, 1993.

92. W. Schöpp, M. Amann, J. Cofala et al. Integrated assessment of European air pollution emission control. *Environmental Modelling and Software* 14: 1–9, 1998.

93. J.P. Shim, M. Warkentin, J.F. Courtney et al. Past, present, and future of decision support technology. *Decision Support Systems* 33(2): 111–126, 2002.

94. D. Snowden and M.E. Boone. A leader's framework for decision-making. *Harvard Business Review* 85: 67–76, 2007.

95. T.J. Stewart. A critical survey on the status of multiple criteria decision-making theory and practice. *Omega* 20(5–6): 569–586, 1992.

96. C. Tang and B. Tomlin. The power of flexibility for mitigating supply chain risks. *International Journal of Production Economics* 116(1): 12–27, 2008.

97. W.B. The. *Natural Hazards, UnNatural Disasters*. Washington, DC: World Bank Publications, 2010.

98. A. Thomas and L. Kopczak. *From Logistics to Supply Chain Management: The Path Forward in the Humanitarian Sector*, pp. 1–15. San Francisco, CA: Fritz Institute, 2005.

99. R.M. Tomasini and L.N. Van Wassenhove. From preparedness to partnerships: Case study research on humanitarian logistics. *International Transactions in Operational Research* 16(5): 549–559, 2009.

100. M. Turoff, R. Hiltz, C. White et al. The past as the future of emergency preparedness and management. *International Journal of Information Systems for Crisis Response and Management* 1(1): 12–28, 2009.

101. S. Ulrich. *Haiti's Emergency Response: An Early Assessment*. London, U.K.: Royal United Service Institute, 2010.

102. UN. Humanitarian assistance and assistance to refugees. http://www.un.org/ha/general.html. New York: United Nations, 1999. Accessed October 27, 2012.

103. UN. UN resident and humanitarian coordinator—Summary. New York: United Nations, 2009.

104. UN, Nutrition information in crisis situation. Report number 21. New York: United Nations Standing Committee on Nutrition, 2010.

105. UNC. *Emergency Health Care Response in Post-Earthquake Haiti*, pp. 1–17. Chapel Hill, NC: The University of North Carolina, 2010.

106. E.M. van Bueren, E.H. Klijn, and J.F.M. Koppenjan. Dealing with wicked problems in networks: Analyzing an environmental debate from a network perspective. *Journal of Public Administration Research and Theory* 13(2): 193–212, 2003.

107. R. Van Dingenen, F.J. Dentener, F. Raes et al. The global impact of ozone on agricultural crop yields under current and future air quality legislation. *Atmospheric Environment* 43(3): 604–618, 2009.

108. L. Van Wassenhove, A.J. Padraza Martinez, and O. Stapleton. An analysis of the relief supply chain in the first week after the Haiti earthquake. Fontainebleau, France: INSEAD Humanitarian Research Group, 2010.

109. P. Vincke. Robust solutions and methods in decision-aid. *Journal of Multi-Criteria Decision Analysis* 8(3): 181–187, 1999.

110. D. von Winterfeldt and W. Edwards. *Decision Analysis and Behavioral Research*, 624pp. Cambridge, U.K.: Cambridge University Press, 1986.

111. S.M. Wagner and C. Bode. Dominant risks and risk management practices in supply chains. In: G.A. Zsidisin and B. Ritchie, Eds., Supply Chain Risk, International Series in Operations Research and Management Science, vol. 124, pp. 271–290. New York: Springer, 2009.

112. S.M. Wagner and N. Neshat. Assessing the vulnerability of supply chains using graph theory. *International Journal of Production Economics* 126(1): 121–129, 2010.

113. D.A. Walton and L.C. Ivers. Responding to cholera in post-earthquake Haiti. *New England Journal of Medicine* 364: 3–5, 2011.

114. W. Watthayu and Y. Peng. A Bayesian network based framework for multi-criteria decision-making. In: *Proceedings of the 17th International Conference on Multiple Criteria Decision Analysis*. Whistler, British Columbia, Canada, 2004.

115. E.P. Weber and A.M. Khademian. Wicked problems, knowledge challenges, and collaborative capacity builders in network settings. *Public Administration Review* 68(2): 334–349, 2008.

116. WHO. *Public Health Risk Assessment and Interventions. Earthquake: Haiti*, pp. 1–33. Geneva, Switzerland: World Health Organization, 2010.

117. World Food Programme, Logistics Cluster, and iMMAP. Haiti—Planned Warehouse Assets. http://epmaps.wfp.org/maps/03967_20100617_HTI_A2_GLCSC_HAITI,_PLANNED_WAREHOUSE_ASSETS,_17_JUNE_2010.pdf. 2010. Accessed August 15, 2012.

118. G. Wright and P. Goodwin. Decision-making and planning under low levels of predictability: Enhancing the scenario method. *International Journal of Forecasting* 25(4): 813–825, 2009.

119. J.L. Wybo and H. Lonka. Emergency management and the information society: How to improve the synergy. *International Journal of Emergency Management* 1(1): 183–190, 2003.

120. D. Yates and S. Paquette. Emergency knowledge management and social media technologies: A case study of the 2010 Haitian earthquake. *International Journal of Information Management* 31(1): 6–13, 2011.

121. L. Zanotti. Cacophonies of aid, failed state building and NGOs in Haiti: Setting the stage for disaster, envisioning the future. *Third World Quarterly* 31(5): 755–771, 2010.

Section III

Culture

8 Construction of an Afghanistan School

Enterprise System Engineering for International Development Assistance in a Foreign Culture

Rob Mitchell

CONTENTS

CASE STUDY ELEMENTS

The study demonstrates that a rigorous enterprise system approach to conceptualizing enterprise operations provides valuable insight into a field (international development assistance) otherwise characterized by *soft* systems, which have not typically been considered from a systems engineering perspective:

- The case addresses the construction of a school catering for 1000 students in a remote province of Afghanistan, as part of a program of community development.
- Its particular relevance is the *respectful partnership* paradigm, its difference from traditional aid approaches, and the strength of the system necessary to interface with the autonomous in-country team.
- The activity domain is international development assistance.
- The country focus is Afghanistan, including the Afghanistan/Australia interface.
- The case will be of interest to the international development community and its associated humanitarian organizations.
- The primary insight is the cost-effectiveness of autonomous, remote operation and the importance of strong system support for its success.

Fundamental essence

The case study is about the operation of an Australian charity, indigo foundation (www.indigofoundation. org), supporting a local community in Afghanistan with funding for the construction and equipping of a school in Borjegai village of Afghanistan. Indigo foundation supports marginalized groups and currently has projects in India, Indonesia, Africa, Solomon Islands, and with the indigenous population in Central Australia.

These projects are typically highly complex due to autonomous independent systems operating, such as, in this case, the Afghan culture, legal system, building practices, the need to coordinate through a very extended supply chain with nontraditional interfaces, and the need to reduce risk in order that good value is obtained with charity-donated funds.

Topical relevance
The project is a good example of complex systems operating that are quite different to the traditional system of systems (SoS), which often involve building on a legacy. This project is quite different although the operation of independent autonomous systems is very evident.

Domain
Charity operating in another culture

Country
Afghanistan and Australia

Stakeholders
The stakeholders are primarily the Australian charity indigo foundation, the Afghan project manager who operates out of Kabul, and the local village elders taking responsibility for the design and construction and operation of the school project for three years.

Primary insights
Effective relationships can be developed through progressively building trust supported by clear agreements that specify milestones and payment for these, cessation of funding if milestones are not achieved, frequent communication along the supply chain emphasizing responsibilities, and a board that remains committed to support of marginalized communities. Given that this project largely depends on relationships, a tool is used to identify the interactions between participants in relationships that are rated on a six-point scale in order to provide focus to the project participants. Imposition of Weltanschauung goals, such as the right to education of girls and women, is possible if enlightened members of the local community are willing to take up the challenge. Community ownership, as an example of self-organization, has been found to be crucial.

KEYWORDS

International development assistance; Respectful partnership paradigm; Case study; Complex systems; Afghanistan school construction; Community ownership; Relationship management assessment tool

ABSTRACT

The case addresses the construction of a school catering for 1000 students in a remote province of Afghanistan, as part of a program of community development. The school building at completion is shown in Figure 8.1. The Australian charity sponsor, indigo foundation, funds the project and manages the development through an extended supply chain. The project is highly complex with the autonomous independent systems being the Afghan culture, the Afghan legal system, building practices in a remote village, and the issues of exerting governance and control, in order that value for money is achieved, through an extended supply chain.

Indigo foundation operates through the *respectful partnership paradigm*—implying genuine community ownership and autonomous in-country conduct—and has arisen in response to the poor success record and cost overruns associated with the more prescriptive, donor-driven traditional approaches involving close, independent oversight and importation of exogenous technologies. The study describes the paradigm by way of an enterprise system that enables autonomous project delivery. The project concerned is the construction and furnishing of a school for 1000 students (boys and girls) in the Hazarajat region of Afghanistan.

The environment of the project is characterized by a diverse group of stakeholders, including

- In-country community-based project operatives and community representatives
- An Australia-based project coordinator and expatriate project advisors
- Executive and governance personnel of indigo foundation, the sponsoring nongovernment organization (NGO)

FIGURE 8.1   School building at completion of construction.

Further, pervading all operations is clarity on the nature of opportunities and aspirations and approaches to identifying and managing the associated risks, many of which are very significant.

GLOSSARY

CBO   Community-based organization, usually established by representatives of a community for a specific purpose, such as community development
DevCo   Development coordinator; executive position responsible for oversight of all *project coordinators (PCs)*
DSC   Development subcommittee, established by the *management committee* to oversee projects
GM   General manager; executive position responsible for operational oversight of all indigo foundation activity
M&E   Monitoring and evaluation; the work done to provide feedback on project delivery
MC   Management committee; the governing body of indigo foundation
NGO   Nongovernment organization, often associated with delivery of aid, community development, etc.
PC   Project coordinator; the (volunteer) role described in detail in Appendix B
SoS   System of systems
ToRs   Terms of reference

BACKGROUND

CONTEXT

The main focus of development assistance provided by indigo foundation (www.indigofoundation. org) is to provide support to *community-based and managed nongovernment groups and organizations for community* development activities. All activities are assessed by the extent to which they can demonstrate: need, expected benefits, community ownership, sustainability, equity and transparency, and over time, results.

SUPPORTING NONGOVERNMENT ORGANIZATIONS

A key focus of this approach is to assist with the establishment and/or sustainable development of new and existing nongovernment organizations (NGOs) and community-based organizations (CBOs). Assistance generally takes the form of catalytic establishment and administrative funding, ongoing support for capacity building and organizational strengthening, and support during periods of critical growth or change. In most instances, the foundation provides core funding to organizations and, preferably, makes a commitment to continue funding for up to three years. This means

- Funds are used according to the community's development priorities
- Critical administrative costs (such as salaries) can be met on an ongoing basis and medium-term planning can be undertaken with a degree of confidence
- An organization is given sufficient support and time to establish itself *and* implement and evaluate activities
- An organization has access to support during times of change or growth enabling ongoing sustainability

By providing a reliable source of funds and management assistance during the establishment phase, indigo foundation's approach means local NGOs and CBOs are supported during one of their most vulnerable periods—a time that is also full of potential. This results in increased organizational confidence and capacity and improved continuity of activities, which lead to a greater likelihood of robust and effective results. Positive results leverage increased support from the community and other funding bodies, improving the chances of long-term sustainability.

Equally, existing organizations sometimes face periods of uncertainty, change, and growth. Small organizations are not always equipped to deal with these periods effectively and may require resources or other support to take the next step forward.

WHY THIS STRATEGY?

Effective Development

Best practice development recognizes long-term positive results are achieved when communities generate and implement their own development solutions. (The notion of *ownership* is articulated in the Paris Declaration on Aid Effectiveness, 2005.)

A community's confidence and capacity to undertake development are enhanced when

- There is *real* community control over decision making; *leadership* is local
- The community *fully participates* in development activities using its own skills, knowledge, and resources

Lessons learned are kept within the community and a sense of pride and independence is facilitated when objectives are achieved. A community-based NGO is one of the most effective mechanisms to achieve this community ownership and build capacity.

The indigo foundation's respectful partnership paradigm mandates community control of priorities, community leadership of in-country project delivery, and genuine community participation in development activities.

Development Needs and Gaps

Established by highly motivated volunteers, NGOs and CBOs often require small amounts of catalytic seed funding, access to management support, and some funds to cover operational expenses in the early stage of development. However, it is often extremely difficult for these organizations to access finance or relevant and appropriate technical assistance: They are too small (and the financial request is too small) for large agencies to consider; they do not have an established track record; and they have no *voice*, having few, if any, national or international networks. Indigo foundation operates in this significant gap in the spectrum of development assistance.

Indigo Foundation Experience and Capacity

Indigo foundation has been established since 2000. Over this time, it has supported a number of NGOs and developed skills and systems to support small, newly established organizations. The foundation has drawn heavily on the experience of its management committee (MC), which has expertise in international development, program management, and capacity building. It is a small organization with limited funds sourced from private donors, and the majority of work is undertaken on a voluntary basis. Providing support to NGOs and CBOs has proven to be an effective and efficient approach to development, which maximizes results from the available resources and skills.

GUIDING PRINCIPLES

Indigo foundation has four *guiding principles* for working with communities.

Community Ownership

Community ownership is vital to the success of community development activities. It is the people in the communities concerned who are in the best position to generate and implement their own development solutions. Project activities (such as design, implementation, monitoring, and evaluation) will be done jointly with the communities. The most vulnerable groups will be identified and included so that they actively participate in project activities. Support for projects will only be provided where participation is voluntary.

Sustainability

Support should be provided to organizations with projects that have a long-term sustainable impact. The design of projects should be flexible so that changing community needs and lessons learnt can be incorporated over the life of the project. Local skills and knowledge will always be used as the first option for solving problems.

Transparency

Projects must operate in a transparent manner. This is particularly important with respect to decision making and financial management. Trust and open communication are essential.

Equity

Projects must operate in an equitable manner. This is particularly important to ensure equitable distribution of benefits as well as child and gender equity. Women, men, and children should have equal opportunities to participate in, and benefit from, community development activities. The foundation commits to follow these principles in its work. In turn, it asks that the communities abide by these principles. If a community cannot follow these principles, the foundation will cease working with it.

OPERATIONAL PRINCIPLES

Indigo foundation projects are chosen, monitored, and evaluated against these four guiding principles. Applying these concepts provides a meaningful framework for communication and negotiation. It avoids a managerial or technocratic approach to development by providing flexible boundaries, which allow for innovation, different processes and timelines, diversity in identity, and ultimately empowerment.

In order to facilitate the implementation of the guiding principles, the type of *operational support* provided to partner communities is critical. This can either enable or undermine efforts to maintain or achieve the guiding principles. Therefore, a set of five operational principles has been developed, which, in many ways, is the practical manifestation of the guiding principles and the core of the respectful partnership paradigm.

Indigo foundation aims to apply the following key operational principles to projects and to its own operations:

1. *Core funding*: Funding should not be tied to specific projects or activities but rather provided for the general use of the foundation. Funds can be used for ongoing costs (such as salaries or recurrent administration expenses) or specific program activities. However, we need to be assured the budget of the foundation aligns to and reflects a robust strategic planning process (including appropriate identification of community needs) and the reporting of funds received and expended is transparent to the community.

     The establishment phase of projects may require some conditionality on use of funds as the relationship (and trust) is established.

2. *Commitment to relationships*: Robust relationships based on equality are at the core of successful community development. Strong relationships provide a foundation for honest and meaningful exchanges. The commitment to relationships is formally expressed through commitment agreements, which should aim to be for 3 years. A preexisting relationship is a key criterion for supporting an organization/community and should be maintained and supported through both formal channels (such as liaison officers, reporting requirements) and informal means (e-mails and telephone conversations).

3. *Partnerships*: Most marginalized communities supported do not have access to national or international support: they do not have adequate access to telecommunications and/or are not large enough to be noticed (or considered administratively efficient). Strengthening and broadening links to the greater development community (and those who can provide higher levels of technical or financial assistance) should be a key component of project management support.

4. *Risk taking*: Empowerment and change do not happen without taking some risks. Calculated risk taking (mitigated by a thorough risk analysis, which includes risk management options) is supported where it has a strong chance of supporting a community's development priorities. Such risk taking may include supporting new or fledging organizations (noting that small CBOs in marginalized communities are inherently risky because of limited capacity and usually a strong reliance on individuals) or providing catalytic funding to test or demonstrate the validity of a development idea.

5. *Reflection*: Quality monitoring and evaluation (M&E) of support is crucial to ongoing development access. Indigo foundation believes that at the core of effective M&E is the ability to critically self-analyze—and an open and *safe* environment for discussions about mistakes and lessons learned should be facilitated. Partners should be strongly encouraged to not only reflect on their own progress but also on the support that has been provided. A process has been developed to encourage partners' self-reflection.

CHARACTERIZATIONS

Indigo foundation depicts its enterprise diagrammatically in Figure 8.2:

A well-defined *enterprise* (Rebovich and White, 2011) overarches systems of integration and delivery, the latter operating autonomously, and with the ultimate aim of sustainable separation from the enterprise. All projects are undertaken on the basis of an exit strategy with ownership and control by the partner community, the aim being that the community development achieved should become sustainable without the ongoing support of indigo foundation (the enterprise).

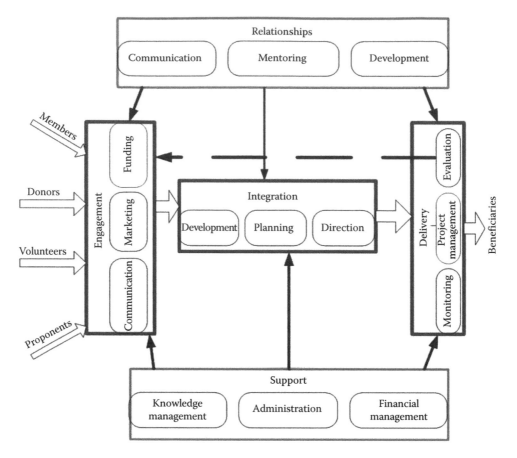

FIGURE 8.2   Enterprise system.

The form of this integrated SoS is designed and centrally *directed* to fulfill specific purposes. It is mature, with little variation over time and different *delivery* projects, but the nature of both integration and delivery varies significantly depending on project context.

The environmental context of the enterprise is characterized by

- A clearly articulated strategic overview around a stable mission and single enterprise
- Stakeholders who often emerge from new relationships but who agree in principle on matters pertinent to enterprise conduct
- System reliability and flexibility
- Widely varying user characteristics across multiple project implementations

This is an enterprise operating in a transitional domain where the influences of autonomous operation would present significant and possibly intolerable challenges for traditional well-bounded behavior in a stable environment.

A distinctive feature of the enterprise is the explicit requirement that one of its major subsystems—delivery—is intended to detach and become self-sustaining for each project. Indeed, the aim of sustainable development is not achieved unless this occurs. While the overarching, guiding architecture begins with the application of an established framework, over the life of the individual programs and projects, the aim is to foster the evolution of delivery capabilities to the point where the partner community is empowered to operate with complete independence from the erstwhile supportive structures.

PROJECT PURPOSE

Initially formulated in broad terms in the enterprise's *engagement* process, project purpose is refined and fully articulated during the development and planning functions of the *integration* process.

The Situation and Its Evolution

The village of Borjegai (population 36,000) is located in a mountainous area of central Hazarajat in Nawor District, Ghazni province. The Hazara population of Borjegai, like the other parts of Hazarajat, has been the victim of institutionalized discrimination by Afghanistan's central governments in the past. The harsh geography and historical discrimination have made it difficult for Borjegai's population to access socially valued resources, such as education. The partly constructed school building shown in Figure 8.3 highlights the harsh surrounding terrain.

The partnership between indigo foundation, the village of Borjegai, and (since 2009) the Rotary Club of Ryde to develop educational opportunities for the village's children began in 2003. After nine years, the Borjegai Schools Program has become a microdevelopment success story with great potential to be used as a model for rural development throughout Afghanistan: it is community managed and focused, results based, and very cost-effective.

For the first time in the village's history, girls are being educated. There are currently nine schools operating in Borjegai that are all registered with the central government's Ministry of Education. The Borjegai School Program has built the buildings for three of them (accommodating 2200 students) and furnished a fourth.

The enterprise, encompassing indigo foundation, the Rotary Club of Ryde, and the Borjegai community, provides project management support and funding for salaries of professional teachers, construction of school buildings, and the purchase of school textbooks and stationery materials. Over 4500 students, of whom 45% are girls (higher than the national average), now attend the network of schools in the village. The early aim of supporting a specific community in need, consistent with guiding principles and areas of emphasis, has evolved to a vision of a wider community committed to the value of education as a key focus for future social improvement.

FIGURE 8.3 Partly constructed school building.

Achievements

Since 2003, approximately $340,000 has been spent in Borjegai to improve education. Of this, almost half has been contributed by the community, in the form of land, labor, and $80,000 cash. Results include the following:

Improved Teaching and Education Outcomes

- Over 4500 students now attending school (an increase of almost 30%), 45% of whom are girls.
- High school graduates increased from 0 to more than 350 since 2003.
- Over 200 students now attend or have graduated from a national university, of which 10% are women. This represents a university entrance exam pass rate of 70%.

How?

- Purchase of 5900 textbooks for students in 2003
- Funding and recruitment of the following:
    - Five professional teachers as well as administrative assistance for the Borjegai High School (the main school in the village) from 2004 to 2011. *This funding has proved to be so effective that Borjegai High School is, in practice, a selective school. It has an annual success rate of more than 90% for university entrance.*
    - Three professional teachers for the girls' high school from 2006 to 2011 enabling girls to attend university and gain employment in the village schools. *This is a significant milestone for achieving gender equity.*
    - As the majority of teachers in Borjegai have limited education, these teachers have been obliged to implement a *teach the teacher* program, thus ensuring there is ongoing improvement in teacher quality.
- Funding for the village's university students, especially underprivileged and female students, to enable them to continue their education in the main cities from 2009 to 2010

Improved Infrastructure

The transfer of 2030 students, both girls and boys, from tents into classrooms in buildings

- Construction of the first girls' high school in the village's history in 2003 for 750 students
- Construction of coed Wali Asr School in 2007 for 750 students
- Construction and furnishing of coed Koshkak High School for 1000 students in 2010, coordinated through the strategic partnership established between the Rotary Club of Ryde and indigo foundation
- Provision of furniture and a well for the fourth school building, built by the local people

Schools have been built at an average cost of $80,000, where the national average is $200,000.

Broader Impact of the School Project

- Increased community understanding and value for education such that families are now encouraging and supporting their children to attend university.
- Increased cultural change toward the education of girls. While there is still a long way to achieve full gender equity, significant improvements have been made. About 45% of the students in the village are girls, representing a much higher rate than that of less than 35% nationwide.
- Increased level of community cooperation and harmony as all three tribes of the village work collectively to take part in the school and their children attend the schools together. The opening ceremony is pictured in Figure 8.4—the enthusiasm of the students and community.
- Cultural shifts in mullahs' attitudes from only supporting religious schooling toward greater understanding of the need of educating of children to be able to respond to the increasingly diverse needs of the community.

FIGURE 8.4   School opening ceremony with students and community members.

- Enhanced capacity of the community to lobby the district and provincial governments not only in matters related to their schools but also in health, security, and legal services. Since 2007, the three tribes have successfully lobbied for a medical clinic in the village, and the governments agreed to build it. The building is built but it still does not have any staff or equipment.
- Lastly, the end of *gun culture* that was crippling the country during the war and the rise of a collective awareness about the importance of peace and education.

Intended Further Stages 2012–2016
- Operational support for the Borjegai High School and girls' high school to fund qualified teachers, until 2014. After this, the community has said this form of support will no longer be needed.
- Indigo foundation and Rotary Club of Ryde completing the furnishing for the Koshkak High School (value of $7000).
- In response to other needs of Borjegai Village and the increasing demand from other neighboring villages, a five-year strategic plan has been developed, which includes the following:
  - In Borjegai, building of one new school building (under way at the time of writing this case study) and completion of three semiconstructed buildings built by the community, as well as provision of furniture for the remaining schools of the village. This stage of the project will involve a financial commitment of well over $200,000.
  - Building another eight schools in the two neighboring villages of Jirghai and Khawat with a capacity of 1000 students for each school. These schools will be fully furnished.
  - Support for improving teacher quality in the new villages.

Achievement of the plan, through the respectful partnership paradigm, will result in the support of 15,000 students in Afghanistan by 2016. A transformation will have occurred from narrow, religion-based teaching to males only, in environmentally constrained circumstances, to broad, secular education of both boys and girls by dedicated teachers in all weather and adequately equipped infrastructure.

ENTERPRISE CONTEXT/ENVIRONMENT

The enterprise has clearly articulated its strategic view, with the following vision:

- Guided always by equity, transparency, sustainability, and community ownership, and through carefully managed growth, indigo foundation will be widely recognized and supported by the Australian community as
  - A benchmark, independent, volunteer-based NGO
  - Achieving the development of small, poor, and marginalized communities worldwide
  - Through genuine and respectful community partnerships

and the following mission:

- With a focus on poor and marginalized communities, we will achieve
  - Community development through respectful, genuine partnerships with project recipients
  - Increasing public recognition of the validity of our *partnership* approach to development
  - Sustainable, productive, and genuine growth
  - Active engagement with our stakeholders
  - Integration and management of available capabilities to match project delivery challenges
  - Robust organizational processes and structures

providing the focus of current effort and the basis for development of a portfolio of projects.

The opportunities identified in the development phase of the integration process represent the practical manifestation of the enterprise mission, while assessing and managing the associated delivery risk has a significant impact on approach to delivery and on staged development.

RISK ASSESSMENT AND MANAGEMENT

A project risk assessment for the current (Salman-e-Fars) stage of the project has been undertaken by indigo foundation resulting in the project carrying some high-risk factors. In spite of the high overall risk, the proven capability and outstanding success achieved by the team and partners in previous work in the Borjegai locality give confidence that the risks can be managed.

As with past work, a key risk management tactic is appropriate phasing of the program to optimize funding availability, value-for-money program deliverables, genuine success milestones as prerequisites for progressive funding, and to give confidence about the longer-term sociopolitical context.

Additionally, building relationships with neighboring communities new to the school project is crucial to maintain the respect/trust themes of the indigo foundation paradigm for possible activity in the years 2014 onward. The existing Afghan team understands this and has already taken action to build these relationships. The scope of the program demands the highest-quality, best resourced project direction team possible. Special attention has been directed at finding and supporting this team and to sustaining it through a multiyear program and schedule.

Close scrutiny continues to be necessary to gauge the sociopolitical environment as the international support for Afghanistan generally changes beyond 2014.

From any perspective, in order to maximize the chance of sustainable success, the program has been structured in phases framed by available funds, judiciously developed relationships with extended communities, and the human resources available in both the Australia-based team and those of indigo foundation's Afghan counterparts.

Summarized Risks and Returns

The aspirational return of this venture is a younger generation capable of making a positive difference for themselves, their families and communities, and their country as a whole. The shorter-term result is good-quality education, improved educational infrastructure, and operational support of the schools themselves as well as the higher quality of teaching.

Success is not guaranteed, but, without the proposed program, the likelihood of an improved future is very small.

Previous *runs on the board* give confidence that the risks, summarized in Table 8.1 (adapted from NSW Government Risk Management Guidelines, 1993), can be managed to deliver an outstanding result and that the potential return does indeed warrant the risk.

The aggregated *risk factor* has been useful as single measure for comparison of projects, particularly in its application to completed projects where it can provide a quantified measure of enterprise risk appetite. Its intent is to provide an indication of probability of experiencing contingencies arising from either or both of likelihood and consequence factors.

The usefulness of engaging in a risk assessment process and of articulating the outcomes in an easily digestible form has been apparent in a number of ways. Importantly, it has provided potential donors with a level of comfort that relevant issues have been considered and that the likelihood of realizing project opportunities warrants the financial outlays required.

Additionally, it has enabled the project team to focus effort on high-scoring aspects. At Borjegai, for example, the logistics of material and key personnel transport between Kabul and Borjegai have resulted in the development of low-profile activities such as multiple small vehicles rather than large (easily targeted) trucks and individual travel by different routes for key personnel rather than group travel.

SCOPE, BOUNDARIES, AND RELATIONSHIPS

The following edited excerpt from the foundation's material used for training of project teams encapsulated these aspects in an overall view of enterprise operation.

Project Team Training Notes

A Process View of Indigo Foundation Operations (see Figure 8.2 for a diagrammatic representation).

Indigo foundation functions to provide a framework whereby interested parties (*stakeholders*) are able to *engage* their resources with us, in a way that enables those resources to be effectively and efficiently *integrated* and applied to the *delivery* of desired outcomes through *respectful partnerships* with beneficiary communities.

These three processes (*engagement*, *integration*, and *delivery*) and the partnership paradigm are the core of our operations. PCs are the key figures in the integration and delivery processes.

Engagement

The process of engaging with stakeholders is one that involves the whole of the indigo foundation team, including PCs.

Much of the activity involved in the foundation's communication, marketing, and funding will be carried out by team members who are not necessarily associated with specific projects.

This does not, however, lessen the significance of engagement for PCs, who are at the *sharp end* of our foundation activities and have a major influence on how we are perceived.

TABLE 8.1

Assessment of Risk (Likelihood and Consequence)

Assessment Summary Sheet

Project: Afghanistan School Project

Project Phase: Expansion Phase							File:	
Risk Category	**Rating (Low to High)**						**Discussion, Key Assumptions, and Responses**	**Score**
Technology	0.1	0.3	0.5	0.7	0.8	0.9	Established, proven processes and tools.	0.1
Capability to deliver	0.1	0.3	0.5	0.7	0.8	0.9	Afghan team being bolstered by training of younger people. In Australia, both funding and project direction capabilities need to be as strong as possible.	0.9
Client quality	0.1	0.3	0.5	0.7	0.8	0.9	New localities requiring new relationships and incremental progress.	0.8
Location	0.1	0.3	0.5	0.7	0.8	0.9	Harsh terrain and weather demanding a high level of local knowledge and experience.	0.8
Project environment	0.1	0.3	0.5	0.7	0.8	0.9	Basically a war zone but a locality, which has been peaceful for a decade and in which the team is experienced.	0.8
Project structure	0.1	0.3	0.5	0.7	0.8	0.9	JV covered by commitment agreement similar to that previously used.	0.5
							Average likelihood score	0.65
Consequence	**Rating (Low to High)**						**Discussion, Key Assumptions, and Responses**	**Score**
Cost	0.1	0.3	0.5	0.7	0.8	0.9	No budget impact. Commitments in fixed AUD and disbursed subject to satisfactory progress. Desired results may suffer due to cost consequences.	0.3
Business development	0.1	0.3	0.5	0.7	0.8	0.9	Significant failure to meet expectations due to major *likelihood* factors would harm relationships, in spite of past success.	0.8
Schedule	0.1	0.3	0.5	0.7	0.8	0.9	Impact potentially severe. Careful phasing necessary to avoid knock-on effect to overall program.	0.8
Our people	0.1	0.3	0.5	0.7	0.8	0.9	Attitudes could be significantly affected, in spite of general awareness that the project carries a high-risk rating.	0.7
Project quality	0.1	0.3	0.5	0.7	0.8	0.9	Validity of paradigm could be adversely impacted by large-scale failure to deliver. Manageable phasing will be very important.	0.7
Core capabilities	0.1	0.3	0.5	0.7	0.8	0.9	Minimum impact. Managing contingencies may actually result in improved competencies.	0.3
							Average consequence score	0.6
Risk factor								0.86
Likelihood score + consequence score − product of scores								

PCs' scope of activity covers coordinated information flows, which enable effective promotion of their projects to audiences likely to be supportive and willing to donate necessary funds. In particular, PCs, whether or not directly involved in fund-raising, are accountable for coordinating project funding in accordance with commitment agreements and project plans. This will entail appropriate communication with both in-country project members and those team indigo members responsible for the stewardship of foundation funds.

The earliest steps of project coordination begin with communication between project proponents and indigo foundation to identify potential projects, assess their suitability for indigo foundation support, and formulate the broad nature of support to be provided.

Integration
Integration is broken into development, planning, and direction.

Development
Project coordination begins in earnest with the development activity of the integration process.

Following confirmation of a project's suitability (compliance with PAC), a sponsorship structure is developed describing the relationship of the stakeholders and the intended mode of project team operation, particularly in relation to leadership and funding.

Project risks are assessed and appropriate risk management approaches developed, and if the project is judged to be feasible in the context of all of the known issues, a broad contractual framework (*commitment agreement*) is developed for endorsement by the partners.

Planning
More detailed arrangements for project execution are added to the commitment agreement during planning activity.
This involves

- *Forecasting* the operating environment over the life of the project
- *Setting objectives* that will serve as project milestones when achieved
- *Programming* (the order of activities) and scheduling (timing) activity necessary to meet objectives
- *Budgeting* resources, particularly financial resources, to match the desired schedule
- *Developing project-specific procedures and policies* to guide planned activity, with a major focus on communication and reporting arrangements

Direction
Our respectful partnership paradigm means that the in-country stakeholder (usually communities and NGOs) carries out the *hands-on* work associated with project delivery. Project coordination provides the very important governance-level oversight for in-country executive action.

Critical to direction is a sound (two-way) communication regime, executed in the context of the relationships defined in the commitment agreement, the style of leadership appropriate to the circumstances, and the technological constraints that may apply. It is very important to understand that relationships and communication among our own team members are every bit as important as relationships with partner communities. The style of leadership required for this aspect must also be well understood and will vary depending particularly on the previous experience of PCs and project officers.

The focus on communications and relationships has been captured for the latest stage of the schools project in the matrix of Figure 8.5, used as an assessment artifact in all projects to clarify the nature of necessary communication among the project team members and stakeholders. Here, a relatively few (10) stakeholders interact through nearly five times that number of relationships, a third of which are rated as either absolutely necessary or exceptionally important, the vast majority of those relating to just three stakeholders.

Typically, the relationship chart is a consensus outcome of a meeting of the project team (often involving teleconferencing) whereby the importance of specific communication links is clarified and reasons are noted. Key links in the communication network emerge (the high-scoring "A" and "E" stakeholders) and their information needs are identified (what, when, how).

Important outcomes of sound direction are assurance of project quality and initiation of timely corrective action should that be necessary.

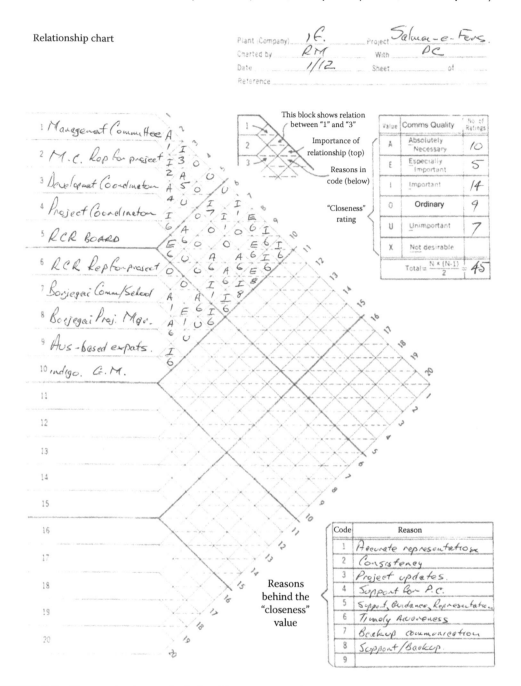

FIGURE 8.5   Communications and relationship assessment chart.

Delivery

The delivery process encompasses the *hands-on* project management task, plus the M&E necessary to satisfy all stakeholders that project purpose is, indeed, being achieved.

Project Management

This is almost exclusively the province of our in-country partners, guided through the oversight and communication exercised in the direction activity.

PCs should understand the challenges faced by in-country project managers and liaison offi-cers so that they are able to offer appropriate support as part of their direction responsibilities. (In Borjegai, e.g., an ever-present challenge is to manage the local community's enthusiasm for the education effort in a wider society where overt demonstrations and promotion of such enthusiasm may attract very negative responses.)

Monitoring and Evaluation

M&E of support provides the crucial *feedback* for the ongoing development success. Quality M&E satisfies the identified information needs of all stakeholders, is reliably formulated against clearly for-mulated guidelines and checklists, and is delivered in a timely manner. It provides an assessment of a community's progress toward the achievement of goals while also providing technical assistance and support. M&E also provides an opportunity for constructive self-reflection and organizational learn-ing for the communities and organizations indigo foundation supports. The aim is to facilitate an open and safe environment for our partners to discuss successes, mistakes, challenges, and lessons learned (including how indigo foundation has performed) and in turn for us to critically analyze our own role.

Monitoring is a continuing function that aims to provide management and the main stakehold-ers of a project or other ongoing support with early indications of progress, or lack thereof, in the achievement of results.

Evaluation is a selective exercise that attempts to systematically and objectively assess progress toward the achievement of an outcome. Evaluation of a project or program should involve assess-ments of differing scope and depth carried out at several points in time in response to evolving needs for evaluative knowledge and learning during the effort to achieve an outcome. An evaluation framework is a planning, implementation, and monitoring tool.

The process of achieving an outcome and how the communities we work with understand and work within our guiding principles (community ownership, sustainability, transparency, and equity) is as important as the outcomes achieved.

M&E guidelines are based on the following principles:

- The value of M&E is directly related to the extent it facilitates empowering stakeholders to reflect on the project and take action to improve it.
- Participation must be as broadly based as possible and include the perspectives of the organization's membership, governing body, the community it serves, other funders (where relevant), and indigo foundation.
- Processes (workshops, interviews, etc.) should be kept as simple as possible while aiming to be useful.
- Evaluations should produce an action plan for improvement, which is a shared document between indigo foundation, the partner organization, and the community.

These guidelines are to assist PCs to develop an M&E framework for projects:

1. Establishing baseline information
   During the assessment stage of the development process, it is useful to establish a simple list of indicators and any baseline information about the economic and social situation of the community. This should be done in consultation with the community. This informa-tion may be a list of priority needs, maps of levels/sources of income, numbers of priority populations, how local community structures operate, etc.
2. An M&E framework
   Stage One: What Are the Goals?
   Good evaluation promotes dialogue and improves cooperation between participants in the partnership process through mutual sharing of experiences at all levels. The aforemen-tioned information will lead into discussions with the key stakeholders about what the community wants to achieve with the support we provide.

While most communities are clear about what they see are the priorities, they are likely to need assistance in defining what is achievable in the time frames and budgets of support. The following key questions will help this process (NB: these are general questions/areas for discussion to be adapted to the community's situation):

- What is the situation now?
- What changes do we want to see as a result of this project/financial support, etc.? Why?
- What are the benefits likely to be?
- What stories passed down to us (even recent day stories) give us strength and hope that things can change?
- What can we learn from these?
- How will we know what the change will look like?
- What resources, knowledge, skills, and relationships/networks already exist in our midst that will contribute to the project? (This way, communities are approaching their improvement from a position of strength [or at least greater confidence] than from a position of need and inability.)
- How are we going to get there?
- Who will be involved? What are the relationships like in the community, between men and women, with the land, etc.? What can we do to mend broken relationships so life for us is more balanced, harmonious, etc.?
- What support do we need to get there?
- What can be achieved in one year? three years?

Discussion of these questions needs to be collaborative between the organization/group and indigo foundation and aims to help stakeholders establish their own goals and processes. Because we support communities that typically have not had donor funding in the past, defining realistic goals and strategies at this stage is critical. Although it probably would not be possible to answer all these questions at this early stage, they need to be kept in mind as they provide the basis for project implementation and ongoing M&E.

Stage Two: Defining the Action Plan

From this information, an M&E framework can be established and agreed on with the community. An M&E framework should include specific goals the community wants to work toward.

Stage Three: Monitoring and Measuring

How progress toward those goals can be measured (what information and documentation will need to be kept by the organization)? The following are some information collection and monitoring tools and processes:

(a) Communication from the field
  (i) Informal reporting
      This can include phone calls and e-mails to and from project partners, the aim of which is to assess progress, find out if there are problems/difficulties, and provide advice/assistance where needed.
          In countries where indigo has a liaison officer, this informal reporting should be a part of that person's responsibilities and will take place on a semiregular basis or as deemed necessary. PCs should aim to encourage liaison officers and project partners to feel comfortable about informally reporting on what progress is being made, impediments hampering progress, and to ask for support where indigo can usefully give it.
  (ii) Formal reporting
      Commitment agreements stipulate the requirement for regular reporting from partners on progress toward goals. The majority of organizations funded have no (or little)

previous experience in producing these reports (as they have not received support from national or international donors). It is crucial that, in early discussions around the M&E framework, all parties should understand what is required in reports, how and when that information is to be communicated. PCs need to ensure that reports are received in a timely manner and to commit to providing feedback to our partners within 4 weeks of receiving a report from the field.

(iii) Information from other NGOs in the field

Where we have an in-country project manager and/or liaison officer, networking with other NGOs will provide a broader perspective on a range of issues including possible sources of funding, network meetings, other groups working on similar projects, potential partnership arrangements, and agreements to undertake monitoring visits. In countries where we have no in-country presence, indigo foundation, through its contacts, may ask a local NGO or contact to assist the project partner as needed.

(iv) M&E visit

An M&E visit should be undertaken yearly and terms of reference (ToRs) agreed on by the project coordinator (PC) and MC representative before the visit. The ToRs will reflect what has been agreed on in the M&E framework and will include specific objectives for the visit. The ToRs should also take into account the findings of the annual project review (or anticipate the review), depending on the timing of the review.

It is important to give the organization/community advance notice of the M&E visit and ask contacts to arrange for you to meet with the key stakeholders. Let them know the aims and length of the visit and whom you wish to meet with.

The M&E visit should provide the opportunity for genuine participation from stakeholders, with the aim of developing the program/project through encouraging self-evaluation processes and practices.

(b) Financial auditing

This is an integral part of every M&E visit to ensure the money donated by indigo is being spent on the agreed project in an appropriate manner. This examination of financial records is, in effect, a mini audit.

Some projects will be keen to show their financial records as they are proud of them and familiar with being audited. Others will be less open as they may worry that it is a test they have to pass or feel that they are not trusted. It is important to explain that as the funds have been donated to indigo foundation, we must ensure that the donors can be confident their money is used properly, and our legal requirements as a charity are met. The *auditor* has a responsibility for the funds in the same way the actual project treasurer does. It is a shared task. We respect the differing ways of handling financial matters in every culture. However, there must be accountability for funds. Questions need to be asked and a critical analysis made. While it is rare for misappropriation to occur, it has happened.

(c) Feedback to the community

Providing feedback to the key stakeholders at the conclusion of a visit is crucial. This provides an opportunity for stakeholders to hear from you about what you have found in your visit and what has been agreed by all parties. If possible, a joint meeting with key stakeholders should be held and actions discussed (and agreed on) for improving the activity. This meeting will provide a shared understanding between stakeholders about the next stage of the project.

Foundation representatives undertaking M&E visits are required to submit a report within 4 weeks of returning from the visit. This report (or relevant parts) should also be shared with the community.

    (d) Communication within indigo foundation
       (i) Project updates for MC meetings

PCs are required to submit project update reports to each MC meeting. These give an overview of the project, progress since the previous update, and bring to the attention of the committee, concerns, issues, and challenges or any decisions that need to be made. Project updates are presented by the MC, and discussion is based on the principles of exception reporting.

       (ii) Annual review

PCs are required to provide to the MC an annual review of the project's progress. This review provides a larger perspective for the MC, indicating the extent to which the project fits strategically with the foundation's overall direction and approach and aligns with the *guiding and operational principles*. The objective of the review is to

- Assess whether the project is implementing the agreed strategies and working toward the project's outcomes
- Assess whether the community/stakeholders understand and are committed to the four guiding principles
- Identify lessons for future engagement, if indicated
- Identify lessons learned for project coordination, in general, and other specific projects, if applicable

The PC reporting schedule is provided as Appendix B.

STAKEHOLDERS

Principal enterprise stakeholder groups are

- Members—financial members of the incorporated association, which is indigo foundation
- Volunteers—constituting the majority of the workforce for project teams
- Service providers—being paid operatives engaged to manage the entity (foundation) created to manage the enterprise
- Donors—ranging from private benevolent entities, through high-wealth private individuals, to regular monthly donors
- Project partners—the organizations and communities and their representatives, with whom a project development and delivery partnership is formed

The governing body of the overall enterprise is the board or MC of indigo foundation. The MC meets quarterly to discharge its governance responsibilities and has established a development subcommittee (DSC) to oversee all projects so as to ensure consistency with enterprise mission and vision and operations within the practical limitations of resources and external constraints.

Before undertaking each new phase in project-specific engagement, relevant stakeholders devote significant effort to explicit formulation of their objectives, roles, and expectations. The enterprise effectively updates its *constitution* with movement through successive phases and with the benefit of mutual understandings, which emerge from the lengthy periods of successful collaboration.

All of these understandings are documented in a *commitment agreement*, of which the most recent (covering construction of a new school building in the Borjegai locality) takes the following summarized form:

- Background
- Purpose

- Agreement
  - Principles
  - Objectives
  - Responsibilities
  - Duration
  - Fund transfer
  - Reporting
  - M&E
  - Relationship of parties
  - Termination

The full document is provided in Appendix A.

Important Project Influences

Most significant factors influencing enterprise operation stem from delivery of a project in a country at war.

The perceptions of stakeholders outside the partner community are heavily influenced by the perceived political, social, and economic risks, which apply to Afghanistan.

The volunteer work ethic on which the enterprise depends almost entirely demands a style of leadership tailored to the strengths of the individual volunteers and sufficiently flexible to accommodate a range of workforce contingencies not normally encountered when dealing with paid employees.

Regulatory compliance within Afghanistan has exercised no significant influence on project development nor delivery, due principally to the competence and political awareness of the in-country team. Compliance with Australian state and federal laws and regulations has not affected project conduct, aside from a level of overhead costs and the ongoing challenge to keep these to a practical minimum.

PROJECT CHALLENGES

The one challenge, which dominates the enterprise and to which most others are related, is securing funding for both the enterprise and the projects.

A great deal of time is spent in framing both the opportunities and approaches to managing associated risks, so that supporters maintain the confidence necessary for sustainable resourcing. The reporting to supporters of successful progress is a major contributor to building that confidence, and the development of communication channels facilitating this kind of information transfer is thus an important focus of effort.

An unanticipated project challenge has been the need to keep the media profile of the considerable success at such a level that it is capable of attracting donors and other supporters, without creating a sentiment within Afghanistan, which may lead to destructive reprisal. In particular, the violent, and sometimes fatal, persecution of women and girls who have engaged in education programs has been prominently featured in the international media. Other NGOs involved in school establishment in this region have had to invest substantial effort in gaining the support of local power groups before attempting projects, which could otherwise be destroyed.

M&E presents significant challenges for the Borjegai schools project. Site access for non-Afghans is effectively prohibited by security concerns. These relate not to the actual project site but to the journey from Kabul to Borjegai.

Communication with the community and its leaders, and with the in-country project leader, can only be achieved in the local language, and direct telecommunication with the project site is impossible.

The project is thus heavily reliant on the abilities of the Australia-based Afghan project advisors. Both have Australian tertiary qualifications in international studies, and their (voluntary) dedication to the project has provided reliable communication between the indigo foundation PC and the in-country management group, as well as M&E visits to the project site, with written and photographic records of progress.

It is through the cultural sensitivity and enterprise comprehension of these two key people that the respectful partnership paradigm has been able to operate.

DEVELOPMENT

A number of issues dominated considerations in the *development* function of the *integration* process.

Overarching all activities were the unknowns arising from working in a country at war. Clearly, with decade free of incidents, the leaders of the community have exercised exceptional political judgment in bringing to their children educational opportunities, which might be judged by some factions as quite contentious.

The wisdom of building both relationships and structures in manageable steps has minimized the likelihood of attracting the potentially very destructive attention of factions opposed to the community's priorities.

Regular communication and sharing of information among stakeholders about project progress were identified as high-priority matters in order to maintain the agreed commitment.

The remote location, poor communications infrastructure, limited physical access, and language barrier were clearly the subject of constant attention to ensure that the relationships defined by the matrix were, indeed, realized.

The strategic core of the relationships was, and continues to be, the respectful partnership paradigm, embodying the autonomy of the Borjegai community, and the trust that its leaders would act appropriately for the long-term beneficial outcomes of the project.

Most prominent in the partnership was the on-the-ground project manager role, filled by a remarkably astute and persuasive local businessman committed to the community's goals for its children. The link between him and the two similarly committed Afghan expatriates based in Australia was pivotal to successful project delivery.

The budgetary aspect of the development process proved to be somewhat unconventional. Plans, specifications, and bills of quantities documented in a manner to which the Western construction industry has become accustomed do not form a prominent part of the working arrangements for local craftsmen. Conveying their vision of a desired structure, and the materials and labor associated with it, was an iterative process commenced with carefully translated verbal descriptions, completed with drawings developed from those words, and budgeted by conversion of traditional measures such as *truckloads* and *barrels* to more familiar metrics.

Scheduling was determined almost solely by climatic constraints and heavy reliance on the ability of the community to judge the requirement for and availability of labor to utilize the time window within which construction is possible.

SYSTEMS ENGINEERING PERSPECTIVE

The diagrammatic depiction of the enterprise (refer to Figure 8.2), while not initially a familiar or easily grasped concept for some in the project team, has proven to be of great value in the framing and delivery of the training necessary to ensure that team members are equipped with the knowledge and capabilities required to participate in their various activities.

One of the more significant aspects of team comprehension proved to be risk assessment. A well-articulated risk factor proved to be valuable in the engagement process, where potential supporters required a level of comfort that the likely results of the project did, indeed, warrant the considerable risks.

The diligent application of *M&E* disciplines has provided confidence that the many uncertainties present in the project environment have been adequately managed. Additionally, this has brought the realization that the project had moved from a *build-a-school* focus to one of *building a community*. This became apparent through approaches from representatives of a number of surrounding localities who had observed the work and outcomes of the project and who also wanted similar benefits for their people. A group of discreet tribal communities had, in fact, coalesced around a strategy of community development through education.

The enterprise response has been to manage this change of emphasis through initiation of multiple, small-scale, first-step engagements with the interested neighboring groups. The life cycle of those new projects will benefit substantially from the lessons learned and trust developed during the original undertaking.

RESULTS

From an enterprise management perspective, the successful delivery of the school construction/furnish/operate project has been an important validation of the respectful partnership paradigm.

Replication is expected as the neighboring start-up projects gain traction.

Confidence in the paradigm and the ability to describe its systemic operation underpin ongoing training and influence the conduct of projects in other countries and communities with which indigo foundation engages.

ANALYSIS

The enterprise system has successfully delivered a project in a complex environment, through autonomous execution. All activity has been on time and on budget.

Engagement of donors through an effectively framed feasibility study—in effect a *project prospectus*—provided a financial foundation without which nothing could have been achieved.

Effective information sharing across a widely dispersed project team, made possible through easily accessible computing and communications technology, rendered achievable a project that could not have been attempted in the absence of those technologies.

The sustained leadership and support of team members, most of whom were volunteers, were critical to success. The inherent level of trust and the paradigm and process awareness were the outcomes of both small-step cooperation over a considerable period and clearly defined system, process, and policies.

By way of comparison with another NGO conceived and operating in similar territory, but within a different paradigm, the international best-selling *Three Cups of Tea* (Mortenson and Relin, 2006) describes the operations of the Central Asia Institute and the schools that resulted from its work in Pakistan and Afghanistan.

The subsequent *Three Cups of Deceit* (Krakauer, 2011) investigates the assertions made in the earlier book and finds many, which it alleges, are overstated or false, as well as operational matters alleged to be improper and schools constructed but not operational.

A supportable conclusion is that articulation of and scrupulous adherence to appropriate guiding principles and an operational paradigm improve the likelihood of successful, sustainable outcomes for both communities and NGOs.

COMPLEX SYSTEM OPERATION

Figure 8.6 depicts a generic complex system and its environment, including external factors, a governing body, and information flows of feedback and governance. The A, B, C, D, E, and F factors for this particular case study enterprise are explained in the following:

A—*Complex system:* The autonomous and independent systems include the Afghan cultural system, the Afghan legal system, Afghan reaction to the education of women, lack of knowledge of Afghan building systems, and the dysfunction of translating English words into Arabic.

B—*External factors:* The external factors included the Afghan legal system, conflicts between Afghan tribes, whether the project may offend the Taliban, and construction in the Afghan winter.

C—*Feedback*: Feedback is very difficult and relies on reporting by the Afghan project manager, relaying this through to the local PC and then to the indigo foundation MC.

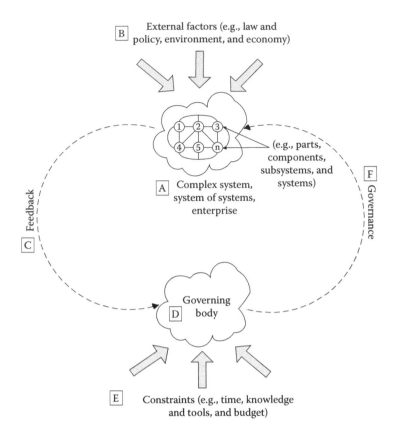

FIGURE 8.6 Generic depiction of a complex system, SoS, or enterprise.

D—*Governing body*: Effectively, the governing body is an MC (board) and members are encouraged to take an active role as a local coordinator, and the on-site project manager appeared to be struggling.

E—*Constraints*: The constraints are somewhat obvious with a long supply line and the need to cope with different cultures and language.

F—*Governance*: Given that traditional governance is difficult, indigo foundation has found a need to use the following measures as part of their governance:

- Gradual buildup of trust through offering low levels of assistance early in the relationship building and increasing support as reliability and trust are established.
- Clear specification of the schedule of achievements against a timeline with cessation of funding if the external party does not achieve the progressive goals.
- Clear exit specification in order to not provide excessive support so that the local community will take responsibility.

RELEVANCE OF SOME KEY ASPECTS OF COMPLEXITY THEORY

1. Recognition of different worldviews
   This issue is extremely important, with interaction between widely differing cultures and expectations.
2. Autonomy of subsystems and processes
   Quite directly related to different worldviews, the need to accommodate autonomy is driven from the realization that more prescriptive approaches are impracticable, unsustainable, and counterproductive.

3. Feedback processes

Feedback, particularly that given through the M&E processes described, has a key role on shaping the timely response of the enterprise to incipient change.

4. Emergence

The emergence of the program from a school building and support exercise to a community-building effort focused on a mutually valued goal was strongly in evidence.

5. Nonlinearity

Nonlinearity was perhaps most observable at the input end of the enterprise, where relatively small increases in support (including financial resources and favorable publicity) can produce very significant increases in observable and tangible outputs.

6. Path dependence

Relevant here is the importance of incremental development as a basis for stakeholders and processes to adjust and familiarize.

7. Self-organization and coevolution

Closely related to autonomy, these aspects are evidenced in the increasingly effective and efficient replication of the delivery process and in the increasing confidence of all stakeholders to trust that mutually understood values will underpin all activities, particularly those necessary to address the inevitable contingencies arising in a turbulent environment. [Refer to Chapter 2]

SUMMATION

An enterprise conceived to support an autonomous subsystem, which is intended for ultimate complete separation, has proven to be highly successful in delivery of education infrastructure and support in a country where high risks apply to both physical and institutional aspects of the undertaking.

Representation of the enterprise in system and process terms has facilitated understanding, analysis, and improvement of operations, as well as training of project team members.

Success of each project stage has brought a broader community development bonus, as long-standing community divisions have been subordinated to a cooperative effort focused on education for the children of the whole community and, potentially, neighboring communities.

The respectful partnership paradigm has been the foundation for highly cost-effective outcomes in a context where more traditional development assistance would not have been possible.

FUTURE WORK

The well-known value of an enterprise system view in enabling a focus on process improvement has resulted in a dedication of research effort to maximize cost-effectiveness of M&E activity.

These aspects of the delivery process constitute by far the largest controllable costs of the enterprise and are expected to benefit significantly from the application of relevant technologies and the development of protocols consistent with the respectful partnership approach.

Accumulation and analysis of outcomes from indigo foundation's other projects will yield insights into those aspects of the paradigm, which require particular attention to match project context.

QUESTIONS FOR DISCUSSION

- Identify the key players in the relationship matrix and indicate with whom they should communicate, how regularly, with what information, and via which media.
- Noting that all project team members, including the PC, are volunteers, consider and describe particular aspects of leadership that you feel are important for project success.
- Given Afghanistan's uncertain and volatile political and social context, consider approaches that might be taken to maximize the likelihood of success of further development of the schools program.

APPENDIX A: COMMITMENT AGREEMENT

1   Background

1.1 General Background

The background of the relationships involving indigo foundation, the Rotary Club of Ryde, the Borjegai community, and the management group established by the community to run *the school* is described in this section. The project support team (PST) comprised of the indigo foundation PC and an MC member, Afghan expatriate project advisor, and Rotary Club project nominee is defined.

1.2 Previous Relationship between PST and the School

The successful outcomes and aspects of the relationship as a basis for continuation of cooperation are detailed.

1.3 Future Relationship between PST and the School

Aspects of the future relationship specific to the next stage of the project are detailed.

1.4 Operation of the PST

Administrative and management arrangements for the PST are detailed.

2   Purpose

The purpose of this agreement is also to outline the basis on which the PST agrees to provide funds to contribute to the building and operation of the school at Salman-e-Fars.

3   Agreement

The parties agree as follows:

3.1 Guiding Principles

By signing this agreement, the school acknowledges indigo foundation's *guiding principles* and, in particular, understands that indigo foundation is committed to the education of both girls and boys.

The school also acknowledges that RCR's participation will be guided by its ethical *four-way test*:

1. Is it the truth?
2. Is it fair to all concerned?
3. Will it build goodwill and better friendships?
4. Will it be beneficial to all concerned?

3.2 Project Objectives

3.2.1 Improvement of the quality of education to be provided to girls and boys by the provision of a new school building

3.2.2 Provision of funds for the operational cost in the school's first year of operation

3.2.3 Financial support of the main school during the first year of operation of the new school

3.3 Principal Responsibilities

3.3.1 The new building will form part of the Borjegai school. It will consist of a school building with eight rooms, plus an amenities block. The building will enclose a large yard to be used as a playground and is expected in due course to provide learning space for up to 500 boys and girls. The school will not charge any fees for the education of its students. The parties acknowledge that the overall cost of the new building will be approximately AUD 80,000, almost half of that amount being contributed in kind and volunteer labor by the community, and AUD 42,535 being provided by the PST for materials, professional services, and support of the first year of operation. The school anticipates that preparation of the materials will commence in June 2012, with building reaching substantial completion by approximately October 2012.

3.3.2 The school has consulted with the three tribes in the wider community and is responsible for the following:

(a) Ensuring that the land on which the building will be constructed is donated.

(b) Ensuring that the labor and tools for constructing the building are provided by the wider community.

(c) Ensuring that the salaries of the necessary professional people (e.g., carpenter, builder, and brick layer) are paid.

(d) Generally paying all the cost of building after taking into account PST's contribution. The parties acknowledge in this context that the school has consulted local business-men and that the local businessmen have agreed to bear the cash costs of the activi-ties, which are not covered by PST's contribution.

(e) Registering the school with the appropriate government authorities if this is necessary to ensure ongoing government support.

(f) Paying the operating costs of running the school each year as is the case now. Some operational costs will also be covered by the government.

(g) Requesting funds in adequate detail from, and reporting to, the third-party financial trustee about the use of the funds (refer to Clause 3.5).

(h) Reporting to and communicating with PST as per Clause 3.6 of this agreement.

(i) Ensuring suitable acknowledgement of indigo foundation's and RCR's contributions to the school as Australian, NGO, and not-for-profit organizations.

3.3.3 PST is responsible for the following:

(a) Providing financial and technical support to the school as detailed in the project feasibility study and the fund transfer and management arrangements in Section 8.3.5 of this agreement. PST will provide the funding (fixed, in AUD) indicated in 3.5, subject to the provision of progressive itemized and costed material and labor requirements.

(b) Indigo foundation will use its established channels to transfer funds to the project. This will be clearly on the basis that 100% of RCR funding will be applied to the building work and operational support. Funds contributed from other rotary sources may be applicable more generally.

(c) Providing the building cost commitment for building trade wages and the purchase of the concrete, wood, plaster, doors and windows, and other building materials neces-sary to construct the primary school.

(d) Ensuring the commitment to the school is provided in a timely and efficient manner in accordance with the fund transfer arrangements in Clause 3.5. Each payment of the funds will be provided subject to PST being satisfied previous funds have been used in accordance with this agreement, *guiding principles*, and the receipt of a satisfac-tory report (and photographs where relevant) outlining the use of such funds (refer Clause 3.6).

(e) Evaluating and providing feedback on the reports and photographs submitted by the school.

(f) Sending a PST representative, security permitting, to Borjegai to meet with the com-munity and the school to monitor and evaluate the use of PST support.

(g) Supporting funds for the school's first year of operation, as detailed in the project feasibility study.

The school and PST will work together in a spirit of trust, respect, and cooperation.

3.4   Duration

This support comprises progress payments as set out in Clause 3.5. It is expected the new building will be completed within 12 months of the first transfer of funds and at the latest within the 18 months for which this agreement is in force.

3.5   Fund Transfer and Management

Details fund transfer and management arrangements.

3.6   Reporting

3.6.1 The school will provide the following to PST:

  (a) Before PST disburses progressive payments, a written report and photographs showing the progress of the construction of the new school building and an estimate of when it will be completed, including budget of remaining costs.

  (b) Following completion of the building, the school will provide a copy of the government registration papers (if necessary), a written report confirming the completion of the building, and photographs and a video showing the completed structure.

  (c) The written report should also contain the following:
   – A financial statement outlining the date funds were received and the use of funds previously received.
   – Any problems encountered with the receipt and use of PST funds.
   – Relevant receipts for use of PST funds.
   – Any concerns or issues that have arisen since the receipt of the previous funds.
   – Identification of the school's ideas and needs for the future and any requests or opportunities where PST might be able to provide further assistance to the school. The report must be signed by at least the principal and the financial trustee.

3.6.2 PST will provide feedback on the aforementioned report within 2 months of receipt.

3.6.3 The school will also communicate with PST by telephone at least every 6 months about the use of funds and any requests or opportunities where PST might be able to provide further assistance to the school.

3.6.4 In addition to the report provided by the school as per Clause 3.6.1(b), the financial trustee will also provide a short written report to indigo foundation after the use of the first part of the building cost commitment and after the completion of the building. The report from the financial trustee must include a copy of all the written requests for funds from the school to the financial trustee and should be endorsed by a representative of each tribe.

3.7   Monitoring and Evaluation

PST's approach to M&E is participatory and is directly related to the extent to which it facilitates empowering stakeholders to reflect on the project and take action on it.

   PST and the school will continue to jointly monitor and evaluate PST support. M&E will include visual documentation of the construction and operation of the school, reporting on the achievements and challenges of the school and the impact of PST support. Processes for M&E will include the following:

  (a) A representative of the PST undertaking a trip to Borjegai within 2 years of the commencement of this agreement. During this visit, the representative will confirm the use of funds as agreed in this commitment agreement through methods such as observation and interviews. The PST and the school will also jointly reflect on the value of the support, the mechanism by which it is delivered, any lessons learned, and how assistance may be improved.

  (b) PST providing feedback on the school's reports in accordance with Clause 3.6.2.

  (c) The school providing feedback on PST's performance.

(d) Where practicable, arranging for another person or agency working in Afghanistan to visit the school and confirm the use of funds as agreed in this commitment agreement through methods such as observation and interviews. In the case of such a visit, the school agrees to cooperate fully with the visiting person or agency.

3.8   RELATIONSHIP OF PARTIES

Details the legal aspect of the relationship, with respect to liabilities, employees, agents, joint venturers, etc.

3.9   TERMINATION OF AGREEMENT

Either the school, RCR, or Indigo Foundation may, at any time by 60-day written notice to the other, terminate its commitments under this agreement. In these circumstances, Indigo Foundation and RCR shall be relieved of their commitment to provide further assistance, and the school shall be responsible for the distribution of any previously advanced funds in accordance with this agreement.

3.10   AUTHORIZED SIGNATORIES

This commitment agreement is to be signed by

(a) The financial trustee
(b) Either or both of the school principal and Mr. Abdullah Fahimi
(c) One or more respected representative of the villages of Borjegai belonging to tribes different from those of the aforementioned signatories so that a representative of each of the three tribes signs this agreement

When the signing of the agreement by these persons is completed and a copy is received by PST, arrangements will be made for transfer of funds as required to meet project requirements and fundraising constraints. A copy signed by indigo foundation and RCR will be sent back in due course.

APPENDIX B: PROJECT COORDINATOR TERMS OF REFERENCE

1. *Role*: Overall responsibility for the *ethical*, *transparent*, and *timely* management of an indigo foundation project
2. *Accountable to* MC (or delegated representative)
   • Project review yearly in conjunction with an M&E trip report where applicable
3. Financial authority: Project budget as agreed by the MC
   • Variations within +/− 10% of overall budget require further approval.
4. Responsibilities
   In general, to manage projects: (a) in the spirit of indigo foundation principles and development philosophy and (b) according to indigo foundation policies and procedures. Specifically,
   (i) General project management
       − Undertake (or collaborate in) any needs assessment.
       − Undertake a risk analysis for new or renewing projects.
       − Submit project proposal for new or renewing projects to the committee for approval.
       − Develop a project framework, as appropriate.
       − In association with MC representative, lead negotiations with partner organizations, including joint donors, draft and finalize commitment agreements for committee approval, and ensure regular reporting mechanisms are installed.
       − Develop and maintain strong active working relationships with partner organizations or communities.
       − Assist the partner organization to achieve its objectives by providing advice, technical expertise, and/or facilitating linkages with other NGOs as appropriate. Renegotiate project objectives when necessary.

- Develop a yearly activity plan and, in collaboration with partners, develop an M&E framework.
- Initiate and manage links with other NGOs, where appropriate.
- Manage project-related travel, including security assessments, debriefing, and trip reporting.
- Ensure all stages in project development and management are adequately documented and project files are maintained.

(ii) Communication and reporting
- Report verbally, on a regular basis, the project activities to the MC representative.
- Report in writing to the committee on project progress yearly, including lessons learned against the yearly activity plan.
- Report against an M&E framework.
- Provide a quarterly project update to MC using the template provided.
- Contribute to indigo foundation newsletters, website, annual report, as requested.

(iii) Finances and fund-raising
- Develop and manage a three-year project budget, with the first year on a quarterly basis.
- Make payments to partner organizations and maintain paperwork associated with these payments, as appropriate and in conjunction with the finance administrator.
- Report against any variations of budget line +/– 10%.
- Initiate and manage, where necessary, any additional fund-raising that may be required by the project.

(iv) Indigo foundation representation
- Represent (and promote) the indigo foundation with partner organizations, other NGOs, and the wider community on project-related matters.

(v) Others
- Provide honest and practical feedback on ways in which the indigo foundation can improve its work.

Project Coordination Checklist

Sticky tape this to the inside of the project folder.

Responsibility	IF Documents
1. Appraisal	• Assessment criteria
	• Assessment commentary/decision sheet
	• MC minutes
2. Implementation	• Commitment agreement
	• Needs assessment
	• Project framework
	• Personnel selection
	• Preparation, briefing, background info, and departure checklist
	• Security plan
	• Field reports
	• Meeting agendas and minutes
	• General correspondence
	• Budget forecast and reporting
	• Debriefing: lessons learned report
	• Lessons learned implementation report
3. Monitoring	• Field reports
	• M&E plan
	• M&E reporting

Please ensure the *project file* has all the aforementioned documents (where relevant) in it. Tick in box when document is in the file.

We must have evidence *of systematic method of delivery and systematic method of documentation. Projects must be clearly identifiable as Australian.*

Activities are undertaken on a partnership basis with indigenous organizations.

Requirement	IF Documents/Response
Evidence demonstrating our role in the project cycle: planning and design, implementation, M&E	• As earlier text for points 2 and 3
Evidence of how partnerships are developed, including how we decide where or not to enter partnership	• Assessment criteria • Assessment commentary/decision sheet • MC minutes

Overseas partners are effective in conducting their activities.

How we monitor and evaluate the work of overseas partners, including example of ongoing communication, reporting documents, and partnership agreement (see preceding text).

PROJECT COORDINATORS' REPORTING SCHEDULE

Quarterly Project Update and Financial Snapshot

Last week of January, April, July, and October.
Use project update template from indigo central.
A very brief outline report that also raises any current issues in the project.

- Forward a version suitable for the website (with any current photos) to the website coordinator and to any project-specific project supporters.

Annual Review:

Conducted after the annual field trip where possible.

Key documents: Trip report, ongoing project assessment criteria, NGO reports, policy and procedures manual.
A detailed project review and assessment.
Guiding notes available on indigo central.
The AR is held as scheduled by the DSC with the PC attending the meeting (a teleconference).

Trip Report:

Four weeks after return from M&E trip, send the report to the development coordinator and, once accepted, to the in-country project partner.
Template available on indigo central.

- Forward a website suitable version to the website coordinator.

New Commitment Agreement:

Due after expiry of existing CA. Key documents as per annual review.

Annual Report:

Article on the project year—due at the end of July.

Newsletter:

Article after M&E visit.

General:

E-mail project financial supporters with new website updates and newsletter articles as appropriate.

ACKNOWLEDGMENTS

The case study has been undertaken with the cooperation of Indigo Foundation and full access to organizational and project documents. I would like to thank the MC of Indigo Foundation and, in particular, Sally Stevenson, the chairperson, who have developed the respectful partnership paradigm and the enabling organizational support; Cynthia Grant, the PC; and Ali Yunespour, the PA.

REFERENCES

Indigo Foundation website. www.indigofoundation.org. Accessed May 29, 2013.

Krakauer, J. 2011. *Three Cups of Deceit: How Greg Mortenson, Humanitarian Hero, Lost His Way*. Norwell, MA: Anchor Press.

Mortenson, G. and D.O. Relin. 2006. *Three Cups of Tea: One Man's Mission to Promote Peace...One School at a Time*. New York: Penguin Group, Viking Press.

NSW Government Risk Management, NSW Public works department, Policy division guidelines. ISBN 0 7310 2704 3, 1993.

Paris Declaration on Aid Effectiveness. OECD 2005. www.oecd.org/dataoecd/11/41/34428351.pdf. Accessed May 29, 2013.

Rebovich, G. Jr. and B.E. White, eds. 2011. *Enterprise Systems Engineering—Advances in the Theory and Practice*. Boca Raton, FL: CRC Press.

Section IV

Environment

9 Utility-Scale Wind Plant System
Value Analysis of Interactions, Dependencies, and Synergies

Lawrence D. Willey

CONTENTS

CASE STUDY ELEMENTS: THE ELECTRICAL POWER GRID AND WIND ENERGY

- Humankind is forever and extraordinarily interdependent on energy and electricity in particular.
- Electricity, which was started as a curiosity and entrepreneurial endeavor, quickly grew into the most complex worldwide enterprise, transcending every sector of humanity's existence.
- The United States was a prime initial adaptor, but the most advanced regions of the world rapidly developed in parallel.
- Sustainable electricity generation coupled with ever-improving distribution, efficiencies, and smart consumption is paramount to meeting man's future needs.
- Large wind turbines hold key advantages that support the future for the electricity generation portion of the grid.
- It matters because humankind's explosive growth and rapid intellectual progress are directly linked to the abundant, reliable, and affordable electricity enterprise system of systems (ESoS).

- Every person on the face of the earth has a stake in the electrical grid.
- Based on the foregoing, it is imperative that more efforts and resources are involved in the electrical grid, the technologies required to grow and sustain it, and the ever-improving complex system engineering (CSE) and ESoS modeling to guide efficient progress.

KEYWORDS

Complex systems engineering, Demand–response, Electrical energy, Electrical load, Enterprise system of systems, Fossil fuel, Grid, Grid energy storage, Human population, Humanity, Life cycle analysis, Macrogrid, Humankind, Microgrid, Power generation, Smart generation, Smart grid, Sustainability, Value analysis, Wind energy, Wind power, Wind turbine

ABSTRACT

Energy leads the list of humanity's top 10 issues. It could very well be humankind's largest and most symbiotic enterprise system above all others. Fossil fuel depletion has now reached or surpassed the midway point relative to even the most optimistic projections of resource stores for the planet. The synergies required to produce, deliver, and maintain electric power production include ingenuity, technology, human development, and economic and political policies. Economics and electricity are interlinked and complex, making power generation the single most important key to sustainable prosperity for humankind.

Wind energy is today's most economical large-scale renewable energy. Lower specific cost and larger wind turbines are possible using the right technology. Cost decisions progress through the system of systems development—consumer, device, grid, power plant, balance-of-plant, turbines, components, parts, and materials. Cost decisions progress throughout the various system levels with continuous interaction that is multidirectional and impacting a large number of uncertainties. Conceptual design using value analysis is a crucial activity, requiring iterations between system requirements and value analysis. Relative value influences are identified early in the conceptual design process, and each aspect is treated accordingly; in the limit, endless cycles of ESoS views that include smart generation and smart grid technologies.

GLOSSARY

AAC	Average Annual Change
AEP	Annual Energy Production (MWh)
BIPS	Building-Integrated Distributed Power Systems
BOP	Balance of Plant
CAES	Compressed Air Energy Storage
CAGR	Cumulative Annual Growth Rate
CCNG	Combined Cycle Natural Gas
CF	Capacity Factor
CIS	Commonwealth of Independent States
CoE	Cost of Electricity ($/MWh)
CS	Complex System
CSE	Complex Systems Engineering
CSI	Coherent Swing Instability
DR	Demand–Response
EPRI	Electric Power Research Institute
EROI	Energy Return on Investment
ESIF	Energy Systems Integration Facility
ESoS	Enterprise System of Systems
FCR	Fixed Charge Rate (%)

FPS	Feet per Second
GHG	Greenhouse Gas
HECO	Hawaiian Electric Company
ISO	Independent System Operator
MDPS	Microgrid Distributed Power System
MIT	Massachusetts Institute of Technology
MMBTU	Million British Thermal Units
MWh	Megawatt-hour
NPV	Net Present Value
NREL	National Renewable Energy Laboratory
POET	Political Operational Economic Technical
PP	Power Plant
PTC	Production Tax Credit
RPS	Renewable Portfolio Standard
SCF	Standard Cubic Feet
SE	Systems Engineering
SoS	System of Systems
T&D	Transmission & Distribution
WPP	Wind Power Plant

BACKGROUND

After World War II, the majority of societies around the globe settled into a prolonged period of growth and prosperity. Up until the 1970s, most hiccups in the economy or spot shortages of energy were short-lived and quickly forgotten. By the end of the 1970s, most big picture thinking revolved around the Three Es—energy, economy, and ecology [1]. From an overall complex system (CS) perspective, the concept of truly long-term sustainability was missing from most assessments of the era. And while some recognized that renewable sources of energy were key to long-term energy sustainability, the mainstream continued as if there was no tomorrow. We now know a better mantra would have been the *Four Es*—energy, economy, ecology, and *endurance*.

> Energy, and in particular electricity, is the single most important factor impacting the continuing existence and welfare of humanity as we know it.

Table 9.1 is a form of Pugh analysis [2] used to illustrate the most important factors supporting humanity. As shown, energy scored the highest in terms of importance across a broad range of considerations. Professor Richard Smalley and a group of colleagues conducted the same exercise during the early 2000s and came up with a similar conclusion [3]—also noting that electricity was the most paramount form of energy for humankind's survival.

As an aside, Dr. Smalley won the Noble Price in Chemistry in 1996, and upon his death in 2005, the US Senate passed a resolution honoring Smalley, crediting him as the *Father of Nanotechnology*.

Not a complete surprise, the overall results of Table 9.1 match the earlier work by Smalley with the exception of education and democracy. Smalley had placed education 8th and democracy 9th, while the current analysis placed democracy 4th and education 5th. All other categories came out the same.

> Potable water is the second most important factor for the welfare of humanity.

Most fossil fuel plants burn fuel to create steam to drive electric generators, using significant quantities of water to cool equipment and generate electricity. Combined cycle coal and natural gas generators can use hundreds or thousands of liters of water per megawatt-hour (MWh) of electricity generated. Nuclear power plants use close to one thousand liters per MWh in a *closed-loop* and one hundred thousand or more liters per MWh in an *open-loop* configuration. Wind energy uses no water in the production of electricity, thus preserving an increasingly scarce resource [4].

TABLE 9.1
Top 10 Issues for Humankind

CTQs (Critical to Quality)	Importance	Baseline	1st Energy	2nd Water	3rd Food	4th Democracy	5th Education	6th Environment	7th Poverty	8th Terrorism and War	9th Disease	10th Population
1 Survive (live)	9	0	1	1	1	0	0	1	−1	−1	−1	−1
2 Intellectual and spiritual growth	9	0	1	0	0	1	1	0	−1	0	−1	−1
3 Happiness	9	0	0	1	1	0	0	0	0	−1	−1	0
4 Harmony	9	0	1	1	1	0	0	1	0	−1	−1	−1
5 Justice	9	0	0	0	0	1	1	0	0	0	0	0
6 Not be picked on (not be targeted)	9	0	1	1	1	0	0	0	0	0	0	−1
7 Accomplishment	9	0	0	0	0	1	1	0	0	0	0	0
8 Good health	3	0	1	1	1	0	0	1	−1	0	−1	−1
9 Comfort	3	0	1	0	0	0	0	1	−1	0	0	−1
10 Ability to solve problems	3	0	1	0	0	0	1	0	−1	0	0	0
11 Least amount of effort	3	0	1	1	0	0	0	0	0	−1	0	0
12 Least amount of money	3	0	0	0	0	1	0	0	−1	−1	0	0
13 Sense of dignity	3	0	0	0	0	1	0	0	−1	0	0	−1
14 No unintended consequences	3	0	0	0	0	0	0	0	0	−1	0	−1
Total weighted score		0	48	42	39	33	30	24	−33	−36	−39	−48

1. Importance: (9) strong or crucial, (3) medium, (1) weak.
2. Scoring: (1) better, yes, or good; (0) same, uncertain, or similar; (−1) worse, no, or bad.

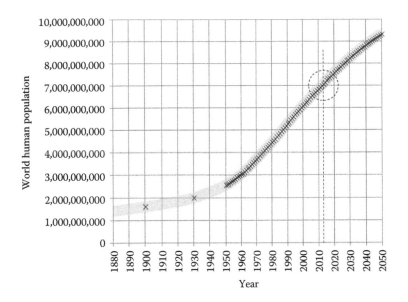

FIGURE 9.1   World human population growth.

Interestingly, population came in 10th in both views. Yet it can very easily be argued that fast-growing numbers of people are actually the root cause for all humankind's shortages and negative pressure on overall quality of life. Figure 9.1 shows the latest listing of recent past and predicted world human population growth [5]. As can be seen, late last year (2011), we passed the 7 billion mark.

Professor Albert Bartlett at the University of Colorado describes sustainability as *development that does not compromise the ability of future generations to meet their needs* [6]. Ultimately, too many people and finite resources are increasingly driving man's challenges, and there is absolutely no such thing—mathematically or logically—as *sustainable growth*. As Professor Bartlett eloquently points out, this phrase is an oxymoron.

Based on the foregoing, we have a chance to meet the challenges for having enough energy by minimizing population, lowering specific consumption, diversifying energy sources, conserving or eliminating less important energy use, and continually reassessing to ensure a truly sustainable system.

With these most important factors for humanity generally agreed, the next overarching view of this most complex system is the physical bounds to contain everything. Figure 9.2 shows a cross-plot for the major world surface areas [7–10], arranged in order of usefulness for human habitation, and estimation for the corresponding total number of people inhabiting these areas. This high-level notional model gives a framework to compare regional and localized population density. It also illustrates the finite limits for the planet (taken in its entirety) to continue to support this model as the earth's human population continues to grow.

Today's most densely populated urban areas range from 20,000 to over 40,000 persons per square kilometer. The present day 7 billion people and the world surface areas shown in Figure 9.2 suggest a gross average of 107 persons for every square kilometer counting all urban lands, agricultural lands, and wetlands. Even adding in all mountains, deserts, and polar regions (all land), this still results in 47 persons/km^2. The actual values are much greater because not all land is accessible or habitable. These values bound the problem and help establish a sober framework as humankind's numbers continue to grow.

Electricity is the flow of electrons from atom to atom after being produced within a generation device, the transport of these electrons to another location, and the collection or consumption of the electrons to produce some form of useful work. Prior to standardized and readily available electricity, people relied on open flame and gas lamps for light and draft animals for overland

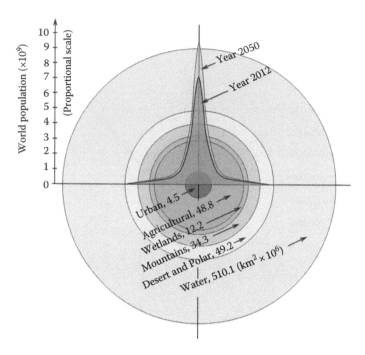

FIGURE 9.2   Human population and relative size of major world surface areas.

transportation. The beginnings of industrialization were powered by falling water or steam from burning wood or coal. Electricity generation and the useful work that is leveraged have permitted humankind to achieve more and more while expending less human energy. In turn, this has driven increased population growth and has created additional need for more energy. This cycle of need and increased demand is endless and represents the trajectory humankind has been on since the early 20th century.

The electrical grid is the largest and most complex machine built by humankind [11,12]. In building the electrical grid, man has produced an ESoS that not only has become the basis for modern life but has also established an environment without which humankind cannot flourish. The electrical grid is a complex system at its outer extent and an ESoS in its technical level of operation.

The electrical grid exhibits features of all of the enterprise models including Directed, Acknowledged, Collaborative, and Virtual; Directed such as electricity transportation and distribution infrastructure; Acknowledged such as individual electricity generating plants; Collaborative such as generation and load balancing; and Virtual in the sense of the invisible influences from all sectors that ensure the growth and stability of the electrical grid.

The electrical grid is a mature complex system that has rapidly evolved to incorporate improvements across all sectors. Technology improvements are characterized by the initial growth and establishment of humankind's dependence on electricity. The grid today continues to benefit from technology improvements, but is shaped more dramatically by government and regulatory policies.

The first electric utilities were highly localized entrepreneurial ventures such as a small urban lighting district. These small regions controlled their production and consumption of electricity very crudely. Interconnection of smaller purpose-built utilities into citywide or regional districts soon followed, greatly improving the ability to meet supply and demand of electricity by sharing more sources of generation and loads. Pursuit of profits, little or no regard for the entire enterprise, or no appreciation of how much more electricity was to become the basis for humankind's survival led to the need for overarching policies to shape further development. The influence of a few early utility leaders eventually required establishing outside authorities to ensure that the ultimate ESoS objectives were promoted and sustained.

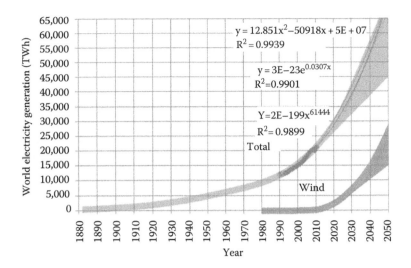

FIGURE 9.3 Total world electricity production and the contribution of wind.

These foregoing phases follow the traditional, transitional, and *messy* domains of the enterprise systems engineering (SE) model. Today's electrical grid is in the full *messy* domain, having to account everything from fossil fuel depletion, potable water consumption, greenhouse gas (GHG) release, environmental pressures, end-user trends, and efficiency impacts—all heavily interactive with humanity's increasing dependence for comfort and survival.

Without regard for the actual wind turbine equipment or installations, the theoretical recoverable wind energy around the world has recently been shown to far exceed humankind's current and projected needs [13]. It was reported that research conducted by Carnegie's Ken Caldeira shows that 20–100 times today's global power demand (all energy categories—not just electricity) could be met by tapping near-surface (like today's 80–100 m hub heights) and high-altitude winds, respectively. Actually achieving some fraction of this extreme case is a more reasonable long-term goal for growing utility-scale wind's share of electricity generation.

Figure 9.3 plots more than 20 years of the most recent world electricity generation data [14] together with the year 1882 beginnings of the electrical grid and projects forward a wide range of increased generation required for the future. Large-scale utility wind growth [14] is also included with a projection of how this renewable resource is expected to gain in percent contribution relative to overall electricity production. The projection of utility wind contribution as a percent of the overall total generation is consistent with the 33%–37% by the year 2050 recently published by the National Renewable Energy Laboratory (NREL) in their latest report [15]. Although this one-third or better wind power contribution prediction is specifically for the United States, it is reasonable to extend this to the world view as most vibrant nations are now advancing on a similar wind power plant (WPP) installation trajectory.

Based on data from a number of sources [16–22], Figure 9.4 takes a closer look at world electricity production for the latest 7-year period. The results include trends for generating capacity, actual generation, and actual consumption. Generating capacity is the summation of all the nameplate ratings across the various forms of electricity production, assuming they are running at rated output for the entire year (8760 h). The actual generation is the true operational time and output across all of the generators. Since more electricity needs to be produced than is consumed due primarily to transmission and distribution losses, the actual consumption is the electricity that is required to service the loads across the entire system. Actual consumption is also influenced by the price being paid by consumers for the power being delivered to them.

Capacity factor (CF) is the amount of electricity actually generated compared to the installed nameplate capability for a given period of time. In the early 2000s, a lot of generation capability was built due to deregulation and lower natural gas prices, and demand did not keep pace—so CFs

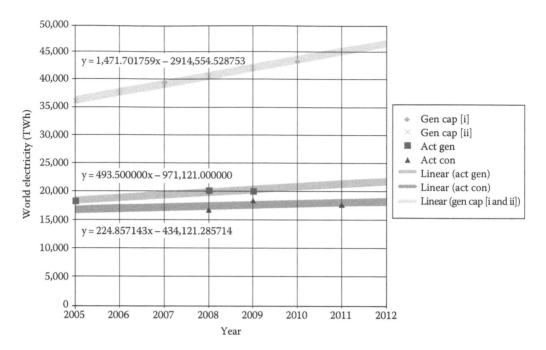

FIGURE 9.4 Recent total world electricity—capacity, generation, and consumption.

fell. Given that it takes years to build large traditional plants and variations of demand are influenced more by short-term factors such as the economy and fuel prices, capacity and generation do not always appear to move in concert. When times are good, we tend to use more of the generating capacity faster than we are adding it, and when times are uncertain, we use less of the generating capacity. In any case, the variation has been relatively small over the past 50 years and around an average of about 48% [23], accounting for all generation sources.

Utilities use base load units that run almost all of the time, except for maintenance, forced outages and refueling for nukes. These plants have a CF of 85% or so. Intermediate size generators and peak load units run significantly less hours in the year and have a CF equal to or lower than a typical wind farm (30%–40%). Peak load units are needed to rapidly increase their output to meet peak demand because base load units normally cannot be quickly started or throttle electricity production.

Latest technology utility-scale WPPs have CFs of 40% or greater. Some wind plants in high wind resource areas achieve 50%–70% CFs. Examples for these project locations include strong northerly winds (winds from the north) frequent in the Isthmus region of Oaxaca, Mexico [24], and northeast-to-southwest trade winds (winds from the northeast quadrant) across the Hawaiian Islands. CFs for wind plants are continually improving due to higher hub heights, improved site planning, technology advancements, and improved specific power (increased swept area for given generator rating).

Table 9.2 was derived from Figure 9.4. This high-level system view is consistent with the beginnings of wind power penetration as a measurable contribution to the overall total electricity supply, that is, ~1% wind-generated electricity circa 2005 and ~3% by 2011 (on our way to ~33% by 2050). The overall decreasing CF for generation and consumption is aligned with higher relative growth in lower CF generation for wind and solar. The increasing trend for losses makes sense too. Losses include transformer, transmission, and power plant parasitic loads. The increase in losses for the period with higher electricity generation is in line with more and more electrical power being transmitted over slower growing transmission infrastructure.

Energy and water are the most crucial factors for the existence of humankind, as we know it. Electrical power is our number one form of energy. Humanity exists as it does today because of electricity. Wind penetration into the electricity supply has been remarkable over the past decade and

TABLE 9.2
Recent Average Annual Change (AAC) for Capacity Factors and Losses

Year	CF Gen' (%)	CF Con' (%)	Losses' (%)
2005	50.67	46.17	8.88
2006	50.00	44.96	10.07
2007	49.38	43.85	11.21
2008	48.81	42.81	12.28
2009	48.27	41.85	13.31
2010	47.78	40.95	14.28
2011	47.31	40.11	15.21
AAC	*−0.56 points*	*−1.01 points*	*1.06 points*

is poised to become one-third of all global electrical power generation by 2050. ESoS and CSE principles will be central to our ability to navigate the coming decades of electrical system growth, while at the same time rapidly transitioning away from burning fossil fuels.

PURPOSE

The discussion thus far supports the need for continued and robust growth in world electricity generation and consumption. This is required for today's modern living. Electricity also supports current population levels, particularly for urban regions around the world. Electric power also contributes to longer lives enabling less manual labor and more leveraged work done by electricity.

US life expectancy has dramatically improved over the last 100 years. Figure 9.5 shows the general trend upward for average life expectancy [25] during the same period electricity generation and the power grid developed. The rapid growth of electricity with the devices and conveniences that came with it appear to be the primary drivers for increased life expectancy. Further increases appear to be difficult and are more likely to result for improvements in the human biological arena. Medical advances will probably not be able to approach the same growth rates for life expectancy as experienced during the first half of the 1900s when the benefits of harnessing electricity made the biggest impact.

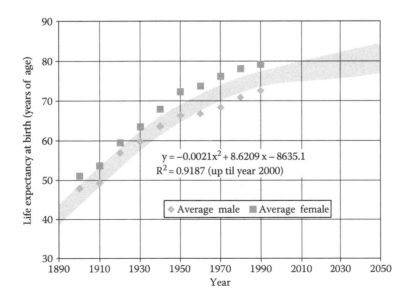

FIGURE 9.5   Average life span increase for the United States.

Figure 9.6 presents the two parallel timelines for (1) the development of the US electric power grid and (2) the five generations that have transpired since the beginning of the electrical grid ESoS (key milestones for the US electrical grid are adapted from [26–28]).

The overall time period of a little more than 100 years and five generations is remarkably short. With respect to present-day middle-aged people (the fifth generation), our great grandfathers living in cities were the second generation to grow up knowing firsthand the benefits and potential for modern electricity. Our grandfathers were the third generation and experienced World War II and the boom years for the growth of the electrical grid and the technology developments that provided more efficient generating capacity. Our fathers were the fourth generation and witnessed the first of the great blackouts, and the realization that improved transmission infrastructure and system reliability was even more important than ever increasing generation.

> The electric power grid cannot continue as we know it.

Our current generation is the first to grapple with the realization that the electric power grid cannot continue as we know it. There are finite fuel sources to burn. Burning fossil fuels cannot continue to add GHGs to our atmosphere. Potable water needs are an important new restriction on operating traditional electric power plants. It is not all about more and more electricity generation—better micro- and macrogrid interconnectivity together with smart handling of loads on the consumer side are just as important. Renewable resources at utility scale together with distributed solutions (e.g., generation, DR, and storage) are a mandate if we are to overcome fossil fuel limitations and electrical grid growth–related issues.

It is now clear that humankind is at a precipice where linear thinking and piecemeal growth cannot sustain the path we are on. The key time to make a difference is now. ESoS and CSE principles are the enabling disciplines to guide our power engineers and policy makers. More and more understanding by a larger pool of cross-functional participants is needed to supplement the traditional engineering-only approach. The perfect result would be electric power that is always available and affordable without burning fossil fuel.

COMPLEX SYSTEM/SYSTEM OF SYSTEMS/ENTERPRISE—THE GRID'S INNER WORKINGS

Electric power sponsors include governments, utility owners, transmission and balancing authorities, infrastructure owners, and consumers, to name a few. Customers (consumers/electrical loads) include all end users in homes, businesses, and industry. Stakeholders are everyone. There is not a person in modern life that is not greatly benefited by electricity. In fact, without electricity, life as we know it would cease to exist. If electricity were to suddenly stop flowing, the most immediate result would be lower life expectancy and the reversal of population growth.

One view for the electrical grid as an ESoS is shown in Figure 9.7. An observer from the governing body toward the grid sees the complex system/SoS comprised of n nodes of generators and loads. An observer from the grid toward the governing body sees n nodes of government administrators, lobbying groups, municipal corporations, and operating policies. Collectively all of these elements come with many interconnections; exchanges and results are part of the overall ultimate ESoS.

Clearly this has become too messy to permit independent and uncontrolled growth to continue. Electricity generation and consumption demands the application of CSE principles. Heuristics and rules of thumb for the electrical grid greatly supported the initial system growth and acceptance by the greater population. Increased economic output came with abundant low-cost electricity. Electricity enables the current high level of human population to exist. No electricity or not enough would cause immediate survival issues and loss of human life.

> An analogy is sunlight shining on a covered petri dish, accelerating light activated bacteria to grow and multiply in symbiotic dependence. Remove the sunlight and it will cause an immediate reversal in the population growth and collapse of the living system. The electrical power system is to people as sunlight is to the bacteria.

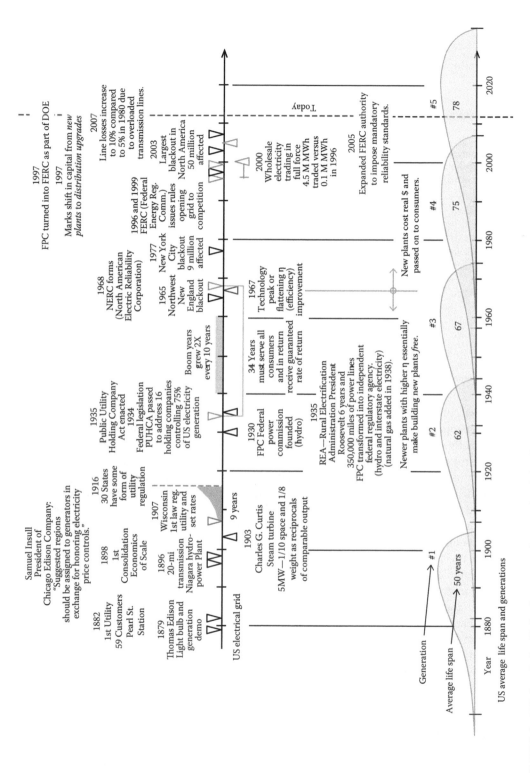

FIGURE 9.6 US electrical grid development and generational timelines.

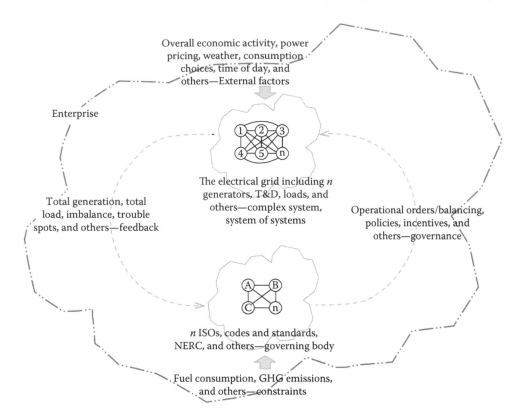

FIGURE 9.7   Context of the grid as a complex system/system of systems/enterprise.

The early growth of electricity was largely driven by the pursuit of financial opportunities that in the end benefit people. Today's electric power grid is *required* to maintain human life and is a very large and interactive environment that contains an enormous number of multidirectional interfaces.

The successful autonomous operation of the grid is dependent on maintaining complete and mutual trust among all of the constituents. Electricity supply and demand is a perfect example of the power of transdisciplines (the ability and desire to bridge across different disciplines) with a broad range of all the people that are an integral part of the ESoS. Government policy, price controls, generation, transmission, consumption, and technology are but some examples of this interplay.

POET (political, operational, economic, and technical) represents key constituents for the success of the electric power grid. The grid contained the spirit of POET from the very beginning, although economic and political goals were sometimes at odds; for example, political pressures and pure profit-motivated movement toward one technology over another were not supported by an ESoS view that contained economic or sustainability considerations. The use of the analysis concept of EROI (energy return on investment) helps to avoid these types of missteps and is explained later in this case study.

The grid provides plenty of opportunity for cross talk, idea sharing, and nurturing discussion. Technology diversity and free exchange have been keys throughout the construction and development of the electric power grids. Critical to the progress has been the sharing of information, establishing standards, and promoting trust among all the participants. This includes individual consumers, communities, local providers, transmission authorities, utilities, and policy makers.

A hallmark of a CS is self-organization among its elements. Self-organization is evident for the generators, transmission, distribution, and end users across the electric power grid system. Achieving balance is also a natural aspect of the grid because creation and consumption of electricity are the basic exchange without accounting for storage or loss. Collaborations are rewarded for achieving balance between generation and consumption. Humility is also a natural aspect of the grid due to the sheer size and complexity of the electricity system. No one organization has the expertise to know the entire system without engaging cross-functional elements and using ESoS and CSE principles.

Of all the initiatives around the world that are working on humanity's greatest issue of energy supply, and in particular electricity, one stands out as a best practice for CSE and ESoS principles. It is known as the Galvin Electricity Initiative.

Robert W. Galvin was the leader and CEO of Motorola for more than 30 years, a business innovator and famous Six Sigma quality management system pioneer [29]. After observing the devastating $10 billion economic loss from the August 2003 northeast US and Canada blackout, the next year Mr. Galvin proposed the ambitious Galvin Electricity Initiative. His vision and astonishing clarity for modernizing the US electrical grid was formulated using the rigorous and illuminating Six Sigma approach applied to the electricity grids' CSE and ESoS issues.

Elements of a vastly improved electric power system included smart technology such as digital communication and control across all system elements, reliability, efficiency, security, distributed generation and storage, cogeneration or combined heat and power, smart meters, smart appliances, and consumer control to ensure a bottom-up mechanism to sustain continuous improvement [30]. While any one of these elements (and the additional ideas that have come along since) taken alone or in small groups may not be completely original, the fact that such a holistic ESoS approach has so eloquently gathered the bigger picture together is the true beauty of the Galvin Electricity Initiative.

Another example of Galvin's truly big systems thinking is the key recognition that government rules will need to be continuously considered and changed accordingly. A perfect example of this is when the US telecommunications rules would not allow Motorola or any other company to access the airways in the days preceding cell phones.

> Motorola built a prototype wireless network in Washington, D.C., and Galvin set about unleashing this secret weapon. Upon being invited to the White House in 1982, he brought along a portable phone prototype. President Reagan asked Galvin what he was carrying around. In his usual modest tone, he said, 'Well, I am so glad you asked. This is a portable, wireless phone.' He then asked President Reagan if he would like to make a wireless phone call. The president called his wife. Recognizing the possibilities, the president next asked Galvin what it would take to make portable phones available to every American. Galvin shook his head and said that the Japanese would likely get to market first as the rules in the U.S. blocked large-scale deployment. President Reagan, with a smile, said he would make a few phone calls and see what he could do. The rest is history. [31]

Without ever-increasing amounts of reliable electric power, life, as we know it, would cease to exist. From the Galvin Electricity Initiative perspective, as well as the ESoS view, construction of increasing amounts of large wind power into the big picture grid is exactly on target. Like the need for humans to have water and air to exist, the stakeholders for a robust electrical power grid are all of humankind.

COMPLEX SYSTEM/SYSTEM OF SYSTEMS/ENTERPRISE—UTILITY-SCALE WIND AND THE GRID

Lowest life-cycle cost to deploy all types of electricity generation makes the overall systems value analysis one of the most important and more difficult challenges. Due to the global scale and large number of uncontrollable interdependencies and uncertainties, the electrical grid is an ultimate ESoS. The electrical power grid contains the greatest amount of interconnectivity as the largest man-made ESoS on the planet.

Where modern man lives, electricity is evident. There is a local environment where end users derive the obvious benefits. Other environments include the interaction of the grid with the weather, the atmospheric impact from burning fossil fuels, and the environments created at the points of fuel extraction, just to name a few. The scope includes all aspects of the world around us. Pick any activity throughout a day and some aspect of the activity will have touched electricity consumption or electrical grid dependence. Even flushing a toilet? Yes—it required electric-driven pumps somewhere in the water system to deliver the water and electricity was used to power pumps, clarifiers, and all the related controls to treat the waste and return clean water back to the environment.

Electricity generation and consumption thrives on diversity. There are many factors influencing electricity generation and consumption. A wide range of generation types and fuels stimulates different perspectives. Numerous different perspectives for all aspects of the system enhance continued progress and optimization of the electricity grid. The grid has continuously adapted and grown as a result of humanity's needs and the profit created to meet those needs.

The electricity grid has never remained static—it has always continually changed and reacted to a wide range of influences that in turn are continually changing and at times seemingly evolving on their own. There are regions within a grid that can remain relatively unchanged with time in terms of physical attributes, but the electricity flow and character is never the same. There is natural growth for technologies that have greater energy efficiency or profit and contraction when there are fuel shortages or high prices. There can be surprises that emerge along the way (e.g., power frequency harmonics).

The number and variety of boundaries for the electrical grid are vast. A crucial economic boundary for wind power is that *cost is king*. The initial and ongoing costs to establish a WPP and operate it for its design life must be competitive with the lowest cost conventional generation. A model by the Electric Power Research Institute, Technical Advisory Group (EPRI-TAG), is commonly used to calculate cost of energy (CoE) for utility-scale wind turbines:

$$\text{CoE} = \frac{\text{FCR} * \text{Cost}_{\text{Capital}}}{\text{AEP}} + \text{Cost}_{\text{O\&M}}$$

where

    FCR is the fixed charge rate (%)
    $\text{Cost}_{\text{Capital}}$ is the total capital cost of the project (\$ or \$/MW × WPP rating in MW)
    AEP is the annual energy production (MWh)
    $\text{Cost}_{\text{O\&M}}$ is the operations and maintenance cost per unit of energy produced (\$/MWh)

From this relationship, FCR, Capital Cost, and O&M must be as low as possible, and at the same time, the AEP should be as high as possible. Using 9% cost of money and assuming installed 2.5 MW turbine example levels of Capital Cost, O&M, and AEP of \$1.43 MM/MW, \$25/MWh, and 8300 MWh, respectively, the resulting CoE is about \$64/MWh. If this example turbine was in an area where retail electricity cost consumers \$80–\$90/MWh, the wind turbine owner would stand to make a reasonable profit, even without government subsidies [32].

Internal and external relationships are key to the grid. Internal relationships include electricity generation, transmission, distribution, electrical losses, and loads (consumption). External examples are government policies, generation technologies, fuel sources, right-of-ways, zoning laws, smart grid technologies, consumer needs, and electricity pricing. Internal and external relationships can exhibit tight or loose couplings. Pairwise interactions with higher frequency, higher intensity, and/ or closer proximity are said to be tight; those with lower frequency, lower intensity, and/or greater distance are said to be loose. The grid acts robustly using many redundant features such as switchyards, fuses, and safety loops.

The grid is sensitive to small effects but incorporates corrective routines. Without these safeguards, sudden small changes in line frequency or voltage could result in much larger disturbances or even system failure. Switchgear and protective devices are used to minimize and even eliminate the chances for small perturbations to grow into larger issues. The openness of generation and load elements throughout the system—primarily in terms of metrics and ancillary impacts—informs the observer and provides for viable interventions. Some of the best improvements to the electrical grid have been made through extensive observations made over a wide variety of technology and operational circumstances.

Table 9.3 presents a collection of 30 key aspects for complex systems and the corresponding elements for utility-scale wind and the electric power grid. These key aspects are presented in general terms. A more detailed treatment is outside the scope of this case study.

Taken together, all of the key aspects of complex systems in Table 9.3 provide rich context to view the electric power grid and the implications for adding more large-scale wind power generation. What also become visible are the many other aspects of the overall system that can be just as important or perhaps even more crucial—energy storage and smart grid technologies.

In the most basic top-level view, the electrical grid is electricity generation inserted into a connective transmission and distribution grid, electricity stored or depleted, and electricity drawn upon (consumed). The overall system is the single most critical machine supporting humankind's existence. The system continues to grow in size and complexity due to population growth and catch-up for many areas of the world relative to the early adopters. All aspects of the ESoS are straining to keep up.

The primary electricity generation has historically been fossil fuel. Fossil fuel is finite in supply, has the undesirable side issue of releasing large amounts of GHG into the earth's atmosphere, and is impossibly unable to keep growing with the overall system. At this pivotal time, large-scale wind represents the best renewable alternative that can inject a significant amount of electricity into the overall generation mix.

Large-scale wind infrastructure can be manufactured and installed relatively quickly. Wind cost of electricity and investment values can be competitive and will only get better as fossil fuel stores are depleted. But most of all, wind is completely renewable and universal, deriving its energy from the sun and the continuous and everlasting differential heating of the earth's surface.

CHALLENGES

Since 1970, the world has more than tripled the production and consumption of electricity. We are on track to triple again by 2050. The United States managed to accompany this growth with a 50% reduction in GHG emissions, but other areas of the world such as China and India added capacity without the same level of emissions reduction. By some estimates, 1.4–2 billion people (20% or more of the earth's total population) still do not have access to electricity. Coal has carried the lion's share of base electricity generation in the past [34], but has begun to slip in recent years yielding to natural gas and renewables. We will need coal for some time to come, but it will not last indefinitely, and continued transition away from coal and all fossil fuels is the biggest challenge for the global electrical power grid ESoS.

Figure 9.8 data is adapted from IEA [35,36]. Coal peaked a few years ago and is expected to slowly decrease generation contribution over the next decades. The most dramatic drops in generation contribution for the past 30 years are for oil and hydro. Increased relative contribution shares are for nuclear, natural gas, and renewables, with wind leading the way for renewables. The world needs to continue having a rich mix of fuel options, and this diversity is key to the transition period needed to move away from burning fossil fuels and toward renewables.

Challenges for the electrical power grid beyond fuel include best practice sharing, technology and development collaboration, affordability, fuel source priority, and efficiency [37]. These issues require government and nongovernment entities and industry to work closer than ever before. This will require collaboration and short-term concessions by some sectors for longer-term prosperity for all participants. If the path can be defined using ESoS principles and data-based system value analytics, then the sacrifice and time needed for redirected efforts will be acceptable to all.

Over the past 30 years, global electricity generation from coal was around 40%. This cannot be sustained and there are overall indications that coal for electricity generation has peaked [38].

TABLE 9.3

Key Aspects for Wind and the Grid as a Messy Complex System

No.	Key Aspects for Complex Systems	Corresponding Elements for Utility-Scale Wind and the Electrical Grid
1	Recognition of wicked problems or messes including different worldviews—total view of a complex system is the summation of as many different view perspectives as possible.	Wind turbine and WPP design. Overall electricity generation growth. How wind contributes. Grid focus shift from increased generation to added storage, DR, and more efficient consumption.
2	Interconnected and interdependent elements and dimensions—different environmental adaptability for tight versus loose coupling within a complex system.	WPP intermittency integration. All generation sources interconnected to all consumers. Government policy influence for all aspects of generation and consumption.
3	Inclusion of autonomous and independent systems—complex system reliability and survivability.	Black start capability required regardless of generation technology. Wind does not have fuel or water delivery dependency.
4	Feedback processes—random nonlinear versus predictable or consistent linear feedback resulting in broad spectrum of system outcomes.	Grid balances generation and consumption. Voltage or frequency deviations from nominal are indicators of unbalance. Balancing authority (Independent System Operator; ISO) continuously weighs infinite solutions.
5	Unclear or uncooperative stakeholders—best results for different worldviews and to understand participative, adversarial, integrative, and managerial stakeholders.	Wind, electricity generation, and the grid work because there is a high level of cooperation among stakeholders. World views locally, regionally, and nationally from generation, T&D, and consumption perspectives.
6	Unclear boundaries—changing and blurred versus obvious or assignable.	Electrical grid boundaries are mostly clear. Transmission, distribution, and operational modeling confirm boundaries and performance within the grid boundaries.
7	Emergence—characteristics, behaviors, or outcomes that are not predicted and occur as a surprise—surprises that cannot be explained even after the occurrence.	Harmonic frequencies in electrical grids are many times a surprise. There are grid code standards to ensure compliance. Determining an exact cause for exceeding allowable resonance is sometimes difficult. Some generators are more susceptible but all can be corrected.
8	Nonlinearity—lack of an independent variable; adaptive behaviors and emergent dynamics are examples causing nonlinearity.	*Small-scale* power grid in which generation and loads are closely coupled within a larger grid, e.g., WPP and microgrid. Stability of short-term swing dynamics for small-scale power grids is called the coherent swing instability (CSI). This is an undesirable nonlinear phenomenon of synchronous machines in a power grid.
9	Sensitivity to initial conditions and path dependence—nonlinearity of relationships and uncertainty buildup are prime drivers.	For a given starting condition, every significant generator or load will to some extent interact differently for changes or growth in an electrical grid. Evolutionary grid modernization is inevitable and positive but path dependent
10	Phase space history and representation—set of all possible states as an analysis approach.	WPP production relative to the project-specific wind *rose*. ISO specified output or ancillary service options.
11	Strange attractors, edge of chaos, and far from equilibrium—underlying patterns showing lack of order or order from phase space mapping.	Time of day or other specific impacts to the electrical grid balancing operations. Other aspects such as specific intermittent generation or load events that can be random.
12	Adaptive agents—elements of a complex system that can perceive what is around them and act on their perceptions.	On-load transformer tap changing to automatically follow local system voltage requirements. Capacitor banks to automatically correct phase shifts and power factor lags. High-voltage switchyard to step up or down transmission voltage level.

TABLE 9.3 (continued)
Key Aspects for Wind and the Grid as a Messy Complex System

No.	Key Aspects for Complex Systems	Corresponding Elements for Utility-Scale Wind and the Electrical Grid
13	Self-organization—spontaneous collection of elements coming together to perform a task or function.	Switchyards, transformer taps, capacitor banks, and the entire range of grid elements interact in a seemingly self-organized way to ensure a balanced supply and demand of electricity.
14	Complexity leadership—disrupt existing patterns, encourage novelty, and help others self-realize good sense.	Burning fossil fuel has environmental downsides and is finite. Large-scale renewables address this. Utility-scale wind is the most scalable and less expensive of today's renewable options.
15	Coevolution—complex system evolution is dependent on the simultaneous and interacting evolution of the system components.	WPP and wind turbine technologies continue to evolve because of lessons learned from operations in different environments and grid conditions
16	Fitness—somewhere between ultrastability and chaos, an entity is best suited to cope with conflict or change.	Ultrastability for a grid would be one generator and one load in perfect equilibrium, and chaos would be multiple generators and loads randomly and rapidly varying electricity production and consumption. Today's electrical grid is somewhere in between.
17	Fitness landscapes—relative fitness of entities mapped to show the state of coevolution.	Wind power versus other forms of electricity generation. There are different types of electrical load and changing ways more work is being produced with less electricity consumption.
18	Recognition of architecture—the embodiment of project strategy in response to system complexity.	Utility-scale WPPs have increased the challenges for managing generation relative to the historical grid of just a few years ago. However, the challenges are being overcome, making the grid better than before.
19	Arrangement of systems into networks versus hierarchy—most SoSs are hybrid.	Power grids are vast networks of interplaying components. There are hierarchical subsystems throughout the grid such as high-, medium-, or low-voltage distribution and single versus three-phase power.
20	Recognition of external environment and project context—strategic, system, stakeholder, and implementation context.	An example of external environment for the grid includes GHG impact for burning fossil fuels and consumption of potable water. The ultimate stakeholders for the electrical grid are all of humankind and the total environment required for sustainable existence.
21	Management of complexity—latency processes, entities' self-interest and reactions, disturbance propagation, and exponential solutions based on entities.	Early large power generation came from hydro and moved to coal and oil. Natural gas is a relative newcomer with significant contribution from wind and solar, the most recent. Fossil fuel production and transportation comes with strong protectionism. A collapse in the electrical grid in one location can rapidly propagate throughout the remainder of the system if protective measures are overcome.
22	Modularity—separating and recombining subsystem elements—entity and function matrix.	Wind turbines offer significant unit power ratings that are combined to reach wind plant power blocks comparable to traditional fossil-fueled generation. Another example of modularity today is new generation that can be inserted into congested areas that no longer have transmission infrastructure capable of bringing more electricity in.

(continued)

TABLE 9.3 (continued)

Key Aspects for Wind and the Grid as a Messy Complex System

No.	Key Aspects for Complex Systems	Corresponding Elements for Utility-Scale Wind and the Electrical Grid
23	Views and viewpoints—perception versus reality.	Perception by most is unlimited electricity supply always available when we need it. Reality is that today, most of the electricity is produced from burning limited fossil fuel resources. Thus far large-scale wind is the best renewable option available.
24	Recognition that there are unknown unknowns—entities do not know what they do not know—variety of solution approaches versus continuous improvement.	Theory of constraints manifests itself in electrical grids when there are brownouts or blackouts. Essentially, the weakest links in the grid are exposed when the system is stressed from either lack of generation, excessive load, or a combination of both.
25	Complexity requires defined completion with the right actions, and creation of a control structure must at least have the range of the system being controlled.	Grid operators have priority order of actions for dispatching more or less generation in response to loads. These actions have the range of impact to cover all load scenarios. Unexpected or catastrophic loads trigger protective actions beyond normal controls.
26	Recognize different organizational structure due to scale and complexity—physical and informational aspects of complex systems become blurred.	Size and complexity of electrical grids results in differences in how they develop and are operated. The repeated process of localized collections of smaller grids into larger and larger interconnected grids is a common theme. The latest example for a super grid is the proposed AC–DC–AC ties to connect the US Eastern, Western, and ERCOT interconnection grids into one super grid servicing the entire continental United States [33].
27	Recognition of the importance of a different leadership style—abstract reasoning, business acumen, ambiguity, and emotional intelligence comfort.	Different and changing leadership adds an important growth and optimization element to the electrical grid. Wind energy is helping to solve problems that otherwise get overlooked; e.g., electricity from wind does not consume dwindling fresh water resources.
28	Governance—SOSs as web of shared interest versus command hierarchy.	The electrical grid elements interact virtually instantaneously by the flow of electrons. This shared interaction is inherent in the physics of electricity generation, supply, and consumption, although command hierarchy plays its part ensuring global aspects, e.g., government policy.
29	Incremental commitment the norm for projects—phase gate or tollgate reviews.	Power generation technologies are complex and expensive to construct and must be developed using an incremental commitment process. This is also true for improved grid or other aspects of the power grid.
30	Systemigram—two-dimensional layer diagram where layers hold families of entities or environments and connections are the interactions between the layers.	The flow of electricity in the grid is well represented by a spaghetti diagram (i.e., Systemigram). There are commercially available software packages that model electrical grids and provide design confirmation for proposed changes or additions.

Historically, the total price for coal has directly determined the cost for coal-fired electricity generation [39]. Total coal price includes mining, processing, storage, and transportation. It requires massive quantities of coal and heavy infrastructure to support coal-fired electricity generation.

A *unit train* of coal is about 2 km long containing 100 rail cars each with about 100 tons of coal. That is about 10,000 tons of coal or the equivalent weight of one Eiffel Tower or 1/36th of the Empire

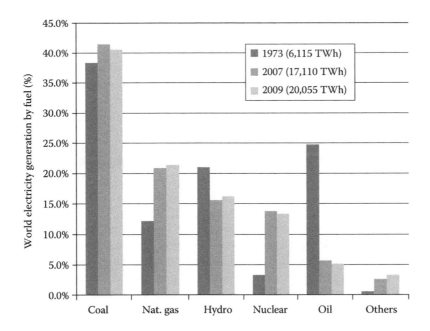

FIGURE 9.8 Share of overall electricity generation by fuel source. *Note:* Others include geothermal, solar, wind, biofuels and waste, and heat.

State Building, inclusive of their respective foundations. A large coal-fired electricity generation plant with 1000 MW rated output would require one of these trains per day with the coal consumed at a rate of 4.3 rail cars or 430 tons per hour.

In recent years, natural gas has begun to supplant some of the coal-fired generation, particularly in the United States where 46% of energy industry executives see natural gas as the most crucial fuel source for expanding the US electricity generation for the coming decade [40]. Nuclear and renewables come in second and third at 26% and 20%, respectively. Only 8% of the executives felt coal was important. But natural gas has its own production and transportation challenges.

A large 1000 MW natural gas–fired electricity generation plant requires a continuous flow of nearly 7.3 million standard cubic feet (SCF) of natural gas per hour. The delivery pipe required and dedicated to the plant for full load operation would be about 30 in. in diameter and flowing over 13 feet per second (FPS) at 600 PSIG.

Taking these two leading types of fossil fuel electricity generation and applying them to the world's current annual electricity generation needs of about 23,000 TWh, humankind is consuming about 9338 TWh/year using coal and 4922 TWh/year from natural gas. Our planet is depleting coal for electricity generation at the rate of 1184 unit trains of coal per day, which is equivalent to the weight of 1172 Eiffel Towers or 32 Empire State Buildings each and every day. Worldwide natural gas depletion for electricity generation is equivalent to 624 individual 30 in. diameter pipes each flowing 175.2 million SCF of natural gas per day, or the equivalent of a single 62 ft diameter pipe flowing 109 billion SCF of natural gas per day.

> Worldwide COAL depletion for electricity generation is equivalent to the weight of 32 Empire State Buildings each and every day.

> Worldwide NATURAL GAS depletion for electricity generation is equivalent to a single 62 feet (18.9 m) diameter pipe flowing 109 billion SCF of natural gas per day (13 FPS at 600 PSIG).

Recent studies for the United States have shown that there is currently the opportunity to supplant some coal electricity generation just by using the existing natural gas–fired power plants [41].

The United States has this option because of our large reserves of domestic natural gas. This could be an early quick hit for lowering GHG emissions without big infrastructure changes. The amount of coal generation reduction possible will depend on the actual gas-fired capacity available, the current and proposed coal and gas operating patterns, the ability for the short-term transmission grid to deliver the electric power flows from the respective plants to consumers, and sufficient gas supplies using existing delivery pipeline and storage capacity.

The US natural gas distribution system is a massive ESoS and system of systems (SoS) in and of itself. This infrastructure covers the entire spectrum from wellhead through processing, transport, storage, liquefied peaking stations, import, export, and consumption. At the start of 2007, there were 55 locations around the country where natural gas was imported and exported. Excluding local distribution company pipeline systems, over 300,000 miles (482,803 km) of major pipeline (>16 in. (0.4 m) diameter) infrastructure interconnects the lower 48 States [42]. That is equivalent to more than 1.25 times the average distance between the earth and the moon.

Years 2008–2011 saw considerable expansion of this US natural gas supply infrastructure averaging about 3300 miles (5311 km) per year or about 1.1% average annual growth [43]. This is equivalent to adding major natural gas pipeline of 1.33 times the distance from New York to Los Angeles every year. This pace of gas infrastructure growth will continue according to BP's most recent world energy outlook report [14], which projects natural gas as the fastest growing fossil fuel globally to 2030 at an average annual rate of 2.1%. This same report shows coal will reduce its contribution to the world's total energy output to 25% and renewable energy (primarily wind) will proceed to grow at an 8% average annual rate to 2030. This level of growth for wind generation is consistent with 33% of all electricity generation from wind by 2050 [15] discussed earlier.

It should also be noted that while this important shift upward for the contribution of natural gas and downward for coal proceeds, oil still plays an especially important role during the next few decades [44]. Oil will continue to contribute less and less to the power generation mix, much more quickly than coal; however, this does not mean that the need for oil infrastructure diminishes. On the contrary, oil demand is still projected to grow nearly 1% per year, although the primary driver for this is in the transport sector and not for electricity generation.

Meeting these challenges of fossil fuel transitions, infrastructure growth, and explosive wind power installations requires ESoS thinking and execution. The trend for larger wind turbines reaching higher into the sky and moving into offshore locations are some of the ways accounting for complex system elements are supporting what needs to happen next.

Take the trend for larger wind turbines for example. Bigger wind turbines are not necessarily better unless they deliver higher value and lower cost of energy for all stakeholders. Value analysis is crucial to the conceptual design process and ongoing reassessments. Mass production of highly reliable, low-cost, big turbines is the Holy Grail. Making larger and larger turbines is doable. The US wind electricity production could be as much as 20% of the total consumption by 2030 [45] and 33% by 2050 [15]. World wind power penetration is poised for similar progress.

It is not just about installing more and more electricity generation of any type, although it is clear that adding wind does away with the pressure on diminishing fuel supplies, less portable water, increased GHG emissions, production, and transportation and delivery headaches. Aside from the obvious conservation and more efficient consumption, electricity storage and smart power routing are increasingly important. Traditional electricity storage uses excess generation at times to power water pumps that returned water to higher head reservoirs.

More recently, battery storage has been used in some limited cases for low-energy applications, but it is very expensive and difficult to maintain. Of all the known storage technologies, hydro pumped storage and compressed air energy storage (CAES) offer reasonably economic grid-scale alternatives. Batteries and flywheel technologies will have their place in the future, but they generally offer smaller scale, lower energy storage capabilities and unfortunately at a much higher cost.

One emergent technology that may change the traditional view of grid-scale batteries is being developed by a Boston company using liquid metal. The company is now called Ambri [46], a takeoff from the word Cambridge—a tribute to the founding technology development by Professor Donald Sadoway at the Massachusetts Institute of Technology (MIT).

Ambri describes their breakthrough as a cost-effective, grid-scale battery technology composed of individual cells where

> Each cell consists of three self-separating liquid layers—two metals and a salt—that float on top of each other based on density differences and immiscibility [substances that cannot undergo mixing or blending]. The system operates at elevated temperature maintained by self-heating during charging and discharging. The result is a low-cost and efficient storage system.

Once low-cost and effective electricity storage can be integrated in large amounts throughout the grid, it will require less and less electricity generating capacity to maintain a reliable system. Less idle generating capacity for more of the time for traditional generation and more variable sources like wind/solar can be more effectively integrated into the grid without the need for installing additional fossil fuel make-up generation.

Economic demand for energy is an indisputable driving force. There are incredible challenges for all of the factors supporting the worldwide generation and consumption of electricity. The continued rapid transition and vibrancy of the fossil fuel mix is an important aspect for the growth of the electricity supply for the foreseeable future.

The continued explosive growth of wind power generation is another. Global free trades of fuel and electricity generation technology, together with product standardization, are crucial features supporting the projected growth. It is more than just electricity generation; there must be *specialization* and efficiency improvements across all grid elements. We also need worldwide per capita electricity consumption to converge to lower and lower global values [14]. The shared belief is that if we can account for all of these ESoS factors and more, humankind will have the electrical energy and make the transition to meet the needs of the earth's population through 2030. If this all goes as planned, or better than planned, the outlook for electricity supply beyond 2030 and to 2050 appears bright.

DEVELOPMENT

The Great Depression primarily drove the US government policy during the 1930s. President Roosevelt recognized that the US electric power grid was at an early juncture in terms of growth and was not available to everyone. It was time to establish transregional and new grid development, especially in rural areas. *Scientific management* (a precursor to SE) was used to execute Roosevelt's vision [47].

Once a larger grid (rural and urban) was established and more people involved, focusing on more electricity generation was the next natural step for rising demand. The demand side was still developing and the initial transmission and distribution infrastructure provided plenty of carrying capacity for the early build-out envisioned at the time. Even today there are still forces that make increased generation a favored solution to addressing growth. But simply adding more generation and buttressing existing transmission infrastructure are not going to get humankind on the track needed to prosper in the second half of this century and beyond.

Recently discussed in a book by Klimstra and Hotakainen [48] was the term *smart power generation*. The three main tenants throughout were that electricity generation must be reliable, sustainable, and affordable. There was some mention of smart grid elements (all other cross-functional grid elements), but the focus was on more efficient and reliable generation. Although this is not ESoS thinking (to only put forward improved generation), the authors did recognize that when it comes to humankind's future condition, business as usual for the advancement of the electric power grid is not an option.

Interestingly, this same book is at best neutral when it comes to large-scale wind, claiming that too much renewable energy in the generation mix requires more backup capacity and reduces all other power plant utilization factors. However, this is not true for either assertion based on newer information, especially when accounting for the bigger picture of a truly smart grid (e.g., Demand Side Management). However, their book does make an excellent point for the narrower view of simply adding conventional fuel-based power generation.

> Parallelized power plants consisting of larger numbers of smaller units do have better flexibility, reliability and fuel efficiency than power plants containing fewer larger units.

A more traditional SE approach takes the view of modular elements. Modular elements lend themselves to limiting scope to the development of subsystems. The next higher-level system is then an assembly of the subsystems. Adding more subsystems to grow the next bigger system is akin to plug and play. This may be acceptable for some circumstances, but not adequate to make bigger and bigger ESoS improvements at the scale needed for the electrical power grid of the future.

Figure 9.9 shows a representation of a regional grid using the traditional modular approach. The electrical power grid is made up of electrical power generators, transmission–distribution infrastructure, and consumers. This is also analogous to sources and sinks in an interconnected network. This modular view reduces all of the system participants into simple elements. Many CSs in nature can be reduced to the simplicity of the participating elements. For the massive size of the electrical grid, today's repetitive elements such as standardized electric power generators, transformers, transmission lines, distribution switching, and end-user consumption devices enable the overall system to grow and thrive. One advantage of this traditional view is that the elemental building blocks are easy to see, evaluate, and relate spatially.

However, this two-dimensional representation gets very complex as the region is expanded to larger geographic areas. Competing power generation technologies are confined to niche comparisons. With further complexity, it becomes more difficult to see and access global considerations such as fuel sources, processing, storage, transport, and consumption. It also becomes much more difficult to foresee unintended consequences.

Figure 9.10 takes two of the Figure 9.9 regional grids and interconnects them as an example of a simple expansion of the traditional modular view. Repeating this process is a confounding aspect that will result in further complications in terms of the original single region. By the time one subdivides the problem, works on optimizing each resulting subsystem, and reassembles the parts,

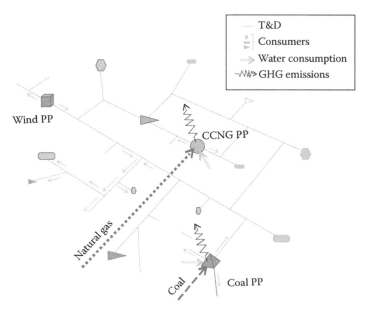

FIGURE 9.9 Modular view of the electrical grid—traditional systems engineering.

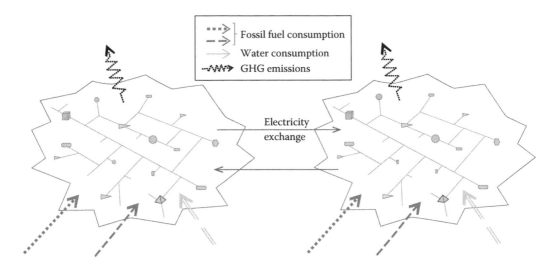

FIGURE 9.10 Modular system of systems view of the electrical grid.

the complex system and its environment have moved on, and likely the larger system will not perform as expected. Fuel inputs, water consumption, and GHG emissions for traditional generation can be represented for each subregion. However, bookkeeping each in terms of their total effect or understanding upstream or downstream impacts and synergies is not easily done.

An alternative analysis approach is to use a *layered view*. This is a much better ESoS mapping approach. A layered view increases visibility to the biggest possible system view, identifies areas of concern has flexibility and is a convenient environment to model and introduce overall system improvements.

Figure 9.11 shows a simplified example to demonstrate the advantages for this approach as applied to the electrical power grid. Each layer provides for a collection of sub-elements. Conveniently, the layers should be a common spatial framework such as a continent, country, or region, for example. In the largest view, the entire world map could be considered. Using this technique, the overall ESoS is more easily visualized, accounted, developed, optimized, and expanded.

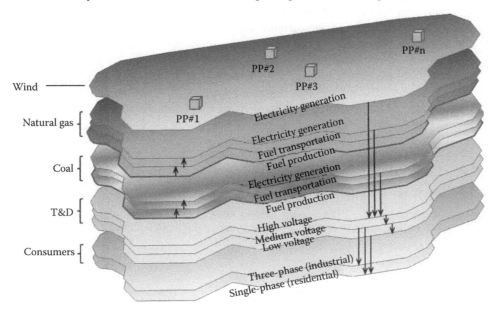

FIGURE 9.11 Layered enterprise system of systems view of the electrical grid.

The range of cross-functional and value stream relationships between layers are more visible using this approach. A clearer understanding of major industry segments and contributions can be seen. There is also clearer line of sight to opportunities for new integration or synergies. There is better overall visibility to subelement technologies, resource transport, and end use (electrical consumption) within the overall multidimensional model. Fuel, water, emissions, electricity and any other transport flows, and total accounting between layers are simplified. Many more layers can be added to improve the model and ensure capturing all of the system interactions without disrupting earlier versions. One can add more layers as new understanding is developed.

Additional layers that could be added to the model in Figure 9.11 include subcategories of electricity consumption and power generation influence characteristics. An example for the latter is the trend in developing taller wind turbines. Larger wind turbines for lower-cost power generation come with the benefit that the free fuel (wind) has higher electricity production potential as the overall turbine height and rotor swept area increase. A wind fuel layer for the United States a few years ago would have been mapped using 50 m height data. Today, that has changed to 80 m high relative to the local topography [49].

The hub heights of the majority of the latest large wind turbine products are 80 m or more. Winds that are 80 m off the ground are higher in speed than the old industry average of 50 m and therefore yield more electricity production per square meter of rotor swept area. This is an example of improved technology (utility-scale wind power) advancing turbine size that in turn influences the quality of the fuel that can be accessed. The improved quality of the fuel (wind) is now simply a new system layer that replaces the previous one. The overall layered ESoS view and the associated analytical model easily accommodate this change, and the corresponding interrelations readily reveal the impacts and opportunities.

Which type of electricity generation has the lowest cost and highest consumer value? Each power generation layer can be further sublayered to highlight other characteristics. Comparing the relative merits and effects on the overall ESoS becomes easier with the addition of more information layers. Table 9.4 shows a comparison of a number of power generation types [50] and shows that even relatively low CF wind power (i.e., 33%) without Production Tax Credits (PTC) can be the lowest cost when the natural gas price is more than $6 per million BTU (MMBTU).

This is not currently true with natural gas costing less than half of this. Natural gas–fired combined cycle gas turbine power plants are currently the lowest-cost electric power generation coming in at about 4.8 $c/kWh. When the price of natural gas goes back up, and it will, utility-scale wind will recapture the lowest-cost position.

The price of natural gas in recent times has ranged from $1.80 to $15.38 per MMBTU, with an average price in August 2011 of $3.89. The continued sluggish economy, increased supplies due to new finds, shale gas, and improved recovery methods such as hydraulic fracturing have resulted in natural gas prices for the electric power generation sector staying in the $3.00–$4.00/MMBTU range.

TABLE 9.4

Cost and Performance Comparison for Electricity Generation

Generation	Capacity Factor (CF or Dependability)	Price Range ($c/kWh)	Nom. Price ($c/kWh)	Total LCoE ($c/kWh)		Fuel		O&M (std)		Capital
Wind	32%–42%	4.4–11.5	8.0	5.7	=	0.0	+	1.3	+	4.4
Nat. gas[a]	70%–90%	6.6–10.9	8.8	6.9	=	4.1	+	0.5	+	2.3
Coal	70%–90%	7.4–13.5	10.5	7.8	=	2.2	+	0.5	+	5.1
Solar	22%–27%	14.1–21.0	17.6	16.0	=	0.0	+	1.1	+	14.9

[a] Using average cost NG of $6.07/MMBTU for years 2000–2010.

As a point of reference, the price for natural gas leading up to the economic downturn in 2008 had reached the $9.00–$13.00/MMBTU range.

Figure 6 in Chapter 6 of the reference book *Design and Development of Megawatt Wind Turbines* [51] compares well with Table 9.4. This figure is shown here for convenience and relabeled as Figure 9.12 relative to this case study.

Traditional types of power generation that burn fossil fuel today are more dependable (reliable) than renewables such as wind and solar. Wind and solar are variable by their very nature, and fossil fuels can be turned on, off, or throttled. But smart grid elements can reduce and even eliminate the impact for variable sources of generation. The most important enabling elements include sensors, communications, and controls.

The explosive growth for wind power penetration and burning less fossil fuel for electric power generation is directly dependent on the rate of smart grid and T&D infrastructure improvements.

There cannot be the required explosive growth of wind and other renewable power generation without the grid infrastructure and smart elements included. There is now a much bigger push than ever before to educate and develop industry interest for tackling this aspect of the electrical grid ESoS. *The super grids market 2012–2022* presents a compelling case for the value of this industry segment for all generation types (renewable and fossil) placing it at $10B globally for 2012 [52]. Seven regional super grids outlined to cover the world are Asia-Pacific, Latin America, North America, Europe, Commonwealth of Independent States (CIS), Africa, and the Middle East. Latin America, the Middle East, Asia-Pacific, and Africa will see the most dramatic super grid opportunities in part due to their current relatively immature starting point and system gains to be realized as national, regional, and continental grids are interconnected.

In a similar report by Pike Research, "Smart grid real progress still to come," the market value of the renewables portion of the overall global super grid is reported as $4B [53]. Comparing this to the $10B in 2012 for all generation types, this means $6B globally is being spent for updating and expanding the smart grid infrastructure for nonrenewable power generators. The 23% Cumulative Annual Growth Rate (CAGR) and $13B in 2018 for renewables reported by Pike Research translate into an all generation smart grid total global market revenue of $35B and $79B for 2018

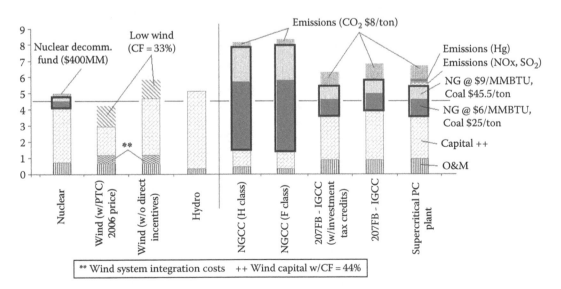

FIGURE 9.12 Generating alternatives—North America 20 year LCoE, $c/kWh. (From Willey, L., Design and development of megawatt wind turbines, in W. Tong, Ed., *Wind Power Generation and Wind Turbine Design*, WIT Press, Billerica, MA, ch. 6, pp. 187–256, 2010.)

and 2022, respectively. With the current global electrical grid infrastructure being valued in trillions of dollars, these relatively modest investments in smart grid elements appear woefully little if we are going to be prepared for humankind's electricity needs for 2050 and beyond.

One of the best examples of a balanced view for the renewable energy future for the grid in the United States is presented in a report released this past summer by NREL [15]. In addition to a comprehensive SoS approach to analyzing all of the cross-functional elements of the electric power grid, there is appropriate treatment of the transmission, distribution, and smart grid elements required to ensure high penetration of renewables overall (80% of all generation) and wind in particular (33% of all generation) by 2050. As mentioned earlier in this case study, projecting these same levels of renewable penetration for the entire world for 2050 seems very plausible based on current trends.

Rather than relying purely on analytical studies or taking chances with the reliability for a particular continent, country, or region of a production grid, these newer renewable technologies and smart grid elements are soon to be evaluated in a laboratory environment. A new US national laboratory is being built in Golden, Colorado [12]. It will be known as the Energy Systems Integration Facility (ESIF) and is colocated at the US DOE National Renewable Energy Laboratory (NREL). ESIF will permit the evaluation of smart grid technologies and intermittency to keep the grid running reliably without involving validation using the production grid.

What was once governed primarily by the variability of demand, the emerging grid will accommodate greater amounts of variable generation such as wind. Industry attention is now beginning to focus on the execution of all aspects of the ESoS with the global electrical power grid properly accounted. CSE principles and evaluation techniques provide a clear view of how humankind can meet electric power requirements at least into the 2050 time frame. This will not be done for the entire world using a government policy similar to what President Roosevelt did for the United States in the 1930s. Instead; SE and the ESoS principles are pointing to a better way for all regions of the world to arrive at the required solution in the most efficient manner possible.

RESULTS

The United States and the world have rapidly transformed such that today 3%–4% of our total electrical energy needs to come from wind power [14]. This compares to less than 1% just a few years ago. All influence factors are driving the growth of wind electricity generation such that about one-third of the world's electricity can realistically come from wind by 2050 [15]. Electrical grids are increasingly more interconnected. Computer control of transmission and distribution has in many cases supplanted manual monitoring.

The Electric Power Research Institute (EPRI) was formed in the aftermath of the 1965 northeastern United States blackout to investigate and promote better generation, distribution, consumption, and overall grid reliability. Prior to this, 99.99% reliability for the grid (equivalent to 53 min per year without power) was considered acceptable. But with the increased reliability demanded by financial, medical, and computational requirements, 99.9999999% or *9-nines* dependability is the target for today's electric power grids [54]. This is equivalent to only three one-hundredths of a second per year without power.

This transition to higher reliability is being accomplished on a number of fronts including installation of wind as the leading new generation (on a nameplate basis), more interconnections between large grids, and decentralization within grids with net metering and smart grid elements such as local storage and smart consumption appliances.

Wind and solar installations continue to increase as utilities and power providers are looking for more sustainable, abundant, and cleaner sources of energy. However, the intermittent nature of wind and solar can create voltage or frequency stability problems between the transmission system and the provider. In response, utilities and regulators are increasingly mandating strict grid interconnection requirements. Renewable energy producers must condition the power produced in order to interconnect with the power grid and not interfere with the grid's overall performance [55].

Having 1/3 of all worldwide electricity from the wind by 2050 is realistic [15]. Having 1/3–1/2 of all worldwide electricity coming from renewables is even more plausible. Getting a glimpse of this future and addressing potential issues early from an ESoS perspective have already been developing for the past few years in places like the Hawaiian Islands.

Demand–response (DR) is the concept that instead of escalating generation to meet a system load surge, one can decrease a corresponding total amount of *least important* loads. The ability for a grid to automatically do this is a perfect companion for variable generation such as wind. Properly implemented DR would also negate the need for fossil fuel backup generation—eliminating one of the arguments used by wind opponents.

In Hawaii, for example, a DR program is already being demonstrated that focuses on automatic Wi-Fi or radio control of commercial and industrial customers having at least 50 kW of discretionary load within their facilities [56]. The express need for this capability is the result of rapid penetration of wind into the Hawaiian Electric Company (HECO) grid. Oahu is poised to easily achieve ahead of schedule its share of the Hawaii Renewable Portfolio Standard (RPS) of 40% by 2030, and proving this technology early is paving the way for others around the world to consider similar SoS electric power solutions.

Other examples of innovative systems approach to the electrical grid are coming from the Galvin Electricity Initiative discussed earlier. An important realization from Galvin is that improved smart grid elements support consumer value (or conversely avoid loss of wealth) for electricity that is at least one or two orders of magnitude greater than the retail price for electricity. This range of consumer value covers the spectrum of end-user electricity consumption from residential and noncritical commercial (10×) to the most valuable critical financial and health-related services (100× or more); that is, electricity that is trouble free and reliable creates value for consumers that is many times the price in cents per kilowatt hour (¢/kWh) that consumers pay for electricity [57].

To gain insight into all the influence factors affecting consumer value, Galvin and others have created a number of analytical models. These will be discussed next.

ANALYSIS

There are far too many electrical power grid modeling software tools to describe or contrast in this case study. However, suffice to say that none known to the author actually account for the true ESoS or CSE elements across all sectors within the grid as it is today, plus distributed generation, plus DR, plus smart meters, plus smart appliances, plus government policies, and so on. This truly holistic analysis would be extremely difficult and computationally intensive. But the Galvin Electricity Initiative is creating analytical models that begin to address some of the design space yet to be explored in a larger systems view.

Two examples are Galvin's building-integrated and microgrid distributed power value models [58]. Building-integrated distributed power systems are also known as *BIPS*, and microgrid distributed power systems as *MDPS*. The models represent just two grid scale views and are provided free at Galvinelectricity.com. What is clear is that by continuing this type of modeling for different grid views, and later computationally coupling this with all of the cross-functional aspects of the ESoS electric power grid, we could be very close to the time when a holistic ESoS analytical model is realized.

In Frank Kreith's article "Bang for the Buck [59]," Energy Return on Energy Investment (EROI) is another key metric for understanding true sustainability for energy-based systems. Money is not the only limiting factor for energy systems. The amount of energy devoted to finding and processing more energy is even more important.

EROI is the amount of energy produced during the life of a system divided by the amount of energy input to find, build, process and run the system.

EROI is defined as the ratio of the amount of energy produced during the life of a system divided by the amount of energy input to find, build, process, and run the system. The larger the EROI ratio, the better it is for the system. In the end, the ESoS is ultimately all of us (humans).

> To put it in terms of an analogy, microorganisms require food input to provide nutrients to run bodily functions with enough calories left over to enable growth. To have the organism make progress in terms of growth, there needs to be easy access to food so that little effort is expended finding food and more of the food nutrients go into growth (this is a relatively high EROI ratio on the order of 100 or more). Worse case is when the effort required to secure food exceeds the energy content of the food (an EROI ratio less than one).

When fossil fuels are viewed this way, EROI started on the order of 100 or more during the early twentieth century. It fell to around 20 by the later part of the century and today is in the low single digits. EROI below 10 is considered problematic and EROI of 3–5 or below does not supply enough surplus energy to sustain modern life, as we know it. This is yet another signal that alternative energy solutions, especially utility-scale wind power, must continue to be rapidly deployed if we are to have a chance at growing the power grid in concert with population growth. This is also a crucial enabler for improving living standards for the 1.4–2.0 billion people that currently do not have access to electricity [36].

Wind, solar, and nuclear power generation currently have EROI's of about 20, 10, and 5, respectively. Wind is the most promising prospect for delivering new alternative energy without fuel depletion issues. Over the next 38 years, the portion of the world's electricity demand provided by wind could go from about 3%–4% to 33%, representing an average CAGR of about 5.7%.

Some reasons that could negatively affect this amount of growth for wind include short-term abundance of reasonably low-cost fossil fuels, erratic or ineffective public policy for wind relative to the historical subsidies given to the fossil fuel companies, and better than expected results for conservation, efficiency gains, and smart grid efforts. With the exception of the impact for improved conservation, efficiency gains, and smart grid elements, we would all do better if large utility-scale wind growth stays on course for 33% of the world's electric power supply by 2050.

Upstream considerations are numerous for supporting the grid, including the resources and materials required to generate, transmit, and consume electricity. An example discussed earlier for resources is water. An example for materials is the amount of steel required to establish an electric power generation facility.

Table 9.5 shows that today's utility-scale wind power is steel intensive on a nameplate MW basis, more than two times the next closet power generation type (nuclear). This becomes further exasperated when accounting for the fact that the WPP is at a CF that is 1/3 to 1/2 as much as a nuclear power plant. But when you take into account that the WPP produces electricity for 20–25 years with zero fuel cost, zero water consumption or GHG emissions, the upfront steel input becomes somewhat negated. Using concrete for the wind turbine tower instead of steel would also eliminate some of the need for large amounts of steel.

Bigger wind turbines (larger rotor diameters and generator ratings) can result in lower-cost electricity production. The physics for sweeping more rotor area for a single foundation can lower energy costs, particularly when combined with new technologies that can reduce the amount of materials in the turbine and improve its reliability.

There is a balance between increasing turbine size and the technology used. To merely scale a particular turbine up in size without improving technology will quickly reach a point where the economics will actually get worse for a larger turbine. This may seem counterintuitive, but consider reliability issues for a smaller turbine translated to a bigger turbine in terms of excessively high costs for failures and it becomes more easily understood.

WPPs of various power block sizes (ratings) have been shown to have better economics using 1 or 2 MW wind turbines versus even larger turbines by Willey et al. [63]. Economic evaluation over a range of technologies throughout the different sizes of wind turbine equipment was used to feed

TABLE 9.5
Amount of Steel Required to Install Power Generation

Power Plant Type	Steel Content (Tons/MW)[a]
Wind	95[b]
Nuclear	40
Coal fired	34
Hydro pumped storage	20
Combined cycle gas turbine	3
Open cycle gas turbine	2

Sources: Mthethwa, M., The impact of power generation projects on the steel industry, October 2006. http://www.cncscs.org/saisc/The%20impact%20of%20power%20generation%20projects%20on%20the%20steel%20industry.pdf [Online]; Peterson, P.F. et al., Metal and concrete inputs for several nuclear power plants, February 2005, http://pb-ahtr.nuc.berkeley.edu/papers/05-001-A_Material_input.pdf [Online]; Meier, P.J. and Kulcinski, G.L., Life-cycle energy cost and greenhouse gas emissions for gas turbine power, December 2000, http://www.ecw.org/prod/202-1.pdf [Online].

[a] Excluding the prime mover (turbine and generator) and building.

[b] Steel tower and rebar in foundation.

an analysis of the overall WPP economic value in terms of Net Present Value (NPV) and CoE. The study accounted for civil works, turbine spacing, wake effects, collection system, balance of plant (BOP), crane costs, and O&M personnel, among others. Larger turbines using today's technologies may only make sense for onshore sites with the most severe land area constraints (& where overall height restrictions allow) or offshore. Bigger turbines must incorporate technology to overcome current limitations and achieve higher reliability with lowest life-cycle CoE.

Offshore wind power is presently not being exploited to the extent it could be, yet the pressure to go offshore, particularly in Europe, is compelling the industry to move quicker than what the present technology and reliability can provide. Offshore turbines demand even better reliability because access to the turbines is so much more expensive and difficult. We must achieve higher reliability for onshore turbines with as much vigor as offshore by developing new technology. Offshore appears to have a chance at becoming practical and cost effective, but making a mistake offshore could have a disastrous economic impact that would then slow or stop technology development progress. Time will show if this mismatch in technology development onshore versus offshore will result in smooth growth in the plan (with the level of collaboration required) or slow things down with costly missteps.

SUMMARY

The electrical power grid has transformed the way humankind flourishes. Affordable and abundant electricity has irreversibly changed virtually every aspect of life for almost 8 of every 10 people on the planet. Transition away from burning fossil fuel for electricity is required as fast as possible. Insist on ESoS analytical modeling and appropriate validation for new grid technologies to ensure compatibility and performance prior to widespread introduction. Renewable sources of energy must swiftly play a greater role for electricity production. Large utility-scale wind can best fit this need for the foreseeable future. More generation is not the leading solution to satisfy growing electricity demand. Better transmission, distribution, lower losses, more efficient consumption, and smart grid elements are crucial to meeting future worldwide electricity requirements.

CONCLUSIONS

1. Humanity's large population today and in the future cannot survive without electricity.
2. Electricity's criticality to man's survival demands that ESoS modeling and guiding principles must be more rigorously applied to support further electric power grid growth and development.
3. Growth and development of the grid cannot compromise the ability for future generations to meet their own needs.
4. Renewable energy sources led by utility-scale wind are key for solving the sustainability problem.
5. Large-scale wind electricity generation is a key enabler to meeting a wider range of electric power grid growth objectives.
6. The transition to renewables requires burning more natural gas in the short term, less coal, and more importantly making system changes to the grid infrastructure and how the grid is operated.
7. Electricity storage, DR, and other smart grid technologies are key enablers for high penetration wind power electricity generation.
8. It is completely plausible that one-third of all worldwide electricity generation could come from wind by 2050.
9. The electrical power grid is the ultimate human-made ESoS.

SUGGESTED FUTURE WORK

This case study put forward a high-level systems overview of the electric power grid and its elements, growth projections, and technology plays with an emphasis on reducing fossil fuel depletion/water consumption/GHG emissions and significantly increasing penetration for the world's most promising renewable energy source—utility-scale wind power. As indicated in the overview, there are some very good studies underway—some of which appear to be using many of the ESoS and CSE principles needed to ensure useful results. However, there are a number of follow-on activities recommended to ensure that the next higher level of ESoS analysis is performed to help understand a number of specific questions and scenarios. This work needs to be a very well–coordinated multimodeling effort for all the different grid views coupled together with all of the cross-functional aspects of the electric power grid—a true ESoS analytical laboratory environment.

DISCUSSION QUESTIONS

1. What would it look like today for (1) the United States and (2) the world if there were an abrupt loss of one or more types of fossil fuel–based electricity generation?
2. Adding more and more electricity generation has been the historical response for meeting increasing electrical demand. What significant changes to the grid, both macro and micro, are needed to decrease the need for more generation for (1) the United States and (2) the world? How fast can these changes be implemented and is the cost/benefit a good trade?
3. It seems that the improving human longevity is at least in part due to better nutrition and improved health care, but is not the real reason for the past 100-year gains rooted in the introduction and growth of electricity? Energy and electricity are the real reasons for higher food yields, the advent of electronic diagnostic medical devices, and comfortable living environments. Is the reason why there is now a flattening trend for human life span the reality of a hard biological limit?
4. In the context of finite fossil fuels and the current reality of mandatory electrical grid growth—are too many people the true root cause of humanity's long-term survivability and quality-of-life issues? What are the most realistic approaches to (1) reducing per capita

electricity consumption, (2) lower population growth, and (3) achieve a long-term balance between lower human population and earth's finite resources?

5. It is been suggested that artificial intelligence (AI) will become so powerful in the next decade or two that it will yield solutions to all of humankind's problems. Does this seem reasonable? Are there potential negative outcomes for applying AI to the problem of electricity supply and humankind's survival?

REFERENCES

1. A.W. Culp Jr. (1979) *Principles of Energy Conversion*. F.J. Cerra, Ed. New York: McGraw-Hill, Inc.
2. Six sigma tools—The Pugh matrix, SIX SIGMA GLOSSARY: Pugh Analysis. (May 2012). http://www.micquality.com/six_sigma_glossary/pugh_analysis.htm. Accessed January 25, 2012.
3. R. Smalley. (September 2003) American energy independence—Our energy challenge. http://americanenergyindependence.com/library/pdf/smalley/OurEnergyChallenge.pdf. Accessed January 6, 2012.
4. NC Offshore Wind Coalition (May 2012). Wind power water consumption versus other energy sources. http://www.ncoffshorewind.org/faqs.html#environment. Accessed December 22, 2012.
5. Negative Population Growth Org & U.S. Census Bureau. World population by year (June 2012). http://www.npg.org/facts/world_pop_year.htm. Accessed May 5, 2012.
6. A.A. Bartlett. (April 2012) Population media center org—The meaning of sustainability. http://www.populationmedia.org/2012/04/04/the-meaning-of-sustainability-by-professor-emeritus-albert-a-bartlett/. Accessed February 24, 2012.
7. C. Korner and M. Ohsawa. (2005) Mountain systems, in *Ecosystems and Human Well-Being: Current State and Trends*, Volume I. R. Hassan, R. Scholes, and N. Ash, Eds. Washington, DC: Island Press, ch. 24, pp. 681–716.
8. E.A. Viau. (2003) World Builders—Primary Productivity Table. http://www.world-builders.org/lessons/less/biomes/primaryP.html. Accessed May 24, 2012.
9. Global Rural Urban Mapping Project (GRUMP). (March 2005) The Earth Institute—Columbia University. http://www.earth.columbia.edu/news/2005/story03-07-05.html. Accessed April 13, 2012.
10. Food and Agriculture Organization of the United Nations. (2010) Yearbook 2010 resources. http://www.fao.org/economic/ess/ess-publications/ess-yearbook/ess-yearbook2010/yearbook2010-reources/en/. Accessed January 9, 2012.
11. The Energy Library. (2009) North American electricity grid. http://theenergylibrary.com/. Accessed January 19, 2012. node/647.
12. H. Lammers. (December 2011) New lab to help utilities "see" grid of the future. *Golden Transcript*, p. 25.
13. K. Marvel, B. Kravitz, and K. Caldeira. (September 2012) Geophysical limits to global wind power. http://www.nature.com/nclimate/journal/vaop/ncurrent/full/nclimate1683.html. Accessed May 13, 2012.
14. C. Ruhl, P. Appleby, J. Fennema, A. Naumov, and M. Schaffer. (January 2012) Economic development and the demand for energy: A historical perspective on the next 20 years. http://www.bp.com/liveassets/bp_internet/globalbp/STAGING/global_assets/downloads/R/reports_and_publications_economic_development_demand_for_energy.pdf. Accessed May 4, 2012.
15. M. Hand and T. Mai. (June 2012) Renewable electricity futures study. http://www.nrel.gov/analysis/re_futures/. Accessed February 5, 2012.
16. Green World Investor. Electric generating capacity by country—world electricity production growth driven by emerging markets (March 2012) http://www.greenworldinvestor.com/2011/03/29/electric-generating-capacity-by-country-world-electricity-production-growth-driven-by-emerging-markets/. Accessed December 29, 2011.
17. CIA World Factbooks 12/18/03 to 3/28/11 Source. (February 2012) Nation master. http://www.nationmaster.com/graph/ene_ele_con-energy-electricity-consumption. Accessed January 16, 2012.
18. OECD/IEA. (2010) International Energy Agency. http://www.iea.org/textbase/nppdf/free/2010/key_stats_2010.pdf. Accessed April 20, 2012.
19. U.S. Energy Information Administration. (October 2011) How much of world energy consumption and electricity generation is from renewable energy? http://www.eia.gov/tools/faqs/faq.cfm?id=527&t=1. Accessed April 4, 2012.
20. REN21. (July 2011) Renewable energy policy network for the 21st century. http://www.ren21.net/REN21Activities/Publications/GlobalStatusReport/GSR2011/tabid/56142/Default.aspx. Accessed June 1, 2012.

21. CIA World Factbook Source. (January 2011) Index mundi. http://www.indexmundi.com/g/g.aspx?c=xx&v=81. Accessed February 22, 2012.

22. IEA Statistics OECD/IEA Source. (February 2012) The World Bank. http://data.worldbank.org/indicator/EG.USE.ELEC.KH. Accessed February 14, 2012.

23. L. Willey. (2009) Gigawatt wind power: Supporting U.S. 20% wind generation by 2030, in *2009 MIT Conference on Systems Thinking for Contemporary Challenges*, Cambridge, U.K., p. 10.

24. D. Elliott et al. (August 2003) Wind energy resource atlas of Oaxaca, NREL/TP-500-34519. http://pdf.usaid.gov/pdf_docs/PNADE741.pdf. Accessed May 26, 2012.

25. A. Noymer. (May 2012) Life expectancy in the USA 1900–1998. http://demog.berkeley.edu/~andrew/1918/figure2.html. Accessed May 17, 2012.

26. National Academy of Engineering. (2012) Greatest engineering achievements of the 20th century—Timeline. http://www.greatachievements.org/?id=2984. Accessed February 14, 2012.

27. Galvin Electricity Initiative. (2007) Power through time: Energy timeline. http://www.galvinpower.org/history/energy-timeline-power-through-time. Accessed January 4, 2012.

28. National Academy of Engineering. (2000) Greatest engineering achievements of the 20th century—Electrification timeline. http://www.greatachievements.org/?id=2971. Accessed December 11, 2011.

29. Galvin Electricity Initiative. (2005) Robert W. Galvin. http://www.galvinpower.org/robert-galvin. Accessed May 1, 2012.

30. Galvin Electricity Initiative. (2005) The elements of the perfect power system. http://www.galvinpower.org/sites/default/files/documents/Perfectpowerforiit.pdf. Accessed February 29, 2012.

31. Galvin Electricity Initiative. (2005) Bob Galvin's recipe for transforming the electricity sector. http://www.galvinpower.org/sites/default/files/Bob_Galvins_Recipe_For_Electricity_Transformation.pdf. Accessed May 16, 2012.

32. L. Willey, R. Budny, and S. Gupta. (August 2012) Challenges & rewards for engineers in wind, in *Global Gas Turbine News, a Supplement to Mechanical Engineering Magazine*, American society of mechanical engineers (ASME), Newyork. pp. 56–57.

33. National Public Radio. (May 2009) Visualizing the U.S. electric grid. http://www.npr.org/templates/story/story.php?storyId=110997398. Accessed May 23, 2012.

34. National Academy of Engineering. (2012) Electrification essay by E. Linn Draper, Jr., past Chairman, President and CEO, American Electric Power Company, Inc. http://www.greatachievements.org/?id=3000. Accessed April 17, 2012.

35. International Energy Statistics. (July 2009) 2009 Key world energy statistics. http://www.iea.org/textbase/nppdf/free/2009/key_stats_2009.pdf. Accessed May 23, 2012.

36. International Energy Agency. (July 2011) 2011 Key world energy statistics. http://www.iea.org/textbase/nppdf/free/2011/key_world_energy_stats.pdf. Accessed May 30, 2012.

37. e7 and its world-wide partners at the request of the United Nations Environment Program. (January 2002) Electricity sector report for the world summit on sustainable development. http://www.globalelectricity.org/upload/File/Electricity_Sector_Report_(E7_Version_-_Full).pdf. Accessed May 12, 2012.

38. E. Glover. (December 2010) Electricity generation in a post-coal world. http://large.stanford.edu/courses/2010/ph240/glover1/.

39. J. McNerney, J.D. Farmer, and J. Trancik. (January 2011) Historical costs of coal-fired electricity and implications for the future. http://www.sciencedirect.com/science/article/pii/S0301421511000474.

40. Forbes Insights in association with CIT. (February 2012) 2012 U.S. Energy Sector Outlook. http://www.rangeresources.com/rangeresources/files/46/46e6cafe-faa3-48b9-9f52-e6a021e3df14.pdf. Accessed February 3, 2012.

41. S. Kaplan. (January 2010) Displacing coal with generation from existing natural gas-fired power plants. http://assets.opencrs.com/rpts/R41027_20100119.pdf. Accessed June 4, 2012.

42. Energy Information Administration. (June 2007) About U.S. natural gas pipelines—Transporting natural gas. http://www.eia.gov/pub/oil_gas/natural_gas/analysis_publications/ngpipeline/fullversion.pdf. Accessed February 28, 2012.

43. Energy Information Administration. (September 2009) Expansion of the U.S. natural gas pipeline network: Additions in 2008 and projects through 2011. http://www.eia.gov/pub/oil_gas/natural_gas/feature_articles/2009/pipelinenetwork/pipelinenetwork.pdf. Accessed May 27, 2012.

44. M. Finley. (December 2011) The oil market to 2030—Implications for investment and policy. http://www.bp.com/liveassets/bp_internet/globalbp/STAGING/global_assets/downloads/R/reports_and_publications_oil_market_to_2030.pdf. Accessed May 14, 2012.

45. Report 20% Wind Energy by 2030. (July 2008) National Renewable Energy Laboratory (NREL). http://www.nrel.gov/wind/pdfs/41869.pdf. Accessed January 15, 2012.

46. Storing Electricity for Our Future. (August 2012) AMBRI. http://www.ambri.com/. Accessed January 9, 2012.
47. National Academy of Engineering. (2012) Electrification history 2—Rural electrification. http://www. greatachievements.org/?id = 2990. Accessed May 1, 2012.
48. J. Klimstra and M. Hotakainen. (2011) *Smart Power Generation: The Future of Electricity Production.* Helsinki, Finland: Avain Publishers.
49. M. Jaffe. (December 2011) Taller towers put wind in turbines' sails, *The Denver Post*, p. 1K & 8K.
50. M. Jaffe. (August 2011) Price of power, *The Denver Post*, p. 1K & 4K.
51. L. Willey. (2010) Design and development of megawatt wind turbines, in *Wind Power Generation and Wind Turbine Design*, W. Tong, Ed. Billerica, MA: WIT Press, ch. 6, pp. 187–256.
52. Visiongain. (2012) The Super Grids Market 2012–2022. Visiongain Ltd, London, Independent business information provider CCR Ref: KD4R6.
53. Pike Research. (2012) Smart Grid Renewables Integration. Pike Research, Navigant Consulting, Inc, Boulder, Independent industry research n/a.
54. National Academy of Engineering. (2012) Electrification history 5—Looking forward. http://www. greatachievements.org/?id=2998. Accessed January 5, 2012.
55. AMSC—American Superconductor. (June 2012) Renewable interconnectivity solutions. http://www. amsc.com/documents/renewable-interconnectivity-solutions-brochure/. Accessed May 22, 2012.
56. A. Beniwal. (June 2012) Demand response could unlock state's renewable energy. http://www. renewgridmag.com/e107_plugins/content/content.php?content.8603. Accessed June 2, 2012.
57. Galvin Electricity Initiative. (2005) The galvin path to perfect power—A technical assessment. http:// www.galvinpower.org/sites/default/files/documents/Perfect_Power_Technical_Assessment.pdf. Accessed April 6, 2012.
58. Galvin Electricity Initiative. (April 2007) Building-integrated and Microgrid distributed power value models. http://www.galvinpower.org/sites/default/files/Models_User_Guide%20[Read-Only]%20[Compatibility%20 Mode].pdf. Accessed May 24, 2012.
59. F. Kreith. (May 2012) Bang for the buck, *Mechanical Engineering*, pp. 24–31.
60. M. Mthethwa. (October 2006) The impact of power generation projects on the steel industry. http:// www.cncscs.org/saisc/The%20impact%20of%20power%20generation%20projects%20on%20the%20 steel%20industry.pdf. Accessed March 19, 2012.
61. P.F. Peterson, H. Zhao, and R. Petroski. (February 2005) Metal and concrete inputs for several nuclear power plants. http://pb-ahtr.nuc.berkeley.edu/papers/05-001-A_Material_input.pdf. Accessed February 25, 2012.
62. P.J. Meier and G.L. Kulcinski. (December 2000) Life-cycle energy cost and greenhouse gas emissions for gas turbine power. http://www.ecw.org/prod/202-1.pdf. Accessed January 4, 2012.
63. L. Willey, J. Nies, and K. Jerwann. (2008) Total customer value in utility scale wind power generation, in *AWEA WindPower 2008 Poster Presentations*, Houston, TX.

Section V

Finance

10 United States Banking System
Systemic Risk and Reliability Assessment

Khaldoun Khashanah

CONTENTS

CASE STUDY ELEMENTS

- The banking system is an example of interconnected sociotechnical complex adaptive systems.
- The banking system is an example of a system of systems (SoS).
- The banking system in a country forms a connected network of enterprises with flows representing cash flows and risk flows. The subjects in this case study are nine systemic banks in the US banking system, which define the *reduced* banking SoS in this case study. We drop the term *reduced* for the rest of the case study and use the banking SoS to refer to the nine systemic banks.
- Systemic risk assessment and indicators are one of the challenges that are faced by today's sociotechnical financial SoS. In this case study, systemic risk assessment of the banking SoS is used as a proxy for systemic risk in the entire financial system.
- The stakeholders in the banking SoS are investors, enterprises, regulators, and ordinary citizens. The public is considered to be a secondary and passive stakeholder, while investors and regulators are active stakeholders. Enterprises depend on the banking system for all aspects of the business stages from funding innovative ideas and start-ups to stable financing of large societal initiatives.

KEYWORDS

anasynthesis, banking system, credit default swaps, financial crisis, financial indicators, sociotechnical systems, systemic risk.

ABSTRACT

Whenever a bank has a large and systemically important role to play in the financial system, it becomes equally important to monitor how much risk it owns as a function of time and how much damage to the system it would cause under scenarios that may lead to its illiquidity or insolvency. The regulator conducts stress tests against each bank to keep capital requirements in check. However, the regulator can only stress test each bank individually. Using the language of SoS, the regulator can stress test components, usually at different snapshots in time, and does not have the tools to stress test the whole SoS through analytic synthesis. The interconnecting links between banks is the missing part of such a systemic stress test. Empirical financial indicators are of great value in providing synthetic information as they act as aggregators of market participants' positions and intelligence. In this case study, we will develop a simple indicator and *heat maps* based on credit default swaps (CDSs) that act as a proxy for banks' credit risk. We will determine how the mutual credit risk between banks can be captured via measuring the relative comovement of CDS spreads expressed in time series of CDS spreads.

GLOSSARY

BIS   Bank of International Settlement
bps   Basis points
CDS   Credit default swap
CFTC   (US) Commodity Futures Trading Commission
DFA   Dodd–Frank Act
EVT   Extreme value theory
FDIC   Federal Deposit Insurance Corporation
FINRA   Financial Industry Regulatory Authority
FSOC   Financial Stability Oversight Council
GDP   Gross domestic product
ISDA   International Swaps and Derivatives Association, Inc.
LIBOR   London interbank offered rate
OFR   Office of Financial Research
OTC   Over-the-counter
P/E   Price to earnings (ratio)
SEC   (US) Securities and Exchange Commission
SIFI   Systemically important financial institution
SoS   System of systems
VaR   Value at risk
VIX   (the ticker symbol for the Chicago Board Options Exchange Market) Volatility Index

BACKGROUND

The aftershocks of Black Monday (October 9, 1987) provided the impetus for an increase in the role of risk management in the enterprise governance. As a result of successive investigations, risk types were identified that included market risk, operational risk, credit risk, reputational risk, management risk, regulatory risk, and, more recently, technology risk and cybersecurity risk. The classification of types of risks helped managers in their day-to-day decisions and also helped regulators in monitoring the banking and investment system in a methodical way. As a result of unprecedented expansion in market types, financial instruments, computational power, financial technology, and intractable mutations of financial innovation and regulation in the early 1990s, it became clear that the old system of conducting business based on name recognition and trust was being challenged. Financial risk management was also being formulated in financial institutions to provide measures of risk. J.P. Morgan developed a measure of risk and made it available to other institutions to use in 1994. A popular measure that emerged was the value at risk (VaR), which also had an unintended consequence to demonstrate the concept of a model risk at the systemic level. Subsequently, extreme value theory (EVT) in risk management was used to account for a more realistic tail risk calculation. The reader who is not familiar with those measures should consult the book by Philippe Jorion (Jorion 2007). It is worth emphasizing that market risk in particular is also referred to as *systematic* risk, which is different from systemic risk. Systematic risk refers to the residual of risk that is nondiversifiable in a portfolio.

Since the 1990s, enterprises and regulators were mostly concerned with classifying types of risks about which an enterprise should be worried in its business and operations *while upholding the fiduciary responsibility to their respective stakeholders in the enterprise*. However, the dominant paradigm in risk assessment in the 1990s was based on a reductionist approach—which neglected the risk transfer and contagion and, most importantly, ignored the risk owned by the SoS and not just its components, as though they were operating independently. Systemic risk was not a focus of attention at that time as it was defined as a property of the financial system as a whole—and it was assumed that the system would be taking care of itself through self-regulation and the effort of existing regulating agencies. Dangerously, it was believed that the system was deemed to be stable if all of its components were individually stable. There was an innocent and implicit assumption of linearity of intercomponent interaction.

Using stress testing, the regulator could assess the probability of collapse of a bank (component) in the SoS. This was satisfactory as a component test. However, the component stress test is not designed to capture the overall SoS stress nor is it designed to produce contingent cash flows under stress scenarios in the entire SoS. Unfortunately, an accurate synthetic aggregation of banking contagion is not yet available. The new generation of stress tests could possibly address such an aggregation if the regulator had data that could aggregate counterparty risks. Thus, putting together the component stress test and the contagion effects under a multitude of scenarios for each component and producing the results in a dynamic setting should result in an effective methodology for an SoS stress test. This test could be a major step in building an early warning system of financial systems that could assess the probability of failure of the SoS due to systemic component failure or due to failure of a number of components simultaneously.

A definition of SoS systemic risk represented by a network of interconnected components must include those (nonlinear) contagion effects. A working definition of the systemic risk of an SoS represented by a network is the probability of a collapse in the operability of the network due to an event or a sequence of events. The reason we choose the definition in terms of probability is simply because systemic risk assesses the *ruin* probability (commonly used in the insurance industry). Systemic risk diverges from the concept of VaR and other risk metrics that seek to quantify the likelihood of sustaining a maximum loss at certain confidence levels. That maximum loss may be sizable or negligible with respect to the enterprise cash flows. The intended operational objective of a banking network is to provide services to customers at the retail level, to provide interbank lending capacity, and to respond to emerging needs of cash flows. Therefore, the dominant sign of operability in a financial network is to provide *network liquidity* characterized by the cash flow rate, which does not happen unless there is a threshold of trust in the network itself. This leads to the questions of how to measure the trust of an SoS and to develop quantitative assessment of component-wise trust and intercomponent trust. Trust, in this context, simply means that when a bank lends to another bank at some interest rate such as the London interbank offered rate (LIBOR), the lending bank has sufficient *faith* in the borrower to pay the loan back. That faith can be broken up into two components: one relates to the borrower only and the other relates to the SoS collectively. The subject of *network trust* is under investigation.

The 2008 financial crisis was the impetus to develop a great number of new and improved techniques in systemic risk measurement. For a survey of the state-of-the-art techniques and analytic tools, the reader is referred to Bisias et al. (2012). In an earlier work on empirical indicators, we implemented an SoS approach (Khashanah and Miao 2011) at the macro level to devise an empirical systemic risk measure in a study that involved systems as asset classes wherein each asset class is represented by an indicator. Financial indicators are of great value in providing information as they act as aggregators of market participants' positions while their comovements convey information on interasset class relative risk flows.

PURPOSE

The purpose of this case study is to provide an empirical model that can convey the following points:

- Explain the concept of empirical systemic risk measurement of an SoS.
- Provide a model that can be improved, expanded, and customized to particular needs or to fit into a systemic risk toolbox.
- Provide a model that can be replicated in other fields of SoS where there are indicators and data.

The advantage of the financial system is the presence of such indicators. For example, the Chicago Board Options volatility index with trademarked ticker symbol VIX reflects the market fear content seen by market participants whereas credit default swaps (CDSs), as contracts, reflect market participants' aggregation of credit risks and counterparty risks dynamically.

BANKING SYSTEM OF SYSTEMS

In order for the case study to be manageable, we chose nine banks that are of sufficient size and connectivity to be systemically important—or in the language of the Dodd–Frank Act (DFA), systemically important financial institutions (SIFIs). Each financial entity is viewed as an enterprise system in its own right. The collection of these system enterprises makes up an SoS of enterprises, which we refer to as the banking SoS. A model for the financial SoS in its entirety does not exist. It will be incumbent on the research to select a subset of the entire system that can most closely mirror and reflect the larger SoS. For this case study, the choice of these banks or investment banks is based on the desire to build a prototype for a small-scale model of a financial SoS. In this exercise, it is important to focus on the scalability of the tools and arguments. A more comprehensive and accurate depiction of the real financial SoS should take more extensive inventory of participating enterprises. The governance, regulatory bodies, and feedback loops are depicted in Figure 10.1. The governance of the banking SoS can be decomposed into two main parts with respect to the boundary of a bank. Each bank is a component in the SoS with operational boundaries known to the bank. In that sense, there is internal governance and external governance. The internal governance of a bank refers to areas such as the board of directors, treasurers, organizational structure, management, compliance, and audit. The external governance refers to the set of rules and regulations that are imposed on the banking SoS to operate in a financial system. Those regulations are intended to safeguard the financial SoS at large, and, at the same time, the banking SoS must not ignore the advice of the experts who work in the field and understand the day-to-day business risks. The distinction between enterprise governance and SoS governance is important for understanding the tension that exists between regulation as an exogenous force and compliance as an endogenous response to the external force, which results in changes in procedures, operations, reporting, accounting, and intercomponent relations.

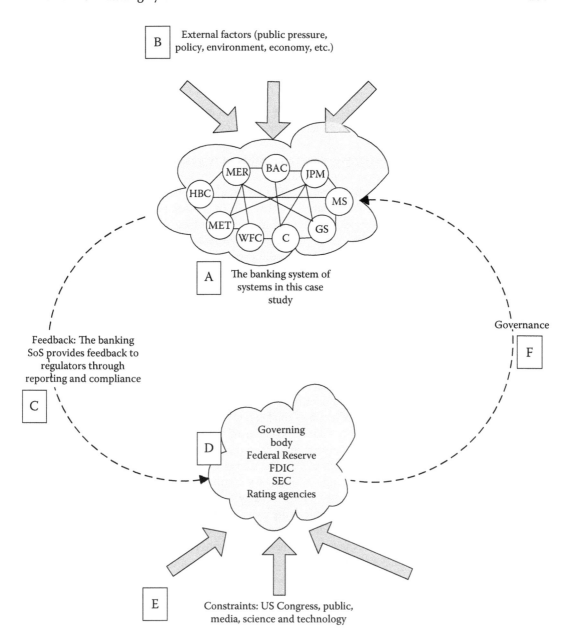

FIGURE 10.1   Schematic shows the banking SoS relational connectivity to its environment and regulators.

A—The component elements of the SoS in this case study are the nine SIFIs: Bank of America (BAC), J.P. Morgan (JPM), Morgan Stanley (MS), Goldman Sachs (GS), Citigroup (C), Wells Fargo (WFC), MetLife (MET), HSBC Finance Corporation (HBC), and Merrill Lynch (MER).

B—Rather obviously, the external factors of this case study consist of public pressure, government policies, the financial environment, the global economy, etc.

C—Feedback: The banking SoS provides feedback to regulators through reporting and compliance.

D—The US governing body consists of the Federal Reserve, the Federal Deposit Insurance Corporation (FDIC), the Securities and Exchange Commission (SEC), and various rating agencies.

E—Primary constraints are the US Congress, the US public, the media, and some limitations of science and technology.

F—The external governance consists of the set of rules and regulations that are imposed on the banking SoS to operate in a financial system.

To make the case study manageable, We chose nine SIFIs to represent our sample of banks that could sufficiently reflect the bigger systemic dynamics of the entire financial system. The SIFIs we chose have a proxy for risk of default. The nine SIFIs are Bank of America, J.P. Morgan, Morgan Stanley, Goldman Sachs, Citigroup, Wells Fargo, MetLife, HSBC Finance Corporation, and Merrill Lynch (MER). We called this set of banks in our case study *the banking SoS*. Ideally, we would like the proxy data to inform the researcher about the credit worthiness of a component system in the banking SoS at a given point in time much like a creditor would like to validate a borrower's credit worthiness. In the individual borrower's case, the lender runs a credit score check and determines the level of credit risk associated with that individual. The rating agencies for financial systems play that role with well-defined parameters that have been developed over the years. The 2008 crisis challenged those rating methodologies as well. Those challenges are being addressed by the rating agencies in an effort to come up with a more efficient and accurate system of rating.

Interinstitutional credit worthiness and solvency questions led to the development of a tool to gauge the level of risk in sovereign and institutional borrowing and lending. A CDS is an agreement (contract) in which the seller of the contract must compensate the buyer based on contingent credit events defined in the contract. Such contingent credit claims can be likened to a default on a loan. The buyer of a CDS *swaps* the credit risk of the borrower and gives it to the seller of the CDS, for the premium paid to the seller of the CDS. The seller of the CDS becomes the new owner of the credit default risk in exchange for the premium. Data on CDSs are recorded in terms of basis points (bps). A basis point is one-hundredth of 1%. The higher the CDS value in bps, the more risk of default associated with bilateral exchange of monetary commitment through contractual agreements. A CDS is not an insurance policy but a form of hedging. The principal difference between a CDS and insurance is in ownership of the underlying asset or instrument. In the case of insurance, one buys an insurance policy for an owned item for which there is a concern regarding loss or damage. A buyer of a CDS, however, does not have to have ownership in the risk of default (e.g., loan). In that case, the buyer has a *naked* CDS describing a position wherein there is a net position in the derivative and no position in the underlying asset of that derivative. To get a feel for the size of this market, the Bank of International Settlement's stable summary of over-the-counter (OTC) derivatives (BIS 2012) cites the total notional amounts outstanding as of December 2011 at $647.762 trillion of which $28.6 trillion are credit default swaps. The gross market values are $27.285 trillion and $1.586 trillion, respectively.

Observing fluctuating bps of CDS provides instantaneous and mostly expert voting results on dynamic credit risk level of a financial entity. This is in contrast to stress testing performed by the rating agencies that provide a static but more thorough snapshot in time for the financial health of the entity. In fact the CDS in an efficient market factors in the information provided by the rating agencies instantaneously. We will not go into a comparison between different types of testing since it would take us outside the scope of this case study.

DEVELOPMENT AND MODEL

In all of our development, we view the nine-bank banking SoS as a dynamic collection of relatively moving components inside the larger banking SoS. Researchers have insightful perspectives on systemic risk albeit not comprehensive because of the presence of deficiencies in quantifying and forecasting attributes of contagion and imminent threats growing in the system. In addition, the mechanisms with which the propagation of risk occurs in financial networks are subjects of current research. The fact that there are numerous papers on financial networks is a good sign that the problem is attracting sufficient attention; however, it should not be assumed that the problem is solved yet. In addition, necessary data to make that assessment of systemic risk accurate are still at the core of these questions and challenges. What type of data would be enough to describe the financial system, and at what granularity are typical questions still unanswered in a definitive way. In order to have reliable systemic risk measures, we must have both reliable data and reliable analytics.

Networks have emerged as useful tools to quantitatively represent interactions among components in an SoS. A financial network is a collection of agents placed in nodes along with connecting links. One can always assume that the graph is complete. Depending on the problem being investigated, the link can represent bidirectional cash flows, relative liabilities, mutual volumes, counterparty risks, return correlation coefficients, or any financially measurable parameter or variable. We can view certain networks as *empirical financial networks* if the variables are *measured* by markets through an exchange or by *marking to market* (a price discovery mechanism) or other valid mechanisms. On the other hand, in some cases, we find it useful to make theoretical assumptions on characteristics of the network flow. For example, imposing some conditions on the probability distributions of cash flows along with assumptions of statistical stationarity should lead to some useful theoretical analysis. In that sense, the network provides a theoretical financial network.

In this case study, we will construct an empirical financial network for nine SIFIs. The collection of SIFIs is the financial SoS in this case study. Each of the nine SIFIs is a component in the financial SoS network. Given a financial network, the importance of an SIFI is dictated by two factors: size and connectivity.

1. The size of an SIFI is determined sometimes in terms of absolute capitalization of that entity (e.g., $50 billion or more). We believe that absolute measures are not useful. A percentage of a moving economic systemic aggregator such as GDP would be a more useful measure.
2. The connectivity of a node can be quantified in terms of the number of links that are incident to it. Mathematically, this is called the node degree in graph theory. The node degree with market information on cash flows in links determines the cash flow volume in a given period of time. Therefore, the relative cash flow size in and out of incident links is an important factor in determining whether or not an enterprise system is becoming a threat to the entire SoS. Depending on the level of transparency in the SoS, it is often difficult to measure relevant cash flows in order to accurately represent asset liability mismatches. As the level of transparency in any SoS is primarily implied by its regulation, the DFA tries, among other things, to remedy this problem by making data available and visible in useful forms for analysis. The Office of Financial Research (OFR) enacted by the DFA as the research arm of the Department of the Treasury is entrusted with advancing the research to see to it that data types and analytics stay commensurate with emerging innovations and complexities in financial systems. All other regulatory agencies, such as the CFTC, SEC, FDIC, ISDA, and FINRA, to mention a few, interact and benefit from the OFR effort.

MANAGEMENT OF UNCERTAINTY

What is new in this case study is expressed in terms of systems thinking in finance. Ultimately, the financial system has to address its architecture, components, jurisdictions, flows, and boundaries. The researcher will need to determine with the defined financial SoS if the financial entity is sizable but not connected, and whether the system would be able to withstand its demise without destabilizing the entire system. Conversely, if the entity has a high node degree with small size capitalization, it might also be liquidated without destroying the entire system. It is rather simple to incorporate the case where a collection of second tier banks can be treated as a major node with SIFI status as a whole.

How does this argument map into the risk space? One way to think about it is by examining the distinction between classical risk management and systemic risk management. The easiest way to see the distinction is to look at the role of a CEO of a SIFI who reports to his or her respective board and explains risk management and compliance in the traditional manner. This distinction was developed during the 1990s for the enterprise system level of risk management, initiating an ongoing debate on how to make it more efficacious. Researchers examined what happened when those SIFIs were put together in one connected system. The observation was made that the whole is greater *or lesser* than the sum of the components. In this case study, the system consists of nine connected enterprise systems each of which generates its own risks—some of which remain internal and some are passed on to other enterprises in one form or another. The fundamental difficulty that distinguishes systemic risk is how to measure the collective risk of the financial SoS, which includes the risks passed on to others through the linkages in addition to *internal* component risks. Our knowledge of individual enterprise system risk should be factored into whatever measure we adopt for systemic risk. The missing link in finance has been the inability to identify the interenterprise risk spillover that leads to contagion and, in effect, makes systemic risk incalculable.

As the industry continues to examine the distinction between classical risk management and systemic risk management, the unanswered question is, who should address and mitigate systemic risk management? The short answer is that it is still not clear how this will happen even after the DFA. However, we can see that the key stakeholders in deflecting the negative implications of excessive systemic risk have been the Department of the Treasury and the Federal Reserve System. The overall unclaimed risks created in making markets, innovative instruments, and conducting the business of financial system enterprises (mainly in the linkages, with CEOs normally doing their jobs) are handed over to the regulators to analyze and identify. Because the SIFIs are indispensable, the regulators have to sell the accumulated overall risk to the public over some reasonable horizon of time, thus realizing the *moral hazard*. This *cleaning up* of the flotsam of systemic risk happens with variable scales every 10 or 15 years—the frequencies are expected to increase. Most importantly, the issue of systemic risk is not just a problem in data collection, interpretation, and tools but is also an issue of SoS architecture, jurisdiction, boundaries, and governance.

DATA

With the emergence of big data in many fields, the general question of what data types are necessary to characterize complex systems becomes a daunting task certainly for a complex sociotechnical and economic system as the financial system. The general question is outside the scope of this case study. The OFR in coordination with other regulatory bodies and specialty organizations is working on this complex problem as part of the overall strategy to better address the shortfall of the previous understanding of systemic risk prior to the 2008 crisis.

For the purpose of this case study, however, the data we use here are taken from daily closing price of published CDS bps on the nine SIFIs: Bank of America, J. P. Morgan, Morgan Stanley, Goldman Sachs, Citigroup, Wells Fargo, MetLife, HSBC Finance Corporation, and Merrill Lynch. The data was collected for the CDS series on five-year senior bonds from November 7, 2002, to August 2, 2012, for a total of 2541 empirical measurements for each of the nine SIFIs. The Bloomberg ticker symbols are given by first identifying the entity to which the CDS belongs and the rest of the symbol describes the contract. For example, BOFA CDS USD SR 5Y Corp refers to Bank of America CDS series on five-year senior bond. The rest of the symbols need only change the entity identifier to JPMCC, MS, CINC, GS, WELLFARGO, METLIFE, HSBCF, MER in the order the entities are listed in this paragraph.

In addition, we used the VIX daily closing value for the same period for almost nine years (similar to the CDS data). The popular VIX has been used often as a gauge for the level of fear in the markets at large, although the VIX itself is calculated as a function of a collection of options on the Standard & Poors (S&P) 500. Therefore, the VIX *value* represents a measure of near-term expectations of the S&P movement. The VIX was introduced by Professor Robert E. Whaley in 1993.

The CDS data on market assessment of credit worthiness of institutions represent an expert opinion, while the VIX represents an aggregate measurement of wide-scope market fear sentiment evolving over time. One data set may be viewed as representing the systemic *analytical* side of data, while the other set can be viewed as a systemic *behavioral* indicator of the financial SoS. Therefore, in general, a useful approach to quantitatively and qualitatively assess the state of an SoS is to look for analytical and behavioral indicators. The two may be, at times, highly correlated and, at other times, not informative about each other but collectively informative about the system. It is worth noting that the VIX level is an observed variable in most decisions on derivatives.

ISDA* has made great strides in setting standards in various financial asset classes.[†] For example, ISDA's standard model for CDSs for fee computations[‡] provides a benchmark for involved parties in a CDS transaction. Markit[§] provides a comprehensive data source of data especially on CDS and plays the administrative role to support and maintain the open source functionality to support data transparency.

ANALYSIS AND SYNTHESIS (ANASYNTHESIS)

In SoS studies, it is not easy to separate analysis from synthesis unless the system is decomposable. The term *anasynthesis* is introduced to mean the process by which analysis and synthesis are continuously applied with iterative multilayered systemic feedback corrections and postulations. See Figure 10.2.

In this case study, the choice of the SoS components is part of the systemic analysis. The fact that the CDS class is chosen as a proxy for component-level intelligence is predicated on both layer-level analysis and component enterprise synthesis reflected by buyers' and sellers' price assessment. In efficient markets, the price of a CDS by a market specialist takes into consideration all available information about the enterprise and its environment at the time of the transaction. Although partly

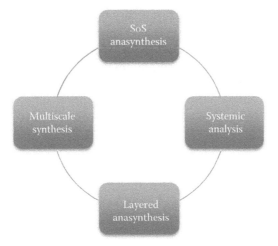

FIGURE 10.2 *Anasynthesis* is the combined nondecomposable process of feedback multilayered analysis and synthesis.

* http://www2.isda.org/
† http://www2.isda.org/asset-classes/credit-derivatives/
‡ http://www.cdsmodel.com/cdsmodel/fee-computations.page?
§ http://www.markit.com/en/

subjective, the specialist includes the environment that, in turn, includes the specialist's assessment of systemic risk levels—in other words, its nondecomposability. The price of a CDS therefore is a component-level analysis and synthesis termed *layered anasynthesis*.

METHODOLOGY

A methodology is a practical framework that affords the confluence of theories and principles with available data measurements toward a usable conclusion. In this case study, we attempt to show that the available transactional data at the level of components can be converted into business intelligence that can tell us something about the state of the financial SoS. We seek empirical measures of interconnectedness and intercomponent risk flows. In order for this conversion to be meaningful, the investigator should have an idea about what he or she is looking for in the face of a tremendous amount of available data at the enterprise level. For example, public data are available on the nine SIFIs in this case study: corporate financial data provided by analysts in terms of key statistics that includes market cap, price-to-earnings (P/E) ratio, price-to-sales ratio, price-to-book ratio, and enterprise value, among many other parameters. Data are also available from public news flows pertaining to the enterprise or the SEC filings, options, competition, management, ownership, insider trades, income and balance sheet, and cash flows. Of all the available data, the corporate financial data are essential for investment-level decision support mechanisms, that is, whether or not an investor or a portfolio manager should buy, hold, or sell the asset in reference to a given portfolio.

Since our intention in this case study is to understand risks internal to an enterprise and risk spillover across an enterprise boundary that is part of the financial SoS, it is important to look for an instrument that summarizes corporate financial data in its pricing on one side and then aggregates markets' propensity to take risk. Market efficiency implies that a buyer or a seller of a CDS must calculate those parameters on both sides of the enterprise boundary. The efficient market hypothesis requires separate investigation but it could in principle be invoked to conclude that expert assessment of CDS is a function of the state of information efficiency in markets.

For this simplified study, our methodology is to construct a financial SoS risk matrix propagating in time alongside the SoS. The matrix that has been traditionally used in portfolio risk assessment is the variance–covariance (var–covar) matrix—the same matrix will be used for an SoS assessment. To construct a dynamic risk matrix propagator, in this case study, we will construct an empirical matrix using 100 data points. The next step will be to move the sliding 100-data window to construct a dynamic var–covar matrix for each day's closing data, resulting in a dynamic correlation matrix. Samples of those matrices will be printed before, during, and after the 2008 crisis in a manner that conveys the change in relative risk allocation to the SoS as it changes in time. The reader is also encouraged to experiment with weighted var–covar correlation matrices and other possible variants of calculations. The results in this case study use the correlation matrix instead of the var–covar matrix. The calculations in this case study are carried out using the R Project for Statistical Computing or simply R.

DISCUSSION OF RESULTS

We extract snapshots of the SoS systemic correlation risk matrix to indicate how this method can inform the state of system components in relation to the systemic risk of the SoS. Figure 10.3 shows a snapshot taken on August 9, 2004, of the CDS-VIX financial SoS of the nine banks along with the VIX. This is one of the data visualization techniques that we employed to automate multiscale and multiasset class systemic risk propagation in a complex system. Figure 10.3 shows a loosely coupled system with a noticeable weak correlation with the behavioral market indicator of fear.

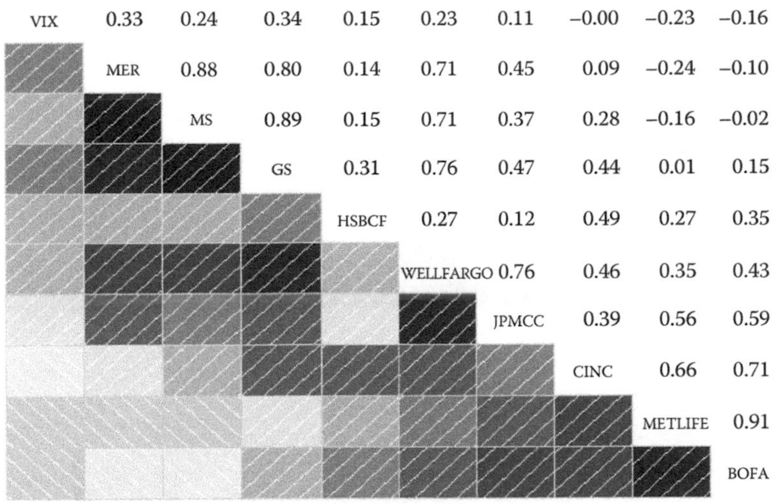

VIX	0.33	0.24	0.34	0.15	0.23	0.11	−0.00	−0.23	−0.16
	MER	0.88	0.80	0.14	0.71	0.45	0.09	−0.24	−0.10
		MS	0.89	0.15	0.71	0.37	0.28	−0.16	−0.02
			GS	0.31	0.76	0.47	0.44	0.01	0.15
				HSBCF	0.27	0.12	0.49	0.27	0.35
					WELLFARGO	0.76	0.46	0.35	0.43
						JPMCC	0.39	0.56	0.59
							CINC	0.66	0.71
								METLIFE	0.91
									BOFA

FIGURE 10.3 This CDS-VIX correlation heat map snapshot is based on market measurements on August 9, 2004.

From this correlation matrix one concludes, for example, that CDS of GS and MS are highly correlated (showing a correlation value of 0.89), indicating some level of relative risk comovement. Similarly, comparing BOFA and Wells Fargo, the matrix shows a negative correlation (a value of −0.10) between BOFA and MER at that time.

Note about Figure 10.3 and the following heat map figures: The color-hatch rectangles correspond to ranges of correlation values. Conceptually, think of flipping the lower left corner of the color-hatched triangular sections to overlay on the upper right corner of the triangular section of correlation values. One finds that the color and hatching of each rectangle is consistent with others of the same range of correlation values.

Snapshots are taken for the successive years on or about the same date. The computations allow all snapshots of the 2351 days to be easily presented. Those snapshots that are of relevance to our study are included. The next snapshot, taken on August 9, 2005 (Figure 10.4), indicates that

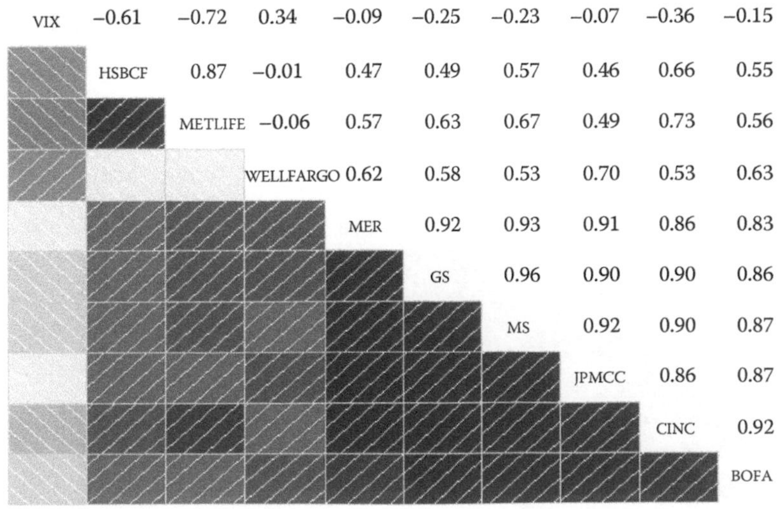

VIX	−0.61	−0.72	0.34	−0.09	−0.25	−0.23	−0.07	−0.36	−0.15
	HSBCF	0.87	−0.01	0.47	0.49	0.57	0.46	0.66	0.55
		METLIFE	−0.06	0.57	0.63	0.67	0.49	0.73	0.56
			WELLFARGO	0.62	0.58	0.53	0.70	0.53	0.63
				MER	0.92	0.93	0.91	0.86	0.83
					GS	0.96	0.90	0.90	0.86
						MS	0.92	0.90	0.87
							JPMCC	0.86	0.87
								CINC	0.92
									BOFA

FIGURE 10.4 This CDS-VIX correlation heat map snapshot is based on market 100 measurements prior to August 9, 2005.

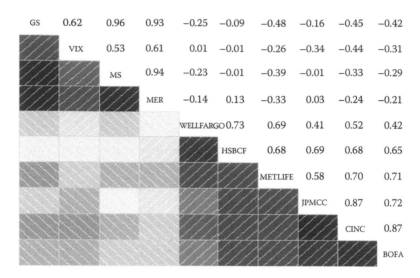

FIGURE 10.5   (See color insert.) This CDS-VIX correlation heat map snapshot is based on market 100 measurements prior to August 9, 2006.

decoupling, in the sense of weak correlations, still persists between SIFIs with negative or weak correlations with the VIX.

A year later as the SoS approaches its crisis, the banks are still oblivious to the gathering threat. Figure 10.5 shows the CDS-VIX correlation heat map snapshot on August 9, 2006. In fact, the system is in maximum decoupling between the SIFIs. The CDS bps prices are only based on the analytic relative credit worthiness of banks.

We now approach the first warning sign of the crisis starting in 2007, which indicated that the financial system was overstretched in its commitments. The CDS-VIX correlation heat map produced by our calculations on August 9, 2007, (Figure 10.6) indicated that the components of the SoS were suddenly tightly coupled and that the CDS correlated with the VIX fear index of the entire market.

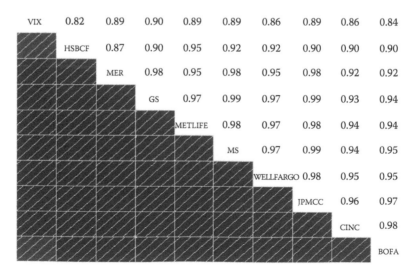

FIGURE 10.6   This CDS-VIX correlation heat map snapshot is based on market 100 measurements prior to August 9, 2007. An alarming pattern!

As the crisis unraveled, the next snapshot of the CDS-VIX correlation heat map is taken on August 7, 2008 (Figure 10.6). The VIX was at the lowest level of 13.27 on August 9, 2007, and the VIX was at 26.62 on August 7, 2008. More importantly, on August 9, 2007, the CDS values of bps were ranging between a minimum of 13 bps and maximum of 34.375 bps, while the values on August 7, 2008, were ranging between a minimum of 111.951 bps and maximum of 249.389 bps. We can express this observation more easily by introducing the sensitivity of CDS relative to the change in VIX. For example, the yearly sensitivity of CDS of BOFA is calculated as

$$S\left(CDS, VIX; \Delta t\right) = \frac{\Delta CDS}{\Delta VIX}$$

over the given period of time. For example, for 2006–2007, the BOFA CDS-VIX sensitivity per year is (15.089–14.5)/(13.27–11.79) = 0.4, while for 2007–2008, it is (124.481–15.089)/(26.62–13.27) = 8.2.

One interesting observation is that a CDS heat map seems to sense a gathering credit default threat even before the VIX is able to reflect such an expectation of the coming crisis. It is possible to interpret this in light of the expert opinion that determines the premium on CDS over a narrower market of options not traded by the public in general. On the other hand, the VIX is affected by the wisdom of the wider crowd that observes the entire equity market and factors in credit markets as they reflect on the components of the S&P through the banking sector. Therefore, with further refinement, the CDS heat map may serve as an early warning indicator of upcoming crises.

Next we examine the CDS-VIX correlation matrix during the crisis to see if the pattern of tight coupling persists. Figures 10.7 and 10.8 show the CDS-VIX correlation heat map up to August 7, 2008, and August 7, 2009, respectively. In Figure 10.8, the pattern persists and the values of CDS range between a minimum of 160.247 bps and maximum of 1002.098 bps, while the VIX reads 45.89 on that day.

Finally, we include a graph of covariance risk propagation in time as implied by CDS components of the SoS. We can produce graphs for all components to see the interactive risk transfer. For the purpose of this study, we include the dynamic CDS covariance for BOFA, showing its intercomponent risk seen from or inflected by BOFA. This is summarized in Figure 10.9.

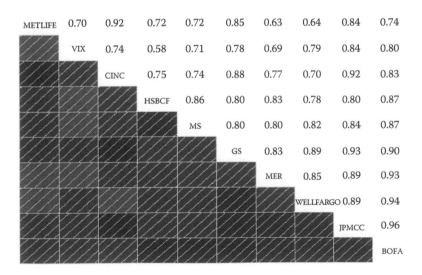

METLIFE	0.70	0.92	0.72	0.72	0.85	0.63	0.64	0.84	0.74
	VIX	0.74	0.58	0.71	0.78	0.69	0.79	0.84	0.80
		CINC	0.75	0.74	0.88	0.77	0.70	0.92	0.83
			HSBCF	0.86	0.80	0.83	0.78	0.80	0.87
				MS	0.80	0.80	0.82	0.84	0.87
					GS	0.83	0.89	0.93	0.90
						MER	0.85	0.89	0.93
							WELLFARGO	0.89	0.94
								JPMCC	0.96
									BOFA

FIGURE 10.7 (See color insert.) This CDS-VIX correlation heat map snapshot is based on market 100 measurements prior to August 7, 2008.

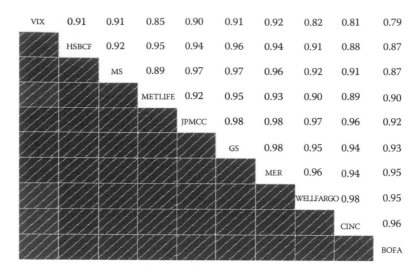

FIGURE 10.8   This CDS-VIX correlation heat map snapshot is based on market 100 measurements prior to August 7, 2009.

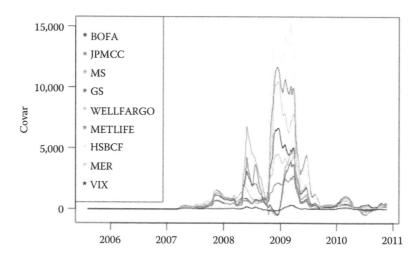

FIGURE 10.9   **(See color insert.)** This graph shows the dynamic CDS-covariance of Bank of America describing its comovement with other components in the financial SoS.

CONCLUSION

The subject of systemic risk requires the development of new analytic tools capable of handling massive data to extract intelligence about the system of enterprises. The method we presented here in this case study is an example of the analytic tools that can be combined into more comprehensive systemic risk analytics software. More refinement of these methods should be the first step in the construction of a systemic risk *toolbox*. Regulators, the Financial Stability Oversight Council (FSOC), and the OFR should encourage the development of these methods in the industry and academia in order to foster a national and international research competence in this critical area.

SUGGESTED FURTHER WORK

Several directions for improving the results in this case study can be undertaken:

1. Using the ideas developed in this study plan, find CDS-VIX correlation heat map snapshot based on market 100-day measurements prior to August 9, 2010.
2. Based on the results in question 1, does the credit freeze persist and, if yes, why? What were the economic and financial conditions that caused the banking system to remain in relative credit freeze in 2010?
3. Following Figure 10.9, produce a graph of covariance risk propagation in time as implied by CDS components of the SoS intercomponent risk seen from or inflected by Wells Fargo.
4. Expand the banking SoS discussed in this case study to include more SIFIs and repeat the steps to produce the corresponding heat maps. Do you observe persistence in patterns of systemic tightening during the crisis with the extended banking SoS?
5. Suggest three more indexes that can be taken to be financial or economic indexes, to include in the case study, and to perform the analysis again to see what excess information one would gain from the inclusion of these indexes.
6. Change the time window from 100 to 50 days and run the calculations again on the entire set of data. Produce a sequence of heat map snapshots in time and make a systemic risk *movie*.
7. Consider a different asset class such as the bond market. What are the suggested empirical data that one can use to extract useful information about the bond market systemic risk content?
8. Discuss the difference between the banking SoS defined in this case study and the corresponding banking system of subsystems. What is lost or gained in the analysis of systemic risk measurement between the two models?

ACKNOWLEDGMENTS

I would like to thank Xugong Li, financial engineering PhD candidate at Stevens Institute of Technology, for his help with the calculations in the R Project for Statistical Computing, but any errors are mine. The data for this case study was provided by the Hanlon Financial System Lab in the Financial Systems Center at Stevens Institute of Technology, http://www.stevens.edu/fsc/content/the-facility/hanlon-financial-systems-lab

REFERENCES

BIS. 2012. *BIS Quarterly Review*. June. http://www.bis.org/statistics/otcder/dt1920a.pdf. Accessed March 8, 2014.

Bisias, D., M. Flood, A.W. Lo, and S. Valavanis. 2012. A survey of systemic risk analytics. U.S. Department of Treasury and OFR No. 0001. January. http://ssrn.com/abstract=1983602 or http://dx.doi.org/10.2139/ssrn.1983602. Accessed March 8, 2014.

Jorion, P. 2007. *Value at Risk, the New Benchmark for Managing Financial Risk*. New York: McGraw-Hill.

Khashanah, K. and L. Miao. 2011. Dynamic structure of the U.S. financial systems. *Studies in Economics and Finance* 28(4), 321–339.

Section VI

Health Care

11 Whole-of-Nation Health Capabilities

A System of Systems Methodology for Assessing Serious and Unusual Emergencies

Richard J. Hodge and Stephen C. Cook

CONTENTS

CASE STUDY ELEMENTS

Fundamental essence: Designing and applying a systems approach to examine health preparedness for serious emergencies—first at local and regional levels and then at a whole-of-nation level.

Topical relevance: Health outcomes delivered as a whole-of-nation system comprising public, private nongovernment organizations, medical associations, not-for-profit, and volunteer organizations. Shaping, directing, and managing that enterprise system to reduce risk, improve readiness and response capabilities, and sustain recovery for serious emergencies. Concurrently, improving governance, planning, management, and delivery of health services from day to day.

Domain: Health and allied services.

Country of focus: New Zealand (NZ), yet relevant to most democratic environments.

Stakeholders: Department of Prime Minister and Cabinet, Ministry of Health, Ministry of Civil Defense and Emergency Management, District Health Boards (especially in Wellington, Auckland, Christchurch, and Dunedin), Public Health Services, public and private hospitals, primary care practitioners, ambulance services, radiological and laboratory services, allied health service providers (pharmaceuticals, psychological services, community services, rest homes), defense health services, NZ Blood Service and other National Health agencies, NZ professional medical associations—over 100 chief executive officer (CEO) or senior executive interviews.

Primary insights: The focal health issues arising at the business enterprise and at the health industry levels are nested and managed by government within the broader national security agenda. While the results of study remain confidential, the design and implementation of a systems approach configured a set of rational, interpretive, and critical systems methodologies that are scalable to support concurrent interventions at the local, regional, and national levels across the enterprise. A process that brings disparate risks down to *rough equivalence* (Helm, 2004) is also demonstrated to help assess and manage risk from a central government's perspective to give it options for balancing risks at a national level where health, however important, is but one consideration in managing a society.

KEYWORDS

Capabilities-based planning, Capability development, Cognitive systems engineering, Complexity, Enterprise planning methodologies, Enterprise risk, Execution, Health planning, Health preparedness, Operations analysis, Option analysis, Risk management, Scenario-based planning, Soft systems methodologies, Strategy, Systems approaches to strategy and execution, Systems engineering

ABSTRACT

In 2004, the NZ government asked a tough question that no other nation at that point had considered: What is the preparedness *of the nation* to deal with serious emergencies at a health services delivery level, at a business level,* and at a health industry level?† All domains are necessarily nested within the national government's broader security agenda. While the results of the broader study remain confidential, the design and implementation of the systems methodology are presented.

A methodology to design and develop an integrated set of rational, interpretive, and adaptive interventions is described, framed on Checkland's soft systems methodology. Systems engineering is adapted as a methodology for synthesizing multiple interventions. A pilot study was used to test the system of interventions at a regional level, the results of which guided the *control* actions required to adapt the methodologies so they could scale to a national-level study. Most interventions were able to scale with little or no modification. However, the modification to the risk analysis was extensive and became a central integrating *force* in the study. The case study concludes with a summary assessment of the extent to which the different rational, interpretive, and adaptive methodologies developed an effective *sweet spot*.

The case study reinforces the methodological application of systems concepts at the business enterprise, industry, and national levels of systems complexity and demonstrates the portability of

---

* As operated across NZ by 21 District Health Boards.
† Incorporating all forms of allied health providers, NGOs, professional and volunteer associations.

this framework of ideas and its adaptability to different areas of concern. The response to the study at the higher levels of government—notably the Department of the Prime Minister and Cabinet and the Treasury—reports a *good outcome*.

GLOSSARY

CATWOE	(mnemonic) Customer, Actor, Transformation, Weltanschauung (Worldview), Owner, Environmental constraints
CEO	Chief Executive Officer
DHB	District Health Board
DPM&C	Department of Prime Minister and Cabinet
NGO	Nongovernmental Organization
NZ	New Zealand
PHO	Primary Health Organizations
RFP	Request for Proposals
SARS	Severe Acute Respiratory Syndrome
SSM	Soft Systems Methodology
UNDP	United Nations Development Program

BACKGROUND

Context: Events in New York on September 11, 2001, in Bali in 2002, in Jakarta at the Marriott in 2003, and in 2004, at the Australian Embassy in Jakarta, contributed to the Australian and New Zealand (NZ) governments establishing a heightened state of preparedness appropriate to their assessment of, and tolerance for, the risks of various kinds.

In NZ, concerns about its national preparedness were brought into a sharp focus in 2003 with threats to the staging of the America's Cup in Auckland, and to the New Zealand Golf Open in Wellington. Concurrently, the global emergence of severe acute respiratory syndrome (SARS) and avian flu emphasized the dangers facing New Zealanders from communicable diseases and the risks—as well as opportunities—inherent in both tourism and terrorism. Tourism is an important part of the relatively small but vibrant NZ economy. How well the nation demonstrated its preparedness to maintain high levels of safety and security in the face of such threats was important to maintain the vibrancy in NZ—among nationals and visitors alike.

NZ made substantial progress early at a whole-of-nation level* in improving its preparedness to the risks of catastrophic terrorism. It also saw the public attention given to counterterrorism as an opportunity to improve national preparedness to other serious emergencies and, in doing so, improve day-to-day operations.

Health systems globally were, and remain, heavily stressed as they expand into a complex social enterprise of public, private, and volunteer elements while responding to populations that are aging in all but 18 countries (UNDP, 2005), while concurrently dealing with catastrophic events borne of wars, terrorism, and natural disasters.

The pressure on a nation's health system therefore demands that every dollar spent increase preparedness to address serious and unusual emergencies and also demands that it add value through concurrent improvements in day-to-day service delivery. Considered in that context, the question asked by the NZ Ministry of Health was therefore highly strategic and deeply operational: *What is the capability and capacity of the NZ Health System to respond to serious and unusual emergencies?*

---

* The NZ government commissioned a number of studies, examining different aspects of the national preparedness *problem space*. Booz Allen Hamilton successfully tendered to review the NZ health system, for which the primary author contributed to all aspects of the project, and led the development of a systems framework for the design of the methodology.

Pertaining theories: The initial approach by the bureaucracy was to commission a number of studies each examining an important part of the problem. The primary focus was one of research and analysis, in parts, by the NZ health and emergency management agencies and/or other contracted firms. An indicative sample of related projects included the following:

- Ministry of Health National Emergency Planning Project (in development)
- Creation of standards for Mental Health Service emergency response (due March 2004)
- Environmental Health Emergency Plan and Operation Links Review (completed)
- National Radiation Laboratories Emergency Plan Review (under review, October 2003)
- Review of Rescue, Triage, and First Responder Paramedical Operations in Hot and Warm Zones (due January 2004)
- Review of DHB capability and capacity to mount a treatment service response to a reemergence of SARS or other infectious diseases (initial report due December 2003, completion due June 2004)
- Research, reviews, and analysis of influenza in NZ in 2003
- Updates to manuals on public health surveillance and public health emergency management in NZ
- Development of a national health emergency infectious disease plan
- Analysis and options for responding to emerging diseases, biosecurity, and chemical, biological, and radiological threats to human health
- Analysis and a report on understanding needs for health system preparedness and capacity for bioterrorist attack
- Updates to minimum standards for intrahospital transport of critically ill patients
- Analysis of workforce capacity in serious and unusual emergencies

In short, the response of the NZ bureaucracy was functionalist and rational, with each part driven by a different agency, conferring widely and backing its options with a strong research and evidence base. The difference we sought to bring was a first-principles approach that combined synthesis and analysis within a whole-of-system (nation) viewpoint.

Research nugget: At the outset, it is prudent to note an important distinction in the use of the term synthesis, as opposed to analysis, which is not always made clear. The distinction is critical to ensuring its full and proper use in strategy formation. For example, the *Australian Oxford Dictionary* (Oxford, 1999) defines analysis as "a detailed examination of the elements or structure of a substance etc. ... [and] ... the act or process of breaking something down into its constituent parts." Synthesis, on the other hand, is defined as "the process or result of building up separate elements, esp. ideas, into a connected whole, esp. into a theory or a system." In this case, analysis and synthesis appear as direct opposites as one (analysis) disconnects the parts of a whole and the other (synthesis) reconnects them. Similar threads of meanings also appear in *Webster's Dictionary* (1989). Based on these words alone, an entirely mechanistic view of synthesis could reasonably be taken: if you understand how to break things down, you will learn how to put them back together. This would be incorrect, as Mintzberg (1994) criticized the planning school (with its emphasis on formal process and rationality) for its "implicit assumption ... that *analysis will produce synthesis*."

Ackoff (1999: 11) offers a useful comparison of analysis and synthesis in an organizational context as a process of three steps, summarized in Table 11.1.

The distinction that Ackoff makes between analysis and synthesis is insightful when considered in the context of strategy formation and execution. The separation of viewpoints between analysis and synthesis aligns them like the two hemispheres of a human brain—the left for analysis, the right for creation. And there is an implicit connection of the two cognitive processes to a *cognitive whole*, where analysis, Ackoff (1999: 12) suggests, produces knowledge and know-how to make a system perform efficiently, but to understand the behavior of an organization, one needs to understand its

TABLE 11.1
Analysis and Synthesis

	Analysis	Synthesis[a]
Step 1	The entity to be understood is taken apart	Identify one or more larger systems of which the entity of interest is a part
Step 2	Understand the behavior of each part of the entity taken separately	Understand the function of the larger system(s) of which the entity is a part
Step 3	Aggregate the knowledge of the parts of the system in an effort to explain the behavior or properties of the whole	Disaggregate the larger containing system to identify and understand the *role* or function of the entity and the nature of its interfaces with the larger system
Focus	*Knowledge* of entity structure, how it works and delivers its *outputs*	*Design* of entity behavior and *outcomes* in the larger system(s) it is part of
Benefit	Analysis provides knowledge to make an entity work *efficiently* and to repair elements of it as they stop working	Synthesis provides systemic understanding of the entity behavior for it to work *effectively* as a whole in a greater system

Source: After Ackoff, R.L., *Re-Creating the Corporation: A Design of Organizations for the 21st Century*, Oxford University Press, New York, 1999.

[a] This is not the meaning of synthesis as used in systems engineering, where it relates to the process of composing a design *from a selection of possible components*, some bought-in, some bespoke. Here, synthesis relates to the design of whole-entity behavior in an organizational setting, typically driven by a selection of measures that are often amended dynamically based on information generated by analysis.

role within a larger system (the enterprise) that it is a part of and design its functions and behaviors appropriately. Synthesis is therefore about design for effectiveness (Blanchard and Fabrycky, 1998) based on an understanding of the relationships among the various elements of the larger system (Ackoff, 1999).

Hart (1992) argues that "over the past three decades, authors have developed scores of different strategy making typologies ... [that have] covered such a wide range of considerations that little cumulative knowledge has resulted. *A conceptualization that is capable of providing a framework for ongoing research is lacking*" (p. 327). With little changing in the following decade, a deep fibrous and fluid connection (Mintzberg et al., 1998; Demos et al., 2001) between the two cognitive functions of analysis and synthesis is yet to be defined in the strategy literature. By extension of Ackoff's metaphor, then, the strategy field lacks an equivalent of a *corpus callosum* connecting the two hemispheres of the brain. The *art* and *science* of the strategy field continue to evolve, while the *expertise* needed to connect them appears missing from the literature.

Hodge (2010) draws on the literature and knowledge of strategy practice in defense and other national security enterprises, from business strategic planning and from systems approaches to management, to synthesize a conceptual model for an integrated strategy framework and an associated methodology that directs practice. The case study *A System of Systems Methodology for Strategic Planning in Complex Defense Enterprises* is a companion study (also in this book) to this case study in the application of the conceptual model for an integrated strategy and execution framework.

Existing practices: In stepping into a new context, it is important to understand the core elements of existing practices and how they govern behavior. The medical fraternity—from research to executive management—places a significant focus on the patient pathway. In practice, this means that all strategy and execution endeavors must be considered in the context of the potential impact on patients as they enter the health system for different diagnostic conditions and the different pathways they would then experience within the health system. Consequently, the medical fraternity places high value on evidence-based decision making—and health policy is no exception.

As Dr. Peter Roberts, an intensive care specialist and clinical adviser to this study, stated, "bad policy was more dangerous to his patients than the flesh-eating bug" (Roberts, 2003).

Guiding principles: One of the medical precepts held by all health and allied health profession-als is one of nonmaleficence or, more colloquially, *first do no harm*.* In medical practice, signifi-cant interventions on a patient take a pathway of assessment (triage), diagnosis (with supporting laboratory evidence), intervention (primary/secondary/tertiary), rehabilitation, and recovery. It is a carefully constructed pathway that takes account of the patient's history, the seriousness of the diag-nostic condition, the options available and their prognoses, and the involvement of multiple people with specialist skills acting collectively in the individual's long-term interest.

Now, when the *patient* is the health system for the whole nation, those operating the health system would not accept a consultant's pathway to be any less well constructed than any medi-cal professional would construct for any of the individuals working within the system or depen-dent upon it for their own health. *Nothing in this case study is to be interpreted to suggest that the NZ health system was in any way "unhealthy." Quite the opposite: it promoted and sponsored this intervention to help it understand—in the face of new, unusual, and serious threats—what more it needed to do, and the options open to government to improve its capac-ity and capabilities.*

PURPOSE

The opening (or *as is*) state of the health system, as viewed by those within the system itself, was defined in the publicly released Request for Proposals (RFP) for a "Review of health workforce capacity and capability to respond to an unusual and serious emergency." The stated objective of the required intervention was to "obtain quality, cost effective and timely information on the gaps in health workforce capacity and capability to respond to an unusual and serious health emergency, and advice on and discussion of options to address these gaps."

While *answering the mail*, it was clear in our initial synthesis of the stated objective that work-force capacity and capability, while important, were not the sole determinant of the national health preparedness for serious emergencies. The Ministry understood this too. And, consequently, sought "the most competitive proposal(s), ... [with which it] may negotiate on a proposal(s), with a view to potential inclusion of key elements or all elements of the proposal in the final contract to be developed."

Our approach was to suggest first a deeper diagnosis of the core issues that would need to be addressed in addition to the workforce question. So we began the intervention (in a Phase 0) with a scoping study to first question what were the right questions to be asked and answered about the health system and its preparedness for serious and unusual emergencies.

The maturity of the NZ health system was such that it agreed. In systems terms, the scoping study had two goals:

1. To assess and manage the cognitive complexity involved in approaching an analysis of a complex adaptive system (of over 8,000 medical practitioners, and more than 45,000 nurses and allied professionals, with an expenditure in excess of NZ$8 billion in public funds per year[†]).
2. To design a systems intervention that would enable the team to assess the health system's capacity to handle a series of unusual and serious emergencies (the prospective situational complexity facing the health system) and develop options for government that integrate with other initiatives under way.

---

* *Primum non nocere* is a Latin phrase that means "first, do no harm." The phrase is sometimes recorded as *primum nil nocere*. (See: en.wikipedia.org/wiki/Primum_non_nocere.)
† 2004 dollars.

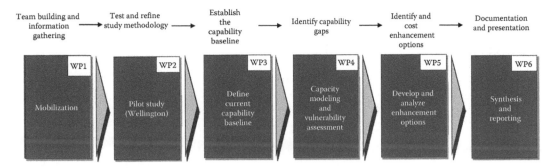

FIGURE 11.1 Linear process diagram of six proposed work packages.

The resultant scope of the intervention was agreed to include the following:

- Strategic issues, including capacity strengths and vulnerabilities, options for improvement in the health system and intersectoral issues (e.g., with the Department of Prime Minister and Cabinet [DPM&C], with Ministry of Civil Defense and Emergency Management, and with other government and nongovernmental agencies)
- Workforce issues
- Equipment, supplier, and infrastructure issues
- Business rules and processes, from policy to tactical processes
- Costs and risks involved in addressing (or accepting) any vulnerability

Subsequent to the scoping study, the objectives of the intervention were revised to be as follows:

- Assess the capacity of the NZ health sector to respond to an unusual and serious health emergency, including assessments of national collaboration, regional coordination, and critical elements of health service delivery.
- Provide advice to the NZ government on options to address gaps in capability.
- Conduct a high-level cost–benefit analysis of the identified options.

Our process for achieving these objectives was first described as a linear six-step process,* as illustrated in Figure 11.1.

COMPLEX ENTERPRISE—COMPRISING MULTIPLE SYSTEMS OF SYSTEMS

The NZ health enterprise is a virtual system that is dispersed across the NZ government and its national security systems, and hence operates in a mixed environment containing aspects that are a public service, aspects that are not-for-profit, and aspects that are for profit service delivery where the constraints are set by government policies and the market economy. The focus of the case study then was an adjunct to the government's system for strategy and execution in the enterprise, illustrated in Figure 11.2.

A. *Health enterprise*
   The first step toward addressing the cognitive complexity involved matching the broad enterprise to Hitchins' five-layer model (Hitchins, 2003), which indicates that a multilevel analysis would be needed across all five layers to truly appreciate the problem:
   - At Level 5, health is viewed as a component of whole-of-nation framework.
   - At Level 4, health is a national enterprise system (an industry in its own right).

---

* Prepared by Les Haines of Booz Allen Hamilton (December 2003) in the firm's proposal to the customer.

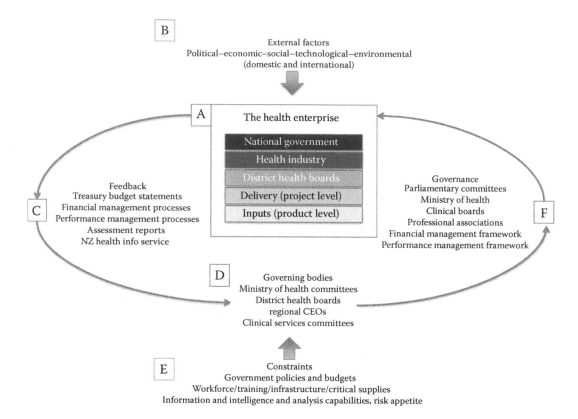

FIGURE 11.2   Health enterprise.

- At Level 3, health is a series of 21 regional systems of health systems, each managed by a District Health Board (DHB), providing service across primary, secondary, tertiary care, mental health, public health, and community health.
- At Level 2, health as a complex capability focused on delivering, or enabling delivery of, a health service; for example, a system of coronary care, a system of surgical treatment, a system of aged care, a system of pharmaceutical supply, a system of engineering services in a hospital—all of which involve people, technology, processes, training, and logistic support.
- At Level 1, health is a range of individual elements of complex capabilities, which are systems in their own right, such as new drug "X," an airworthy neonatal crib, surgical training, nurse education, and people (skills, expertise).

B. *External factors*

The environment for the study was the full health enterprise embedded in the national security architecture for NZ. This meant that the scenarios forming the planning basis for the study required inputs from over 100 CEOs or senior executives across the enterprise, as summarized in the stakeholder listing at the beginning of this chapter. The external factors also involved an appreciation of the wide range of all hazards facing NZ. While the study was initiated by terrorist concerns, the DPM&C and other ministries involved were intently driven to ensure that the intervention considered terrorism within a much wider context of hazards. Approval of the scenario-planning base therefore required sign-off by the National Steering Committee cochaired by the DPM&C. Concerns at this level extended beyond the government sector because external factors such as the health industry base, emergency management workforce, and volunteers also needed to be considered. Tackling a system problem at this level entailed traversing the full five tiers of Figure 11.2

because the outputs from the intervention would entail changes at the lowest level, that is, new equipment, doctrine, etc., to achieve an outcome at the whole-of-nation level.

C. *Feedback*

Formal feedback from the NZ Health Enterprise to the governing bodies was provided in the form of a number of statements and reports and through the financial and performance management processes. It is important to point out that this study was aiming to obtain insight into the performance and behavior of the enterprise during eventualities outside of its normal operating conditions, and, as such, in common with defense planning and other disaster-related enterprises, the normal steady-state feedback mechanisms were not seen as entirely appropriate in informing the governance bodies on the potential of the enterprise to deal with serious emergencies.

In a similar fashion to the defense case study in this book, problem boundary definition was indicated not just to contain the scope of the problem (using the usual sources of feedback within the enterprise—as shown in Figure 11.2) but also to identify the stakeholders and to improve the systemic intervention in the social system of interest (Ulrich, 1983). Boundary critique at the early stages of an intervention helps surface boundary definitions that might change the feedback loops required as the debate goes along to address, for example, microeconomic concerns at a quantitative level and then bring the results for synthesis into a whole of enterprise dialogue.

D. *Governing bodies*

Governance of the NZ Health Enterprise is overseen by the Minister of Health and policy is developed by the Ministry of Health. Provision of health and disability services within defined geographical areas is the responsibility of 21 DHBs that provide funding to Primary Health Organizations (PHO) that are funded on a capitation basis. PHOs are usually set up as not-for-profit trusts and have as their goal the improvement of their population's health, and their governance comprises both community and provider representation. It is the PHOs that provide health and disability services ranging from GP clinics to specialist medical, dental, and allied health services delivered through a range of clinics and hospitals. Thus, the NZ Health Enterprise has many layers of governance as is common in similar public health systems.

Governing bodies for our study began with the RFP issued by the Ministry of Health, with the senior executive responsible for governance and risk being the accountable officer, with day-to-day responsibilities held by an executive responsible for emergency response and communicable disease controls. These two executives brought significant expertise in management across the health network. Following the scoping study, they established a Steering Committee cochaired by a senior executive in the DPM&C who was also a senior adviser to the prime minister of the day. This added significant value when it came to the issue of addressing health and emergency management risks in whole-of-government and whole-of-nation contexts. With these governing bodies for the study, the team was then able to operate with broad boundaries for strategy and execution across the enterprise.

E. *Constraints*

The constraints on the NZ Health Enterprise are as one would expect for a national public health system and are listed in Figure 11.2.

The biggest constraints on delivering a quality whole-of-nation intervention are cost and time. The subconstraint within the time domain is who, of everyone you could speak to, you choose to invite to be involved in such an intervention. We found everyone exceptionally willing to be involved and exceptionally generous with helping us to understand the import of their contribution and its impact (to second and third order) in the wider enterprise. The other constraint is to manage within the limits of one's own cognitive limitations for taking in new information. How we structured our approach to manage the cognitive complexity, using Soft Systems Methodology (SSM), is discussed in the forthcoming sections.

F. *Governance*

The governance of the NZ Health Enterprise is achieved through the entities and framework listed in Figure 11.2. Note that this comprises medical practice governance in addition to enterprise governance.

Governance of the study was considered at two levels. Governance of the study's design and implementation was conducted, as mentioned earlier, with the executive oversight of the Steering Group. The DPM&C and Ministry of Health cochairs of that group also assisted in building our understanding of, and interaction with, the national government's governance apparatus—especially the interface at the end of the study with the NZ Treasury to present our findings on the options available to the NZ government.

CHALLENGES

Clearly, the concerns arising from the advent of new threats in the strategic environment were a great concern to the NZ government, its bureaucracies, and its people. They were prepared to face that by exploring their preparedness by virtue of this intervention. The main challenge was to make the intervention count; noting that, in a complex whole-of-nation enterprise three challenges arose:

- *Scale*: How to understand and define the systems complexity of the *problem space*?
- *Scope*: How to establish boundaries on the scope of the investigation by defining subordinate research questions?
- *Target*: How to identify the *what*, *where*, and *why* of each layer in the problem space and so establish a basis for designing the systems interventions that uncover the *how* and the *who*?

Scale: While a prime tenet of the systems concept is that we can hide some complexity through abstraction at higher layers (after Boulding, 1956), the study would be conducted extensively at the higher levels of abstraction, yet require a number of *deep dives* into the detail of selected elements at the lower levels.

Conceptually, Hitchins' five-layer model (Hitchins, 2003) was used to depict the scale of systems complexity, as illustrated in Figure 11.3, where the length of the bars illustrates the increasing level of ambiguity and scope for disagreement, and indicates an initial estimate of the division of our consulting time in order to manage the participant stakeholders, who were many.

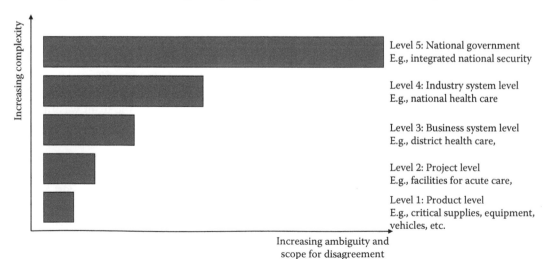

FIGURE 11.3 Indicative layers of systems complexity. (After Hitchins, D.K., *Advanced Systems Thinking, Engineering and Management*, Artech House, Inc., Norwood, MA, 2003.)

TABLE 11.2

Research Questions—Setting Boundaries and Targeting the Intervention into the Health Enterprise

Issues/Research Questions	Complexity Level	Methods
Overall capacity, priorities, and options		
What overall strengths in response capacity have been identified through this study? For example: skills? Planning? Flexibility?	Level 4	Convene National Steering Group
What overall vulnerabilities have been identified? For example: specific choke points? Specific skill gaps? Weaknesses in strategies/plans?	Level 4	Conduct National Survey
What are the options and indicative costs for addressing the gaps?	Level 4	Conduct Regional Focus Groups
What are the costs additional to those incurred through the response that arise due to deferring *business as usual*? For example: increased morbidity and mortality from deferred elective procedures?	Level 4	Conduct stakeholder interviews—National Health, Intersectoral, and Professional Medical and Nursing Associations
What residual risks may not be addressed through these options?	Levels 1–5	
Clinical and operational management		
How well do current DHB emergency plans and arrangements for regional agency cooperation and overall national coordination service the requirements of responding to a serious or unusual health emergency?	Levels 4–5	Convene National Steering Group
Has this analysis (within the health sector) identified any support or coordination issues at a regional and intersectoral level?	Level 5	Conduct National Survey
What are the options for addressing identified gaps in coordination, e.g., in planning, interagency arrangements, exercises?	Levels 4–5	Conduct Regional Focus Groups Conduct stakeholder interviews—national, regional, and district level, and intersectoral emergency service agencies Conduct visits to representatives of eight DHBs Conduct scenario-based analysis
Workforce		
Are there shortfalls in the availability of specific health occupations relevant to the emergency response? (Note: overall recruitment and retention of the health workforce is a background consideration)	Levels 1–3	Conduct National Survey
Are there gaps in skills related to basic professional training, specialist professional training, or training through participation in emergency response exercises?	Levels 1–3	Conduct Regional Focus Groups
What are the options to address identified gaps, e.g., use of specific training courses, improvements to emergency planning and exercising, targeted recruitment and retention strategies (specific to emergency preparedness), sourcing personnel from other regions to meet response requirements?	Levels 1–3	Conduct stakeholder interviews at district and hospital levels Conduct facility visits to selection of hospitals in specific areas identified in the scenarios Review nationally sourced data

(*continued*)

TABLE 11.2 (continued)

Research Questions—Setting Boundaries and Targeting the Intervention into the Health Enterprise

Issues/Research Questions	Complexity Level	Methods
		Conduct systems dynamic modeling of public hospital capacity
		Conduct scenario-based analysis
Infrastructure, equipment, and supplies		
Are there shortfalls in infrastructure, critical equipment, or critical supplies that would impede the progress of an effective health emergency response?	Levels 1–3	Conduct National Survey
What are the options to strengthen infrastructure and improve the availability of critical equipment and supplies?	Levels 1–3	Conduct Regional Focus Group
		Conduct stakeholder interviews at stores, hospitals, and districts
		Conduct facility visits to selection of supply and infrastructure managers in specific areas identified in the scenarios
		Review nationally sourced data

Scope: In view of the broad, outcome-focused question the study set out to address, the potential existed to overwhelm the study with considerations of every issue that might arise across the five layers of systems complexity. To help manage the cognitive complexity, and scope out the boundaries of the study, the team developed a set of focal research questions in consultation with the Ministry of Health, as summarized in the first column of Table 11.2.

The detailed scope of the problem was cross-matched to the scale of the problem as modeled in Figure 11.3. The identified systems layers were aligned with the research questions, as shown in the second column of Table 11.2, to generate a guide to the choice of methodologies for collecting the data. Questions aligning with higher levels of systems complexity would require more emphasis on pluralist approaches. Those at the lower levels were likely to be more open to rational approaches or a mix of rational and pluralist approaches.

Target: With the scope of the systemic intervention now defined, the study team was able to define those elements of the national health infrastructure that needed to be targeted and involved in this study. A mapping was created to guide those parts of the study to be addressed against the different elements of the NZ health enterprise. This mapping, referred to as the information architecture for the study, is shown in Figure 11.4. The architecture relates the focal areas of analysis with those functional areas of health capability with which the study team needed to interact in order to acquire adequate information and organizational perspectives for the study. The interventions also addressed issues at all four phases of emergency management: reduction, readiness, response, and recovery (known colloquially in NZ as the *4 Rs*).

With such a wide spectrum of agencies in this information architecture, it was decided to define a thin *vertical slice* and exercise the intervention first at a local and regional scale before taking the step of conducting the intervention at a national level across the full information architecture. This was seen as a risk reduction measure (any initial problems would not be experienced on a national scale) and as a confidence-building measure once the methodology was established on a smaller scale.

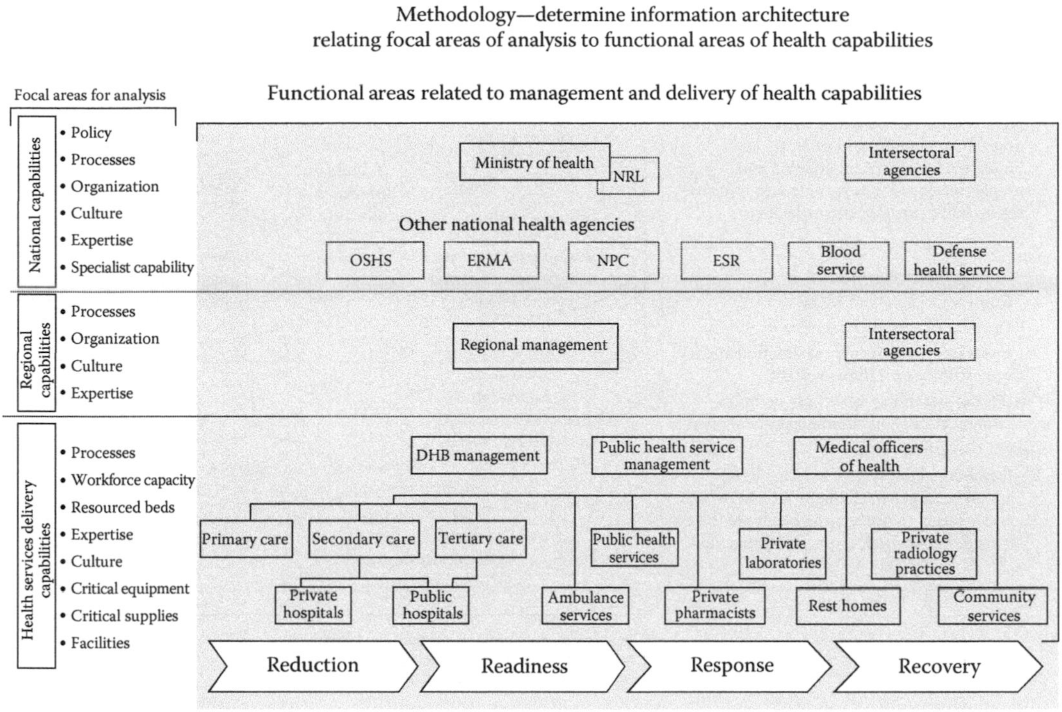

FIGURE 11.4 Information architecture.

DEVELOPMENT

In the first stage of managing the cognitive complexity, as early as in the scoping study (Work Package 0), the review and control processes involved extensive consultation and negotiation with the customer and the national steering group before the integrated methodology design (as theory) could be applied in practice. Using the Checkland and Scholes (1990) version of SSM illustrated in Figure 11.5, root definitions and CATWOE assessments helped focus the systemic intervention, while also identifying critical elements in the design and control of the systems intervention.

Seven elements were involved in the process of managing the cognitive complexity of designing a methodology for the whole-of-nation intervention. The study team sought to do the following:

1. Appreciate the cognitive complexity of the *problem space* as discussed around Figure 11.3
2. Define the problem space using research questions as discussed around Table 11.2
3. Create an architecture of those agencies involved and with which we had to seek and exchange information as discussed around Figure 11.4
4. Appreciate the available rational, adaptive, and interpretive methodologies
5. Synthesize a methodology to combine the selected methods/methodologies
6. Appreciate the situational complexity in managing the systemic intervention in a complex area of concern
7. Define performance criteria for the conduct of the implementation of the systemic intervention
8. a. Implement the methodology
   b. While evaluating the performance of the methodology, take control action to ensure the systemic intervention is achieving a *good result*

SSM 1—Generic concept for the process of managing cognitive complexity

Root definition: A government-owned and contractor-supported system of enquiry to investigate the capacity and capability of the NZ Health system to respond to serious and unusual emergencies, in order to understand the current situation and prepare options for government to improve the system's capacity and capability.

C—Ministry of Health (MOH)
A—Contractor team and MOH
T—Disparate knowledge sources→
    integrated view of current situation and its opportunities for improvement
W—*Wicked* systemic problems in complex situations can be systematically addressed
O—NZ Government
E—Elements of the system external to NZ are taken as given, elements internal to NZ and external to the health system are to be considered where it impacts health action

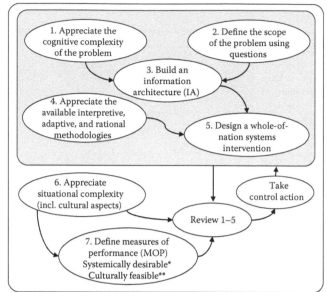

* Fit to suit scope of problem efficiently and effectively
** Balance of interpretive, adaptive, rational methodologies
    to suit the culture of the organization

FIGURE 11.5 SSM (round 1) to manage cognitive complexity of the methodology. (From Hodge, R.J., A systems approach to strategy and execution in national security enterprises, PhD thesis, University of South Australia, Adelaide, Australia, 2010.)

The following sections now continue the case study by discussing Steps 4–8 in turn.

Step 4: Appreciate the available interpretive, adaptive, and rational methodologies
The next step involved consideration of the modes of intervention that the study team would engage in. There were several, comprising interpretive, adaptive, and rational modes.

The principal *interpretive modes of intervention* included the following:

- Interviews with stakeholders with responsibilities represented across all five levels of systems complexity, to identify user requirements, health system capacities and interdependencies, capability gaps and vulnerabilities, and the potential initiatives to address the gaps
- Scenario-based analysis using three scenarios to support focused *deep dives* into the health system and its capacity to respond, noting that each incident begins locally (at the health services delivery level at one location) and rapidly escalates to a regional and national contingency. The three scenarios covered[*,†]:
    - A flu pandemic that was also representative of a possible terrorist or naturally occurring biological incident
    - A building collapse that was representative of a possible terrorist incident or a possible natural disaster from an earthquake
    - A chemical poisoning incident that was representative of a major industrial accident and a possible terrorist incident including a radiological device (or *dirty* bomb)

---

* These scenarios were developed by the Ministry of Health in consultation with the Booz Allen team to ensure that they provide adequate context and information to support all relevant modes of intervention.
† The study team remains indebted to the work of Dr. Nick Wilson and Jane Allison of the Ministry of Health for their medical research and preparation of the scenario descriptions.

- Facility visits to the principal hospital facilities, public health, and primary care capabilities represented in the principal locations of each of the three scenarios to determine local response capacities
- Vulnerability analysis of the issues arising to identify the criticality of each issue and its interdependencies with other causal factors

The principal *adaptive modes of intervention* included the following:

- The National Steering Group guided the direction of the study. This group included representatives with operational responsibilities at different levels of systems complexity—including Level 5 (DPM&C), Level 4 (Ministry of Health), Level 3 (DHB representation), Level 2 (General Manager of a major hospital), and Level 1 (Clinical Advisers).
- Regional focus group meetings to validate the vulnerability assessments and explore feasible options to enhance the health system.

The principal *rational modes of intervention* included the following:

- National survey of all 21 DHBs on emergency management preparedness issues
- Review of nationally sourced data on emergency management preparedness plans, processes, and constituent data (e.g., workforce and supplies)
- Modeling of health demand to provide plausible upper and lower bounds of cases of mortality and morbidity by number and type
- Systems dynamic modeling of health supply capacity* to assess the impact on the health system of major health demands arising from serious emergencies, taking account of the normal load on the system, the beds, the people, and the resources available, and the average length of stay for each condition
- Risk modeling and analysis of the vulnerabilities arising in the system

The cognitive framework of questions (Table 11.2) constrained and focused the study across the five levels of systems complexity. The team anticipated that answers to these questions would yield a mix of outputs based on calculation, expertise, and judgment, and that the outputs would be supportive of a mix of rational, adaptive, and interpretive strategy formation modes needed to address the situational complexity.

The multimethodological design, likewise, had to span all five levels. The principal constraints of the study were time (limited to nine months for delivery of the major elements of the study to ensure that the outcomes are fed into the government decision cycle) and costs (it was conducted under contract as a fixed-price commercial venture).

The initial assessment of the prospective methods of investigation identified in Table 11.3 reflected a mix of rational, adaptive, and interpretive modes.

Step 5—Synthesize a methodology to combine the selected methods/methodologies

The designed approach drew on systems engineering principles, whereby the three modes of intervention were interwoven across four main types of activity:

- Information collection and interpretation
- Analysis of the information to identify the potential weaknesses in, and requirements of, the health system
- Synthesis/design of the initiatives to address the requirements and the weaknesses
- Evaluation of the designs to ensure that, as potential solutions, they provided an appropriate level of fit, balance, and compromise to deliver a sustainable outcome for NZ health

---

* The primary author acknowledges the contribution of the Booz Allen Hamilton team led by Joan Bishop in the United States and her team's persistence in applying systems dynamic modeling as one of the *rational* interventions in the study.

TABLE 11.3
Modes of Investigation

Rational Modes	Adaptive Modes	Interpretive Modes
Systems dynamic modeling of workforce, beds, patient load, requirements, bottlenecks	Systems dynamic modeling— sensitivity analysis	Scenario analysis
Pandemic spreadsheet model	Focus groups	CEO and stakeholder interviews
Risk modeling	Scenario-based seminars/workshops	Facility visits
Cost–benefit analysis	Steering Group meetings	Mission needs analysis
National survey		Capability baseline assessments
National data sources		Vulnerability analysis
		Design of mitigation measures

The application of the systems engineering model to manage the situational complexity incorporated the mix of rational, adaptive, and interpretive methods for intervention in the health system, and is discussed later in Step 8a.

Step 6—Appreciate the situational complexity in managing the systemic intervention in a complex area of concern
The mix of rational, interpretive, and adaptive methods addressed all five levels of systems complexity as shown indicatively in Figure 11.6.

This initial appreciation also suggested that, while interpretive modes of strategic analysis conceivably covered the full scale of the problem, they could not adequately address the full scope of the problem. Other means were better suited to investigate the detailed complexity (in operating a hospital, for example) and the adaptive nature of different communities of practice and their ability to work together. With these *pieces* of the intervention identified, the next step sought to integrate them into a workable whole.

Step 7—Define performance criteria for implementation of the systemic intervention
From the outset, the study team recognized that its initial approach would need to be flexible to a reasonable degree to include new people, new knowledge, and new or revised methods/methodologies if these were judged to be desirable.

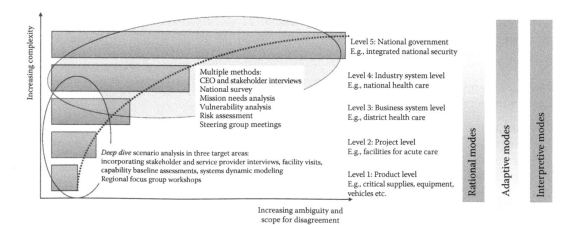

FIGURE 11.6   Hitchins' five-layer model guides methodology design.

Three measures of performance were used. Consistent with Checkland (1981), they are as follows:

1. The *fitness-for-purpose* of the methodology to deliver a systemically desirable outcome
2. The extent to which the study achieved a balance of interpretive, adaptive, rational methodologies
3. The extent to which it would be culturally feasible for the process and the content to be addressed by the health system

These criteria shaped many of the conversations with the Ministry staff and the judgments made by the study team in designing and conducting the intervention.

Step 8a: Implement the methodology

The four principal types of activities presented in Step 5 formed the basic approach taken to the study. To manage the situational complexity involved in conducting a whole-of-nation study, it was felt prudent to test the approach in local and regional contexts first before rolling it out nation-wide. These four main activities form the quadrants of a spiral development path that incrementally worked from the conception of the study to its trial in a pilot study and then to a whole-of-nation level involving multiple scenarios. This spiral path is shown in Figure 11.7.

This approach, common to systems engineering, enables the following functions:

- The methodology and the mix of component methods to be tested and validated
- The study to evolve through a series of scenario-based investigations that allow issues to be identified, tested, and evaluated in other scenarios for their commonality and criticality to the NZ health system as a whole

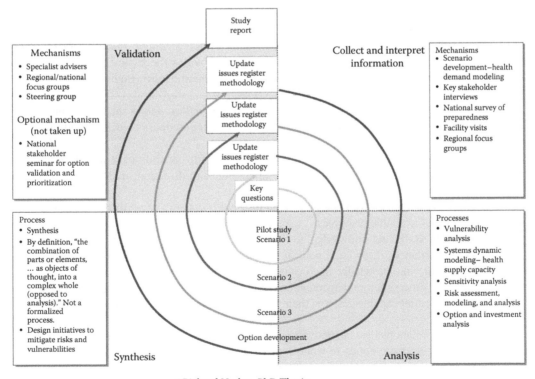

©Richard Hodge–PhD Thesis

FIGURE 11.7   Spiral development path for the designed methodology.

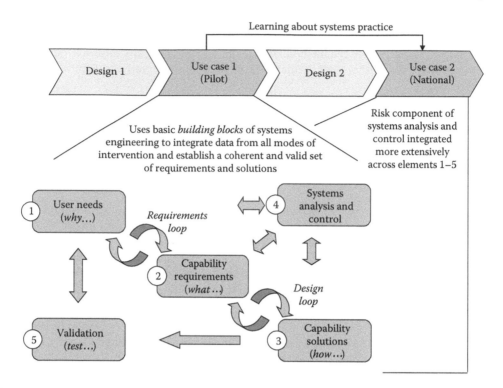

FIGURE 11.8 *Building blocks* of an integrated process and a rational product.

The use of a systems engineering framework in the pilot study and in the national study is summarized in Figure 11.8.

As in the case study *A System of Systems Methodology for Strategic Planning in Complex Defense Enterprises*, the systems engineering framework provides the basic building blocks for integrating the data outputs of each of the interventions to define the following:

- *The user needs* derived from the health system operating in the contexts set by the scenario-planning base and expressed as a set of five high-level preparedness objectives addressing the abilities for
  - Planning and direction, to develop the critical leadership capabilities required for unity of purpose, coherent planning, resource allocation and direction, maintaining strong engagement with stakeholders
  - Organization and management, to develop the health system's capability to optimize its utilization of available resources
  - Developing skills and knowledge, to raise, train, and sustain the skill base and team-work that balances directional control with adaptivity, and to provide the teams with rich feeds of vital knowledge in constant flow
  - Developing emergency management capability and support, to raise, prepare, and sustain the essential capabilities, and their enabling support systems, required in response/recovery
  - Developing relationships and networks, to ensure consistent engagement with the health sector from strategic policy to front-line health services delivery, and to establish vital links to critical resources outside the health sector
- *The health system capability requirements*, in this case expressed as a set of functional capabilities that have been derived through the vulnerability assessment, risk analysis, and risk treatment prioritization processes

- *The design of health system capability solutions*, which were appropriately packaged for consideration by government in cost bands matched to the achievement of the user needs and objectives
- *The conduct of systems analysis and control functions*, particularly the integration of risk extensively throughout the analysis and the *configuration management* of the design solutions to achieve the preparedness objectives, cover the vulnerability set, and establish a culturally feasible set of options for government
- *The validation of the solution set*, through interaction with the National Steering Group and in presentation of the option packages to the Treasury, noting that the customer chose not to pick up the analysis option of testing the option packages with the broader health community

As noted in the discussion of the user needs, the input to the framework was shaped by the context of the planning scenarios. The output of the process was ultimately a report to the customer addressing the user needs, the current capability baseline, the extant and prospective vulnerabilities and risks, the options for enhancement, and a set of recommendations for government—integrated in its analysis to demonstrate the cost–benefit–risk of the user needs and objectives.

IMPLEMENTATION

With the benefit of the initial insights into prospective methods to be applied in the study, the team implemented its approach in four distinct components:

- A scoping study, during which the boundaries of the system under study were defined, research questions were scoped out, and an initial methodology was designed and planned
- A pilot study across two DHBs to test and refine the methodology before applying it nationally
- An updated methodology to elicit and collect information through multiple methods across the full 21 DHBs and private entities comprising the health sector, to analyze the vulnerabilities in the sector, and to determine risk treatment priorities
- A synthesis approach to generate four options for government, to be presented as packages of initiatives across the health *value chain* to mitigate the assessed risks, including a cost–benefit analysis of the proposed investments to improve health system preparedness

A linear overview of the approach is presented in Figure 11.9. Taken at face value, this representation *hides* the inherent system components of the design. In reality, this linear path tracked along a spiral development path to ensure that the methodology design remained adaptive to the complexity arising in the problem situation. A linear representation proves useful in managing a conversation with the client to explain in a stepwise manner the approach being taken during the course of the study and assists in assuring the client that a rational mode of strategy formation with rational outputs linked to a budget will follow.

TESTING THE METHODOLOGY DESIGN IN A PILOT STUDY

The pilot study tested the methodology using one scenario (the building collapse scenario) operating across multiple stakeholders at the national, regional, and local levels of operations covering two major public hospitals, three private hospitals, two DHBs, the Ministry of Health and the intersectoral response and recovery agencies (Police, Emergency Services, Ministry of Civil Defense and Emergency Management, etc.). It sought to validate two primary issues:

- That the research questions would sufficiently address the problem driving the whole-of-nation study
- That the modes of intervention would scale from a pilot study to a national study

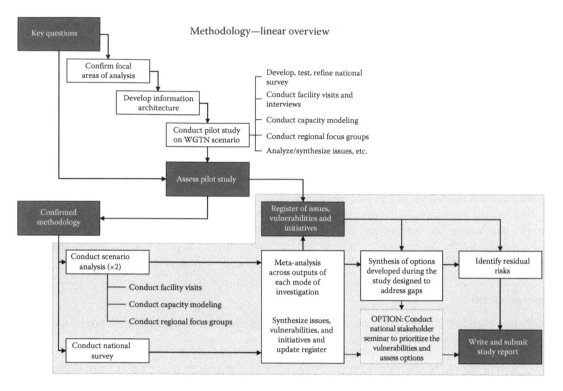

FIGURE 11.9   Linear overview of the methodology.

The question set: The critical outcome sought in the pilot study was the establishment and validation of the partitioning of the problem into the research questions that would be sufficiently coherent as a set so that the data they uncover would address the problem driving the study. It was also required that the pilot study demonstrate that the methodology and the associated research questions had the ability to scale in a practical way to address the primary question at a national level. This certification—by the Ministry of Health, not the team engaged to conduct the study—that the system-of-systems intervention would be both satisfactory and feasible is the natural conclusion of a pilot study before widening the scope of the problem to the national level (Maier and Rechtin, 2002).*

The data gathered from the various methods of intervention listed in Table 11.2 were captured in a central repository or register of relevant issues, critical vulnerabilities in the health system emergency management preparedness, and potential enhancement initiatives. These data remain confidential with the Ministry of Health.

The focal areas of analysis were derived from the research questions, which are illustrated in Figure 11.10. Several other issues that required exploration in order to assess the effectiveness of NZ's health emergency management capabilities were added. In particular, issues of process, organization, and culture are important indicators of effectiveness that needed to be analyzed. These tend to be significant drivers of other vulnerabilities and they lead to including initiatives impacting leadership, communication, and business processes that otherwise might not be considered.

It was quite natural for the interventions in the pilot study to address the 4 Rs (reduction, readiness, response, and recovery) as the philosophy is embedded in NZ's emergency management legislation

---

* Adapted from Maier and Rechtin's (2002, p. 179) discussion of the goal of scoping, which is to "form a concept of what the system [of intervention] will do, how effectively it will do it, and how it will interact with the outside world. The level of detail required is the level required to gain customer acceptance first of the continued development [in this case, during the pilot study] and ultimately of the built system [in this case, when considering the final outcomes on the NZ Health System]."

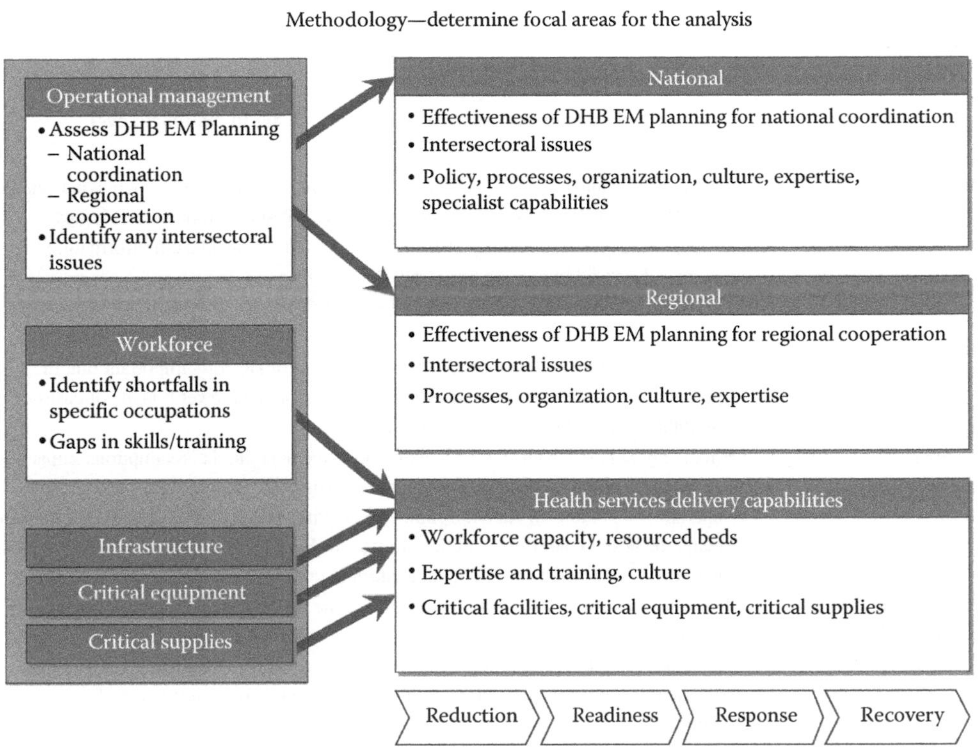

Methodology—determine focal areas for the analysis

FIGURE 11.10 Relationship between key questions and focal areas for analysis.

and its culture. It also resonated well in terms of outcome with the systems-based approach to the study intended to advise on the balance of improvements required in preparedness as well as response and recovery activities.

The pilot study validated the research questions to be sufficiently complete as a set, and that with the addition of a few subordinate issues, would adequately address the problem driving the whole-of-nation study.

Ability to scale the modes of intervention: In the pilot study, a vertical *slice* of the information architecture was taken by restricting the scope to 2 of the 21 DHBs and all of the national-level agencies. In that way, a sample of the full set of bilateral and multilateral relationships was tested among the national, regional (district), and local entities. This approach anticipated that scaling up to a national-level study would add complexity in the detailed content to be analyzed, without adding significantly to the types of relationships and interfaces shared by the DHBs and each of these interfaces was tested in the pilot study.

The information architecture was a key element in the design and conduct of this multimethodological study as it identified the multiple organizational and functional boundaries and critical interfaces where—most likely—the greatest opportunities for improvement and the greatest vulnerabilities in the health system reside (Rechtin, 1991, 2000). The information architecture provided a summary integrated view of the focal areas of analysis with the actual entities of a multitiered health system. In doing so, it framed the cognitive complexity of the study and guided the integration of the different interventions to address the key study questions. It also guided the integration of issues arising in the assessment of the situational complexity facing the health system in its capacity to handle a series of unusual and serious emergencies, in mapping the data from the interventions, and the prospective options for government to improve the health system, back to the problem space.

The study team applied all modes of intervention in the pilot study. The summary assessment of the changes required to scale up to a national-level study is indicated in Table 11.4.

TABLE 11.4

Assessment of Modes of Intervention in the Pilot Study

Mode of Intervention	Summary Assessment of Ability to Scale to National Level
Rational modes of intervention	
National survey	Pilot questionnaire required clarification. Question set simplified to operate nationally to suit 21 District Health Boards on emergency management preparedness issues.
Nationally sourced data	Increased the type and quantity of data sourced from national health information sources. Reduced burden on the 21 DHBs.
Review preparedness plans and constituent data	No adjustment required to scale to a national level.
Modeling of health demand	No adjustment required. Conducted by the Ministry of Health—important that the primary stakeholder set and agree the plausible upper and lower bounds of cases of mortality and morbidity by number and type.
Systems dynamic modeling of capacity	Pilot study required extensive review of, and agreement on, the assumptions impacting *every* entity in the model and how it interacts with other entities to represent the operating conditions of each hospital. The activity was resource-intensive for the study team, the hospital advisers, and the Ministry of Health, but without this effort, it could not have been applied acceptably at a national level.
Risk modeling and analysis of vulnerabilities	The risk assessment in the pilot study applied the AS/NZ Standard 4360 to assess the likelihood, consequence, and risk criticality of each of the vulnerabilities. To scale to a national level, the methodology required extension of the risk standard to address (1) the relative importance of the vulnerability relative to the full range of hazards facing NZ, (2) ease of addressing the vulnerability, (3) prospective cost, (4) potential for mitigation action to have wider benefits, and (5) the impact of different judgments arising in different scenarios for any one vulnerability. This proved a significant integrating feature of the different modes of intervention.
Adaptive modes of intervention	
National Steering Group	The National Steering Group provided a strong unifying force at the national level to review and validate the judgments of the utility of the study to address the primary question and the scalability of the methodology to provide an integrated national assessment. The role of the steering group in the national study shifted to review, challenge, and validate the outcomes at a national, whole-of-government level.
Regional focus groups	The pilot study demonstrated the importance of these interventions in validating the vulnerability assessments and the potential initiatives to improve the system. In scaling to a national level, care was exercised to ensure that a full range of stakeholders were included in the focus groups.
Interpretive modes of intervention	
Interviews	The interviews were highly productive in identifying systemic issues, particularly at the higher levels of systems complexity (Hitchins Levels 3–5). The team found it important to tailor the generic research question set to the specific interests of the interviewee. It was also imperative for these to be conducted *anonymously* without attribution to interviewee.
	The value of these interventions during the pilot study encouraged the study team to interview a broad range of stakeholders including different professional medical associations and the Ministry of Transport.
Scenario-based analysis	The pilot study demonstrated the importance of scenarios as a tool to bound the national problem and enable extrapolation to an *all hazards, all assets* approach. In scaling to a national study, adjustment was required to integrate the interpretive scenario analysis into the quantitative risk assessments.

TABLE 11.4 (continued)
Assessment of Modes of Intervention in the Pilot Study

Mode of Intervention	Summary Assessment of Ability to Scale to National Level
Facility visits	The pilot study demonstrated the importance of validating data, assessments, and assumptions in the physical environment to ensure the practical limits in each major facility and how they might manage any given scenario was satisfactorily understood. In addition, the visits often demonstrated the importance of extending *ownership* in the study as examples of practical innovative thinking—e.g., in creating additional facilities for infectious patients—which could be extended through the study to other DHBs. In scaling to a national study, more time was scheduled to allow extensive visits to key facilities in combination with a wide series of interviews.
Vulnerability analysis	No change required to scale from a pilot to national level.

In summary, most modes of intervention were scaled with little or no modification from the pilot study to the national-level study, and this enabled the customer of the study to accept the results of the pilot study and to support it being fully scaled to a national study—subject to the modifications indicated in Table 11.4.

There were two elements requiring greatest modification:

1. The national survey instrument of health emergency preparedness to refine the required level of detail to that agreed by the Ministry of Health as being appropriate with a view to reducing the effort required in each DHB to complete the survey.
2. The risk assessment and risk treatment prioritization methodologies. Noting the constraints of the available guidance (then AS/NZ 4360:2004, 2004), the Ministry required that scenario-based judgments be integrated for each of the identified vulnerabilities, and other issues had to be addressed that were important for integration into a whole-of-government context.*

VALUE OF INTEGRATING RISK

How risk is handled and presented in the systems approach is justified by the theoretical argument underpinning the development of the approach (Hodge, 2010). While there are many possible candidate risk assessment approaches available globally, the study team was required by the customer to comply with the requirements of the Australian/New Zealand Standard for risk management (AS/NZS 4360:2004, 2004).†

In summary, the value of the risk model lies in several factors:

- Risk modeling may be adapted to match the government's tolerance for risk through the national steering group and accepted as an informed basis for developing options for government.
- Risk became a central element in the analysis process and in the process to integrate security within the health system's strategy formation process.
- The outputs of, and insights from, the model were validated in the regional focus groups and the national steering group and, while imperfect, provided the level of quantification required by the customer.
- The resultant communication of risk-related issues had an integration role, and improved the balance between the interpretive, adaptive, and rational modes of intervention to support decision making, direction setting, and resource allocation.

---

* See Hodge (2010)—Appendix 2, for details of the changes made to the risk assessment methodology to meet the customer requirements for scenario-based analysis.
† This Standard has an essential companion handbook, *HB 436 Risk Management Guidelines—Companion to AS/NZS 4360*. The intention is for the two documents to be used together, with the handbook providing important commentary, guidance, and examples on the implementation of the Standard.

The study team sought to enable progressive improvement during the study by placing risk assessment and risk management as a central systems analysis capability in the health strategy formation process, which included the pursuit of opportunities for improvement. Ultimately, this change has been adapted to establish a model for integrating security issues within strategy formation studies to assist in the design of enterprises operating at the firm, industry, and national levels.

In a systems sense, the pilot study achieved its desired outcomes of modeling the methodology, testing it for feasibility and scalability, and refining it to better meet the needs of a national, whole-of-government assessment. In the process, a suitable fit, balance, and compromise (after Rechtin, 1991)* is achieved between the interpretive, adaptive, and rational modes of intervention.

SYNTHESIS AND ANALYSIS OF OPTIONS FOR GOVERNMENT

To conclude the national study, options for government were synthesized and then analyzed for cost, benefit, and risk.

Synthesis of options: Fifty-one high-level initiatives were developed *top-down* across the four phases of preparedness (reduction, readiness, response, and recovery) to address the complete set of prioritized vulnerabilities. These were mapped against the prioritized list of vulnerabilities to identify their effectiveness in reducing one or more vulnerabilities in the context of one or more of the representative scenarios. This effectively linked the top-down and *bottom-up* analysis to provide confidence in the completeness of the set of initiatives.

The 51 candidate initiatives were then assessed and ranked using a weighting process that assessed each of the initiatives against five criteria:

1. Its risk treatment priority to prioritize initiatives addressing risks with the highest priority for treatment
2. Its *footprint* over the complete set of vulnerabilities to prioritize initiatives offering wider effects on reducing vulnerabilities
3. Its impact on representative planning scenarios—to prioritize initiatives addressing scenarios where intervention should offer higher expected benefits
4. Its indicative cost—to prioritize those that might be relatively cost efficient
5. Its ease of implementation—to prioritize those that may be easier to implement in terms of cost–benefit

In several respects, this prioritization process is an extension of the risk treatment prioritization process. The first criterion is the product of the previous process. The remaining four criteria revisit the three factors of the implementation index for risk prioritization—ease, cost, and benefit—as improved judgments on cost and implementation of the initiatives can now be made, given the detail in the description of the initiatives. The benefit assessment is made more tangible by addressing both the extent to which each of the initiatives impacts the wider range of vulnerabilities and the more specific benefit each of the initiatives offers to each of the scenario events.

The resultant weighted ranking of the initiatives guided the packaging of four option sets for government:

1. Option A—Establishes the policy framework
2. Option B—Establishes the operational foundations
3. Option C—Establishes a systemic capability
4. Option D—Establishes an *All Hazards* national capability

---

* Rechtin (1991, p.156) summarizes "pairs of competing factors pulling in opposite directions, held together by fit, balance and compromise—the essence of systems architecting." Similarly, architecting interventions of complex adaptive systems such as a national health system requires an appropriate fit, balance, and compromise to be found between the competing modes of intervention.

As each of the option sets incorporates the previous option(s), the aim is to give advice to government on different levels of improvements, at different cost bands, with different levels of residual risk.

Cost, benefit, and risk analysis of the options: The approach to cost, benefit, and risk analysis used a composite of well-known methods* involving the following:

- "Developing a Base Case that includes the indicative costs of mortality and morbidity associated with each of the three representative emergency scenarios
- Assessing the indicative capital, operating, and maintenance costs of each option
- Assessing the indicative quantitative benefits of each option in terms of potential cost savings through reduced mortality and morbidity in each scenario
- Assessing the qualitative benefits of each option in terms of
  - Direct benefits to reducing vulnerabilities and improving health sector EM preparedness
  - Indirect benefits to national health outcomes, national security, and national EM preparedness
- Assessing the costs and benefits of each option against the base case to derive net benefits
- Assessing the implementation risks of each option
- Providing an integrated assessment that compares the indicative costs, benefits, and risks of each option in both quantitative and qualitative terms Booz Allen Hamilton (2004)."

As a mark of the customer's acceptance of the results of this analysis, the customer invited the team to present the cost–benefit–risk analysis to senior Treasury officials.

Step 8b: While evaluating the performance of the methodologies, take control action to ensure that the systemic intervention is achieving a *good result*.

In applying the *model* of the designed methodology in the complex situation of the pilot and whole-of-nation elements of the study, the monitor and review functions were now able to operate with the benefit of data from the operation of the model in the real world.

The high level of interaction with the customer group during the study drove the discussions to extend the two criteria of systemic desirability and cultural feasibility (after Checkland, 1981) to ensure adjustments in the design could be made as the study progressed. The SSM's *3 Es* (Checkland and Scholes, 1990)[†] form the basis for an extended set of criteria:

1. Does this methodology, as applied, comprehensively address the primary question?
2. Is the use of resources efficient in conducting the study?
3. Do the outcomes offer effective improvements to the enterprise that are systemically desirable and culturally feasible?

A further application of SSM was under way during the complex systems intervention to ensure the designed methodology remained efficacious, efficient, and effective as it went along. An interpretation of the processes that were under way is given in Figure 11.11.

The considerations and judgments of the *Actors* in the study, the *Customer*, and the *Owner* (all of whom were represented in the national steering group) guided improvements in the cognitive framework to enable an improved design in scaling the methodology to a national study, keeping systems theory and systems practice close together, as illustrated in Figure 11.12.

In this manner, the cognitive complexity (of the methodological approach) and the situational complexity (of changing methodologies to suit the consideration of the health enterprise operating

---

* The author contributed significantly to the costing of the individual initiatives, the packaging of the options for government, and the risk analysis in this work step. Andrew Tessler (Booz Allen Hamilton) conducted the economic analysis (EV and CBR) and contributed with Les Haines (Booz Allen Hamilton) and the author to the qualitative analysis.
† Checkland and Scholes (1999, p. 39): "3 Es": efficacy (for "does the means work?"), efficiency (for "amount of output divided by amount of resources used"), and effectiveness (for "is T [Transformation element of CATWOE mnemonic] meeting the longer-term aim?").

SSM 2—Generic concept for
the process of managing the
situational complexity

Root definition: A government-owned and
contractor-supported system of enquiry to
investigate the capacity and capability of the
NZ health system to respond to serious and
unusual emergencies, in order to
understand the current situation and
prepare options for government to improve
the system's capacity and capability.

C—Ministry of Health (MOH)
A—Contractor team and MOH
T—Disparate knowledge sources →
    Integrated view of current situation and its
    opportunities for improvement
W—*Wicked* systemic problems in complex
    situations can be systematically addressed
O—NZ Government
E—Elements of the system external to NZ
    are taken as given, elements internal to NZ
    and external to the health system are to be
    considered where it impacts health action

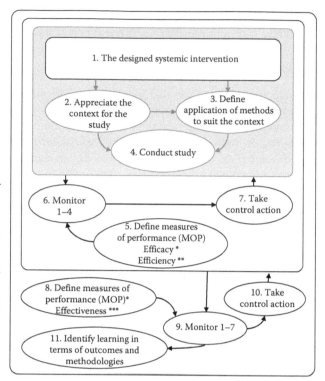

1. The designed systemic intervention

2. Appreciate the context for the study

3. Define application of methods to suit the context

4. Conduct study

6. Monitor 1–4

7. Take control action

5. Define measures of performance (MOP)
Efficacy *
Efficiency **

8. Define measures of performance (MOP)*
Effectiveness ***

10. Take control action

9. Monitor 1–7

11. Identify learning in terms of outcomes and methodologies

* Does this methodology comprehensively address the primary question?
** Is the use of resources efficient in conducting the study?
*** Do the outcomes offer improvement to the enterprise that are systemically desirable
and culturally feasible?

FIGURE 11.11 Managing the situational complexity of dynamically changing the mix of methodologies. (From Hodge, R.J., A systems approach to strategy and execution in national security enterprises, PhD thesis, University of South Australia, Adelaide, Australia, 2010.)

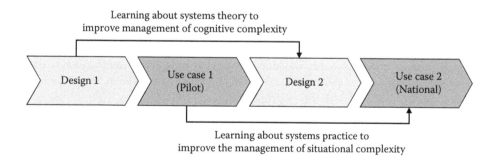

Learning about systems theory to
improve management of cognitive complexity

Design 1

Use case 1 (Pilot)

Design 2

Use case 2 (National)

Learning about systems practice to
improve the management of situational complexity

FIGURE 11.12 Close interaction of theory and practice.

in unusual and serious emergencies) were both handled in an evolutionary manner. In a sense, it is fair to say that the cognitive frameworks (the theory) evolved in close step with the practice of discovery and the application of analytical processes (the practice of applying the theory of investigation of a health enterprise). Checkland's SSM (Checkland and Scholes, 1990) provided an invaluable framework for managing the cognitive complexity and framing the whole study.

RESULTS

As a result of this study, the NZ Ministry of Health developed a deeper—yet still imperfect—understanding of the issues addressing the question: *What is the capacity of the nation's health system to respond to serious and unusual emergencies and what strategy should government consider to enhance its preparedness?* With the assistance of this complex and unique whole-of-nation assessment of health capabilities, the customer developed its agenda for improving the health system operated by 21 DHBs across the nation. Its agenda was subsequently endorsed and funded.

ANALYSIS

The multiple methodologies in Case Study 2 are assessed from a systems' viewpoint and from the viewpoint of strategy formation as a management activity.

VIEW ON THE DOMINANT SYSTEMS METHODOLOGIES

The *surprise* contributor in this study—more so than in the authors' other case study, *A System-of-Systems Methodology for Strategic Planning in Complex Defense Enterprises*, of this book—was the influence of SSM (1990 version) as a dominant methodology. It illustrated the value it can provide to the strategy formation process by first assisting to manage the cognitive complexity of the intervention by structuring the thought processes and presenting key criteria needed to ensure a good result at the methodology level; and then—although not by initial design—the thought processes and criteria at work in managing the situational complexity (after Warfield, 1999) of changing the piecewise methodologies or methods during an intervention.

Being able to cycle dynamically between the applications of SSM at two levels (as shown in Figure 11.13) enhanced the attempts at continuously improving the design of the methodology and its application in complex situations.

In summary, the value of the systems engineering methodology was found in the rational organization of main processes to derive a basis for *hard* decisions by government. In particular, in this case study, the systems engineering processes developed a strong basis for the design of health system capability solutions, appropriately packaged for consideration by government in cost bands matched to the achievement of the user needs and objectives. The conduct of systems analysis and control functions provided this foundation. In particular, the integration of risk was pervasive throughout the analysis. It was assisted by the configuration management of the design solutions to cover the set of vulnerabilities of greatest concern and frame the solutions to achieve the Ministry's preparedness objectives.

Working with both dominant methodologies in combination was considered a significant contributor to the success of the study. In particular, because the solution set was *validated* through interaction with the National Steering Group and in the presentation of the option packages to the Treasury, the customer chose not to pick up the analysis option of testing the option packages with the broader health community. For these reasons, the combination of the dominant methodologies (SSM and systems engineering) was a significant factor in establishing a culturally feasible set of options for government.

VIEW ON THE COMPREHENSIVENESS OF THE STRATEGY FORMATION

From a strategy perspective, improvements were made on all counts, as summarized in the *scorecard* shown in Table 11.5, which draws on the guidance of Mintzberg et al. (1998) on what is needed for better strategy formation.

A subjective and summary view of the *whole beast* of strategy formation in case study two is presented in Figure 11.14, illustrating the perceived extent of integration of the three modes of strategy

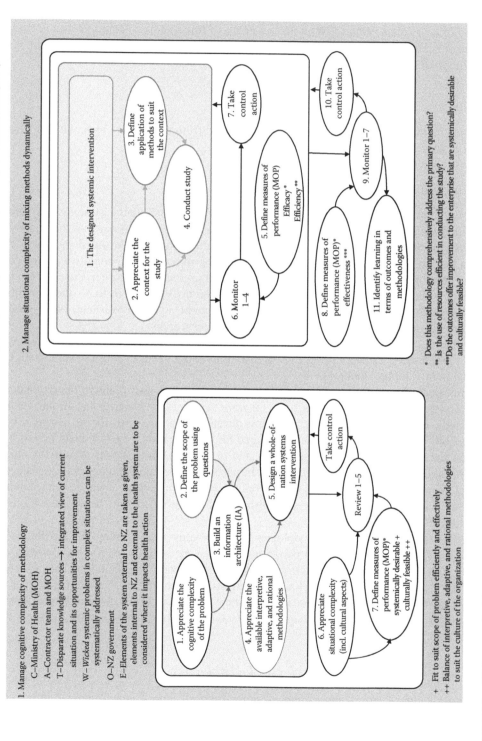

Root definition: A government-owned and contractor-supported system of enquiry to investigate the capacity and capability of the NZ health system to respond to serious and unusual emergencies, in order to understand the current situation and prepare options for government to improve the system's capacity and capability.

1. Manage cognitive complexity of methodology

    C–Ministry of Health (MOH)

    A–Contractor team and MOH

    T–Disparate knowledge sources → integrated view of current situation and its opportunities for improvement

    W–*Wicked* systemic problems in complex situations can be systematically addressed

    O–NZ government

    E–Elements of the system external to NZ are taken as given, elements internal to NZ and external to the health system are to be considered where it impacts health action

2. Manage situational complexity of mixing methods dynamically

*   Does this methodology comprehensively address the primary question?

** Is the use of resources efficient in conducting the study?

*** Do the outcomes offer improvement to the enterprise that are systemically desirable and culturally feasible?

+   Fit to suit scope of problem efficiently and effectively

++ Balance of interpretive, adaptive, and rational methodologies to suit the culture of the organization

FIGURE 11.13 SSM as a methodology to manage cognitive and situational complexities. (From Hodge, R.J., A systems approach to strategy and execution in national security enterprises, PhD thesis, University of South Australia, Adelaide, Australia, 2010.)

TABLE 11.5

Strategy Scorecard of Health Preparedness Case Study

Item	Summary Assessment
Better strategy formation needs to	
Ask better questions and fewer hypotheses	The study began with a primary question from which the research questions were devised and then tested in a pilot study for completeness and scalability.
Be pulled by concerns *out there* not pushed by strategy concepts	The concerns of the NZ government as customer that were major drivers of the content and process were as follows:
	• Concern for the *affordable* security of the people of NZ and that the government exercised its duty of care
	• Coverage of intersectoral/whole-of-government issues
	• Wide stakeholder engagement in interviews, visits, and focus groups
	• Need to bring analysis to a quantitative measure to guide resource allocation
	Other concerns were reflected in the range and content of the research questions driving the interventions.
Design of interventions should be more comprehensive and concerned with	
Process and content	The content and the process were framed principally by the information architecture agreed with the customer.
	Process and content were modeled in the scoping study and tested in the pilot study and then remodeled before being scaled to conduct the interventions at a national level.
Statics and dynamics	The study developed a baseline understanding of the health system as it was before it could begin to understand the dynamics of the system capabilities as they both evolve from day to day and would respond/recover to emergencies in future.
	Strategy analysis also incorporated spreadsheet modeling of static data and systems dynamic modeling of the dynamic complexities, where they could be modeled.
Constraint and inspiration	As a first-world nation of *c*.4 million people, government in NZ is necessarily one of constraint and inspiration to sustain high-quality services from a limited resource base. The strategy interventions (process) and outcomes (content) were driven by affordability, timeliness, and wider concerns and benefits.
Cognitive and collective	Strategy as a *mental* process began with managing the cognitive complexity of the study through careful question definition and a systems approach (discussed later).
	The collective processes involved a wide range of interviews, visits, focus groups, and steering groups to help manage the cultural perspectives across the 21 DHBs in the health system and to cater for the cultural diversity in NZ, including its Maori and Islander populations.
Planned and the learned	While formal health and civil defense and emergency planning is well advanced in NZ, the study itself planned a formal strategy process—through its methodology description, to conduct the study. The study also integrated adaptive methods to generate learning and validate assessments, and gain insights into initiatives to address the situation.
Economic and political	The study concluded with a rigorous economic analysis of expected value and cost–benefit ratio to ensure that the interpretive/political perspectives of the outcomes of the study could be balanced with and informed by the bounded rationality of the cost–benefit–risk analysis. This enabled strategic conversations to be at all times constructive in bringing the study to a conclusion at a whole-of-government level. Personal conversations (Helm, 2005) indicate that the study continued to be referenced by the customer in its broader dealings with central government and other government departments.

Source: After Mintzberg, H. et al., *Strategy Safari—A Guided Tour through the Wilds of Strategic Management*, The Free Press, New York, 1998.

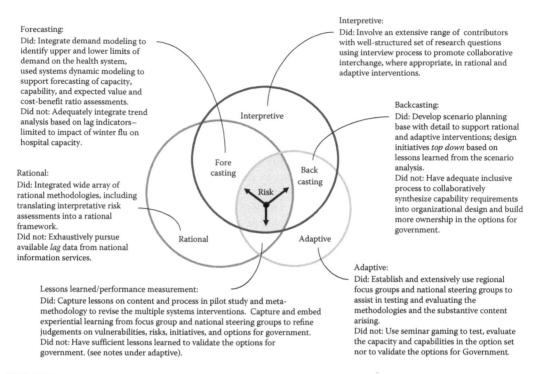

Forecasting:
Did: Integrate demand modeling to identify upper and lower limits of demand on the health system, used systems dynamic modeling to support forecasting of capacity, capability, and expected value and cost-benefit ratio assessments.
Did not: Adequately integrate trend analysis based on lag indicators– limited to impact of winter flu on hospital capacity.

Rational:
Did: Integrated wide array of rational methodologies, including translating interpretative risk assessments into a rational framework.
Did not: Exhaustively pursue available *lag* data from national information services.

Interpretive:
Did: Involve an extensive range of contributors with well-structured set of research questions using interview process to promote collaborative interchange, where appropriate, in rational and adaptive interventions.

Backcasting:
Did: Develop scenario planning base with detail to support rational and adaptive interventions; design initiatives *top down* based on lessons learned from the scenario analysis.
Did not: Have adequate inclusive process to collaboratively synthesize capability requirements into organizational design and build more ownership in the options for government.

Adaptive:
Did: Establish and extensively use regional focus groups and national steering groups to assist in testing and evaluating the methodologies and the substantive content arising.
Did not: Use seminar gaming to test, evaluate the capacity and capabilities in the option set nor to validate the options for Government.

Lessons learned/performance measurement:
Did: Capture lessons on content and process in pilot study and meta-methodology to revise the multiple systems interventions. Capture and embed experiential learning from focus group and national steering groups to refine judgements on vulnerabilities, risks, initiatives, and options for government.
Did not: Have sufficient lessons learned to validate the options for government. (see notes under adaptive).

FIGURE 11.14 Perspectives on developing a *sweet spot* among the mix of methodologies. (From Hodge, R.J., A systems approach to strategy and execution in national security enterprises, PhD thesis, University of South Australia, Adelaide, Australia, 2010.)

formation and the extent to which their integration improved the formation of a sweet spot. While Figure 11.14 summarizes the main opportunities taken to integrate the three modes, it also provides some insights into where opportunities were missed or not taken up due to customer choice.

Incorporating risk was a critical element in the integration of the three modes of strategy formation. The use of risk, first to assess the identified vulnerabilities in the health enterprise *as a complete set of interdependent systems* and then to assess the opportunities for enhancement, drove the expansion of the sweet spot from the *middle out*.

Of the interventions that were not adopted in this process, the lack of seminar gaming was the most significant omission. It presents an opportunity to include an evaluation process to test the options for government against the requirements defined in the set of representative scenarios. In engineering a systems approach to strategy formation, this is a highly desirable inclusion because it offers a means to validate the options and the assessment of residual risk. While it was systemically desirable to the strategy formation process—for the opportunity it gives to expand the adaptive mechanisms in use and enhance the sweet spot—it was not culturally feasible at the time to expose the set of vulnerabilities beyond the national steering group. The application of judgment, therefore, prevailed as the dominant interpretive means of validating the options and the residual risks.

SUMMARY

Situation: In the face of terrorism threats and SARS, the NZ Ministry of Health responsibly asked the question: What is the capacity of the nation's health system to respond to serious and unusual emergencies and what strategy should government consider to enhance its preparedness? The Ministry of Health sought assistance to help develop its agenda for improving the health system that is operated by 21 DHBs across the nation.

Contribution of the case study: In support, the study assessed the baseline capabilities and analyzed the vulnerabilities, risks, and risk treatment priorities. The focal areas of analysis covered all issues impacting health preparedness—policy (including legislative authority), business processes, culture, organization, expertise and specialist capabilities, workforce capacity and bed capacity, critical supplies, equipment, and facilities. The study analyzed the baseline capabilities of the primary, secondary, and tertiary care, as well as links with private hospitals, laboratories, pharmacies, rest homes, and community services. We assessed issues arising in district management across the 21 DHBs and their capacities for regional coordination. We also assessed issues of national collaboration and decision making, particularly leadership issues at the Ministry and those needing national direction such as stockholding policies, health surveillance, and intelligence, and enabling information systems to support health system response in a serious emergency.

The study team developed a strategy that contained four options for systemically enhancing the high levels of reliability in the health system. We conducted a cost–benefit analysis that integrated quantitative and qualitative measures and, at the request of the Ministry of Health, we presented to the NZ Treasury our cost–benefit analysis of those options.

Result: The NZ Ministry of Health was able to deliver against these challenging outcomes by addressing a complex and unique whole-of-nation assessment of health capabilities. The agenda that was subsequently set gained endorsement and funding from government. The Ministry is also well positioned to help NZ understand the requirements such emergencies place on their systems and guide courses of action they might take to continuously improve the agility of the health system and the delivery of health outcomes for the NZ people.

CONCLUSIONS

In outcome terms, the Ministry is well positioned to help NZ understand the requirements that serious or catastrophic discontinuities place on their systems and guide courses of action they might take to continuously improve the agility of the health system in such events. In the process, the day-to-day delivery of health outcomes for the NZ people will also be improved.

QUESTIONS FOR DISCUSSION

1. What makes a whole-of-nation study so difficult?
2. Why was a multimethodological approach selected to tackle this problem?
3. The case study explicitly describes the principal interpretive (SSM) and functionalist (systems engineering) methodologies employed. It has been mentioned that SSM was the biggest *surprise* in how it helped to manage the cognitive and situational complexities in near-real time.
   a. Explain the main issues in using SSM to manage cognitive complexity.
   b. Explain the main issues in using SSM to manage situational complexity.
   c. In bringing a study team together for a whole-of-nation study, what suite of competencies would you bring into your team to assure success? If your core team is restricted to two, three, or four people, what competencies would you ensure they hold, and why?
4. If your brain hurts, that is an expected condition in whole-of-nation studies. Worse still is the hurt to the reputation and *bottom lines* of the individual and the agencies involved if your study fails to deliver a quality outcome within the constraints of time and cost. What confidence do you have that the selected multimethodological approach would actually deliver a *good result*? Explain your reasoning from the different perspectives of
   a. The cochairs of the Steering Group
   b. A DHB or a CEO of a major hospital
   c. A professional medical association
   d. The head of an intensive care department in a major hospital
   e. The lead of the study team

5. If you were an adviser to government, what would you expect to see in an options paper on ways to improve the capacity and capability of a nation's health system to respond to new, unusual, and serious emergencies? What would you expect from the outputs of a multimethodological approach to support your options paper? How would you improve the control of the study to give a government the assurance it needs?

6. If, as the same adviser to government, you were told at the outset that you are to be made accountable for implementing the study outcomes—would this change your approach? Please explain.

REFERENCES

Ackoff, R.L. (1999) *Re-Creating the Corporation: A Design of Organizations for the 21st Century.* New York: Oxford University Press.

AS/NZS 4360:2004. (2004) *Risk Management.* Melbourne, Victoria, Australia: Standards Australia.

Blanchard, B.S. and Fabrycky, W.J. (1998) *Systems Engineering and Analysis*, 3rd edn. Upper Saddle River, NJ: Prentice Hall. ISBN 0-13-095062-9.

Boulding, K.E. (1956) General systems theory—The skeleton of science. *Journal of the Institute for Operations Research and the Management Sciences* 2(3): 197–208.

Booz Allen Hamilton (2004)—Internal correspondence to the Ministry of Health, Wellington, New Zealand.

Checkland, P.B. (1981) *Systems Thinking, Systems Practice.* Chichester, U.K.: Wiley & Sons.

Checkland, P.B. and Scholes, J. (1990) *Soft Systems Methodology in Action.* Chichester, U.K.: Wiley & Sons.

Demos, N., Chung, S., and Beck, M. (2001) The new strategy and why it is new. *Strategy+Business.* Winter (25): 15–19. New York: Booz Allen Hamilton. http://www.strategy-business.com/article/14254?pg=all. Accessed February 25, 2014.

Hart, S.L. (1992) An integrative framework for strategy-making processes. *Academy of Management Review* 17(2): 327–351.

Helm, P. (2004) Email correspondence on *risk management*, 17 November 2004.

Helm, P. (2005) Personal conversations with the author on the outcomes of the study, Department of Prime Minister & Cabinet, Wellington, April 2005.

Hitchins, D.K. (2003) *Advanced Systems Thinking, Engineering and Management.* Norwood, MA: Artech House, Inc.

Hodge, R.J. (2010) A systems approach to strategy and execution in national security enterprises. PhD thesis, University of South Australia, Adelaide, Australia.

Hodge, R.J. and Cook, S.C. (2005) *Human Development Report 2005—International Cooperation at a Cross-Road—Aid, Trade and Security in an Unequal World.* United Nations Development Program. http://hdr.undp.org/sites/default/files/reports/266/hdr05_complete.pdf. Accessed February 25, 2014.

Maier, M.W. and Rechtin, E. (2002) *The Art of Systems Architecting*, 2nd edn. Boca Raton, FL: CRC Press.

Mintzberg, H. (1994) *The Rise and Fall of Strategic Planning.* New York: The Free Press.

Mintzberg, H., Ahlstrand, B., and Lampel, J. (1998) *Strategy Safari—A Guided Tour through the Wilds of Strategic Management.* New York: The Free Press.

Oxford. (1999) *Australian Oxford Dictionary.* South Melbourne, New South Wales, Australia: Oxford University Press.

Rechtin, E. (1991) *Systems Architecting—Creating and Building Complex Systems.* Upper Saddle River, NJ: Prentice Hall PTR.

Rechtin, E. (2000) *Systems Architecting of Organizations—Why Eagles Can't Swim.* Boca Raton, FL: CRC Press.

Roberts, P.R. (2003) *Snakes and Ladders: The Pursuit of a Safety Culture in New Zealand Public Hospitals.* Wellington, New Zealand: Institute of Policy Studies and Health Services Research Centre, Victoria University.

Ulrich, W. (1983) *Critical Heuristics of Social Planning.* Bern, Switzerland: Haupt.

Warfield, J.N. (1999) Twenty laws of complexity: Science applicable in organizations. *Systems Research and Behavioral Science* 16: 3–40.

12 Hospitals and Health-Care Systems

Methods for Reducing Errors in Medical Systems

Bustamante Brathwaite, Eranga Gamage, Shawn Hall,
Karunya Rajagopalan, Mariusz Tybinski, and Danny Kopec

CONTENTS

OVERVIEW: HOSPITALS AS COMPLEX SYSTEMS

A hospital is a collection of systems with many subsystems. However, many of the systems are effectively complex due to their operational independence and relative autonomy. The prime example is the role of medical practitioners; however, nurses and other hospital staff operate with some degrees of autonomy and independence. Certainly the whole hospital-based care system does not operate as a single system. Coordination, synchronization, and integration of systems of personnel, information, and other systems occur in order to deliver health care safely. In health-care settings, the challenges entail a set of entities (patients, nurses, physicians, other hospital staff, electronic equipment, to name some) executing a set of functions whose interactions comprise a global system behavior (Kopach-Konrad et al. 2007). There are many systems in hospitals employing and depending on information technology including a computerized physician's order entry system, intensive care unit (ICU) monitoring systems, nursing stations, and various life-sign monitoring systems, just to name a few. Furthermore, new complicated software and protocols for operation are frequently introduced to operatives who are under pressures of intensive workloads and, to some extent, operate on the borders of experimentation, in order to deliver possible cures for life-threatening conditions. For this reason, the health-care system can be thought of as an integrated and adaptive set of people, processes, and products—consequently, we have a system of systems with the objective of enhancing efficiency and effectiveness (improved health).

Such systems are often also complex in nature, due to the uncertainties related to the human element, which interacts and depends on the work of technology (Milstein et al. 2000). Michael Lissack, in his book *The Interaction of Complexity and Management*, defines such a setup as "...complex adaptive systems are composed of a diversity of agents that interact with each other, mutually effect each other and in so doing generate novel behavior for the system as a whole.... But the pattern of behavior ...is not constant, because when a system's environment changes, so does the behavior of its agents, and, as a result, so does the behavior of the system as a whole" (Lissack 2002). Interestingly, this also means the system will constantly adapt to the conditions and setup around it.

According to Kopach-Konrad, large hospitals are extremely complex systems that are not well understood, very expensive, and frequently inefficient. The consequences of this complexity are that there are often hospital systems for which there are no detailed models for simulating and capturing the overall operations of all the individual systems and their operation as an overall complex system. William Rouse, National Academics, based on many studies, has concluded that a major problem with the health-care system is that it is a complex adaptive system (Rouse 2008). Rouse argues that most of us understand a traditional system in terms of examples such as vehicles, utilities, airports, or businesses such as Walmart. Additionally, we tend to believe that a system may be improved by decomposing individual components (e.g., propulsion, suspension, and electronics) and then by recomposing them with individual improved components, thus improving the overall system design. This approach worked well with generic traditional systems such as those used in the design of laptops, cell phones, retail systems, or automobiles. Noticeably the success of such systems greatly depends on the ability to decompose and recompose the individual improved elements. This generally does *not* work with a complex system because it and its environment will have changed while one is trying to fix the decomposed components!

Thus, not all problems can be resolved by addressing individual components' issues and recompiling them back to the system. For example, decomposition could result in the loss of information about how individual components work or integrate. Another significant problem for complex systems such as health care is that there is not a single *entity* in charge and none has the authority to alter or design the overall system. Fortunately, those complex adaptive systems are sufficiently intelligent to *game the system* by finding work around and resourcefully identifying ways to serve for their self-benefit. Whether this scenario truly benefits the overall health-care system remains to be seen.

KEYWORDS

Accelerators, Checklist, Cobalt-60, Complex systems, Complications, Functional requirements, Health care, Medical errors, Mishaps, Normal accident, Overdose, Patient safety, Radiation therapy, Requirements engineering, Safety culture, Therac-25, UML

ABSTRACT

Doctors, nurses, physicians, and qualified administrators are part of the medical field, one of the 20 largest industries in the United States. Yet, in recent years, ongoing medical reports show that yearly, somewhere between 44,000 and 98,000 people die in hospitals due to "errors committed by medical professionals" (Kopec et al. 2003).

Related studies conducted by the Agency for Healthcare Research and Quality (AHCRQ) report that those mistakes and related complications are preventable. Furthermore, costs are estimated to be in excess of $9 billion per annum.

Rapid developments have indeed increased the complexity of health-care systems. Significant involvement in mechanical and computer systems for health care can be identified to date back to the 1950s. However, unanticipated interactions of these complex systems can result in *normal accidents* (Perrow 1999). Complexity alone should not be a cause for accidents in a system.

In the first case study of this chapter, two massive accidents in health care (the Therac-25 and the case of the Instituto Oncologico Nacional of Panama) are studied to identify their root causes. Both these cases involved radiation therapy machines and the malfunction of a computer system. The Therac-25 machine overdosed cancer patients and caused a number of fatal injuries in the United States and Canada. The Cobalt-60 machine treated cancer patients at the Panama Cancer Institute where a massive overdosage of radiation caused many deaths.

For these reasons, the number and frequency of medical errors have become a larger concern for the medical industry in the past two decades. Any systems or methods that have proven effective in reducing errors are highly valued. Identification of safe practices/universal procedures, checklists, and various computerized systems has proven effective in reducing medical errors. More recent studies show that the medical field has improved greatly since the 1999 IOM Report (Brennan 2000), but current error rates can still be improved upon. There is still considerable room for improvement given that peoples' lives are at stake and hospitals' financial burdens need to be reduced in the near future. In industrial countries, complications occur with inpatient surgical procedures in 3%–16% of cases, with that figure fluctuating by 0.4%–0.8% with regard to permanent disability or death rates. In the event of trauma surgery, the rate of serious complication is significantly higher and estimated at 1 per 100 surgical exposures.

Hence, the second case study focuses on a surgical safety checklist.

Sources of medical error–related information presented are Kopec et al. (2003, 2004); Haynes et al. (2009); Tybinski et al. (2012); and Panesar et al. (2011).

Complex systems design, implementation, and use have become major issues in health care, as there is a significant concern to reduce medical errors linked to health-care information systems. The enormous potential of systems to improve processes in advanced medicine and patient care requires a critical need for the relevant stakeholders to master the skill of understanding how to adopt and maintain technologies that ensure patient safety.

The third case addresses *requirements engineering*, which is a fundamental process in the successful development, adoption, and maintenance of health-care information systems. The results of this process are often unsuccessful, which is frequently reflected in the lack of software quality leading to medical errors. Within the scope of health-care information systems, this case proposes to (1) reintroduce and summarize the impact of the requirements engineering process in the development and support of health systems and (2) shed light on a relatively new and efficient requirements engineering model—User Requirement Notation.

GLOSSARY

AECL	Atomic Energy of Canada
AHCRO	Agency for Healthcare Research and Quality
CRPB	Canadian Radiation Protection Bureau
DSR	Department of Radiological Health
FDA	Food and Drug Administration
GRL	Goal-oriented requirements language
HCIS	Health-care information system(s)
IAEA	International Atomic Energy Agency
ICU	Intensive care unit
IOM	Institute of Medicine
ION	Instituto Oncologico Nacional
NFR	Nonfunctional requirements
RE	Requirements engineering
RPB	Radiation Protection Bureau
TPS	Treatment planning system
UCM	Use Case Map
UML	Unified Modeling Language
URN	User requirements notation
WHO	World Health Organization

Figure 12.1 and the following labeled notes characterize the health-care complex system being discussed, which provides a context for all studies.

A—*Complex system, system of systems (SoS), or enterprise.* This complex system includes a hospital care system. Examples of its individual systems would be physicians, nurses, radiotherapy machines and systems, computer systems, operators, patients, procedures, checklists, etc.

B—*External factors* impacting this complex system include rules and regulations given by the Food and Drug Administration (FDA) and the Department of Radiological Health (DSR). Other environmental influences involve World Health Organization (WHO)'s recommendations and protocols for patient safety.

C—*Feedback* from the complex system would include case study analyses of procedures, implemented checklists, and requests for evaluating of machines and their errors.

D—*Governing body* includes the FDA, an agency responsible for protecting and promoting public health through regulation such as supervision of feed safety, prescriptions, and over-the-counter medications, vaccines, blood transfusions, and medical devices; the WHO of

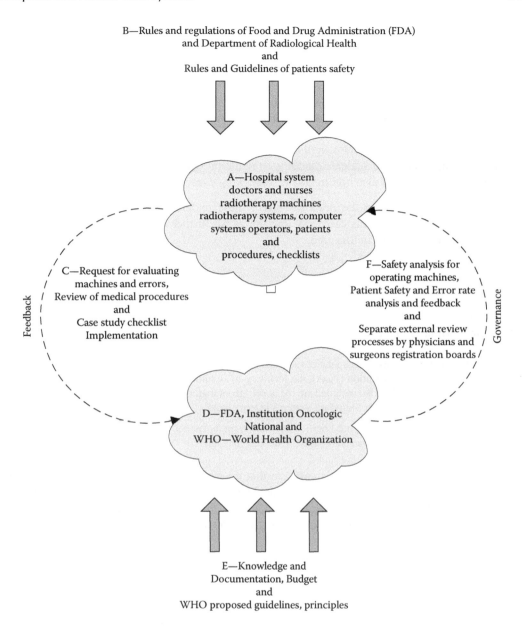

FIGURE 12.1 Context of complex system/system of systems/enterprise.

the United Nations that is concerned with international public health; and the National Oncologic Institute or ION (Spanish: Instituto Oncologico Nacional) of Panama.

E—*Constraints*. Even though the FDA and WHO are the primary governing bodies for patient safety and public health overall, there are other constraining factors such as available budgets, existing knowledge, principles, and documentation (or the lacks thereof).

F—*Governance*. The governing body provided patient safety, patient error rate, and operating machinery analysis for review.

CASE 1: THERAC/PANAMA CASE

Bustamante Brathwaite, Eranga Gamage, Karunya Rajagopalan, and Danny Kopec

CASE STUDY ELEMENTS

Fundamental essence—Software is widely used in health care for diagnostic devices, the management of patient records, patient admittance and scheduling, and accounting, among other purposes. The fundamental essence of this case revolves around two well-known accidents that occurred with the use of software for radiation therapy. The infamous case of the Therac-25 accident that caused the deaths of six patients has garnered much attention to the need for proper systems engineering and user monitoring. However, as we know, 15 years later in the case of the ION of Panama, the development and use of such complex systems in health care was the underlying cause for an even more tragic event whereby 17 patients died due to radiation overdosage. Not only does this case explore the two events, but it also highlights the underlying causes.

Topic relevance—Recognizing the need to correct these underlying causes within the critical domain of health care makes this case very relevant as we further develop complex systems and systems engineering practices in this area.

Domain—Health care.

Countries of focus—United States and Panama.

Stakeholders
   Active and primary—Patients who received radiation treatment and physicist operators.

   Active and secondary—Governing agencies such as the Canadian Radiation Protection Bureau (CRPB), the Food and Drug Administration (FDA), the Ministry of Health in Panama, and the Department of Radiological Health (DSR). Also included are the accelerators manufacturers such as CGR, the Atomic Energy of Canada (AECL), and Multidata Systems International.

   Passive and secondary—The hospital boards that oversee the hospitals where overdosages occurred in places such as the ION in Panama and the East Texas Cancer Center in Tyler, Texas, in the United States.

Insights—The primary insights of this case are that despite advances in technology and public documentation of past accidents such as the Therac-25, the possibility and probability of even worse future accidents in the same domain is still relatively high. Furthermore, there is a significant need to address and correct the underlying causes for such accidents.

BACKGROUND

The AECL and a French company called CGR worked together to build linear accelerators in the 1970s (Leveson 1995). Linear accelerators create high-energy beams of charged atomic particles that are used to treat cancer patients. The two businesses also created the previous Therac-6 and Therac-20 versions. The Therac-6 was a 6 MeV (million electron volt) accelerator producing x-rays only, and the Therac-20 was a 20 MeV dual-mode accelerator producing x-rays and electrons.

In the mid-1970s, AECL designed and developed the Therac-25 that was a dual-mode linear accelerator that could deliver 25 MeV photons or electrons at different energy levels. This machine was supposedly easier to use. It is worth noting that before the Therac-25 design was started, the business partnership of AECL and CGR ceased. CGR had developed the computer control for these machines and the same base software routines were reused in Therac-6 and Therac-20. Hardware safety features and interlocks in the two machines were untouched.

The ION in Panama provided treatment for cancer patients. The ION used blocks of shielding material to modify shapes of the radiation beams to protect normal tissue during treatment (IAEA 2001). A computerized treatment planning system (TPS) was used by ION to calculate dose distribution and treatment times. The data entered are done separately for each shielding block. The TPS allowed the operator to enter up to four shielding blocks per field at a time.

According to the International Atomic Energy Agency's (IAEA) investigation report, the patient starts his or her treatment process by making an appointment with a radiation oncologist who then prescribes a radiation dose for the patient's cancer treatment. This radiation is then delivered to the patient's tumor by a beam of gamma rays from the radiotherapy machine. The dose is divided into fractions and this fraction is given over a period of time on a regular basis. Each fraction may involve different *fields* in which the radiation beam is pointed at the tumor from a different direction (IAEA 2001). The prescribed radiation dose is then noted in the patient's chart and a physicist enters data into the TPS. The TPS then calculates dosage and dosage times to be delivered to the patient.

COMPLEX SYSTEMS

A degree of overconfidence in any system can lead to significant consequences as was the case in the Therac-25 accidents, which were considered by J. A. Rawlinson as the worst in the 35-year history of medical accelerators (Rawlinson 1987). The Therac-25 machine was introduced in 1982 and was an upgrade with significant changes from the previous versions. It introduced a new concept of *double pass* where one accelerator was capable of delivering photons at 25 MeV or electrons at various energy levels. Another significant change came about from the higher dependence on the embedded software of the system for safety maintenance as opposed to an independent machine used in the previous versions.

In Panama, the Cobalt-60 radiotherapy machine was used by ION to provide treatment for cancer patients. The computerized TPS used within their radiotherapy machine was the RTP/2 Multidata System, and like the Therac-25, it had the capability to administer both photons and electrons. The TPS of this system had four computing options for calculating prescriptions—the dose chart calculator, which calculates the treatment times needed to deliver a given dose to a prescription point; Irreg, which can calculate the treatment times needed to deliver a given dose to selected points, specifically for complicated, irregular shaped fields; the external beam when one intends to generate isodose distributions together with the calculation of the treatment time; and brachytherapy, which calculates isodose distributions when using brachytherapy sources (IAEA 2001).

As with any other complex system used in the highly critical domain of health care, certain legal and economical governance factors will play a role in its operation. The CRPB and the FDA of the United States were both external entities that governed the pre and post-installation activities of the Thereac-25. Before AECL's product was allowed to be legally used on cancer patients, AECL had to obtain approval from the FDA. Furthermore, any accidents that occurred during operation had to be reported to the FDA and the CRPB. Additionally, there was the Radiation-Emitting Devices (REDs) Act that detailed the standards of different classes of REDs' design, construction, functionality, and components (Leveson and Turner 1993). For the Therac-25, one significant economic factor was gained by patient users due to a new *double-pass* feature where both photons and electrons could be delivered with the same machine.

The AECL company along with the hospital clients' board of directors and machine operators made up the governing body of the Therac-25. AECL provided training, maintenance, and troubleshooting support through its engineers. Feedback was received by AECL and its engineers through the reports generated by its hospital client users. These hospital users were the actual operators who interacted with the cancer patients and operated the machine. And it is through the operator DEC VT100 terminal interface that the operator would enter the patients' information such as treatment prescription, energy dosage level, and beam type and receive feedback about the status of the entered instructions (Leveson 1995).

According to the Executive Decree No. 1194 that is dated December 3, 1992, the Ministry of Health in Panama has executive regulatory control over the operation and administration of radiation therapy. The DSR of the Social Security Agency also plays a role in regulatory control as the technical unit of consultants. Some of the responsibilities of the DSR are conduct of regulatory inspections, maintenance of instruments, management of nuclear waste, operator training, and consultation

(IAEA 2001). In order to gain legal authorization to operate any radiation equipment in Panama, an application had to be submitted for review by both regulatory parties. As a result of their roles, the Ministry of Health and the DSR along with the operators (medical physicists) who use the accelerator all made up the governing body of the radiation therapy system in the Panama case.

CHALLENGES

When considering the challenges that surfaced due to the Therac-25 accidents, it is important to note that this system (the Therac-25 and its previous versions) operated without an accident for several years. Even with the key changes that were made to the Therac-25 version, it operated incident-free for a long period of time. For these reasons, it was believed that the original investigations of the first accidents led to the conclusion that errors occurred in other areas and not the embedded software. This belief was further solidified when the AECL engineers were unable to reproduce the error messages that the operators claimed to have seen.

However, it was not until the second accident that occurred in the East Texas Cancer Center on March 21, 1986, that the operator was able to replicate the conditions that led to the accident. The key factor in producing the error came about due to the speed of data entry during the editing process of the prescription data (Leveson 1995). It is interesting to note that after the Therac-25 accidents, a University of Chicago physicist discovered that this same error potential existed in the Therac-20. However, it never surfaced due to the independent protective machine that monitored the electron beam scanning and blocks the beam from operating in the event of an error. And as stated before, this independent machine was not present in the Therac-25 (Leveson 1995). Like the Therac-25, the radiation overdosage that occurred in the Panama case in the year 2000 was initially not attributed to the system's software.

After observing specific symptoms of its patients, the ION physicians began reviewing for all possible causes. The system calculations were assumed to be correct. This was evident from the practice of not conducting a review of these calculations on the patients' charts. However, it was not until 5 months after the first report of the symptoms that the calculation of treatment times came into question. It was discovered that the adjustment made in the use of five shields as opposed to four, which had been the original procedure, was the source of the problem. When this cause was confirmed, all patients who displayed this symptom were identified. Furthermore, all other patients who had been treated using the alternative procedure that the physicians had developed could be identified. These patients were then examined to determine the level of overdosage in each case.

DEVELOPMENT

The AECL and CGR worked together for the production of medical linear accelerators and the Therac-6 and Therac-20 machines based on older CGR machines. The Therac-6 was capable of producing x-rays only, while the Therac-20 was capable of both x-rays and electrons (dual mode). Both machines had limited software functions. CGR developed the initial software that was used in Therac-6 and reused in Therac-20. The Therac-25 was a dual-mode linear accelerator based on previous versions and its commercial version was introduced in late 1982.

All Therac machines, including the Therac-25, were controlled by a PDP-11 computer. However, both the Therac-6 and Therac-20 could also work without computer control. The turntable design in Therac-25 was a major cause of the accidents. The upper turntable rotated the accessory equipment into the beam path to produce two therapeutic modes called the electron mode and photon mode. The other mode available was the field light position and it was not a treatment mode. All three modes were controlled by three microswitches. The accelerator beam is harmful to living tissues, and the scanning magnets in the turntable are used to spread the beam safely. The computer program is used to control the beam energy and moves the scanning magnets into their proper positions. Electromechanical interlocks were used in the Therac-20 to ensure that the turntable and scanning magnets are in correct position when treatment started. The Therac-20 had an independent protective circuit for monitoring electron beam work with the mechanical interlocks. However, in the Therac-25, the hardware interlock was replaced by a software program.

The software used in Therac-25 evolved from Therac-20 software and was developed by a single person using PDP-11 assembly-level language. The software was responsible for data input as required by the desired treatment dosage, setting the machine for treatment and operator function and monitoring of machine status. The software developed for the Therac-25 was modified just after it was first built because of complaints from operators that the process time of inputting treatment plans via the keyboard was too lengthy. The software was changed to simply use carriage returns to copy the treatment site data. The error messages and malfunction numbers that occurred thereafter were cryptic to operators, and the operator and maintenance manuals did not provide explanation for the malfunction (Leveson 1995). In addition, the amount of software testing conducted was minimal. Some testing was conducted on a simulator and most of testing was done on the whole system (Leveson and Turner 1993).

The Old Gorgas Hospital was previously a US Army-controlled hospital that was transferred to Panamanian government in 1997. With the exception of its radiation therapy service, ION transferred all its services from its old facility on Justo Arosemena Avenue in Panama City to the Gorgas Hospital in 1999. The Gorgas Hospital had a Cobalt therapy room but the structural shielding in the treatment room did not meet the Panamanian radiological protection regulation 1995. The structural changes needed were too difficult and costly. The Ministry of Health of Panama decided to keep external beam therapy treatments under the Justo Arosemena Avenue facility and brachytherapy (internal radiotherapy with sealed source) treatments in the Gorgas Hospital (Borrás 2006).

ION (on Justo Arosemena Avenue) used the Theratron 780C-type Cobalt-60 unit model for external beam therapy treatments. This hospital also used ATC/9 Picker and a Stabilipan orthovoltage unit (Siemens). Both were decommissioned in April 2000, while a new radiotherapy facility was being built in Gorgas Hospital. Only the Theratron 780 Cobalt-60 was used for treatments during the accidents. The RTP/2 Multidata System Version 2.11 was created by International Corp. and was used as the TPS for the Cobalt-60 accelerator. The gamma ray beam data were entered and verified for both 60 Co units (Theratron 780C and Picker ATC C/9) in the first installation in 1993.

The TPS User and Reference Guide manual was used to operate the machines. However, the instructions in the manual about the blocked fields were not complete. The section of the manual entitled Beam Handling and Field Setup, subsection Block, stated that "this function allows for entry of up to 4 blocks of 32 points per beam" and "once the block is nearly finished, strike enter to close the block contour." However, there were no instructions about how to digitize contours. The section in the manual for computational methods only described the algorithm for block calculation. In the section Block Calculations, it stated that "the calculation is symmetric with respect to whether the outline goes clockwise or counter clockwise." From these sample examples, it is evident that the manual did not give clear instructions to users on how to enter the blocks.

The TPS only allowed the entry of data for four blocks per field under the options. Usually four blocks were used for treatments. While all five blocks were actually used during patient treatment, data for one of the blocks were not entered and therefore not considered in the calculation. ION physicists thought that the new computer calculations were valid in the cases tested and thus introduced a new method to enter data for the 5th block. From August 2000 onward, this method was used to enter data for multiple blocks (IAEA 2001).

Timeline of the Therac and Panamanian Cases

Eleven accidents were reported by Therac-25 from 1985 to 1987 and six of them involved massive overdoses to patients (see Table 12.1).

The patient who was involved in the first accident in Kennestone Regional Oncology Center, Georgia, filed a lawsuit in October 1985 listing the hospital, manufacturer, and service organization

TABLE 12.1

Summary of Six Major Cases of Massive Therac-25 Overdoses

Major Accident Date and Place	Patient and Incident	Result of Injury	Action Taken by Responsible Parties
June 3, 1985 Kennestone Regional Oncology Center, Georgia	61-year-old woman treated for nearby lymph nodes Despite claims by the patient that she had been injured during treatment.	Radiation burns the patient and caused death.	No careful investigation.
July 26, 1985 Hamilton, Ontario Cancer Foundation, Canada	40-year-old patient treated for carcinoma of the cervix. Therac shut down with error message. Patient complained of burning, hip pain.	Total hip replacement was needed. Patient died, but the reported reason was cancer.	The machine was taken out of service and modified.
December 1985 Yakima Valley Memorial Hospital, Yakima, Washington	A woman treated for cancer.	Developed erythema on her hip after one of the treatments.	Staff at Yakima sent letters to AECL and AECL; technical support supervisor response was that it could not happen from a Therac-25 error or operator error.
March 21, 1986 Tyler, East Texas Cancer Center	Male patient treated for removal of tumor from back. Malfunction 54 displayed but the error was not explained in manual.	Patient got electric shock and numerous complications that led to death.	AECL engineers then suggested that an electrical problem might have caused this accident.
April 11, 1986 East Texas Cancer Center, Tyler	Male patient treated for skin cancer in face. Malfunction 54 error displayed again.	Patient's face burned and overdose caused death.	Machine taken out of service and AECL was called. An engineer from AECL response was "the ion chambers being saturated in the machine."
January 17, 1987 Yakima Valley Memorial Hospital, Yakima, Washington	Patient treated for carcinoma was to receive two film-verification exposures of 4 and 3 rads, plus a 79 rad photon treatment (for a total exposure of 86 rads).	Patient death caused by overdose of radiation.	AECL discovered a flawed software function that causes error.

responsible for the machine. The AECL received an official notification of the lawsuit in November 1985 (Leveson and Turner 1993). After that the CRPB asked AECL to make changes, AECL made the changes to microswitches (Leveson 1995).

In November 2000, the radiation oncologists in Panama started to observe prolonged diarrhea in some patients. By December 2000, similar abnormal symptoms were observed in other patients. Finally, the source of the abnormal symptoms was identified and the cause stated was that 28 patients were overdosed.

ANALYSIS

The following brief timeline for both events was adapted from the investigation reports of the Therac-25 accidents and accidental exposure of radiotherapy patients in Panama.

The first Therac-25 accident was reported on June 3, 1985, from the Kennestone regional Oncology Center, Marietta, Georgia. A 61-year-old woman was overdosed during a follow-up radiation treatment after removal of a malignant breast tumor. This incident was not investigated. In November, AECL received the official notification about a lawsuit filed by the patient. Ten accidents were reported thereafter from 1985 to 1987 and six of them involved massive overdoses to patients (Leveson 1995).

In November 2000, the radiation oncologists in ION Panama started to observe diarrhea in some patients that was unusually prolonged. In March 2001, physicists identified the reason for the problem. In December 2000, similar abnormal symptoms were observed in other patients. Finally, they identified the abnormal symptoms and the 28 patients that were overdosed. By May 2002, 17 of the 28 overexposed patients had died in times ranging from 35 days to 21 months after being treated (Borrás 2006).

As one would expect, the biggest challenge was trying to identify and replicate the events of the error that produced the accidents. In both cases, the computer system being the source of the fault was initially regarded as impossible. However, further analysis proved otherwise as the true reason was discovered through careful replication and review of the methods of operation. In the case of the Therac-25, the physicians at the East Texas Cancer Center were the first to identify and replicate the steps that led to the overdosage of two of its patients. In all the previous cases, AECL attributed the accidents to other causes such as electrical failure and operator error.

The overdosage case that occurred in Panama follows a similar pattern where there was an interval of time that had elapsed before operators pinpointed the computer calculations as the reason for the accident. The systems used and their operator interaction in both cases have two common key characteristics. The first deals with the mindset of members of the governing body where the creators (the companies) of the systems were adamant that their machines could not be the source or the error and that the fault lay in the operators' negligence in the process of entering and reviewing data. The second characteristic is the lack of a safeguard component to prevent the overdosage. The independent machine of the Therac-20 that regulated the radiation output was removed in the Therac-25 version and the TPS of the Cobolt-60 accelerator allowed for the entering of a fifth shielding block when it was designed to only do calculations for a maximum of four.

The manufacturers and operators were both responsible for the safety culture and their actions should have been checked by a regulatory board. This analysis is identified by those manufacturers who were responsible for providing full service in both cases. User manuals were incomplete and not well explained. The operators of the Therac-25 seem to be overly concerned about the difficulties they encountered in entering data into the system. However, the fact that the user manual was unclear did not receive the same level of consideration. We also see the need for a user-friendly interface. The operators had little to no understanding about the underlying reasons for the error messages that were displayed on the interface screen.

The physicists at ION Panama developed their own pattern for solving problems that would take less time to implement. The completed user manual was a partial certificate for all functions were on functionality rather than a booklet. Both user manuals confused the users and led to faulty decisions. As users, operators and physicists in the institutes should have become aware of the safety issues related to unknown malfunctions and unclear manual instructions.

Both cases also emphasize the importance of testing. Basically, both systems were not sufficiently tested. Thousands of successful operations of the system could still be followed by a faulty operation by changing a single component or command in the software program. The Therac-25 software used software that had worked satisfactorily in early versions of the machine, but with some modification. It was not tested thoroughly before its first use. A new method introduced by Panamanian physicists for entering data for fifth block was tested. But it is unclear how much they met quality assurance requirements. This analysis report also highlights the need of testing of every single change performed on a system.

In Therac-25 case, AECL ignored ethics in systems engineering and relied on their own limited point of view. The role played by the AECL during accidents also demonstrated poor organizational structure. The AECL relied on their software developers without further analysis. It was unable to find the fault of the system for two years, while six people were literally burned to death by the overdosages. This situation demonstrated that the AECL performed superficial technical activities rather than effective measures. Software quality is usually considered *good*, but what exactly is the definition of *good*, and how do we measure this?

In summary, the errors occurred due to the complexity of autonomy in operator and administrator judgment of eliminating the independent checking of dosage and elimination of the use of the original procedure of use of five shields and errors in the software coding.

RESULTS

As one would expect, the aftermath of such accidents would garner a lot of attention and once again call into question the use of computers in critical areas of health care. Given the US FDA responsibility as an external entity governing safety in this area, their methods of approving software-controlled devices before they can be used legally were reviewed and changed after these accidents. Prior to that change in policy, however, the FDA only required premarket notification, which is a process whereby the new product is proven to be equivalent in safety and process to another product that is currently in public use (Leveson 1995).

Following the report of the accidents to the FDA, the manufacturers of the Therac-25 were required to complete a corrective action plan (CAP). As the name suggests, this was a detailed plan that featured the changes that were to be made to the accelerator so as to increase the safety margin of the machine. Until these changes were reviewed and deemed as suitable enough to reduce the probability of future accidents, the Therac-25 was ordered to be decommissioned from all further use (Leveson 1995). Changes that were detailed in the CAP included software operation improvements, replacement of error messages with more meaningful ones, changes in the system documentation, and the installation of an independent single-pulse hardware and software configuration to shut down the machine (Leveson 1995). Several CAP revisions were made and the fifth version was finally accepted in 1987. It is interesting to note that it was through these revisions that a number of the underlying causes for the faults in the system were exposed. Since then, a number of reports and publications have since been released, which went a step further to suggest and explain the underlying reasons for the overdosages. In Sara Baase's book *A Gift of Fire* (Baase 2008), the Therac-25 case was featured and some of the reasons highlighted for the failure were design flaws, lack of proper and extensive testing, software bugs, overconfidence, carelessness, and poor training. This case has even been used in the curriculum of university courses as an example of failures in a complex system and real-world lessons (for both hardware and software designs) that can be and should be applied.

As in the case of the Therac-25 accidents, the Panamanian radiation therapy accidents were also a combination of human factors and machine/computer factors. Regulatory authorities in Panama and the United States started investigations to look into the root causes of the accidents and also documented lessons learned to prevent such future accidents. In 2004, a court trial was initiated for the physicists who were part of the cancer treatment in Panama. Under Panama's law, the three physicists were charged for irresponsible and untested *introduction of changes into the software*. In late 2004, the courts announced that one physicist was acquitted but the other two were found guilty. Panama authorities also sued multidata in the United States and in Panama. However, both the cases were dismissed. The US case was dismissed on the grounds that the case should be filed in the same location as that of the accidents. The Panamanian case was dismissed citing that multidata could not be sued in two different locations simultaneously (Borrás 2006). As part of the US investigation (Kansas), investigators found that it should have been good practice to manually recheck the calculation times before starting the radiation treatment (USCA 2008).

Since these accidents and investigations, there have been a few changes in Panama's Cancer Institute (ION). New treatment rooms have been built with inputs from regulatory bodies. A new set of machines are now used at the ION for cancer treatment. The quality and safety of the machines are continuously monitored by the Inter-institutional Committee on Radiological Protection and Quality Control. This body oversees and manages all treatments and activities that use radiation.

Conclusion

After the accidents of both the Therac-25 and the Panamanian overdoses, detailed reports were generated from the investigations that were carried out by the governing bodies for each case. Prior to the last accident in 1987, the FDA mandated the manufacturers of the Therac-25 to complete a CAP so as to implement corrective measures. The FDA reviewed the safety changes that were made and concluded that these changes should reduce the probability of such future accidents. The aftermath of the overdose cases in Panama saw a number of improvements in the system and the procedures of using that system, for example, the ability to correctly enter calculations into the TPS for more than four blocks and then manually checking those computed calculations.

CASE 2: SURGICAL SAFETY CHECKLIST

Mariusz Tybinski and Danny Kopec

Case Study Elements

Fundamental essence—Medical complications are devastating to patients and costly to the health-care system but often preventable. Examining existing protocols, checklists, and various computerized systems helps in determining the cause for various mishaps and in turn reducing medical errors.

The case study describes the use of checklists, which illustrate how teams of medical individuals, who all demonstrate autonomy and individuality in their actions and judgments, can be moved toward becoming a single coherent system, which does not induce emergence into complicated medical procedures. The checklist also addresses the patient, whose autonomous reactions can interfere with medical procedures. Equipment functioning is also addressed, another aspect of autonomy if not constantly coordinated with the patient's needs.

Topic relevance—Analyzing the checklist program demonstrates how this relatively small and simple step has proven effective in medical error analysis in a diverse group of institutions around the world.

Domain—Health care.

Country of focus—Global.

Stakeholders

- Active and primary: patients
- Active and secondary: hospitals, doctor, and nurses
- Passive and secondary: hospital management and the researchers

Insights—The introduction of the checklist program in the hospitals demonstrated a significant improvement of medical error ratios in operating rooms.

Background

Surgical complications are particularly devastating to patients and costly to the health-care system overall and are often preventable. Typically, for prevention of errors to be possible, significant changes in procedures and individual behaviors are necessary. In this case study, we will present and analyze the *checklist program*, which has facilitated a decline in the rate of mishaps in surgical procedures and corresponding complications. This relatively small and simple step has proven effective in a diverse group of institutions around the world (Haynes et al. 2009). The study spanned about a year between October 2007 and September 2008 in eight hospitals and in eight different cities. The researchers attempted to collect data about processes and outcomes from 3733 consecutively enrolled patients of 16 years of age or older who were undergoing a noncardiac surgery. After the introduction of the *Surgical Safety Checklist*, they were able to collect data from 3955 consecutively enrolled patients. The primary goal and focus of the study was the rate of complications (including death) when hospitalized in the first 30 days after the procedure/operation (Haynes et al. 2009).

Complex System

> In 1935, the U.S. Army Air Corps held a flight competition for airplane manufacturers vying to build its next generation long-range bomber. In early evaluations, the Boeing plane had trounced other designs... With the most technically gifted test pilot in the army on board,... it stalled, turned on one wing, and crashed in a fiery explosion. Two of the five crew members died, including the pilot. ... The pilot had forgotten to release the new locking mechanism on the elevator and rudder controls. A few months later, army pilots were convinced the plane could fly and invented something that would be used on the few planes that had been purchased... A checklist, with step-by-step checks for takeoff, flight, landing, and taxiing. With the checklist in hand, the pilots went on to fly the model (B-17) a total of 1.8 million miles through several conflicts without one accident. (Panesar et al. 2011)

Clearly, a checklist benefits aviation by ensuring consistency and completeness. Today, it is a standard for preflight checks, quality assurance of software engineering, and various operations procedures to name a few (Gawande 2007). This event has been the key milestone in the birth of the checklist in general. In particular, the hospital (and health care overall) like most systems is complex, and integrated structures are easily prone to error typically due to human factors and/or system failures and more recently the combination of the two (system failures or human errors). In a study published in 2009 by *the New England Journal of Medicine*, it was reported that implementation of the surgical safety checklist (based on the first edition of the World Health Organization Guidelines for Safe Surgery) helped eliminate some surgical complications and mishaps.

Development

In 2008, the WHO published recommended guidelines, principles, and recommended practice to improve the safety of patients. Based on the WHO's guidelines, a 19-item checklist was designed with the intention of reducing the rate of major surgical complications and mishaps (Haynes et al. 2009). The implementation involved a two-step process. First, the initial baseline data were collected, and then each local investigator was advised of the analyzed deficiencies and asked to implement the 19-item checklist (Table 12.2). The checklist is based on the first edition of the WHO guidelines for safe surgery (also available with the full text of this article at NEJM.org). This study was conducted in eight participating hospitals within the WHO region (Table 12.3). St. Mary's Hospital has since been renamed St. Mary's Hospital—Imperial College National Health Service Trust (Haynes et al. 2009). For each hospital, a local data collector was chosen and trained accordingly (Table 12.4). Sites 1 through 4 are located in high-income countries; sites 5 through 8 are located in low- or middle-income countries (Haynes et al. 2009).

This employee worked full time on the conducted study and was not assigned any other clinical or hospital duties at the time. The list was introduced to the staff by lectures, written material, or direct training. Each hospital designated between 1 and 4 operating rooms for the study and all patients of 16 years of age and older were consecutively enrolled (Haynes et al. 2009).

TABLE 12.2
Elements of the Surgical Safety Checklist

>> Sign In

1. Before induction of anesthesia, members of the team (at least the nurse and an anesthesia professional) orally confirm that the following applies:
   - The patient has verified his or her identity, the surgical site and procedure, and consent.
   - The surgical site is marked or site marking is not applicable.
   - The pulse oximeter is on the patient and functioning.
   - All members of the team are aware of whether the patient has a known allergy.
   - The patient's airway and risk of aspiration have been evaluated and appropriate equipment and assistance are available.
   - If there is a risk of blood loss of at least 500 mL (or 7 mL/kg of body weight, in children), appropriate access and fluids are available.

>> Time Out

2. Before skin incision, the entire team (nurses, surgeons, anesthesia professionals, and any others participating in the care of the patient) orally performs the following:
   - Confirms that all team members have been introduced by name and role.
   - Confirms the patient's identity, surgical site, and procedure.
   - Reviews the anticipated critical events.
   - Surgeon reviews critical and unexpected steps, operative duration, and anticipated blood loss.
   - Anesthesia staff review concerns specific to the patient.
   - Nursing staff review confirmation of sterility, equipment availability, and other concerns.
   - Confirms that prophylactic antibiotics have been administered ≤60 min before incision is made or that antibiotics are not indicated.
   - Confirms that all essential imaging results for the correct patient are displayed in the operating room.

>> Sign Out

3. Before the patient leaves the operating room,
   - The nurse reviews items aloud with the team
   - The name of the procedure is recorded
   - The needle, sponge, and instrument counts are checked if complete (or not applicable)
   - The specimen (if any) is correctly labeled, including the patient's name
   - Any issues with equipment are addressed
   - The surgeon, nurse, and anesthesia professional review aloud the key concerns for the recovery and care of the patient

TABLE 12.3
Characteristics of Participating Hospitals

Site	Location	No. of Beds	No. of Operating Rooms	Type
Prince Hamzah Hospital	Amman, Jordan	500	13	Public, urban
St. Stephen's Hospital	New Delhi, India	733	15	Charity, urban
University of Washington Medical Center	Seattle, Washington	410	24	Public, urban
St. Francis Designated District Hospital	Ifakara, Tanzania	371	3	District, rural
Philippine General Hospital	Manila, Philippines	1S00	39	Public, urban
Toronto General Hospital	Toronto, Canada	744	19	Public, urban
St. Mary's Hospital	London, England	541	16	Public, urban
Auckland City Hospital	Auckland, New Zealand	710	31	Public, urban

Source: Tybinski, M. et al., *Health*, 4, 165, 2012.

TABLE 12.4

Surgical Safety Policies in Place at Participating Hospitals before the Study

Site No.	Routine Intraoperative Monitoring Pitch Pulse Oximetry	Oral Confirmation of Patient's Identity and Surgical Site III Operating Room	Routine Administration of Prophylactic Antibiotics in Operating Room	Standard Plan for Intravenous Access for Cases of High Blood Loss	Formal Team Briefing	
					Preoperative	Postoperative
1	Yes	Yes	Yes	No	No	No
2	Yes	No	Yes	No	No	No
3	Yes	No	Yes	No	No	No
4	Yes	Yes	Yes	No	No	No
5	No	No	No	No	No	No
6	No	No	Yes	No	No	No
7	Yes	No	No	No	No	No
8	Yes	No	No	No	No	No

Source: Tybinski, M. et al., *Health*, 4, 165, 2012.

The study data were collected from the individual local collectors and/or from the clinical teams involved. The information gathered from surgery recovery period included the demographics of patients, procedural safety, and type of the anesthetic data. The information was collected primarily from patients' first 30 days or until discharge. The collection of data aimed for 500 consecutively enrolled patients (Haynes et al. 2009).

The researchers' defined complications included "acute renal failure, bleeding requiring the transfusion of 4 or more units of red cells within the first 72 hours after surgery, cardiac arrest requiring cardiopulmonary resuscitation, coma of 24 hours duration or more, deep-vein thrombosis, myocardial infarction, unplanned intubation, ventilator use for 48 hours or more, pneumonia, pulmonary embolism, stroke, major disruption of wound, infection of surgical site, sepsis, septic shock, the systemic inflammatory response syndrome, unplanned return to the operating room, vascular graft failure, and death" (Haynes et al. 2009).

RESULTS

There were 3733 enrolled patients during the baseline period and 3955 patients after the introduction of the 19-item checklist (Table 12.5). Plus–minus values are means ± SD. Urgent cases were those in which surgery within 24 h was deemed necessary by the clinical team. Outpatient procedures were those for which discharge from the hospital occurred on the same day as the operation. P values are shown for the comparison of the total value after checklist implementation with the total value before implementation (Haynes et al. 2009).

ANALYSIS

In this study, a 20% reduction in complications was found by the researchers after the checklist was implemented with a statistical power of 80%.

The error complications for any of the sites dropped from 11% to 7% (at the initial baseline) on the average of every site, and the total in-hospital rate of death dropped to.08% from 1.5% (Table 12.6).

P values are shown for the comparison of the total value after checklist implementation as compared with the total value before implementation (Haynes et al. 2009). The surgical complications and death rate fell on average by 36%. Every hospital had a decrease of major postoperative

TABLE 12.5

Characteristics of the Patients and Procedures before and after Checklist Implementation, According to Site

Site No.	No. of Patients Enrolled		Age		Female Sex		Urgent Case		Outpatient Procedure		General Anesthetic	
Column I	Before	After	Before 2	After 3	Before 4	After 5	Before 6	After 7	Before 8	After 9	Before 10	After 11
1	524	598	51.9 ± 15.3	51.4 ± 14.7	58.2	62.7	7.4	8	31.7	31.8	95	95.2
2	357	351	53.5 ± 18.4	54.0 ± 18.3	54.1	56.7	18.8	14.5	23.5	20.5	92.7	93.5
3	497	486	51.9 ± 21.5	53.0 ± 20.3	44.3	49.8	17.9	22.4	6.4	9.3	91.2	94
4	520	545	57.0 ± 14.9	56.1 ± 15.0	48.1	49.6	6.9	1.8	14.4	11	96.9	97.8
5	370	330	34.3 ± 15.0	31.5 ± 14.2	78.3	78.4	46.1	65.4	0	0	17	10
6	496	476	44.5 ± 15.9	46.0 ± 15.5	45	46.6	28.4	22.5	1.4	1,1	61.7	59.9
7	525	585	37.4 ± 14.0	39.6 ± 14.9	69.1	68.6	45.7	41	0	0	49.1	55.9
8	444	584	41.9 ± 15.8	39.7 ± 16.2	57	52.7	13.5	21.9	0.9	0.2	97.5	94.7
Total	3733	3955	46.S ± 18,1	46.7 ± 17.9	56.2	57.6	22.3	23.3	9.9	9.4	77	77.3
P value			0.63		0.21		0.26		0.4		0.6	8

Source: Tybinski, M. et al., *Health*, 4, 165, 2012.

TABLE 12.6

Outcomes before and after Checklist Implementation, According to Site

Site No.	No. of Patients Enrolled		Surgical Site Infection		Unplanned Return to the Operating Room		Pneumonia		Death		Any Complication	
	Before	After	Before 2	After 3	Before 4	After 5	Before 6	After 7	Before 8	After 9	Before 10	After 11
1	524	59$	4	2	4.6	1.8	0.6	1.2	1	0	11.6	7
2	357	351	2	1.7	0.6	1.1	3.6	3.7	1.1	03	7.8	6.3
3	497	486	5 8	4.3	4.6	2.7	1.6	1.7	O.S	1.4	13.5	9.7
4	520	545	3.1	2.6	2.5	2.2	0.6	0.9	1	0.6	7.5	5.5
	370	330	20.5	3.6	1.4	1.8	0.3	0	1.4	0	21.4	5.5
6	496	476	4	4	3	3.2	2	19	3.6	17	10.1	9.7
7	525	585	9.5	5.S	1.3	0.2	1	1.7	2.1	17	12.4	8
S	444	584	4.1	2.4	0.5	1.2	0	0	1.4	03	6.1	3.6
Total	3733	3955	6.2	3.4	2.4	1.8	1.1	1.3	15	O.S	11	7
P value				<0.001		0.047		0.46		0.003		<0.001

Source: Tybinski, M. et al., *Health*, 4, 165, 2012.

complications with a significant reduction at three sites (one in a high-income location and two in lower-income locations). Even though in some hospitals a few effects of the intervention were stronger than others, no single hospital was responsible for the overall rate change, nor was the rate limited exclusively to a high- or low-income hospital.

CONCLUSION

The introduction of the checklist in those eight hospitals was clearly an improvement of medical error ratios in operating rooms. From these findings, it was concluded that the checklist program is an improvement among diverse clinical and economical environments. Clearly, implementation of more of such checklists for surgery and other medical practices will be helpful in reducing errors.

CASE 3: REQUIREMENTS ENGINEERING IN THE DESIGN AND MANAGEMENT OF HEALTH-CARE INFORMATION SYSTEMS—IMPLICATIONS FOR COMPLEX SYSTEMS ENGINEERING

Shawn Hall and Danny Kopec

CASE STUDY ELEMENTS

Fundamental essence—Information systems have the potential to significantly improve quality, efficiency, and safety in health-care delivery. However, technology characterizes complex systems that require extensive investigation in order to enable all these potential benefits in health care. This case study applies requirements engineering (RE) to the whole complex system, in an attempt to integrate the separate autonomous elements of the medical system within a hospital, including the operatives and the equipment they use in providing medical services to patients who each behave in independent manners. Such an approach offers far-reaching benefits to the engineering and implementation of complex systems, and hence, the application of these principles to produce more efficient and safe health-care systems is the core of this case study.

Topic relevance—The critical relationship between quality of human life and health care is reason enough to investigate an engineering principle that will produce more quality and safety in health care through health-care information systems (HCISs).

Domain—Health care.

Country of focus—United States.

Stakeholders

- Active and primary: patients
- Active and secondary: health-care providers and organizations such as hospitals, clinics, physicians, and specialized health-care facilities (e.g., nursing homes, rehabilitation/treatment facilities)
- Passive and secondary: federal policy makers in health care

Insights—RE as a comprehensive and systematic approach to minimize errors and lapses in systems engineering is recognized to be a crucial part of the complex system engineering process. To this end, the case elevates the design, engineering, and implementation of HCISs as parts of an extensive RE process in a model-driven development approach. RE should be taken into consideration by modern inventiveness in the area of complex systems engineering. We believe that this understanding is precisely the main driver of safety critical complex systems.

Background

Thousands of lives are lost each year as a result of medical errors. Thus, medical errors leading to severe injury or death are a serious concern. This means that medical errors are now a leading cause of death and a significant problem in health care worldwide. Medical errors are increasingly related and linked to the computer systems used in patient care. The medical errors that are linked or related to computer systems are unintended consequences of technology in health care. Thus, these unintended consequences define errors made in patient care where computer systems are involved.

The number of medical errors is a dire international problem, and the current huge number needs to be reduced. The Institute of Medicine's (IOM) report on medical errors focused attention to dangers inherent in the US medical care system that might cause up to 98,000 deaths in hospitals and cost approximately \$38 billion per year (Kopec et al. 2003). The United States is one among the many countries struggling with the profound and dire consequences of medical errors in its health-care systems.

The pervasive nature of technology has inevitably molded the expectation that technology will improve health-care efficiency, patient quality, and safety. However, many fail to consider that these technologies may also introduce errors and adverse events that compromise the mission of quality and safety in patient care. Considering the significant use and dependence on technology in health care, errors, lapses, and mishaps as a result of these complex systems is inevitable (Nadzam and Mackles 2001). In light of the vast advances, accomplishments, and potential of complex technology and systems in medicine, the efficiency of these systems is compromised (Institute of Medicine 2004), as a result of the following:

- Poor design that does not adhere to human factors—this relates to the functional features of a system that affects how humans understand and interact with the system or technology. This may induce user error.
- Poor technology interface with the patient or environment—some complex systems lack thoroughly designed interfaces between the system and patients that are inefficient and increase the probability of errors.
- Inadequate plan for implementing a new technology into practice.
- Inadequate maintenance plan.
- Errors as a result of poor interface between multiple technologies—miscommunication of information transmitting between devices or systems.

Technologies in health care offer the means for preventing errors but present some unintended side effects as well, such as medication errors, miscommunications, and delays in treatment (Kuehl 2001).

Reduction of medical errors through the design or development and maintenance of safety critical health-care systems is a direction to be pursued in health care because if better planning is involved, fewer errors would be triggered. Therefore, focusing on the design of complex HCISs through the RE process will invariably reduce the frequency and magnitude of these medical errors.

RE in the context of assessing, selecting, and developing quality HCISs as complex systems is the focus in this study. RE is the practice used to first identify and then translate stakeholder needs to system requirements (Zave 1997). The concern for the constantly increasing frequency of medical errors in medical institutions, a situation that is highly dependent on the efficiency of these complex medical technologies, needs to shift focus to solving the problem from the perspective of implementing quality RE processes. High-quality RE in HCISs can be achieved by continuously measuring RE processes and, if they are found to be lacking, improving them by eliminating the process problems or gaps (which are defined as the difference between the desired and existing states for a process), which result in poor quality (Zave 1997; Castro et al. 2002). High-quality RE processes will ensure that the quality of requirements that are developed is good.

Therac-25 is a major example of how inconsistencies in the requirements of health information systems can lead to devastating outcomes. Therac-25 was a radiation therapy system produced by AECL Limited. As covered in Case Study 1 of this chapter, the system was involved in several grave accidents between 1985 and 1987, in which patients were given huge overdoses of radiation, in the region of 100 times the intended dose. The root causes of these errors were due to bad software and development practices (Leveson 1995). Therac-25 is investigated further here. The failure of such a safety critical health system highlights the importance of clean and clear RE in the process of adopting, developing, and maintaining safe complex systems in health care.

RE is an extremely important activity, considering the measure of success of information systems is the degree to which they satisfy their requirements. Many of the existing software engineering methods have concentrated on the design of the system and devoted relatively little attention to RE (Nuseibeh and Easterbrook 2000, Gawande 2007).

Therefore, implementing the principles of RE in the development and maintenance of complex health systems is a significant addition to facilitate the following:

- Utilizing standard-quality control procedures, such as surgical checklists (this model is explored further in Case Study 2 in this chapter), to diminish risk and ensure efficiency in patient care (Amyot and Mussbacher 2001).
- Designing user interfaces for systems that are clear and functional and reflect the thorough characteristics of the environment in which medicine is practiced.
- Facilitating the chances of errors that might be introduced by new IT systems—RE enables a more meticulous approach to simplifying the design process of complex health systems that will in turn minimize anticipated errors through clear system requirements.

PURPOSE

It is widely accepted that the development of information systems is a complex task, involving technical, human, and organizational issues. In the evaluation of RE and software methodology, the major concern in software and information systems development pointed to the lack of quality. The lack of quality is a significant component that prevented the implementation of the most adequate technological solution for solving problems. However, if quality is adequately improved, it can significantly decrease the frequency of errors in the implementation of information systems.

Considering the prominence of information systems as a means to generally improve patient health and health-care processes, there is a major concern for the responsibility of information systems in the health-care industry. On this basis, health-care administrations are presented with a vast number of complex information systems and technology to solve their problems and satisfy objectives. The inability to choose the most efficient and effective systems has become a major problem for health-care organizations because the ultimate issue of choosing safety critical systems needs to be accomplished.

RE principles have been included as part of the development approaches in different categories of systems of various sizes. These systems are becoming increasingly complex in architectures and functionalities because they are becoming more dynamic in nature and evolve at an exponential rate (Goguen 1994).

The need for patient safety and health-care quality through HCISs is the core motivation. Therefore, it is the aim of this RE to focus on presenting principles along with the narrative of historic events that will shed light on the need for safety critical complex systems and technology in health care.

DEVELOPMENT

RE modeling: RE is an important phase in the development of complex systems. The concept of RE is a relatively new discipline in the software engineering discipline and is intended for capturing,

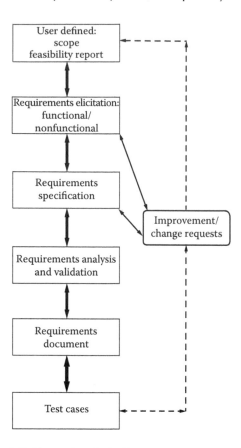

FIGURE 12.2 Requirements definition and analysis diagram.

analyzing, specifying, and managing system requirements. RE can essentially be defined as the process of discovering the software purpose, identifying who are the affected stakeholders and their needs, and then documenting such knowledge in a useful way, so it can be used in later phases during the software development process (Gawande 2007).

One of the greatest challenges in the software process is to identify the client's real needs, that is, what the client expects from the system, hence the system requirements. Some researchers have shown that a large number of big projects fail because of invalid or misunderstood requirements (Sommerville 2001; Brooks 2005). These researchers explicitly state that the central measure for a project's success is to what extent the client's needs have been fulfilled, which in the case of health care, the safety and quality of care of patients is critical.

Due to the inherent subjectivity in RE, researchers have failed to identify a standard model to solve the RE activities issue. Therefore, a standard format has been developed to evaluate the execution of the RE activities. The following illustration (Figure 12.2) shows the details of the interaction between the RE phases and their objects. Solid arrows indicate the iterative process of requirements definition and elicitation. These arrows represent the continual feedback from the user and improvement process. It is important to observe the iterative progression of requirements definitions, indicative of an imperfect and repetitive process. Broken arrows indicate a cycle of change/improvement requests.

REQUIREMENTS ELICITATION AND ANALYSIS

Requirements elicitation is a process in RE where software/system requirements are elicited from a variety of sources, while requirements analysis focuses on analyzing and refining the requirements gathered from requirements elicitation (Pressman 2005). Essentially, this phase is concerned with the production of a set of system models that should contain high-level descriptions of the client's needs. Possible sources of information for this task are the customer, end user, any other stakeholder, books, existing applications, domain knowledge, organizational standards or constraints, and any other rules or regulations.

REQUIREMENTS SPECIFICATION

This phase is focused on building requirements models. These models are detailed representations of the system models, that is, they have a low abstraction level because of the great detail associated with depicting the design of the system. The system models are its input artifacts and the requirements model is its output object (Pressman 2005).

The term specification has a myriad of associated meanings in the software development and engineering context. Each is related to ways of representing requirements acquired during the elicitation phase. Accordingly, one can conclude that there are a large number of notations to represent requirements model objects (Chung et al. 2000).

These notation choices include written documents, graphical models, mathematical models, and interface prototypes. Essentially, requirements specification provides solid objectives for the developer during product development. It also serves as a means against which product features can be validated.

REQUIREMENTS VALIDATION

After the requirements elicitation/analysis and specification phases, the following phase is used to validate the requirements for their relevance, completeness, and precision with regard to stakeholders. Requirements validation is concerned with checking the requirements for omissions, conflicts, and ambiguities and for ensuring that the requirements follow quality standards (Chung et al. 2000).

RESULTS AND ANALYSIS

This section covers the primary RE techniques relevant to this study.

USER REQUIREMENTS NOTATION

In the development of complex systems in health care, certain critical qualities are required for the engineering process. These qualities are imperative to safety, the prevention of medical errors and the level of success of user-centered or functional requirements, which focus on functionalities provided to the end user. Second, there are context-centered requirements, which may represent system, process, and human requirements. Context-centered types of requirements are referred to as nonfunctional requirements (NFRs) (ITU 2003).

NFRs are mainly system constraints, which may significantly affect the operational environment and design choices a developer may pursue during the development of the system. In the context of HCISs, NFRs have become increasingly crucial. NFRs include operational environment, which deals with hardware and software interfaces; accuracy; and performance, which handles timeliness and storage requirements, security, reliability, maintainability, portability, robustness, and usability (ITU 2003).

Functional requirements mainly describe the visible and external input and output interactions with the system being developed, whereas NFRs are those that impose special conditions and qualities on the system. Consequently, system acceptance testing is based on both functional and nonfunctional system requirements, which offer great potential as a preventive approach to medical errors.

Considering that functional requirements and NFRs are the two crucial groups of software or system requirements, the ability to adequately model them within the same RE language is an advantage. In view of that, the capacity to model both functional requirements and NFRs is a major advantage of user requirements notation (URN).

URN is the model that we chose to highlight as the RE framework in this study. URN provides a notation for the representation of NFRs such as performance, cost, security, and usability, in addition to a complementary scenario notation for functional requirements (ITU 2002). As a basis for a framework for complex systems engineering of HCISs, URN proposes a more comprehensive capture of requirements to improve the safety of these systems. URN proposes two languages: Use Case Maps (UCMs) and goal-oriented requirements language (GRL) to describe functional and NFRs, respectively. The proposed NFR notation—GRL—is the first attempt for the explicit capture and representation of NFRs. In addition, UCMs are used as scenarios that help to describe and understand the complex functional behavior of software systems.

URN is a modeling language or methodology used for eliciting requirements and using them in the software development process (ITU 2002). GRL and UCM are components of URN. GRL is used as a notation for URN; GRL is concerned with the modeling of NFRs, such as the performance and usability of a system. The other component of URN is UCM, a notation for functional requirements, which relates to the functionalities provided to the system's users. Both GRL and UCM utilize graphical models in the notation of NFR and functional requirements, respectively.

UNIFIED MODELING LANGUAGE

The Unified Modeling Language (UML) is a standard developed by the Object Management Group (a group of software companies) to deal specifically with software systems and systems in general (Goguen 1994). Recently, the increase in failure of systems to satisfy their NFR has established the need for developing standards for formalizing the description and capture of these requirements.

Different approaches to a process can be presented with nine different types of UML diagrams. In HCIS modeling, some of the most applicable diagrams are use case; activity and class collaboration diagrams are also very useful.

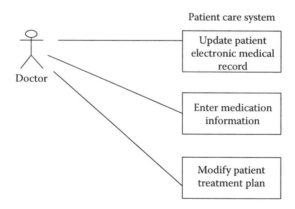

An example of a simple use case diagram relevant to a medical scenario.

GOAL-ORIENTED REQUIREMENTS LANGUAGE

The GRL represents a complementary component of URN that is needed as a graphical notation for describing nonfunctional user requirements, business goals or objectives, alternatives, and rationales. GRL proposes fundamental capabilities in the RE process by supporting goal- and agent-oriented modeling and reasoning about requirements. These include usability, user interface, performance, operational, maintainability, security, and social context, among cultural and political and legal requirements (ITU 2002).

In the initial stages of RE, the stakeholders will provide requirements in terms of objects and desired goals. Therefore, the representation of these goals in GRL enables developers to analyze various alternatives and trade-offs in building the most efficient and effective system. Subsequently, the initial and critical decisions can be made wisely based on the early analysis optimization that GRL provides.

The following descriptions illustrate the many different properties of GRL. All the properties are described in the succeeding text, but only goals, soft goals, and actors are used for illustration purposes in our case study.

Actor: An actor is an active entity within the system that carries out actions to achieve certain goals. Graphically, an actor is represented by a circle.

Goal: A goal refers to a real-world condition that the users of the system would like to achieve, satisfy, or meet. A goal can represent a number of things, for example, a business goal or a system goal (normally functional) to be achieved by the business and the system, respectively. A goal is graphically represented by a rounded rectangle as shown in the following.

Soft goal: A soft goal is similar to a goal, but it is normally difficult to verify if a softgoal has been achieved in an implemented system because of its qualitative nature. The requirements of the system proposed in this study will be represented by softgoals, as the safety of systems is largely dependent on NFRs. A soft goal is graphically represented by an irregular curvilinear shape.

Task: A task refers to a process or a set of activities. Tasks are used to address goals and softgoals. Tasks are closely linked to requirements. Therefore, tasks can be used to provide processes, data representations, and constraints required to meet the system's goals and softgoals. A task is graphically represented by a hexagon.

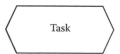

Resource: A resource refers to a physical or data entity of a system. In the design process of a system, the requirements model should specify if the resource is available or not. A resource is graphically represented by a rectangle.

Resource

Belief: A belief is used to represent a design rationale of the system. Beliefs are significant in the design of complex systems to assist in tracking and tracing actions and decisions taken at the design and implementation stages. A belief is graphically represented by an ellipse.

Contribution and correlation relationship: Contribution relationships allow the illustration of how goals, softgoals, tasks, and other elements in a system contribute to other elements within the system. The graphical illustration of a contribution describes how an element contributes to the satisfaction of another element.

Correlation relationships illustrate the interactions and relationships between actors, goals, softgoals, and other elements of a system (ITU URN 2002). The following types of contribution relationships exist:

(**Note:** Arrows are used to model contribution relationships—the element that the arrow points to is affected by the contribution; the element at the starting end of the arrow contributes to the element pointed to.)

- The MAKE contribution is used to illustrate a positive and sufficient contribution of an element.

- The BREAK contribution is used to illustrate a negative and sufficient contribution of an element.

- The HELP contribution is used to illustrate a positive but not sufficient contribution of an element.

- The HURT contribution is used to illustrate a negative but not sufficient contribution of an element.

- The SOME + contribution is used to illustrate a positive contribution of an element with an unknown extent.

- The SOME − contribution is used to illustrate a negative contribution of an element with an unknown extent.

- The UNKNOWN contribution is used to illustrate a contribution the extent and direction of which is unknown at the modeling phase.

To facilitate the basic idea of developing a GRL to represent system requirements, the following illustrates the model as it may be applied to the Therac-25 system. Specifically, actors, goals, soft-goals, help, and some positive are applied.

The illustration in Figure 12.3 represents a simple GRL model for an HCIS process. The model shows the goals, softgoals, and tasks required to enable safe and quality patient care in a patient

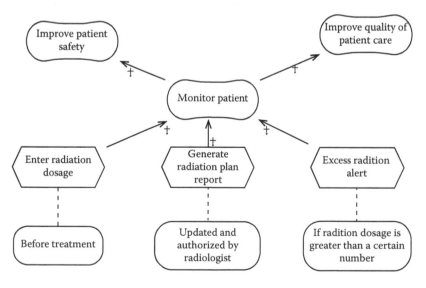

FIGURE 12.3 GRL illustration of attempting to develop a requirements analysis of the whole integrated system addressing patient needs in the administration of Therac-25.

care environment. Here, we provide three processes as examples. A radiologist is required to enter a radiation dosage into the HCIS. A radiation plan report is created, based on the radiation dosage from the radiologist, and sent to a doctor for review. If the radiologist failed to enter a dosage amount within the set safety level, the system will raise an alert. The preceding illustration offers significant implications for the engineering of safety critical complex systems. The Therac-25 was one such complex system that would have benefited greatly from the principles of RE.

Further Analysis and Contributions

A major contribution of URN in the RE process of complex reactive systems is its ability to graphically model and measure the impact of goals and requirements on systems. GRL goal models can be used to graphically show the impact of stakeholders' goals on systems. Physicians and nurses are primary stakeholders in the use of HCIS(s) to deliver care to patients. As a result, goals and requirements need to be clearly illustrated in order to ensure clarity and quality in HCISs. Although the process of measuring the impact of goals and requirements on system performance is subjective, URN provides a clear outlet to model the impact of requirements on system safety.

URN is very useful in organizing requirements to model the processes of HCIS. This ability enables the illustration of system processes and their impact on patient care. UCMs are used to show this relationship and impact. URN effectively supports the modeling of health-care processes, stakeholder goals, and requirements to operate efficient and safe HCISs. Fortunately, the identification and linking of stakeholders' goals to system selection and development can create a significant structure for evaluating system integrity and performance. This structure is important in systems such as HCIS where traceability is imperative to ensure that health-care processes support quality in patient care.

The RE approach proposed provides a systematic, iterative RE approach for the selection and development safe, efficient, and effective HCISs. The greatest advantage is the use of goal-oriented and scenario requirements concepts to model health-care processes through systems that focus on quality patient care.

MAIN CONTRIBUTIONS

- The case presented highlights the impact of RE on the safety of systems. One main contribution and motivation of this study was to bring renewed light to RE and the impact the process has on the efficiency of complex reactive systems. RE is a relatively new discipline in software engineering that focuses on the iterative and detailed process of requirements management. The study identified and outlined the impact of RE on the quality and safety of systems, especially in preventing errors in health-care delivery by a medical team of individual experts.
- Application of a new RE modeling technique to complex HCISs. URN is a requirements modeling language that uses two complementary components, GRL and UCMs, to model the processes and goals of systems. GRL enables the modeling of goals: goals relating to stakeholders, which are the foundation for creating clear and strong requirements. UCM enables the graphical representation of scenarios that relates to the processes of the system. The capabilities of URN and its components can significantly contribute to the RE process to develop and maintain safe HCIS. URN is particularly helpful, as it is one of the only modeling languages to model goals and processes at the same time and has the ability to integrate the two.

Future Work

The impact of URN as a requirements modeling language presents immense advantages and opportunities for safer complex information systems in health care. URN is particularly valuable in the campaign toward patient safety, as it offers great components to model goals and processes. Goals and processes are extremely significant in establishing quality and focusing on the efficiency of these systems. Yet one of the biggest drawbacks of URN is its lack of a clear policy to manage continuous refinement of requirements. URN needs to establish a clear iterative approach to refinement, which supports the constant cycle of validation in RE. Nonetheless, the capability exists in URN but has not yet been clearly defined. There appears to be no solid model of tracking requirements as they progress through their iterative cycle.

Future research must also consider the high level of abstraction of the URN component—UCMs. The capabilities of UCM notations should be evaluated from a resource perspective in order to gain a more complete overview of the applicability of UCMs from scenario descriptions in general.

Research would serve great purpose to focus on continuous refinement of requirements and the general application of UCMs as an excellent way to validate systems. This would be beneficial in evaluating the impact of systems on the processes and goals of health-care organizations, which aim to constantly improve quality and safety in patient care.

SUGGESTIONS FOR FUTURE WORK AND QUESTIONS FOR DISCUSSION

As tragic as the cases discussed in the Therac/Panama case were, we have learned many lessons about how to avert future accidents. Some questions to be discussed are as follows:

- What would define a system to be safe?
- What are our responsibilities as system builders and system designers?
- How are we going to evaluate and measure safety for our critical systems?
- What must be considered when a working system is adapted for other purposes?
- What steps have we taken since the Therac-25? Why did we have another similar incident in Panama?
- What should be the role of software testing in a project?

Further questions to be discussed for the surgical safety checklist include the following:

- How might the implementation of such a checklist be improved when one needs to accommodate changes in both the systems protocol and in the behavior of individual surgical teams?
- To what extent will medical practitioners be more attuned to the tasks at hand when they are cognizant of and responsible for following a checklist, especially when they know they are being observed?
- How can the effectiveness of a checklist be studied in a controlled manner when it is impractical for researchers to choose every operating room?

RE questions include the following:

- How can URN be efficiently integrated with other graphical RE models such as UML in developing a requirements analysis of the whole integrated system addressing patient needs in the administration of drugs by a medical team?
- How can URN better assess the continuous development of requirements in addressing overall medical care by a team using equipment?

DANNY KOPEC'S ACKNOWLEDGMENTS

As an instructor, the greatest joy one can have is to see that your students are motivated and involved and do something productive related to the course subject that goes beyond the classroom. As I have said, most student projects "die in my office."

I take great satisfaction in the fact that in September 2012, these five master's students Karunya Rajagopalan, Bustamante Brathwaite, Eranga Gamage, Shawn Hall, and Mariusz Tybinski accepted my offer to make a contribution to this book. We were just studying complex systems failures, and the students were preparing their individual case studies for my course in software methodology. Furthermore, a year earlier, Shawn Hall had completed a master's thesis related to requirements for hospital systems. As it turned out, the next year, Shawn found a job whereby the ideas and skills he developed in his thesis are directly employed. Shawn also happens to write very well.

Mariusz Tybinski had already published in the area of medical errors and was pursuing a thesis in this area when this project came about. So it was a natural fit for him and it helped him progress on his thesis. Bustamante Brathwaite is a superb student who has assisted me in many ways, and Eranga Gamage took on many unpleasant tasks to help the medical errors' team finish its work.

And then there is Karunya Rajagopalan. Karunya not only led the development of the Miracle on the Hudson Case Study (see Chapter 26) but also made a full contribution to the Medical Errors Case Study (see Chapter 12). This also led to her writing and completing a thesis in the area of mobile medical systems.

There were many meetings our team had during the course of the 2012–2013 academic year. I enjoyed working with all of them and feel they can be proud of what was produced.

I have been publishing well (and poorly) for nearly 40 years. Never in the course of those years have I been through so many editing cycles. The publishers can rest assured that the editors of this book made every effort to achieve *total quality management* in the delivery of our material. I would particularly like to thank Brian White for being kind, approachable, supportive, and responsive throughout this project. I learned we have a number of interests in common. I would also like to thank Professor Vernon Ireland for his thoughts and contributions, as well as Professor Jimmy Gandhi for his suggestions.

REFERENCES

Amyot, D and Mussbacher, G.(2001) Bridging the Requirements/Design Gap in Dynamic Systems with Use Case Maps (UCMs). International Conference on Software Engineering. pages 743–744,

Arora, A., Dutta, P., Bapat, S., Kulathumani, V., Zhang, H., Naik, V., Mittal, V., Cao, H., Gouda, M., Choi, Y., Herman, T., Kulkarni, S., Arumugam, U., Nesterenko, M., Vora, A. and Miyashita, M., A line in the sand: A wireless sensor network for target detection, classification, and tracking, In: Computer Networks (Elsevier), 2004, volume 46, pages 605–634.

Baase, S.B. (2008). *Gift of Fire: A Social, Legal, and Ethical Issues for Computing and the Internet*, 3rd edition. San Diego, CA: Prentice Hall.

Borrás, C. (2006). Overexposure of radiation therapy patients in Panama: Problem recognition and follow-up measures (Article and Special report).

Brennan, T.A. (2000) The Institute of Medicine Report on medical errors could it do harm? *The New England Journal of Medicine*, April 13, 342:1123–1125.

Brooks, F., Jr. (2005). *The Mythical Man-Month: Essays on Software Engineering*. Reading, MA: Addison-Wesley.

Castro, J., M. Kolp, and J. Mylopoulos. (2002). Towards requirements-driven information systems engineering: The tropos project. *Information Systems*, 27(6): 365–389.

Chung, L., B.A. Nixon, E. Yu, and J. Mylopoulos. (2000). *Non-Functional Requirements in Software Engineering*. Dordrecht, the Netherlands: Kluwer Academic Publishers.

Gawande, A. (2007). The checklist. *New Yorker*, December 10, 1994. http://www.newyorker.com/reporting/2007/12/10/071210fa_fact_gawande. Accessed November 17, 2012.

Goguen, J., Jirotka, M., Monk, A.F. et al. (1994). *Requirements Engineering: Technical and Social Issues (Computers and People)*. Waltham, MA: Academic Press.

Haynes, A.B., T.G. Weiser, W.R. Berry, S.R. Lipsitz, A.-H.S. Breizat, P. Dellinger, et al. (2009). A surgical safety checklist to reduce morbidity and mortality in a global population. *New England Journal of Medicine* 360: 491–499. doi:10.1056/NEJMsa0810119.

Institute of Medicine. (2004). *Keeping Patients Safe: Transforming the Work Environment of Nurses*. Washington, DC: The National Academies Press.

International Atomic Energy Agency (IAEA). (2001). Investigation of an Accidental exposure of radiotherapy patients in Panama, June.

ITU-T (2002). *Workshop on the "Use of Description Techniques."* User Requirements Notation (URN). Retrieved from www.UseCaseMaps.org/

ITU-T, Draft Recommendations Z. (2002). 151—Goal-oriented Requirements Language (GRL), International Telecommunication Union, Geneva, Switzerland.

ITU-T, Recommendation Z.150 (2003). User Requirements Notation (URN)—Language requirements and framework, Geneva, Switzerland.

Kopec, D., M. Kabir, D. Reinharth, O. Rothschild, and J. C. Castiglione. (2003). Human errors in medical practice: Systematic classification and reduction with automated information systems. *Journal of Medical Systems*, August, 27(4): 297–313.

Kopec, D., G. Shagas, M. Kabir, D. Reinharth, J.C. Castiglione, and S. Tamang. (2004). Errors in medical practice: Identification, classification and steps towards reduction. medical and care compunetics 1. Ed. Bos, L., S. Laxminarayan, and A. Marsh, *Proceedings of the 1st ICMCC (International Congress on Medical and Care Compunetics)*, The Hague, the Netherlands, June 2–4, pp. 126–134.

Kopach-Konrad, R., M. Lawley, M. Criswell, I. Hasan, S. Chakraborty, J. Pekny, and B.N. Doebbeling. (2007). Applying systems engineering principles in improving health care delivery, December.

Kossiakoff, A. and W. Sweet. 2003. *Systems Engineering Principles and Practice*. New York: Wiley.

Kuehl, C.S. (2001). Improving system requirements quality through application of an operational concept process: An essential element in system sustainment. *4th Annual Systems Engineering Conference*. http://www.dtic.mil/ndia. Accessed November 17, 2012.

Leveson, N. (1995). *Safeware: System Safety and Computers*. Reading, MA: Addison-Wesley, Pages 18–41.

Leveson, N.G. and C.S. Turner. (1993). An investigation of the Therac-25 accidents. *IEEE Computer*, July, 26.

Lissack, M. (2002). *The Interaction of Complexity and Management*. Westport, CT: Greenwood Publishing Group, p. 26.

Milstein, S., R.S. Galvin, S.F. Delbanco, P. Salber, and C.R. Buck. (2000). Improving the safety of health care: The Leapfrog initiative. *Effective Clinical Practice*, December, 3(6): 313–316.

Nadzam, D.M. and R.M. Mackles. (2001). Promoting patient safety: Is technology the solution? *Joint Commission Journal on Quality Improvement*, 27: 430–436.

Nuseibeh, B. and S. Easterbrook. (2000). Requirements engineering: A roadmap. In *ICSE—Future of SE Track*, pp. 35–46. New York, NY: ACM.

Panesar, S., D. Noble, S. Mirza, B. Patel, B. Mann, M. Emerton, K. Cleary, A. Sheikh, and M. Bhandari. (2011). Can the surgical checklist reduce the risk of wrong site surgery in orthopaedics?—Can the checklist help? Supporting evidence from analysis of a national patient incident reporting system. *Journal of Orthopaedic Surgery and Research*, 6:18. doi:10.1186/1749-799X-6-18, Page 2, http://www.ncbi.nlm.nih.gov/pmc/articles/PMC3101645/pdf/1749-799X-6-18.pdf.

Perrow, C. (1999). *Normal Accidents*. Princeton, NJ: Princeton University Press.

Pressman, R.S. (2005). *Software Engineering: A Practitioner's Approach*. Boston, MA: McGraw-Hill.

Rawlinson, J.A. (1987). Report on the Therac-25. In *OCTRF/OCI Physicists Meetings*, Kingston, Ontario, Canada.

Rouse, W.B. (2008). Health care as a complex: Adaptive system: Implications for design and management. *The Bridge*, Spring, Page 1.

Sommerville, I. (2001). *Software Engineering*, 6th edition. Harlow, U.K.: Addison-Wesley.

Tybinski, M., P. Lyovkin, V. Sniegirova, and D. Kopec. (2012). Medical errors and their prevention. *Health*, 4: 165–172. doi: 10.4236/health.2012.44025

U.S. Court of Appeals. (2008). *Johnston v. Multidata Systems International Corp. MDS* (2008). United States Court of Appeals. http://www.ca5.uscourts.gov/., Accessed January 31, 2013.

Zave, P. (1997). Classification of research efforts in RE. *ACM Computer Surveys*, vol 29, no 4: 315–321. New York, NY.

Section VII

Homeland Security

13 Disaster Response System of System Case Study Outline

Maite Irigoyen, Eusebio Bernabeu, and Jose Luís Tercero

CONTENTS

CASE STUDY ELEMENTS

The management of an emergency situation depends strongly on the type of emergency that occurs and the protocol or strategy to be followed for faster resolution. In this overall case study, we propose four emergency situations and the ways to solve them. The following summarizes the four situations:

1. Human rescue through robotic and unmanned vehicles
    - The use of unmanned and robotic vehicles to save human lives in hostile situations or areas.
    - Some characteristics of the equipment: overhead obstacles detector, real-time remote location system, a communication system, a command recognizer module, a leader–follower system.
    - Integration of several cameras to manage the vehicle as far as needed to be safe.
2. Visibility management: smart sensors
    - The use of sensors and emergency lighting to manage dangerous situations.
    - Expand the use of smart sensors toward smart cities.
    - Proper distribution and structuring of sensors to cover all the area controlled.
    - Analysis and communication among sensors to share information.
    - Sensor operation under bad weather conditions.
3. Flood control
    - Protect the areas most vulnerable to flooding.
    - Design a quick action protocol or plan to maintain flood control and to secure the safety of the residents of that area.
    - Implementation of new architectures and materials that minimize both floods and the possibility that these will occur.
    - The use of modular retaining walls that are easy and quick to build.
4. Crowd management
    - Develop and build an alert system to rapidly and efficiently manage a crisis situation like a fire or a terrorist attack.
    - Design a complete alert system in enclosed buildings or open spaces to control the crowd.
    - Improved utilities from sensors responsible for capturing the sound, video, or another parameter, indicating that an emergency is occurring.

Also, we propose a methodology to follow in emergency cases, a strategy structured in seven steps capturing the tasks needed to control the situation. These steps are as follows:

1. *Pre-event stage*: before the disaster occurs. Sensors and all the control technologies are making measurements of critical parameters.
2. *Alert stage*: apply the best emergency protocol; warn emergency bodies and provide all known information.
3. *Preventive stage*: the actors start the preventive actions to avoid the disaster or minimize the damage.

4. *Action stage*: the main stage, in which the emergency situation is resolved by the emergency bodies (army personnel, policemen, health staff, firemen, civil guardians, etc.).
5. *Restoration stage*: immediately after the emergency, begin working to restore normality.
6. *Maintenance stage*: the continuation of the previous phase to repair all damage, material and human, until predisaster conditions are achieved.
7. *Learning stage*: time to learn and correct errors, to improve future actions and decision making.

We describe the disaster management strategy in more detail in the *Scenario Development* section.

KEYWORDS

Crowd, Decision making, Emergency situation, Explosive disposal, Flood, GIS, Human rescue, Network sensor, Prevention, Rainwater, Risk, River, Robots, Smart grid, SoS, Surveillance systems, Unmanned, Visibility sensor

ABSTRACT

There is a large type of emergencies that are caused by events of a very different nature. May be due to natural disasters (fires, floods, earthquakes, volcanoes, etc.), accidents (derailment of a train or subway accident of traffic chain, leakage of hazardous gases, etc.), or situations (such as a terrorist attack, war, or any situation of violence or danger generated by man). The resolution and decisions to manage any of the aforementioned situations are very different. But the concept in the information management to anticipate or prevent a catastrophe is the same in all emergencies. To be able to correctly solve a hazardous situation, it is of vital importance to have all the information necessary to know the cause of the situation and of course handle all parameters that enable us to find a solution. This information must be available in real time for all the parties involved in solving the crisis: firefighters, police, health, military, etc.

The proper way to address and manage a crisis or emergency situation is a challenge for which it is critical to be well prepared. Only proper planning and equipment will allow us to achieve a favorable outcome. The integration of system of systems (SoS) or distributed systems has turned into an essential approach to effectively manage or drive any emergency situation.

A SoS involves a group of independently operating subsystems that collaborate to face a global objective. In order to emergency management, this common goal is the efficient resolution of crisis and emergencies, and the available resources or subsystems can range from surveillance systems, information/communication platforms, and intervention vehicles and units to intelligent sensing systems—even emergency management agents that can be coordinated and managed from a central system.

DISASTER MANAGEMENT STRATEGY

In recent years, there has been an increased demand for automated systems for the management of emergencies that result in increased competition and a wider range of solutions. This also leads to an increased investment of resources in the training of specialized employees. The development of new training tools to manage emerging technologies in the field of emergency management is a reality, and it will be a growing sector in the coming years; some of the existing and currently developing tools are as follows: training tools for emergency control of unmanned vehicles, training tools for information platform coordination in the emergency management field, and training tools for automated surveillance and security systems.

At some point in a disaster situation, there needs to be an assessment of the capacity of the community to cope so that the appropriate level of emergency relief can be determined.

For the purposes of examining the ingredients of disaster management strategies in more detail, it is useful to look at the frameworks that have been used to describe the stages in response to disasters at the community level. The most interesting one that we have found is drafted by Fink (1986). The different stages in a community's response to a disaster are the following:

1. *Pre-event*: where action can be taken to prevent disasters (e.g., growth management planning or plans aimed at mitigating the effects of potential disasters).
2. *Prodromal stage*: when it becomes apparent that the crisis is inevitable.
3. *Emergency-acute stage*: the point of no return when the crisis has hit and damage limitation is the main objective.
4. *Intermediate phase*: when the short-term needs of the people affected must be dealt with restoring utilities and essential services. The objective at this point is to restore the community to normality as quickly as possible.
5. *Long-term (recovery) stage*: cleanup, postmortem, self-analysis, and healing.
6. *Resolution*: routine restored or new improved state.

Disaster strategies clearly need to articulate a set of appropriate actions for each of the stages described earlier. For this reason, it is necessary to create a roadmap or strategy to follow in a disaster situation. Suggest the strategy developed by Turner (1994), which is compounded by the steps shown in Table 13.1, and the implementation suggested by Quarantelli (1984) in Table 13.2.

TABLE 13.1
Strategy Development for a Disaster Survival

Strategy Development

Form disaster recovery committee and convening meetings for the purpose of sharing information.

Risk assessment. Identify potential threats/disasters and prioritize in terms of probability of occurrence: real, likely, and historical threats.

Analysis of anticipated short- and long-term impacts.

Identification of strategies for avoiding/*minimizing* impacts, critical actions necessary, chain of command for coordination, responsibilities, and resources.

Prepare and disseminate manual and secure commitment from responsible parties and relevant agencies. Relevant contact information must be included.

TABLE 13.2
Strategy Implementation for a Disaster Survival

Implementation

Holding disaster drills, rehearsals, and simulations

Developing techniques for training, knowledge transfer, and assessments

Formulating memoranda of understanding and mutual aid agreements

Educating the public and others involved in the planning process

Obtaining, positioning, and maintaining relevant material resources

Undertaking public educational activities

Establishing informal linkages between involved groups

Thinking and communicating information about future dangers and hazards

Drawing up organizational disaster plans and integrating them with overall community mass emergency plans

Continually updating obsolete materials/strategies

In an emergency or crisis situation, it is very important to have predefined preventive tasks and the needed resources and qualified staff available. But also it is very important to have a strategy to follow in each disaster to facilitate and expedite the actions to manage the situation. An alternative approach to organize the steps in crisis events in six phases is given by Buchanan (2012):

1. *Precrisis or incubation*: It is the period during which combinations of *slow-moving causes*, including cumulative and threshold effects, and outcomes from past events contribute to the next incident and outcomes.
2. *Event*: Multilevel is needed, sociotechnical explanations for extreme events, and such modes of reasoning have become central in accident modeling.
3. *Crisis response management*: Studies have also focused on *the four Cs* of emergency management: communication, coordination, control, and the cognitive processes of risk detection, recognition, and interpretation that initiate the emergency response process.
4. *Investigation*: Investigations into extreme events can be pivotal to understanding what happened and why, to *learning lessons,* and to identifying recommendations for change.
5. *Organizational learning*: This phase is dominated by an organizational learning perspective, and most commentary concerns failures to learn.
6. *Implementation*: Active learning involves implementing change after an event. A learning perspective alone is inadequate; a complementary change management perspective is required.

We have designed roadmaps to manage four different emergency situations. In the following text, we describe the different technologies that can be used in different situations to manage and solve several types of crises:

1. Human rescue through robotic and unmanned vehicles
   Semiautonomous and unmanned robots can play important intelligent and technological roles that support or replace first responder equipment and personnel in unreachable, harsh, or dangerous environments. Robotics solutions that are well adapted to local conditions of an unstructured and unknown environment can greatly improve safety and security of personnel as well as work efficiency, productivity, and flexibility. Solving and fulfilling such needs present challenges in robotic mechanical structure and mobility, sensors and sensor fusion, autonomous and semiautonomous control, planning and navigation, and machine intelligence.

   Robotic technologies can be used for disaster prevention or early warning, intervention, and recovery efforts for various missions while ensuring quality of service and safety of human beings. Some of these missions may include the following: demining, search and rescue, surveillance, reconnaissance and risk assessment, evacuation assistance, and intrusion/victim detection and assessment.
2. Visibility management: smart sensors
   One visibility sensor aims to control the power on/off streetlights using high-power light-emitting diodes (LEDs). These will be used in urban routes to control the lighting and the visibility levels in streets as much for pedestrians as for vehicles. Also, this visibility sensor tries to normalize the illumination levels independent of adverse weather like rain, snow, haze, and fog. The sensor network will be easy to install, considering that we can reuse the existing system of the public streetlights. The philosophy of the smart grid includes reducing operating costs by exploiting the existing networks. In addition, public lighting is an infrastructure that is widespread in all types of urban environments. For this reason, a solution applied to public lighting network can have a universal application. With the intention of carrying out the control of several vital parameters for the functioning of communications within the city, we propose the integration of different sensors. We can control a lot of parameters, thus forming a system of systems (SoS) formed by fire sensors, smoke sensors, visibility sensors, temperature sensors, etc.

3. Flood control

The flood risk management has been studied extensively in areas that have suffered a real risk of flooding. The study of the topography of the areas close to rivers and how to build cities and their drains is important in preventing these events. But keep in mind in these studies the actors involved in the process. Due to the increasing number of public gatherings, safety measures and efficient management are necessary; thus, public administrations and other entities involved in issues such as traffic, civil defense, mass event management, or local festivals management will be especially interested in the possibilities offered by this system.

4. Crowd management

A fast, modular, and mobile crowd alert and management system can be developed in which social behavior patterns can be embedded in real-time prevention models of crowd behavior and evacuation (Helbing et al., 2002). Fast deployment is based, on the one hand, on the use of new low-energy consumption and autonomous sensors that will reduce current reaction time and, on the other hand, on the optimization of real-time model processing. Spatial analysis within a geographic information system (GIS) will ensure the modularity of the system through the optimization of the number, size, and location of sensor and alerting devices. The post of control post is designed and dimensioned to fit in a vehicle.

Due to the increasing number of large public gatherings, safety measures and efficient management are necessary. Public administrations and other entities involved in issues such as traffic, civil defense, crowd event management, or local festivals management will be especially interested in the possibilities offered by this system.

GLOSSARY

ARA	Applied Research Associates
CMOS	Complementary metal oxide semiconductor
EOD	Explosive ordnance disposal
GDC	Global decision center
GIS	Geographic information system
GPS	Global Positioning System
GSM	Group Spécial Mobile or Global System for Mobile Communications
IT	Information technology
LIBS	Laser-induced breakdown spectroscopy
MAV	Micro air vehicle
MDE	Model-driven engineering
RFID	Radio Frequency IDentification
SCADA	Supervisory control and data acquisition
SDR	Software-defined radio
SoS	System of systems
SoSE	System of systems engineering
UAS	Unmanned aircraft system
UAV	Unmanned aerial vehicle

BACKGROUND

In recent times, systems of systems (SoSs) or distributed systems are increasingly present in modern society. As new fields of economic activity emerge daily, a SoS-based solution is then developed to meet business needs more efficiently and in an automated way. The following describes some of the most relevant projects that are being currently developed on the use of SoSs in the field of emergency management.

EUROPEAN PROJECTS

Today, there are a variety of industrial devices for emergency response and crisis situations; mostly, automated systems consist of a dedicated and interconnected subsystem.

The function of these devices in automated emergency intervention can range from rescuing victims in situations of natural disasters (fires, earthquakes, etc.) to counterterrorism, surveillance, and crime prevention tasks. In France, the Groupe INTRA (Intervenion Robotique sur Accident) has developed, operated, and maintained a fleet of specific remote-controlled equipment, able to intervene instead of human beings, in the case of an accident in one of its members' nuclear site. This technology has proved its value in the tragic earthquake and nuclear power plant disaster in Fukushima where robots and unmanned vehicles capable of operating in highly radioactive conditions have been employed. In addition, some emergency management projects aim to develop a coordination platform for the emergency actors:

- CHORIST proposes solutions to increase rapidity and effectiveness of interventions following a major natural and/or industrial disaster in order to enhance citizens' safety and communications between rescue actors (CHORIST, 2009). Tools developed in Integrating Communications for enHanced envirOnmental RISk management and citizens safety (CHORIST) aim at providing more information to authorities and to the population. Three modules were developed and tested in the frame of the project:
- Module 1 (situation awareness) provides an overall real-time picture of events with an assessment of the consequences on the population and on property. This information helps authorities to take decisions.
- Module 2 (warning the population) allows authorities to warn the population quickly and through several media routes simultaneously.
- Module 3 (rapidly deployable PMR systems) allows both field rescue and support teams in control rooms to get more information on the situation.

The CHORIST project is a 38-month project (June 2006–July 2009), although some of the modules that make up the project (module 1) will be available in about five to ten years.

- The Experimental Update Less Evolutive Routing (EULER) project proposal gathers major players in Europe in the field of wireless systems communication integration and software-defined radio (SDR); it is supported by a strong group of end users and aims to define and actually demonstrate how the benefits of SDR can be leveraged in order to drastically enhance interoperability and fast deployment in case that crisis needs to be jointly resolved. The proposed activities span the following topics: proposal for a new high-data-rate waveform for homeland security, strengthening and maturing ongoing efforts in Europe in the field of SDR standardization, implementation of SDR platforms, associated assessment of the proposal for high-data-rate waveform for security, and realization of an integrated demonstrator targeted toward end users. Significant interaction with EU stakeholders in the field of security forces management will contribute to shape a European vision for interoperability in joint operations for restoring safety after crisis (EULER, 2012).
- Project objectives: Communication systems used in the field by security organizations constitute major elements enabling restoring security and safety after crisis in an efficient manner. Large-scale events require the cooperation between security organizations of different nature and different nations. In connection with a strong group of end users in Europe, EULER will contribute in proposing a more agile, interoperable, robust communication system supporting a new range of services to its users. In order to achieve these goals, three main components will be combined: a reference high-data-rate radio technique, a communication system architecture allowing integration of heterogeneous radio standards, and SDR as a key enabler for this.

- Seamless Communication for Crisis Management (SECRICOM) is a European project that will create an infrastructure of secure operative communication, a new generation of networks utilization. SECRICOM is intended to solve problems of contemporary crisis communication infrastructures. Some of the SECRICOM's objectives and characteristics are as follows:
  - Seamless and secure interoperability of many hundred thousand mobile devices already deployed
  - Smooth, simple, converging interface from systems currently deployed to systems of new SDR generation
  - Creation of pervasive and trusted communication infrastructure; offer interconnectivity between different networks
  - Provide true collaboration and interworking of emergency responders
  - Seamlessly support different user traffic over different communication bearers
  - Add new smart functions using distributed information technology (IT) systems based on an SDR secure agent's infrastructure
  - Easier instant information gathering and processing focusing on the emergency responder's main task—saving lives

The SECRICOM will develop and demonstrate a secure communication infrastructure for public safety organizations and their users. Achievements will include the following:

- The exploitation of existing public available communication network infrastructure with interface toward emerging SDR systems
- Interoperability between heterogeneous secure communication systems
- A parallel distributed mobile agent-based transaction system for effective procurement
- Infrastructure based on custom chip-level security
- SiCoSSys is a European project about agent-based modeling and simulation of complex social systems, involving the Complutense University of Madrid (UCM) and the INSISOC research group (Group of Social Systems Engineering) from the University of Valladolid (UVA) (GRASIA, 2011). The goal of this project is to provide a well-sound methodological framework for the treatment of complexity by policy makers and social scientists, endowed with an updated theoretical body of knowledge, a set of tools that enable scenario simulation, and a collection of case studies to guide and demonstrate the applicability of the framework. This framework will be based on agent-oriented modeling and simulation methods and tools: its bases are on model-driven engineering (MDE) to facilitate adaptation and integration of different profiles in a system of systems engineering (SoSE) process.

The SiCoSSys approach considers that models are the main artifact for the development of SoS. Models are conceived or already exist for each system. And there should be models as well to specify the SoS. Models are therefore heterogeneous, in principle, so different modeling languages should coexist in the SoSE process. Models are useful not only for the specification of SoS. They can be processed by tools for analysis and validation. This is important as the cost of deployment of SoS is usually high. Once models are validated, they can be implemented and deployed. It should be noticed at this point that the target platforms can be heterogeneous devices and computer infrastructure. At this point, new tools can assist the process by facilitating code generation for the diversity of target platforms.

Another issue that should be taken into account is maintenance of the SoS. With this respect, self-management capabilities can help. This is quite relevant as the number of elements in a SoS can considerably grow and their configurations change through time. These are the main considerations that make the SiCoSSys approach. They are developed from different viewpoints in the following subsections: which are the roles in a model-driven SoSE process, which are their main activities in the SoSE process, and how to organize all together.

INTERNATIONAL PROJECTS

Applied Research Associates (ARA) is a US company specializing on a variety of tracked and wheeled commercial construction unmanned vehicles designed for explosive ordnance disposal (EOD), reconnaissance, and surveillance functions. One of the projects and products is shown in the following:

- The Nighthawk micro air vehicle (MAV) is a rugged, simple, and affordable unmanned aerial vehicle (UAV) system that allows an operator with minimal training to conduct intelligence, surveillance, and reconnaissance (ISR) missions (ARA, 2013). Waypoint navigation, touch screen controls, and automated takeoff, flight, and landing modes make Nighthawk one of the simplest UAV systems to fly. With a range of more than 10 km and flight time of more than 60 min, the 1.6 lb Nighthawk system is the smallest, most capable ISR platform available today. The vehicle carries forward- and side-looking electro-optical (EO) cameras and a side- or forward-looking thermal image in a removable pod providing real-time situational awareness and targeting information to the operator. The ground station incorporates PC-based graphical user interface (GUI) technology to provide real-time visual feedback and mode control. *Point & click* waypoint navigation makes operation extremely user friendly.

RELEVANT DEFINITIONS

Complementary metal oxide semiconductor (CMOS) sensor: The CMOS are the newest technology in 2D detectors' record and represent an excellent novel alternative to CCD cameras, which are already fully exploited. This technology provides higher resolution and lower cost and energy consumption than other technologies, for example, CCD, and are easily included or embedded. Another advantage of the CMOS technology when compared to traditional devices is that CMOS technology can work using Fourier transform, thus absolutely preserving personal privacy while allowing the extraction of information of each pixel.

GIS: GIS is a type of information system that integrates tools (software), hardware, spatial datasets, and technical knowledge (i.e., specialized staff) that allow the visualization, analysis, and modeling of georeferenced thematic data. These processes result in valuable information in which the spatial component is the main added value.

UAVs: According to the US Department of Defense (DoD), UAVs are powered aerial vehicles that do not carry a human operator, use aerodynamic forces to provide vehicle lift, can fly autonomously or be piloted remotely, can be expendable or recoverable, and can carry a lethal or nonlethal payload. (Ballistic and semiballistic vehicles, cruise missiles, and artillery projectiles are not considered UAVs.)

PURPOSE

The purpose is to design a SoS that allows the interconnection and the transport of the information related with a danger situation. The correct use of this information can be very important in the prevention of catastrophes and very useful to the emergency forces. As previously advanced, systems-oriented systems of emergency management can be classified into the following:

SYSTEMS OF SYSTEMS DESIGNED TO ACT DIRECTLY ON RISK AND INCIDENTS (EMERGENCY RESPONDERS)

The primary functions directly related to crisis and emergency situations can be classified into monitoring/observation and intervention tasks. This implies that the majority of systems developed for the direct management of emergencies consist in automated surveillance and monitoring systems of crisis (or prone to become a crisis) scenarios or otherwise intervention unmanned vehicles designed for emergency scenarios and disasters.

SoS that operates directly at the stage of the crisis has a similar structure and subsystems due to the common functionality required to manage such situations. Therefore, a surveillance system or emergency response includes the following: an artificial vision subsystem or unit of pattern recognition, a subsystem of decision making or artificial intelligence, a transport system, a communication system, and finally, in some cases (such as intervention vehicles), an actuator subsystem able to operate in catastrophe stages and intervene on behalf of the victims and reduce the damage caused.

SYSTEMS OF SYSTEMS DESIGNED TO ENSURE INTEROPERABILITY, COMMUNICATION, AND COOPERATION AMONG THE VARIOUS EMERGENCY SERVICES AND AGENTS

In the case of disaster vigilance and mitigation, several public organizations (not only disaster prevention agencies) must be coordinated to provide emergency services such as rescue and replacement of critical infrastructure. How to efficiently develop a SoS at societal scale beyond the boundary of individual organizations is a very important research challenge that provides large benefit to society.

Such SoS architecture, normally consisting of a hardware/software network, is linked to the local area headquarters, fire stations, intervention unit, and communication platforms to enable coordination between agents and emergency services. Consequently, the information remains updated and may be contrasted at the time of intervening in a crisis, and relevant information can flow between those on site and those involved in command, control, and coordination.

The combined and coordinated actions of the parts of the system achieve more than all of the parts acting independently. This concept known as *synergy* is critical to the field of emergency management. These systems have three main tasks to be addressed: first, the communication system between the actors involved; second, the data processing; and finally, the decision-making unit.

The work performed by those responsible people for the coordination and emergency management is very intense and responsible; the work capacity ranges from 14 to 24 h, and their decisions may cost lives so consequently they want the best possible timely information from both humans and sensors to give them accurate assessments of circumstances. Integration of SoSs for the management and coordination of intervention units supports workers who may entrust part of their decision to the SoS decision-making unit and finally increase the autonomy of the emergency services.

EMERGENCY MANAGEMENT SYSTEM

The core aim is to develop a new, highly integrated crowd control system during short temporary events of mass transit, such as political meetings, sport events, shows, or demonstrations. This crowd control system will be designed to help event managers and other involved entities to manage the whole event life cycle in a safe and efficient way.

To achieve that goal, we will combine the examination of the sociopsychological background of such gatherings and of general human behavior with the development of innovative methods, technologies, and tools designed to advance and optimize today's way of managing complex events. The context of our emergency management system is shown in Figure 13.1. By way of explanation,

A—This part is composed of the control central unit. It is responsible for managing the information provided by a smart grid of sensors. This grid is embedded along the system under surveillance and where the emergency situation could occur. To have diverse information is necessary to install several and diverse types of sensors:
- A visibility sensor with its software algorithm, it is able to regulate the lighting for the best local illumination and minimize accidents on the roads due to darkness, fog, rain, etc.
- Network of interconnected smart cameras, information of the entire scene, like some place where a sport or musical event is celebrated or simply a crowded place, can be managed. This information together with the location information (provided by

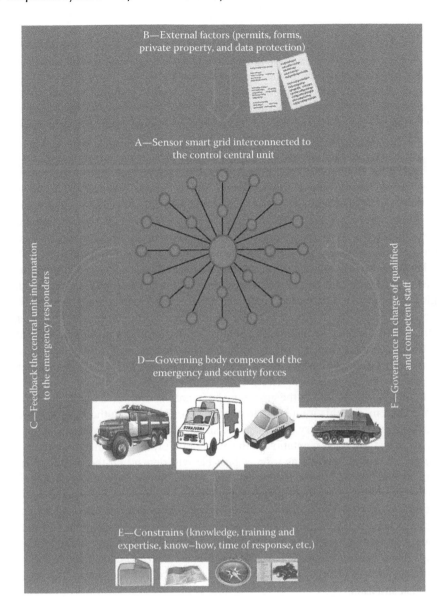

FIGURE 13.1 **(See color insert.)** Outline of the components of an emergency management system.

the Global Positioning System [GPS]) could be used to manage the flow and to prevent a possible emergency situation by a crowd.

With more types of sensors measuring humidity, noise, temperature, smoke, chemical composition, pressure, caudal of water, etc., embedded in the smart grid, one could better manage the situation and get it under control, keeping the conditions between the normal values.

B—There are several external factors that have to be managed for the functioning of a complete SoS. Depending on kind of system, their components will be a different external factor to overcome. To install the Smart Streetlight in the streets, some permission must be given by city authorities. In the installation of smart cameras, an important and delicate issue can be solved: the protection of the identity of persons recorded. This problem is removed by programming an algorithm that does not show the faces of the people. It is possible to blur faces or simply takes the persons like blobs. In the placement of external devices or control elements in a situation of risk, the environmental control laws and private property rights in areas in which it is necessary to put a control element like a dike in a flow-sensitive area, or the mounting of a field hospital in a catastrophic situation, or the assembly of a control sensor are also taken into account.

C—All the information collected by the control devices will be stored in hard drives. If an emergency situation takes place, the information will be available for the emergency services and security forces and all the bodies (political, legal, surveillance, etc.) in charge of managing the situation. Knowing the location, the conditions, the background, the victims, etc., in the risk situation could make a difference between minimizing and anticipating the threats.

D—The people in charge of the complex system could be experts in the use of the devices that compound the system and also could be persons trained in emergency management. If there is some problems, these persons could be able to manage the information and make the best decision to solve the emergency situation, coordinating the emergency services. Teamwork, coordination, and rapid and accurate decision making will make the difference between the resolution of the crisis in real time and with no loss of human life, minimal economic cost, and few, if any, operational failures. The organization and decision making must be under the purview and control of military forces, not politicians, because of the special expertise required to mitigate situations in dangerous and hostile environments. The civil authorities composing emergency management include health staff, doctors, firefighters, policemen/policewomen, Special Operations Group (SOGs), Técnicos Especialistas en Desactivación de Artefactos Explosivos (TEDAX [Spanish technician EOD]), and military personnel.

E—Constraints that would affect the governing body are the unknown use of the devices in a complex system, and this is an important task due to the fact that those sensors are where all the information necessary to manage the situation is. The need to know all the factors that may affect the implementation of the evacuation protocol of the area or the strategy in the situation management as soon as possible will depend on the coordination of the governing body members and the availability of the relevant individuals or organizations involved in providing required information.

F—One of the most striking qualities of military and emergency professionals is discipline. And hierarchies in this institution are clearly marked. Nevertheless, it will still be difficult to have a governing body capable of dealing with an emergency in the shortest time possible and without human error. For this reason, the decision making has to be managed by an emergency or military body and only for qualified and competent staff able to deal with a risk situation.

CHALLENGES OVERCOME

There are still a lot of improvements related to the management of emergencies and the systems that support this type of management. For this reason, we want to divide this section in the four main case studies that we proposed and analyze individually the technologies, architectures, scenarios, protocols, or decision making we have to develop to get the needed improvements in all the problematic situations:

1. Human rescue through robotic and unmanned vehicles
   The design and implementation of a robot capable of carrying and moving people is a challenge already overcome. Thus, a robot would be useful in moving injured or dead soldiers in the battlefield, transporting transmission equipment or heavy luggage of soldiers, or moving sick or old persons with limited mobility. The time of intervention can be reduced significantly in emergency situations with limited access, as robotic devices are

capable of carrying about 200 kilos at 15 miles/h in high mountain areas, rocky terrain, or collapsed buildings and rubble (DARPA, 2013). They can also automatically follow a leader using computer vision or travel to designated locations using appropriate optical sensors and GPS.

Currently, US military technology has revealed its latest effort to improve a robotic beast of burden (Dynamics, 2009). This device can carry significant squad member burdens and follow through rugged terrain and interact with them in a natural way, similar to the way a trained animal and its handler interact (Figure 13.2a).

Another challenge already overcome is the improvements in the UAVs due to the better technologies in image acquisition and the resolution and definition in the images. All these improvements have been able to manage emergency or surveillance situations providing information of the danger area. The small unmanned aircraft system (UAS), a UAV (Scientist, 2012), will likely play an increased role in search and rescue (Geiger, 2011). This was demonstrated by the use of UAS during the 2008 hurricanes that struck Louisiana and Texas (Dempsey, 2010). Micro-UASs, such as the Aeryon Scout (Aeryons, 2013), have been used to perform search and rescue activities on a smaller scale, such as the search for missing persons. UASs provide military and public safety users with immediate high-quality aerial intelligence, controlled directly by ground personnel. Micro-UASs can be broadly deployed with minimal training and operate reliably in extreme environments—not just calm sunny days.

For many applications, collection of aerial imagery from the micro-UAS is faster, more accurate, and more cost-effective than satellites, manned aircraft, or ground-based alternatives. UASs provide an immediate eye in the sky for law, fire, and emergency management personnel—and is the ideal small UAV system for public safety agencies. We can see an image of an aerial vehicle in Figure 13.2b.

There are several robots developed by the army or companies working for them, designed for spying, rescuing people, or carrying heavy-load transmission equipment (Dynamics, 2009). One of these examples is *The Spider*, robotic insects aimed at enhancing the military's situational awareness capabilities, capable of operating in places too inaccessible or dangerous for humans. They will be able to travel over very rough terrain, do things on their own autonomously, and carry lots of sensors and equipment for gathering useful information. Their main task is spying. Another very useful robot in the battlefield is the Battlefield Extraction-Assist Robot (*BEAR*). The BEAR can be controlled

(a)

(b)

FIGURE 13.2 (a) LS3 robot images courtesy of Boston Dynamics and (b) small UAS Aeryon Scout, A Lightwave, Easy-to-Use System. (From George Wong, http://www.ubergizmo.com/2011/05/videozoom10x-aeryon-scout/. Accessed June 5, 2011.)

remotely by a motion-capture glove or specially equipped rifle grip. A warfighter could use the equipment to guide the robot to recover a wounded soldier and bring him or her back to where a combat medic could safely conduct an initial assessment. The BEAR is a multi-modal, high-degree-of-freedom robot that can reach out with its hydraulic arms to lift and carry up to 226 kg, complete fine motor tasks with its hands and fingers, maneuver with a dual-track system, stand up and balance, and use cameras and sensors.

2. Visibility management: smart sensors

The main goal is to reduce the number of accidents and emergencies due to poor visibility. Therefore, the most important challenge is to develop a visibility sensor able to work almost autonomously and autoregulate the luminosity in response to weather variations. The perspective is to design an *artificial eye* able to see like a human eye and adapt the lighting conditions for a better visibility for a better driving and higher security on streets and roads. Each integrated sensor and processing device in the final design will contribute to increasing safety in adverse weather conditions. With today's vision technologies, overcoming these challenges is feasible, and improvements in this technology indicate that performance will improve considerably. These visibility sensors may be integrated into each car and lamppost if the costs of the various components are optimized, as we can see in Figure 13.3. It is composed of a camera, a processor, and a Ronchi test. The Ronchi test is a picture with stripes (black and white). The camera is always focusing this picture, and the variations in the luminosity of the image obtained by the camera are detected by the processor.

Visibility sensors can be attached to street lamps but may not be needed on every lamp. We place a visibility sensor on a streetlight every 500 m. The visibility information obtained by each such sensor is extrapolated to be the same for the other streetlights in line within that 500 m because it is assumed that the visibility conditions do not vary significantly in this area. In this area, we have the information necessary, for example, the luminosity and the weather conditions, with only one sensor. Thus, fewer sensors can cover a large area. In this way, cost is considerably reduced and network saturation with irrelevant data is avoided. Although the CMOS technology is cheap and the visibility sensor is easy to build again, it is not necessary to configure a network with one visibility sensor per streetlight. We show in Figure 13.4 two pictures taken during abundant sunshine (a) and normal (b) conditions. In picture (b), we can distinguish white and black stripes, that is, the so-called Ronchi test can be seen properly, in normal conditions of sunshine. However, in picture (a), we cannot distinguish the stripes, so the Ronchi test can't be seen properly by excessive sunshine. This is acceptable because the camera processor takes the Ronchi test information, which is continuously fed back, and appropriately adjusts the effective amount of light

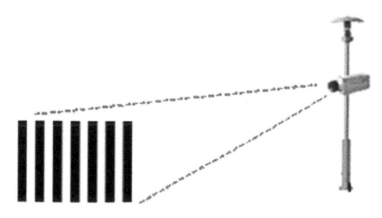

FIGURE 13.3   Qualitative description visibility sensor.

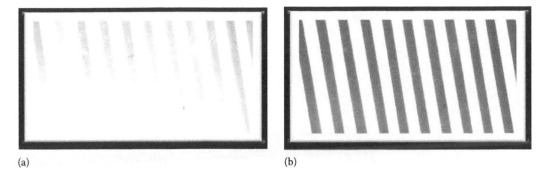

(a)   (b)

FIGURE 13.4   Pictures taken in conditions of (a) abundant sunshine and (b) normal.

supplied by the streetlight, that is, the processor decreases the luminosity derived from the streetlight, when a given level of artificial illumination is unnecessary.

3. Flood control

Prevention efforts should include a rigorous study of the topography of the area, moisture, soil porosity, and average rainfall per year. Depending on the results of the analysis, the buildings in the area must be fit for their purpose, and one should not build homes in riverbeds, if possible, and make solid retaining walls (of concrete) in the areas most sensitive to flooding or where there has already been some flooding in the past. These walled structures are responsible, in case of emergency, for containing rainwater and protecting buildings and people living in nearby villages.

But it is necessary to make an emergency management plan for floods and overflows of the river banks in the rainy season even if this occurs in an area not usually expected to experience this type of disaster. In planning for and managing prudent actions in placing containment systems for new building, one can avoid situations that endanger people in the affected areas.

A novel containment system can be rapidly constructed of lightweight preassembled panels that are joined together to build booms, barriers, or retaining walls for rerouting or diverting water. We show in Figure 13.5 the water containment system.

FIGURE 13.5   Water containment system. (From SolucionesEspeciales.Net, http://www.solucionesespeciales. net/Index/Noticias/374752-La-mano-fuerte-de-las-defensas-de-costas-Los-gaviones.aspx.)

FIGURE 13.6 Oosterschelde barrier in the Netherlands. (From AbsolutViajes, http://www.absolutholanda. com/estuarios-en-holanda-oosterschelde/. Accessed June 22, 2012.)

Such installations allow increases in the level of rivers and streams and/or help prevent oil contaminant spills. The system does not require heavy machinery or technology deployments for placement and provides a durable and sustainable solution for the control and prevention of floods. Such a system can work even if the permanent containment systems have failed.

This solution can be considered temporary or permanent. Such containment systems have been put in place in the United States to prevent flooding in South Sioux City, Iowa, due to flooding of the Missouri River, in the Valle de Sula in Honduras, on Colombia's Bogotá River, and on the River Thames in London, which has the second-largest mobile containment barrier in the world after the Oosterschelde barrier in the Netherlands (Figure 13.6).

4. Crowd management

To fulfill its challenging objectives, the project research will focus on the following:

- The analysis of psychological, social, and cultural behavior, on both individual and group bases
- The development of a new simulation framework based on multiagent models fed by the social behavior patterns and historical and real-time affluence data
- Integration of the prevention and real-time models into a GIS, which will help in positioning the sensors and the intervening emergency forces on the field and visualizing the predicted reactions of the crowd
- The development of CMOS, acoustic, and video sensors with new capabilities like embedded pattern analysis and autonomous decision for alarm triggering
- The optimization of the adaptive streaming protocols for obtaining more efficient transmissions
- The adaptation of the already established management protocols to the new management techniques, improving the information flow for the intervening agents (the police, fire departments, event organizers, etc.), which would bring clear benefits for the decision-making process
- The integration of all information received from sensors, social networks, GIS, etc., in a global decision center (GDC), where the new management protocols are set and modeling will be available
- The optimization of the communication channels for them to be secure enough so as to guarantee privacy and independent enough so as to keep on working even if the

traditional systems (Group Spécial Mobile or Global System for Mobile Communications [GSM] or Internet) collapse
- The observance of the highest ethics and privacy standards, as well as their application to the development of the tools and devices mentioned on this project

The advanced overall system will have new real-time information capabilities, and for the first time, the event managers will be able to efficiently understand the event dynamics and to fully coordinate with the intervening forces.

To sum it up, the aims are to solve the current problems in event management and to provide an efficient protection system for event goers.

DEVELOPMENT

In the case of the crowd management, we are in front of a complicated scene and more difficult resolution. There are many people involved in the situation and there is a need to evacuate without causing damage to somebody. This type of crowded places is usually closed. And it is easy to position a series of sensors for control of the critical parameters like the crowd (through the camera images). Below, we show the model of this emergency situation: the protocols to follow for managing the project, the architecture needed to have the place controlled, and the scenario development where to focus the crowd.

PROJECT MANAGEMENT

Crowd management in any scenario will be achieved with the combination of two different modes of management that can be applied to any emergency situation:

- *Preventive Management*: study of the event venue and deployment of an early alarm system, designed for the detection of a wide range of potentially dangerous circumstances, endogenous or exogenous, caused by human factors or by natural phenomena.
- *Conflict Management*: this kind of management would start to act in case the preventive system fails and the danger materializes. The conflict management system would provide real-time critical data, which, together with already established protocols, would be extremely useful for the emergency management bodies to take the best decisions possible.

ARCHITECTURE

In the case of an emergency situation in a mass event (Still, 2012), an adequate crowd-management (Abbott and Geddie, 2000) system enabling early evacuation of the crowd is essential to prevent crowd disasters.

Social networking contributes to an increasing number of mass events and their overcrowding, because it largely facilitates both the creation of events and its spread over a large and diverse community.

In the first moments of a crowd disaster, a quick intervention with the best available information (Helbing and Mukerji, 2012) is basic for a successful management of the tragedy; current communications and data acquisition do not meet the requirements of usability, simplicity, and support of the emergency corps.

In summary, there are three factors affecting the provision of adequate responses to protect crowd gatherings (Richardson, 1993): scarce data availability, limited crowd-management-trained staff, and restricted or nonexistent integrated communications and coordination center where all the relevant information is displayed, thus facilitating decision making while minimizing uncertainty and errors.

The goal is to design a general architecture that tackles the three aforementioned weaknesses of current crowd disaster prevention systems: data availability, skilled personnel, and integrated communications. We propose a GDC in which an early warning system based on a set of latest technologies will be integrated with a simulation model, spatial (GIS) data, and nonstructured data in order to directly communicate with the crowd and to support the *outside* authorities in the mobilization of the emergency services. Apart from research centers, universities, and end users (inclusive of security forces), the consortium counts with the participation of several small and medium enterprises (SMEs).

The control system design mass is shown in Figure 13.7. It relies heavily on a network of sensors and devices that provide video and location information. A management center to process this information and provide guidance to emergency services if their intervention was necessary in an emergency situtation.

1. The scenario information is obtained from a series of devices:
   - Audio sensors, which capture noise levels above critical values considered normal depending on number of attendees. If the noise level exceeds this estimated threshold, then an alert will be sent to a Management centre.
   - From CMOS sensors, which are obtained by Fourier transform of the captured images of the scene. This image processing allows us to constantly compare the images obtained with a test pattern that is considered within normal. If the result of image processing differs considerably from the test pattern, it will trigger an alert.
   - A video system that allows us to visualize the scene to control and check in real time if there is an emergency situation.
   - GIS: information system capable of integrating, storing, editing, analyzing, sharing, and displaying geographically referenced information. In a more generic sense, GISs are tools that allow users to create interactive queries, analyze spatial information, edit data and maps, and present the results of all these operations.
2. The GDC is the central system responsible for managing information from the devices described previously. Information from the warning signs are analysed, discarding false alerts and dissecting real alerts. In an emergency, it collects information from all devices to reconstruct the situation that generated the crisis. Depending on the facts of the situation, it will alert emergency services to resolve the crisis. Ultimately, the GDC is the center of decision making.

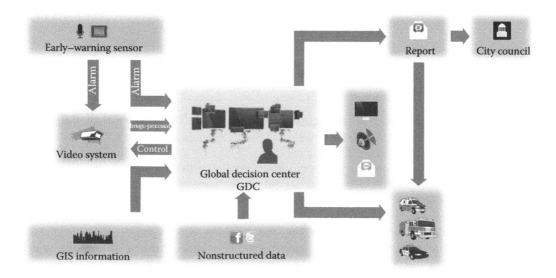

FIGURE 13.7   **(See color insert.)** Control mass system layout.

3. The fate of the information managed by the center of decisions is conveyed via satellite and television messages to the heads of government of the affected area as emergency services. The real-time communication with those responsible for resolving and managing an emergency situation is of vital importance in the resolution rate.

SCENARIO DEVELOPMENT

Figure 13.8 shows the virtual model that the event manager can see on its screen. It reflects the current situation, based on the following:

- Camera systems measuring the flow (vertical yellow planes).
- Cameras measuring the density of the market area (horizontal yellow planes).
- Cell phone GPS data of visitors and emergency forces (forces: orange spheres), localization.
- Spatial GIS data (buildings, road network, and so on).
- The simulation's core will assure to reflect agents that are not covered by real sensors in the best way possible.

Having a look at the virtual model at the screen, the event manager knows where the participating agents are located (real-time mode). Introducing a measurement object into the simulation (*prestage occupation measurement*), he will find out about the real density—the simulation scenario can be investigated in real time. He detects also that the overall occupation of the market place is average, but the distribution is heterogeneous. The event manager develops a plan:

- Shut down the flow at the stage entries.
- Reduce the flows at the rear entries by half (still space left).
- Make the visitors in the rear area step back—starting with an App message to the very rear area, proceeding to the front after reaction time, and so on until the front area becomes free (message: "step back, and tell your neighbors"—expected perception around 10%).
- Delegate half of all forces to the stage in order to help the casualties.

The event manager's simulation engine allows testing the actions in fast time before really applying them (fast-time mode). *Average* agent behavior is assumed here, developed based on behavioral models regarding perception of messages and flow behavior. The simulation runs fast into the future and shows what would happen if the actions were chosen: the persons now seem to be distributed.

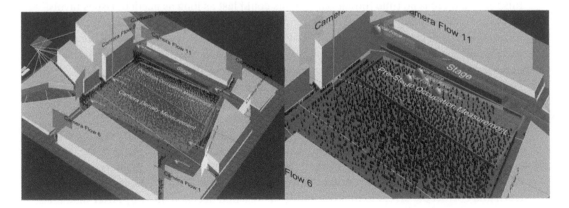

FIGURE 13.8 (See color insert.) CAST simulation, Scenario event management. (From Airport Research Center GmbH. Control mass scenario, http://www.airport-consultants.com/index.php?option=com_contnet& view=section&id=8&Itemid=74.)

DISASTER STRATEGY

In our opinion, the best way to manage the strategy of an emergency situation is to consider the emergency as a whole: pre, during, and post. For this reason, we consider the following steps in the strategy:

1. *Pre-event stage*: Before the disaster takes place. During all this time (between emergencies), all the sensors and control technologies are making measurements of critical parameters. At this stage, it is important to continually observe the measurements and initialize the protocol or strategy when there is an alarm and when any parameter is outside the normal range of values.

2. *Alert stage*: In the order of the critical parameters that have given the alarm, we apply the best protocol of emergencies. The first action is to warn emergency bodies involved in the emergency: policemen, firemen, army personnel, health staff, etc. With this notice, the following information should be provided: the kind of emergency, location of the best road to arrive in function of the traffic jams, time of the disaster, people involved, known victims, and the other emergency bodies that will give support.

3. *Preventive stage*: If we can anticipate the emergency, the actors start the preventive actions to avoid the disaster or minimize the damage. For example, if the emergency is a flood, they will put movable retaining walls on the limits of the buildings or areas to conserve.

4. *Action stage*: The main stage, in which the emergency situation is resolved by the emergency bodies (army personnel, policemen, health staff, firemen, civil guardians, etc.). Depending on the type of emergency, the way to act will be different, and maybe the security forces, too. In this stage, the main goal(s) is eliminating the danger and/or minimizing the damages.

5. *Restoration stage*: It is the phase immediately after the emergency. It is the time to work to restore normality. Both emergency bodies and social organizations must try to restore public services as soon as possible, to fix all things damaged, and to coordinate the work of the professionals needed to achieve this. And the most important action is to rescue surviving victims and recover any dead. This could be a difficult work due to the instability of the situation and the fragility of the human lives.

6. *Maintenance stage*: It is the time to repair all damage, material and human, until predisaster conditions are achieved. It is like the continuation of the previous phase but with less urgency because the most important things have been solved in the restoration stage. For this reason, there is no time limitation, and this could last as long as needed. Issues involving streets, the buildings, the public lightings, telephones, other public lines, etc., can be restored and, if necessary, improved in design. And if there is loss of human life, one must exercise all the protocols for identifying the bodies, finding their families, and giving all necessary help to the families and other victims of the tragedy.

7. *Learning stage*: When its over, that is, the emergency is resolved and the main damages are repaired, it is time to analyze the entire situation—starting from the working of the sensors in the pre-event stage, then the decision making and the coordination of the actors involved, through the way to act in the preventive management of the disaster and the result of the preventive elements, the performance of the different bodies involved in the action stage, the results of these actions, the time spent in resolving the main dangers, the number of victims and dead, and the restoration and maintenance of operations. It is the time to learn and correct the errors to improve future actions and decision making. After that, it is necessary to feed back strategies for improvements, building a new protocol and/or adding new concepts to each step in which there were failures or deficiencies. In this manner, the disaster management strategy will be continually improved.

In applying this strategy to each emergency, the only differences among them are the resources used in the management of the emergency:

- *Human rescue through robotic and unmanned vehicles*: This type of technology is used during the *action stage* in situations like explosive situations, shaky ground, and the need of carrying people and ultimately in situations where the lives of members of the emergency bodies are in danger.
- *Smart management sensors*: These technologies could be included in the *pre-event stage* because their main task is to change the luminosity and visibility conditions in streets and roads to avoid accidents or danger situations due to the bad weather.
- *Flood control*: In the control of floods, the *pre-event stage* and the *alarm stage* are very important. In these stages, there is the difference between a disaster occurring and controlling the situation and with only some damage. Movable retaining walls can be used in several kinds of situations to help prevent spillage, flooding, or land collapse.
- *Crowd management*: The tools used in these situations include smart cameras for surveillance that allow managing crowd situations before problems occur (this is part of the *pre-event stage*), lighting devices to show emergency exits and ways to follow to the exits (this belongs to the *action stage*), devices to make communication possible among all staff in charge of controlling the event (this belongs to the *pre-event, alarm, action, restoration, and maintenance stages*), and all the surveillance and security forces that control the event in real time in the *action stage*.

Of course, in all the cases, there are professional crews responsible to give the alarm, to coordinate and make decisions in the *alarm* stage, to resolve the situation in the *action stage*, to restore the normal order in the *restoration stage*, and to perform all the tasks required in the *maintenance stage*.

RESULTS

The employment of SoS within the emergency field has the following advantages:

- Successful and efficient address/management of the situation ensuring a reliable and failsafe service.
- Reduce the economic impact caused by the emergency and avoid the potential victims.
- Avoid overflow conditions due to the rapidity with which events can happen, due to lack of foresight, lack of organization, or by an inadequate response.
- Enhance the versatility of the emergency system by providing a flexible and adaptive SoS according to the nature of the event risk to deal with.
- Streamlined and efficient use of energy and emergency services.
- Increase the coverage and the autonomy of the emergency and intervention systems as well as guarantee an automatic answer to give priority to the situations.

ANALYSIS

In this section, we discuss some of the most important applications in the field of emergency resolution. Devices and architectures for crisis resolution that exist today are useful and can help to prevent many crises.

With the advance of new technologies, there are many tasks to prevent emergency situations that have been greatly simplified: fire prevention, preventing floods, make people buried alive in the rubble, restoring a telecommunication network after an earthquake that allows people to connect with isolated, locate injured persons in remote areas.

Natural Disasters

The first step in avoiding a disaster is knowing how to anticipate the same.

Among them, we can name the most important natural disasters (tsunamis, earthquakes, floods, fires, and hurricanes), terrorist attacks, traffic accidents chain, etc.

We know that in most cases, natural disasters are inevitable, but we also know the damage they cause will be lower the greater our commitment to monitor its progress and the more we prepare for its arrival. It is also vital to the reaction after a natural disaster, since the speed in mitigating its effects and the organization will influence the number of lives saved.

In applications of *fire prevention*, it is the use of sophisticated sensors that are placed at key points in an area and continually send data on different parameters, so that its evolution can inform on the risk of disaster. In the coastal range of Collserola (Spain) should be controlled in the rear of the city of Barcelona, numerous sensors have been installed, scattered among the pine and oak forests in order to draw a temperature map of the area. Any abrupt rise in temperature in a given area will be drawn immediately on the map as an anomaly (a principle of fire) and can be controlled effectively and quickly. *Autonomous nodes for sensor networks* and the nodes are basically miniature computers with sensors that can be connected together as a network of sensors for different measurements (temperature, humidity, motion, sound) and provide useful information in various settings. This sensor acquires values from environmental parameters: temperature, humidity, lightness, presence, pressure, or (almost) whatever you can sense and operate with these values when required. The values are transmitted using a low power, making their installation economically feasible (SQUIDBEE, s.f.).

In a devastated area by an *earthquake* or a *tsunami*, for example, it is easier than just mobile infrastructures are in place, so you have to rebuild a communication network from scratch and with extremely moving. One of the alternatives recommended are WIMAX antennas, which emit an Internet signal over a wide area and can thus reach isolated areas where there are connected computers.

The company Albentia Systems SL has successfully launched a draft implementation of WIMAX technology in an area of Sierra Norte de Guadalajara (Spain), enabling broadband access to many villages. We used base stations with a capacity of up to 34.4 Mbps with a typical coverage radius of 20–30 km. This model increases the capillary deployment of broadband service to remote areas, reducing network cost.

But in a natural disaster, in which telephone systems and other communications are useless, the best option and safest communication seems to be via satellite. In the deployment of US troops in Haiti, there has been widespread use of tactical communications radio AN/PRC-117F Harris (Harris, 2005). This is one of the most advanced military radios on the market to operate in the VHF and UHF, particularly in the segment between 30 and 512 MHz. The radio is equipped with secure communications capabilities (COMSEC) and secure transmission (TRANSEC) and supports several encryption systems, both owners of Harris as standards in the US military or NATO itself.

Security

Scientists from the US Army Research Laboratory have developed a portable laser that allows detecting chemicals, explosives, and biological threats instantly. It is the laser-induced breakdown spectroscopy (LIBS): a series of laboratory experiments have been performed highlighting the potential of LIBS as a versatile sensor for the detection of terrorist threats. LIBS has multiple attributes that provide the promise of unprecedented performance for hazardous material detection and identification. We can see the experimental setup in Figure 13.9. These include the following: (1) real-time analysis, (2) high sensitivity, (3) no sample preparation, and (4) the ability to detect all elements and virtually all hazards, both molecular and biological (Delucia, 2005).

The laboratory studies we have conducted for a variety of potential threats suggest that LIBS has the potential to become an important tool for security-related applications once it will be field validated under different environmental conditions. The development of remote sensors based on

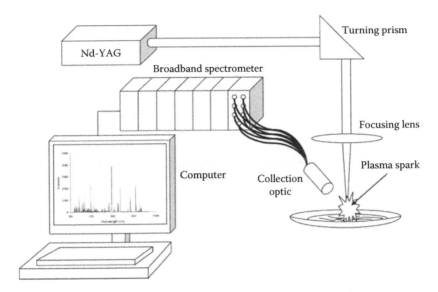

FIGURE 13.9 LIBS experimental setup. (From DeLucia, F.C. et al., *IEEE Sensor J.*, 5(4), 681, 2005.)

LIBS offers the ability to detect traces of explosives at significant distances from the target, while the transportability of these systems enables inspection of samples in a real environment.

One embodiment of the system used to identify traces of explosive vapor has the following building blocks: air sampling (with filter elements and trap), the biosensor, and the hardware and software required to process the signals. Thus, we generally suggest three steps in the process of detection:

- First phase: sample collection and concentration
- Second phase: detection (by the biomolecules and the transducer)
- Third phase: measured signal analysis and decision making (positive or negative identification)

Its main application is the detection of traces of explosives, with a simple and quick method.

SUMMARY

In summary, the implantation of SoS in emergency management is an important factor within the current trend toward the establishment of smart cities where most social services and citizenship-oriented functions are automated and the environment becomes increasingly autonomous. Smart buildings are an example of this transformation where emergency management is prevented and managed in an automated way.

SUGGESTED FUTURE WORK

There are a variety of applications that can be carried out in the context of SoS and more specifically in the management of emergencies and crises. Depending on your difficulty, the resources available today, and the most immediate needs, future applications can be classified into two groups: short-term and long-term developments.

SHORT-TERM DEVELOPMENT IN THE EMERGENCY MANAGEMENT DOMAIN

The field of emergency management has undergone an important development due to the introduction of new technology that provides a more comprehensive, reliable, and functional service. A clear example is the implementation of SoSs and distributed systems that automate the process of crisis management and provide a degree of interoperability unknown so far. Trends and short-term strategies can be summarized in the following points:

Implantation of Nanotechnology in SoS Field

Nowadays, the scientific sector is crossing the threshold of the third revolution in semiconductor packaging. Most of the efforts of R&D are dedicated to the development of new concepts, a mix of conventional assembly and innovative processes. The trend leads us to 3D packaging that promises to break the limits of miniaturization. This potential will be applied to the building blocks of SoSs thus achieving a higher rate of compaction and robustness, resulting in more reliable, compact, and improved autonomy systems that can respond under extreme conditions. The implantation of nanotechnology will also improve the integration of SoS in the operation area and the velocity of response under emergency scenarios.

An example of this technology is the nanotransducers applied to a sensor network. There are several kinds of sensors that carry out different tasks; therefore, there will be many ways to integrate it in a SoS. New technologies are facilitating the integration of more and smaller sensors around us, capable of processing huge amounts of data to help improve the operation of factories and cities and the control of production processes or even to detect earthquakes. Some of the applications of this sensor network in a SoS are the following:

- Climatic management: temperature sensors, thermostats, temperature probes for immersion, pipes or ducts, humidity sensors, pressure, etc.
- Fire management: optical sensors, infrared, optical barrier, expansion, smokes, etc.
- Management of robberies and intrusions: infrared presence sensors, microwave or ultrasonic presence sensors, sensors of opening doors or windows, microphonic sensors, etc.
- Presence control: card readers, physical identifiers, etc.
- Control of lighting: light sensors

Improved Algorithms for Pattern Recognition, Computer Vision, and Decision Making

There are significant barriers in the disciplines of pattern recognition and computer vision; these problems hinder the use of automated systems in the performance of tasks of direct intervention and disaster crisis resolution. An improved system for pattern recognition and identification of objects allows the use of automated systems with a lower rate of errors in decision making. It takes a global approach in the design of automated systems that take into account the performance of sensors and processors available and the control software together with the decision-making algorithms (neural networks, AI, genetic algorithms, etc.).

New Information and Communication Platforms

The introduction of new information platforms and network infrastructure can represent a revolution for the world of communications and therefore play a key role in the collaboration between agents in the management of emergencies. Perhaps the most representative model with a greater potential is the well-known *cloud computing*, which offers information services and computing resources over the network.

This concept is ideal for collaboration between different actors in the emergency management as it ensures automated updated information and a fast response and reliability due to the large amount of computing resources provided. This model of collaboration based on cloud computing is absolutely necessary and will be a benchmark in the coming years. Today, there are already emerging companies and research projects that develop these applications:

- Community-Based Cloud for Emergency Management
- EMI SIG (Emergency Management Issues Special Interest Group)

LONG-TERM DEVELOPMENT IN THE EMERGENCY MANAGEMENT DOMAIN

The long-term prospects for the development of SoSs applied to emergency management are difficult to predict; in particular, we will focus on industry trends over the next five years. Previously, a number of threads for short-term development have been summarized; these trends are valid long term, taking into account the degree of improvement of technology and the needs of society in a period of five years.

Intelligent Buildings and Smart Cities

A technological breakthrough that will grow and become standard in the next five years consists on the rise of intelligent buildings and even smart cities. Through the improvement and cheapening of new sensors, it can be measured each brick temperature outside the building, how are the oxygen and CO_2 levels within each office, or even the and thus more efficiently manage situations of emergency. It also gives the building a preventive element that allows active monitoring of such buildings and avoids accidents, for example, someone leaving a tap running in the sink or a gas spigot.

As an extension of the intelligent buildings, intelligent cities are gaining prominence. At this moment, many urban services have been digitized. Presently there are monitoring stations and air quality camera operating unit of traffic control and fast-speed communication networks that cover the whole urban area. Overall, the city has a wire mesh controlled digitally; the trend in the near future is to develop a centralized control unit that can take advantage of these data to generate synergies, optimize traffic, prevent crime, and manage crises and disaster scenarios. For example, if a fire breaks out in an intelligent city, the technology will automatically detect the fire and notify the software that controls the fire and ambulance systems. We gain nothing by automating and digitizing all city services if there is no integrated planning that allows each of them to interact.

Development of Nanotechnology and Surveillance Networks

Within five years, the rate of miniaturization in technology systems of all kinds will take a big step forward; the integration of electronic components in 3D wafer can allow the nanotechnology to improve existing sensors and integrate them into everyday objects such as clothing items, urban infrastructure, and power grids. This degree of integration may allow monitoring such a big area never reached before at a much lower cost and higher efficiency and reliability. Emergency management can take advantage of this breakthrough to detect earlier and possible security crisis and better coordinate the resources used to solve them.

The networks and surveillance systems are implemented in the urban facilities as part of urban infrastructure and allow for the detection and control of abnormal situations while also ensuring the flow of information between the emergency management intervention units.

Some of the most important applications in the field of emergency management by the security forces and emergencies will be the virtual helmet. The speed of decision making and coordination between the teams involved in the management and resolution of emergency situations is vital. It is therefore necessary that emergency personnel have a range of information on the disaster area both real-time status as the area before the emergency occurs (maps, plans, existing infrastructure, etc.). That allowing for a composition of place at the situation and also an interoperable communications to enable communication between all bodies involved in emergency themselves. The main goal is to develop assistive equipment for emergency personnel, enabling the virtualization of the state and control of the situation in vivo, through the connection of multiple technological devices.

Another important application would be a sensor grid. The intention is to develop a system that enables the rapid deployment of specific sensors for treating emergencies or disasters. This system allows the collection of information about the area and the evolution, if any, of the condition. In turn, this identifies robust communication platform to support the sensors and networks, integrated individually, and the different agents. The main task will be the analysis and development of an interoperable system that allows communication between all sensors involved in the management of emergency lighting and the actors and assets involved in the communication network.

QUESTIONS

1. What is the best generalized strategy to manage an emergency situation of different nature?
2. Why is a sensor or device network the best way to manage the information involved in an emergency?
3. What is the way to handle the feedback from the network and how to decide what is important and what is not?
4. Who composes the control central unit and what must be the hierarchy?
5. What is the difference in the modus operandi, the information needed, the strategy designed, and the bodies involved if the situation is a natural disaster or a security emergency?
6. What are the main constraints in the management and subsequent resolution of a disaster or emergency situation?
7. What is the main driver needed to facilitate the emergency resolution?

CONCLUSIONS

Today, emergency management is a complex function that involves many factors such as public safety and security, management information systems, communications technology, science, mapping and modeling risks, legal issues, and coordination with many other organizations. In view of this diverse set of functions and activities, emergency management systems have evolved into SoSs or distributed systems that can handle all these elements.

Since the 1970s, most advances in emergency management came from more/better improved monitoring, instrumentation, data collection, and data processing. Some of these have resulted from advances in theories and models, but no radical theoretical breakthroughs have occurred in the past 20 years.

On the one hand, the way to solve a crisis and the protocol or strategy that the security and emergency forces have to follow are fundamental. This aspect is easily improved with the help of the technological advances that occur every so often. An emergency situation is composed of several stages, and the strategy to follow will be divided in several steps. The first one is the formation of a disaster recovery committee with the end to share information from all the possible sources. This action will allow knowing all the factors involved in the emergency. The second action should be the risk assessment; identify the threads and prioritize it depending on the probability. After this, the short- and long-term impacts due to the disaster must be analyzed, and try to minimize it with the design of the best strategy. This strategy has to include the hierarchy in the chain of command for coordination, the decision making, the resources needed and the resources available, and the modus operandi. And last but not least, it is necessary to prepare and disseminate manual and secure commitment from responsible parties and relevant agencies, with relevant information.

On the other hand, the integration of SoS, and technological advances they represent, has significantly affected several fields of emergency management domain as automated surveillance systems, emergency intervention vehicles, warning systems and monitoring, and hazard modeling and prediction of disasters.

The evolution of integrated circuits propitiated/caused by advances in the manufacture of silicon wafers together with the advance of hardware/software design and the great improvement of communication platforms has caused a technological leap that allows the development of innovative automated monitoring and intervention and organization systems that drastically change the strategy of dealing with disasters and crisis situations. These technological advances result in several emergency management improvements:

- Shorter decision making in emergency systems automated and less error rate in the same
- Increased robustness and autonomy of surveillance and intervention systems
- Improved sensing technology
- Improvement of computer vision and pattern recognition techniques
- Higher levels of interoperability between heterogeneous emergency actors
- Integrating data acquisition and real-time updating automated systems (GIS and supervisory control and data acquisition [SCADA])
- The use of new communication/service platforms (cloud computing) and target location systems (Radio Frequency IDentification [RFID], GPS)

Some advances have been also made in predicting, detecting, and forecasting floods, tornadoes, volcanoes, landslides, and chemical accidents, but these improvements have to still be fully integrated into warning dissemination systems.

SoSs have also revolutionized the traditional concept of building, stepping beyond the home automation, building monitoring systems, control, actuators, and notifications and alarms that prevent and eliminate crisis situations and emergencies.

REFERENCES

Abbott, J.A.L. and Geddie, M.W., 2000. Event and venue management. Minimizing liability though effective crowd management techniques. *Event Management*, 6(4), 259.

AbsolutViajes. http://www.absolutholanda.com/estuarios-en-holanda-oosterschelde/. Accessed June 22, 2012.

Aeryon, 2013. Small Unmanned Aerial Systems (sUAS). Available at: http://www.aeryon.com/

Airport Research Center GmbH. Control mass scenario. http://www.airport-consultants.com/index.php?option=com_contnet&view=section&id=8&Itemid=74.

ARA, 2013. Applied Research Associates, Nighthawk Micro Air Vehicle (MAV). Available at: http://www.ara.com/robotics/Nighthawk.html

Buchanan, 2012. Researching tomorrow's crisis: Methodological innovations and wider implications. *International Journal of Management Reviews*, 15, 205–224.

CHORIST, 2009. CHORIST Project. Available at: http://www.chorist.eu/index.php?page=11&sel=11&sel=11

CIVIBASTION by Arquitecsa. http://www.milibastions.com/milibastions/.

DARPA, 2013. Available at: http://www.darpa.mil/Our_Work/TTO/Programs/Legged_Squad_Support_System_%28LS3%29.aspx

DeLucia, F.C. et al., 2005. Laser-Induced Breakdown Spectroscopy (LIBS): A promising versatile chemical sensor technology for hazardous material detection. *IEEE Sensor Journal* 5(4), 681–689.

Dempsey, M. E., 2010. Eyes of the army. Available at: http://www-rucker.army.mil/usaace/uas/US%20Army%20UAS%20RoadMap%202010%202035.pdf. Accessed April 14, 2014.

Dynamics, 2009. Available at: http://www.bostondynamics.com/robot_bigdog.html

EULER, 2012. EULER consortium, European SDR for wireless in joint security operations. Available at: http://www.cwc.oulu.fi/euler/Euler_general_presentation.pdf

Fink, 1986. *Crisis Management*. American Association of Management, New York.

Geiger, H., 2011. The drones are coming. Available at: https://www.cdt.org/blogs/harley-geiger/2112drones-are-coming

George Wong. http://www.ubergizmo.com/2011/05/videozoom10x-aeryon-scout/. Accessed June 5, 2011.

GRASIA, Pavón, J. and Gómez-Sanz, J., 2011. GRASIA Research Group, Universidad Complutense de Madrid, Adolfo López Paredes, INSISOC, Universidad de Valladolid. *The SiCoSSyS approach to SoS Engineering. 6th International Conference on System of Systems Engineering*, Albuquerque, NM.

Harris, 2005. AN/PRC-117F(C) Multiband multimission radio. Available at: http://www.armyproperty.com/Resources/Catalogs/Harris-AN-PRC-117F%28C%29.pdf

Helbing, D., Farkas, F., and Vicsek, T., 2002. Panic: A quantitative analysis. Available at: http://angel.elte.hu/panic/

Helbing, D. and Mukerji, P., 2012. Crowd disasters as systemic failures: Analysis of the love parade disaster. *EPJ Data Science*, 1(7).

INTRA, s.f. Groupe INTRA, Intervention Robotique sur Accident. Available at: http://www.groupe-intra.com/pages2/presentation/historique1.htm

Joshua Topolsky. http://www.engadget.com/2008/05/01/bea-systems-working-on-spider-bots-other-ways-to-scare-you-to-death. Accessed in 2008.

Pouted Online Magazine. 7 Newest Robot Generations and Their Uses. http://www.pouted.com/7-newest-robots-generations-uses/.

Quarantelli, E.L., 1984. Organisational behaviour in disasters and implications for disaster planning. *Monographs of the National Emergency Training Center*, 1(2), 1–31.

Richardson, W., 1993. Identifying the cultural causes of disasters: An analysis of the Hillsborough Football Stadium Disaster. *Journal of Contingencies and Crisis Management*, 1(1), 27–35.

Scientist, F.O.A., 2012. *Excerpts on unmanned aircraft systems*, 158(16), 230–304 (SPAIN, 2009) Spain Ministry of Defense, 2009, *Security and defense: Safety from explosive devices*, Top center of national defense studies.

SQUIDBEE, s.f. Available at: http://www.libelium.com/squidbee/index.php?title = Main_Page

Still, K., 2012. Crowd disasters. Available at: http://www.gkstill.com/ExpertWitness/CrowdDisasters.html. Last accessed on November 16, 2012.

Turner, D., 1994. Resources for disaster recovery. *Security Management*, 57–61.

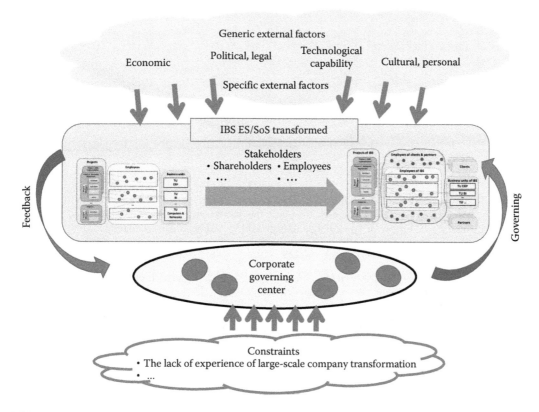

FIGURE 5.9 Context of ES/SoS.

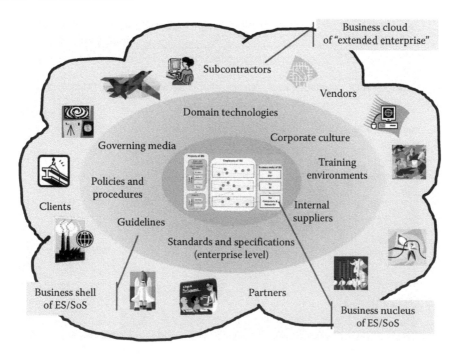

FIGURE 5.13 ES/SoS nuclear model.

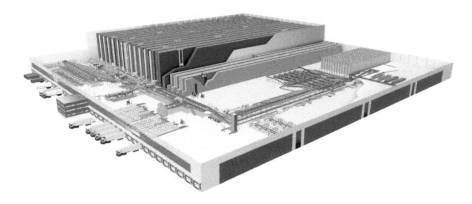

FIGURE 6.3 Artist impression of an entire DC. The blue area is the main pallet store, transport, and pick and place; the green area is a tote store for pick and place; the orange area is storage of oddly sized goods that are handled differently. (Courtesy of Vanderlande Industries B.V., Veghel, the Netherlands.)

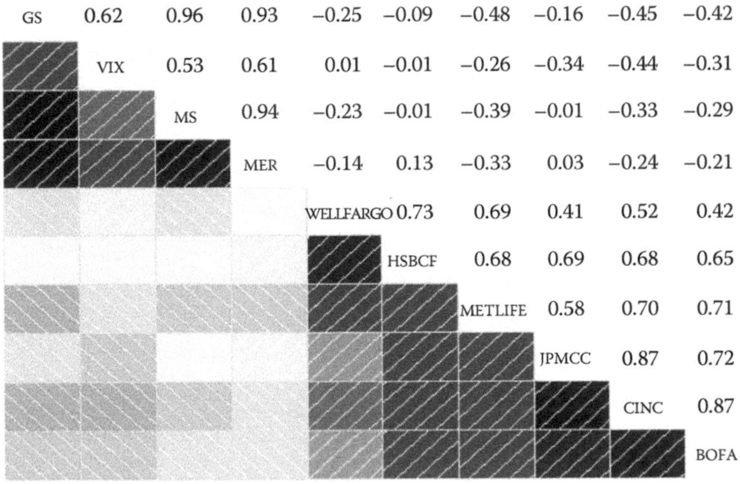

GS	0.62	0.96	0.93	−0.25	−0.09	−0.48	−0.16	−0.45	−0.42
	VIX	0.53	0.61	0.01	−0.01	−0.26	−0.34	−0.44	−0.31
		MS	0.94	−0.23	−0.01	−0.39	−0.01	−0.33	−0.29
			MER	−0.14	0.13	−0.33	0.03	−0.24	−0.21
				WELLFARGO	0.73	0.69	0.41	0.52	0.42
					HSBCF	0.68	0.69	0.68	0.65
						METLIFE	0.58	0.70	0.71
							JPMCC	0.87	0.72
								CINC	0.87
									BOFA

FIGURE 10.5 This CDS-VIX correlation heat map snapshot is based on market 100 measurements prior to August 9, 2006.

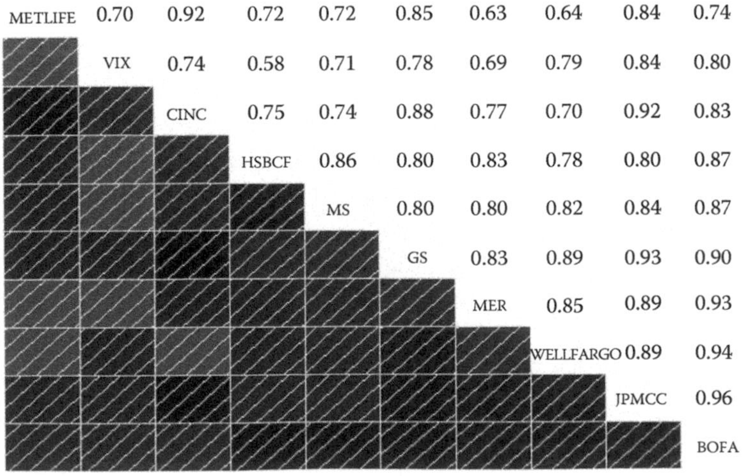

METLIFE	0.70	0.92	0.72	0.72	0.85	0.63	0.64	0.84	0.74
	VIX	0.74	0.58	0.71	0.78	0.69	0.79	0.84	0.80
		CINC	0.75	0.74	0.88	0.77	0.70	0.92	0.83
			HSBCF	0.86	0.80	0.83	0.78	0.80	0.87
				MS	0.80	0.80	0.82	0.84	0.87
					GS	0.83	0.89	0.93	0.90
						MER	0.85	0.89	0.93
							WELLFARGO	0.89	0.94
								JPMCC	0.96
									BOFA

FIGURE 10.7 This CDS-VIX correlation heat map snapshot is based on market 100 measurements prior to August 7, 2008.

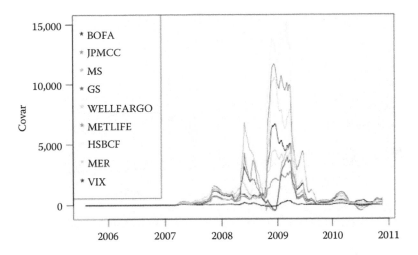

FIGURE 10.9 This graph shows the dynamic CDS-covariance of Bank of America describing its comovement with other components in the financial SoS.

FIGURE 13.1 Outline of the components of an emergency management system.

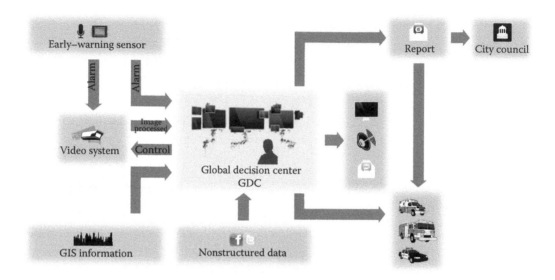

FIGURE 13.7 Control mass system layout.

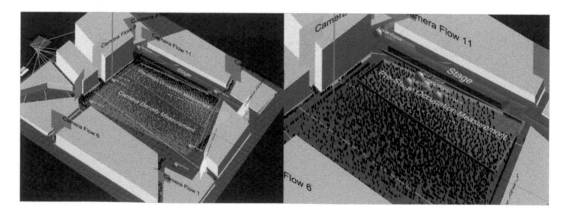

FIGURE 13.8 CAST simulation, Scenario event management. (From Airport Research Center GmbH. Control mass scenario, http://www.airport-consultants.com/index.php?option=com_contnet&view=section&id=8&Itemid=74.)

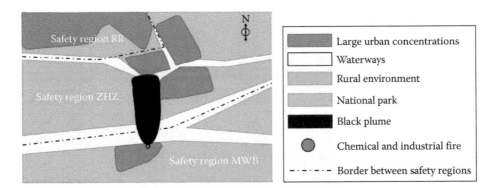

FIGURE 15.3 Location of the incident and effect of the black plume of smoke crossing the border of the safety region MWB into the neighboring safety region ZHZ and threatening urban population.

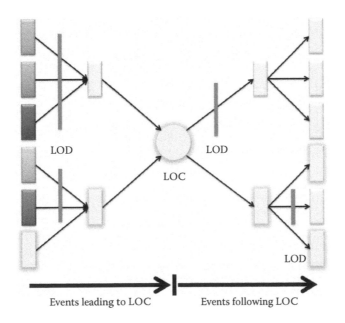

FIGURE 15.4 Bow tie model. LOC, LOD.

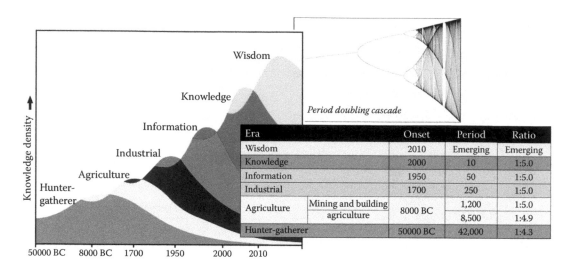

FIGURE 16.1 Waves of change, duration, and increasing frequency.

Era		Onset	Period	Ratio
Wisdom		2010	Emerging	Emerging
Knowledge		2000	10	1:5.0
Information		1950	50	1:5.0
Industrial		1700	250	1:5.0
Agriculture	Mining and building agriculture	8000 BC	1,200	1:5.0
			8,500	1:4.9
Hunter-gatherer		50000 BC	42,000	1:4.3

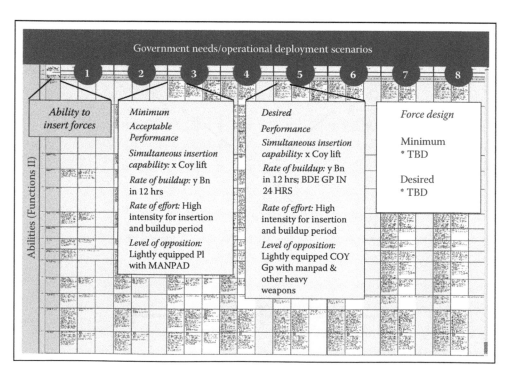

FIGURE 17.20 Strategic functional architecture for defense showing minimum and desirable parameters to guide design.

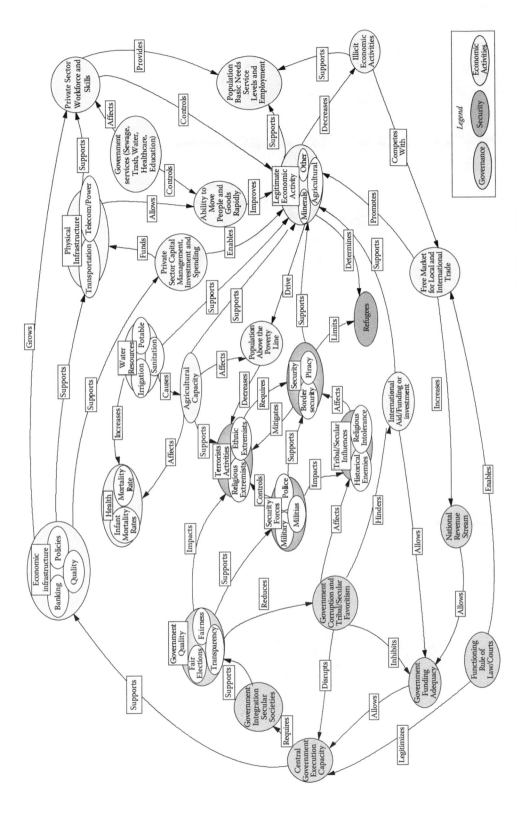

FIGURE 18.3 Systemigram showing some of the essential elements of country capacity for a country in sub-Saharan Africa.

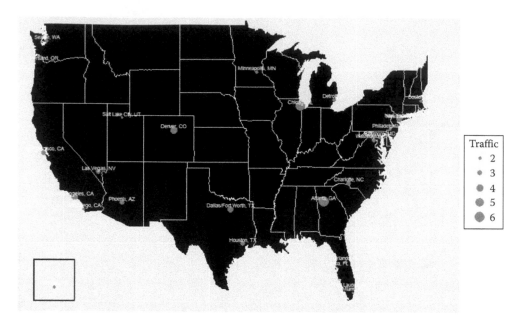

FIGURE 22.1   Major hubs in the United States and the percentage of total network traffic.

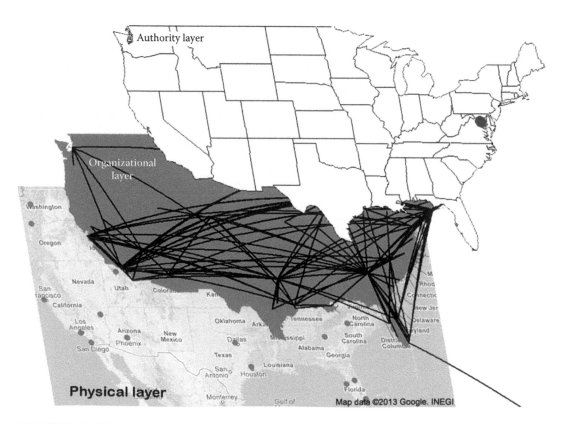

FIGURE 22.6   Three-layered framework for analyzing network industries.

14 Improvised Explosive Devices in Asymmetric Conflicts
Multisource Data Fusion for Providing Situational Information

Jürgo-Sören Preden, Leo Motus, James Llinas, Raido Pahtma, Raul Savimaa, Merik Meriste, and Sergei Astapov

CONTENTS

CASE STUDY ELEMENTS

- *Fundamental essence*: The case study describes an SoS solution development for improvised explosive device (IED) detection.
- *Topical relevance*: IEDs are one of the major threats in the operational theater, and there are no effective methods for IED detection. Therefore, new approaches for IED detection and defeat are sought.
- *Domain*: The case study falls in the military domain.
- *Country of focus*: Estonia.
- Stakeholders
  - European Defense Agency project ATHENA—countries involved—the Netherlands, France, Norway, Poland, Finland, and Estonia.
  - Primary/secondary institutions:
    Research Laboratory for Proactive Technologies, Tallinn University of Technology, Tallinn, Estonia
    Nederlandse Organisatie voor toegepastnatuurwetenschappelijk onderzoek TNO, Delft, the Netherlands
    EADS Defence and Communications Systems/Defence & Security Systems, Elancourt, France
    Forsvarets forskninginstitutt, Kjeller, Norway
    ITTI Sp., Poznan, Poland
    Military University of Technology, Warsaw, Poland
    VTT Technical Research Centre, VTT, Finland
    Center for Multisource Information Fusion, University at Buffalo, New York, USA
- *Primary insights*: It was shown in simulation that the nominated SoS architecture for the IED threat detection is viable. The experimental validation (and deployment) of the prototype architecture assumes development of the individual case study systems.

As the functionality (data processing capabilities, dynamic establishment of communication links, etc.) of these systems deviates from conventional functionality, substantial development effort is required.

KEYWORDS

Asymmetric threat, IED detection, Multisource data fusion, Proactive middleware

ABSTRACT

The system of systems (SoS) case study on improvised explosive device (IED) detection presents the problem of defeating IED threat in asymmetric conflicts. The Case Study describes an SoS solution for detecting the IED threat by applying multisource data fusion using data from a set of diverse information sources for identifying the locations with high probabilities of IED threats. The presented solution is an SoS where the individual systems act both as providers of data and consumers of data. In order to ensure correctness of data for the consumer in a dynamically changing system configuration, ProWare proactive middleware is applied. ProWare realizes a subscription-based information exchange model that makes use of validity information associated with the data to ensure data correctness. The SoS architecture concept is validated in a simulation of the IED detection scenario.

GLOSSARY

BML	Battle Management Language
C4ISR	Command, Control, Communications, Computers, Intelligence, Surveillance, and Reconnaissance
DIF	Data and information fusion
DoD	Department of Defence
HCI	Human Computer Interface
HUMINT	Human intelligence
IED	Improvised explosive device
ISR	Intelligence, Surveillance, and Reconnaissance
JDL	Joint Directors of Laboratories
JIEDDO	Joint Improvised Explosive Device Defeat Organization
KF	Kalman filter
MRAP	Mine Resistant Ambush Protected
NATO	North Atlantic Treaty Organization
ProWare	Proactive middleware
RCP	Route clearance patrol
REMBASS	Remotely Monitored Battlefield Sensor System
SA	Situation awareness
SoS	System of systems
UAV	Unmanned aerial vehicle
UGS	Unmanned ground sensor
UGV	Unmanned ground vehicle
XML	eXtensible Markup Language

BACKGROUND

Current and future conflicts are characterized by asymmetry—there is no clear battlefield any more, instead untraditional tactics and weapons are used in unexpected locations. An important weapon used in these conflicts is the improvised explosive device (IED)—an unconventional explosive weapon.

An IED is fabricated using destructive, lethal, noxious, pyrotechnic, or incendiary chemicals and designed to destroy or incapacitate personnel or vehicles. An IED is composed of a switch (activator), an initiator (fuse), container (body), charge (explosive), and a power source (battery). The types of IEDs and modes of activation range from very crude to rather sophisticated—radio detonated (using battery-powered doorbell devices, pagers, cell phones as senders and receiving units), wire detonated (using electrical wires between devices and/or leading away from the bomb), locally detonated (e.g., applying pressure onto a wooden plunger to detonate the device), with and without metallic parts—so developing one detection solution that works for all types of IEDs has not been possible and is still being researched.

There are several approaches for dealing with the IED threat. Typically, IEDs have a long supply chain, involving first the procurement of chemicals and the rest of the components needed to assemble these devices. As the life cycle of every IED device from assembly of the device to delivery for deployment is rather long, one can use different strategies for countering the IED threat at every step of the life cycle. One of the strategies is to collect intelligence on the entire network of suppliers, designers, developers, those who emplace the devices, etc., which is of course a huge undertaking requiring substantial resources. Realizing this, we deal in the context of the current case study with the bounded problem of IED detection and mitigation after the device has been deployed, which can be classified as a tactical problem.

IED Detection

IEDs have become a major threat and they will continue to be part of the operating environment for future NATO military operations. Therefore, technologies for countering IEDs have become and are becoming more relevant. IEDs are one of the main causes of casualties among troops the current conflicts and also take a heavy toll on local population in the areas of conflict.

Low-tech measures have usually proven more effective at detecting and defeating IEDs before they cause harm, according to published reports: trained dogs, local informants, and a trained marine or soldier's eye. No high-tech tactic has undisputedly emerged as a reliable means to consistently detect and defeat IED attacks. For example, dogs that are trained to recognize fertilizer smells can help detect IEDs in Afghanistan. Joint Improvised Explosive Device Defeat Organization (JIEDDO) officials estimate that 83% of IEDs used in Afghanistan are made with calcium ammonium nitrate fertilizer smuggled in from Pakistan.

The IED threat has evolved over time—in the United States—Iraq war mainly military-grade ordnance was used to construct IEDs, while in Afghanistan very simple methods and technologies are used to construct the devices (Murray, 2012). Officials have also noted that the 2011 IED attacks in Kenya, Nigeria, and Somalia conducted by al-Qaida-affiliated groups have shown an increased sophistication, including trying to create greater force through an explosion to penetrate armors.

Methods for Defeating IED Threats

Several methods for IED detection and defeat are being employed, involving various low- and high-tech technologies from dogs to conventional metal detectors to ground penetrating radar technology (Sprungle, 2008).

Specialized route clearance patrols (RCPs) are used in areas where IED probability is high and where troops and materials need to be moved, regardless of the threat. RCPs are able to detect and neutralize threats they encounter, clearing routes designated by the commanding officers in order to facilitate freedom of movement of coalition forces. With the new and emerging technology, all patrols and units have increased IED detection and defeat capabilities.

In addition to detecting IED threats directly by detecting the presence of chemicals or metals, indirect methods are also being employed, as noted by the US military personnel—one example is that IED detection can be performed by identifying changes in the environment (Murray, 2012), for example, artificial changes in vegetation and traces of digging possibly indicating a buried IED. Changes in the environment can be detected by various means, including unmanned aerial vehicles (UAVs) and other airborne assets. For example, detectors on aircraft, first used in Iraq, have successfully assisted troops in locating wires attached to bombs, which enable them to be defused.

In addition to the active countermeasures described earlier, passive countermeasures in the form of heavily protected vehicles are also used. An example of this technology is the Mine Resistant Ambush Protected (MRAP) vehicles, for which active development was started in 2007. It was known at the outset that MRAPs would not be a final solution to the IED problem due to several problems such as high cost, potential logistical difficulties (due to high fuel consumption and varied designs), and an increased disconnection between military personnel and the local population due to their huge size and threatening looks. While the protection offered by MRAPs is passive and limited (the adversary is always coming up with greater threats reacting to the protective actions), MRAP vehicles certainly form one of the building blocks aiding operations of countering the asymmetric threat—but they are by no means a complete solution to the asymmetric threat posed by IEDs.

Even in cases where MRAPs are used, the routes still need to be cleared for convoys. Very often the adversary tactics are such that the lead protected vehicles are allowed to pass the IED, and the IED is detonated when an unprotected vehicle passes it. Such tactics immobilizes the entire convoy and enables a direct attack against the immobile convoy by the adversary.

Considering the hurdles involved in IED detection and defeat and the low effectiveness of existing countermeasures, improved methods for IED detection are clearly needed. Especially desirable are methods that would offer greater probability of detection with reasonable requirements on resources and minimal human involvement.

PURPOSE

As established in the previous section, IEDs pose a serious threat to lives in asymmetric hostile scenarios. As IEDs vary substantially in the materials used, triggering, and concealment methods, there is no fitting universal method for countering IED threats. However, the need remains to provide the war-fighter with up-to-date situational information on possible IED threats. In order to accomplish this objective, information from a range of data sources available at the area of operations must be acquired, communicated, and fused to form a situational picture reflecting the IED threat. As the individual data sources are autonomous systems, the resulting information acquisition and processing system is an SoS. In this chapter, we present a case study as a framework for discussing an architectural solution for information exchange and processing in such an SoS.

IED THREAT EVALUATION

Information from a range of sources is required to effectively detect the IED threat. The information sources that are used in the case study are human intelligence (HUMINT), unmanned ground sensor (UGS), unmanned ground vehicle (UGV), and unmanned aerial vehicle (UAV). Information from every one of these complex systems is useful but not individually sufficient to identify an IED threat successfully. Instead, the (possibly contradicting) information from the individual sources must be fused to obtain a threat estimate. In addition to information fusion performed between the various information sources at a high level, information is also fused at a low level. The information fusion issues are discussed in more detail in "Data and Information Fusion" section. The information generated by the individual sources could be also evaluated individually either by a human or a machine. However, such an approach is not efficient as the total amount of information, if not collected in a discriminative manner, is quite high and will very likely result in overload of the data receiver. In the following paragraphs, we describe various characteristics of each of these data sources.

HUMINT

HUMINT sources (human observers) provide both human judgments as well as observations. This means that HUMINT sources can also provide information on the *hearts and minds* of the local population, the activities of people, and their opinions and views, which may contain information on the probability of an IED threat. This is an important distinction compared to conventional sensors that do not provide such judgmental insight. HUMINT data are typically received in an unstructured form—as natural language reports from the area of operation. Therefore, the unstructured data must be converted to a formal representation, which can be interpreted by a machine, before the information can be used in a data fusion process as an input. All HUMINT data must be also augmented with spatial and temporal validity labels, that is, when and for what area the information is valid. Previous work has shown (Rein and Schade, 2012) that converting reports in natural language to a formalized representation (e.g., Battle Management Language (BML), a constrained vocabulary evolving NATO standard command and control language; see Simulation Interoperability Standards Organization (2006)) is possible with automated tools.

This formalized information can then be further communicated to other systems to be used as an input. The system providing formalized HUMINT information can be viewed as one of the data sources in an information fusion system, although in reality, it may combine information from a range of individual sources.

UGS

UGS systems have been used for decades and they can be (and are being) used to detect a range of phenomena, depending on the type of sensors and processing capabilities available in a UGS device. A UGS device is essentially an autonomous computer-based system, equipped with a power supply, a set of sensors, and (typically) a wireless communication interface. The operational effectiveness of UGS can be potentially limited by a high false positive alarm rate. This problem can be alleviated by the use of efficient target detection algorithms executed in situ, which are able to discriminate humans from animals, mobile equipment by their type, etc. These objectives can be reached even better if the sensors are not used in a stand-alone mode (where decisions are made with limited information) but instead by performing intersensor data fusion. A substantial amount of work exists on this topic (Goodman, 1999; Arora et al., 2004; Li et al., 2006). Systems such as the Remotely Monitored Battlefield Sensor System (REMBASS (Hewish, 2001)) or Battlefield Anti-Intrusion System (BAIS) (Stockel et al., 2004) offer advanced remote monitoring capabilities (utilizing acoustic, seismic, and magnetic sensors). However, these systems are targeted for monitoring a dedicated area and have limited data fusion capabilities, having not been designed to be used as data sources for data fusion purposes. These systems provide early warning, intrusion detection, and threat classification of vehicles and personnel. However, UGS systems are not typically considered autonomous systems that are part of a greater data fusion network; instead, the UGS systems are viewed as mere data sources for tactical data from a limited area. In the context of the current case study, the UGS systems are considered autonomous systems that collaborate in the data fusion tasks. This makes the UGS systems rather complex (when compared to conventional UGS) as they have to perform a range of complicated and time-sensitive tasks, for example, sensor data processing, feature extraction, multisource fusion, and communication of data. The use of a proactive middleware as proposed in the case study description makes it possible to isolate different aspects of communication and computation, making it possible to realize the objectives described here.

UAV

UAV systems have seen a rapid development in the past decades. UAVs have been considered as communication nodes—range extenders for UGS deployments (Nemeroff et al., 2001)—but integrated systems where UAVs are part of a larger system have not been considered very much. UAVs have become bigger, and they are able to and do carry complex payloads—sensor packages on board of modern UAVs surpass sensor systems on manned aircraft of a decade ago (which is mainly due to the fact that the certification for systems to be used on board a manned aircraft is complex and time consuming, thereby hindering the quick proliferation of the latest technology for use on manned aircraft). So utilizing the sensor information available on board an unmanned platform, quite complex information can be inferred about the environment. Attempts in this domain are also being made in civil applications (Hardin et al., 2011). The NATO standardization agreement STANAG 4586 (NATO, 2005) defines the architectures, interfaces, communication protocols, data elements, and message formats for operating and managing legacy and future UAVs in an operational environment. As STANAG 4586 also specifies interfaces for direct receipt of Intelligence, Surveillance, and Reconnaissance (ISR) and other data, this also opens a path for UAVs to be used as data sources in a fusion process.

Data Fusion Operations

As noted in Figure 14.1, the IED detection/defeat network (only a portion of what could be a much larger network is shown) includes various data fusion operations. Local fusion can be occurring at the various sensing subsystems such as UGS and UAV nodes, as well as at designated fusion centers. In a truly decentralized topology, no overarching fusion center is specified, since these are single-point failures nodes. The nature of such possible fusion operations is discussed more fully in "Data and Information Fusion" section.

SoS Considerations

The considered IED threat detection system must be able to cope with a dynamically changing configuration of the system due to inherent uncertainties and the nature of the operations. One cannot assume that the system configuration is static, that is, the system components are fixed and that the interactions between the components are fixed. One does not know the configuration of the system beforehand; that is, it is not known what assets are in place, nor what are the exact configurations and capabilities of the individual systems that form the SoS. The resulting SoS must be also able to function if only some of the components or only parts of their capabilities are available.

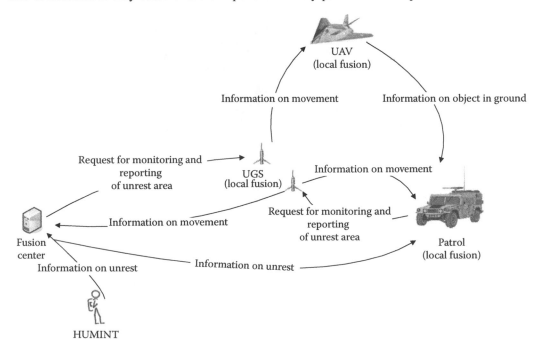

FIGURE 14.1   IED detection scenario.

This same functional requirement applies for the interactions among the components—the delays and interaction patterns are not known precisely, which means that the systems must be able to adapt according to the evolving interaction scenarios.

OPERATIONAL SCENARIO

The asymmetric threat detection scenario that will be discussed herein is based on the work done in the scope of the European Defence Agency's program "Defence R&T Joint Investment Programme on Force Protection" project (contract no A-0937-RT-GC) "Asymmetric Threat Environment Analysis (through Asymmetric Engagement Modelling, Modelling of Impacts on Hearts & Minds and Threat Scenario Generation from environmental factors)—ATHENA." In the Athena project, the Research Laboratory for Proactive Technologies at Tallinn University of Technology in Estonia was responsible for creating a threat detection simulation using a set of diverse information sources and making use of the proactive middleware (which we have called *ProWare*) concept. The simulation and demonstration done in the context of the project was focused on visualization of the operation of the middleware. The simulation has been extended later to illustrate the benefits of the chosen SoS engineering approach for communicating situational information. An example screenshot of the Athena third vignette (scenario) demonstrator is depicted in Figure 14.2.

The scenario describes a patrol vehicle starting from Base 1, moving along the track highlighted on the figure, past an airfield, and through an urban area past many threat zones. During the passage, the patrol vehicle obtains information from HUMINT sources on an elevated probability of an IED in an area where IEDs had been identified before. This information is used to cue the patrol vehicle, which then also subscribes to data from the UGS in the area and also

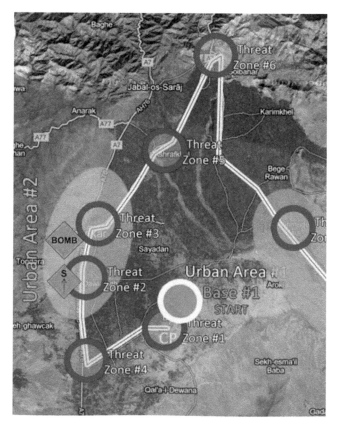

FIGURE 14.2   Project Athena vignette 3 demonstrator screen shot.

makes a request to a UAV for information on the same area. These different systems form an SoS in the course of the scenario and maintain the interaction patterns over the scenario period.

The UGS in the area of interest (there are also other UGS in the scenario) start providing sensor information to the patrol vehicle right away. The UAV starts providing information once it has reached the area of interest, and it is able to provide up-to-date situational information on the area. Using the information from these information sources, it is possible to identify the probable location of the IED and the potential IED can be neutralized.

SITUATION AWARENESS

As the concept of situation awareness is in the core of the case study, the basics of situation awareness (SA) are introduced. A general definition for SA is "the perception of the elements in the environment within a volume of time and space, the comprehension of the meaning and the projection of their status in the near future" (Endsley, 1988); notice that this is an inherently human capability. A human needs good SA in order to perform his or her tasks effectively. For most of history, acquiring good SA has been a responsibility of humans, gained by experience—what aspects to consider in a specific situation in order to form an adequate understanding of the situation and act accordingly. However, with advancement of technology, the tasks humans have to perform have become far more complex, and also the amount of information required for performing the tasks has increased.

While in the past a person may have lacked information required to produce good SA, in the present, the situation is the opposite—there is a huge amount of information. In many cases, the problem is also the presence of incoherent or contradicting data, which cause degradation of situation awareness, which in turn may cause suboptimal and even possibly dangerous decision making and action taking. The reason for degradation of SA in the presence of contradicting data lies also in the fact that people are usually not trained for such occurrences and therefore are unable to cope with such situations. With the vast amount of information that modern systems are able to provide, the recipient of information may be actually less informed than he was with less information because there is a gap between the amount of information disseminated and the recipient's ability to process the incoming information and combine the different bits of information for an objective assessment of the situation. The same principles apply both in case of biological (humans) and artificial systems—overloading a system with too much information will exceed the processing capabilities of the system, actually reducing the SA of the system.

Therefore, the information provided to the war-fighter must be carefully analyzed, and importantly and ideally, only the information relevant to a specific situation should be provided to the war-fighter. The objective is to provide the human situational information at a high level of abstraction required for fast decision making. Because in military operations the ability to obtain better SA could be a question of life and death (not only for the given war-fighter), enhancement of war-fighter SA is of high interest and has been studied quite thoroughly (Endsley et al., 2000; Strater et al., 2001).

The definition of a good automated system supporting the development of situational awareness is one that is providing the required information where one needs it and in the format one needs it. Good SA in an area of operations is required in order to enable the entities to perceive the situation, project the possible future scenarios, and select the optimal course of action that will result in the desired outcome. One of the purposes of an automated system for generating and delivering situational information to the war-fighter is to reduce the possible overload of the human caused by excessive amounts of data. By fusing the information automatically to an abstract level where it conveys more integrated situational components, the human is rescued from interpretation and low-level information fusion tasks.

While the classic approach to SA defines a three-level model (perception, comprehension, and projection [Endsley et al., 2000]) for generating the situational information, no established models exist for distributed generation of situational information. In order to generate these data in a distributed manner from distributed sources, we need an architecture that supports the exchange and processing of situational information at different levels of abstraction, while ensuring data validity and integrity and minimizing traffic.

DATA AND INFORMATION FUSION

Early developments in the field of *fusion* centered on inputs from electronic sensor systems and so the early characterizations of this estimation process were labeled as *sensor fusion* or *data fusion*. More recently, the spectrum of possible input data has grown dramatically to include so-called *open source* information such as web-based information, as well as field and intelligence reports (HUMINT, as described earlier), a priori database information and the sensor data as well. The modern label for this process is *data and information fusion (DIF)*. As just noted (and is important to understand), DIF is an estimation process that uses these different inputs and a priori knowledge to develop *best* estimates of a situation or of an event, a behavior, or the attributes of some entity. The notion of *best* or optimal is usually couched in a statistical sense where the fusion process exploits all available information about the errors, reliability, or uncertainty in the contributing data or information. DIF can thus also be seen as an estimation process focused on uncertainty management and minimization.

The data fusion concept was first introduced in the literature in the 1960s, as mathematical models for data manipulation. First implementations were started in the 1970s in the United States in the fields of robotics and defense. In 1986, the Data Fusion Sub-Panel of the Joint Directors of Laboratories (JDL) was established by the US Department of Defense to tackle some of the main issues in data fusion and map the new field in an effort to unify the procedures and terminology. The applications of data fusion have been spanning a wide range of areas for decades: maintenance engineering (Edwards et al., 1994), robotics (Ayari and Haton, 1995), pattern recognition and radar tracking (Linn and Hall, 1991), other military applications (Harris et al., 1998), remote sensing (Bruzzone et al., 1998), and many other walks of life (e.g., traffic control, aerospace systems, law enforcement, medicine, finance, metrology, and geoscience). More specifically, the data fusion methods are applied in military contexts to solve the following tasks: automated identification of objects (targets), object (target) tracking, guidance for autonomous vehicles, remote sensing, battlefield surveillance, and automated threat recognition systems, such as identification-friend-foe-neutral (IFFN) systems (Hall et al., 1991). These applications and a structured characterization of the DIF process have been framed in a process/pedagogical model about DIF that was developed by the US JDL organization in what is now the well-known *JDL fusion process model*.

JDL Fusion Process Model

The JDL fusion process model serves as a reference model for DIF processing operations. It should be understood that it is not an architecture but really a conceptual and instructional model of the major functions of a DIF process. Formal taxonomies of DIF functional architectures were first documented by Waltz and Llinas* in what is generally considered the first integrated book on DIF (Waltz and Llinas, 1990). The JDL Data Fusion Working Group also developed a Data Fusion Lexicon, and the process model for data fusion was created (Kessler et al., 1991) in an effort to formalize the terminology related to data fusion. One version of the JDL model is shown in Figure 14.3. It can be seen that the inputs (on the left side) are generally labeled as *Sources* since, as noted previously, inputs may not only come from sensor systems. Note that there can be both a priori and/or online database information such as contextual data (e.g., terrain information) in a support database and various types of deductive a priori knowledge about the domain in a fusion database.

The JDL model basically is framed around a *divide and conquer* philosophy in which complex problems are broken down into conceptually understandable subproblems. This approach led to the naming of *levels* shown here that are abstractions of the major functional goals of DIF processing. Note that in each case the term *refinement* is used, to emphasize that most DIF applications deal

---

* James Llinas was a long-time member of the JDL.

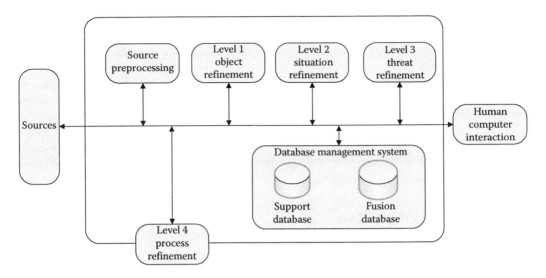

FIGURE 14.3   Nominal JDL fusion process model.

with streaming data and dynamic environments wherein all operations involve temporal dynamics and fusion estimates are being continually refined. An initial function is source preprocessing that reflects various operations that may have to be done on the raw input data to condition it for later operations. The goals or levels can be briefly described as follows:

- Level 1: Object Refinement—Following the divide and conquer idea, this is the level that focuses on the individual *piece-parts* of a situation such as the individual physical objects or more broadly the individual events and behaviors. The fusion goal here is to assemble the identity and attributes/features of these entities.
- Level 2: Situation Refinement—It can be argued that a situation is a collection of entities in a set of interrelationships; thus, Level 2 involves developing estimates of the interrelationships among the Level 1 entities. This can involve aggregation type operations such as estimating organizational affiliations among the Level 1 objects as well as many other relational structures or patterns or clusters. Situation refinement in this step focuses on relational information (i.e., physical proximity, communications, causal, temporal, and other relations) in order to determine the meaning of the Level 1 collection of entities.
- Level 3: Threat Refinement—Since this DIF model is defense oriented, the model includes the general military concern of estimating a threatening condition. The notion of threat is usually partitioned into the Boolean triple of capability–opportunity–intent, where these terms are associated to the adversary. That is, a threat arises if an adversary has a capability (to do harm), the opportunity to use that capability, and the intent in fact to use it. These estimates are developed from, for example, knowledge of adversarial weapon systems and tactics, hostile value systems, and various type of intelligence information.
- Level 4: Process Refinement—It characterizes what could be called the *control law* for the DIF process. This function may, for example, have a source management logic to dynamically alter the flow or characteristics of the source inputs in order to improve the DIF estimates or it may have threshold control/adaptation logic to alter certain thresholds within the DIF process itself (such as detection thresholds). One could also envision a knowledge management function that alters the knowledge base in accordance with the estimated situations being developed.

On the right side of the JDL data fusion framework diagram, the *human–computer interaction (HCI)* element is depicted. The HCI allows the human to input commands and requests for information to the system and also to communicate the outputs of the system, such as alerts and positional and identity information, to the user. For many modern DIF applications, no small part of the process design can involve architecting the human role (and human intelligence) into the process.

The *data management* element is an equally important support function required for supporting data fusion processing. In reality, it is a collection of functions that provide access to, and management of, data fusion databases, including data retrieval, storage, archiving, compression, relational queries, and data protection.

The JDL model was developed to be generic and is intended to serve as a basis for common understanding and discussion. The separation of processes into several levels is to some extent an artificial partition. An implementation of a real data fusion system may integrate and interleave these functions in an overall processing flow.

COMPLEX SYSTEM OF SYSTEMS

CONTEXT OF THE SOS

An overview of the complex system of this case study is depicted in Figure 14.4.

A—The SoS of the case study must serve the objective of IED detection in a tactical military environment. The selection of components of the SoS is dictated by the objective—individual systems must contribute to the enhancement of SA for entities

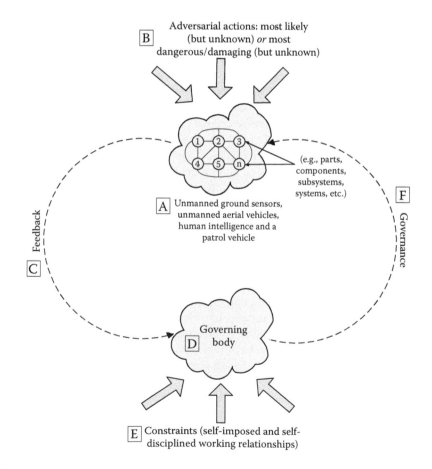

FIGURE 14.4 Context of complex system of systems.

in the IED detection scenario. The sensing systems involved are UGSs, UAVs, human intelligence, and a patrol vehicle. The SoS is able to serve its purpose by correct data exchange between the individual systems.

B—Certainly the main external factors in the application environment we are addressing here are the consequences of (at least partly unknown) adversarial actions; self-surveillance can reveal some but not all of these factors and effects, so SoS-level operations will be guided by decisions being made under this risk. Strategies to deal with these effects can include means to address those unknown behaviors that can be labeled as *most likely (but unknown)* or *most dangerous/damaging (but unknown)*.

C—Feedback to the governing body is largely provided in performance and quality of service assessments that are ongoing for the SoS. Measures include performance measures, quality measures, and security measures.

D—As the SoS is created at the level of a nation or a coalition, the specific governing body in the creation of similar systems depends on the specific national and/or coalition restrictions and requirements. The governing body for this project may be the *Systems Engineering Guide for System of Systems* [A] that identifies seven core SoS system engineering elements needed to evolve and sustain SoS capabilities. Of the four elements identified in [A], we see the following as guiding the design and development of our SoS application:
- Understanding the constituent systems and their relationships over time
- Assessing extent to which SoS performance meets capability objectives over time
- Monitoring and assessing potential impacts of changes on SoS performance
- Orchestrating upgrades to SoS
  These elements are associated with the evolution of the SoS in the sense of assuring specified performance and incorporating necessary changes and upgrades in this regard; they implicitly address SoS scalability.

E—Our working relationship with the governing body is essentially self-imposed and self-disciplined. The constraints mentioned here are those on the performing team that will experience the types of constraints listed.

F—As noted in E, as this is a self-disciplined process, the effects of the constraints have resulted in a weighted priority–based approach to achieve the overall SoS goals. That is, we have developed a weighting scale for performance, quality, and security measures and respond to any degradations in the SoS when the thresholds of these measures have been exceeded.

Typically, when an SoS solution is considered, a fixed system structure is assumed. A fixed system structure with defined subsystems and interaction patterns makes it possible to design a system that behaves in a predictable manner and is able to deliver repeatable results. However, such a rigid approach from the system structure point of view has several drawbacks and will also introduce issues related to performance, scalability, and robustness, which are difficult to overcome. Scaling a system with a fixed structure—adding new components to such a system—also involves the redesign of a system as processing of the data must be handled and the interactions between the systems defined. A system with a fixed structure is typically not robust (unless redundancy is built into the system) as it is difficult to facilitate automatic replacement of system components that are not identical to the original system components.

Alternatively, the SoS architecture proposed in the case study is a loose collection of systems. The systems form an unstructured network, where the availability, location, and capabilities of the individual systems are not known a priori (not at design time nor at deployment time). For performing a specific function, the availability of a certain set of systems is required. As different capabilities from individual systems are required for performing different functions, the set of systems that form an SoS may vary, thus the functionality of the SoS may also vary over time.

The composition of the SoS may be subject to availability of resources (e.g., a UAV may not be always available; thus, one cannot rely on the information from that system to be present every time the collection of systems is performing a specific task). Also in a tactical scenario, the resources may become unavailable due to a range of reasons (e.g., actions of the adversary, a higher priority demand for the same resource from a cooperative entity, end of service time, etc.); thus, also the performance of the system may vary. One can argue that if the system performance cannot be guaranteed then such a system is not very useful as it cannot be relied on. However, one must understand that the same applies for a system with a fixed structure in such an operational environment—if all the system components are not available, the system is also not functional. While by design the performance of an adaptive SoS (able to cope with changing structure) will degrade if some subsystems are not available, the system with a fixed structure will most probably fail completely.

In case of a loose system configuration as described earlier, we run into the problem of optimizing the system resources and configuration. The optimization questions being what data the individual systems should provide, to what systems the data should be delivered, what is the required fidelity of the data to be delivered, etc. One cannot assume that there is global information on the system and the individual systems do not have complete knowledge of the surrounding network configuration, the available systems, or their capabilities. So one can only apply a local optimization approach (handled by ProWare in the case study), where every system makes decisions based on the locally available information. In order to improve this quality of the decisions, the nodes must try to obtain all the information required for this. Part of the optimization problem involves the interaction patterns between the systems and the validation of data before consumption. As the system structure is not fixed, the systems must make sure that they are able to obtain the type of data required for fusion and that the data delivered also satisfies the (temporal, spatial, reliability, and confidence) constraints set by the given process.

Yet another issue in the described self-organizing SoS is the synchronization of clocks and other data alignment issues. Such a distributed SoS is not synchronized and cannot be realistically synchronized. The reasons for this are the intermittent connections between individual systems, the nondeterministic communication delays in the geographically distributed system, as well as clock jitter of individual systems. In addition, the time counting systems of the individual systems may be different; the systems must cater for clock drift, jitter, and offsets of the systems themselves and other systems.

To handle the problems that arise in building a loosely coupled SoS, we suggest the use of a distributed proactive middleware—ProWare. The proactive middleware has a role in intersystem communication, and in principle, it can also be applied in intrasystem communication. However, since intrasystem communication is defined more strictly, the uncertainties—therefore also the applicability of the middleware—are much smaller. ProWare handles the following tasks:

- Discovery of information providers for the individual systems
- Agreeing on data exchange
- Checking that the produced data satisfy the consumer requirements
- Delivering the data
- Validating that the delivered data are fit for consumption

The middleware makes it possible to decouple the data processing and fusion from the communication aspects, ensuring that only data satisfying the requirements of the algorithms are input to the algorithms. The middleware can be used to cater for the exchange of data at any level of abstraction and any validity requirements.

Using the JDL fusion model as a reference, we can say that ProWare makes it possible to distribute processing at all levels of the fusion process (and also between the different data fusion levels). Naturally, at different levels of the process, the data types, constraints, and metadata exchanged are different.

In the context of the case study, we view the individual systems as information consumers and/or information providers—any system can assume any role and also be in two roles at the same time, that is, consume information (e.g., from a sensor system) and provide information to other systems (e.g., after fusing information received from a sensor system with local sensor data).

In order to make it easier to understand what is the functionality of ProWare, how it is employed in the case study, and what are the concepts used, the individual issues are tackled one by one in the following sections.

SYSTEMS IN THE CASE STUDY

As stated in "Methods for Defeating IED Threats" section (IED threat evaluation), information is required from a range of sources to effectively identify and detect the IED threat. Information sources used in the SoS solution for IED threat detection are HUMINT, UGS, and UAV. Information from every one of these complex systems is useful but not sufficient to identify an IED threat successfully.

No fixed relationships or interactions patterns are foreseen in the SoS; that is, any system can interact with any other system, and the interaction patterns are formed based on the need of information that may arise dynamically. The system requiring information from the other systems needs to discover those systems that are able to provide the required information. This means that direct interaction between any two nodes (at the logical level, on network level the connections can span several nodes) in the network of systems is possible.

The information providers in the scenario are able to provide the data listed in Table 14.1, although they may be also able to provide other types of data, which is not relevant from the perspective of the current application.

The data sources must be able to describe the information they are able to provide, process source data, perform fusion locally, and be able to provide the data upon request.

The sensors are located on the systems providing information, and the entire sensor signal processing and conditioning is performed locally at the sensor system. Sensor types in the case study are the ones listed in Table 14.2.

TABLE 14.1
Types of Data by Provider

Parameter	Data Type
HUMINT	Elevated possibility of IED in area
UGS	Movement of vehicles in area
	Movement of people in area
UAV	Ground changed in area
	Unknown object in ground

TABLE 14.2
Sensors and Their Outputs in the Case Study

Sensor	High-Level Output	Sensor Location
Microwave (ground penetrating radar)	(1) Unknown object in ground	(1) UAV (2) Ground vehicle
Optical	(1) Unusual movement in the area (2) Unusual change of ground/area compared to stored	(1) UAV
Audio	(1) Movement of vehicles in the area	(1) UGS
Movement sensor (PIR)	(1) Movement in the area	(1) UGS
HUMINT	(1) IED danger in area	(1) HUMINT

SITUATION PARAMETERS

In order to have a common representation of the exchanged situational data, the concept of situation parameters (introduced in Motus et al., 2009) is employed. The value of a situation parameter can be computed from a single sensor signal source, fused from several sensor signal sources or fused from several situation parameter values. The concept of situation parameters enables hierarchical distributed detection of situations and in the estimation of high-level situations, exchange of intermediate processing results (situation parameters) between the individual systems. A situation parameter is the result of a single computation or a series of computations that can be interpreted autonomously and that can be associated with validity information, that is, temporal and spatial intervals where the parameter value is valid. So one can say that a situation parameter characterizes an aspect of a situation, is computed from data acquired from the environment on the situation of interest, and reflects a property of interest. As the computation of situation parameters is an ongoing process, a single parameter value is valid only for a limited amount of time. The computation must operate on a data stream and the processing chain is depicted in Figure 14.5.

The granularity of situation parameters depends on the nature of the computations, the sources of information, and requirements of the application.

The hierarchy of the situation parameters is depicted in Figure 14.6. One can map elements on the situation parameter diagram in Figure 14.6 to the JDL data fusion framework diagram presented in sections "Data and Information Fusion" and "JDL Fusion Process Model". The *source preprocessing* step in the JDL process can be matched to the perception step, the *object refinement* step and the *situation refinement* step can be matched to the comprehension step. Mapping the elements makes it simple to correlate between the two models and create a logical data flow, which can be interpreted in either domain. The use of the concept of situation parameters, which are exchanged with the validity metadata, makes it easier to achieve a deterministic result in case the processing is distributed.

Examples of situation parameters are features extracted from periodic signals, (intermediate) results of pattern matching for the purpose of detecting an object of interest, and detected objects or states of detected objects. Many of the situation parameters form quite clear hierarchies, for example, features extracted from a periodic signal are clearly an input to a pattern matching algorithm and the output of the pattern matching algorithm has clearly a higher level of abstraction than the output of the feature extraction algorithm. Other situation parameters may not form such clear hierarchies, and the fusion of these parameters may be highly application, algorithm, and situation dependent. For example, pattern matching results for identifying an object may be obtained using information from the visual domain or the electromagnetic domain (e.g., using synthetic aperture radar). So depending on the situation (e.g., presence of fog), the data fusion may use the situational information obtained from a camera or from an SAR or the information from both of these sources may be fused.

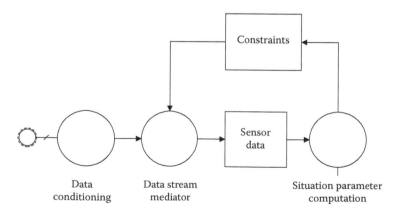

FIGURE 14.5   Data processing chain for situation parameter computation.

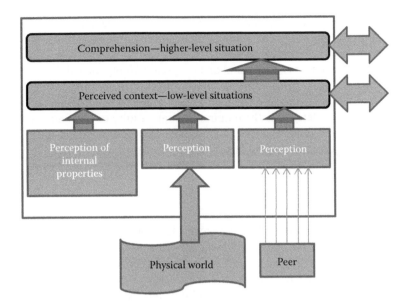

FIGURE 14.6 Hierarchical buildup of situation parameters.

The higher-level situation parameters can also be used as cues for the data processing routines to notify the system of events in the environment that need further attention.

In the context of the case study, we assume that every system that is used as a data source provides data already fused to situation parameter(s) locally using the data available from the local sensors. All situation parameters are augmented with validity information to ensure that only valid data are used in data fusion. This makes it possible for the information consumers to receive situation parameters. Consumer and provider roles make sense in the context of an individual interaction.

COMMUNICATION OF SITUATION PARAMETERS

As stated earlier, the main reasons for using the situation parameter concept lies in the need to communicate these data across the network using nonfixed system architecture, ensuring the usability (i.e., the validity) of the data for the data fusion process. Every situation parameter is augmented with validity information as its value is computed, where the computation of the validity of the parameter is based on the validity of the source information. The validity information describes different properties of data, for example, validity can be expressed in temporal, spatial, accuracy, or uncertainty domain. In the temporal and spatial domains, it is described for what area and in what time interval the data are valid, for example, for an object detected on a visual image, the validity information describes the coordinates of the objects and for what time interval that information can be expected to be valid (e.g., an object to be located in that area based on the known dynamics of the object). Clearly, the information is valid at the moment when the image was captured, but if there are identical subsequent images over a certain time interval, then that information is also valid for the entire time interval. Also, if there is additional information that there has not been any movement in that area for a certain time interval, the information may also be considered valid for that time interval. Different validity intervals depending on the type of the data are, for example, an increased IED threat, which may be valid for a few days, while information on a person or a vehicle moving in a certain area is valid only for the period of time when the sensor data indicate this. The validity depends both on the consumer and the provider of the information. The consumer may specify the interval when the data must be valid; the provider has the information on when the data are actually valid.

The systems computing only certain situation parameter values request only the data needed in the current fusion step, which helps to reduce bandwidth requirements for the overall system.

Other systems may require a wide range of situation parameters for data mining purposes. In either case, the specific source of the parameter is not of importance, but instead, the parameter type and its spatial and temporal validity intervals must satisfy the constraints set by the data consumer.

The properties of the data, that is, the spatial and temporal validity may also depend on the location and speed of the data consumer (in case the consumer is a mobile entity). Given the location of the consumer, the consumer sets constraints on the data it subscribes to. The location and direction of movement can be used to determine the required spatial validity of the source data, that is, the area where the information is valid. The speed of movement may dictate the desired temporal properties of the situational information—for a fast mover, information with a shorter validity period is important compared to a slower mover.

SUBSCRIPTION-BASED DATA EXCHANGE MODEL

A data subscription model is used in ProWare, where the data consumer subscribes to data (restricted by the constraints specified by the consumer) from the provider that is required by the consumer. In a subscription-based model, the data consumer sends out a subscription or request that specifies the type of data the consumer needs, but also the constraints on that data. The potential providers in turn check if they are able to provide that data and if so they respond to the subscription request with a positive acknowledgement. The message exchange with the subscription message and the following messages is depicted in Figure 14.7.

So in the model, the consumer specifies the constraints on the source data in the subscription, and the data provider makes sure it is able to produce data that satisfies the constraints. After the provider has agreed to provide the consumer with the requested information, it will either start producing the information (if it is something to be processed and can be done in a way that satisfies required criteria) or provide it directly to the consumer if it is data that are immediately available at the producer node. In either case, the validity information in the requested validity domains is also provided with the data.

As stated in the case study description, it is assumed that all data providers are able to provide data already abstracted to a situation parameter to reduce required communication bandwidth. This means, for example, that the UAV platform (or possibly its ground station) is equipped with advanced processing capabilities and is able to provide higher-level situation parameter values

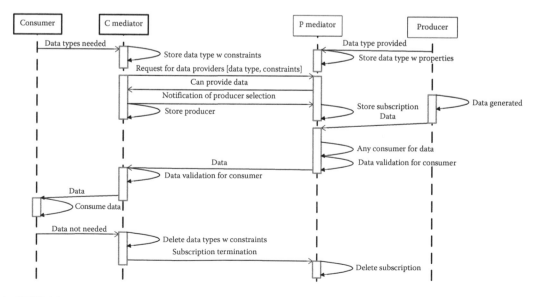

FIGURE 14.7 Message exchange in the subscription-based model.

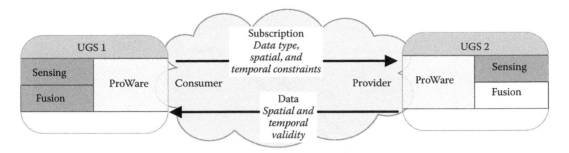

FIGURE 14.8   Data exchange between consumer and provider.

to the consumer. These situation parameter values may be the result of either on-board or off-board data fusion. The situation parameter value provided to the consumer system is in the form of changes of ground condition (which can be related to a possible IED) in a specific area. This information may be either derived from comparing historic visual data with current visual data from the area of interest or from information acquired with a ground-penetrating radar (where also current and historic information may be combined).

The concepts of data consumer and data provider have a meaning only in the context of an interaction. Any system can assume the role of the consumer or a provider, and a single system may have one role in one interaction and another role in another interaction. In Figure 14.7, the interaction between two UGS systems is shown, one being an information provider and the other a consumer. The cloud designates the interaction and the green highlights show what functionality is being used by the UGS system.

As stated earlier, a single system can have the role of an information consumer in one interaction and the role of an information provider in another interaction. In Figure 14.8, such a case is depicted. *UGS 2* on that figure is just providing information to *UGS 1* based on the constraints specified in the subscription. *UGS 1* is both obtaining information locally and also fusing locally obtained information with information provided by *UGS 2*. *UGS 1* is also providing the fused information to the patrol vehicle, assuming the role of a provider in that interaction.

The concepts of data consumer and data provider have a meaning only in the context of an interaction. Any system can assume the role of the consumer or a provider, and a single system may have one role in one interaction and another role in another interaction. In Figure 14.8, the interaction between two UGS systems is shown, one being an information provider and the other a consumer. The cloud designates the interaction and the green highlights show what functionality is being used by the UGS system.

As stated earlier, a single system can have the role of an information consumer in one interaction and the role of an information provider in another interaction. In Figure 14.9, such a case is depicted. *UGS 2* on that figure is just providing information to *UGS 1* based on the constraints specified in the subscription. *UGS 1* is both obtaining information locally and also fusing locally obtained information with information provided by *UGS 2*. *UGS 1* is also providing the fused information to the patrol vehicle, assuming the role of a provider in that interaction.

So the UGS sources are able to provide situation parameters, indicating the movement of both people and vehicles in the monitored area based on information acquired locally and information received from other systems. It has been shown in practice that the computation of such situation parameter values is feasible on embedded computing nodes (Astapov et al., 2012).

MIDDLEWARE FOR SITUATIONAL INFORMATION EXCHANGE

As stated in the SoS solution introduction, we assume an SoS with a nonfixed structure where the system configuration is determined at runtime. We have conceptualized the ProWare middleware

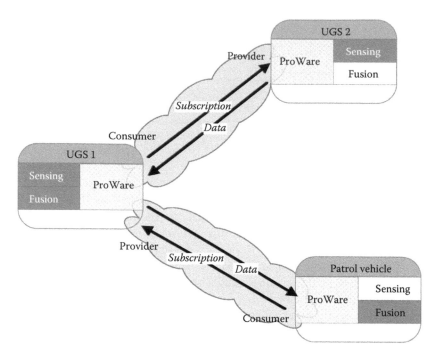

FIGURE 14.9   Data exchange with multiple roles.

structure to handle the tasks involved in making such a system work. ProWare must perform a variety of functional operations that lie in the interfacing domain between the elements of an SoS. Table 14.3 shows a list of various ProWare functional tasks in the SoS interaction domain.

As any network and system is limited by its resources and also by rules of operation, the problem of resource management according to constraints must be also addressed. As no node in the network has global information (attempting to acquire global information is not realistic in a nontrivial SoS), the system must be built in a manner where local optimization is sufficient for realizing the system objectives. So ProWare must optimize system operation at a local level using the provided constraints (Table 14.4).

TABLE 14.3
Middleware Tasks and Activities

Middleware (ProWare) Task	Middleware Role in the Interaction	Functional Activity
Locate information provider	Consumer	Send subscription request
Check if data with specified constraints is available	Provider	Check constraints against system capabilities
Responding to information requests	Provider	Check data properties against consumer constraints on data
Validate data before transmission	Provider	Check data properties against consumer constraints on data
Transmit information with validity data	Provider	Send data with metadata on validity
Align data to local representation	Consumer	Align temporal, spatial, engineering units
Validate data before consumption	Consumer	Check data properties against local constraints on data

TABLE 14.4
Optimization Tasks Performed by Middleware

Middleware (ProWare) Task	Middleware Role in the Interaction	Optimization Activity
Bandwidth	Provider	Data are sent only when needed
Processing power/feasibility	Provider	System evaluates if it is able to service the subscription requests, rejects if it is not able to
Data access rights	Provider	Authenticate consumer, access rights based on policies

TABLE 14.5
Types of Data by Provider

Parameter	Semantic Content of Data
HUMINT	Elevated possibility of IED in area
UGS	Movement of vehicles in area
	Movement of people in area
UAV	Ground changed in area
	Unknown object in ground

As ProWare must also handle the discovery of information providers for the consumers, this functionality in ProWare can be discussed (as one example) in the context of the contract net protocol (CNP) (Smith, 1980). The CNP defines a model of contractor and manager, where a manager is a network node that has a task that must be executed. The manager sends out bid specifications for the task, to which the contractors respond—the subscription model used in ProWare is similar to the CNP model. ProWare makes however some explicit additions to the CNP model by enabling also temporal constraints for the data requests. Of course, with the advent of more capable computing equipment, modern sensing systems are able to handle more than one information request in parallel, so the limitation of a single task per system is not valid. The authors of CNP suggest that once a contract has been made between a contractor (provider) and a manager (consumer), any application-specific protocol can be used for communication. ProWare extends that concept by explicitly including metadata with the application level data exchanged between systems, thereby enabling data validation and other functionality.

Validity constraints are specified on the source data by the data consumer in order to guarantee the validity of the situation assessment. The provider of the data must adhere to the constraints set by the consumer but in order to deliver the data, which satisfies the constraints, the concept of mediated interaction (Motus et al., 2009) must be used. The mediator in the current case is comparable to the theoretical concept of the channel function in the Q model (Motus and Rodd, 1994) as it forwards only data, which satisfies the constraints set by the final consumer of the data (i.e., the algorithm used for processing of situation parameters) (Table 14.5).

MIDDLEWARE IMPLICATIONS IN SUPPORT OF DIF TRACKING

From the perspective of the specific task of DIF tracking, ProWare can provide support in data alignment (temporal, spatial, and also type of data to some extent), relieving the fusion part of the system from these tasks. The alignment functions are embedded in the middleware, although the extent of them is limited.

ProWare can be also applied for realizing information sharing strategy (ISS) functionality, providing functionality to enforce access rights and manage bandwidth and loads of individual systems according to predefined policies and rules. Middleware can ensure that the capacity of the internodal communication bandwidth that is available is not exceeded as well as the computational capacity of the individual nodes.

DATA PROCESSING ON THE UGS NODE

As an example of the systems that make up the SoS, the operation and architecture of a UGS node is described in this section. Data processing on the UGS node considers a method of mobile vehicle identification based on acoustic signal analysis and the implementation of the method as a specific embedded system. The algorithm is designed as a multistage decision-making scheme, which involves frequency domain feature extraction, fuzzy classification, correlation analysis, and signal dynamics monitoring.

The state-of-the-art methods of vehicle identification based on audio signals implemented on resource-constrained systems consist of cluster analysis (Malhotra et al., 2008), processing stationary interval estimation using hidden Markov models (HMM) (Aljaafreh and Dong, 2010), application of artificial neural networks (ANNs) (Eng-Han et al., 2009), etc. However, the problem of unwanted noise separation is regarded rarely and superficially, thus the robustness of existing algorithms needs to be increased.

We make use of the recently developed algorithm in Astapov et al. (2012) on an embedded device that is to function as a subsystem of a larger system. The classification algorithm is specifically developed for time-critical operation; thus, its performance has been tested and the resulting statistics analyzed.

The algorithm operates frame-by-frame on the incoming signal. The acquired signal is stored in an input buffer, and frames of length N are input successively to the algorithm. A decision on the state of the environment is generated on the basis of every frame (Figure 14.10).

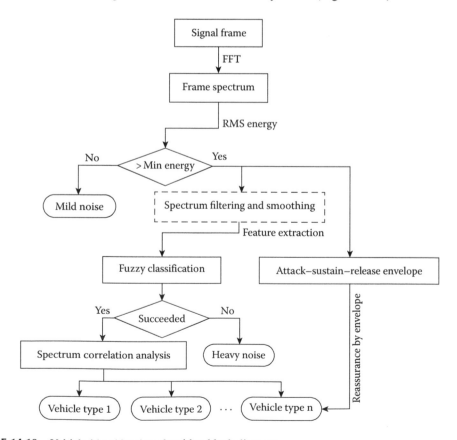

FIGURE 14.10   Vehicle identification algorithm block diagram.

If a feature with a sufficiently high energy is detected in the signal, the signal is analyzed for specific parameters that are relevant in establishing the vehicle noise pattern. A fuzzy classification algorithm is used to determine if the feature vector belongs to the L-dimensional feature space, in which the vehicle pattern features reside; the process is described in greater detail in Astapov et al. (2012).

Potentially, a large amount of information may be generated on movement of people and vehicles if it is acquired in a nondiscriminative manner from all available sources. As suggested in the case study (and made possible by the middleware), the UGS data sources are requested for information by the data consumer that estimates possible IED threats (e.g., a data fusion engine on a patrol vehicle) only if there is an indication that there is a higher probability of an IED in that area. Only after the indication for that fact has been received do the UGS sources start processing, storing, and forwarding the data. Only the UGS data sources that are able to provide the requested information from the area of interest respond to the request and start providing the data to the data consumer. This does not preclude the use of data from these UGS sources for other purposes, such as collecting information on general movement patterns of people in the area (which in turn may be used to estimate other parameters of the situation).

CHALLENGES

It can be appreciated that IED detection/defeat systems-of-systems involve complex components and complex algorithmic operations. In this section, we discuss the various challenges involved in the system of system solution design and development.

SYSTEM INTEROPERABILITY, INFORMATION EXCHANGE, AND KNOWLEDGE REPRESENTATION

Individual systems that form an SoS must be able to exchange data, both data for control and information management purposes and data for fusion purposes. An SoS is a set of cooperating autonomous systems, which may have been developed at different times and by different system providers. However, these diverse systems must still be able to interact in order to form an SoS and the interoperability must be addressed at different levels—syntactical, semantic, and ontological levels. As the systems are also distributed in the environment, the interoperability at a practical level also depends heavily on SA about peers and the environment.

The issues of information exchange between different architectures and systems have been a massive challenge for the last three decades. Problems typically arise when a perfect model that deals with the precise exchange of data among different systems and architectures is required for a successful operation of the system. In the context of military frameworks on a common battle field, exchange of efficient and precise information is the key for correct decision making. Data exchange within heterogeneous nodes has been and still is a challenge.

The interoperability of systems applied in the military domain is defined by the US DOD as the ability of different systems, units, or forces to provide services to and accept services from other systems, units, or forces and to use these services and to enable them to operate effectively together. In this context, the C4I interoperability includes both the technical exchange of information and the end-to-end operational effectiveness of the exchange of information as required for battle-mission accomplishment (Department of Defense, 2000). In order to understand the semantic interoperability issues within C4I systems, it is imperative to understand syntactic interoperability first.

The capability of two or more systems to communicate and exchange data is called syntactic interoperability, which can be enabled, for example, by specified data formats and communication protocols. Generally speaking, XML, SQL standards, or CASE Data Interchange Formats are standardized data formats that provide syntactic data interoperability. Syntactic interoperability is required for any efforts of further interoperability (Department of Defense, 2000).

Semantic interoperability is the ability to automatically understand the information exchanged meaningfully and correctly in order to produce useful results as defined by the end users of both systems. To achieve semantic interoperability, both sides must put into practice a common information exchange reference model. The contents of the information exchange needs are clearly defined: "what is sent is the same as what is understood" (Department of Defense, 2000). An alternative to using a common ontology across systems is also to use different ontologies at individual systems and to perform conversions between ontologies at interaction points.

In order to convey the information in a system, the system components must all use a common ontology for representing the concepts used in the system, or possibly disparate ontologies need somehow to be normalized. In order to overcome the interoperability challenge of mixed semantics, methods for connecting or integrating various ontologies through known preexisting ontology databases have already been developed (Collard, 2005).

A classification of ontologies is presented in Zhang et al. (2010). The definition of ontology can be borrowed from computer science and information science: ontology is a *formal, explicit specification of a shared conceptualization* (Gruber, 1993). The ontology system can be partitioned into four types of ontology that are domain ontology, mission ontology, task ontology, and service ontology.

The top-level ontology is the domain ontology that is used to define the C4ISR capability concepts, such as goal, behavior, and information. The mission ontology defines the processes typical for a mission. A mission associates a group of tasks that contribute to achieve the mission, and a task associates sequences of functions that can be described as services. The task ontology in turn specifies the tasks related to the domain and the associations between the mission and services. The service ontology describes the services that refer to the functions of a military unit. All of the ontologies are often described using XML.

High-level interoperability remains a crucial open problem—a reasonable solution is far beyond the current interoperability standards DIS and HLA. Establishing conceptual, semantic, as well as dynamic consistency between integrated models and simulations is a complex task, lacking common approaches and standards. In addition to information exchange between federated autonomous entities, we also have to solve the problem of conceptual alignment between the underlying data models in the context of particular information fusion tasks. Pragmatically, while looking for potentially ready-to-use components for an information fusion task, the decisions ought to be made based on knowledge about the interoperability of these components or assuming that the code can be easily modified to meet the interoperability requirements.

VALIDITY OF DATA

A critical issue in the distributed system is the validity of data—when the value of a data item is computed or when a situation is identified, it must be possible to estimate where and when that data item or situation is valid. This estimation must be performed based on the validity of the source data. Validity estimation starts from the raw sensor data, and a data item or situation inferred from this sensor data is valid only in the region where the sensor data are valid and also for the period of time when the sensor data are valid. So the properties of the sensor data that are monitored determine the validity of the data items and situations inferred based on these parameter values. A trivial example of this is weather monitoring—for most regions, we are able to predict the dynamics of temperature change and the area where the temperature is homogeneous quite well, so based on temperature measured in one spot at one specific moment in time, we can quite well estimate the temperature for the adjacent regions for some period of time.

An important factor of data validity lies in the fact that the use of data that are not valid (i.e., do not satisfy the temporal or spatial constraints of the data consumer) is in some cases even worse than not having the data if the consumer of the data is not notified. If any of the constraints on the source data is not met for the computed data items, these items should be associated with low-confidence estimations. It is up to the consumer of the data to decide if the data can be used or not.

The data items determining the situations are measured or computed by systems in the network, and the output information provided by each system is considered to be a stream. Each element of this stream must be tagged with the information, enabling cross-checking and validity assessment of its value, which enables the correct interpretation of the value of data items.

The tagged information accompanying data items is derived separately from the data item itself. Specific sensors, additional supporting hardware, algorithms, and computing power—for example, for time measurement and synchronization, for positioning, for estimating stress in humans, and for online estimation of communication delays—are often required.

FUSION CHALLENGES

Brief Fundamentals of Distributed Information Fusion

The typical architectural setting for a distributed information fusion (we use *DIF* here again but the *D* signifies distributed for this discussion) system or process involves a number of important design parameters that can greatly influence the nature of and complexities in various information handling and processing operations. It should again be emphasized that *fusion* is an (imperfect) estimation process that attempts to estimate either some attributes of an entity in a *world* of interest or the nature of possible events and activities or of the existence of certain organizations in that world. These estimates are formed from all available and relevant information that is often but not always dominated by multisensor-derived observational data on that world of interest and is also imperfect and perhaps contaminated by adversarial actions such as jamming and deception. As also pointed out before, other information such as contextual information or knowledge of policies and tactics can aid in forming these fused estimates. For noncooperative defense/military applications involving an intelligent adversary, the *true* world is strictly unknown. The typical task for any fusion process then is to form some type of *best* estimate regarding these entities, attributes, events, etc., and ideally to include in those estimates an estimate of the correctness of such estimates.

An important property of any DIF architecture is its topology. DIF systems can be designed in various graphically based topologies such as tree structures, star structures, and fully connected structures; all these topologies have various benefits and also disadvantages regarding DIF from different points of view. A crucial parameter in any DIF architectural design is the link or internodal communication bandwidth that is available. In an era of *big data*, there is a serious question as to what bandwidth may be required to exchange *raw* data across nodes in DIF architecture; this may also be an issue even for fused or processed data. Another design factor is the degree of knowledge any nodes have about their overall environment. In many applications, it is unreasonable for any node to have knowledge of the overall surrounding network and all the nodal properties and operations.

Fundamentals of any Fusion Process

As described originally by Llinas et al. (2009), any fusion process can be described by a *fusion node*, and any fusion architecture can notionally be described by a network of such nodes. The nominal fusion node is shown in Figure 14.11.

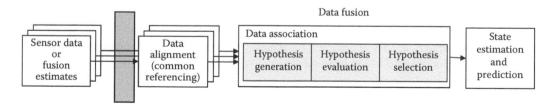

FIGURE 14.11 Nominal data fusion node.

Here, the inputs (in the DIF case, these can be inputs from a local or *organic* [belonging to the node] input, that is, a *within-node* input, or a communicated input from another node) are either sensor observational data or estimates formed at some predecessor node or process. These inputs typically need to be aligned or normalized, that is, put into a common referential framework, such as common coordinates, common time basis, and common units, among other such alignment operations. The next functional operation is called data association (DA). It is easiest to explain DA if we assume for the moment that some recursive fusion algorithm is estimating the states of various entities at some time t0. Assume too for the moment that all observational processes are synchronous and that the next batch of measurements is expected at t0. We now have a situation where there are a number of estimated entity states and a number of measurements to be used for state updating to time t0+. The question is: which measurements go with which entities that should be used for the next-time fusion-based state updating? DA techniques (there are many) are designed to be statistically optimal methods to determine the *best* allocation of measurements to state-estimating fusion algorithms so that the best set of next-cycle estimates can be formed. DA is characterized by a three-step operation:

- Hypothesis Generation: This is usually an offline (design-time) function that enumerates the feasible states to which measurements can be allocated or assigned (note that in this discussion the term *hypothesis* means an association hypothesis).
- Hypothesis Evaluation: Of the feasible possible assignments of the measurements to an entity (note that this really means the assignment of the measurements to the fusion algorithm for that entity), this step poses a *score* (typically a statistical score accounting for known sensor errors that produce any measurement) for all feasible assignments as a way to gauge their likelihood of being the best measurement to use.
- Hypothesis Selection: This is an algorithmic step involving combinatorial optimization methods to select, from all feasible scores of candidate measurements, the best measurements to give to the fusion algorithms for next-time-step updating.

The final step in this fusion node process is state estimation, where the fusion algorithm for any given entity or event, etc., is updated, using the assigned measurements coming out of the DA process. These state estimates are then forwarded to some recipient node in the notional architecture. Notice that these operations can be applied to any fusion level as in the JDL process model; the state of interest here can be a Level 1, 2, or 3 type state.

It should also be noted that additional types of information (besides observations/measurements) can be employed in the fusion process, such as contextual information. Such inputs still need to go through the common referencing and DA processes, but can be factored into modern fusion methods.

Some Complexities in Distributed Fusion Processes

There is one fundamental tenet about DIF that should be understood: any fusion node in a DIF system can only fuse two things: (1) the information that it gathers and manages within its own operational authority (usually called *organic* information, meaning that it is information that is an integral part of that node) and (2) any information that arrives at or is sent to the node from other nodes in the DIF system. Regarding the second item, information arrives at a node only in accordance with some type of what we will call here *information sharing strategy (ISS)*, that is, some inter-nodal data sharing policy that determines how information flows in the network (e.g., what node sends what information to specified nodes, in what form, how often, etc.); these are crucial issues that impact what fusion is happening at what nodes but is a design issue separate from designing the techniques for specific fusion processes.

Complexities in Distributed, Multisensor, Multitarget Tracking

Tracking of an entity—meaning the fusion-based estimation of its kinematic properties—is a Level 1 fusion process. Clearly, this is an important function in any military problem environment as

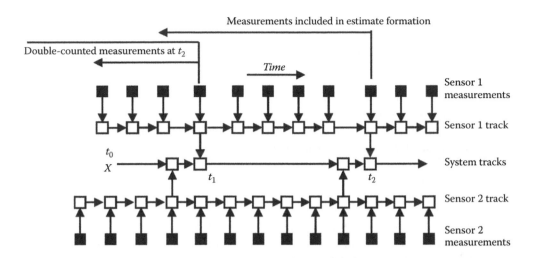

FIGURE 14.12 Graph for a 3-node DIF tracking process. (From Chong, C.-Y. et al., Distributed tracking in distributed sensor networks, in *American Control Conference*, pp. 1863–1868, 1986.)

it provides answers to a fundamental aspect of such problems, to include the IED problem, of where an object is at any moment (position), what direction it is traveling in (velocity), and whether it is maneuvering (acceleration). The tracking function can be related to the steps of Figure 14.11, where the inputs could be multiple sensor data from such sensors as radars on various UAVs in the IED problem situation. Assuming there are multiple moving objects on the ground of interest (multiple targets), each trying to be tracked, we have the classical fusion problem of multisensor, multitarget tracking to be solved. As most if not all military problems are temporally dynamic, fusion algorithms ideally need to be designed to be recursive so that they have an update-projection capability, that is, to provide an estimate at any given time while being able to project the future estimated state of an entity forward in time. The classical recursive algorithmic method used in tracking is the Kalman filter (KF), of which there are many problem and data-specific variants (Rong Li and Jilkov, 2003; Chong et al., 1986). Being recursive, the KF at any time has a memory of the effects of all prior inputs (sensor measurements). Given these remarks, let us pose a simple DIF tracking problem situation; consider the following *information graph* diagram (Chong et al., 1986) of Figure 14.12.

Shown here is a simple 3-node DIF architecture: the upper line depicts the multisensor fusion-based tracking at Node 1, the lower line measurements and fusion at Node 2, and the center line fusion of communicated track estimates from Nodes 1 and 2 at Node 3 (system tracks). Black squares depict local-node/organic sensor measurements; white squares depict the (3-step/Figure 14.12) fusion process that locally estimates the state of an object of interest. We assume all nodes are using KF-type algorithms. We assume too that there is some ISS policy that is governing the communication rate of track estimates from Nodes 1 and 2 to Node 3. Regarding this, we can see that there is a fundamental problem at Node 3 in that, at the second communication of track estimates from Nodes 1 and 2 to Node 3 at t2, there is a double-counting of the effects of prior measurements from the t0 to t1 interval. That is, the estimates communicated at t1 include the information related to the measurements over the interval t0–t1, and the estimates communicated at t2 include the information in the measurement interval t0–t2, thus redundantly communicating the effects of the t0–t1 interval at t2. Problems of this type can occur in any DIF topology that includes cycles or topologies in which loops between nodes are possible. This double-counting problem is one type of DIF issue for the Level 1 tracking case.

Notice that the case described here is what is usually called the *track fusion* or state estimate fusion case, where estimates are being communicated between nodes (i.e., not measurements), because it is usually assumed that internodal communication of raw measurement data would be prohibitive from a bandwidth-consumption point of view.

Another type of tracking complexity can arise in the case of DIF when performing track fusion. Again assuming the use of KF methods at any node, the so-called *process noise* in each contributing KF from each node (assuming they are indeed tracking the same target) is statistically correlated. To form the mathematically correct fused estimate at Node 3 in the previous example from the estimates from Nodes 1 and 2, it requires that Nodes 1 and 2 also communicate their respective covariance estimates for each state estimate. This again leads to the bandwidth issue, since now even to fuse estimates the contributing nodes need to send these matrices. These issues have given rise to the development of methods that can form reasonable fused estimates without accounting for the covariance information (e.g., *covariance intersection* methods (Julier and Uhlmann, 2009)). Studies have also been done to assess the effects of ignoring the covariance information in order to avoid the bandwidth issue.

The problems described earlier are just a few of the issues that can arise in DIF systems for target tracking as might be applied in the IED case study. These problems give rise to careful consideration of what type of metadata needs to be communicated between nodes to support correct fusion operations. Extensive discussions of these DIF tracking issues are included in Julier and Uhlmann (2009). There are a number of implications for the services and capabilities of a middleware system that can support correct DIF tracking operations; these are discussed in the following.

Complexities in Distributed, Multisensor, Multitarget Object Classification/Identification

Object classification is the process of providing some level of identification of an object, whether it is at a very specific level or at a general level. In a defense or security environment, such classes can range from distinguishing people from vehicles, a type of vehicle, or in more specific cases a particular object or entity, even to the *serial number*. By and large, an object's class is discerned by an examination of its features or attributes where we distinguish a feature as an observed characteristic of an object and an attribute as an inherent characteristic. Regarding information fusion processing, automated classification techniques are placed in the *Level 1* category of the JDL data fusion process model. In conjunction with tracking and kinematic estimation operations, they help to answer two fundamental questions "Where is it?" and "What is it?" Identity and/or class membership is determined from features and attributes (F&A henceforth) by some type of pattern recognition process that is able to assign a semantic label to the object according to the F&A estimates and perhaps some ontology or taxonomy of object types and classes.

In a broad sense, the basic processing steps for classification involve sensor-dependent preprocessing that in the multisensor fusion case includes the nominal standard fusion node processes of Figure 14.11: common referencing or alignment (for imaging sensors often used for classification operations, this is typically called coregistration), F&A extraction, F&A association, and class estimation. There are many complexities in attempting to automate this process since the mapping between F&A and the ability to assign an object label is a very complex relationship, which is complicated by the usual observation errors and, in the military domain, by efforts of the adversary to incorporate decoys and deception techniques to create ambiguous F&A characteristics.

Again because of bandwidth consumption/limitation concerns, it is typical in the classification/identification (C/I henceforth) case that nodes in a DIF architecture share locally developed (fused) C/I estimates* rather than raw data such as raw imagery or F&A data, etc. In these cases, we can identify various potentially problematical situations that need to be considered in DIF design for C/I.

Some Pathological Cases in Distributed Classification/Identification
Explicit Double-Counting

In this case, as can also happen in tracking, a node's estimate (here a C/I estimate) is sent out onto the network and it returns to the originally sending node as seemingly new information.

---

* Similar to the Track Fusion operations for kinematics; the term used for C/I operations is usually "Declaration Fusion" to signify that what is being fused are multiple, locally-developed declarations of a given object's C/I.

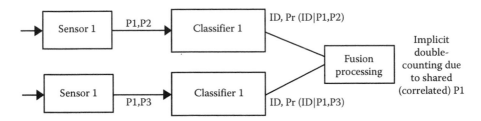

FIGURE 14.13   Notional idea of implicit double-counting.

This problem can be managed notionally by appropriate metadata tagging by all nodes as to the originating source node for any data but does need to be prevented.

Implicit Double-Counting

In this case, the local/sending node uses sensor-specific classifiers, and the performance of the individual classifiers is quantified using confusion matrices, resulting in the statistical quantification as shown (Prob [ID|Features]) in Figure 14.13, which are sent to the fusion node along with the declarations. Note that the sensors provide a common feature/attribute P1 that is used by both sensor-specific classifiers, resulting in the individual classifier outputs being correlated. The receiving fusion node operates on the probabilities in a Bayesian way to develop the fused estimate but may erroneously assume independence of the received soft declarations, and thereby double-count the effects of Feature F1, upon which both individual declarations are based. Pedigree tags would signify the features upon which the decisions are based.

Note that this problem is similar to the problem of correlated variances in the track fusion case.

DEVELOPMENT

While the objective of the case study was to validate the architectural solution proposed for the IED detection scenario, it is clear that showing the feasibility of the architecture using a real system would be too time and resource consuming. Instead, a simulation-based approach was selected, as many of the SoS aspects can be refined and validated in simulation. As an SoS can be viewed as a multiagent system from a simulation standpoint (every system in the simulation is autonomous, preserves an internal state, and changes its state based on information in the interaction streams), a multiagent simulator must be considered for simulating the behavior of an SoS successfully. While one can argue that a myriad of simulation tools and environments exist, among which there are many multiagent simulation environments that can be used for simulating SoS, the matter of the fact is that very few simulation environments are well suited for simulating dynamic SoS.

SYSTEMS ENGINEERING

Architecture

The crucial aspect of SoS engineering is the communication between the systems—while one can show the correct behavior of an individual system, the behavior of the communicating systems will be different. As the SoS under simulation will contain many different types of autonomous systems, the simulation environment must be able to support the simulation of autonomous systems and provide good control and observability of the communication between the systems.

Simulator Alternative Analysis

As stated earlier, a multitude of multiagent simulation environments exist (which can be used for simulating the autonomous behavior of the individual systems), yet none of them are well suited for simulating the communication aspects of the SoS. The main reason why these alternatives were not found to be fit for our purposes is that none of them are particularly focused on simulating communication or, more specifically, mediated communication. The other factor was that many use Java or less-known domain-specific languages and neither was an option for us. A short overview of some of the simulation environments considered is given in the following.

NetLogo

Perhaps the most famous member of this list, NetLogo is a programmable modeling environment for simulating natural and social phenomena. It is particularly well suited for modeling complex systems developing over time [4]. NetLogo is sufficiently advanced to be used on some of the real-world projects, but it was mainly developed to be used as a learning tool for students.

It uses a dialect of Logo programming language with added support for agent-based models to create and run the simulations. The application itself runs on Java. It has a nice set of tools to visualize the simulations. The core of NetLogo is free and open software; licenses for the numerous extensions by third parties vary.

Repast Simphony

This is the youngest member of the Repast agent-based modeling and simulation toolkits family (preceded by Repast for Java, Repast for the Microsoft.NET, and Repast for Python Scripting) [5]. Of all the alternative solutions, Repast might be the most advanced. Repast is really a general platform of modeling and simulation of multiagent systems and can be modified to do almost anything the user can imagine. In some cases, it might be also considered a weakness—the general approach oftentimes results in more work on fulfilling specific tasks.

Three of the most important Repast Simphony abstractions are (proto-)agents, contexts, and projections. Protoagents are modeling entities that maintain a set of properties and behaviors but need not exhibit learning behavior. If protoagents gain learning behavior, they become agents. These represent roughly the same ideas, albeit on a more abstract level, as agents, environments, and connections in MACE (description in the next section).

Repast runs on all of the modern computing platforms and uses Java and/or Groovy (an agile and dynamic language for the Java Virtual Machine) as the languages for writing the software of the agents. Repast is licensed under the *New BSD* style license.

MASON

Multiagent simulator of neighborhoods: MASON is a fast discrete-event multiagent simulation library core in Java, designed to be the foundation for large custom purpose Java simulations and also to provide more than sufficient functionality for many lightweight simulation needs. MASON contains both a model library and an optional suite of visualization tools for 2D and 3D [6].

MASON is, strictly speaking, an application programming interface (single-process discrete-event simulation core and visualization library), not an application itself. MASON is widely used in academia but the documentation of the library is quite minimal.

Some of the MASON features are visualization of independent models, check-pointed and recoverable models, guaranteed repeatability of simulations given the same parameters, self-contained models runnable inside other Java frameworks and applications, and possibility to take snapshots or video clips, charts, and graphs from the simulation.

Ascape

Ascape is yet another tool for developing and exploring general-purpose agent-based models. As many others, it is written in and uses Java as its main programming language. The agent is the basic structure in Ascape. All of the Ascape objects are formally Ascape agents. Ascape emphasizes (because of historical reasons?) that agent systems are a generalization of an earlier approach to representing interacting objects (e.g., particles, molecules) known as the cellular automata [7].

User must first write an agent class (inheriting from agent class) in Java and a model class (extending scape class) that acts as base (environment) of the system model. Agent scape must contain at least a description of what kind of agents it contains and in what order the actions of these agents are executed (can be both synchronous and asynchronous). While the model is written extending Ascape classes, the compiled outcome is not meant to be run as a standalone application. Instead, the user starts the Ascape application and opens the model for analysis and executing the simulation. Ascape is released under a BSD standard open source license—it is free to use and redistribute.

MASS

Multiagent simulation suite: MASS consists of four major components, which are centered around simulation core, largely based on Repast J (an ancestor of Repast Simphony). Even though the core uses Repast, MASS adds notable features of its own [8]. FABLES, a Functional Agent-Based Language for Simulations is an easy-to-use programming language and its integrated environment designed for creating agent-based simulations. Almost no programming skills are required to set up simulations, as it uses a lot of mathematical formalism instead of a scripting language. FABLES can be used to generate Java code for the simulation. MEME, Model Exploration Module of MASS is a tool to handle data gathered by running experiments on models. It is able to run Repast J 3.1 (FABLES), NetLogo 4.0.4, and pure Java simulations. It can be used to organize, merge, and split raw data in database. User can filter and transform data and get aggregated results over the data sets. Also featured are statistics and charting utilities. The participatory extension (PET) enables multiple users to run and participate in simulations (as nonartificial agents) by using a web interface while VISU (the Visualization Module), a (largely undocumented) part of the suite, supposedly helps to visualize the simulations and the results (charts, etc.).

MACE (Multiagent Communication Environment)

The MACE (Multiagent communication environment) simulator developed at the Research Laboratory for Proactive Technologies (Preden et al., 2013) specifically for the purpose of simulating system of systems behavior and communication aspects was used for the simulations. In the context of the simulator, the individual systems are also called agents as the simulator was developed using the agent-based modeling perspective. The agents are placed in the simulation environment, being invoked, controlled, and destroyed by the simulation environment.

The MACE framework allows setting up a virtual environment, where agents with software that is generated independently from the simulator environment may communicate with each other and the simulated environment via predefined interfaces. The mediators (middleware components) for the agents are designed, implemented, and deployed in the simulation as individual autonomous software components.

The simulation platform consists of a simulator application and a set of interface functions (an application programming interface or API) used by the agents in the simulation to communicate with the simulation environment. The simulator makes it possible to execute individual agents and mediators (in the form of C++ code) exhibiting unique behavior and to orchestrate their activities in a well-controlled environment, monitoring the communication between the entities in a great detail. All communication to and from the agents is controlled by the simulator, as well as the execution opportunities for agent internal actions. The simulator provides also clock events information to the agents, but no central clock is used—every agent is responsible for maintaining a local clock based on the clock events received from the environment.

The MACE simulator application is implemented in C++, relying heavily on Boost library. The MACE simulation environment has been developed on Windows (MinGW environment), but it can be ported to Linux fairly easily if needed. What is more important—the agent and mediator software can be developed in any programming language or combination of languages that allow the final result to be compiled as a shared library (in case of Windows, it is DLL—dynamic-link library). This includes practically all of the well-known programming languages, allowing the user a lot of freedom in the choice of tools.

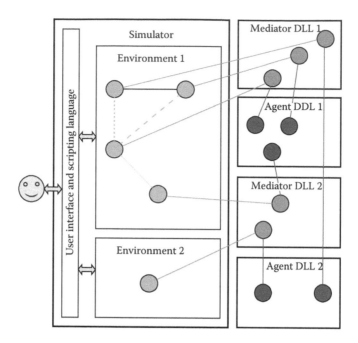

FIGURE 14.14   Relationship between agents and dynamic link libraries.

The main objects of the simulator of Figure 14.14 are as follows:

- Environment: the space and time in which agents act
- Agent: an autonomous entity in environment
- Mediator: an independent software layer between the agent, environment, and other agents
- Connection: a direct link between agents that describes the communication channel

Design

As every system in the case study was simulated as an independent agent, the behavior of every agent was specified—the input and output data types and the output data transformation rules were described. This means that for each system in the simulation—each agent (e.g., UGS)—a single DLL was designed, which enables the instantiation of the agents from that DLL (e.g., all the UGS systems in the simulation). In the same fashion, the mediators for the agents were designed and implemented as individual DLLs. This approach makes it possible to maintain an internal state for every agent and mediator in the simulation, making the behavior of the agents situation dependent.

Due to the fact that the simulation was focused on the communication aspects of the system, the data created by individual agents were prespecified as well as the data generation intervals. In the same manner, the temporal constraints (i.e., the maximum age of data) on consumption of data were specified at design time.

The design phase also incorporated the design of the interfaces and the semantics to be used in the communication, which is required to ensure compatibility between individual systems and what makes later integration possible.

Implementation and Integration

Once the behavior of the agents had been specified, the functionality of the agents was implemented in the C++ programming language in DLLs. Agent and mediator DLL templates were used in development to reduce the workload involved. The behavior of individual entities was validated to confirm to the specification before the agents were integrated to form a system of systems.

The agents were integrated into a system in the simulation by instantiating the agents and enabling communication between the agents. As the interface definitions of the individual agents were well defined and the interfaces were implemented correctly, the integration did not pose serious problems.

Testing

The testing of the agent and mediator behavior was performed in simulations in the MACE environment. The behavior, internal states, and communication patterns of the agents and mediators was observed via the logging information stored in text-based log files created at each execution of the simulation. Several iterations of testing and bug fixing were performed.

RESULTS

The case study system architecture, including ProWare, was validated in the MACE simulated environment once the behavior of all the systems involved in the simulation confirmed to their specifications.

BANDWIDTH REDUCTION

The computation and communication intervals of data provided by distinct data providers are listed in Table 14.6. In order to ensure correctness of data acquisition (regular sensor sample intervals, sensor data preprocessing), the data providers generate data regardless if the data have been subscribed to or not, and the mediator of the data provider decides if the data are to be forwarded to the data consumer.

The systems in the simulation were able to communicate with each other directly. Direct communication ability may not be realistic in a real application on a physical level; however, direct communication in higher networking layers is feasible and the details of network routing are not of importance in the context of the current simulation, as the purpose of the current simulation is to demonstrate the issues related to the operation of the middleware in an SoS context. The configuration of the network is depicted on the simulator screenshot in Figure 14.15, each node in the graph designating one system and each edge designating a link between the systems. Such an abstract representation of the system is appropriate in the current simulation as the simulation is focused on the interaction aspects, not the specifics of the individual systems.

The simulation was executed for 100, 500, 1,000, 5,000, and 10,000 time units. Each simulation featured a single high-level data consumer in the network. At the start of the simulation the data consumer made a subscription to the HUMINT data source. When in the course of the simulation information on a possible IED in an area was received, also subscriptions to other data sources were made. In the subsequent information requests the consumer specified that it required data from the same area where the HUMINT source had indicated the possibility of an IED.

For every situation parameter value generated by a data provider, a data packet was generated with the validity information; this data packet was evaluated by the mediator of the data provider to

TABLE 14.6
Types of Data by Provider

Provider	Reporting Interval	Number of Providers	Data Type
HUMINT	5	1	Possible IED
UGS	5	10	Movement of vehicles
			Movement of people
UAV	5	1	Ground changed in area

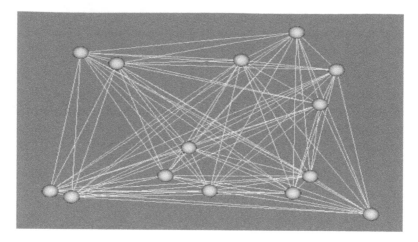

FIGURE 14.15   Visualization of the network configuration in the simulation.

evaluate if the data satisfied the consumer's validity constraints. If the validity constraints were not satisfied, the data packet was dropped at the mediator.

In the evaluation of the simulation results, the total number of data packets generated by the data providers was compared to the number of data packets that actually satisfied the constraints specified by the data consumer and that were forwarded to the data consumer to be stored there for use in data fusion (Figure 14.16).

The left vertical axis and the solid line depict the total number of packets, while the right vertical axis and the dashed line depict the number of packets that were forwarded to the consumer.

As one can deduce from Figure 14.16, the number of data packets communicated from the providers to the consumers was reduced ten-fold by using the mediated interaction concept realized in the form of a proactive middleware. Naturally, the bandwidth requirement reduction depends also on the configuration of the system, the constraints set by the data consumers, and other aspects.

The simulation results show that the proposed architecture is applicable to the problem domain.

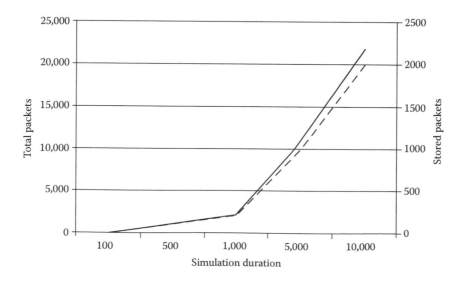

FIGURE 14.16   Total number of packets versus number of packets that reached and were stored by the consumer. The left vertical axis and the solid line depict the total number of packets while the right vertical axis and the dashed line depicts the number of packets that were forwarded to the consumer.

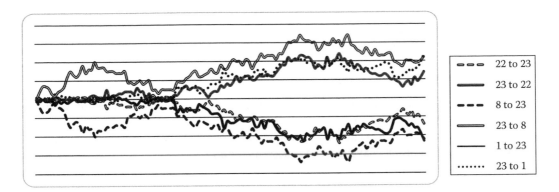

- - -	22 to 23
——	23 to 22
- - -	8 to 23
——	23 to 8
——	1 to 23
······	23 to 1

FIGURE 14.17 Diagram of clock differences observed and stored at mediators.

TEMPORAL ALIGNMENT OF DATA

It is clear that it is extremely difficult (if not impossible) to synchronize systems in a distributed SoS with dynamically changing configuration, the problem is especially acute in case of two SoS merging into one system due to mobility of individual systems or one system being part of two SoSs.

ProWare can be used to overcome the challenges that result from these problems by measuring and storing the clock offsets between individual systems (compensating for clock drift and jitter) and adjusting timestamps of received information based on known clock offsets before the information is delivered to the algorithm using the information as an input. The mediator at each communication entity is responsible for determining the clock offset between the current system and the system with which information is exchanged. See Figure 14.17.

In the simulation targeted for validating the principles of temporal alignment of data in an SoS, the clock offsets were determined by the individual mediators. The simulation spans over 2000 time units, which was sufficient for establishing the viability of the approach. All the individual systems that made up the SoS used local clocks, and the update events for the clocks were introduced by the simulation environment. Clock jitter of the individual systems was introduced in the simulation by adding a pseudorandom variable value (normalized to a value in the range of −1 to 1) calculated at each clock update to the current clock value of the individual system. Current clock value was communicated to the local mediator at each clock update.

Every line on the graph on Figure 14.17 illustrates the clock difference observed between two communicating systems. The clock difference was computed by the mediator associated with the system identified by the ID in the first position (i.e., in the graph *22 to 23* illustrates the clock offset between systems with IDs 22 and 23 as observed by the mediator of the system with the ID 22). Every mediator determined (and stored locally) the clock offsets with the other systems that communication was established with. Every time information was received from the system to which clock offset was known, the timestamps of the information were modified by the mediator to correspond to the clock of the local system. As the simulation results show, it is possible to determine and compensate for the clock jitter and channel delays in an SoS by using a proactive middleware. This makes it possible to design and run an SoS where individual systems are truly autonomous, that is, not synchronized to a central entity.

SUMMARY

Future SoS must be able to cope with dynamically changing system configurations, especially in tactical applications, such as IED detection. As one cannot expect that the future SoS will be designed as complete systems, rather the SoS will be formed in the field at runtime from the systems available at a given time in a given area, the SoS must be adaptive and be able to cope with dynamic configuration changes at system run time. The case study presented a self-organizing adaptive SoS solution where the use of the ProWare middleware enabled multisource data fusion between autonomous systems in a deterministic manner. ProWare enables information source discovery, data validation, and data alignment between individual systems. The presented approach enables creation of SoSs that are able to cope with dynamically changing system configuration in a deterministic manner.

CLASS DISCUSSION QUESTIONS

1. What are the benefits/drawbacks of SoS with a fixed structure, compared to an SoS with a dynamically changing structure?
2. What are the drawbacks of a centrally managed data fusion system in a tactical environment?
3. What are the data fusion challenges in the described SoS?
4. Why is data validation important?
5. What constraints besides temporal could the systems specify for the data?
6. What is the benefit of employing mediators in the SoS architecture?
7. Why is a simulation-based approach appropriate for validating the properties of the SoS architecture?
8. What is (are) the crucial aspect(s) in the design and implementation of a complex SoS?

REFERENCES

Alghamdi A.S., Siddiqui Z., and Quadri S S.A. A common information exchange model for multiple C4I architectures. In: *Proceedings of the 12th International Conference on Computer Modelling and Simulation.* Washington, DC: IEEE Computer Society, 2010, pp. 538–542.

Aljaafreh A. and Dong L. Ground vehicle classification based on hierarchical hidden Markov models and Gaussian mixture model using wireless sensor networks. *IEEE International Conference on Electro/Information Technology.* [s.l.]: IEEE, 2010, pp. 1–4.

Astapov S., Preden J.-S., and Suurjaak E. A method of real-time mobile vehicle identification by means of acoustic noise analysis implemented on an embedded device. In: *BEC 2012: 13th Biennial Baltic Electronics Conference: Proceedings of the 13th Biennial Baltic Electronics Conference.* Tallinn, Estonia: Tallinn University of Technology, 2012, pp. 283–286.

Astapov S. and Riid A. A hierarchical algorithm for moving vehicle identification based on acoustic noise analysis. *Mixed Design of Integrated Circuits and Systems*, 2012.

Ayari I. and Haton, J.P. A framework for multi-sensor data fusion. In: *IEEE Symposium on Emerging Technologies and Factory Automation.* [s.l.]: IEEE, 1995, pp. 710–713.

Bruzzone L., Fernandez, D., and Vernazza G. Data fusion experience: From industrial visual inspection to space remote-sensing application. In: *Academic and Industrial Cooperation in Space Research.* Vienna, Austria: ESA, 1998, pp. 147–151.

Cevher V., Chellappa R., and McClellan J.H. Vehicle speed estimation using acoustic wave. *IEEE Transactions on Signal Processing.* [s.l.]: IEEE, 2009, 57(1), 30–47.

Chong C.-Y., Chang K.-C., and Mori S. Distributed tracking in distributed sensor networks. In: *American Control Conference*, 1986. pp. 1863–1868.

Collard M. Ontologies-based databases and information systems. *Lecture Notes in Computer Science, Theoretical Computer Science and General Issues.* [s.l.]: Springer, 2005, 4623.

Edwards I., Gross, X.E., Lowden, D.W., and Strachan, P. Fusion of NDT data. *British Journal of NDT.* Mar 1994.

Endsley M.R. Design and evaluation for situation awareness enhancement. In: *Proceedings of the Human Factors Society 32nd Annual Meeting*, Vol. 1. Santa Monica, CA: Human Factors Society, 1988.

Eng-Han N., Tan S.-L., and Guzman J.G. Road traffic monitoring using a wireless vehicle sensor network. In: *International Symposium on Intelligent Signal Processing and Communications Systems*, 2009.

Frigo M. and Johnson S.G. FFTW: An adaptive software architecture for the FFT. In: *IEEE International Conference on Acoustics, Speech and Signal Processing*, 1998, pp. 1381–1384.

Goodman G.L. Detection and classification for unattended ground sensors. *Information, Decision and Control.* [s.l.]: IEEE, 1999, 419–424.

Gruber T. A translation approach to portable ontology specifications. *Knowledge Acquisition.* [s.l.]: Academic Press Ltd, 1993, 5, 199–220.

Hall D.L. and Linn R.J. Survey of commercial software for multisensor data fusion. In: *SPIE Conference on Sensor Fusion and Aerospace Applications*, Orlando, FL: [s.n.], 1991.

Hall D.L., Linn R.J., and Llinas J. A survey of data fusion systems. In: *SPIE Conference on Data Structure and Target Classification*, Orlando, FL: [s.n.], 1991, 1470, 13–36.

Hall D.L. and Llinas J. A challenge for the data fusion community I: Research imperatives for improved processing. In: *Seventh National Symposium on Sensor Fusion*, Albuquerque, NM: [s.n.], 1994.

Hall D.L. and Llinas J. An introduction to multisensor data fusion. *Proceedings of the IEEE*. [s.l.]: IEEE, Jan 1997, 85, 6–23.

Harris C.J. Bailey A., and Dodd T.J. Multi-sensor data fusion in defence and aerospace [Journal]. *Aeronautical Journal*, 1998, 102, 229–244.

Julier S. and Uhlmann J.K. General decentralized data fusion with covariance intersection. In: *Handbook of Multisensor Data Fusion*, eds. Liggins M., Hall, D., and Llinas, J. [s.l.]: CRC Press, 2009.

Kessler O. et al. *Functional Description of the Data Fusion Process*. Warminster, PA: Office of Naval Technol., Naval Air Development Ctr., 1991.

Li M., Lu Y., and Wee L. Target detection and identification with a heterogeneous sensor network by strategic resource allocation and coordination. In: *Sixth International Conference on ITS Telecommunications*, Chegdus, China, 2006, pp. 992–995.

Linn R.J. and Hall D.L. A survey of data fusion systems. In: *SPIE Conference on Data Structure and Target Classification*, 1991, pp. 13–36.

Llinas J. et al. Studies and analyses within project correlation: An in-depth assessment of correlation problems and solution techniques. In: eds. Liggins M., Hall D., and Llinas J. [s.l.]: CRC Press, 2009.

Malhotra B., Nikolaidis I., and Nascimento M.A. Distributed and efficient classifiers for audio-sensor networks. In: *Fifth International Conference on Networked Sensing Systems*, 2008, pp. 203–206.

Milner R.A. *Communicating and Mobile Systems: The π-Calculus*. Cambridge, U.K.: Cambridge University Press, 1999.

Motus L., Meriste M., and Preden J. Towards middleware based situation awareness. In: *Fifth IEEE Workshop on Situation Management, Military Communications Conference*, 2009. Boston, MA: IEEE Operations Center, 2009.

Motus L. and Rodd M.G. *Timing Analysis of Real-Time Software*. Oxford, U.K.: Elsevier, 1994.

Murray W. IED detection/defeat, new anti-IED technologies emerge to counter cunning enemy. SOTECH—Special operations technology. [s.l.]: KMI Media Group, 2012, Vol. 10(5).

NATO STANAG 4586 Edition 2. Standard Interfaces of UAV Control System for NATO UAV Interoperability. [s.l.]: North Atlantic Treaty Organization, 2005.

Nemeroff J. et al. Application of sensor network communications. In: *Military Communications Conference, 2001(MILCOM 2001), Communications for Network-Centric Operations: Creating the Information Force*. [s.l.]: IEEE. 2001, Vol. 1, pp. 336–341.

Peeters G. A large set of audio features for sound description (similarity and classification) in the CUIDADO project. [s.l.]: CUIDADO I.S.T., 2004.

Rein K. and Schade U. Battle management language as a "Lingua Franca" for situation awareness. In: *2012 IEEE International Multi-Disciplinary Conference on Cognitive Methods in Situation Awareness and Decision Support (CogSIMA)*. New Orleans, LA: IEEE, 2012, pp. 15–21.

Riid A. and Rustern e. An integrated approach for the identification of compact, interpretable and accurate fuzzy rule-based classifiers from data. In: *15th IEEE International Conference on Intelligent Engineering Systems*. [s.l.]: IEEE, 2011. pp. 101–107.

Rong Li X. and Jilkov V.P. Survey of maneuvering target tracking. Part I. Dynamic models. *IEEE Transactions on Aerospace and Electronic Systems*. Oct 2003, 39(4), 1333–1364.

Smith R.G. The contract net protocol: High-level communication and control in a distributed problem solver. *IEEE Transactions on Computers*. [s.l.]: IEEE, 1980, C-29, 1104–1113.

Sprungle R.J. Forward-looking lateral wave radar for ied detection and classification. Department of Electrical and Computer Engineering; Ohio State University. [s.l.]: Ohio State University, 2008.

Van Trees H.L. *Detection, Estimation, and Modulation Theory, Part I: Detection, Estimation, and Linear Modulation Theory*. New York: Wiley, 1968.

Van Trees H.L. *Detection, Estimation, and Modulation Theory, Part II: Radar-Sonar Signal Processing and Gaussian Signals in Noise*. New York: Wiley, 1971.

Waltz E. and Llinas J. *Multisensor Data Fusion*. Norwood, MA: Artech House, 1990.

Wright F. The fusion of multi-source data. *Signal*. 1980, 39–43.

Zhang Y. et al. C4ISR capability analysis based on service-oriented architecture. In: *The 2010 Fifth IEEE International Symposium on Service Oriented System Engineering*. Washington, DC: IEEE Computer Society, 2010.

15 Dutch Emergency Response Organization

Safety Region—Quantitative Approach to Operational Resilience

John van Trijp and M. Ulieru

CONTENTS

CASE STUDY ELEMENTS

- *Fundamental essence*: This case study is about elements making up operational resilience of a Dutch emergency response organization, a safety region. As an example, an industrial blaze at an industrial plant is examined.
- *Topical relevance*: To date, a quantitative approach for operational resilience in case of an emergency response organization has not been used before. This case study presents insight on a new technique that defines the way the organization behaves.
- *Domain*: Government.
- *Country of focus*: The Netherlands.
- *Stakeholders*: Passive primary stakeholders are (inter)national administrations and emergency response organizations of all kind: fire services, crisis management services, and services like US Homeland Security and Federal Emergency Management Agency (FEMA). Active stakeholders are directly found in the Netherlands itself as they are directly confronted with the results of the research performed. One may think of primary stakeholders like the Dutch safety regions and the Dutch local, provincial, and national administrations. The Dutch citizens are secondary stakeholders as they are confronted with any actions performed, without having a direct influence on the quality of the actions.
- *Primary insights*: A very important insight is that only very resilient organizations are capable of dealing with all the risks and hazards in an efficient way. Very resilient organizations should be excellent in managing awareness, keystone vulnerabilities, adaptive capacity, and quality.

KEYWORDS

Adaptive capacity, Awareness, Blaze, Emergency response organization, Keystone vulnerabilities, Management, The Netherlands, Threats, Operational resilience, Safety region, System of systems, Total of systems

ABSTRACT

Currently, emergency response organizations face tough challenges caused by nature, technology, or man. Many threats are fully capable of disrupting society on a local, national, or global scale.

To cope adequately with these challenges and threats, emergency response organizations should possess the right structure, so a minimum amount of (long-term) disruption of society may be achieved. This structure may be defined by resilience or more precisely operational resilience also known as organizational resilience. This chapter defines and quantifies operational resilience in relation to a Dutch Emergency Response Service: a *Safety Region* where the safety region acts during a crisis as a system of systems (SoS).

The case study involves a major incident in the Netherlands in 2011 (an industrial and chemical blaze) to determine the operational resilience of the safety region of middle and West Brabant (MWB) battling the blaze. The case study is based on data provided by official and independent incident reports.

In this particular case study, the operational resilience of MWB as an SoS was determined; the MWB operated at 18.10% of its capacity in the midst of a crisis.

GLOSSARY

A full glossary of abbreviations and acronyms is presented in the Appendix.

BACKGROUND

Emergency response organizations face tough challenges caused by nature, technology, or mankind. Some well-known challenges are, for example, hurricanes, pandemics, urban riots, food and water shortages, major power failures, terrorist threats, cyber security attacks on the Internet, earthquakes, large wildfires, flooding or drought, and industrial fires. Emergency response organizations can be seen as a system of systems (SoS) as they are a collection of systems that functions to achieve a purpose not generally achievable by the individual systems when acting on their own. Some examples from crises of recent history are hurricanes, for example, Sandy that hits the east coast of the United States (October 2012) and Katrina in the US Gulf of Mexico (2005); a 2005 industrial petrochemical blaze in Buncefield, United Kingdom; wild fires in British Columbia, Canada (2009), and California, United States (2010); global-scale cyber attack *Ghostnet* (Delbert and Rohozinski, 2009); large-scale urban riots in London, United Kingdom (2011); earthquake and the related tsunami in Japan (2011); and the earthquake in Christchurch, New Zealand (2011). All such threats are fully capable of disrupting society on a local, national, or global scale.

To adequately cope with these challenges and threats, all emergency response organizations should be capable of mitigating the degree and duration of disruption to society to the extent possible. This depends strongly on the way an emergency response organization is designed and operated that is mainly dependent on its structure as an SoS. This design and operation is defined by the operational resilience (Van Trijp et al., 2011, 2012) building blocks espoused in this case study.

PURPOSE

In this chapter, we analyze a major incident in the Netherlands in 2011 that involved a chemical and industrial blaze and how well the operational resilience of the Dutch emergency response organization called *safety region* was at the time. We will present a time line and other relevant facts of the incident for full understanding of some theoretical background and definitions on operational resilience. We also present an overview of what is understood by a safety region, how it works, its history, and the present-day challenges and role of operational resilience in relation to performance and needed capabilities. The results are of particular interest for fire services and related emergency response organizations.

The objective was to determine the operational resilience of the safety region battling the blaze in a so-called *quick scan* mode. This quick scan mode was chosen as the preferred method due to its relative simplicity in comparison to the full scan mode. The calculated results from this quick scan multiplied by two equal the results of a full scan complex method. Hence, using the quick scan method simplifies the amount of work to be done and still presents valuable results. This case study is based on the data provided through both official and independent incident reports.

PRINCIPLE OF A DUTCH SAFETY REGION

Due to new legislation in the Netherlands that came into effect on October 1, 2010, local fire departments, municipal medical departments, medical emergency services, etc., will be working together in a new entity called a safety region or security region. This complex organization finds itself in the midst of a great variety of stakeholders varying from media (including social media) and citizens on one side to influences from the environment like risks, crises, and public opinion on the other side (see Figure 15.1).

It should be clear from Figure 15.1 that there are numerous internal and external relationships: the internal relationships (inside the oval) are mainly those that are defined by law and based on influence, either directly like the board of a safety region or indirectly like the board of national police.

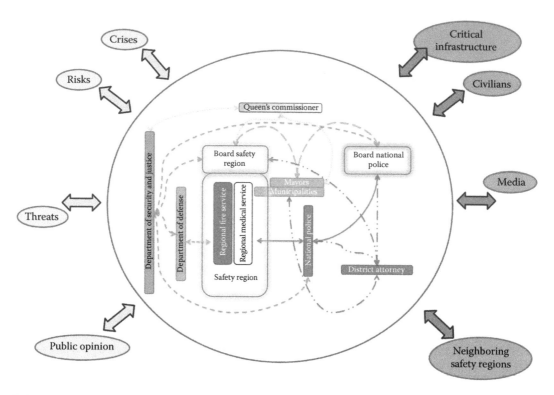

FIGURE 15.1   Internal and external relationships of a Dutch safety region.

Inside the safety region, one finds at least the regional fire service and the regional medical service, often supplemented with specific local staff of municipalities. The members of the board of the safety region are the individual (lord) mayors of the municipalities. The chair is normally held by the mayor of the so-called *center* municipality, which is usually the largest and most influential municipality in the region. This specific role is named *coordinating mayor*. In case of a crisis or emergency, liaisons provided by the police, the department of defense, and other vital service organizations may join at different levels of the crisis emergency staff in order to optimize the response (Van Trijp et al., 2011, 2012). Outside the safety region (outside the oval), one finds any possible kind of influence categorized as *risks*, *threats*, *crises*, and, last but not least, *public opinion*. Some examples of risks are changes in the environment that enhance the danger in this environment, that is, a new chemical plant or transports of hazardous materials. Threats may be anything that is not a risk but is still of influence on the integrity of the safety region. Good examples are a lack of funds or material. The meaning of crises is rather self-explanatory: think of a large-scale incident like the one in this case study—industrial blaze over a longer period of time with great impact on all stakeholders, the safety region included. The public opinion is of great importance as members of the public use new social media with instant messaging and services creating opinion that have the potential to go viral and but are not necessarily conform the current status of the crisis hence creating an alternative reality. The safety region has to deal with all the inner and outer influences and stakeholders before, during, and after a crisis.

By Dutch law, safety regions have to provide better protection of civilians from risks; offer better emergency management and aftercare during disasters and crises; act during emergencies as one administrative organization that coordinates and addresses the fire service, medical service, disaster and crisis control service, and the operational use of police; and enhance the administrative and operational mitigation capabilities (Anonymous, 2010). To meet these criteria and to deal with all these stakeholders and influences, the system needs operational resilience.

(COMPLEX) SYSTEM/SYSTEM OF SYSTEMS/ENTERPRISE

To combine Figure 15.1 and operational resilience (see later on in this chapter) into an SoS, we use Figure 15.2.

Figure 15.2 represents the following:

A: Our safety region is in fact the SoS in the midst of the crisis. Inside the upper cloud are the separate systems visible (1, ..., *n*) that make up the SoS. The separate systems consist of the fire fighting units, support units, command and control units, and communication. Next to these internal systems, there are also external systems, for example, research institutions, the national disaster support service, and the environmental service, which are added to the SoS on a temporary basis to act as part of the internal system. All the systems are visualized as being interconnected in the cloud. The descriptions and values of those connections are described in full in the section that covers organizational resilience.

B: The SoS is influenced by many external factors that are also valid when it is *peacetime* or in the absence of a crisis. Some of those factors are the Netherlands Act of Safety Regions,

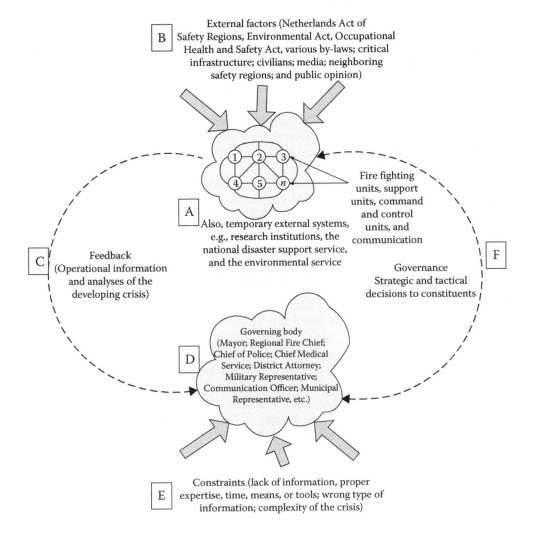

FIGURE 15.2   (Complex) system/SoS/enterprise. (Adapted from Gandhi, S. Jimmy, Alex Gorod, Brian White, Vernon Ireland, and Brian Sauser. 2013. Personal communication.)

Environmental Act, Occupational Health and Safety Act, by-laws regarding operability of fire services, agreements with other safety regions, and the Netherlands Department of Defense concerning assistance and support. Furthermore, extra factors are depicted in Figure 15.1: the presence of critical infrastructure (like power plants) and the presence of civilians (urban centers), media (TV, newspapers), neighboring safety regions (may be influenced by what an SoS does), and public opinion. One may argue the external factors are also system adding more complexity to the current SoS.

C: In the midst of a crisis, our SoS provides feedback to the governing body. The feedback comprises of operational information and analyses of the developing crisis. In this case study, the development of the blaze and the impact of it on the first responders, the environment, urban areas, and neighboring safety regions are studied.

D: As sketched, the executive leader of the governing body is the coordinating mayor, assisted by the regional fire chief, chief of police, chief medical service, district attorney, military representative, communication officer, municipal representative, and any extra representatives from other relevant organizations when called for.

E: This governing body has to deal with constraints that may influence the proper functioning. Some constraints are the lack of information, lack of proper expertise, wrong type of information, complexity of the crisis, lack of time, and lack of means or tools. Nevertheless, along the chain of command, strategic decisions are transformed into operational commands.

F: The operational data are received by the governing body upon which strategic and tactical decisions are made that are funneled down through the chain of command to the first responders on the scene or the support units, communication and information staff, and up the chain to national institutions and departments. Again, a very complex system that may be regarded as an SoS as they need to work together from their own expertise to produce new insights needed to battle the blaze and its consequences.

In this case study, the total of systems (ToS) as shown in Figure 15.2 is governed and described by organizational resilience and its attributes. The sections in which the organizational resilience of the SoS (safety region Midden-en West Brabant [MWB]) is analyzed give detailed information about the way the ToS functioned during the crisis.

EMERGENCY RESPONSE AND HANDLING OF AN INDUSTRIAL CHEMICAL FIRE

This case study is with permission by the respective organizations based on official reports generated by the Dutch Public Order and Safety Inspectorate of the Ministry of Security and Justice (IOOV) (2011, 2012), the Dutch Safety Board (OOV) (2012), the Netherlands Society of Fire Service and Crisis Management (2012), and the safety region MWB (2011). Some of the information of the Dutch Safety Board is also available in English at their website (www.onderzoeksraad.nl/en).

The reports were drawn up to evaluate the handling of a large blaze that occurred at an industrial chemical plant (Chemie–Pack NV) on January 5, 2011, in the industrial zone of the town of Moerdijk located in the south of the Netherlands. This location is part of the territory of the safety region MWB that is responsible for the emergency response in Moerdijk.

INTRODUCTION

At January 5, 2011, a large industrial chemical fire broke out at the site of Chemie–Pack NV. This plant was located at the site since 1982 and specialized in handling and packaging of large amounts of hazardous chemical products (see Table 15.1).

TABLE 15.1

Licensed Amount and Nature of Hazardous Substances Present at the Site of Chemie–Pack

Substance	Licensed Amounts
(Highly) flammable liquids	750 tons in packaging
	75 tons in storage tanks
Flammable solids	94 tons
Poisonous substances	750 tons in packaging
	75 tons in storage tanks
Corrosive substances	1100 tons in packaging
	75 tons in bulk
Miscellaneous hazardous substances	400 tons in packaging
	75 tons in bulk
Hazardous waste	35 tons in bulk

OVERVIEW AND LOCATION OF THE INCIDENT

The town of Moerdijk, where the incident took place, is located in the safety region MWB, on the banks of a major waterway and directly bordering the safety region, Zuid-Holland Zuid (ZHZ). The fire caused a large back plume of smoke that quickly left the service area of MWB entering the service area of ZHZ subsequently threatening major urban areas. See Figure 15.3.

The local fire service that is part of MWB was alerted at 14:26 h. The fire incident scene officer of MWB raised the coordinated threat level (GRIP) to "2," which was raised again to level "3" at 16:52 h, and to level "4" at 21:43 h, and lowered to level "3" at 02:19 h the next day. A fine explanation of the GRIP is presented by Wolff (2011). The minimum coordinated threat level "0" equals day-to-day operations while the maximum level "4" equals the situation where more than one safety regions are involved to cope with the incident. It may readily be understood level "4" has the highest complexity on all operational and strategic levels including the civil command structure supervising the fire operational command structure.

The neighboring safety region ZHZ, as the recipient of the black plume, was raised to level "2" at 14:50 h, immediately followed by level "4" at 15:43 h, and lowered to level "2" again at 02:19 h the next day. At a national decision level, the National Coordination Center (NCC) and the interministerial committee crisis management (ICCB) were involved. In the weeks following the incident, numerous institutes and institutions all over the country were involved in interpreting the incident.

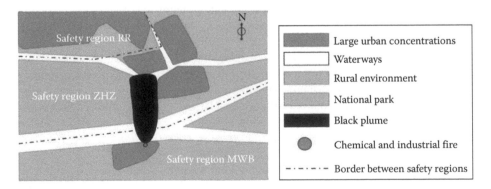

FIGURE 15.3 (**See color insert.**) Location of the incident and effect of the black plume of smoke crossing the border of the safety region MWB into the neighboring safety region ZHZ and threatening urban population.

At 00:15 h the next day, the fire was finally contained as no new outbursts of fire were expected. The aftermath of the incident lingered on until this case study was prepared (March–May 2012). Fortunately, no personal casualties were reported.

The entire plant of Chemie–Pack, including part of a plant of a neighboring maritime service provider, was destroyed in the blaze.

According to a fact sheet of MWB, the following figures can be presented to indicate the severity of the situation (2011):

The size of the fire pool was about 6500 m². The plant consisted of five big warehouses with hazardous chemicals. Outside in the yard, several hundred intermediate bulk containers (IBCs), each with 1000 L of flammable liquids, were stored. Also placed there was a shipping container with 80 drums totaling 16,000 L of acetone and a tanker truck with an additional 33,000 L of flammable liquid. The list documenting the great variety of hazardous substances comprised a total of 52 pages.

Four hundred fire fighters with 54 fire trucks battled the blaze, and the total number of emergency service personnel in action was 700. They were assisted by three crash trucks from a neighboring military air base, one fire service vessel, and a police helicopter. They used 14,000,000 L of water complemented with 18,850 L of foam-generating liquid.

The blaze took 10 h to contain, and the total cost of the entire operation is estimated to be $100 M.

OPERATIONAL RESILIENCE

This section introduces operational resilience and the (quantitative) status of operational resilience of the safety region MWB during the incident.

DEFINITIONS ACCORDING TO LITERATURE

In the literature, many resilience features are described. Some of those features are used to construct the survey that underlies this study. Brouns et al. (2009) present the following definition for resilience in relation to a network: "The social structure of a network determines resilience." In centralized networks, activity revolves around a small core group of people. Te Brake et al. (2008) describe as a major characteristic for resilience (for man) "to sustain normal development despite long-term stress or adversity." Wildavsky (1988) presented the following description: "The capacity to cope with unexpected dangers after they become manifest." Rutter (1985) states "Resilience is the potential (of organizations and individuals) to adapt to changing circumstances in the face of adversity, and the ability to recover after a disaster or other traumatic event."

Stolker (2008) describes operational resilience as "The capabilities of operational resilience in an organization are defined as: The ability of an organization to prevent disruptions in the operational process from occurring; When struck by a disruption, being able to quickly respond to and recover from a disruption in operational processes."

McManus et al. (2007) and Seville (2009) state "Resilience is a function of an organizations': Situation Awareness; Management of Keystone Vulnerabilities and Adaptive Capacity."

The authors finally conclude "An organization with heightened resilience is able to quickly identify and respond to those situations that present potentially negative consequences and find solutions to minimize these impacts. Furthermore, resilience enables an organization to see opportunities in even the most difficult circumstances that may allow it to move forward even in times of adversity."

Vargo and Seville (2008) combine the data from *Resilience is a function of...* into a diagram that bears a strong similarity to a bow tie model (see Figure 15.4).

The bow tie model obviously derives its name from the way it looks and is widely used in safety and risk analysis. The model was originally introduced by the hazard analysis department of Imperial Chemical Industries (ICI) in 1979 (Ale, 2009). The model shows on the left-hand side the causes (in fact a fault tree) and the consequences on the right-hand side (in fact a consequence tree) combined with lines of defense (LOD) that act as barriers and either interrupt the progression

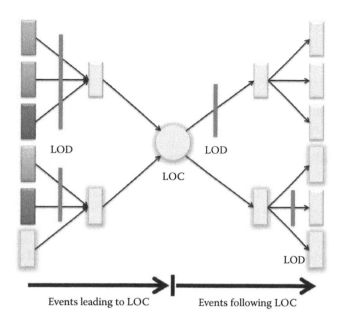

LOD

LOD

LOC

LOD

Events leading to LOC   Events following LOC

FIGURE 15.4   (See color insert.) Bow tie model. LOC, LOD.

of the causes leading to the loss of containment (LOC) or interrupt the consequences after LOC. In daily life, an emergency response organization working on the left-hand side of the bow tie model is applying processes like *prevention* and *risk control*. However, on the right-hand side, some known processes are *suppression* and *search and rescue*.

On the left side of the diagram (that represents *reduction and readiness*) prior to the event (LOC), Vargo and Seville position and rank (with weight-factor values from 0.00 to 1.00) factors like *situation awareness, management of key vulnerabilities,* and *adaptive capacity.* After the event on the right side of the diagram (that represents *response and recovery*), they rank factors based on organization culture and leadership.

According to Hollnagel et al. (2006), resilience may be found on the left and right side of the undesirable event in the bow tie model. Van Trijp (2012) postulates resilience as part of a learning feedback loop, which starts from the right side of the bow tie diagram and connects to the left side, offering the opportunity to learn from crises or emergencies.

In general, it may be concluded from literature that the concept of *resilience* can be best described by the generic approach *operational resilience.*
The generic capabilities of operational resilience in an organization are defined by Van Trijp et al. (2011, 2012) as

- The ability of an organization to prevent disruptions in the operational process from occurring
- When struck by a disruption, being able to quickly respond to and recover from a disruption in operational processes

To obtain and sustain these capabilities, the following four items from literature are considered of prime importance and a function of an organization's operational resilience (Van Trijp et al., 2011, 2012):

- Situation awareness
- Management of keystone vulnerabilities
- Adaptive capacity
- Quality

MODELING OPERATIONAL RESILIENCE

In 2009–2010, an extensive survey (Van Trijp, 2010) was conducted among major stakeholders of a Dutch safety region (see Table 15.2) to determine the intrinsic value *resilience* in case of an emergency response organization like a safety region.

In total, 454 (100%) requests (total subset) to fill out the survey were sent by regular mail, and 112 (24.7%) respondents (starter subset) started filling out the survey, and 84 (18.5%) made it through the entire survey (final subset).

The following preliminary objectives for the survey were formulated (see Table 15.3):

From the survey, all identified attributes (see Table 15.1 and Appendix for a full description) were ranked and sorted in a value tree where the most prevalent score receives the highest ranking or weight factor (1.00). Other attributes received lower scores and thus rankings or weight factors between 0.00 and 1.00. A method described by Goodwin and Wright (2004), based on the multiattribute utility theory (MAUT), is utilized in Figure 15.5.

In this value tree, the quick scan method uses all the attributes shown under *objectives* and the first two attributes shown under *performance measures*. The full scan method uses all the attributes instead.

According to the results of the survey, the identified attributes (these attributes make up the separate items of operational resilience as identified in literature) describing resilience R_{ero} are for the left side of the bow tie (see Table 15.4) and for the right side of the bow tie (see Table 15.5).

When we combine these results according to the value tree (see the right-most column of Figure 15.5) for operational resilience R_{ero}, we get Equation 15.1:

$$R_{ero} = \left(1.00c + 0.20a + 0.10d\right)_{Reduction+Readiness} + \left(0.70b + 0.30e\right)_{Response+Recovery} \tag{15.1}$$

See also Appendix for a glossary of used symbols for Equations 15.1 through 15.10.

TABLE 15.2
Major Stakeholders of a Dutch Safety Region Used as Expert Judgment Panel

Coordinating mayor/chair safety region
Managing director/chief executive officer safety region
Regional fire chief regional fire service
Chief medical officer regional safety service
Chief of regional police
District fire chief regional fire service
(Deputy) fire chief municipal/local fire department
Manager
Other functional titles

TABLE 15.3
Preliminary Objectives to Determine the Intrinsic Value *Resilience* by Means of a Survey

In what way is the theoretical concept from literature (see Paragraph 2.1) valid for an emergency response organization in general and a Dutch safety region in particular?
What are relevant key aspects determining *resilience*?
Is a quantitative measure of *resilience* possible/feasible?

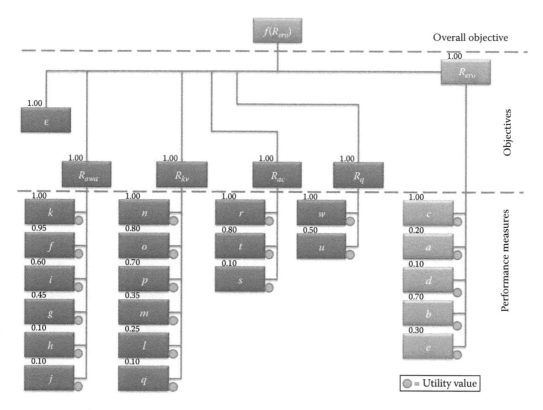

FIGURE 15.5 Value tree describing dynamic operational resilience $f(R_{ero})$ with weight factors (figures) and undetermined UVs (spheres). (Reproduced from Van Trijp, J.M.P., Ulieru, M., and van Gelder, P.H.A.J.M. Quantitative modeling of organizational resilience in case of a Dutch emergency response safety region. *Proc. Inst. Mech. Eng. Part O J. Risk Reliab.* 266, 666, 2012. With permission.)

TABLE 15.4
Reduction + Readiness (Variables in Arbitrary RU)

a The sustenance of normal development despite long-term stress or adversity
c The potential (of organizations and individuals) to adapt to changing circumstances in
     the face of adversity and the ability to recover after a disaster or other traumatic event
d The readiness of an organization before the shock or disruptive event

---

TABLE 15.5
Response + Recovery (Variables in Arbitrary RU)

b The capacity to cope with unexpected dangers after they become manifest
e The response of the organization after the disruption has struck

---

Equation 15.1 is an additive function as both sides of the bow tie are regarded to be of equal weight according to the concept of resilience as described by Vargo and Seville (2008). It should also be noted all attributes have as dimension arbitrary resilience units (RU). Quantification occurs by multiplication of the attributes with the weight factors and the utility values (UVs). The weight factors are shown as values on the top left-hand side of each square under the lower dotted line in the value tree (performance measures). The UVs are visible as the small spheres on the bottom right-hand corner of each square, and they can only be determined by conducting a full assessment or audit of the emergency response organization. These UVs may range between 0 and 1 (or any other acceptable scale like 0–100). In this case study, we use a range of 0–1.

In a similar way, the identified attributes *f*, *g*, *h*, *i*, *j*, and *k* describing awareness as a function of resilience (R_{awa}) are presented by the second column from the left that result in Equation 15.2 (Van Trijp et al., 2011, 2012):

$$R_{awa} = (1.00k + 0.95f + 0.60i + 0.45g + 0.10h)_{Reduction + Readiness} + (0.10j)_{Response + Recovery} \quad (15.2)$$

See Table 15.6.

The identified attributes *m*, *n*, *o*, *p*, and *q* describing keystone vulnerabilities as a function of resilience (R_{kv}) are presented by the third column from the left that result in Equation 15.3 (Van Trijp et al., 2011, 2012):

$$R_{kv} = (1.00n + 0.80o + 0.70p + 0.35m + 0.25l + 0.10q)_{Reduction + Readiness} \quad (15.3)$$

See Table 15.7.

The identified attributes *r*, *s*, and *t* describing adaptive capacity as a function of resilience (R_{ac}) are presented by the fourth column from the left that result in Equation 15.4 (Van Trijp et al., 2011, 2012):

$$R_{ac} = (1.00r + 0.80t + 0.10s)_{Reduction + Readiness} \quad (15.4)$$

See Table 15.8.

TABLE 15.6

Reduction + Readiness and Response + Recovery Variables (in Arbitrary RU)

f	The ability to look forward for opportunities as well as potential crises
g	The ability to identify crises and their consequences accurately
h	The level of enhanced understanding of the trigger factors for crises
i	The level of increased awareness of the resources available both internally and externally
j	The level of better understanding of minimum operating requirements from a recovery perspective
k	The level of enhanced awareness of expectations, obligations, and limitations in relation to the community of stakeholders, both internally (staff) and externally (customers, suppliers, consultants, etc.)

TABLE 15.7

Reduction + Readiness Variables (in Arbitrary RU)

l	The level of importance of buildings, structures, and critical supplies
m	The level of importance of computers, services, and specialized equipment
n	The level of importance of individual managers, decision makers, and subject matter experts
o	The level of relationships between key groups internally and externally
p	The level of importance of communication structures
q	The level of perception of the organizational strategic vision

TABLE 15.8
Reduction + Readiness Variables (in Arbitrary RU)

r	The level of importance of leadership and decision-making structures
s	The level of importance of the acquisition, dissemination, and retention of information and knowledge
t	The degree of creativity and flexibility that the organization promotes or tolerates

TABLE 15.9
Reduction + Readiness Variables (in Arbitrary RU)

u	The level of greater awareness of itself, its key holders, and the environment with which it conducts business
w	The level of ability to adapt to changed situations with new and innovative solutions and/or the ability to adapt the tools that it already has to cope with new and unforeseen situations

TABLE 15.10
Overall Object Function [$f(R_{ero})$] Variables (in Arbitrary RU)

$f(R_{ero})$	Dynamic operational resilience of an emergency response organization (i.e., a Dutch safety region)
R_{ac}	The level of adaptive capacity of an emergency response organization
R_{awa}	The level of awareness of an emergency response organization
R_{ero}	The level of resilience of an emergency response organization
R_{kv}	The level of importance of keystone vulnerabilities of an emergency response organization
R_q	The level of quality of an emergency response organization
ε	The level of unspecified data and items that are also a function of resilience

The identified attributes u and w, reduction and readiness variables, describing quality as a function of resilience (R_q) are presented by the fifth column from the left that result in Equation 15.5 (Van Trijp et al., 2011, 2012):

$$R_q = \left(1.00w + 0.50u\right) \tag{15.5}$$

See Table 15.9.

By another look at the value tree, we notice a set of variables between the dotted lines called *objectives*. These variables are each the sum of the five individual equations we have just calculated (15.1)–(15.5) plus the variable ε that stands for the level of unspecified data and items that are also a function of resilience.

We already know

- Situation awareness
- Management of keystone vulnerabilities
- Adaptive capacity
- Quality

are a function of an organization's operational resilience (see that section). In the value tree, this is expressed by multiplying the right-most column (R_{ero}) with the remaining columns on the left-hand side as is described by Equation 15.6 (Van Trijp et al., 2011, 2012):

$$f\left(R_{ero}\right) = R_{ero}\left(R_{awa} + R_{kv} + R_{ac} + R_q + \varepsilon\right) \tag{15.6}$$

See Table 15.10.

Maximum resilience $f(R_{ero})_{max}$ is achieved when R_{awa}, R_{kv}, R_{ac}, R_q, ε, and R_{ero} are all as large as possible. A high score for R_{ero} alone is no guarantee that the resilience of an emergency response organization like a Dutch safety region is good as well. Equation 15.6 shows us this is also dependent on good scores for awareness, keystone vulnerabilities, adaptive capacity, and quality that are all part of reduction and readiness before the event (LOC) takes place (Vargo and Seville, 2008).

Hence, $f(R_{ero})$ is defined as dynamic operational resilience of a Dutch safety region as it dynamically describes the organizations' actual state of resilience.

We also have to take into account the UVs (remember the little green spheres in the value tree) and include those into Equation 15.6 giving a *unique dynamic operational resilience* $f(R_{ero})_{UV}$ factor that is made up of all relevant key aspects (15.7) (Van Trijp et al., 2011, 2012):

$$f(R_{ero})_{UV} = (R_{ero})_{UV}\left(R_{awa} + R_{kv} + R_{ac} + R_q + \varepsilon\right)_{UV} \qquad (15.7)$$

It is assumed R_{ero}, R_{awa}, R_{kv}, R_{ac}, R_q, and ε are all of the same importance and as a result have a weight factor equal to 1.00. Hence, $f(R_{ero})_{UV}$ has a maximum value of 22.54 **RU** ($= f(R_{ero})_{max}$) and a minimum of 0.00 **RU** when calculated using the value tree based on the MAUT developed by Goodwin and Wright (2004) as cited in Stolker (2008). To calculate the maximum value, all **UVs** have to equal to 1. For practical purposes, ε will not be taken into account ($=$ nullified) as we like to concentrate on the key aspects we can measure.

The undetermined **UVs** can be assessed for each safety region by auditing this organization. In general when an attribute is fully implemented, an operational score of 100% is assessed, and the related UV equals 1.00. A score of 80% results in a UV of 0.80, like a score of 35% results in a UV of 0.35.

It is also possible to use a simplified approach for $f(R_{ero})_{UV}$ by using a quick scan method, $f(R_{ero})_{QS}$. This quick scan contains only the two most important attributes from each column of the value tree, which are part of $f(R_{ero})_{UV}$. In this case, the two attributes with the highest scores result in Equation 15.8:

$$f(R_{ero})_{QS} = \left(1.00c + 0.70b\right)_{UV} \cdot$$

$$\left((1.00k + 0.95f) + (1.00n + 0.80o) + (1.00r + 0.80t) + (1.00w + 0.50u) + \varepsilon)_{UV}\right) \qquad (15.8)$$

Thus, reducing the amount of work to be done by almost 50%! (10 vs. 22 attributes!).

Again in a similar fashion as we described for Equation 15.8, we nullify ε and set UV = 1.00. We do this as we like to calculate the highest value possible. As a result, the maximum achievable dynamic operational resilience $f(R_{ero})_{QSmax}$ using a quick scan totals

$$f(R_{ero})_{QSmax} = 11.99\textbf{RU} \qquad (15.9)$$

which is 53.19% of $f(R_{ero})_{max}$.

When taken all uncertainties into account, it is postulated that the result from the quick scan method is about half the score possible. So we conclude (15.10)

$$f(R_{ero})_{max} = 2.0 f(R_{ero})_{QSmax} \qquad (15.10)$$

Based on the results presented, it is concluded it is possible to quantify resilience of an emergency response organization like a Dutch safety region by using the quick scan method and by assessing the unique **UVs** of the attributes of this emergency response organization.

ANALYSIS

OPERATIONAL RESILIENCE OF THE SAFETY REGION MWB DURING THE INCIDENT

In this section, we look at the operational resilience of the safety region MWB around and during the incident at Chemie–Pack. For convenience, we use the quick scan method $f(R_{ero})_{QS}$ (15.8). The range of the UVs is 0–1. The reader may notice the absence of any values between 0 and 1. We use 0 and 1 only as this case study is not based on an extensive audit that should provide intermediate values but solely on the interpretation of external reports regarding the incident that leaves no other room than to use 0 or 1. This remark applies for the rest of this section.

UV OF c

According to Table 15.4, c equals "the potential (of organizations and individuals) to adapt to changing circumstances in the face of adversity and the ability to recover after a disaster or other traumatic event." In this case, it may be translated into the potential of MWB to adapt to changing circumstances and recover. According to the reports, MWB was slow in determining the correct coordinating threat level; instead of level 3, which was needed, MWB remained too long at level 2, hence lacking the possibility of proper coordination at the public order level (mayor is involved as coordinating authority from level 3). Adaptation to changing circumstances like suppression of the blaze lacked proper coordination and control. When we look at these important findings, we should conclude on a scale of 0–1, the UV for c equals 0. In reality, it is not 0 but close to 0 as the blaze was of such a complex nature that any possibility of saving the plant was negligible. For this case study, $c = 0$.

UTILITY VALUE (UV) OF b

According to Table 15.5, b equals "the capacity to cope with unexpected dangers after they become manifest." In this case, it may be translated into the capacity of MWB to cope with unexpected dangers after they become manifest. The reports state MWB failed to react adequately to the large amounts of polluted water that were generated in the process of firefighting: subsequent decontamination of personnel and vehicles was not carried out. This danger was underestimated in the command and control chain, and as such, we should conclude, on a scale of 0–1, the UV for b should equal 0. As other more immediate dangers were met when needed (like removing fire fighting crew when walls became unstable or the quick suppression of a large pool fire that moved toward a crash truck), it is decided for this case study not to use the value "0," but $b = 1$.

UTILITY VALUE (UV) OF k

According to Table 15.6, k equals "the level of enhanced awareness of expectations, obligations, and limitations in relation to the community of stakeholders, both internally (staff) and externally (customers, suppliers, consultants, etc.)." In this case, it may be translated into "the level of enhanced awareness of expectations, obligations, and limitations in relation to the municipality of Moerdijk and its administration." The reports state at the time of the blaze the establishment of a special dedicated plant fire service as was advised by MWB was not yet enforced by the responsible authorities of the town of Moerdijk, which was then responsible for enforcing this dedicated fire service. According to the reports, this was still in the process of communication. In October 2010, the law in the Netherlands was changed, and after this change, MWB became responsible for enforcement of the dedicated plant fire service instead of the administration of Moerdijk. MWB failed to do so. MWB is not alone in this as still two-third of 539 similar plants in the Netherlands have to be evaluated by the respective safety regions whether a dedicated plant fire service is needed (2012).

When we look at these findings, we should conclude, on a scale of 0–1, the UV $k = 0$.

UTILITY VALUE (UV) OF *f*

According to Table 15.6, *f* equals "the ability to look forward for opportunities as well as potential crises." In this case, it may be translated in the ability of MWB to look forward for opportunities as well as potential crises. The reports state MWB did not possess the compulsory coordinating multidisciplinary and monodisciplinary exercise and training plan following the year 2009. This plan should have comprised data how and when MWB would practice and drill emergency response in case of major crisis at plants like Chemie–Pack. Practice training for senior fire officers that did occur did not include major incidents involving hazardous chemicals although such type of incidents could be expected due to the nature of the risks involved. Hence, it is concluded the UV *f* = 0.

UTILITY VALUE (UV) OF *n*

According to Table 15.7, *n* equals "the level of importance of individual managers, decision makers, and subject matter experts." In this case, it is interpreted as "the level of importance of the senior officers involved in handling the incident." The reports state severe shortcomings were noted in leadership, command structure, and coordination during the incident. The result was the emergence of an informal command structure next to the official command structure that clouds the effort of the organization of controlling the blaze. It is also noted leadership and coordination at the administrative level of Moerdijk was flawed contrary to leadership and coordination at the regional administrative level, which did function properly. Hence, it is concluded the UV of *n* = 0.

UTILITY VALUE (UV) OF *o*

According to Table 15.7, *o* equals "the level of relationships between key groups internally and externally." This is interpreted in this case as "the level of relationships between MWB key groups internally and between key groups externally like other safety regions and administrative bodies." The reports concluded the relationships between MWB and the neighboring safety region ZHZ (see also Figure 15.2) were adequate in contrast to the relationships with other safety regions and national partners that were mainly carried out at informal levels creating a lot of noise. This was not due to MWB but to the system in general and a troubled perspective on the role each partner has. Recommendations were made by the investigating parties to all partners involved to improve the relationships in case of a crisis. Hence, it is concluded the UV of *o* = 1.

UTILITY VALUE (UV) OF *r*

According to Table 15.8, *r* equals "the level of importance of leadership and decision-making structures." For MWB, these are very important factors to deal with as they make up a substantial part of the operational structure and striking capability. The reports state shortcomings in the leadership function and decision-making structures concerning getting an overall picture of the incident at hand. In this particular incident, no negative effects could be noted as the blaze was too large to be influenced by the shortcomings. Nevertheless recommendations were made to improve leadership and decision-making structures to cope with future major crises. Hence, it is concluded the UV of *r* = 0.

UTILITY VALUE (UV) OF *t*

According to Table 15.8, *t* equals "the degree of creativity and flexibility that the organization promotes or tolerates." This is interpreted as the degree of creativity and flexibility MWB promotes or tolerates while battling the blaze. MWB showed great skill in creativity and flexibility in fighting the blaze using nonconventional techniques like deploying crash trucks from a nearby military air base, a police helicopter, and unprecedented large amounts of foam. The incident was of such a magnitude that out-of-the-box thinking was needed to contain the blaze and its effects.

Although the reports have noted shortcomings that influence the other described attributes, enough flexibility and creativity was shown to be effective. Hence, it is concluded $t = 1$.

Utility Value (UV) of w

According to Table 15.9, w equals "the level of ability to adapt to changed situations with new and innovative solutions and/or the ability to adapt the tools that it already has to cope with new and unforeseen situations." This attribute is closely related to attribute t, and it is assumed in this case (i.e., battling the blaze) $w = t$. Hence, it is concluded $w = 1$.

Utility Value (V) of u

According to Table 15.9, u equals "the level of greater awareness of itself, its key holders, and the environment with which it conducts business." It is interpreted as the level of greater awareness of MWB, its key holders, and the environment with which it conducts business prior and during the incident. According to the reports, MWB failed to communicate properly with its prime key holders the media (including social media) and citizens. MWB also failed to carry out surprise inspections at the site of Chemie–Pack during operational hours. The dates of the inspections that MWB did perform were shared in advance with Chemie–Pack hence facilitating masking any shortcomings prior to the actual moment of inspection. MWB was closely involved in the process of the needed permits but was not able yet to enforce the establishment of a dedicated plant fire service at the time of the incident. It is also noted communication between MWB and key holders was flawed and in time unidirectional. The relationship with prime key holder the administration of Moerdijk was such that in the years 2009 and 2010, no multidisciplinary exercises at the proper level including the administration of Moerdijk were conducted involving large amounts of hazardous chemicals like the incident at Chemie–Pack. In addition, MWB did not have a training and exercise plan for the years following 2009 by which it was in violation of the gentlemen's agreement MWB had with the minister of the Dutch Ministry of the Interior and Kingdom Relations. Based on these data, it is concluded the UV of $u = 0$.

Calculating Operational Resilience

When all UVs are entered in Equation 15.8 and again nullifying ε, the result is (15.11)

$$f(R_{ero})_{QS} = 2.17\mathbf{RU} \tag{15.11}$$

To equate the percentage, we divide 2.17 by 11.99 (from Equation 15.9) and multiply by 100%: $f(R_{ero})_{QS}$ is 18.10% of $f(R_{ero})_{QSmax}$ at the time of the incident and just under one-fifth of the possible operational resilience! Hence, we can conclude the SoS operated in the midst of a crises at less than one-fifth of its potential.

Lessons Learned

When we combine the findings of the time line of the incident and the characteristics of operational resilience as shown in the paragraphs dealing with the UVs, we can conclude the challenges were met by means of improvisation during the unfolding events. This was necessary as proper preincident planning was not adequate enough, and the improvisation led to confusion between operational units in and between the fire service, neighboring safety regions, communication staff, and regional and national administrations. When we take a look at Figure 15.1 again, we notice a high complexity of relationships the safety region has to deal with. These relationships exist in a preincident situation but are also active during and after an incident. It was recommended afterward by the investigating authorities the way these relationships work during an incident or crises should be reevaluated as the law involved (act concerning the safety regions). With specific attention to the competences, authority figures like the coordinating mayor should need to work with issues on transboundary incidents between safety regions, the competences of all departments involved (in fact who is in command and when) and who is in operational command and in control of related and relevant national services concerning the environment, and a large array of diverse advisory boards.

Furthermore, it was recommended to provide a national communication center to assist safety regions and municipalities during a major incident or crisis.

The type of communication and its contents should directly link up with the needs of the general public—especially today where social media provide instant news beyond control of the authorities.

Basically, it all boils down to the right type of communication between people, parties, and the subsequent follow-up whether before, during, or after the incident. Failing to do so and trusting paper structures beyond reasonable belief creates a nonexisting world for which no safety region can prepare itself. The calculated operational resilience attributes in this case study show the safety region MWB was not prepared and resilient enough to deal with this particular type of incident.

This raises the question when MWB would have been prepared enough. In the current stage of our research, this can only be answered by looking at the outcome the way this major incident was handled. In an ideal world, the incident either never occurred or when it does occur is quickly brought under control. For this to happen, MWB needs to be resilient in relation to the prospective incident(s) to happen. In other words, MWB should be aware of the risk profile (i.e., type of industries, critical infrastructures, urban and rural population, presence of nature, possible types of weather, possibility of flooding) present in the safety region and adjust the organization accordingly to deal with all possible risks based on an elaborate risk analysis of each hazard. For those hazards that may be beyond the capability of MWB to control or contain, MWB has the possibility to join forces with neighboring safety regions and the department of defense for assistance. For this to happen, MWB needs to make solid agreements and in case of a crisis needs to live up to them or *to be resilient*.

A scale defining when a safety region is or is not resilient cannot be presented as throughout the Netherlands, the risk profile of each safety region varies greatly. We can only state resilience should meet the respective risk profile. Further research is planned to define and explore the limits of this statement.

BEST PRACTICES

Other safety regions and safety-related structures in the Netherlands are intensely studying the outcome of the different types of investigations of these incidents. The results are used to enhance the command, communication, and control structures. Also extra courses are provided for fire fighters on junior and senior levels with regard to this type of incident. These courses are in the field of suppression techniques, how to inspect similar type of plants, risk management and preparedness. One of the immediate results in the Netherlands was safety regions gave a closer look to the need of assigning private dedicated plant fire services. These plant fire services are based on location and paid for by the plant owners itself but are under direct command of the safety region. The use of this type of fire service has the possible advantage of immediately responding to the first sign of a possible incident. This improves the potential for containment before the fire service of the safety region arrives on the scene.

EPILOGUE

In October 2012, the eastern seaboard of the United States was hit by Hurricane Sandy. This giant made landfall on October 29 and created coastal havoc in the states of New Jersey, Connecticut, Rhode Island, and New York, including Staten Island and the south tip of Manhattan of New York City. Many houses alongside the coastline were swept away, major power outages occurred, and the financial heart of the United States and the world, Wall Street, was unable to function for a period of time. Subways were submerged with water and local activities came to a full stop. By March 27, 2013, the FEMA reported (FEMA, 2013)

- FEMA personnel deployed: 2884
- Assistance registrations: 527,089
- Approved in assistance: $1.31 billion

while damage is estimated to range in the billions of dollars.

Although help and assistance was provided and a state of emergency was declared by the federal administration, the general question was how this may have happened.

One of the key aspects in this was the absence of any flood and/or tidal barriers that may have protected the mainland from flooding. Over time, many residents built their houses close to the shoreline without any protection from tidal surges. The same can be said about lower Manhattan where street level is just a few feet above the water table. Entrances of the subway could not be closed as flood doors were not present causing major flooding of the transport system. It seems residents and the administration were not prepared for what was to come. It can readily be recognized that operational resilience (look again at the attributes defined!) was substandard as especially attributes at the left side of the bow tie model were not adequately dealt with by the authorities.

Hurricane Sandy will undoubtedly induce a steep learning curve for everyone involved. One aspect of this learning curve will surely be increased operational resilience of emergency response organizations.

Increased operational resilience of the emergency response organizations will demand a different type of strategy in the future: not wait and see if an incident happens and subsequently response but be actively involved in urban planning and zoning and use risk analysis as a standard tool to provide answers to questions like "what are the hazards involved, what is the possibility an incident happens, and what will occur when an incident happens?" Where do we want the population to live? Near a hazard like a plant or an unprotected seaboard—or is it a good idea to relocate the population and/or install special safeguards like flood protection? How do we inform the public? What can we do to make the population more self-sustainable? A sound risk analysis followed by the policy decision-making process creates more resilient emergency response services (and societies!). To do so, emergency response organizations should be actively involved in these types of processes. If this was the case in NY, one can imagine flood defense systems would have been in place at the time Sandy struck presenting a whole different outcome with less economic damage and less casualties.

When we look at Figure 15.2, where the (complex) system/(SoS)/enterprise of the Dutch situation is described, we see a lot of similarities with this case. The separate systems involved in the governing body (D) and the external factors (B) in the United States have different names, but the problem is exactly the same: how to respond to an ongoing crisis that transgresses state borders like in the Dutch situation, a crises that transgresses safety region borders complicating the response by the governing body. Of course, the scale of the US Hurricane Sandy-induced crisis is much larger than the scale of the Dutch industrial blaze, but the (complex) system/SoS/enterprise relationships are similar.

SUMMARY AND CONCLUSIONS

Based on the findings of the cited reports, the Netherlands Branch Organization of Fire Services (NVBR) has started a program titled *Learning Arena Moerdijk* (2011) to improve the learning capabilities of all the Netherlands regional fire services, which are part of the respective safety regions. It is expected that this program will enhance operational resilience at future incidents. In addition at the national level, all procedures are carefully reexamined to be either enhanced or abolished. The Public Order and Safety Inspectorate of the Dutch Ministry of Security and Justice has released a report in February 2012 with the results of a survey among the Dutch safety regions concerning the allocation of dedicated plant fire services at high-risk locations. The results show a substantial backlog in evaluations and allocations. Recommendations are made by Dutch authorities to improve the situation.

Resilience can only be achieved when a number of criteria are met: be aware of the risk profile; make sound policy decisions on the basis of this risk profile; get the communication structures in order; be aware of the (im-)possibilities of the emergency response organization and report those to the responsible authorities; invest actively in the relationship with key stakeholders (internally and externally); be proactive instead of reactive and be ahead of any possible incident; invest actively in leadership; and finally make sure the emergency response organization has eyes and a voice in society before, during, and after an incident or crisis.

Failing to do so and trusting paper structures beyond reasonable belief create a nonexisting world for which no safety region (or any other emergency response organization—SoS) can prepare itself.

SUGGESTED FUTURE WORK

QUESTIONS FOR DISCUSSION

Consider the following questions:

- Based on the presented data, what may be needed to reach maximum operational resilience for an emergency response organization or SoS? To what extend do you believe this may be possible?
- This case study presents a model to calculate operational resilience. Should in your opinion all emergency response organizations strive for the highest score (22.54 RU), or do you see other possible solution(s) that work? If so, present the solution(s) with a sound argumentation. Take into account the different high-/low-risk environment emergency response organizations have to work in.
- Study the case of Hurricane Sandy (see Epilogue) and construct a (complex) system/SoS/enterprise diagram during the crises similar to Figure 15.2. What part of the (complex) system/SoS/enterprise would you change so emergency response organizations can cope better with a similar kind of crisis in the future and why? Can you think of any aspect or aspects how operational resilience of the emergency response organizations can be enhanced regarding super storms like Sandy?
- Can you think of other applications of operational resilience other than optimizing the performance of an emergency response organization? You may consider in your answer the stakeholders to whom the organization is accountable.

FUTURE WORK

Future research may focus on establishing correct UVs (validating the model) by auditing a representative set of emergency response organizations and checking the findings from this case study to the situation in North America (United States, Canada) and compare those to the Dutch results.

Research is planned to define and explore the limits of resilience in combination with the respective risk profile, and research is planned to study the resilience relationships of a self-organizing security network (SOSN) (Ulieru, 2007, 2008, 2009). An SOSN is a network of security or safety stakeholders (internal and external) related to each other by meta-organizational decision-making structures, practices, and processes.

APPENDIX

Glossary of Used Symbols

a	The sustenance of normal development despite long-term stress or adversity
b	The capacity to cope with unexpected dangers after they become manifest
c	The potential (of organizations and individuals) to adapt to changing circumstances in the face of adversity and the ability to recover after a disaster or other traumatic event
d	The readiness of an organization before the shock or disruptive event
e	The response of the organization after the disruption has struck
ε	The level of unspecified data and items that are also a function of resilience
f	The ability to look forward for opportunities as well as potential crises
$f(R_{ero})$	Dynamic operational resilience of an emergency response organization (i.e., a Dutch safety region)
$f(R_{ero})_{max}$	Maximum achievable dynamic operational resilience of an emergency response organization (i.e., a Dutch safety region)
$f(R_{ero})_{QS}$	Dynamic operational resilience of an emergency response organization (i.e., a Dutch safety region) using the quick scan method
$f(R_{ero})_{QSmax}$	Maximum achievable dynamic operational resilience of an emergency response organization (i.e., a Dutch safety region) using the quick scan method
$f(R_{ero})_{UV}$	Unique dynamic operational resilience of an emergency response organization (i.e., a Dutch safety region) dependent on UVs
g	The ability to identify crises and their consequences accurately
GRIP	Coordinated threat level
h	The level of enhanced understanding of the trigger factors for crises
i	The level of increased awareness of the resources available both internally and externally
IBC	Intermediate bulk container
ICCB	Interministerial committee crisis management
ICI	Imperial Chemical Industries
IOOV	Inspectorate of the ministry of security and justice
j	The level of better understanding of minimum operating requirements from a recovery perspective
k	The level of enhanced awareness of expectations, obligations, and limitations in relation to the community of stakeholders, both internally (staff) and externally (customers, suppliers, consultants, etc.)
l	The level of importance of buildings, structures, and critical supplies
LOC	Loss of containment
LOD	Line of defense
m	The level of importance of computers, services, and specialized equipment
MAUT	Multiattribute utility theory
MWB	Safety region Midden-en West Brabant
n	The level of importance of individual managers, decision makers, and subject matter experts
NCC	National Coordination Center
o	The level of relationships between key groups internally and externally
OOV	Dutch Safety Board
p	The level of importance of communication structures
q	The level of perception of the organizational strategic vision
r	The level of importance of leadership and decision-making structures
R_{ac}	The level of adaptive capacity of an emergency response organization

R_{awa}	The level of awareness of an emergency response organization
R_{ero}	The level of resilience of an emergency response organization
R_{kv}	The level of importance of keystone vulnerabilities of an emergency response organization
R_q	The level of quality of an emergency response organization
RU	Resilience units
s	The level of importance of the acquisition, dissemination, and retention of information and knowledge
SoS	System of systems: a collection of systems to achieve a purpose not generally achievable by the individual acting independently
SOSN	Self-organizing security network
t	The degree of creativity and flexibility that the organization promotes or tolerates
ToS	Total of systems: a system of systems plus the governing body relationship including external factors, feedback, governance, and constraints
u	The level of greater awareness of itself, its key holders, and the environment with which it conducts business
UV	The utility value of an attribute in a value tree
v	The level of increased knowledge of its keystone vulnerabilities and the impacts that those vulnerabilities could have on the organization: both negative and positive
w	The level of ability to adapt to changed situations with new and innovative solutions and/or the ability to adapt the tools that it already has to cope with new and unforeseen situations
x	The level of importance of individual aspects of resilience
y	The level of importance of keystone vulnerabilities and adaptive capacity for resilience
ZHZ	Safety region ZHZ

ACKNOWLEDGMENTS

The authors wish to thank Mrs. Arlette van de Kolk, Mrs. Annabelle van Roosmalen, Mrs. Annemarie van Daalen, and Mr. Noud Bruinincx for reading the manuscript. We like to thank the editors Dr. Jimmy Gandhi, Dr. Alex Gorod, Dr. Vernon Ireland, Dr. Brian Sauser, and Dr. Brian White for their continuous support and patience.

REFERENCES

Ale, B.J.M. 2009. *Risk: An Introduction—The Concepts of Risk, Danger and Chance*, 1st ed., pp. 48. New York: Routledge.

Anonymous. 2010. Dutch Security Regions Act Official Consolidated Version. The Hague, the Netherlands: Ministry of the Interior and Kingdom Relations. Retrieved March 7, 2012, from: http://www.government.nl/documents-and-publications/decrees/2011/09/29/dutch-security-regions-act-official-consolidated-version.html.

Brouns, F., Berlanga, A.J., Van Rosmalen, P., Bitter-Rijpkema, M.E., Sloep, P.B., Kester, L., Fetter, S., and Nadeem, D. 2009. ID8.16—Policies to stimulate self-organisation and the feeling of autonomy in a network. Heerlen, the Netherlands: Open University of the Netherlands, TENCompetence. Retrieved March 7, 2012, from: http://hdl.handle.net/1820/1944.

Delbert, R. and Rohozinski, R. 2009. Tracking Ghostnet: Investigating a Cyber Espionage Network. Toronto, Ontario, Canada: Information Warfare Monitor. Retrieved April 28, 2012, from: http://www.scribd.com/doc/13731776/Tracking-GhostNet-Investigating-a-Cyber-Espionage-Network.

Dutch Safety Board. 2012. Brand bij Chemie–Pack te Moerdijk (Fire at Chemie–Pack in Moerdijk). The Hague, the Netherlands, Report in Dutch. Retrieved March 23, 2012, from: http://www.onderzoeksraad.nl/index.php/onderzoeken/onderzoeksraad-start-onderzoek-naar-brand-in-moerdijk/.

Goodwin, P. and Wright, G. 2004. *Decision Analysis for Management Judgment*. Chichester, U.K.: John Wiley & Sons.

FEMA. 2013. Hurricane Sandy. Retrieved March 30, 2013, from: http://www.fema.gov/hurricane-sandy.

Fire Safety Region Midden-en West Brabant (MWB). 2011. Fact Sheet. Breda, the Netherlands, Fact Sheet in Dutch. Retrieved March 23, 2012, from: http://www.veiligheidsregiomwb.nl/Organisatie/Onderzoeksrapporten-Brand-Moerdijk.aspx.

Hollnagel, E., Woods, D.D., and Leveson, N. 2006. *Resilience Engineering, Concepts and Precepts*. Farnham, U.K.: Ashgate Publishing Limited.

McManus, S., Seville, E., Brunsdon, D., and Vargo, J. 2007. Resilience management, Resilient Organizations Research Report 2007/01. Retrieved September 25, 2012, from: http://www.resorgs.org.nz/pubs/Resilience%20Management%20Research%20Report%20ResOrgs%2007-01.pdf.

Netherlands Branch Organization of Fire Services NVBR. 2011. Leerarena Moerdijk (Learning Arena Moerdijk), Arnhem, the Netherlands, Report in Dutch. Retrieved March 31, 2012, from: http://www.nvbr.nl/publish/pages/18638/leerarena.pdf.

Public Order and Safety Inspectorate, Dutch Ministry of Security and Justice. 2011. Brand Chemie–Pack Moerdijk (Fire at Chemie–Pack Moerdijk), The Hague, the Netherlands, Report in Dutch. Retrieved March 28, 2012, from: http://www.ioov.nl/aspx/download.aspx?file =/contents/pages/106258/rapportagechemie-packmoerdijk.pdf; Summary, conclusions and recommendations in English retrieved March 28, 2012, from: http://www.ioov.nl/aspx/download.aspx?file =/contents/pages/106291/chemi-packmoerdijk.pdf.

Public Order and Safety Inspectorate, Dutch Ministry of Security and Justice. 2012. Aanwijzing bedrijfsbrandweer risicobedrijven (Allocation of dedicated plant fire service for high-risk plants). The Hague, the Netherlands, Report in Dutch. Retrieved April 28, 2012, from: http://www.rijksoverheid.nl/bestanden/documenten-en-publicaties/rapporten/2012/02/10/rapport-aanwijzing-bedrijfsbrandweer-risicobedrijven/aanwijzing-bedrijfsbrandweer-risicobedrijven.pdf.

Rutter, M. 1985. Resilience in the face of adversity: Protective factors and resistance to psychiatric disorder. *British Journal of Psychiatry* 147: 598–611.

Seville, E. 2009. The Goal of Resilient Organizations. Keynote presentation—Business Continuity Institute Summit, Brisbane, Queensland, Australia. Retrieved September 25, 2012, from: http://www.resorgs.org.nz.

Stolker, R.J.M. 2008. A generic approach to assess operational resilience. Technische Universiteit Eindhoven (TUE). Capaciteitsgroep Quality and Reliability Engineering (QRE), Eindhoven, the Netherlands. Retrieved March 7, 2012, from: http://library.tue.nl/catalog/LinkToVubis.csp?DataBib=6:639658.

Te Brake, H., van de Post, M., and de Ruijter, A. 2008. Resilience from concept to practice—The balance between awareness and fear; citizens and resilience; impact, Dutch knowledge and Advice Centre for Post-Disaster Psychosocial Care, Amsterdam, the Netherlands.

Ulieru, M. 2007. A complex systems approach to the design and evaluation of holistic security ecosystems. In: *International Conference of Holistic Security Ecosystems*, Boston, MA. Retrieved September 25, 2012, from: http://www.theimpactinstitute.org/Publications/Boston-submitted.

Ulieru, M. 2008. Enabling the SOS network. In: *Proceedings of the IEEE SMC 2008 Conference*, Singapore, Singapore. Retrieved September 25, 2012, from: http://www.theimpactinstitute.org/Publications/Ulieru-Formatted-Final-sos.pdf.

Ulieru, M. 2009. Towards holistic security ecosystems, opening keynote address and invited tutorial lecture. In: *3rd IEEE International Conference on Digital Ecosystems and Technologies*, Istanbul, Turkey. Retrieved September 25, 2012, from: http://www.theimpactinstitute.org/Publications/Keynote-Reformatted.pdf.

Van Trijp, J.M.P. 2010. An attempt to quantify resilience of emergency response organizations—Results from a large scale survey among safety stakeholders in the Netherlands, Master of Public Safety thesis Delft TopTech/Delft University of Technology, Delft, the Netherlands.

Van Trijp, J.M.P. 2012. Propelling beyond the Bow Tie: An emergent dynamic risk—Resilience model. In: *Proceedings of the PSAM11/ESREL2012 11th International Probabilistic Safety Assessment and Management Conference and The Annual European Safety and Reliability Conference*, Helsinki, Finland.

Van Trijp, J.M.P., Ulieru, M., and van Gelder, P.H.A.J.M. 2011. Quantitative approach of organizational resilience for a Dutch emergency response safety region. In: *Advances in Safety, Reliability and Risk Management*, eds. Bérenguer, C., Grall, A., and Guedes Soares, C., pp. 173–180. Taylor & Francis Group, London, U.K.

Van Trijp, J.M.P., Ulieru, M., and van Gelder, P.H.A.J.M. 2012. Quantitative modeling of organizational resilience in case of a Dutch emergency response safety region. *Proceedings of the Institution of Mechanical Engineers, Part O, Journal of Risk and Reliability* 266(6): 666–676.

Wildavsky, A. 1988. Playing it safe is dangerous. *Regulatory Toxicology and Pharmacology* 8(3): 283–287.

Wolff, C. 2011. Procedures regarding major waterway blockades. The Hague, the Netherlands. Retrieved March 23, 2012, from: http://www.ivr.nl/fileupload/publications/congres2011/ivrcongres2011/WS-7-WOLFF-ENGELS.pdf.

Section VIII

Military

16 Complex Adaptive Operating System
Creating Methods for Complex Project Management

John Findlay and Abby Straus

CONTENTS

FUNDAMENTAL ESSENCE

- The majority of complex major projects and programs are not completed on time or budget nor meet the requirements.
- Traditional project management methods no longer work reliably in complex or rapidly changing situations.
- Most projects, no matter how large or small, can now be considered complex, due to the accelerating rate of social and technological change, the complex intersection of systems, and conflicts at the boundary of world views, cultures, and economic or political interests.
- New methods based on the science of complexity and the features of complex adaptive systems (CASs) offer a promising alternative for dealing with changing requirements, technologies, and stakeholder interests.
- An organization that is seeking to identify and develop more reliable solutions to the multiplicity of problems facing the complex project management community is the International Centre for Complex Project Management (ICCPM). ICCPM represents project and program leaders across government, corporate, consulting, and academic sectors around the world.
- One of these initiatives is a series of international roundtables attended by the leaders of ICCPM member organizations as well as associate partners including consultants and academics.
- This case study reports on a suite of new tools, methods, and frameworks known collectively as the complex adaptive operating system (CAOS), based on the laws of complexity and the features of CASs that has been developed by the authors in response to the needs identified by the leaders of the complex project management community around the world.
- Three of the tools are used to frame or facilitate the roundtable conversations and have proved useful in dealing with complex environments, the Zing complex adaptive meeting environment for knowledge co-creation (Findlay, 2009); polarity thinking, which helps manage ongoing issues that have both–and solutions (Johnson, 1996); and the complexity model of change (Findlay and Straus, 2011), which offers a robust model of the changes that are occurring in the socio-technological system.

TOPICAL RELEVANCE

The world of complex project management is in an acute state of flux. Contributing factors include the speed of change, the increasing complexity of and interactivity between systems, and the need for a higher level of cognitive complexity to understand the context and actions required in the current environment. In order to address this situation effectively, we need to be able to create and apply new knowledge quickly and wisely across the boundaries of worldviews, disciplines, and cultures.

Many of the methods and models currently in use—including some of the most innovative—are based on an understanding of human systems that is neither sufficient to explain what is happening nor capable of being the sole basis of sustainable activity. While many traditional approaches may have worked well when transformational change occurred in tens of generations, today's *new normal* of constant, disruptive change requires an entirely new approach.

With transformational change now occurring in less than a human generation, there is insufficient time for new tools to develop a pedigree. The current generation of project management tools, although so new that they are unproven, may offer a much better chance of success than the tools that are no longer consistently reliable.

In its discussion paper for the 2012 roundtable series, ICCPM says the following about what is required for success:

> Today's leaders of complex major projects need a set of capabilities that enable them to deal creatively, adaptively and successfully with emergence, collaboration and cross-cultural/sector issues, all prevalent features of the new economy. Flexible models are required to support a richer, more complex approach to benefits realization of cost, schedule (time), scope, quality and risk. (ICCPM, 2012, p. 16)

This case study demonstrates how ICCPM and its community are developing a richer understanding of the issues and new frameworks, models, and method to achieve project success.

DOMAINS

- Complex project management
- Defense acquisitions
- Aerospace projects
- Infrastructure projects
- Large-scale government programs
- Information technology (IT) and telecommunications
- Supply chain management

COUNTRY OF FOCUS

International: primarily Australia, North and South America, and Europe.

INTERESTED CONSTITUENCY

This case study is of particular interest to those whose job it is to successfully design, undertake, or operate complex major projects including

- The leadership and staff of defense and other government agencies responsible for procuring and operating defense systems and infrastructure, airports, railways, freeways, telecommunications networks, IT systems, and large programs of any kind
- Private sector owners/operators of IT systems; chemical and process plants; aerospace; manufacturing production lines; infrastructure such as airports, freeways, and railways; and supply chains

- The engineering profession generally
- IT professionals, including systems analysts and programmers
- Professions that engage with political leaders and stakeholders generally in order to secure and/or sustain support for major projects

Primary Insights

1. The developers of complex major projects are struggling to deal with accelerating change, increasing complexity and greater uncertainty and ambiguity.
2. Disruptive transformational change is due to the creation and application of new theories, methods, models, tools, and processes that have developed as a consequence of feedback loops in the human knowledge creation system.
3. At the transition from one stage of socio-technological development to the next, new constellations of technologies, structure, roles, rules of interaction, and skills emerge or are required.
4. The dynamics of the changes are experienced by the developers and managers of major projects as *wicked problems*, which manifest as constellations of evolving issues including changing requirements, stakeholder activism, inconsistent or ineffective governance, escalating costs, and suboptimal results.
5. Linear project management methods are failing in this new context, and SE/SOS methods, although generally effective, have proven difficult for some practitioners to implement.
6. Human systems—our brains, communities, projects, and organizations—are CASs and need to be managed as such.
7. A promising way to deal with the new context is to create and use tools and methods based on the laws of complexity and the features of CASs to design, manage, commission, and operate projects.
8. In order to successfully approach the challenges facing complex project management, we need a new operating system on which to run our major projects and, ultimately, our society. To do this, we need to get everyone in the room who has a role in how we govern, create new knowledge together, make decisions, and implement new ideas. There is no single owner of this larger system.

GLOSSARY

CAOS   Complex adaptive operating system
CASs   Complex adaptive systems
CASE   Complex Adaptive Systems Engineering
CEO   Chief executive officer
COO   Chief operating officer
CSIS   Center for Strategic & International Studies
EEE   Enlightened evolutionary engineering
GAO   Government Accountability Office (United States)
ICCPM   International Centre for Complex Project Management
IT   Information technology
NAO   National Audit Office (United Kingdom)
PM   Project/program manager
STEM   Science, technology, engineering, and math
SWOT   Strengths, weaknesses, opportunities, and threats

BACKGROUND

CONTEXT

In 2007, the International Centre for Complex Project Management (ICCPM) was established as a not-for-profit company to improve "the international community's ability to successfully deliver very complex projects and manage complexity across all industry and government sectors" (2012). The company grew out of an initiative begun with the support of Australian, US, and UK government agencies and defense sector corporations, previously known as the College of Complex Project Managers.

ICCPM is supported by a growing number of international partners who undertake major complex projects, such as BAE Systems, Lockheed Martin, Booz Allen Hamilton, and Thales, and has thus far appointed representatives in Australia, South America, North America, Europe, Russia, and the United Kingdom.

ICCPM also has an associate partner network of consultants to the complex project sector. It offers a Masters in Complex Project Management through the business school of the Queensland University of Technology. ICCPM also sponsors research into complex project management issues.

One of its major initiatives is the biannual international roundtable series, which brings leaders in the complex project management community together to better understand the reasons why complex projects fail and to develop better methods, processes, and tools to ensure their future success.

This case study focuses on how the roundtable process has played a critical role in the creation of new knowledge to help its members more successfully manage major complex projects; build new relationships between partners, industry professionals, and practitioners with experience in dealing with complex project issues; and develop a research agenda with regard to the issues facing program/project managers.

The first ICCPM roundtable series, conducted in Canberra, Australia, in May 2009 and Washington DC, United States, in October 2009, focused on "The Conspiracy of Optimism: Why Megaprojects Fail." It was sponsored by participating organizations such as the Defence Materiel Organization in Australia, which is the acquisitions arm of the Australian Department of Defence, the National Audit Office of the United Kingdom, Lockheed Martin, Raytheon, Business Executives for National Security, and Center for Strategic & International Studies (CSIS) from the United States.

The workshops explored the megaprojects paradox, where industry and defense purchasers and suppliers have a propensity to "strike agreements that are so optimistic as to be unsustainable in terms of cost, timescale or performance" (Hayes, 2013).

A series of studies characterize the problem of complex project failures:

- An IBM (2008) survey of 1500 change managers found that only 40% of projects finished on time and budget. Barriers to success included the inability to change attitudes and mind-sets (58%), dysfunctional culture (49%), and lack of senior management support (32%). Managers underestimated the degree of complexity in 35% of projects.
- A Logica Management Consulting (2008) of 380 European executives found that a third of companies abandoned a project in the past 3 years. Some 37% of *process change* projects failed to deliver the required benefits.
- A KPMG survey (2008) of 600 global IT project management firms found that almost half suffered a project failure in the past year. Only one in 50 had achieved 100% project success.
- A Government Accountability Office (June, 2009) review of US government–funded technology projects found that 49% were poorly planned or performed badly.
- The United Kingdom spent £4 billion on a succession of failed IT schemes between 2000 and 2008. Only 30% of projects were successful (Johnson and Hencke, 2008).

PERTAINING THEORIES

The case study is informed by six broad theoretical frameworks, each of which pertains to the nature and behavior of complex systems. We show how each of these frameworks offer an explanation of the issues addressed by the ICCPM roundtables and how they contribute to the process of new knowledge creation and its conversion into tools, frameworks, and methods:

- Systems thinking
- Complexity theory and complex adaptive systems (CASs)
- The complexity model of change
- Wicked problems
- Polarity thinking
- Cognitive complexity and adult development theory

Systems Thinking

Systems thinking is a broad and complex field, much of which is beyond the scope of this case study to define. There are, however, some key definitions and principles, which are central to our work including the following:

- *Definition of the term "system"*: A set of things, organisms, and people that, as a result of their interconnection, produce their own patterns of behavior over time. A system may experience outside influences, but it responds in characteristic ways, which, in the case of living systems, are almost always complex (Meadows, 2008).
- *Living systems is more than the sum of its parts*: You cannot understand the system by taking it apart and studying the bits (Wheatley, 1999). It must be understood as an interconnected whole. Frequently, critical discussions are to be had across boundaries of parts of a system, making it necessary to expand our understanding of where the boundaries lie.
- *Only one system*: In reality, the world is one large system, a continuum where everything is interconnected. We draw *boundaries* around parts of this continuum in order to have particular discussions, for example, about our organization or our country's political system. It is critical to be able to draw the boundary big enough to have a robust model of the system (Conant and Ashby, 1970) while also making it small enough to be able to "get our minds around" and managed (Meadows, 2008).
- *The structure of a system influences its behavior*: Different people placed in the same system tend to produce similar results. Therefore, the idea of simply changing personnel without changing the system within which they operate is like rearranging the deck chairs on the Titanic, that is, it leads to the same predictable behaviors over time.
- *We are the system and the system is us*: The very people and events we tend to blame for disruption in our systems—and consider *external influences*—are actually integral parts of those systems, including competitors, the press, markets, and vocal stakeholder populations. The key to solving problems frequently lies in cultivating productive relationships across the boundaries to other parts of the system that we have created, thus including critical elements of the system that have an effect on the whole.
- *There are many places to intervene in a system*: Some intervention points are more effective than others for creating sustained change. Meadows (2008) has identified 12 places to intervene and has categorized them in ascending order of effectiveness, changing numbers and parameters being the least and transcending paradigms being the most effective. Lower-level interventions may be deceptively attractive, because many of them seem to work in the short run. They do not necessarily, however, produce changes to the structure of the system, and so do not change the system's behavior over time. Intervening at the highest levels of a system may be said to be more difficult, because higher levels of cognitive complexity—and the accompanying capabilities—are required to do so; but they are more likely to produce systemic change over time.

Leverage Points to Intervene in a System

Meadows (2008) describes 12 leverage points that offer ways to successfully intervene in the operations of a system (in increasing order of effectiveness):

12. Numbers: Constants and parameters (such as subsidies, taxes, standards)
11. The size of buffers and other stabilizing stocks, relative to their flows
10. Structure of material stocks and flows (such as transport network, population age, structures)
9. Delays: Lengths of time relative to the rate of system changes
8. Balancing feedback loops: The strength of negative feedback loops, relative to the effect they are trying to correct against
7. Reinforcing feedback loops: The strength of gain of driving loops
6. Structure of information flow: Who does and does not have access to what kinds of information
5. Rules of the system (such as incentives, punishment, constraints)
4. Self-organization: The power to add, change, or evolve system structure
3. Goals: The purpose of the system
2. Mind-set or paradigm from which the system arises, its goals, structure, rules, delays, and parameters
1. Transcending paradigms: The ability to view multiple paradigms and integrate or choose between them
    - *Team learning is vital*: In order to tackle complex issues at organization and greater systems levels, it is essential that teams be able to learn together. Senge (2006) points out that teams, not individuals, are the fundamental learning unit of organization systems. In order to successfully navigate today's complex environments, people need to be able to let go of their accustomed patterns of advocating for what they believe and looking for one *right* answer in favor of collective inquiry and *thinking together*. Harvard's Chris Argyris (1991) argues that this is very difficult for many managers, because they tend to be rewarded for the former behavior and not the latter. The result is what Argyris calls *skilled incompetence*, a proficiency in keeping teams and organization from learning (Senge, 2006). While learning at a systems level this may be difficult to achieve, it is vital to the process of affecting sustainable change.
    - *Systems are fractal*: As Senge (2006) puts it, "We are the seed carriers of the whole in that we carry the mental models that pervade the larger system" (p. 348). As individuals, teams, and organizations start to change the structure of their thought and the ways they organize, they affect the greater system of which they are a part.

COMPLEXITY THEORY AND COMPLEX ADAPTIVE SYSTEMS

For much of the past century, with its intense focus on scientific management, managers have attempted to control human systems in a linear way, as if they were machines; but this metaphor no longer represents what we are experiencing. It is now more useful to regard complex projects and organizations operating in rapidly changing environments as CASs, complex systems that evolve over time as the result of the interaction of the parts.

In order to do this, we need to take into account three fundamental laws of complexity:

- *Requisite variety*: Ashby's law (Ashby, 1956) contends that any control system must have an equal or greater number of states than the system to be controlled. In other words, complexity can only be dealt with by equal or greater complexity. For example, one human brain is no longer complex enough to match the complexity of the challenges we are facing in many areas worldwide, including that of major projects.
- *Robust model of the system*: Conant's theorem (Conant and Ashby, 1970) states that "every good regulator of a system must be a model of that system." This means that

no one can effectively influence a system until they have a thorough understanding of its scope and the connections and interdependencies or an algorithm that explains the development of the system. An example of a model of a CAS is the logistics map (Gleick, 2008), which offers an explanation of the periodicity (and chaos) of booms and busts in a predator prey system. The complexity model of change is the same type of model (see period-doubling cascade).

- *Scale*: Highly complex situations can best be addressed by greater degrees of freedom at the local scale so that innovation and adaptability are maximized (Bar-Yam, 2004). An example of this is the inability of traditional infantry-type warfighter systems to deal with conflict in places such as Afghanistan, where the terrain and the culture of the conflict are highly complex, ambiguous, and rapidly changing.

A key point to be made is that the appropriate tools for the *management* of CAS are tools that are themselves complex adaptive and that have the following features:

- *Nonlinearity*: CASs evolve over time in a nonlinear fashion. "Feedback processes among interconnected elements and dimensions lead to relationships that exhibit dynamic, nonlinear and unpredictable change" (Stacey, 1996).
- *Emergence*: Emergence is a quality of a system where new patterns of activity arise as the result of the interactions between the parts. Unlike a machine, where the future state of the system can be predicted by understanding the relationship between the parts, it is not possible to *control* CAS in the classical sense, because they are inherently unpredictable.
- *Simple rules of interaction*: Simple rules of interaction between the agents or participants in a system can lead to complex patterns of activity or behavior, such as occurs when fish shoal or birds flock, by following rules of interaction like swimming or flying a fixed distance from others and at a specific angle.
- *Feedback*: Feedback is a process by which information flows from one part of the system to another and influences the development of other parts of the system over time (Sterman, 1994).
- *Self-organization*: The local interactions between the parts of an initially disordered human, economic, physical, chemical, or biological system lead to new patterns of organization at a global level, which in turn influence the pattern of the local interactions.
- *Attractors*: Emergent patterns of activity in CAS develop as attractor basins. They include limit cycle (circular loop) and strange (infinite loop) attractors. These patterns have their analogues in human systems. For example, many either/or problems, or polarities (see Figure 16.2), are infinite loop attractors that persist over time, and which are major sources of conflict when dealt with in a linear, either/or way by project leaders, managers, or meeting facilitators.
- *Sensitivity to initial conditions*: Small differences in the starting point of a system—the initial conditions—can result in widely differing behavior of the system. For example, minor changes in requirements for a complex project, particularly those involving multiple, interdependent systems, can result in unanticipated or exponentially large changes to schedule, cost, and performance.
- *Autocatalysis*: A process where a set of things—agents, molecules, or concepts—interact and create more of themselves (Kauffman, 2008), for example, when written or spoken words stimulate related concepts.
- *Phase transitions*: Systems undergo transformations from one state of organization to another when they reach a critical point in their development, for example, from ice to liquid water, to water vapor to plasma, or in reverse. Examples of phase transitions that occur in human systems include the stages of child and adult development, the process of a group becoming a team, or the reorganization of an entire system undergoing disruptive technological change.

- *Period-doubling cascade*: Phase transitions usually occur in a succession of shifts according to a predictable mathematical principle to more complex states (See period doubling cascade, Figure 16.1). This aspect of CAS explains how children develop in stages and how society undergoes sociotechnological transformations in which the roles people play, the rules of interaction and relationships that follow, are transformed to a higher level of order, transcending and including that which has gone before. Often, the technology gets ahead of the community's ability to wisely and effectively apply new scientific discoveries on which a project depends, or conversely, where public knowledge of complex issues is in advance of a project owner's preparedness to embrace change.
- *Fitness peaks*: Fitness peaks are optimal states of the system. Many of the tools, methods, and processes that complex project managers use today are artifacts of old fitness peaks, states of the system that were optimal in earlier socio-technological stages of development and may no longer be so.

Complexity Model of Change

At the core of the CAOS approach is a model (Figure 16.1) of socio-technological change (Findlay, 2009; Findlay and Straus, 2011), which provides a longitudinal explanation of the large-scale transformations that have occurred over the past 50,000 years but with particular emphasis on the last 200 years. The model also provides a general indication of future trends.

The past waves have familiar names: the Industrial Age (1700–1950), the Information Age (1950–2000), and the Knowledge Age (2000–2010); recently, there is the Wisdom Age (2010–2012) (Table 16.1). With the arrival of each wave, the technologies, products, and services of the new era tend to change. The tools of the new era automate and/or eliminate the dominant work of the immediate previous era and are as much as 4–10 times more efficient and/or productive than the predecessor they displace in a process of what the Austrian economist, Joseph Schumpeter, calls *creative destruction* (Schumpeter, 2003).

Successive waves of new technologies require new constellations of work roles, skills required, rules of interaction, and culture (Table 16.2). The model also anticipates the changes that are likely to arise in the future, especially in terms of new technologies; the skills required to design, build, and operate the more complex systems; and the shifts in stakeholder needs and interests that might be expected.

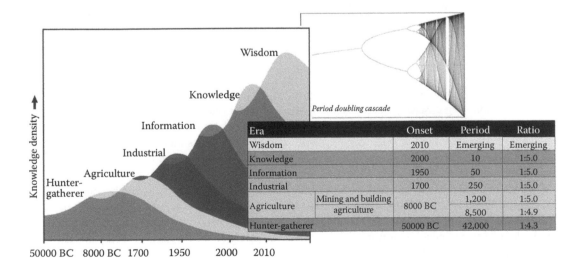

FIGURE 16.1   (See color insert.) Waves of change, duration, and increasing frequency.

TABLE 16.1

Features of the Stages of the Waves of Change

Era	Industrial	Information	Knowledge	Wisdom	?
Relationship to change	Reactive: transitions occurred beyond our understanding and therefore our capacity to deliberately direct thought and action.				Creative and deliberate direction of thought and action
Span	1700–1950	1950–2000	2000–2010	2010–2012	2012+
Metaphor	Machine	Computer	Network	Ecology	Hologram
Character	Liner	Algorithmic	Connected	Interconnected	Interpenetrated
Rate of change	12 generations	2 generations	<1	<0.1	Constant transformational change
Interaction	Monologue	Discussion	Dialogue	Dialectical	Ethical dialectical
Knowledge transfer	Telling	Reproduction	Creation	Wise application	Intrinsic intuition (reading the system)
Productivity gain: automates	Farm and artisanal work	Clerical and process work	Expert work	Governance work	Ability to leverage the whole system
Tools	Hardware	Software	Systems	Complex adaptive systems	Consciousness/quality of attention
Roles	Dependent	Independent	Shared purpose	Interdependent	Ability to wisely use all as appropriate
Coordination	Instructions, orders	Expert methods, procedures, and know-how	Facilitation, teams, expert systems, and databases	Self-organization via simple rules of interaction	Anticipatory awareness drawing on all previous methods

TABLE 16.2

Restructuring of the System between Paradigms

Technologies	More knowledge dense: the new tools automate the dominant work of the previous era.
Methods	Faster and more efficient.
Skills	More complex and cross-boundary.
Roles	More interdependent and integrative.
Rules of interaction	More combinations and more powerful.

The waves of change follow a predictive pattern, which obeys the laws of complexity. The shifts are a period-doubling cascade, where each new wave is about one-fifth the length of its predecessor (Findlay and Straus, 2012), approaching a limit, the Feigenbaum number, 4.669 (Gleick, 1988). When a shift to a new kind of order occurs, some species of products and services die out (record players, carbon paper, and 8 mm movies are now just distant memories) and are displaced by new products and services.

Corporations such as Google and Amazon are already gearing up for the world beyond the current era, the Wisdom Age (2010–?) characterized by the "wise application of knowledge." The two Internet behemoths are moving into the *anticipatory awareness* business using big data methods that track your every message or transaction sent via the Internet or IT system (Findlay and Straus, 2013) and that translate into actively *thinking about* how they can meet your needs, before you even realize it yourself. This will help set a new standard for stakeholder relationships management that the developers of complex projects and programs will be compelled to consider.

WICKED PROBLEMS

Many of the problems facing the complex project management community are known as *wicked problems*: those that are very complex, hard to define, ongoing, and constantly evolving, becoming *moving targets*. Wicked problems may have many interconnected parts and contributing factors. They have no clear solutions and seem to evade all attempts at resolution. Bringing a complex project in on time, on budget, and meeting requirements is itself a wicked problem, since every aspect of the endeavor may be subject to *forces* seemingly beyond the control of those charged with delivery, such as the invention of new technologies, dealing with politically active stakeholder groups and keeping track of shifting requirements due to unanticipated events.

One of the main issues experienced by project managers is that there is an extensive list of problems to be solved, many of them systemic. Each one depends on the resolution of others, which depend on the resolution of others; all of which are outside of the project/program manager's (PM's) direct control. The authors propose that we can begin to *tame* many wicked problems by regarding them as CASs and working with them using the laws of complexity and the features of CAS.

POLARITY THINKING

Polarity thinking is a framework developed by Johnson (1996) that helps practitioners proactively deal with the class of issues or problems that cannot be *solved* in the traditional sense—as either/or problems—but rather must be managed as a both/and relationship to achieve success over time. The complex project management space is rife with polarities, including the tension between centralizing operations to achieve coordination across a system and decentralizing to achieve autonomy of the parts and keep subsystems flourishing; the need to minimize cost while delivering a quality product; taking both a strategic, long-term view and a tactical, short-term view; building relationships; and getting results and mitigating risk while also reducing the probability.

We can identify benefits of focusing on each side of the equation, or pole (Figure 16.2). But when we focus on one pole to the neglect of the other, we eventually experience negative results. So we want to maximize the benefits while minimizing the negative results of focusing on one and not the other. In this way, we can realize high performance over time.

Many issues that we experience as *wicked problems* have at their core polarities, which need to be recognized as such and managed well. Trying to *solve* them for one right answer will only exacerbate the issue, causing conflict and seesawing between *solutions*, both of which result in suboptimal performance. The reason for this is that, while each perspective is *right*, it is only one half of the story.

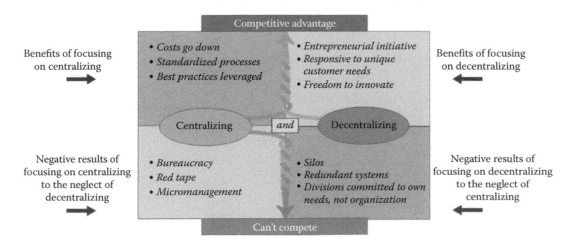

FIGURE 16.2   Polarity map.

It is important to distinguish between problems to solve and polarities to manage (which is itself a polarity) and to use each appropriately. We would not, for example, want to develop the specifications for an aircraft using a both/and approach because these kinds of problems have one unique, optimal solution. And, by the same token, an either/or approach frequently fails in stakeholder affairs where people align on opposite sides of issues such as new energy generation and responsible care for the environment, where both are *right* and neither political position is the sole answer.

COGNITIVE COMPLEXITY AND ADULT DEVELOPMENT THEORY

There is a considerable body of research indicating that adults have the capacity to continue their cognitive development (Cook-Greuter, 2004; Rooke and Torbert, 2005) long after childhood. Each level of development, or action logic (Rooke and Torbert, 2005), makes sense of the world in a different way. Every successive stage embodies greater cognitive complexity. This is the ability to *open the aperture* to perceive a greater web of systemic connections and relate to others and to the world in ever more sophisticated ways.

Research suggests that only a small percentage of the population operates from the highest action logics—strategist (4%) and alchemist (1%)—which are associated with exceptional emotional intelligence (Goleman, 1995) and the capacity to lead transformational change (Rooke and Torbert, 2005).

A key feature of later stages of adult development is the capacity to apply *triple-loop learning* (Gragert, 2013) (Figure 16.3). Figure 16.3 illustrates the orientation toward getting results of from the three different logics.

Single-loop learning is primarily oriented toward action and assumes that cause and effect are related closely to each other in time and so can be understood sequentially. The person at this stage is concerned with doing things right, frequently calling on rules and procedures as guides. A useful metaphor is that of a thermostat. At single-loop learning, the question is "Am I at 68 degrees or not? If not, I will adjust myself until I am."

Double-loop learning begins to examine our actions in the context of our beliefs and assumptions. We look for patterns and are more attuned to structures and how the decisions we make affect them. At this level, people are involved in organizational structures and function. People might ask: Are we doing the right things? The thermostat would ask, "Is 68 degrees the right temperature?"

Triple-loop learning looks at context and the principles that inform it. At this level, we can see that cause and effect may be separated widely in time and space, and we have a greater sense of the interdependencies between our thoughts and behaviors and how they interact with the world around us. From this perspective, the questions are "How do we know what is right? What principles are we using to decide?" The thermostat asks, "Given my concern for the environment, is heating with coal the best choice? What other choices might there be that would be more in alignment with my values?"

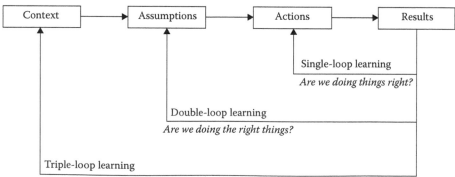

FIGURE 16.3   Single-, double-, and triple-loop learning model.

TABLE 16.3

Principles for Complex Project Management Based on Polarities

1. Develop the field of complex project management as a whole.	*and*	Develop the capabilities of individual people and organization systems.
2. Look for and embrace new models, methods, and tools that will help us be successful in the *new normal*.	*and*	Respect and leverage our existing expertise.
3. Know the outside of our organization system.	*and*	Know the inside of our organization system.
4. Be visionary.	*and*	Be grounded in current reality.
5. Invite participation.	*and*	Provide direction.
6. Be adaptive and flexible to accommodate change.	*and*	Provide structure.

GUIDING PRINCIPLES

The authors have identified six pairs of principles (Table 16.3), which are at the heart of the activities of ICCPM in general and the roundtable process in particular. These are arranged as polarities (Johnson, 1996), which, when managed well, contribute to the successful navigation of complexity and rapid change.

CHARACTERIZATIONS

TYPE OF SYSTEM

The ICCPM roundtable participants, and the organizations they represent, form a CAS with constituent parts operating at multiple stages of socio-technological development.

The participating organizations employ a wide range of organization structures (hierarchical, networked, matrix, innovative, and complex adaptive) and cultural types (entrepreneurial, machine bureaucracy, and scientific).

SYSTEM MATURITY

ICCPM is a young entrepreneurial organization.

ENVIRONMENT

The environment in which the ICCPM roundtables operate is highly political, dynamic, and rapidly evolving, with significant levels of uncertainty and ambiguity.

SYSTEMS ENGINEERING ACTIVITIES

As Is System Description

- The system is a confederation of individuals and organizations interested in addressing the most pressing challenges and leveraging the most promising opportunities in the field of project management.
- The system is not completely aware of the nature of itself as a system and how its structure and behaviors affect desired outcomes.
- There is a tendency in the system to place blame on *outside influences* such as government regulation, governance failures, or stakeholder activism for the difficulties it encounters.

To Be System Description

- The system is more aware of itself and its dynamics, thus empowering people and their organizations to engage with the system to create desired outcomes.
- People and organizations in the system consciously expand boundaries to incorporate more groups whose voices and capabilities are essential to solving problems or dealing with evolving issues.
- The system is aware of and adopts/creates the tools and methods that will help make interaction with the system more reliable and produce desirable outcomes.
- The system itself is more readily able to learn and adapt.

PURPOSE

The primary purpose of the ICCPM roundtables is to provide a forum for leaders in the complex project management sector from both the supply side and the buy side, and some of their key advisors, to come together to learn from each other and, in so doing, develop more effective ways to lead complex major projects.

In continuing to pursue this goal, ICCPM embarked on a second and much larger roundtable series in 2010, building on the 2009 report, with a focus on developing a research agenda that could be undertaken by ICCPM in association with its partner universities and consultants.

HISTORY

The focus by ICCPM initially on one principal issue—the *Conspiracy of Optimism*—soon came to be seen as related to many other issues involving a much broader view of the system. This led ICCPM to expand the boundaries of the system to be considered during the second and third roundtable series.

As we will show, each subsequent roundtable series has considered a much broader system than the earlier roundtable, which is tacit recognition that the dynamics of a much larger system are at play. This larger system takes into account market forces, especially socio-technological dynamics, government regulation, policy setting, and decision making as well as a broad range of community stakeholders, all having an interest in the outcome and contributing to the systems dynamics.

LEGACY OF THE 2009 ROUNDTABLE SERIES

The approach adopted by ICCPM for the 2009 roundtables was to develop a discussion paper to circulate in advance, which set the boundaries of the discussion and raised awareness of the issues.

The briefing notes (ICCPM, 2009a) cited a list of possible causes of complex project failings identified by Flyvbjerg (2005) including technical issues such as poor forecasting techniques, inadequate data, and general difficulty in predicting the future. The notes also identified psychological issues such as making decisions with "an optimistic bias rather than a rational evaluation of the risks and returns" and political issues where "planners and promoters purposefully misrepresent costs and benefits for either political or economic advantage" (ICCPM, 2009a).

The paper argued that the desire to equip defense forces with the best and latest in a world of rapid technological change leads to efforts to contract for *developing* rather than *proven* solutions, to win contracts at any cost, to ignore the unknown risks that might emerge over long implementation times, or assume wrongly they can be controlled or managed.

Along the way, the success of other more urgent projects is jeopardized. There is often a huge waste or misallocation of public funds and political and commercial reputations are frequently tarnished. Potential solutions offered for consideration included greater transparency/accountability on the buy and sell side; ensuring that aspirations/requirements are trimmed to match what is possible; improving cost estimation including greater use of parametric techniques; having the political/commercial will to cancel failing programs; lowering the risk of technological interfaces by better testing; abandoning fixed-price development contracting; using of joint supply-and-buy-side teams; managing risk holistically; approaching risk from a problem exploration, option generation, and analysis before decisions are made; and better training.

At the conclusion of the series, project management consultant Michael Cavanagh, who wrote the report of the roundtable discussions (ICCPM, 2009b), expanded the list of issues facing complex projects to include many that could be characterized as *wicked problems*, further developing the understanding of the issues cited in the discussion paper.

Cavanagh identified 10 issues (ICCPM, 2009b) that continue to haunt complex projects. His report pointed out that, paradoxically, some projects that finish way behind time or grossly over budget can often, in hindsight, be a big success for key stakeholder groups, such as the Channel Tunnel between the United Kingdom and France and the now much-loved Sydney Opera House, which is both a world-famous tourist attraction and a lively arts complex for Sydney's theater, music, and opera goers. The issues were as follows:

1. Unmet or misaligned customer expectations.
2. Changes in the political climate, such as a change of government with radically different policies or a preparedness to blame their predecessors or scapegoat a project for political expedience.
3. Ignoring the interests of seemingly minor external stakeholders whose influence increases when the political winds change and who can stop projects in their tracks or halt their commissioning.
4. Projects that are underbid to win or underestimated to gain political support.
5. Shortage of leadership capacity, especially experienced project managers with both exemplary technical and soft skills, and as projects become more complex, the ability to comprehend and work with people from a very broad range of both project-oriented and non project management disciplines.
6. Poor coordination/alignment across the boundaries of disciplines, teams, and subprojects.
7. Inflexible, institutionalized procurement strategies on both the supply side and buy side, which limit the agility/flexibility to deal with change, and unanticipated events, both black swans and unknown unknowns (black swans are events that can be imagined but have not been observed before; unknown unknowns are events that do not fit any current reality).
8. Culturally bound expectations that suppliers and purchasers will conduct the relationship in transparent, independent manner, but few opportunities to align supply-and-buy-side goals.
9. Dependence of a project on scientific or engineering breakthroughs, or estimates based on wishful thinking, guesstimates or hope, can result in suboptimal results, failure, or scope creep.
10. Most current management methods/models, risk management tools, and decision processes deal poorly with uncertainty.

2010 ROUNDTABLE SERIES

Before embarking on the roundtable series, ICCPM established an international task force comprising partners, academics, and consultants who agreed to contribute chapters or working papers to the report on specific issues. The task force report, titled *Global Perspectives and the Strategic Agenda to 2025*, was published in 2011 jointly by ICCPM and Global Access Partners (GAP).

The report points out that

> ... most large capital investments come in late and over budget, never living up to expectations. More than 70% of new manufacturing plants in North America, for example, close within their first decade of operation. Approximately three quarters of mergers and acquisitions never pay-off ... And efforts to enter new markets fare no better. (Lovallo and Kahneman, 2003)

The report then makes recommendations about strategies and tactics that complex project leaders can apply under the following headings:

- Delivery leadership
- Collaboration and competition
- Benefits realization
- Risks
- Opportunity and resilience
- Culture
- Communication and relations
- Sustainability and education
- Recommendations for further research

The report also includes an appendix, a set of working papers prepared by thought headers in the complex project management field.

The present authors were invited to facilitate the 2010 roundtable series and facilitated two of the sessions—in Canberra, Australia, in August and Washington, DC, in September 2010. A colleague agreed to facilitate the UK roundtable at Ashridge University in October using the same methodology and an identical set of questions.

The outputs from the three roundtables were consolidated into a narrative, which informed the report as published, the recommendations, and several of the working papers (ICCPM, 2011). Although the final report incorporated many of the ideas generated at the roundtables, the tenor of the task force report tended to reflect the views of the experts tasked to contribute working papers, and the world views of the consultants tasked with crafting the report, rather than truly synthesizing the ideas of all participants into a new, jointly held perspective. Nevertheless, there was a significant shift in emphasis away from the *symptomatic* issues that dominated 2009 toward a larger, more systemic, *causal* view of the issues.

The report proposed 9 broad policy approaches, 60 targeted recommendations, and 31 suggestions for further research by a global collaboration between ICCPM partners and practitioners, theorists, and academicians.

These recommendations have been consolidated into five main research areas under an initiative to establish a Cooperative Research Centre into Complex Projects and Programs (CRC for MCPP, 2013):

1. *Managing risk and uncertainty*: this program is concerned with the systemic and holistic identification, quantification, classification, and mitigation of the risks and uncertainties associated with complex projects and programs. It also considers in a systemic way how individual risks and uncertainties interact and how project managers make sense of and act on risks and uncertainties.
2. *Delivery and dynamic tensions in complex projects*: this program is concerned with the social, symbolic, and cultural dimensions of projects as soft systems of project and stakeholder relationships. It investigates how project managers make, give, take, and communicate meaning and sense in the process of delivering projects, navigating issues of organizational creation, maintenance and change, organizational learning and knowledge management, and power relationships.
3. *Leadership of complex projects and programs*: this program investigates the issues associated with the leadership of complex projects and programs and the (distributed) agency of project managers within governance structures as social systems. Associated themes include the delivery of benefits to stakeholders and project/program outcomes.
4. *Analysis and modeling of projects as complex systems*: this program aims to investigate projects/programs as complex (adaptive) systems and provide models/archetypes of configurations dependent on context, a mapping of project complexities and associated strategic responses.

5. *Complex project/program design*: this program investigates how project managers design complex projects/programs, which evolutionary pathways project take from design types, and how project methodologies can be mobilized to match the design and strategy (p. 11).

2012 ROUNDTABLE SERIES

The 2012 roundtable became an exercise in working out how to deal with the major issues of complex project management, not by focusing on the parts alone, as the industry had tried to do in the past, but by taking a broad, systems-wide view, not only examining the politics of the issue but also asking whether the larger system in which we are all operating is also in need of some kind of transformation.

The theme chosen was *Complexity in a Time of Global Financial Change: Program Delivery for the New Economy.*

The Asia Pacific event was held in Canberra, Australia, in March. The European event was held in Lille, France, in August, in conjunction with the SKEMA business school research conference, an annual event in which ICCPM has participated since 2009. The United Kingdom event was held in London in November at the Government Audit Office, sponsored by the Major Projects Authority, which is part of the Cabinet Office. The North American event was held in November in Washington DC, sponsored by Booz Allen Hamilton.

Prior to the first formal roundtable meeting, 16 members of the ICCPM community participated in a preliminary online workshop to understand the issues in order to prepare the questions for the series and to generate ideas from which to prepare a discussion paper as stimulus/pre-reading. The online version of the Zing meeting environment was used to conduct the sessions with participants joining in from Australia, the United States, Brazil, and Europe. Skype was also used to discuss the issues. The preliminary workshop was attended by representatives of ICCPM (the chief executive officer [CEO] and chief operating officer [COO]), academics, partners (NASA, EADS), associate partners and consultants (the present authors), and affiliated organizations (CSIS and the International Association for Contract and Commercial Management).

The questions asked during this session, shown in Appendix B, generally dealt with the current context, particularly the big economic/financial issues facing people, organizations, nations, and the world, how the education/learning system was contributing to the system, what research was required to deal with a more complex mix of politics and economics, how the political system was responding to the situation, and the extent to which globalization and innovation was playing a role.

During the session, supplementary questions were asked to deal with issues that had not previously been considered important, such as the implications for complex project management of the World Economic Forum Risks Report Seventh Edition (2012), one of several documents introduced into the conversation by members of the ICCPM community.

Volunteers from among the group accepted responsibility for crafting the discussion paper. Three members of the group shared the majority of the task of reviewing the session outputs, assembling the ideas into a narrative, undertaking additional research to add factual information to support the arguments. Once a near-final draft was complete, the document was circulated to all the members of the group for comment, additions, deletions, and improvements. The main authors were located in Australia and the United States, and other significant contributors resided in Brazil, the United States, and Europe.

CURRENT SITUATION

From an analysis of the results of the preliminary workshop, it was determined that the issues confronting the complex project management community are systemic, connected, and evolving/transforming. It was clear that the current political environment is no longer working, that the governance of the system no longer works well, and that a new kind of *operating system* is required.

The key question was what new kind of operating system do we need to put in place? What new system(s) might we create that would leverage what we know about complex systems and the *rules* that govern them and that, by transitioning to the new and simpler state, would resolve many of the problems that occur as a consequence of the highly cross-connected interdependences in the current system, and what would be the features of the higher-level system?

KNOWN PROBLEMS

The participants identified 16 issues that would impact on complex projects in the immediate future that were important to discuss at the roundtables (ICCPM, 2012). These were as follows:

1. Structural reorganization of the global economy, where investments in new arrangements/ ecologies of physical and soft systems infrastructure are unlikely to be made, because business and the government do not understand the size and transformative nature of the socio-technological shifts under way.
2. Foresight capacity, especially the inability of politicians, educators, and business and community leaders to make sense of the emerging future.
3. There is now insufficient time between the introduction of a new technology (system, product, service, tool, method, or technique) and its deployment to develop a deep understanding of all the consequences of its use and to get buy-in from stakeholders uncertain of its impact.
4. Growing political complexity and partisanship resulting in the lack of political support for new projects or projects and programs caught in the crossfire between feuding political interests.
5. Capital investment risk devolving from declining government revenues and rising costs, resulting in less investment in projects normally funded or underpinned by the government sector.
6. Project complexity requires ever higher levels of specialized leadership, engineering, scientific, and technical competence.
7. The need for new tools to simplify complexity and uncertainty especially management, coordination, interests integration, and knowledge co-creation processes.
8. Learning from chaos and complexity science in order to develop the capacity to run projects as CASs.
9. The urgent need for project leadership capacity at every level of organizations, especially to deal with uncertainty, paradox, politics, stakeholder and supplier conflicts, and unexpected events.
10. Dealing with competition from new entrants to the complex project management field who are more flexible and adaptive than the current major players.
11. Skills shortages, especially in the critical math, science, and engineering fields, especially in combination with leadership skills.
12. Stakeholder activism, particularly if the interests of stakeholders are ignored or subjugated to the client's interests, resulting in opposition or delays to major projects.
13. Cross-cultural complexity, especially the difficulties faced by operating across numerous boundaries, nations, time zones, continents, suppliers, professional disciplines, organization departments, and a myriad of community stakeholders.
14. The impact of natural events on complex systems, which become more vulnerable the more complex and interdependent they become.
15. Resource constraints as a result of world population growth and economic expansion, especially for critical resources such as energy and rare earths.
16. Regulatory risk, especially the lack of uniformity across nations, in relation to the stock market, the trade in commodities, taxation, subsidies, financial controls, labor prices, and different approaches to climate change or farm and industrial production and concerns that arbitragers might artificially distort these markets.

MISSION AND DESIRED OR EXPECTED CAPABILITIES

The primary mission of ICCPM with respect to the roundtables is to grow a network of practitioners, professionals, researchers, and educators who can deliver leading edge complex project management solutions to client organizations and partners around the world.

TRANSFORMATION NEEDED AND WHY

The complexity and uncertainty of the task are described in the 2012 ICCPM roundtable series discussion paper (ICCPM, 2012):

> The issues now facing complex project leaders are beyond the borders of their projects and exceed what might be expected from increasing complexity and accelerating change. They now have to contend with the downstream by-products of the change—fiscal chaos, bureaucratic inertia, vested interests unwilling to give up hard won privileges, critical skills gaps, looming resources shortages, social unrest, political partisanship and reluctant, nervous capital. New unanticipated issues emerge every day.
>
> The issue has been characterized as learning to fly a plane, while the plane is already in the air, and being re-assembled into another kind of transportation technology altogether. And, at the same time, the current passengers are disembarking and another group is boarding that demands a better quality of service or experience at lower cost than ever before. (p. 21)

CONSTITUENTS

The ICCPM is a company registered in Australia, with a headquarters in Canberra, and it represents an international community keenly interested in the issues facing complex project management leadership.

Many of the members of the community are affiliated organizations with no formal partnership relationships, such as the National Audit Office and the Major Projects Authority in the United Kingdom, but who nevertheless have a vital interest in better understanding the reasons why the projects over which they provide oversight continue to fail in some way.

The company also has customers in the form of partners that fund the business, who are seeking to acquire leading edge insights into how to operate differently, as well as associate partners, such as consultants, whose primary interest is doing business with the partner organizations. The associate partner network provides much of the intellectual input as well as on-the-ground, practical knowledge around implementation of new models, methods, and processes. The universities in the network have an interest in undertaking research and providing training services to the staff of the member organizations.

There are also other organizations in the network who partner with ICCPM, sometimes as customers or partners and at other times as knowledge co-creators. Some of these organizations have developed new kinds of tools they are hoping to trial in real-world settings, or to have more widely adopted by ICCPM partner organizations.

The ICCPM community works in many different combinations, ranging from small groups engaged in research, delivering courses in some aspect of complex project management, or undertaking consulting assignments together, to larger gatherings at the roundtables and the research conferences.

(COMPLEX) SYSTEM/SYSTEM OF SYSTEMS/ENTERPRISE

An overview of our CAS is depicted in Figure 16.4.

Scope (A of Figure 16.4)
A series of roundtables to be conducted around the world in multiple locations.

Structure (A of Figure 16.4)
A membership organization with affiliates.

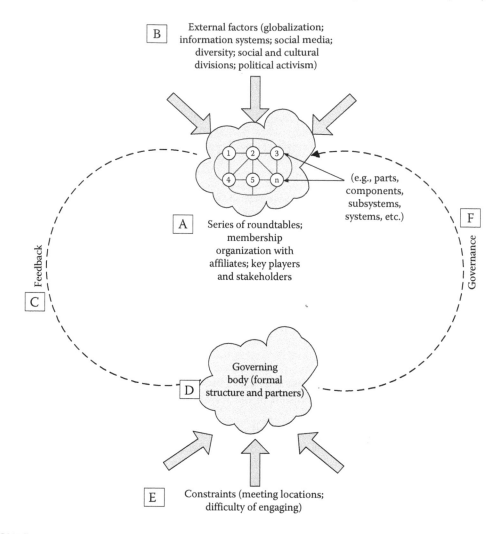

FIGURE 16.4 CAS overview.

Boundaries (**A** of Figure 16.4)
The system comprises the leaders and staff of organizations involved in the design, delivery, ownership, and operation of complex major projects and academics and consultants who support the sector. However, as the concerns of the participants in the roundtable series have made clear, the problems they are dealing with seem intractable and incapable of resolution by this group alone. Even industry and government leaders feel powerless to act on the system, citing others as having *control* or causing the problem.

Internal Relationships (**A** of Figure 16.4)
Key stakeholders in this project management community include leadership, project management, and acquisitions staff of government agencies (Department of Defence [Australia], NASA [United States], Major Projects Authority [Cabinet Office], and National Audit Office [United Kingdom]), prime contractors (BAE Systems, Boeing, Lockheed Martin, EADS, and Thales), universities (Central Queensland University and University of Technology [Australia], University of Hull [United Kingdom], and the Defense Acquisition University [United States]), and providers to the sector (Booz Allen Hamilton, Human Systems, Association for Project Management, the International Association for Contract and Commercial Management), and

more than 30 other partners, think tanks, and associate partners who consult to the sector and their customers or stakeholders.

External Factors (B of Figure 16.4)
Participants in the roundtables have identified numerous global trends having an impact on the complex project management community, including the following:

- Globalization of everything: business, entertainment, travel, communication, and personal networks
- Exponential increase of the collection and exchange of information
- The growing interdependence of complex technological, economic, and information systems including logistics, transportation, energy, communication, and distribution systems, where a minor glitch in one part of a system can bring the entire system to a standstill
- A new period of transparency and engaged debate challenges the power structure and spheres of influence in society. With social media no one can hide anymore
- The divergence and diversity of religious, professional, activist, community, and cultural interests and how they interfere or intersect with other systems
- The need to address the social and cultural divides that have become more obvious than ever before
- The desire for or expectation of instant gratification, the *must have it now* generation
- Growth in political activism with increasing education and dissolving boundaries between countries and communities

Constraints (E of Figure 16.4)
ICCPM is international, so the roundtables were required to be conducted where there are concentrations of partners and associate partners.

However, the more critical constraint is the difficulty in persuading stakeholders outside ICCPM's network to participate in such a broad and seemingly wide-ranging discussion, particularly political leaders, regulatory agencies, and the broadest range of community stakeholders who perceive their interests to lie in only narrow issues.

Governing Body (D of Figure 16.4)
Although ICCPM has a formal structure and existence with a board, a CEO, and operational staff, it relies for the delivery of many of its services on the talents and skills of organizations with which it partners, who come together from time to time to deliver courses, provide consulting services, present at conferences, undertake research, publish e-books, or organize and facilitate roundtables.

Decision-Making Process (**C** and **F** of Figure 16.4)
Decisions are made consultatively. Most contributors to the roundtable project operated autonomously, providing feedback from time to time to the CEO or the operations manager.

The CEO sought suggestions or recommendations from contributing organizations for the role they might play and the broad strategy they might follow. Each organization then did its part and kept the CEO and each other informed of progress via Skype conversations and emails.

The authors, John Findlay and Abby Straus, were responsible for the design of the main roundtable process, the facilitation of the events, and writing and editing the discussion paper. They traveled to each country, as required, to facilitate the events.

The task of writing a series of questions to guide a pre-workshop to gather ideas for the 2013 roundtable series was as a joint effort of the CEO and the facilitators.

Sponsorship for the London roundtable was provided by the Major Projects Authority and the National Audit office who hosted the activity. In Lille, France, the event was held in parallel with the SKEMA university project management research conference. In the United States, the event was organized by the US ICCPM President Fred Payne with sponsorship from Booz Allen Hamilton, who also provided the venue in Reston, Virginia.

The preparation of the discussion paper was also a collaborative effort, with editing undertaken by the authors with support from Professor Carolyn Hatcher of Queensland University of Technology. Other members of the ICCPM community from the United States, Europe, and Brazil offered comments and suggestions on the final drafts.

The CEO and operations manager of ICCPM were responsible for making arrangements with all of the parties for the provision of a venue, the catering, promotion, and printing of materials and circulation of reports.

At the time of writing this case study, the final report was being edited by Michael Cavanagh of Ireland, who was the author of the first roundtable report in 2007. The results will be published as part of the ICCPM e-book series and launched at the annual research conference, which is being held in London in October 2013.

METHODS/TOOLS/PRACTICES/PRINCIPLES USED

The technology used to support the roundtables was the Zing complex adaptive meeting environment (Figure 16.5). Its purpose is to provide a container for a suitably representative sample of the people in the system to meet and conceptualize a robust model of the system and develop strategies for how to leverage the system.

The tool can also be used with many kinds of decision, thinking, and learning methods to intervene in a system at different intervention points and at any scale, from small work groups to large community meetings.

The conversation at the roundtables was guided by an etiquette, talk–type–read–review, a simple rule of interaction, which, when adopted by the whole group, results in a phase transition to a higher level of order and orchestrated performance.

Rich open-ended questions stimulate discussion in small groups or pairs and elicit a diverse range of participant responses that exhaustively explores a specific issue. The question sequences guide the conversation and facilitate the self-organization of shared knowledge.

The ideas are read aloud and the participants look for the patterns in the ideas. This is a sense-making step, which leads to some kind of emergent conclusion (a plan, decision, new model theory, feedback, etc.). The themes are recorded by the facilitator at the conclusion of each question.

Ideas generated by the participants also catalyze further ideas. Kauffman (2008) shows that ideas/concepts are autocatalytic, because they stimulate the expression of other related ideas, concepts, memories, or schema in the brain. These are often stored as what Wittgenstein (2009) called *language games*, which are families of related concepts.

No single human brain is complex enough anymore to deal with the level of complexity we are facing in many communities and industry sectors worldwide, including that of major projects (Bar-Yam, 2006). By ensuring that stakeholders representing multiple interests and disciplines are in the room for crucial conversations, the required amount of complexity thereby assembled is usually significantly greater than that of the system to which the new shared model and solutions will be applied.

Another way of satisfying Ashby's law is to reduce complexity. In a CAS decision-making or learning setting, this is achieved by increasing the variety or richness of the ways in which we explore an issue, usually by breaking the problem down into sequences of rich, open-ended questions and presenting one question at a time. It reduces the complexity of the learning process by progressively widening the aperture of the lens through which the participants see and deconstruct the issue, so the process explores the entire space in an overlapping way.

This process has the added benefit of increasing the overall *cognitive* complexity of the group, thus enabling more sophisticated thinking to take place and more robust understanding to be reached.

During the course of a complete question sequence of six or seven questions, participants who began the meeting with diverse viewpoints about an issue generally reach similar, and often radical, conclusions. This occurs as a consequence of continual feedback from other participants during the reading aloud and theming steps.

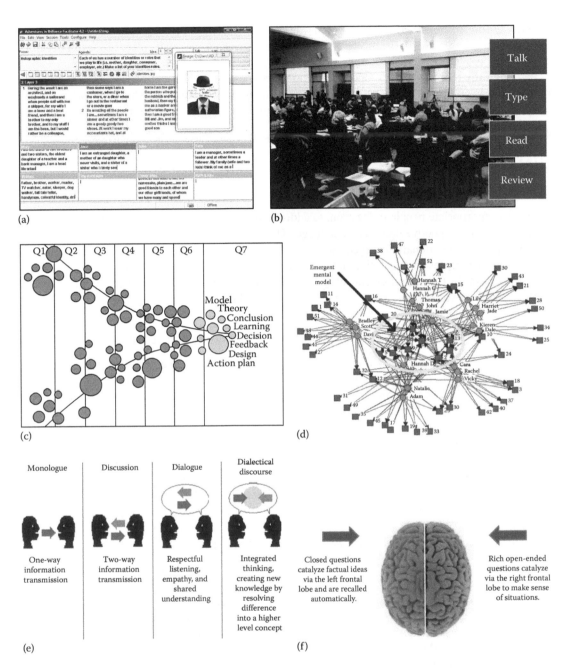

FIGURE 16.5 CAS features of the Zing meeting environment. Zing is tool that helps groups deal productively with multiple viewpoints and interests/facilitates the integration of multiple viewpoints and interests. (a) The Zing user-interface has personal spaces where ideas are generated and a group space where they collect. (b) Simple rules of interaction help people think/work together more effectively. (c) Rich question sequences facilitate and funnel the self-organization of shared knowledge. (d) New mental models and shared understanding emerge. (e) The appropriate conversation type is chosen to match the complexity of the issue. (f) The autocatalytic features of words and concepts help generate new possibilities.

When separate meetings are held in different locations at different times with a similar community, and using the same question sequence—as occurred with the roundtables—all groups generally produce similar responses and reach similar conclusions. Multiple meetings are thus structurally equivalent to a single very large meeting. This means that very large-scale interventions in organizations can be reliably carried out using tools such as Zing.

Although the tool can also be operated using older socio-technological logics, for example, as an information collection device or to vote on issues, as some facilitators do, this is not what it was designed for, and the results achieved in this manner are nowhere near as effective as those we have been describing.

CHALLENGES

GETTING THE WHOLE SYSTEM IN THE ROOM

Getting the appropriate people in the room to address the issues of complex project managers from a whole-system perspective presents a challenge. This is particularly the case when the stakes are high, the issue is emotionally charged, and the entities and individuals involved are diverse and believe that they represent competing interests. However, having everyone in the room is essential in order to meet the twin requirements of Ashby and Conant for dealing successfully with complex systems, requisite variety, and the potential to develop a robust model of the system.

STEERING PROJECTS THROUGH MULTIPLE DISRUPTIVE SOCIO-TECHNOLOGICAL SHIFTS

The current socio-technological environment of accelerating change and greater complexity involving rapid transitions from one major phase to another is particularly problematic for complex projects. People, their organizations, and their projects need to be capable of reorganizing into new forms, which are a better fit with the new context.

Iansiti (2009) makes the point that "megaprojects face considerably greater complexity and uncertainty in phase transitions because of their long duration and overlap of design, production and fielding, and so are the most vulnerable they have ever been, at this point in time" (p. 97). Any model of the system has to take into account successive disruptive technological and social shifts. Many people see it, instead, as a single shift or undifferentiated continuum, neither of which is a sufficient description of the system.

DEVELOPING NEW WAYS OF THINKING, ACTING, AND INTERACTING

Many of the tools, methods, and processes that complex project managers use today are stuck on old fitness peaks, optimal states from an earlier socio-technological paradigm. While traditional project management skills are still vital, a variety of new skills is now also required in order to successfully navigate the new terrain. White (2009) subscribes to the view that "most acquisition program failures result more from ineffective interactions of people and inappropriate processes than from technology shortfalls." A related challenge is to develop a culture that acknowledges and rewards these skills, and one where today's project managers—and the next generation of PMs—may acquire them with as much speed and ease as possible.

ANALYSIS

The ICCPM roundtables have provided a productive environment for examining and addressing the challenges facing the complex project management community.

Getting everyone in the room: All of the people who needed to be in the room for the much broader systems discussion were not present. During the closing comments at the London 2012 roundtable, several of the participants argued that the complex projects community could not resolve the issues on its own, because the community does not represent enough of the system. These issues include

- Developing adequate technical and leadership education systems necessary to train those who will design, build, and operate future generations of ultra-complex projects
- Making sure we have an economic system that allows the adequate funding and management of complex projects that support our current level of security and safety
- Evolving a political system that is able to make wise choices based not on purely self-interest but also for the betterment of society as a whole

Ensuring that representatives of the whole system will be in the same place at the same time is difficult; but a reasonable approximation can be achieved by having many types of groups with sufficient diversity equipped with same information doing the same type of thinking.

When ICCPM began in 2007 to address a localized systems problem within a government acquisition community, to ensure that major projects could be undertaken on time and on budget and meet requirements, it set what seemed at the time to be a broad boundary. The boundary of the system was to include key players on the supply side and the buy side in an effort to come to grips with the problem of project failure in terms of cost, time, and meeting requirements. Over the ensuing 6 years, the community expanded the boundary of the system to include the project management community worldwide. This, also, provided too narrow a view of the system, and the aim for the next round is to invite greater participation from education, government, and regulatory agencies, as well as community stakeholders.

Steering projects through multiple disruptive socio-technological shifts: In the past, it was possible for the life cycle of a major project to exist within one socio-technological period or paradigm, the Industrial Age, for example. However, as the rate of change has increased, almost all projects are conceived and executed in an environment of disruptive change (Figure 16.6). Roundtable participants identified the need to rapidly adapt multiple aspects of a project or program to keep pace with the change, including leadership and coordination, governance, stakeholder relations, and technological innovation, testing, and deployment.

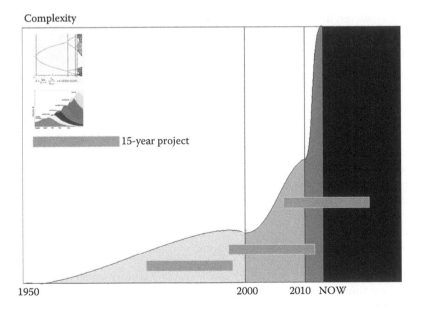

FIGURE 16.6 Major projects now span one or more paradigms.

Although the participants were able to identify individual aspects of the system, and some of the necessary shifts, they were generally unable to develop a comprehensive model of the system as a whole, especially the successive transformations. To invoke a much-used metaphor, they could name the individual parts of the elephant, but not the elephant as a whole.

The complexity model of change was developed by the authors to explain the observations and experience of the community and to provide an overall view of the system that can be used as a reliable map of the terrain. The authors have subsequently tested the model with members of the complex project management community worldwide, who have found it useful for understanding their current circumstances and plotting their route forward.

A common issue that the model explains is the embedding of old/earlier logics in systems, especially IT systems, which limits the ability of the larger system to adapt. Another is when different parts of an organization system are operating from the perspectives of different socio-technological paradigms. When this happens, conflicts arise about strategy, coordination, and what methods and tools to use, which reduces the capacity of the organization to perform. In addition, the model helps identify potential productivity gains that can be achieved by choosing the appropriate alignment of technologies skills, roles, and rules of interaction. For example, directive leadership—which has its center of gravity in the Industrial Age—may have been broadly appropriate for coordinating the activities of a minimally educated workforce, but today's highly educated teams require a much more flexible and distributed style of leadership, characterized by what we see in the Knowledge Age, and moving into the Wisdom Age.

This is not to say that all aspects of previous eras should be jettisoned, however. There may be times when a directive leadership approach is appropriate, for example, during emergencies. Leaders must have the ability to apply both/and polarity thinking in order to transcend old paradigms *and* include the best traditional models and methods.

By and large, the complex project management community has taken a unipolar approach to addressing its challenges. Its singular focus on technical competence to the neglect of *soft skills* results in lack of trust and poor coordination. Conflicts may also arise when strategic choices are made between two poles, for example, undertaking only continuous improvement and not participating in cycles of breakthrough innovation, which can easily result in an organization falling behind its competitors.

This is not surprising when one considers that the prevalent culture displays a preference for either/or problem solving, clearly a necessary approach for many aspects of engineering and technical management. However, this approach alone no longer suffices in environments of high complexity, ambiguity, and rapid change where high-touch relationships management is just as important to a project as high-tech tools and processes.

Developing new ways of thinking, acting, and interacting: In addition to dealing with paradox and dilemmas using both/and thinking, roundtable participants have identified numerous new or developing skills (Straus and Findlay, 2011) that are now needed including the ability to

- Take an integrative, multi-systems perspective
- Embrace/accept multiple viewpoints
- Be self-aware
- Lead well and have the confidence to support others in leading
- Communicate to multiple levels of understanding and ability
- Reach a good balance between focusing on relationships and getting results
- Facilitate conversations across multiple disciplinary, functional, and cultural boundaries
- Be aware of and work with multiple paradigms

As people acquire these skills, they develop greater levels of cognitive complexity and the ability to engage in triple-loop thinking and learning. This allows them to take a systemic view, comprehend emerging paradigms, develop new mental models (Senge, 2006), and make fundamental changes to systems at the highest place of leverage (Meadows, 2006).

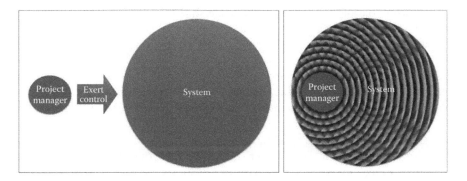

FIGURE 16.7   The shift from outside *control* to inside influence of the system.

When a project manager sets out to deal with changing requirements, technological transformations, and stakeholder issues from a *control* or outside-the-system perspective, these issues may appear to be *wicked problems* that keep evolving and never go away.

But when a project manager regards himself or herself as a participant in a CAS, a new perspective emerges. The manager is now capable of co-creating high-leverage actions with other participants in the system, thus influencing it, but not with the precise predictability of systems with linear and algorithmic features, such as machines, processes, or networks (Figure 16.7).

The good news for project managers seeking to achieve some kind of control over a system is that CASs have features and patterns of behavior that can be relied upon. And some of the unique states of a CAS are in fact linear or algorithmic and can continue to be dealt with using existing methods when they show up.

MediaCityUK: Complex Construction in a Cooperative Framework

The MediaCityUK case study contributed by Barbara Chomicka is an example of a very complex project that was successfully undertaken using an evolutionary, complex, adaptive-style approach. The project team worked well together as a team, constantly collaborating and brainstorming solutions as they arose. They were able to complete the fixed-price contract on time and on budget by responding rapidly and flexibly to changes in *scope, program, budget, and strategy* and unanticipated events such as the global financial crisis and by shifting resources and talents around when needed.

DEVELOPMENT

The authors have taken advantage of their participation as associate partners of ICCPM to proactively develop and trial the new kinds of tools that members said they needed and to collaborate with other associate partners in their development.

From these collaborations has emerged a suite of interdependent tools, methods, and processes, known as the complex adaptive operating system (CAOS), which together can be used by people anywhere in the system to lead organizations and projects.

The methods have similarities to the Complex Adaptive Systems Engineering (CASE) approach of White (2009/2010) and the Enlightened Evolutionary Engineering (EEE) approach proposed by Bar-Yam (2004).

The methods have been developed by applying abductive (metaphorical) reasoning (Prawat, 1999), which involves starting with a metaphor or conceptual principle and applying it to each aspect of another conceptual system. In this case, the authors applied the laws of complexity and the features of CASs to the latest in management, leadership, thinking, decision-making, and coordination methods. They compared the results with the emergent patterns in the complexity model of change. The process resulted in a well-defined family of concepts that is consistent across six paradigms of transformational change.

CASE, CAOS, and EEE all acknowledge that we must start by identifying and involving the *whole system*. Bar-Yam (2006) points out that the ideas central to the discussion of complex systems have to do with the relationships between the parts of the system and how these relationships contribute to the behavior of the system as a whole. We come to understand that, rather than residing in the parts of a system, the solutions to systems failure can often be found in the relationships of the parts to one another. This is equally true whether we're talking about the interoperability of technological components or the ability of humans to draft contracts that either smooth the way for successful project delivery or bog the process down in a quagmire of litigation and political disputes.

The first CASE activity is to *create a climate for change*, which is to set up an ongoing process involving key stakeholders who accept the principle of using a self-organizing approach to create solutions (at least for the time being), to collectively identify customer and other stakeholder needs and how the organization needs to change to do this, and to harmonize interests and propose solutions.

This is similar to the first steps in the CAOS process at the ICCPM roundtables, which begins with a context-setting activity where participants are asked to share their knowledge or perceptions of the major trends in the world that have (or could have) an effect on the issue at hand. This step helps them identify the broadest possible set of causal connections (Senge, 2006), thus opening the aperture to define the *system* as a larger entity than it may have at first appeared.

Other steps in the CAOS strategic planning process include the dynamic equivalent of strengths, weaknesses, opportunities, and threats (SWOT) to help participants sort through the complexities of an issue to see what elements may influence either success or failure of various parts of the system. This process, one of many developed for the Zing tool asks as a single question: Which aspects of the organization/project are working that one would want to *keep*? Which ones are no longer working or necessary that it would make sense to *abandon*? What aspects have possibilities that it would be profitable to *reinvent*? Ultimately, it will be changes in behavior—what people commit to doing—that change the system, and this is a first step toward determining what these might be.

The CAOS approach offers a way of relating to and mediating activity between stakeholders of all kinds using simple rules of interaction. These rules are complex adaptive interpretations of the best-of-breed leadership, discourse, coordination, learning, and decision-making process.

The CAOS methods and approaches developed for complex project management include the following:

- The complexity model of change offers a robust model of the larger socio-technological system that affects a project and provides a way for project managers to understand the kinds of shifts in technology they might anticipate and what new tools, skills, methods, and systems might be required. It can also be used to analyze an organization's stage of development and to identify opportunities to realize productivity gains by the adoption of more advanced socio-technological configurations.
- The complex adaptive meeting environment supports collective knowledge creation with stakeholders of all kinds—owners, suppliers, cross-functional teams, and community leaders. It provides a container for the level of complexity suitable for addressing even the most complex problems. This tool is also a platform for the development of question sequences to perform new learning, model creating, plan conceptualization, feedback or relationship development functions, and rules of interaction to guide the conversation.

- Stakeholder and cross-boundary team interests and activities integration methods that involve internal and external stakeholder from the outset in the design of a project so they develop shared interest and commitment, as well as involving them throughout the project in critical decisions about design changes. This approach replaces the current practice of *selling* fully scoped out projects to stakeholders in order to get buy-in, which often leads to resistance.
- Distributed or *fractal* leadership that is self-similar at every scale, allowing leadership to emerge wherever it is needed in the system. Fractal leadership for complex project management combines technical (STEM) skills with leadership/coordination skills that, together, allow personnel to effectively navigate complex, ambiguous, and rapidly changing environments.
- Iterative cycles of breakthrough and continuous innovation that help projects adapt to cycles of disruptive transformation. This paradigm-jumping process focuses not only on product and service innovation, but also on re-inventing leadership, governance, policy, coordination, business model, decision making, and learning systems. It also recognizes that the creators of new product/service concepts in complex, rapidly changing environments need to simultaneously create ecologies of companion products and services and develop people with a higher-level mix of capacities, particularly STEM skills + leadership/coordination skills to undertake the work and operate the new system.
- Polarity (both/and) thinking to assist in managing the paradoxes, dilemmas, messes, and wicked problems that arise at the boundaries of value systems and the intersection of socio-technological paradigms. This method supports the development of strategies for leveraging the benefit of apparently conflicting points of view, such as breakthrough and continuous innovation or the need to provide stability in the midst of change.
- Other methods have also been developed for specific functions of project management including evolutionary contracting, process improvement and transformation, requirements gathering, and project and program reviews.

RESULTS

The roundtable process has proven to be a source of many new opportunities for participants. For ICCPM, it has resulted in the creation of an organization that, although still in its infancy, is growing internationally.

Some of the new initiatives include the following:

- A series of short-term research assignments to better understand complex project management issues including a cross-cultural complex project management study in Germany, a research project supported by the National Audit Office in the United Kingdom focused on measuring project success, the creation of a digital library in a focused on complex project management issues, and a study of best practice acquisition and sustainment strategies
- The establishment of a Cooperative Research Centre in Australia as a long-term approach to the extensive research agenda identified during the 2010 roundtable series
- The creation of an associate partner network, as the consulting arm of the center, to deliver short courses in complex adaptive and systems thinking approaches to leadership, management, process improvement, risk management, and innovation and undertake reviews of projects or programs that are experiencing difficulties
- The creation of suites of new tools such as Michael Cavanagh's complexity assessment tool and Booz Allen Hamilton's Polaris real-time risk management tool

- In a cooperative effort between ICCPM team members in Australia, the United Kingdom, and the United States, the development of a new approach to project and program review that combines traditional, well-established tools and methods with new CAS approaches
- The publication of an e-book series on complex project management issues
- Cooperation between ICCPM, its partners, and associate partners in new countries around the world

For the authors, the roundtables have been a source of insight and inspiration for the CAOS suite of complex adaptive methods, which are being created in collaboration with other members of the ICCPM community. Many of these methods have been trialed in Australia, the United States, South America, and Asia and continue to be further enhanced. New methods continued to be developed for application at different scales and for different types of communities (whole of government, whole states, regions, professional communities) in fields as diverse as defense, economic and community development, and international development health.

In particular, project managers are finding polarity thinking to be useful in helping them recognize the both/and nature of many issues that they had previously viewed as insoluble choices, paradoxes, or dilemmas. The framework helps them understand the issues from the perspective of each pole and allows them to create strategies that leverage the benefits of both. The authors have facilitated polarity thinking workshops for PMs in North and South America, Australia, and Europe and have received consistently positive responses from the ICCPM community indicating this as a much-needed tool for resolving complex, wicked problems.

SUMMARY

- The ICCPM roundtable series has enabled leaders in the complex project management community to discuss, explore, and develop a better common understanding of the challenges and opportunities in the current environment and to identify solutions.
- New frameworks, methods, and approaches are being developed by members of the community to deal with complexity and uncertainty, including an approach centered on the laws of complexity and the features of CASs.
- However, new approaches will need to continue to be developed to deal with phase transitions yet to emerge. These new paradigms will require new constellations of roles, rules, and tools.
- The complex project management community will need to develop better mechanisms for reliably applying our new knowledge to our organizations and projects to produce desired behaviors and to do this repeatedly in ever shorter time frames.

CONCLUSIONS

The roundtables have played a vital role in the development of ICCPM and its community in an emergent, rather than a planned, way, providing opportunities for its partners, associate partners, and affiliates to deal with a very complex international problem from a global, systemic perspective. The cooperative effort to organize and facilitate the roundtables has parlayed into new business relationships between ICCPM, its members, and associate partners and the launch of new products and services.

The case study highlights the urgent need for a real-time, systems-wide approach to the development of the methods and tools for managing complex projects and the new skills that are needed and the new roles people will play.

It also demonstrates the importance of engaging the entire system surrounding the development of major projects rather than just the project management community.

The conversation needs to embrace the communities in which the projects will operate and the political and business leaders who will commission, own, or regulate the projects—in order to successfully compete and operate the projects in the context of increasingly shorter successive waves of transformational change.

Those charged with the responsibility for investing in, owning, designing, and developing projects will not only need regularly updated models of their projects in order to successfully *control* them, they will also need more robust models of the broader social and political systems in which the projects will be developed and operated to replace traditional models of political decision making that are no longer working as well as they might.

There is also a need for project managers to continuously anticipate the skills, leadership and coordination roles, technologies, methods, and processes that will be required to successfully surf the waves of change that are coming our way.

A broad representation of the internal stakeholders of project management system is therefore also needed, so people from across the entire organization system and not just the leadership or the project management professionals are able to engage in real-time methods, processes, learning, decision making, and structure innovation.

Political, community, and educational leaders should be also involved in the conversation as well as the owners and investors in complex projects.

A form of the roundtables should be extended to include the staff on the supply and buy sides of major projects who have responsibility for the day-to-day implementation. They have an important contribution to make to the development and trial of tools that combines the best of the complex adaptive approach with the best of the linear or algorithmic.

SUGGESTED FUTURE WORK

Further work needs to be done to apply the CAS approach to the other aspects of project management. In particular, we need to marry the best of the complex adaptive approach to the best of proven technical systems in a well-integrated and seamless manner. Many discussion questions that have been, and could continue to be, asked are listed in Appendices A, B, and C.

More sophisticated tools will be required in the future that are the CAS analogues of risk management, work breakdown structures, etc. They will need to take into account the basic features of CAS especially polarities/strange loop attractors (which explain the dynamic of two poles of values that need to be managed as a pair over time) and period-doubling cascades (which explain the shifts to new configurations of the system at higher fitness peaks).

Project managers and project owners will also need to take informed risks on seemingly unproven but promising emergent methods rather than relying solely on those that no longer work.

QUESTIONS FOR DISCUSSION

1. What are the design features of a project capable of shape-shifting in order to survive one or more socio-technological phase transitions that occur in less time than a human generation of 25 years? What might you progressively *keep, abandon, and reinvent*?
2. How would you apply the laws of complexity and the features of CASs to the following aspects of a project: requirements gathering, the work breakdown structure, risk management, scope management, resource management, and contract and commercial management? How is a project emergent and self-organizing? What are the simple rules of interaction? How are the technologies, methods, and skills evolving or transforming? What are the highest leverage actions that are needed? How much of the system should you engage?

3. Thinking about the features of polarities, pairs of values that depend on each other over time, what are some of the polarities you might find in managing or leading a project? Contracting? Dealing with risk and uncertainty? Cross-boundary team coordination? Soft systems innovation? Technological innovation?

4. Thinking about what you have learned from the case study, describe the ideal set of skills that people in your organization might begin to develop, starting today, that would help you design, build, and operate complex projects more successfully.

5. How might our political and governance systems be reinvented in order to deal more successfully with the greater complexity, diversity, and uncertainty we have today? What role might project managers play in this?

APPENDIX A: QUESTION SEQUENCE FOR THE 2010 ROUNDTABLES

The question sequences that guided the 2010 roundtables comprised the following:

1. Chapter 1—*Baseline trends*. What for you is complex and how do you deal with it? Compared with what? What are the big trends/discontinuities emerging?

2. Chapter 2—What kinds of *executive behaviors, leadership qualities, or culture* are essential for great projects or lead to project failures? What qualities do you need to develop?

3. Chapter 2—Describe in 25–50 words a real example of a project failure, cost overrun, cancellation, etc., where issues of *executive behaviors/leadership qualities/culture* was a factor. (Your initials—example of the success/failure)

4. Chapter 3—What is *risk* for the private sector? What is risk for the public sector?

5. Chapter 3—Give examples of projects where *risk* was poorly or well managed and the factors that led to these outcomes.

6. Chapter 4—*Commercial management*. What do you see as the main contracting challenges in the future?

7. Chapter 4—Describe a project where *commercial management* was either poor or brilliant and the factors that led to these outcomes.

8. Chapter 5—In what ways do *stakeholder needs/aspirations* contribute to the failure/cost or time overrun of complex projects or resulted in great projects. How do stakeholder influences affect your program/initiative?

9. Chapter 5—Give examples of how the management of *stakeholder needs/aspirations* got in the way of a successful project. Ensured a project was very successful despite the complexity *or* reduced conflict and was beneficial.

10. Chapter 6—Thinking about successful/difficult projects in which you have been involved, what role does *knowledge management* play/need to play? Is knowledge management internal or external to the function of human resources?

11. Chapter 6—Describe in 25–50 words one or more fundamental lessons learned through the application of *knowledge management* processes, how was the knowledge accumulated, developed, dissipated, communicated, lost, added to, etc.

12. Chapter 7—What kinds of *tools* are useful/not useful in rapidly changing/complex environments? Where are the gaps in our tool sets? What new tools are needed? In what areas are there no tool solutions? What would you endorse?

13. Chapter 7—Describe in 25–30 words examples of existing *tools* that did not work/got in the way of a successful project, limited the availability of necessary information, innovation, etc., or ensured a project was very successful despite the complexity.

14. Chapter 8—What kinds of *education* programs do you need to help deal with complex projects in increasingly ambiguous or fluid environments?

15. Chapter 8—Give examples of exemplary *education* programs that have resulted in the transfer from the classroom to the workplace, or from practice to theory, or fail to equip the industry with the methods, models, or mind-sets we need.
16. Chapter 9—Describe an aspect of complex projects you believe needs to be explored, better understood, solution developed, etc., through *research.*
17. Chapter 9—Describe a research project that is vital to help manage/implement complex major projects in the future, where we don't have the answers. Create a snazzy 4–5-word title and 25–50-word description. Describe the research hypothesis and how the research might be conducted, with what benefit.

APPENDIX B: QUESTION SEQUENCE FOR THE PRELIMINARY WORKSHOPS FOR THE 2012 ROUNDTABLE SERIES

1. Context: What is happening in the world economy/financial system? What are the big economic/financial issues facing people, organizations, nations, and the world? How is it changing?
2. Education/learning: In what ways are our education systems around the world supporting or acting as barriers to economic/financial success?
3. Research: What research is required to deal with a more complex mix of politics, economics, technology, etc., in order to create a better approach to economic/financial success of projects and programs?
4. Politics: What are the cultural, resource, and ideological differences/realities/complexities at local, state, national, and international levels that influence economic/financial success?
5. Globalization: What are the financial/economic complexities that have developed as a result of globalization and how can we better deal with it?
6. Innovation: What new kinds of innovation (products, services, methods, models, etc.) are needed for the new economy/financial system that is emerging?
7. Analysis: What does all this mean?
8. Title: Craft a snazzy 5–10-word title for this discussion paper.
9. Supplementary question: Describe an issue identified in the World Economic Forum risks report that would be important to consider/explore further in the 2012 roundtable series.

APPENDIX C: QUESTION SEQUENCE FOR THE 2012 ROUNDTABLE SERIES

1. What big social, technological, economic, and cultural shifts or structural reorganizations are under way in the world for which the fallout from the global financial crisis may be merely a symptom?
2. In this new context, what kinds of government and private sector activities must now be considered to be complex major projects? Give detailed examples.
3. What are the major issues/difficulties you and your organization are encountering with designing, winning support for, and delivering complex major projects and programs that you lead/manage or are involved?
4. Give examples of how we might turn economic risk and uncertainty into possibilities and new solutions. What is fabulous about what we do or can do?
5. In the parts of the world with which you are familiar, what impact has the changing economic/social world order profile had on complex major projects or programs delivered by the government, the not-for-profit sector, and the private enterprise?
6. Thinking about how major complex projects/programs are currently led/organized/implemented, what should we *keep* that's working, *abandon* that is no longer useful, and *reinvent* what needs to be transformed?

7. If we could invent a new/ideal socioeconomic–political *operating system,* what features would it have in which society could thrive? For example, insight, foresight, and oversight. Make a list of 10 features, that is, 1, 2, and 3.

8. What are the top five changes we feel it is critical to make to the way we run complex major projects in this new environment that we or our organizations might research/explore in more detail? Describe it as a project: 3–5-word title and 25-word description.

REFERENCES

Argyris, C. (1991). *Teaching Smart People How to Learn* (pp. 99–109). Harvard Business Review, May–June.

Ashby, R. (1956). *An Introduction to Cybernetics.* London, U.K.: Chapman & Hall.

Bar-Yam, Y. (2004). *Making Things Work: Solving Complex Problems in a Complex World.* Cambridge, U.K.: Knowledge Press.

Conant, R. and Ashby, R. (1970). Every good regulator of a system must be a model of that system. *International Journal of System Sciences* 1(2): 89–97.

Cook-Greuter, S. (2004). Making the case for a development perspective. White paper. *Industrial and Commercial Training* 36(7).

CRC for CMPP. (2013). Proposed flagship projects. Slide presentation. Cooperative Research Centre for Complex Major Projects and Programs, February 6 and 7.

Findlay, J. (2009). Learning as a game. Exploring cultural differences between teachers and learners using a team learning system. Doctoral dissertation, University of Wollongong, Wollongong, New South Wales, Australia.

Findlay, J. and Straus, A. (2011). A shift from systems to complex adaptive systems thinking. In O. Bodrova and N. Mallory (Eds.), *Complex Project Management Task Force Report: Compendium of Working Papers* (pp. 24–26). Canberra, Australian Capital Territory, Australia: International Centre for Complex Project Management.

Findlay, J. and Straus, A. (2012). Navigating uncertainty with a complexity map. Slide Presentation. *International Centre for Complex Project Management, Research Conference,* Lille, France.

Findlay, J. and Straus, A. (2013). *The Wisdom Economy: Opportunities in the Chaos.* Occasional series. Boston, MA: Maverick & Boutique.

Flyvbjerg, B. (2005). Policy and planning for large infrastructure projects: Problems, causes, cures, World Bank Policy Research Working Paper 3781. Washington DC: World Bank.

Gleick, J. (1988). *Chaos: Making a New Science.* London, U.K.: Cardinal.

Goleman, D. (1995). *Emotional Intelligence.* New York: Bantam Press.

Gragert. T. (2013). Triple loop learning. Thorsten's site. http://www.thorsten.org/wiki/index.php?title = Triple_Loop_Learning. Accessed June 14, 2013.

Hayes, S. (2013). Conspiracy of optimism: Why megaprojects fail. Slide show. http://www.kcpm.com.cn/2010iflcpm/notice/10.pdf. Accessed June 6, 2013.

Iansiti, I. (2009). Managing megaprojects: Lessons for future combat systems. In G. Ben-Ari and P. Chao (Eds.), *Organizing for a Complex World: Developing Tomorrow's Defense and Net-Centric Systems* (pp. 88–110). Washington, DC: Center for Strategic and International Studies.

IBM. (2008). *Making Change Work.* New York: IBM.

ICCPM. (2009a). Roundtable Discussion Paper. Canberra, Australian Capital Territory, Australia—May 27, 2009, Washington, DC—October 13, 2009.

ICCPM. (2009b). Complex project management: The conspiracy of optimism. A Position Paper. In M. Cavanagh (Ed.). Canberra, Australian Capital Territory, Australia: International Centre for Complex Project Management.

ICCPM. (2011). Complex Project Management; global perspectives and the strategic agenda to 2025. In O. Bodrova and N. Mallory (Eds.), *Complex Project Management Task Force Report.* Canberra, Australian Capital Territory, Australia: International Centre for Complex Project Management.

ICCPM. (2012). Roundtable discussion paper. In J. Findlay, A. Straus, and C. Hatcher (Eds.), *Complexity in a Time of Global Financial Change: Program Delivery for the New Economy.* Canberra, Australian Capital Territory, Australia: International Centre for Complex Project Management.

ICCPM. (2013). Associate Partner network. http://www.iccpm.com/content/associate-partner-network. Accessed February 24, 2013.

Johnson, B. (1996). *Polarity Management: Identifying and Managing Unsolvable Problems.* Amherst, MA: HRD Press.

Johnson, B. and Hencke, D. (2008). Not fit for purpose: £2bn cost of government's IT blunders. *The Guardian*, January 5.

Kauffman, S. (2008). *Reinventing the Sacred: A New View of Science, Reason, and Religion*. New York: Basic Books.

KPMG. (2008). Global IT Project Management Survey. Hong Kong: KPMG International. https://www.kpmg.com/CN/en/IssuesAndInsights/ArticlesPublications/Documents/Global-IT-Project-Management-Survey-0508.pdf. Accessed June, 14 2013.

Logica Management Consulting. (2008). Failing business process change projects substantially impact financial performance of UK business. http://www.consultantnews.com/article_display.aspx?p = adp&id = 5170 October 2008. Accessed June 14, 2013.

Lovallo, D. and Kahneman, D. (2003). *Delusions of Success—How Optimism Undermines Executives' Decisions*. Harvard Business Review, July.

Meadows, D. (2008). *Thinking in Systems: A Primer*. Chelsea, VT: Green Publishing.

Prawat, R.S. (1999). Dewey, Peirce, and the learning paradox. *American Educational Research Journal* 36(1), 47–76.

Rooke, D. and Torbert, W. (2005). *Seven Transformations of Leadership*. Harvard Business Review, April.

Schumepter, J. (2003). *Capitalism, Socialism and Democracy*. London, U.K.: Taylor & Francis e-Library.

Senge, P. (2006). *The Fifth Discipline: The Art and Practice of the Learning Organization*. New York: Doubleday.

Stacey, R. (1996). *Strategic Management and Organizational Dynamics*. 2nd ed. London, U.K.: Pitman.

Sterman, J. (1994). Learning in and about complex systems. *System Dynamics Review* 10(2–3): 91–330.

Straus, A. and Findlay, J. (2011). ICCPM CPM international facilitated workshops: Global feedback on leadership for complex projects. In O. Bodrova and N. Mallory (Eds.), *Complex Project Management Task Force Report: Compendium of Working Papers* (pp. 43–47). Canberra, Australian Capital Territory, Australia: International Centre for Complex Project Management.

Wheatley, M. (1999). *Leadership and the New Science: Discovering Order an a Chaotic World*. San Francisco, CA: Berrett-Koehler.

White, B.E. (2009). Complex Adaptive Systems Engineering (CASE). *Systems Conference, 2009 3rd Annual IEEE*, pp. 70,75, March 23–26, 2009; also, B.E. White, "Complex Adaptive Systems Engineering (CASE)." *IEEE Aerospace and Electronic Systems Magazine*, 25(12). December 16–22, 2010. ISSN 0885-8985.

Wittgenstein, L. (2009). *Philosophical Investigations*, 4th ed., In P.M.S. Hacker and J. Schulte (eds. and trans.). Oxford, U.K.: Wiley-Blackwell.

World Economic Forum. (2012). *Global Risks Report,* 7th ed. Geneva, Switzerland: World Economic Forum.

17 Australian National Security and the Australian Department of Defence

Framework to Enhance Strategic Planning and Capability Development in Defense Acquisition Organizations

Richard J. Hodge and Stephen C. Cook

CONTENTS

CASE STUDY ELEMENTS

- *Fundamental essence*: planning the evolution of a medium-sized defense force.
- *Topical relevance*: force planning and evolution is an archetypal complex systems problem requiring a multidimensional mix of *hard*, *soft*, and critical systems approaches.
- *Domain*: government—defense and national security.
- *Country of focus*: Australia, yet relevant to most democratic environments.
- *Stakeholders*: government in the national security context, the department of defense,* national intelligence agencies, defense industry, public policy administrations, and private enterprises delivering for the common good. Defense is also of interest to the general population, and as such, few people within the country are not stakeholders. The most active stakeholders are within defense: the strategic planning community, the capability development community, the military, and the defense science community.
- *Primary insights*: Using a systems approach, as described, it is possible to tackle the vexing problem of strategic planning and execution in a socially and culturally aware fashion and produce a useful outcome that is supportable through evidence, readily updated in a dynamic evolutionary process, is cost-effective, and imbued with inherent stakeholder buy-in. In short, strategy and execution work in harmony and the people driving them coproduce defense outcomes with a good understanding of the *why*, *what*, and *how* across multiple tiers and communities in a complex enterprise.

KEYWORDS

Defense strategic planning; Systems of systems (SoS) planning methodologies; Systems approaches to strategy and execution; Strategy; Execution; Capability development; Capabilities-based planning

ABSTRACT

In a climate of reducing defense budgets and government demands for high levels of transparency in the expenditure of public monies, defense planning needs to produce capability outcomes that, as a set, are traceable to policy objectives.

This case study describes a systems framework that was devised from the strategy and complexity literatures to enhance the preexisting planning processes in a medium-sized defense department. The application of the methodologies and tools resulting from the framework will be discussed along with an illustration of how the framework was tested within the Australian National Security and the Australian Department of Defence (ADoD) planning environments.

This framework was seen to be successful by the stakeholder community, and its use has subsequently changed the way strategic planning is conducted in that enterprise from being unitary (and exclusive to a small group of strategists) to being more inclusive of the subcommunities involved and more rigorous and traceable to meet the government needs for transparency.

---

* Henceforth, the Australian Department of Defence is abbreviated to *ADoD*. The generic business of defense will be differentiated by the use of US spelling and the absence of capitalization.

GLOSSARY

ADoD	Australian Department of Defence
ADHQ	Australian Defence Headquarters
AEW&C	Airborne early warning and control
CDF	Chief of Defence Force
CONOPS	Concept of operations
DSTO	Defence Science and Technology Organisation
FIC	Fundamental Inputs to Capability
HQ	Headquarters
SEA	Systems Engineering Assessment
SME	Subject Matter Expert
SSM	Soft Systems Methodology
VCDF	Vice Chief of the Defence Force

BACKGROUND

Strategic planning and strategy execution (hereinafter referred to as strategy and execution) is a challenging activity for all nations. The goal is to determine what preparations should be made and what military capabilities should be developed or enhanced to meet often unforeseeable, politically defined needs some 15 years or more into the future. The defense of a nation in common with disaster relief is an unusual enterprise in that it does not conduct the principal activities for which it exists on a routine basis as would say air traffic control, banking systems, civil infrastructure, etc. Thus, deficiencies are often hard to identify, and the potential upgrade solutions are hard to validate given the changing nature of how governments choose to use their military capabilities across a wide range of combat and noncombat operations.

In terms of systems types, defense can be considered a complex adaptive system that interacts strongly with its environment and must adapt to suit the imperatives of deployments (Grisogono and Ryan, 2003). Thus, defense strategic planning needs to determine the nature of the future force and identify the most desirable set of sociotechnical elements that are best able to synergize to provide desired military effects. The process needs to align these capabilities with changing government defense policies, the external national security environment, and the state of current and future defense capabilities. It also needs to reflect changing social norms, advances in technology, and international legal frameworks. It is not surprising, therefore, that defense strategic planning in Australia is considered a challenging and important problem.

Assessed with the Enterprise Systems Engineering Profiler (Stevens, 2011), the problem would occupy the outermost ring in every attribute, and using the System Engineering Application (SEA) Profiler (Gorod et al., 2008), it would align with the right-hand column in every attribute. There are four main characteristics of the case study problem of improving the effectiveness of strategy and execution in a defense environment:

1. It is a sociotechnical problem of high complexity with many dynamic interfaces across the political, bureaucratic, economic, environmental, and military strategic domains. There are many *fingers in the defense pie*.
2. Military strategic planning is itself a *wicked* problem, the meaning of which will become clearer, noting the root cause is that planning often takes 10–15 years for new military capabilities to move from concept to operations where they remain in service for up to 30 or more years. The planning basis for investments is uncertain and ambiguous. Once the planner gets below the general acceptance of making national investments in security, the question of *why specific choices should be made relative to others* does not have a simple answer.

3. Military operational requirements are clear when there is present danger. However, the clarity in the requirements diminishes as a function of time when planning for future crises or contingencies because ambiguity increases in the definitions of what will constitute operational success for a perceived event over which there is no certainty. The question of *what specific operational requirements should drive military preparedness* is difficult to address as the planner is traditionally working in a vacuum with no feedback.

4. Militaries do what they do with what they have available at the time a government authorizes a military intervention. So it is perhaps as a result of the foregoing characteristics of the problem that the military strategist and planner faces the question of *how to achieve the right balance between investment in the force-in-being (to do what we are now doing) and in the future force (to be prepared for uncertain contingencies).*

At the time this case study originated, these characteristics of military strategic planning were well known (Hodge, 2010). The strategic planning framework (*as-it-was*) operated roughly as illustrated in Figure 17.1.

The greatest challenges were how to improve the future thinking and then make that thinking impact on decisions that shape both the programmed force (where investment decisions had already been made and were often remade) and the preparedness of the current force. The box labeled *capability planning* in Figure 17.1 is where the main elements of integration occurred—in the early stages of the work, this box lacked definition and the team regularly regarded this as *the magic pudding.* This was the focus of the work: to define the integration of future and current force imperatives and determine force structure priorities. It is worth stating that Australia's efforts were generally on par with other first-world nations at the time.

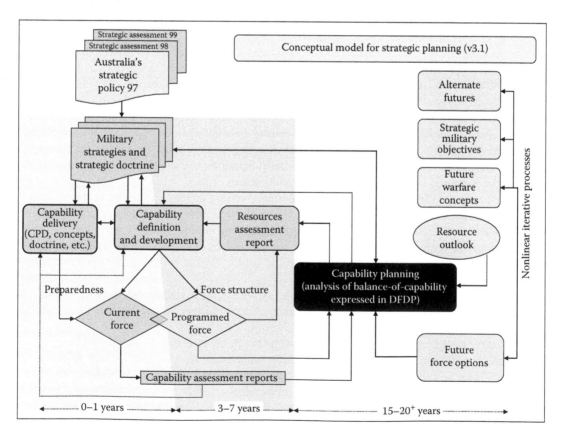

FIGURE 17.1   *As-is* planning framework c. 1999.

To approach the problem of building an integrated approach to strategy and execution, the methodology of systems engineering was seen as desirable and adaptable from the outset. The pragmatic reasons were that the first logical phase of implementing defense planning happens with acquisition and logistics elements in the ongoing acquisition and support of defense capabilities. These agencies applied systems engineering extensively in practice. So, systems engineering language—requirements analysis, functional architecture, system design, test and evaluation, analysis, and control—became structural elements of the conceptual thinking around the project.

Additional elements of the conceptual framework were synthesized following a review of the nondefense strategy and management literature. Firstly, it was found that the need for an integrative framework was well documented (Chaffee, 1985; Hart, 1992; Hart and Banbury, 1994; Mintzberg, 1994) but remained elusive (Mintzberg et al., 1998). Nonetheless, this literature provided some relevant insights for the problem in hand. For example, Hart (1992) identified five typologies of strategy formation—command and control (*Just do it*), symbolic (strategy based on *I have a vision*), rational (strategy driven by mission, vision, goals, etc.), adaptive (strategy directed from learning by doing), and generative (strategy directed by bright ideas, mostly generated bottom-up). Importantly, Hart's analysis indicated performance improvements across enterprises correlated with the increased use of a wide range of strategy formation processes. In other words, using a self-designed mix of methods across the five process typologies gave greater performance improvement than any one alone. This contrasts with his finding that most firms employ traditional, top-down approaches in a rationalist hierarchy.

Hart's contribution relates well to the work of Chaffee (1985) who describes the strategy field using three modes of strategy: linear, adaptive, and interpretive. Chaffee describes the modes as interrelated and able to be applied singularly to an issue or in combination to address more complex issues with complex enterprises requiring all three modes operating together.

Chaffee's constructs are summarized in Table 17.1.

Chaffee's review of the literature led her to the conclusion that "interpretive strategy ignores linear and adaptive strategy" (p. 95) and foresaw that "the full value of strategy cannot be realized in

---

TABLE 17.1

Three Modes of Strategy Making

	Linear Model	Adaptive Model	Interpretive Model
Primary focus	Methodical, directed sequential action in planning	Development of a viable match between external and internal conditions.	"Orienting metaphors or frames of reference" allowing organization and its environment to be understood.
Strategy construct	Integrated decisions, actions, plans, goals, and means of achieving them	Monitoring environment and making changes are simultaneous and continuous functions.	Strategy is socially constructed and multifaceted "relying on individuals to cooperate in mutually beneficial exchange."
Top managers	Hold considerable capacity to change the organization	Attend to means/process, where goals are driven by the "co-alignment of the organization and the environment."	Shape the attitudes of participants toward the organization and its outputs; they do not make physical changes in the outputs.
Organization	Composed of tightly coupled parts	Open to the environment and structurally focused.	Open, emphasizing attitudinal and cognitive complexity.
Environment	"A necessary 'nuisance' composed mainly of competitors"	"A complex organizational life support system."	The organization and the environment constitute an open system.

Source: Data from Chaffee, E.E., *Acad. Manage. Rev.*, 10(1), 89, 1985.

practical terms until theorists expand the construct to reflect the real complexities of organizations" (p. 96). Yet, in the midst of her thesis, Chaffee has added considerable value to the theoretical construct by correlating the three models of strategy with Boulding's hierarchy of systems complexity (Boulding, 1956), to illustrate how strategy has "evolved to a level of complexity almost matching that of the organizations themselves" (p. 89).

An examination of the complexity literature followed, and Warfield's (1999) "Twenty laws of complexity: Science applicable in organizations" and Warfield's interpretation of these laws are insightful and helped Hodge (2010) to identify three insights for strategy making (see Table 17.2).

Considering the characterizations of the problem in the light of these research *nuggets*, it becomes quickly apparent that systems engineering, due to its rationalist essence, would be a necessary part of the final approach but not sufficient alone to address the *why*, *what*, and *how* characteristics of the problem in a complex social, economic, and political context.

Hence, concepts were developed that incorporated other methodologies such as the following:

- Soft systems methodology (SSM) to *orchestrate the debate* (Checkland, 1981).
- Critical systems thinking approaches to address issues of power and obtaining permission to act—including permission to involve more stakeholders in an orchestrated strategy process than had previously been the norm.
- Strategy methods drawn from corporate practices and management literatures because many people know something about strategy and have familiarity with management methods that work for certain circumstances—for example, there are many more methods in the management field to address power (manifested in politics, hidden agendas, and strategic

---

TABLE 17.2
Interpretation of the Laws of Complexity

Interpretation of Laws (Warfield, 1999)	Insight for Strategy Making (Hodge, 2010)
"The limitation [of the human mind] should also persuade individuals that intuitive decision making, carried out without careful analysis, is likely to produce bad decisions and bad outcomes when complexity is present."	There is benefit in designing a component of the strategy-making process to focus on analyzing and controlling the strategy process and on developing effective products of the process. Likely benefits include improved management of complexity in the strategy products and processes and improved understanding of the potential risk of bad decisions or outcomes arising.
"The limitation to interactions among three items suggests a very serious limitation on creative ability as might be reflected in the design of complex systems.... Many situations in life have been approached as though there were a dichotomy involved. Instead of allowing our thinking to be limited to dichotomies, we should be encouraged to move to trichotomies in a way of becoming more flexible in thought and action, wherever appropriate."	Designing the strategy process (i.e., organizing the management of the cognitive complexity) with tension among three domains of the strategy-making processes will engender greater flexibility in the thought and actions within the designed process. Implicitly, one could expect the resultant strategy to better reflect and manage the situational complexity that the designed strategy process is contending with. Designing a trichotomy into the strategy-making process is consistent with a tenet of multidisciplinary thinking that a solution to a complex problem requires a minimum of three different perspectives of that problem to be taken (Kline, 1995).
"Ad hoc designs, arrived at in ordinary conversational modes (as for example, in government bodies or committees) might be looked upon as unlikely to be of high quality, and likely to produce bad outcomes."	Strategy making by an appointed (or self-appointed) *inner circle* invoking interpretive strategy models in an unorchestrated, dichotomous debate is likely to produce bad outcomes. Strategy-making processes that are innovative and designed to be inclusive of other strategy models (and therefore more demanding in orchestrating analysis and synthesis) are a hallmark in high-performing organizations (Hamel, 2000; Hart, 1992).

Source: After Warfield, J.N., *Syst. Res. Behav. Sci.*, 16, 3, 1999.

alliances among other ways) than exist in the systems field. Many of these methods in the strategy literature are reviewed in Mintzberg et al. (1998). They focus on strategy as a process of negotiation and include, for example,

- Political gaming—line management vs. staff; strategic candidates and rival candidates games; and *whistle-blowing* games among others
- Use of classical political tools—satisficing; anticipation of coalition behaviors; equifinality to drive outcomes rather than specific means
- Stakeholder analysis and strategic maneuvering
- Cooperative strategy making in a range of ways including, for example, future forums, seminar gaming, alliancing, joint ventures, and/or cooperative contracting.

For the purposes of segmenting the cognitive problem of developing an integrated approach that covered the main characteristics of the substantive problem (of improving strategy and execution outcomes), the systems engineering process illustrated in Figures 17.1 through 17.3 of DSMC (2001) was abstracted into Figure 17.2, which retains all seven main features of the original.

In many ways, the concept in Figure 17.2 has similarities with Flood and Jackson's Total Systems Intervention (1991) for selecting and applying multiple methodologies to richly address complex organizational problems. Where it differs is that it seeks to identify specific methodologies for specific aspects of the problem.

SSM was used to assist in the selection and integration of the methodologies and adapt the application of the systems engineering process using the three Es as criteria:

- *Efficacy*—was this process adaptation right for the purpose?
- *Effectiveness*—was this process adaptation effective for the purpose?
- *Efficiency*—did this process adaptation achieve an effective outcome for minimum resource utilization?

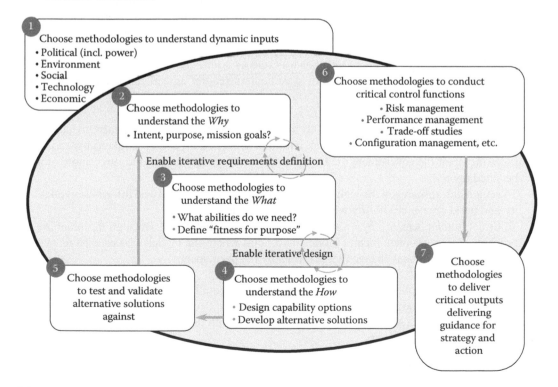

FIGURE 17.2   Systems engineering as an approach to integrating methodologies.

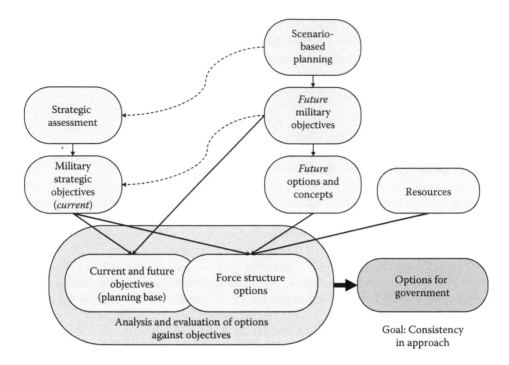

FIGURE 17.3 Integrating current and future thinking.

SSM also set the criteria for success of the *to be* process architecture to ensure outputs in process design and substantive content were systemically desirable and culturally feasible. The major variables in the *to be* planning process, whether in a current or future time domain, were seen to be the strategic objectives and the shape of force structures. The relationships between these major inputs are depicted in Figure 17.3.

How these elements were to come together was sold as a *little bit of magic happens here (and you can help us create it)*. However, senior stakeholders readily agreed the goals of the *to be* architecture:

- To bring the organization through the torrents of ambiguity and uncertainty that are endemic in defense planning (and stick with that process until action is taken) instead of thrusting upon the government the best plan of several bright minds and directing that it be acted upon
- To establish transparency throughout the process to give auditability of information quality and build greater credibility with defense stakeholders
- To direct human behavior by a shared image of reality that would, through the planning process, change peoples' mental models to a point where the models coalesce to form a basis for organizational change that is *owned* by the participants

Agreement on the goals of the *to be* architecture conveyed to all that a transformative undertaking in strategy and execution now had the sanction of the then Secretary of Defence* and Chief of Australian Defence Force (ADF),† who gave their approval with the explicit understanding that the journey would take at least 3 years.

---

* Mr. Paul Barrett, AO.
† Admiral Chris Barrie, AO.

PURPOSE

At this stage, it is useful to state that the purpose of the system transformation documented in this case study was to produce a defense in-house capability planning system that could repeatedly produce a convincing future-force structure (i.e., a set of military capabilities comprising current assets and those to be acquired) that would be a good fit against strategic guidance, be affordable, and be broadly accepted by the stakeholders.

At the time of the study, there was a general lack of ease about the profoundness of the approach used and the quality of the outputs produced and the lack of a methodology that could be consistently applied. This is exactly the sort of circumstances that Checkland (1981) identifies as the trigger for a systemic intervention: the wide recognition that the current situation is unsatisfactory but also that it is hard to define the problem situation and what would constitute a good solution.

ENTERPRISE

The Australian defense enterprise is a virtual system that is contained within the Australian Department of Defence (ADoD) and hence operates within a public service not-for-profit environment where the constraints are set by government policies. The focus of the case study then was on transformation of the system for strategy and execution in the enterprise.

A. *Defense Enterprise*

In its simplest representation, ADoD is an enterprise with an annual resourcing of over $20bn (Defence, 2004) that is structured to contribute to national power through the synergistic output of two *products*: *influence* and *combat power* (Hodge, 1998 after Hitchins, 1992).

Influence: Defense forces provide nations first and foremost with the ability to exert their will and influence over other state and nonstate actors just by their presence. The effects of any influences by a nation state and those of the counterinfluences upon it will change as the strategic environment changes (von Clausewitz, 1832/1984). In Australia's case, a stable state has rarely been perceived (if ever) and much less achieved. Consequently, the spectra of influences and counterinfluences are potentially very broad in scope and comprehensive in their effects. The balance of power issues involved becomes a complex equation to comprehend, yet they remain an important component in the exercise of national power.

Combat Power: Defense's number-two *product* is combat power, which is not an end in itself but supports the application of deterrence as one element of the national capability to influence other actors. Not surprisingly, the terms *complexity* and *uncertainty* feature strongly in the business and planning vocabulary of defense decision makers.

The heart of many military capability studies involves the analyst determining how to measure, assess, and report military capability. For such studies to remain focused, it is important to clarify the level of the system of interest. Table 17.3 that was constructed by abstracting Hitchins' five-layer model (Hitchins, 2000, 2003) to the defense context has been found useful for that purpose, and it also illustrates the degree of complexity that must be tackled at each level.

B. *External Factors*

The environment for the study was the full defense enterprise embedded in the national security architecture for Australia. This meant that the contingencies forming the planning basis for strategy and execution required inputs from national security agencies and sign off at the Deputy Secretary level. Concerns at this level extend beyond the government sector because external factors such as the workflow to the national defense industry base also need to be considered. Tackling a system problem at this level entailed traversing

TABLE 17.3

Hierarchy of Complexity of a Defense Enterprise

Fundamental inputs to capability (e.g., equipment, people, training, infrastructure, logistics support)
- The basic *building blocks* that can be acquired and/or developed as inputs that must be integrated to develop a defense capability.
- These elements can usually be quantified and costed directly.

Capability (e.g., the ability to conduct air combat, maritime surveillance, counterterrorism)
- Functional expression of defense operational outputs.
- The actual capability is more than the sum of the physical elements; all FIC need to be considered.
- The actual value of the capability cannot be directly costed; judgment is required on doctrine, tactics, morale, leadership, command intent, etc.

Capability domain (e.g., the ability to achieve control of the air, control of the sea, special operations, including key enablers)
- Functional ability comprising a group of capabilities (e.g., control of the air comprises offensive counter-air, air defense, with key enablers command and control, intelligence, logistics)
- Produces operational level outcomes

Component Force (air force, land force, or maritime force)
- Functional expressions of the enterprise's ability to deliver armed force for a range of desired effects in one or more operational environments
- Includes raise, train, and sustain functions
- Delivers strategic level outputs

National Defense Force
- Functional expressions of the national ability to deliver military strategic outcomes supporting national interests in concert with the nation's political and economic objectives
- Integrates strategic support functions of strategic and capability planning, development and acquisition, science and technology, national industry support, and corporate governance functions
- Defense is costed as a government line item

---

the full five tiers of Table 17.3 because the outputs from the intervention are changes at the lowest level, that is, new equipment and doctrine to achieve an effect at the national defense level.

C. *Feedback*

Problem boundary definition was indicated not just to contain the scope of the problem (using the usual sources of feedback within the enterprise—as shown in Figure 17.4) but also to identify the stakeholders and to improve the systemic intervention in the social system of interest (Ulrich, 1983). Boundary critique at the early stages of an intervention helps surface boundary definitions that might change the feedback loops required as the debate goes along to address, for example, microeconomic concerns at a quantitative level and then bring the results for synthesis into a whole of enterprise dialogue.

D. *Governing Bodies* (*For the Study*)

Our work began with the sanction of the Secretary and Chief of Defence Force following a presentation to the Defence Committee. The Deputy Secretary (Strategy) was the accountable officer, and day-to-day management responsibility was held by the Division Head for Strategic Policy and Planning. With these governing bodies for the study, the team was then able to operate with broad boundaries for strategy and execution across the enterprise.

E. *Constraints*

The Australian Defence White Paper (Australian Department of Defence, 2000) that articulated the then government's defense policy was the key source of direction for strategy development and described Australia's defense strategic environment, role of the ADF, its priorities, and level of funding, thus also articulating additional constraints on the defense

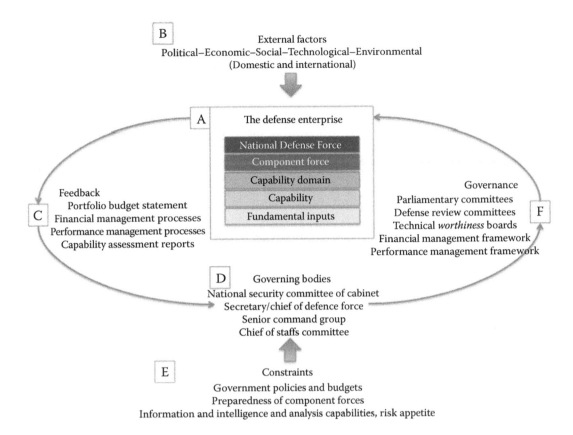

FIGURE 17.4   The defense enterprise.

enterprise. The constraints on the system of strategy and execution were to conduct its transformation as leanly as possible—small numbers of competent people, small budget, and tight timeframes.

F. *Governance*

For the pragmatist operating at the enterprise level, the advice of Checkland (1981) is profound: it matters less where you start and it matters more that you start and adjust the boundaries and the focus as you proceed guided by the learning achieved in the process. The boundaries were defined using a hierarchy of measures built around the recursion of *why*, *what*, and *how* (after Rasmussen, 1998) at each of the identified focal areas:

1. At the strategy level to build a coherent view of the portfolio and dissolve the basis for any future *warring* among the *defense tribes* (McIntosh et al., 1997)
2. At the capability strategy level to align different program interests and close the *gap in explanation* between strategy and capability development
3. At the execution level to assure a shared understanding of priorities across the force structure and assure capabilities were developed and employed in a manner that was fit for purpose and balanced to mission requirements and trade-offs were made in a manner traceable to strategic priorities

A generalized version of the established hierarchy of measures is illustrated in Figure 17.5. The main insight this provides is the coherence achieved across the three focal areas, as the performance of strategic policy is dependent on the effectiveness of the capability strategies and the success of their execution (Hodge, 2010).

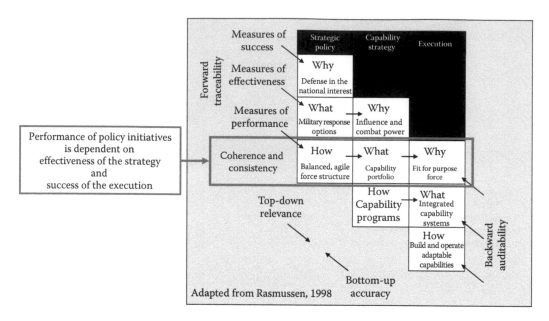

FIGURE 17.5   A generalized hierarchy of measures.

These boundaries then helped to further refine the judgments about who was invited to participate in the work. At a later stage, it would also raise contentious discussion over who should *own* the work—the strategy group or the capability development group.

CHALLENGES

The foremost challenges to the strategic planning and execution system had been identified back in the 1970s by the then Secretary of Defence, 1970–1979, Sir Arthur Tange, who sought to improve the level of integration in strategic policy, which he saw as being hampered by interservice rivalry and stereotyped thinking (Edwards, 2006). These were once again spelled out in the Defence Efficiency Review (DER) of 1996–1997 that reported publicly to the Australian Minister for Defence, noting the *warring* nature of the tribes (McIntosh et al., 1997). The effect of this tribalism on the strategic planning function was made clear in the Secretariat Papers to the DER:

> In the absence of a sufficiently articulated strategic direction, there remains considerable scope to dispute guidance on parochial grounds or to continue to dispute decisions once taken. ... There are too many organisations competing in establishing capability priorities. ... In attempting to set capability priorities, the focus has tended to be more on the platforms required than the holistic consideration of capabilities, being not only the platform, but also the manner in which that platform will be employed, supported and staffed. The end result is that it is difficult to agree priorities for force capability development, with those priorities which have been established in the Five Year Development Program (FYDP) having been constructed largely on the basis of resources available in the capital investment program. ...
>
> **McIntosh et al. (1997)**

The challenges can be summarized thus:

1. Strategy and execution is one of the more complex systems problems and one beset by inherently conflicting priorities, worldviews, and tribal behaviors.
2. Strategic guidance, which is the input to strategy and execution process, will never be precise, comprehensive, nor entirely explicit.

3. The nature of the strategy and execution task is such that there is a natural tendency of adversely affected groups to dispute and undermine capability decisions, historically with some success.*

4. Much conventional equipment planning followed a bottom-up equipment replacement paradigm that had strong support in some areas as it was seen to be serving the needs of the services and *looking after our boys in uniform.*

5. Defense departments are process-oriented organizations whose people solve problems based from a predominantly functionalist worldview. In this Australian case, there was a need to engender an understanding in the stakeholder group at both executive and working levels that something more than a better process would be needed and that better defense outcomes would be achieved through cultural and behavioral changes.

DEVELOPMENT

From the problem analysis, it was clear that to succeed in tackling the defense strategy and execution problem, it would be necessary to

1. Work with the stakeholder to help them appreciate the cognitive complexity of the problem (Checkland, 1981; Warfield, 1994) and have them understand the need for a revised approach and secure buy-in for the form it might take

2. Design a multimethodological approach (Midgley, 2000) guided by (1), to manage the cognitive complexity of the problem in tandem with changes in the situational complexity (Warfield, 1999)

3. Apply the methodology in a manner inclusive of all stakeholders using the approach designed in (2) without losing sight of systems thinking and systems engineering principles to produce the desired planning products and organizational change

Each of these components will now be described in turn.

APPRECIATION OF THE COGNITIVE COMPLEXITY

To address the complexity set out in the previous section, the primary author employed Checkland's SSM because it provides a framework for dealing with large, unstructured problems; for conceptualizing a solution to a problem; and for comparing the cognitive output/concepts with the real-world problem before taking actions to improve the problem situation (Checkland, 1981). The SSM also enables a *learn by doing* approach to managing the cognitive complexity of designing a new strategic planning framework for defense. The approach uses SSM to adjust the mix of planning methods designed to address the situational complexity of warfighting. To this end, the normal seven-step methodology, shown in Figure 17.6, is presented in four main elements.

The four main elements are

1. Clarifying the problem definition (steps 1 and 2)
2. A systems definitional view of the problem space, introducing relevant conceptual models (steps 3 and 4)
3. A comparison of the models with the *real* world of defense strategic planning and a discussion of the appropriate changes that were designed (steps 5 and 6)
4. The systems-based actions that were implemented to improve the problem situation (step 7)

---

* The Chief Defence Scientist advised: "If people do not like the answer you arrive at, they will attack your methodology. So, test your methodology to make sure it will withstand such an attack" (Brabin-Smith, 1999). Similarly, the then Deputy Secretary (Strategy) advised: "The critical vulnerability in the methodology is the people you engage in it. If any of the Service Chiefs and Defence Group heads do not hold trust in their capacity to speak for their organization, they will immediately dismiss the input as invalid and then throw out the findings" (White, 1999).

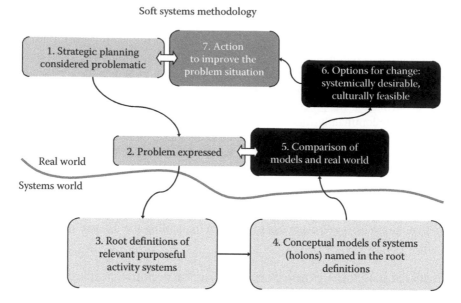

FIGURE 17.6 Checkland's SSM. (Modified from Checkland, P.B., *Systems Thinking, Systems Practice*, Wiley & Sons, Chichester, U.K., 1981.)

The first author led the study team through these four steps in a naïve manner, which, on reflection at the end of the process, is far from the intended process outlined by Checkland (1981). The approach outlined here may, therefore, not be recognizable to an SSM specialist. However naïve, these actions were, at the time, the application of SSM proved sufficiently useful to get started in our job of managing the cognitive complexity of the problem and using that understanding to guide the application of systems engineering in managing the situational complexity.

In attempting to express the problem, Argyris's advice was noted: "…defining and solving problems can be a source of problems in its own right" (Argyris, 1991). In defining a system that is considered problematic, Checkland considers six focal issues: the *C*ustomers of the system, the *A*ctors that carry out the main activities of the system, the *T*ransformation process that is occurring, the *W*orldview (or image) of the system that makes the overall definition meaningful, the *O*wnership of the system, and the *E*nvironmental constraints acting on the system. Consequently, Checkland uses the mnemonic *CATWOE* to assist in remembering the elements of a well-formed definition (Checkland, 1981; Checkland and Holwell, 1998). In the case of the ADoD, the discussion pertains to the worldview held in 1999 by the (then) Strategic Policy and Planning Division of the ADoD headquarters and is summarized here using Checkland's mnemonic CATWOE:

- The primary *customer* is the National Security Committee of Cabinet as represented by the Minister for Defence.
- The key *actors* in the process are the members of the ADoD leadership team or their representatives.
- The *transformation* that is desired through the appreciation process is to continually adjust the mental models of the customers and actors directly and indirectly the mental models of those in receipt of the products of the strategic planning process (i.e., the defense staffs who use them in capability planning and analysis). The focus of the transformation is to improve the cognitive processes of those directly involved in strategic decision making and to have them accept the need for an improved approach: one that can gap between strategic assessments and capabilities to be acquired.

- The *worldview* that the primary author and his team, as leaders of the transformation, is that a skillful engagement process that gradually expands participation in the strategic debate will improve strategic planning outcomes (after van der Heijden, 1996).
- The *owner* of the planning process and its products is, ultimately, the Minister for Defence. Of recent times, the ADoD looks to the government through two perspectives: as owner and as customer.
- The *environmental constraints* have been defined in large part previously and may be seen as common to many organizations that are an amalgamation of several large *stovepipes*.

Strategic Planning Considered Problematic

Within this context where strategic planning is considered problematic, the reasons for it are now explored with a view to expressing the problem more specifically within the context of the larger *system* of planning in ADoD covering the spectrum from strategic planning, capability development, to defense acquisition.

The *why*, *what*, and *how* for capability development starts with an assessment of the strategic environment, the output of which provides the *hooks* for candidate systems to contest a place in the schedule of unapproved projects (then colloquially known in ADoD as the *Pink* Book). Unapproved projects were often little more than a reviewed set of documented capability options. As depicted in Figure 17.7, an extreme *gap* existed between the strategic assessments that form the beginning of a strategic planning process and the formation of capability systems designs (Baker, 2000). *This is the heart of the problem being addressed by this case study.*

The following decision-making processes of the defense committee structures focused on refining capability options to transform them into approved equipment acquisition strategies. However, the decision-making foundations relied on to select and transform these options were not based on solid ground because of the strategic *gap in explanation* (Baker, 2000).

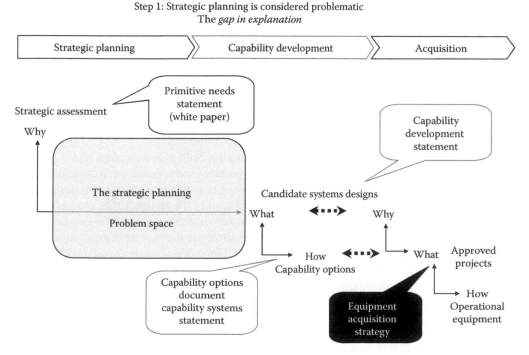

FIGURE 17.7   The *gap* in explanation.

The main points of reference on this spectrum, as they stood in 1999–2000, are listed in the following:

- The White Papers that the government issues periodically to outline their forward policies for defense that provided the initial *hooks* for capability options. (The White Paper typically offers no more guidance than a primitive needs statement. Prior to 2000, it had typically said little about what capability areas would be needed or how much of each would be enough or why.)
- The capability development statement that canvasses the need for new or replacement systems.
- The capability options documents and capability systems statements through which unapproved projects move into the *Pink* Book as candidates for approval by the government. These pave the way for a project to be included in a budget submission for approval and then to the schedule of approved projects (the *White* Book).
- The equipment acquisition strategy then translates the concept into specific systems design language for the project to proceed to its ultimate in-service capability.

Problem Expressed

The *strategic planning problem space* depicted in Figure 17.7 defines the defense strategic planning problem dealt with in this case study. Thus, the aim of the case study can be alternatively stated to produce the missing links between the primitive needs statement presented in the White Paper and the *Pink* Book of unapproved projects. Some might argue that the *missing links* were never missing and are implicit in the White Paper. Without opening that debate, the aim of the case study is to make the argument offered in such planning documents more transparent. In other words, defense needs to be able to find a way of uncovering the *what* and the *how* at the strategic planning level and use this understanding to define a range of joint functional abilities expressing what effects are needed in which environments and to what ends. Further, it then needs to examine what priorities attach to these abilities, given the range of strategic contexts that the government defines as important.

It is also necessary to consider the standards that should apply in meeting these high-level needs. For this understanding to impact across the development spectrum, capability priorities need to be clearly communicated and some high-level measures of effectiveness (MoE) must be articulated in a capability priorities paper that can influence and shape the capability development and acquisition program of activities.

That type of formative guidance—diagrammatically represented by the wide horizontal band in Figure 17.8—did not exist in 2000, and while the overall framework is better in 2013, there still remains room for significant improvement.

Root Definitions and Conceptual Models for Strategic Planning

With the problem thus expressed, the next phase in applying SSM is to articulate what Checkland calls *root definitions* of systems that are relevant to the problem space and develop conceptual models of these systems (Checkland, 1981). The reader may or may not agree with the root definitions used here. These are the ones that our team settled on. Those of the reader, if different, are equally valid. It is worth remembering that there are no *right* answers because these expressions and models exist only in the cognitive domains of those who were involved in this transformation activity as plausible descriptions (Beer, 1980; Maturana and Varela, 1980). These expressions and models represent *our* images of the real-world human activity of strategic planning in the ADoD.

For the strategic planning problem area, the study team developed root definitions of three relevant systems to close the *gap* in explanation shown earlier in Figure 17.7:

- [Root definition 1] System to sustain learning about the development, growth, and survival of the ADoD that uses that learning to design and build a defense enterprise for the government

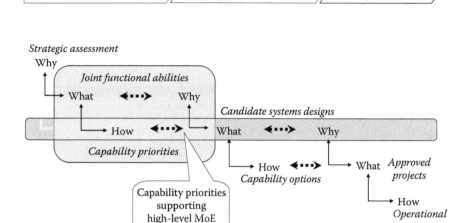

FIGURE 17.8 The problem expressed.

- [Root definition 2] System to guide the design and development of a holistic strategic architecture for the ADoD
- [Root definition 3] System to design and mentor cultural and behavioral change to ensure that future actions and decisions are appropriate and culturally acceptable to fully achieving the evolving strategic intent

For each of these definitions, conceptual models of the systems were in the form of parsimonious accounts of the human activity systems describing what the systems must directly carry out (after Checkland, 1981; p. 313).

The conceptual model for root definition 1 is an adaptation of a systems engineering model, shown in Figure 17.9. In particular, the role of the test and evaluation element of systems engineering is explicitly included as it was seen as an ideal concept for testing potential force structure designs against multiple scenarios to generate learning about whether the force structure option under test would deliver the functionality and capacity to execute the scenarios that capture government needs. The processes of evaluation and systems analysis were also seen to offer an ideal conceptual basis for making the strategic planning process more inclusive of the *warring tribes* (after McIntosh et al., 1997).

The conceptual model for root definition 2 is Boulding's hierarchy of systems complexity, appropriately adapted to the context. As previously illustrated in Hodge (2010, p. 140), an adaptation of Boulding's hierarchy of systems complexity to defense planning provides a multidisciplinary framework of methodologies across nine systems levels. Ideally, an adaptation of Boulding's hierarchy of systems complexity would guide the development of a strategic architecture for defense and help decision makers not accept a level of analysis inappropriate for the level of the problem being studied. This model is explained further in Hodge and Walpole (1999).

The conceptual model for root definition 3 is a performance management system based on an understanding of the intrinsic motivators for human behavior. This is an important third component in the overall concept and is still under development.*

---

* See Black (2011) review of the Defence Accountability Framework.

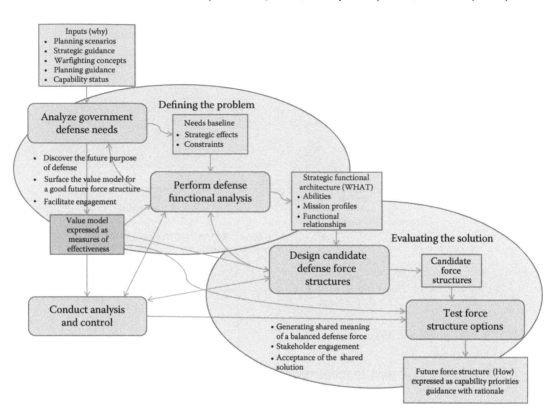

FIGURE 17.9 The conceptual model for the system that sustains learning about the defense organization and uses that learning to design and build a defense enterprise for the government.

Importantly, when dealing with an enterprise-wide problem in a political, social, and technical environment as complex as defense, it is extremely difficult to be precise when first applying SSM. The root definitions used here were later refined and supported with more detailed conceptual models that look more like the conceptual models that Checkland and experienced users of SSM might recognize. Checkland's own guidance was again taken: it matters less where you start, it matters more that you start.

This case study concentrates on describing the activities shown in Figure 17.9 that constitute the conceptual model for a system to sustain learning in which systems engineering represents the *why*, *what*, and *how* of a strategic planning process. Starting with a statement of what government wants defense to achieve, the *why* is expressed in a range of scenarios, which provide the strategic context and the national security outcomes required from defense. An iterative process of requirements analysis then leads to a statement of the *what*—those abilities defense must develop in order to achieve the government's needs for a defined national security outcome. A design activity then leads to exploring the *how*—the particular, physical instantiation of how the Defense Organization can be structured to in order to provide the abilities needed to achieve the government needs. Remaining true to the systems engineering model, these force structures or capability priorities can then be tested back against the strategic scenarios (the *why*) to ensure that they do meet the government needs with which the process started. It is important to note the highly iterative nature of this process that allows activities to occur concurrently in each part of the system.

For such a process to be successful, it was recognized early in the intervention that the senior defense leadership needs to play a controlling role in all activities, comparing risks and trade-offs, deciding on the capability options that go to government, and interpreting the guidance that emerges from their customer/owner.

Comparison of the Model with the Real World

A qualitative and theoretical comparison of the model with the real world was conducted through discussions with the highest levels of senior leadership in the DoD. The problem of interest—the *gap in explanation* and the need to address it—was recognized. The changes suggested by this application of systems theory and practice as captured in the model were seen as highly desirable, with the secretary (then Mr. Paul Barrett) noting that it would take some years to mature. It was recognized by the leadership that the conceptual model fits the *problem space* described earlier in a manner that was seen to be consistent with ADoD's traditional *top-down* approach to strategic planning—the feedback and testing loops being the main difference along with decision traceability to provide rationale. The leadership also anticipated that the process could provide a basis for discussing with the government its needs statement.

Action to Improve the Problem Situation

The senior leadership accepted the need to take action. They funded a trial of a prototype of the proposed process. They set up a group to oversee and approve the inputs to the process and they participated in the development test and evaluation. So, the final step in applying SSM began with the design, development, and implementation firstly of the testing process to manage the situational complexity involved in applying military power in the national interests, both now and in the future. Three teams were established:

1. A scenario-development team led by the Director Force Structure Guidance to devise a complement of 14 different vignettes written around a selection of classified military strategic objectives. These were to incorporate a threat statement of probable adversary capabilities and a suite of planning assumptions.
2. A force-option development team led by military personnel to design options for force structures around three different resource profiles. All three services and the HQ staff of the Australian Theatre were intimately involved in creating these artifacts.
3. A third team led by the first author to develop a testing methodology within a systems engineering framework. Tools such as databases and collaborative computing techniques to collect and assess the required data were to be developed. This team was also allocated responsibility to evaluate the process to ensure the analysis provided the transparency needed to develop a capability priorities paper for defense.

Following the success of the prototype, the Deputy Secretary (Strategy) also agreed to fund the establishment of an in-house capability and move the project to operational test and evaluation and then to full-scale implementation.

Thus, Checkland's SSM was successful in dealing with the cognitive complexity of identifying and expressing the strategy and execution problem, developing subtle definitions of the transformations an improved process would entail, using conceptual models to identify an alternative approach that would be culturally feasible, and convincing the stakeholders that the alternative approach to this important problem should be trialed.

It is worthwhile to note that SSM continued to operate in the background to ensure that the activity retained its stakeholder buy-in and to surface changes required to ensure efficiency, effectiveness, efficacy, and stakeholder acceptance. Given the complexity of the enterprise, SSM was used to surface the nature of the deep issues considered most problematic and most inhibiting to the development of an integrated framework for strategy and execution. These are illustrated with *crossed-swords* icons in the rich picture (after Checkland and Scholes, 1990) shown in Figure 17.10; they were *hinge* issues for the concept of strategy and execution.

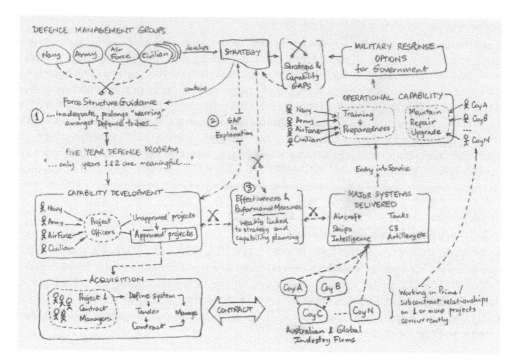

FIGURE 17.10 Rich picture of the issues *considered problematic* in the defense strategy and execution process in 2000.

The three key deep issues are stated in the following, paraphrased in the words of defense leaders of the time:

1. Inadequate force structure guidance and inadequate force structure development process. This deficiency was seen to promote and prolong the *warring* among the defense tribes (McIntosh et al., 1997).
2. The *gap in explanation* between strategy and capability development. This deficiency meant that there was no basis for understanding or determining relative priorities for the development of current and future capabilities (Baker, 2000).
3. The available measures used to assess the effectiveness and performance of new or upgraded capabilities were at best weakly linked to capability planning, the further consequence being that the military response options available to the government of the day presented both strategic and capability gaps (McIntosh et al., 1997).

Figure 17.10 illustrates how the problem space evolved from the beginning of the research program through discussion with the major actor communities. This depiction shows that there are three military and several civilian defense groups involved in strategy and execution activities as well as the military operational community that needs to use the resulting capabilities. In a sense, both sets of military stakeholders can be considered the customers (indeed in the United Kingdom, they are labeled as such). Another principal actor community is the capability development group who are collectively tasked with interpreting the strategic guidance and the five-year defense program (which embodies the future-force structure) and from it defining the projects to fill identified or perceived capability gaps. The remaining actor communities are the acquisition community who execute the projects defined by capability development and the defense industry community that delivers the material to fulfill the equipment component of projects.

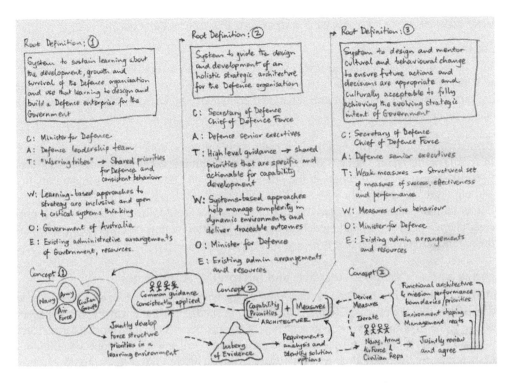

FIGURE 17.11   Root definitions aid in identifying customers of the transformation.

In the same SSM session that saw the development of the rich picture, the problem domain was explored to produce root definitions (after Checkland and Scholes, 1990) using the CATWOE* mnemonic to identify

- How the systemic issues were nested within one another
- Who the customers, actors, and owners of the transformation were
- What the prevailing worldview driving each part of the transformation was

These root definitions are shown in Figure 17.11. The derivation of the root definitions for the problem clarified the insight that the customers and stakeholders vary depending on the part of the problem under investigation. It should be noted that while Figures 17.4 and 17.5 show the main actors, many other constituents existed but are now shown for clarity. Additional players in the strategy execution process included the intelligence and defense science communities who had strong hands to play, respectively, in the content development and the methodology development. International policy, finance, quality assurance, logistics, personnel and training, and facilities were among other groups from defense who have defined roles at various stages of the strategy and execution. Such groups were invited to participate whenever it was perceived they had a legitimate contribution to make.

Our attention now turns to the detailed design of the process.

---

* CATWOE: *C*ustomers, *A*ctors, *T*ransformation, Weltanschauung, *O*wner, *E*nvironment (constraints)—this is described in greater detail later.

IMPLEMENTATION OF A SOCIALLY AWARE STRATEGY AND EXECUTION PROCESS

Goals of the Strategy Process

To ensure the results from a defense strategy process were seen as credible to major stakeholders like the Chief of the Defence Force and VCDF and the chiefs of the three services, the strategic planning process needed to manage the situational complexity of warfare. Therefore, the goals of the strategy process were to

- Bring the organization through the torrents of ambiguity and uncertainty that are endemic in ADoD planning (and stick with that process until action is taken)
- Establish transparency throughout the process to give auditability of information quality and build greater credibility with ADoD stakeholders
- Direct human behavior by a shared image of reality that would, through the planning process, change peoples' mental models to a point where the models coalesce to form a basis for organizational change that is *owned* by the participants

Configuring a Mix of Rational, Adaptive, and Interpretive Approaches

To achieve our aim of bringing the organization along with planning decisions, the study team needed to ensure that the process offered consistency above all else in the way in which it assessed the strategic value of options for force structuring. In any process where high-priced force structure trade-offs are made, there will always be some people who will not like a proposed solution and respond by attacking the methodology that delivered such an answer (Brabin-Smith, 1999).

The major variables in the planning process, whether in a current or future time domain, are the strategic objectives and the shape of force structures. The relationships between these major inputs are depicted in Figure 17.3.

The need for consistency in the analysis and evaluation of options against objectives was recognized early as paramount when presenting options for the government. The apparent risk was mitigated using a conventional systems engineering process as a methodology to guide the integration and implementation of appropriate subordinate methods and tools. In this problem context, the systems engineering model was seen to offer the capability to

- Manage a process in a dynamic environment where the government needs and system requirements are ill defined and are changing and/or changeable—thus, supporting the development of various *use cases* in multiple scenarios
- Develop and assess multiple force structure solutions at different levels of interoperability
- Test and evaluate potential solutions to examine their fitness for purpose in delivering the government needs for joint and/or combined operations
- Capture and analyze the data in each of the preceding process elements to examine the common and critical issues affecting the fitness for purpose of each force structure option—whether at the technical, conceptual, or behavioral levels of interoperability

The decision to focus first on the testing component of the systems engineering methodology was a good place to start a transformation of a defense strategic planning process because it satisfied two criteria from Checkland's SSM. It was systemically desirable because it enabled the author to design and integrate a mix of interpretive, adaptive, and rational methodologies into the planning process. It was also culturally feasible because the major stakeholders retained ownership and control of the major inputs—namely, the development of the scenarios and the force structure options. This brought in a large number of senior and middle-ranking people from the strategy group, the operations community, and the capability development group. It proved to be a catalyst for them to participate in the testing process, then to assess the efficacy, effectiveness, and efficiency of the force structures, and then to own the outputs of the test and evaluation process to which they would contribute.

The interplay between managing the cognitive complexity of transformation and managing the situational complexity of developing and implementing new processes for achieving better clarity in decision making on warfighting incorporated Checkland's SSM and systems engineering into direct and symbiotic interplay throughout the case study. While SSM helped to define the problem and select a mix of methods by which to take action, systems engineering provided a structured framework of processes by which the mix of methods could be integrated. Monitoring the effectiveness and efficiency of the mix of methods continued as a background SSM activity and control action was taken to adjust the mix to suit the evolving planning conditions.

Strategy Process Model

The conceptual model of the system of interest (described in Figure 17.10) formed archetype for the process implementation. Figure 17.12 illustrates the process steps in a way that reinforces this systems approach as a cyclic process of inquiry (which operates on roughly a two-year basis). The drawing takes the form of the *wave model* of systems of systems (SoS) development that has been formulated to tackle SoS development in the US Department of Defense (Dahmann et al., 2011). This model recognizes that the environment changes with time as does the capability and preparedness of the force-in-being.

The process is best viewed as an action-learning approach melded with a systems engineering-based activity model. Action learning is a cyclic process of inquiry that usefully surfaces organizational issues and change agendas in pluralistic environments, while promoting stakeholder buy-in to the solution, whereas systems engineering brings the methods and discipline needed to produce solutions and artifacts to agreed schedules, budgets, and performance levels.

The mature application of cyclic methods of inquiry recognizes that the cycle can start anywhere within the process. In this case study, it was at the synthesis end of the process and the description that follows extends to the end of the first complete iteration.

Development of the Scenario Planning Base (Expression of the *Why*)

A set of 14 real-world scenarios was developed to represent the type of operations that the Australian Government would expect the ADF to successfully prosecute. They included a range of geographical locations and various adversaries and cover many potential points along the operational intensity spectrum from support to national operations to defense of Australia. The scenarios also included information such as the events leading up to the crisis, the political climate, the geographical location, time of year, rules of engagement as directed by the government, and timing imperatives. They also indicated other players, such as coalition forces and what Australian forces were currently tasked elsewhere and therefore may not be available for use in this crisis. Each scenario was accompanied by a concept of operations (CONOPS) describing a possible way to tackle the problem, including indicative objectives and missions. This was used as a strawman during the testing, allowing the players to modify it and in that way gain ownership of the military solution to a real-world problem and—importantly—ownership of the data gathered in the seminar gaming process.

Development of the Force Structure Options (Expression of the *How*)

The development of force structure options, that is, the overall packages of military capabilities, was performed by a team led by military personnel that included input from all three services and from the staff of HQ Australian Theatre.* For the first run with the methodology, three capability options were designed to three different cost profiles relative to the existing budget: 0%, 3%, and 6% real growth over seven years.

The inclusion of a broad cross section of personnel in the force-option generation process produced fairly conservative capability options, but they were culturally acceptable within ADoD. More radical testing artifacts (e.g., a force structure without airborne early warning and

---

* Now known as Joint Operations Command.

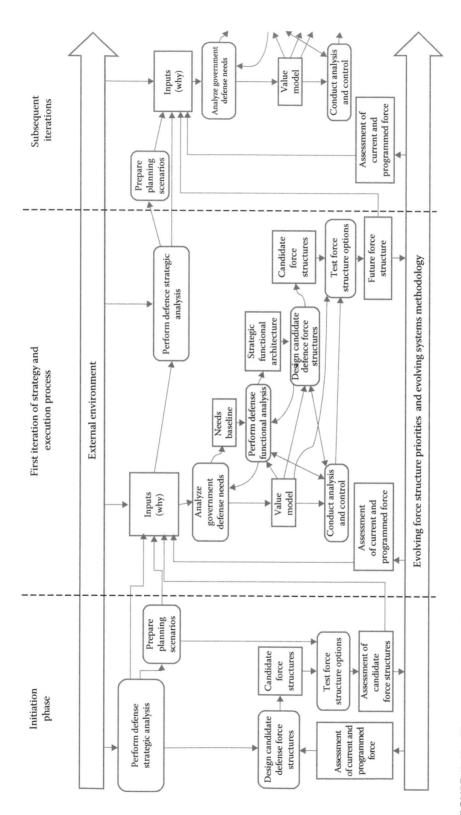

FIGURE 17.12　The strategy and execution process employed.

control [AEW&C] aircraft) would have stressed the force options in different ways. While this would have been an interesting analytical technique to triangulate on the value of the AEW&C contribution, the players found it broke the bounds of plausibility and such options were generally not accepted.

All capability options were costed, albeit approximately, to ensure that they did not exceed the realms of plausibility. This allows ADoD to give the government guidance along the lines of the statement "If you structure the ADF this way, and spend approximately this much money, these are the military response options available to government, and these are the things that you cannot achieve." It was recognized that the need for costing tends to make the capability options rather platform focused as these are easy to cost. Ideally though, capability options should contain information on the readiness levels of the force elements and their sustainability to allow these aspects to be experimented with and tested. In addition, softer issues should be included if possible such as the general organization, doctrine, and people that together with platforms make capability.

It is seen as important that as the process matures, more radical capability options that do not include the traditional means (e.g., fighter aircraft) for a particular effect (such as strike in the land environment) be employed by force participants to consider how essential those means are to successful achievement of an end state. Being unable to complete a mission without a particular platform or *capability* in a strategic simulation of warfighting is stronger evidence for its necessity than being able to complete a mission with ease using it.

Development of the Testing Methodology (Expression of the *So What*)

In the process of implementing the conceptual model of systems engineering, a seminar gaming approach was selected to bring together the senior stakeholders in an orchestrated debate that was aided by several software tools. A two-part process was designed. In the first part, the assembled subject matter experts (SMEs) debated and agreed the efficacy of the objectives and missions to achieve a military strategic objective. The approach is based on a modified military appreciation process, which connects strategy to tasks and resources. In the second part, the assembled SMEs debated the fitness for purpose of the allocated forces to achieve the hierarchy of objectives and activities, using the key criteria of likelihood of success (effectiveness) and capacity (efficiency). This process is outlined in Figure 17.13 and in the following text.

The process involves eight main steps as follows. To expedite the process, a *strawman* answer at each step of the process was previously prepared. The group however had complete discretion on the final data entered into the database:

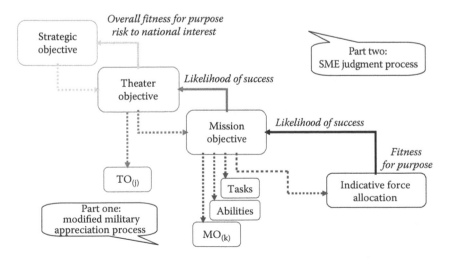

FIGURE 17.13   Summary of test methodology and measures.

Step 1: An assembly of 12–20 military expert representatives from the single services, ADHQ, intelligence staffs, and Defence Science and Technology Organisation (DSTO) was convened and briefed on a real-world scenario and a force structure option of a particular cost and design profile. The group was asked to develop a CONOPS from the inputs and develop an agreed military strategic objective, noting that the ultimate aim of the seminar game was to assess the fitness for purpose of the given force structure option to achieve the strategic objective—it was not about getting the best military plan of attack.

Step 2: The strategic objective is reduced to several theater objectives, followed by several mission objectives for each of the theater objectives. The essential aspect was to ensure the objectives clearly state what effect is to be achieved in which environment. A range of indicative tasks was then listed for each mission objective. These are mainly used as aides-memoire when assigning forces to each mission objective. A list of abilities essential to achieving the mission objective was also collected. Finally, force elements were assigned to each mission objective, with these assignments being restricted by the number and type available in the order of battle (ORBAT) that was being played. Multiple assignments of force elements were tracked throughout the process so that the degree of overuse of force elements could be gauged. Note that although there were loose similarities to the military appreciation process, deliberate joint force planning was not employed.

Step 3: The players then assumed the mantle of *military inspectors general* and worked in reverse order up the hierarchy of objectives, making judgments about the likelihood of success of the allocated forces to achieve each objective. First, they considered the fitness for purpose of the allocated force elements by considering the functionality and the capacity of the allocated force to achieve the objective. The likelihood of success of the mission objective was assessed with an indication of the impact of failure of this mission objective on the outcome of the strategic objective to give some sense of the strategic criticality of each objective. These judgments were made for all the mission objectives.

Step 4: One bridging judgment is made at the theater objective level: the likelihood of theater objective success. This was assessed for each theater objective from the judgments arrived at for the mission objectives that underlay it, bearing in mind their previous judgments on the fitness for purpose of the forces allocated.

Step 5: Before making the final strategic judgments of risk and overall fitness for purpose, the participants were asked to identify key strategic issues relating to the ADoD's fundamental inputs to capability (FIC) and preparedness. This brings out issues such as there not being adequate doctrine for a particular way of doing things or a lack of support in particular areas. It also allows the participants to focus on the broad spectrum of issues in making their judgment, rather than just on the underlying judgments obtained during the process to that point.

Step 6: At the strategic level, the likelihood of success was assessed based on the hierarchy of judgments previously made. The participants are also asked to assess the risk to the national interest. This is defined as the product of the likelihood of failing to achieve the strategic objective and the impact of such a failure.

While the scenario-based planning and evaluation to this point keeps the two major inputs, the scenario and the force structure option constant, it is important at the end of the process to give the participants an opportunity to review how sensitive their collective decision making was to any changes in either or both of the inputs. Thus, steps 7 and 8 in the process were included.

Step 7: The participants were asked to judge the sensitivity of the outcome to changes in the scenario by asking the following: what small change, if any, in adversary capabilities would significantly reduce the judged likelihood of success at the strategic level? This allows the participants to highlight issues such as if the adversary had slightly more capable platforms, the force option would have been unlikely to succeed.

Step 8: The participants are then asked to judge the sensitivity of the outcome to
- The given course of action by estimating the relative value of other friendly courses of action
- The given CONOPS by identifying changes that would significantly improve the campaign plan
- The given force structure by identifying changes in capabilities that would significantly improve the effectiveness of the force

The purpose of step 8 is to prompt the participants to suggest elements that would increase the overall capability of the force structure option being played. The participants were free to think *outside the square* and include futuristic suggestions. However, for each suggestion made, an offset had to be indicated in an attempt to keep the enhanced force structure option affordable within its broad cost profile. Not surprisingly, the participants found it much easier to suggest additions to the force structure rather than find offsets.

Confidence in the outcomes of force-option testing is directly proportional to the people selected to make the judgments—this represents a critical vulnerability with the process itself. To mitigate this vulnerability, we worked with the primary stakeholders at the chief of service level to carefully select participants with appropriate expertise. While it can be said that expertise brings its own bias, the process requires all participants to *vote* to achieve a collective judgment. An onus therefore falls on the experts to convince their peers of the strength of their position so that overall, the *wisdom of the crowd* prevails.

A *gamebook* developed to orchestrate the seminar gaming process is included in Hodge (2010). The gamebook lists the key roles of the exercise control team members, the players, the questions they are to address and where they iterate, and the definitions that apply on each judgment scale.

Reporting of these trials took two forms: scenario reports, termed *phase one reports*, that were likened to *lab* reports that detailed what happened in each seminar game and identified the principal issues that became evident and effects-based reports, termed *phase two reports*, that analyzed multiple phase one reports to identify what effects are delivered in which environments while noting the commonalities and criticalities across all scenarios.

Tools for Data Capture and Knowledge Elicitation

A range of tools was developed to support the elicitation and capture of the vast quantities of data produced. The designed gaming process was applied over 40 times in the first implementation where each run addressed one combination from a matrix of 14 scenarios with over six options for force structuring (at least four future-force structure designs, the force programmed to exist in three years' time, and the current force structure).

The tools that were designed and used included

- Map-based software to situate the scenario and support the military appreciation process
- Electronic meeting support equipment to capture individual inputs, voting on the values for all of the measures and categorization of key issues
- A relational database of the agreed hierarchy of objectives, force allocations, and the agreed judgments on the measures
- A visual reference of the decision tree and the judgments recorded in the primary database to allow the players to contextualize any one judgment within a view of the whole framework of decisions they have made

The tool environment is schematically represented in Figure 17.14 projecting four sets of information concurrently across four screens. Those interested in the details of the tools are referred to Hodge (2010).

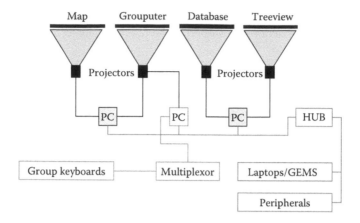

FIGURE 17.14   Seminar gaming tools for force structure analysis.

Initial Analysis

Phase One Analysis: Production of Scenario Reports

Once each scenario had been tested against all the force structure options relevant for a trial, it was analyzed by a desk officer. There are various inputs to this process, including the information collected in the gaming database and the coded data collected by the analysts throughout the gaming process. The initial output of the phase one analysis is a set of reports, one for each seminar game assessing the performance of the force structure option as it was tested in each scenario. The analyst then collected all of the initial reports for a given scenario thus having all the data at hand for all of the capability options tested in that scenario. The analyst then drew out significant capabilities for each scenario as well as highlighting any assumptions made, the CONOPS used, and the success, or otherwise, of the force structure options tested. The results recorded in the scenario reports that were known as the *phase one reports* were sent out to the players for comment.

Phase Two Analysis: Production of Effects-Based Reports

The phase one reports of the scenario-based analysis were then analyzed across broad environmental areas to understand the key strategic functionality required to structure a force for military success across the scenario planning space. In the 1999–2001 period, the analysis addressed eight areas: strategic issues (e.g., preparedness), combat-strike effects, combat-maritime effects, combat-land effects, combat-air effects, cyber effects, support enablers, and knowledge enablers.

Although the phase one reports were a major input, the desk officer also considered the data collected during gaming and used their military knowledge and their intimate knowledge of the scenarios to inform the analysis process. The effects-based reports were then *workshopped* to ensure all the analysts present during the testing had an opportunity to comment and validate the analysis. The output was a set of reports that draw information from all scenarios and all capability options and indicate the common and critical issues as well as the range of strategic needs and indicative forces required to meet these needs. The phase two reports were also sent out to the players for review and comment.

The production of these reports brought the process to the end of the initiation phase in Figure 17.12. The concept of using force structure options testing against an agreed scenario set had been demonstrated and had been found to be culturally feasible as evidenced by positive feedback and high degree of acceptance of the results. The process now started on its first full iteration of systems analysis to provide additional rigor and to compile the capability priorities guidance for the government.

Analyzing Government Defense Needs

From the previous phase, the team had to hand the key artifacts that defined the government's needs for defense such as the Defence White Paper, Warfighting Concepts, Defence Planning Guidance, and knowledge of the current and programmed force structure. The team also had the 14 validated capability planning scenarios that had been prepared earlier. From this information in hand, the needs analysis began by reviewing the traditional approach to defining *user requirements* described by Blanchard and Fabrycky (1998) and matching that with appropriate sources of data. The main process elements in defining user requirements are shown in Table 17.4 together with the source material that was adapted for each process.

Using this adaptation of the traditional systems engineering requirements analysis process, the primary author led the team through the process outlined in the following to synthesize the first-order summary of the government's needs illustrated in Figure 17.15. Firstly, the 14 scenarios where represented in the first two columns of Figure 17.15 grouped into three strategy bands corresponding to broad geographic regions. Secondly, the need for military effects in the air, land, and maritime environments was ascertained for each scenario against a scale defined in capability and constraint terms. The assessment was recorded using a coarse scale employing 4-point gray scale coding; nil (black), low (dark), medium, and high (light), respectively. This was followed by similar judgments for the key capability areas of strike, strategic support, and knowledge (the last category including effects in the cyber environment). Thirdly, the key capability conclusions for each scenario were recorded in written form. Lastly, conclusions about the key constraints that will limit the candidate forces structure solutions were recorded, which usually addressed adversary capability, and own force readiness and sustainability.

While the picture of the government needs is coarse, and far too coarse to guide force structure design, a picture nonetheless begins to emerge. For example, if scenario group 2 (central band) were the prime determinant for force structure design, then indications are that high demands for capability (shown in light shaded) appear in only two of the six environments, with nil demand emerging in different scenarios for other environments. The challenge in the final analysis, of course, is to integrate the judgments across the three scenario bands because all three scenario sets drive government needs for defense and each will impact in varying degrees on the final design solution.

TABLE 17.4

Adaptation of the User Requirements Process

Traditional Systems Engineering *User Requirements* Process Elements	Systems Engineering for Strategic Planning Adaptations for the Defense Case Study
Analyze mission and environment.	Analyze the scenarios—these were well defined and were easily grouped by geographic region, mission, and environment.
Identify functional requirements.	Identify requirements for strategic level effects, using judgment to synthesize the data within the phase two reports to provide a useful starting point for scenario-based functional demands.
Define or refine performance and design parameters.	Define the key demands for defense capability in each scenario by a process of synthesis of the data from the phase one and phase two reports.
Define the constraints that will limit the solutions.	Define the constraints on the application of the defense capability set in each scenario, where the principal constraints were the adversary capability, and own force readiness, reach, and sustainability.

Source: After Blanchard, B.S. and Fabrycky, W.J., in: W.J. Fabrycky and J.H. Mize (eds.), *Systems Engineering and Analysis*, 3rd edn., Prentice Hall, Upper Saddle River, NJ, 1998.

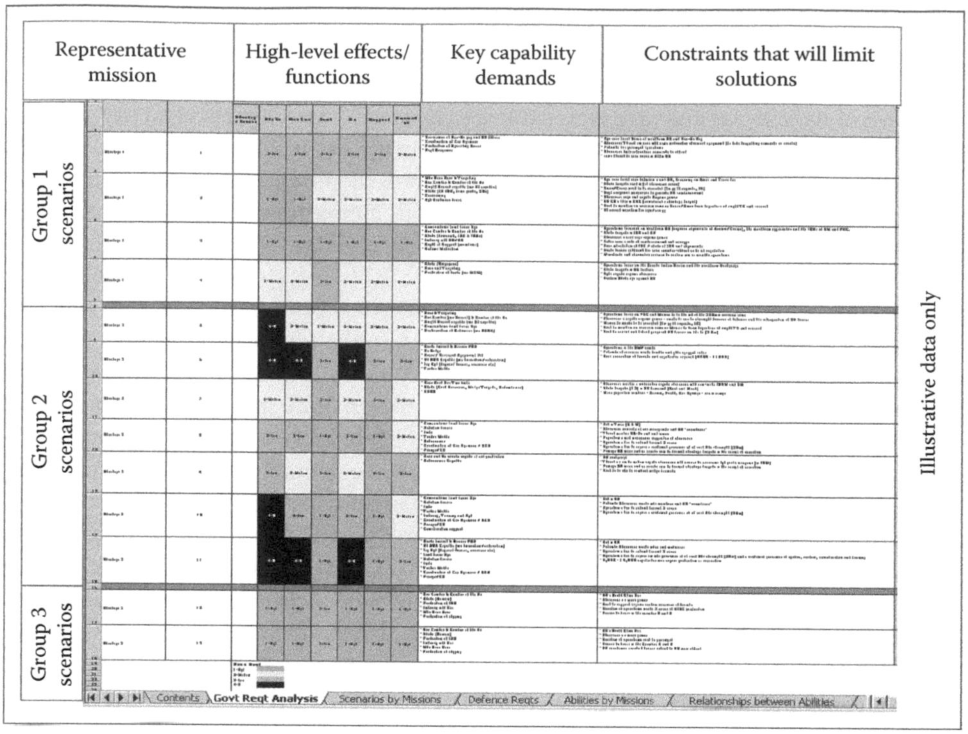

FIGURE 17.15   First-order assessment of government needs. (Adapted from Blanchard, B.S. and Fabrycky, W.J. *Systems Engineering and Analysis*, Third Edition, Edited by W.J. Fabrycky and J.H. Mize, Prentice Hall, Upper Saddle River, 1998.)

Having captured the government needs, the next step is to conduct a requirements analysis process to identify the functional requirements for defense and complete the needs baseline. Guided again by Blanchard and Fabrycky (1998), requirements at the system level are traditionally driven by those issues listed in the left-hand column of Table 17.5—noting that they wrote this primarily focusing on project and product-based systems engineering. Their guidance was adapted to defense strategic planning as indicated in the table.

Using this adaptation of the traditional systems engineering process, the primary author led the team through the following activities to synthesize a first-order summary of the defense *system* requirements.

TABLE 17.5
Adaptation of the System Requirements Analysis Process

Traditional Systems Engineering	Systems Engineering for Strategic Planning
System Requirements Analysis Process Elements	**Adaptations for the Defense Case Study**
Define operational distribution/deployment (geographic considerations).	Sort scenarios into a number of groups ordered by operational demand and reach from Australia.
Define mission profiles to identify what the system must accomplish.	Define functions at a primary level (strategic functions indicating mission profiles) and at a secondary level (first-order statement of abilities needed by defense to achieve the missions).
Identify • Performance parameters • Utilization requirements • Effectiveness requirements • Maintenance and support requirements	Identify key performance parameters and qualify these with statements of utilization, effectiveness, maintenance, and support requirements.

Defining Defense Functions: Operational Distribution/Deployment

The analysis of the *system* requirements for defense considered first the range of strategic effects/objectives needed to cover the government needs and some of the higher-level operational or theater objectives. Eight different categories of strategic deployments were synthesized based on the geographic reach from Australia and level of warfighting intensity as follows:

- Support to domestic emergencies and peacekeeping
- Defending Australia (denial missions)
- Defending Australia (deterrent missions)
- Maritime operations in the *inner arc*
- Land operations in the *inner arc*
- Comprehensive joint operations in the *inner arc*
- Maritime operations in the *nearer region*
- Comprehensive Joint operations *nearer region*

In general, these strategic functions become more challenging as they progress down the list. For example, designing a force structure to achieve the one mission profile of peacekeeping would have little capability to do more demanding tasks. At the time of developing these strategic functions, the government's intentions (later expressed in the 2000 White Paper) were not clear. However, traceability in the systems analysis was assured because the analysis drew on the agreed statements of strategic effects recorded in the primary database from the testing of the 14 scenarios, each of which could be shown to be directly related to one of the eight strategic functions in the defense systems requirements.

Defining Defense Functions: Mission Profiles

The second part of defining defense system requirements involved deriving a set of mission profiles to address the question: *what must the system accomplish?*

In order to meet the operational deployments described in the first part, a list of 11 mission profiles were developed as follows*:

1. Basic services protected evacuation and counterterrorist options (SPE/CT)
2. Basic maritime zone security
3. Secure bases
4. Defeat incursion (I)
5. Defeat incursion (II)—more demanding than (I)
6. Assault
7. Dominate a land area
8. Attack adversary homeland (I)
9. Attack adversary homeland (II)—more demanding than (I)
10. Strategic command, control, communications, and intelligence (C3I)
11. Strategic support

The study team developed these first by working with the most challenging deployment and worked down from there, appropriately adjusting the deployment definition to determine other less demanding deployments (e.g., defeat incursions I and II vary according to adversary capability level, distance from Australia, and likely duration of operation; similarly, for *attacking adversary homeland*). Traceability in the systems analysis was assured because the analysis drew on the agreed statements of strategic effects recorded in the primary database from the testing of the 14 scenarios, each of which were related to one of the eight mission profiles in the defense systems requirements. To validate the analysis, the team checked the extent to which one or more of these mission profiles,

---

* The authors acknowledge the contribution of Martin Dunn in designing this list of mission profiles and to the study team for validating them.

		Government needs							
		Peacekeeping	DAA - Denial	DAA - Deterrent	Inner Arc - Maritime	Inner Arc - Land	Inner Arc - Comprehensive	Nearer Region - Maritime	Nearer Region - Comprehensive
Missions (functions I)	Basic SPE and CT	✓	✓	✓	✓	✓	✓	✓	✓
	Basic Maritime Zone Security	✓	✓	✓	✓	✓	✓	✓	✓
	Secure Bases		✓	✓	✓	✓	✓	✓	✓
	Defeat Incursion (I)		✓	✓		✓			
	Defeat Incursion (II)				✓		✓	✓	✓
	Assault				✓	✓	✓	✓	✓
	Dominate a land area	✓				✓	✓		✓
	Attack adversary homeland (I)			✓		✓			
	Attack adversary homeland (II)				✓		✓	✓	✓
	Strategic C3I	✓	✓	✓	✓	✓	✓	✓	✓
	Strategic Support	✓	✓	✓	✓	✓	✓	✓	✓

FIGURE 17.16   Mapping of mission profiles against strategic functions.

singly or in combination, would adequately describe each of the operational deployments defined by the strategic functions, as shown in Figure 17.16.

This set of mission profiles informed the system requirements; however, it was still at too high a level of granularity to enable us to think sufficiently deeply about design issues. To achieve further refinement of the activity model, a secondary level was developed to describe the mission profiles, which we referred to as the *abilities* required in defense that singly or in combination will achieve the missions.

Based on the CONOPS developed by the participants during scenario gaming, the study team synthesized a set of secondary mission profiles as follows:

Attack adversary strategic targets	Command and control
Conduct mine countermeasures (MCM)	Conduct surveillance and reconnaissance operations (SRO)
Control the air	Control the sea
Degrade adversary	Insert forces
Organize the population	Protect shipping
Provide strategic lift	Respond to incidents
Secure land areas	Sustain (forces in theater)

The list of secondary mission profiles traced directly to the list of important *capabilities* recorded in the primary database complied during the initial force structure analysis and in the phase one reports. With this two-tiered definition of mission profiles and defense system functions, the team was then in a position to examine the relationships between the primary and secondary functions in order to support the definition of defense system requirements and their respective performance parameters.

Defining Defense Functions: Performance Parameters (Generic)

The primary author then led the study team through a process that mapped the relationship between the primary and secondary functions to build a taxonomy of the mission profiles that ADoD (as a systemic enterprise) must accomplish. The questions guiding the process were the following:

- Is there a relationship between the two functions?
- If so, what performance parameters define the relationship? Here, the team aimed to describe the degree to which the secondary *ability* function was required in order to successfully achieve the primary function defining the mission profile.

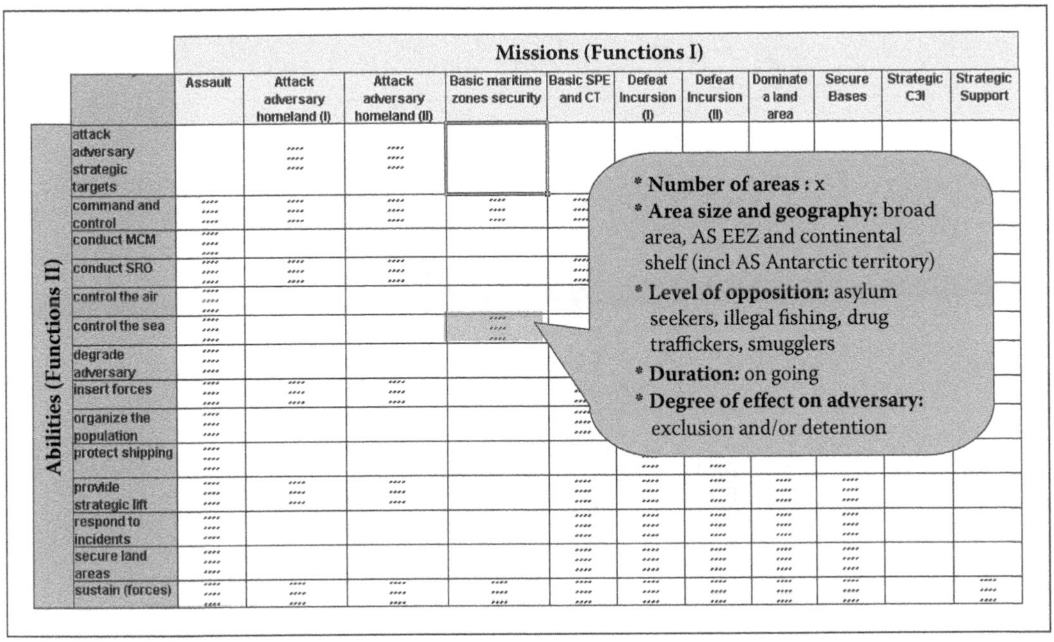

Missions (Functions I)

Abilities (Functions II)	Assault	Attack adversary homeland (I)	Attack adversary homeland (II)	Basic maritime zones security	Basic SPE and CT	Defeat Incursion (I)	Defeat Incursion (II)	Dominate a land area	Secure Bases	Strategic C3I	Strategic Support
attack adversary strategic targets		▨	▨								
command and control	▨	▨	▨	▨	▨						
conduct MCM	▨										
conduct SRO	▨	▨	▨		▨						
control the air	▨										
control the sea	▨			▨							
degrade adversary	▨										
insert forces	▨	▨	▨								
organize the population	▨										
protect shipping	▨										
provide strategic lift	▨	▨	▨			▨	▨	▨	▨		
respond to incidents	▨					▨	▨	▨	▨		
secure land areas	▨					▨	▨	▨	▨		
sustain (forces)	▨	▨	▨	▨	▨	▨	▨	▨	▨		▨

Call-out box:

* **Number of areas :** x
* **Area size and geography:** broad area, AS EEZ and continental shelf (incl AS Antarctic territory)
* **Level of opposition:** asylum seekers, illegal fishing, drug traffickers, smugglers
* **Duration:** on going
* **Degree of effect on adversary:** exclusion and/or detention

FIGURE 17.17 Generic performance parameters between defense functional decompositions.

The resultant chart, shown in Figure 17.17, was created quickly *once-over-lightly* without being specific to any particular needs of the government expressed in the operational deployments. In this way, this chart provided a generic building block for later analysis against the government needs for defense.

An example is shown in the call-out box in Figure 17.17 illustrating the definition of generic performance parameters for the ability *control of the sea* as it might be applied to the mission profile *basic maritime zones security*. These performance parameters often differed from cell to cell, both in the parameters that were defined and in the values ascribed to them. Traceability in the definition of performance parameters is maintained by drawing on the evidence in the primary database that captured many relationships between theater-level and mission-level effects and the fitness for purpose of a given force to deliver the performance effectiveness needed for mission success.

Defining Defense Functions: Performance Parameters (Specific)

The aim now was to complete the definition of defense system requirements expressed as a function of needs (operational deployments), mission profiles (primary functions), and *abilities* (secondary functions). To ease the complexity of this problem, the analysis began by limiting the needs profile to the operational deployment for peacekeeping noting that the analysis for the government need for defense only relates to five primary functions, as shown in Figure 17.18.

The goal of the analysts was to derive a single set of design criteria for ADoD to meet the government's needs for peacekeeping deployments—shown as the right-hand column in Figure 17.19 To achieve this, the primary author conducted a workshop comprising the planning team to simplify the chart further by creating a summary of the performance parameters for each of the abilities or secondary functions. Taking a workshop approach also provided a means for validating the analyst's evaluation of the data from the relevant phase one reports and phase two reports that they had individually undertaken. In some cases, this exercise was trivial, for example, when the ability was not required, or was required only in one circumstance, in which case, the existing statement *rolls up* to become the summary design criteria for an ability (secondary function) to meet the government's need for peacekeeping. In other cases where there was more than one set of performance parameters specified, some synthesis was required to achieve a *rolled up* set of criteria. This synthesis

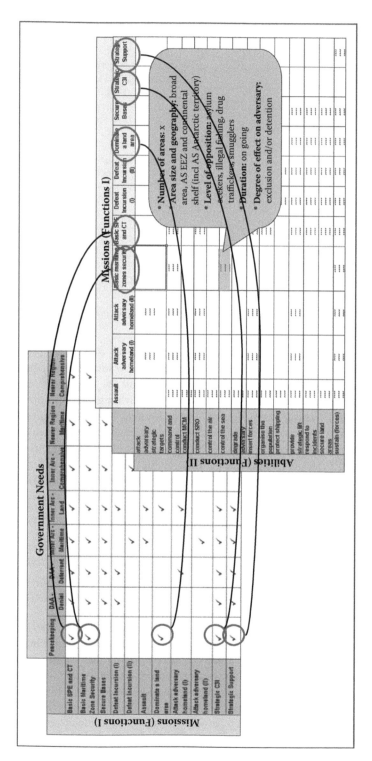

FIGURE 17.18   Relating operational deployments to mission profiles and abilities.

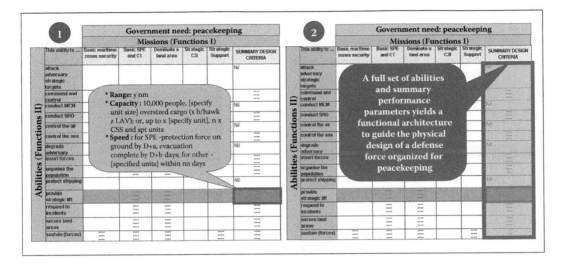

FIGURE 17.19 Developing summary design criteria for peacekeeping.

took different forms, for example, for parameters like range, this was set to be the maximum value for sensing systems. In contrast, for parameters like transport capacity, the parameter was set to the sum of individual platform capacities. In most cases, the synthesis involved creating design criteria by trading off the multiplicity of demands (e.g., to set a duration to complete an SPE, a protection force might be specified to be required on ground by $D + a$ days and have the evacuation complete by $D + b$ days, while sustainment and reconstruction [if required] be completed within nn days).

The workshop environment proved critical to achieve consensus in the final set of design criteria. So that, in the end, the design team established a set of abilities with performance parameters synthesized specifically to assist in designing a defense force structured, in this example, for peacekeeping.

To extend the summary design criteria to form a basis for designing force structures for all government needs, the same process is repeated for all other operational deployments listed in Figure 17.16. This is necessarily time consuming. In this case study, four operations analysts were engaged for 2 weeks in their analysis and synthesis of eight tables similar to Figure 17.19. Further four people were involved in the workshops over 3 days to validate the eight sets of summary design criteria for the eight representative operational deployment scenarios.

The eight sets of design criteria were then compiled into one large table in the form of Figure 17.20 representing all government needs of ADoD at the time and the abilities (secondary functions) required in a defense *system* to deliver on the needs. Working from this sheet, ADoD was in a better position organizationally to engage the minister in a conversation to refine the extant statement of government needs. The key question discussed was: *what are the minimum and desirable levels of operational deployment the government expects ADoD to be able to achieve?* In 2000, the senior executive was able to draw from that discussion a lower and an upper bound; a representative bound is shown overlaid on Figure 17.20.

DESIGNING SOLUTIONS

This section addresses the final step in applying systems engineering processes to the design and development of force structure options. It discusses first the design process element for force structure options and then addresses the issue of balance that is required across the capabilities within the force structure.

Designing Force Structure Options

In view of the *wish list* approach that had been experienced in the first attempt at designing force structure solutions based on functional requirements, the primary author reviewed the work of

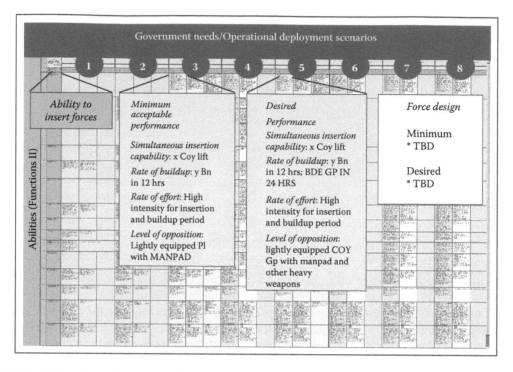

FIGURE 17.20   (See color insert.) Strategic functional architecture for defense showing minimum and desirable parameters to guide design.

TABLE 17.6
Key Mission Components

Conceptual Capability	Forecast Deficiency	Necessary Requirement
Military task to be performed, but without stating how	Assessed deficiency in light of the estimated threat	Draws on military judgment, operational analysis studies, and exercise data to indicate appropriate solutions
Worked Example		
Detect mines in harbor approaches	Perfect capability—so no deficiency	Perfect capability—so no requirements
Clear mines from harbor approaches	Slow clearance of conventional mines	Increase numbers of mine clearance vessels
	Inability to locate modern high-technology mines	Develop new methods of mine hunting

Coyle (2001) for a conceptual planning framework* of key mission components (KMC), which incorporates a three element table for identifying necessary design components, as illustrated in Table 17.6 (Coyle, 2001).

The primary author extended the work of Coyle (2001) to develop a design worksheet of six columns that includes the minimum and desirable functional requirements as a basis for design as well as perspectival information relating to forecast deficiencies, resulting in Table 17.7.

Coyle's array (Coyle, 2001, 2004) developed from a traditional military appreciation process of *gap* identification and analysis, from which options are designed to close the *gap*. However, as government requirements change in a dynamic strategic environment, so too does the definition

---

* Coyle (2001) extends the discussion of the conceptual planning framework from the military context to a broader range of complex social problems. Coyle (2001) suggests the Delphi approach (or other satisfactory method or balance of methods) might be used with a view to finding good candidate solutions to remove the deficiency.

TABLE 17.7
Options Design Based on Functional Requirements and *Gap* Analysis

(1)	(2)	(3)	(4)	(5)	(6)
Conceptual Capability	*Backcast* Minimum Parameters	*Backcast* Desirable Parameters	Forecast Deficiency	Options for Architecture/ Design	Necessary Requirement
Military capability needed without stating how	Functional performance parameters	Options for force structure/ architectural design	Assessed deficiency in light of strategic context	a. Forecast options b. *Backcast options*	Indicate appropriate solutions, use judgment, OA studies, EX data, etc.

of the *gap*. While *gap analysis* is also a traditional planning component in Australia, the primary author was also interested in introducing the minimum and desirable functional and performance parameters from the strategic functional architecture as a basis for capability design. In doing so, the approach has shown the value in addressing capability design from a strategy-led, top-down and a deficiency-driven, bottom-up push on capability development. In effect, this provided an avenue for bringing the future-force capability development group and the current force operations group together in designing the next-generation force structure.

Using illustrative data in Table 17.8, the team gained significant benefit in an orchestrated discussion about design options from both *ends* and then struck a balanced solution that was systemically desirable and culturally feasible.

TABLE 17.8
Worked Example of Options Design

Conceptual Capability	Minimum Performance (Backcasting)	Desired Performance (Backcasting)	Forecast Deficiency	Design Options (Forecast) (Backcast)	Appropriate Solutions
Ability to insert forces	*Simultaneous insertion capability*: x Coy lift *Rate of build up*: y Bn in 12 h *Rate of effort*: high intensity for insertion and buildup period *Level of opposition*: lightly equipped Pl with Man-portable air defence capabilities (MANPAD)	*Simultaneous insertion capability*: x Coy lift *Rate of build up*: y Bn in 12 h; BDE GP in 24 h *Rate of effort*: high intensity for insertion and buildup period *Level of opposition*: lightly equipped COY GP with MANPAD and *other heavy weapons*	Inability to provide lift capacity appropriate to achieve desired rate of buildup (currently programmed for low– medium intensity)	Forecast options: Replace landing ship heavy (LSH) with multi-role auxillary vessel (MRA) in 2015 Backcast options Desired level: 40 (30) battlefield RW 1 Para Bn capability 108 vehicles 8 C130 Amphib platforms (MRA) Light tactical airlift capability (LTAC)	Bring forward: 1. LSH replacement 2. additional mobility assets (RW and vehicles)

The only new data appearing on this chart are that in columns five and six. The collective goal was to ensure a wide variety of options were canvassed and the solution was fit for purpose and deliverable within a given cost profile.

Working from the left, the task involved stating the conceptual capability, defining the minimum and desired performance parameters, defining the forecast deficiency, collating the options from each *end*, and orchestrating the debate around the information before the team with the communities of interest.

Changes to data captured in the earlier needs and functional analysis phases were only permitted in the light of traceable and sustainable arguments. In this way, the design process provided a functional basis for new force structure options that met the strategic requirements within a cost profile. These newly designed force structure options were then tested back against the same, or new, scenarios giving further insights and some validation.

The performance parameters and their appropriate solution sets were then assembled into one *common sense* arrangement—not cast in stone—rather held in a living document that was adjusted in keeping with agreed implications of changes in either the strategic environment or in the resources available to ADoD. Doing this for the minimum and desired performance parameters provided the defense committees information to assist in decision making regarding capabilities to develop, maintain, or retire. The intention was not to provide a *follow-this-plan* instruction for capability development but a means of dynamically planning defense capability that reduces the uncertainty currently faced by committees in judging each proposal for its fitness for purpose in the government's strategic context. The process survives on contestability. What is before them needs to be open to challenge and changes need to maintain traceability to assumptions and requirements.

Balancing Capabilities

In performing their deliberations, the committees needed to address one final issue: the balance of capabilities. They recognized that it is fine to design the physical elements to fit acceptably within the boundaries of functional requirements. However, they risked skewing the overall structure if they did not complete the analysis of balance, taking into account how important each ability is relative to other abilities and then make certain the force structure is appropriately balanced. This issue is important because warfighting requires interdependent, coordinated action across the sensor, command, information, and engagement grids in keeping with the commander's intent. As no capability acts alone, no capability should be developed without a full appreciation of the interdependencies among sibling capabilities. The key question to be answered then is: *in performing a given ability, how important is each of the other abilities?*

The primary author therefore led an orchestrated debate in a seminar format that listed the [then] 14 abilities (secondary functions) in a table and addressed the key question in relation to the other abilities. A judgment was made using a five-point scale ranging from no importance (white) to critical importance (dark gray). Analyst comments or explanatory notes were captured on most judgments. The context for the discussion was the summary backdrop of over 40 runs of force-option testing, the phase one and phase two reports, the analysis of government requirements, and the design of the strategic functional architecture. Those invited to the seminar included the analysts involved to date and a small group of SMEs from the three services and the public service. A completed interdependency matrix, showing only illustrative data, is depicted in Figure 17.21.

The resultant matrix is itself a complex picture. The importance of command and control and sustainability to a balanced force structure should not be a surprise. Where a range of strong interdependencies occur (e.g., the rows to control the air, control the sea, and secure land areas in particular), the proposed force structure solution set would need to be tested for the desired level of flexibility in the capabilities to deliver adequate support to the range of interdependent abilities. Time did not permit this reiterative testing during the case study. However, the general point

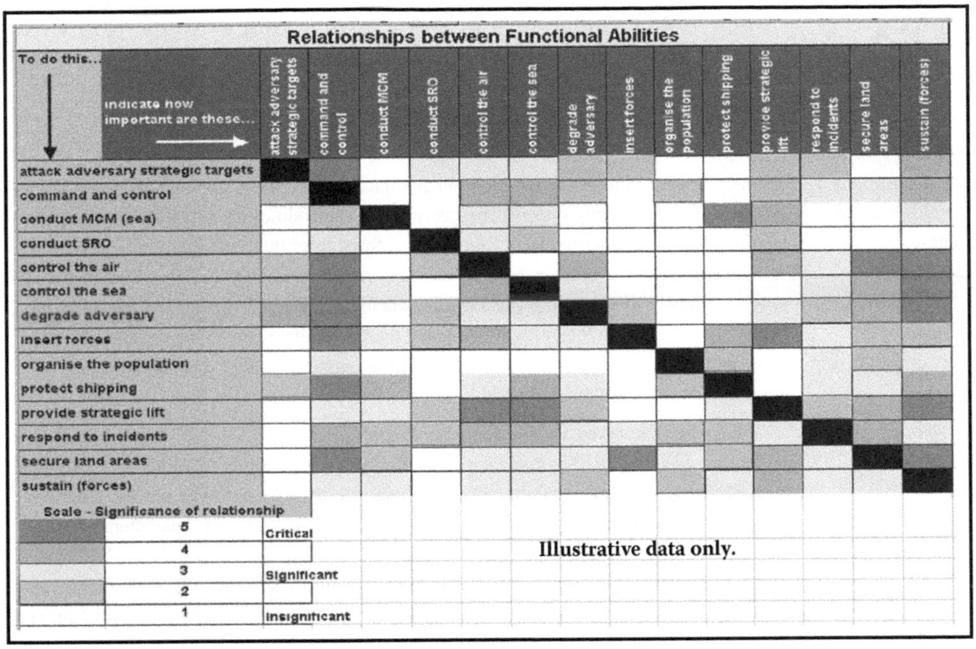

FIGURE 17.21   Illustrative matrix of interdependent capabilities.

illustrated is that strategy does not have an end point: rather, there is a constant to need to continue to develop options relevant to changes in the strategic environment and test them:

> …from the moment of acknowledgement of uncertainty the key to success moves from the idea of one-time development of "best strategy" to the most effective strategy process.
>
> **Michael (1973, cited in van der Heijden, 1996)**

The process challenge is to make it the most skillful, to deliver meaningful content over time and guidance at any point in time, and to address the key strategic issues with efficiency with respect to time and human resources.

RESULTS

It is worth remembering the guidance of Mintzberg et al. (1998) on what is needed for better strategy formation:

> We need to ask better questions and generate fewer hypotheses – to allow ourselves to be pulled by the concerns out there rather than the being pushed by the concepts in here. And we need to be more comprehensive – to concern ourselves with process *and* content, statics *and* dynamics, constraint *and* inspiration, the cognitive *and* the collective, the planned *and* the learned, the economic *and* the political. In other words, in addition to probing its parts, we must give more attention to the whole beast of strategy formation. We shall never find it, never really see it all. But we can certainly see it better.
>
> **Mintzberg et al. (1998, p. 373)**

From a strategy perspective, improvements were made on all counts, as summarized in the *scorecard* shown in Table 17.9.

TABLE 17.9

Scorecard of Case Study 1

Item	Summary Assessment
Better strategy formation needs to:	
Ask better questions and produce fewer hypotheses	The study used SSM to define the problem and the root definitions upon which a conceptual system was used to apply systems engineering as a methodology guiding the strategic planning process to address the key question of force priorities.
Be pulled by concerns *out there* not pushed by strategy concepts	The concerns of the government as customer and owner were major drivers of the content and process. Namely, an extreme *gap* existed between the strategic assessments that form the beginning of a strategic planning process and the formation of capability systems designs.
	It was important to establish a systemic connection between strategy and capability development that could be sustained.
Design of interventions should be more comprehensive and concerned with:	
… process *and* content	The process was developed in consultation with senior leadership in ADoD in a manner that ensured the strategy and capability inputs remained in their control. The process aimed to bring the organization along with it and contribute to the evolution of the content of the outcomes.
	Process and content were modeled in an initial trial of two weeks before a further series of force structure testing and analyses were approved by the [then] Deputy Secretary (Strategy).
…statics *and* dynamics	The study enabled a static understanding of defense capabilities at any one point in time for a given force addressing a given contingency. Through taking many *static* viewpoints across a wide range of contingencies, it also enabled—through the meta-analysis—an understanding of the dynamics of the system capabilities as they both evolve day to day and how they might offer the government a range of military response options in the future.
…constraint *and* inspiration	As a middle-ranking power, Australia aspires to maintain a capability to defend its sovereign territory and contribute to regional and wider security interests. As each force structure option was constrained to a given cost profile, the inspiration of the participants in force structure design and in their concepts for how to employ such a force drove the process.
…cognitive *and* the collective	Strategy as a *mental* process began with managing the cognitive complexity of the study using SSM and employing systems intervention methods such as seminar gaming to ensure the collective wisdom was captured.
…the planned *and* the learned	Defense force structure planning recognizes that much of the current capability will be enhanced by new capabilities from previous planning decisions. These form legacy elements in the force structure and reduce the scope for investment in additional new capabilities. Introducing adaptive strategy interventions improve the capacity to learn new ways of designing and employing different force structure options to improve the range of military response options for the government.
…economic *and* the political	The construction of a strategic functional architecture that would form a major component of a capability priorities statement enables ADoD to improve its capacity to communicate with other economic departments of state (e.g., Treasury and Finance) and with the government itself as force structure decisions can be related to economic implications for employing the force and political implications on the military response options available to the government.

Source: After Mintzberg, H. et al., *Strategy Safari—A Guided Tour through the Wilds of Strategic Management*, The Free Press, New York, 1998.

ANALYSIS

A subjective summary view of the *whole beast* of strategy formation in the case study is presented in the comparison between Figure 17.22, which illustrates at the start of the study the degree of integration of three modes of strategy intervention, and Figure 17.23, which illustrates the perceived extent of integration of the three modes of strategy formation during the study and the extent to

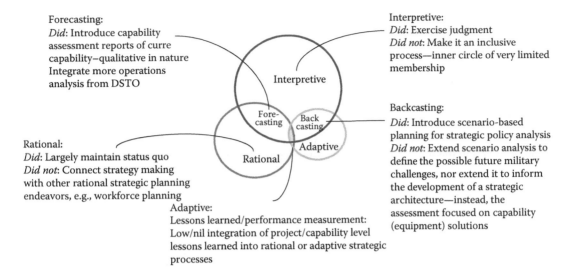

Forecasting:
Did: Introduce capability assessment reports of curre capability–qualitative in nature Integrate more operations analysis from DSTO

Interpretive:
Did: Exercise judgment
Did not: Make it an inclusive process—inner circle of very limited membership

Rational:
Did: Largely maintain status quo
Did not: Connect strategy making with other rational strategic planning endeavors, e.g., workforce planning

Backcasting:
Did: Introduce scenario-based planning for strategic policy analysis
Did not: Extend scenario analysis to define the possible future military challenges, nor extend it to inform the development of a strategic architecture—instead, the assessment focused on capability (equipment) solutions

Adaptive:
Lessons learned/performance measurement: Low/nil integration of project/capability level lessons learned into rational or adaptive strategic processes

FIGURE 17.22 Interpretive viewpoint of defense strategic planning (c. 1999).

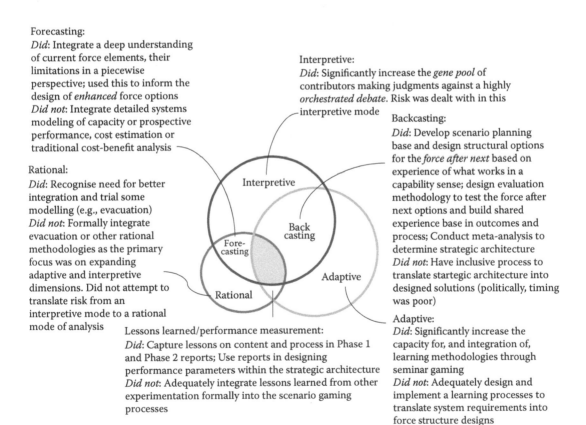

Forecasting:
Did: Integrate a deep understanding of current force elements, their limitations in a piecewise perspective; used this to inform the design of *enhanced* force options
Did not: Integrate detailed systems modeling of capacity or prospective performance, cost estimation or traditional cost-benefit analysis

Interpretive:
Did: Significantly increase the *gene pool* of contributors making judgments against a highly *orchestrated debate*. Risk was dealt with in this interpretive mode

Rational:
Did: Recognise need for better integration and trial some modelling (e.g., evacuation)
Did not: Formally integrate evacuation or other rational methodologies as the primary focus was on expanding adaptive and interpretive dimensions. Did not attempt to translate risk from an interpretive mode to a rational mode of analysis

Backcasting:
Did: Develop scenario planning base and design structural options for the *force after next* based on experience of what works in a capability sense; design evaluation methodology to test the force after next options and build shared experience base in outcomes and process; Conduct meta-analysis to determine strategic architecture
Did not: Have inclusive process to translate startegic architecture into designed solutions (politically, timing was poor)

Lessons learned/performance measurement:
Did: Capture lessons on content and process in Phase 1 and Phase 2 reports; Use reports in designing performance parameters within the strategic architecture
Did not: Adequately integrate lessons learned from other experimentation formally into the scenario gaming processes

Adaptive:
Did: Significantly increase the capacity for, and integration of, learning methodologies through seminar gaming
Did not: Adequately design and implement a learning processes to translate system requirements into force structure designs

FIGURE 17.23 Interpretive viewpoint of defense strategic planning after the intervention.

which they and their integration improved through the formation of a *sweet spot* from the conditions at the start of the study in 1999.

While Figure 17.23 summarizes the main opportunities taken to synthesize the three modes, it also provides some insights into where opportunities were missed or not taken up due to choices made by ADoD. These judgments are the authors' own, made after they had left employment in ADoD.

EPILOGUE

The approach described in this case study made an enduring change to the ADoD strategy and execution process and the organization continues to conduct force-option testing to examine capability gaps in the force structure. This unusual level of endurance for an intervention approach is remarkable and, in this case, is made more so because both authors no longer work for the organization and have no direct influence on the selection of methods, tools, or processes. Thus, we would claim that the approach has merit and indeed it has been applied elsewhere to useful effect.

In reflection, there are a few aspects that we would like to have included or done differently. Of the interventions that were not adopted in this process, the lack of engagement by the organization in the design process was the most significant omission. In engineering a systems approach to strategy formation, this is a highly desirable inclusion because it offers a means to move force structure design from a platform replacement program to one that consolidates the trade-offs between function, performance, and cost that are needed to build an agile, responsive, and sustainable defense force structure capable for a given cost.

While it was systemically desirable to the strategy formation process—for the opportunity it gives to expand the rational mechanisms in use and enhance the *sweet spot*—it was not culturally feasible at the time to engage the services in an agenda-free discussion at a time when a highly political White Paper process was underway. Consequently, the judgments of a small group of analysts in Strategic Policy and Planning Division prevailed as the dominant interpretive means of force design. Validation of the design was limited at the time to one-on-one discussions of the outcomes with senior executives across the services. With this event arising, the study team considered applying modeling and simulation to help validate the new design; however, time and resources acted against this.

Finally, we would like to have implemented a structured and documented after action review to provide reviewable evidence to support later judgments for the continuous improvement of the systems-based, multimethodological approach.

SUGGESTED FUTURE WORK

Notwithstanding the actions at the time, modeling and simulation offers significant capabilities to increase the contribution of rational modes of strategy intervention. The effectiveness and efficiency of the testing and evaluation component of the process could have been enhanced with modeling and simulation methodologies. The perceived benefits accruing would have been as follows:

- Visualization of an Australian Illustrative Planning Scenario (AIPS) to aid the immersion of SMEs into each AIPS to help them internalize the start-state disposition of blue, red, and other forces. This would be a significant step forward from the tools used to support scenario briefings during case study 1, which included a documentary version of the AIPS, a simple map, and a briefing at the commencement of each scenario exercise. Simulation-based tools ensure consistency in the briefing and enable reuse.
- Simulation could be extended to all of the AIPS and used prior to the scenario game, to assist with
    - The definition of a scenario, including a red-force CONOPS and new adversary capabilities
    - The definition of a force option and the introduction of new technologies
    - The definition of an operational concept

- Simulation could be applied after the scenario exercise to explore the following:
  - Key judgments where the assessed risks were high and/or the confidence in the judgments were low.
  - The sensitivities identified during the end stages of the testing process to validate the judgments made by SMEs and/or identify any further combat resolution issues that need to be considered. In these ways, simulation could provide strategic analysts with a form of after action review.

CONCLUSIONS

The strategy formation process described in this case study uses two dominant methodologies:

1. The SSM as a methodology to manage the cognitive complexity
2. Concurrently, an adaptation of systems engineering as a framework for harnessing the multiple methods involved in requirements analysis, enterprise design, solution validation, and systems analysis and, as a methodology, to integrate the products of those processes into a tangible set of assessments and options for government

The case study describes the first iteration of the process, and much was achieved in terms of engaging a wide range of stakeholder in the strategy process both to harness their knowledge and to achieve buy-in with the outputs prepared by the processes. The process also succeeded in producing the needed outputs: capability priorities and guidance with rationale for the government to consider in the succeeding defense White Paper that outlined the revised Australian Defence Strategic Policy.

The integration of soft systems approaches from the strategy and systems thinking communities has demonstrated a capacity to help manage the cognitive complexity of defining one or more problems and surfacing an appropriate mix of methods (which are systemically desirable and culturally feasible), enabling their conceptualization as a *cognitive process* in a strategy and execution system where the selected methods are executable within a systems engineering framework.

Further case studies within the fields of health, transport, infrastructure, and education have since been conducted. In response to the initial work (reviewed in this case study), ADoD recognized the first author's contribution with a research award for *Best Advice to Defence in Force Structure Analysis*. ADoD continues to conduct force-option testing to examine capability gaps in the force structure. This is ongoing nine years after its initial trials and eight years since the first author left the organization. In the transport sector, the work for security risk management has been informally recognized as setting the current *gold standard* in the State of Victoria under its Counter Terrorism (Community Protection) Act.

QUESTIONS

1. What makes the problem of strategic planning and execution so difficult?
2. Why was a multimethodological approach selected to tackle this problem?
3. The case study explicitly describes the principal interpretive (SSM) and functionalist (systems engineering) methodologies employed. Mention is made that practice was also informed by critical systems thinking approaches. What evidence is there that this is the case? How did it add value? What more could have been done to increase the contribution of critical systems thinking?
4. If you were asked to create a business case to apply a multimethodology of the type described in this case study in order to effect some major change across a complex enterprise, what would be your major arguments?
5. If you were told that your business case has been accepted subject to you being able to articulate and report against success criteria, what would these be and how would you provide the evidence needed to satisfy this requirement?

REFERENCES

Argyris, C. (1991). Teaching smart people how to learn. *Harvard Business Review*, May–June 1991, pp. 99–109, Harvard Business School Press, Boston, MA.

Australian Department of Defence. (2000). *Defence 2000 Our Future Defence Force*. Defence Publishing Service: Canberra, Australian Capital Territory, Australia.

Baker, J. (2000). Australia's defence posture. *Australian Defence Force Journal*, 143(July/August): 15–18.

Beer, S. (1980). Introduction to two essays published, in: Maturana, H. and Varela, F. (eds.), *Autopoeisis and Cognition: The Realization of the Living*. D. Reidel: Dordrecht, the Netherlands, pp. 11–30.

Blanchard, B.S. and Fabrycky, W.J. (1998). In: Fabrycky W.J. and Mize, J.H. (eds.), *Systems Engineering and Analysis*, 3rd edn. Prentice Hall: Upper Saddle River, NJ, pp. 45–72.

Boulding, K.E. (April 1956). General systems theory—The skeleton of science. *Management Science Journal of the Institute for Operations Research and the Management Sciences*, 2(3): 197–208.

Brabin-Smith, R. (1999). Discussion between the study team led by the author and Richard Brabin-Smith in his role as Chief Defence Scientist on the importance of ensuring rigour in designing methodologies for Defence strategic planning and capability analysis.

Chaffee, E.E. (1985). Three models of strategy. *Academy of Management Review*, 10(1): 89–98.

Checkland, P.B. (1981). *Systems Thinking, Systems Practice*. Wiley & Sons: Chichester, U.K.

Checkland, P.B. and Holwell, S. (1998). *Information, Systems and Information Systems—Making Sense of the Field*. Wiley & Sons: Chichester, U.K.

Checkland, P.B. and Scholes, J. (1990). *Soft Systems Methodology in Action*. Wiley & Sons: Chichester, U.K.

Coyle, R.G. (2001). The conceptual military framework. Course notes issued by Professor Coyle at his Strategic Analysis Course/Seminar, Australian Defence Force Academy, Campbell, Australia, February 14–20, 2001.

Coyle, R.G. (2004). *Practical Strategy, Structured Tools and Techniques*. Pearson Education Limited, Prentice Hall: Harlow, U.K.

Dahmann, J. et al. (2011). An implementers' view of systems engineering for systems of systems, in: *Systems Conference (SysCon), 2011 IEEE International*, Montreal, Quebec, Canada, pp. 212–217.

Defence, Australian Department of _. (2004). Department of Defence Annual Report 2003-2004, Defence Publishing Service: Canberra.

Defence, Australian Department of _. (2011). Review of the Defence Accountability Framework, Defence Publishing Service: Canberra (see: http://www.defence.gov.au/oscdf/BlackReview/).

DSMC. (2001). *Systems Engineering Fundamentals*. Defence System Management College, US Government Printing Office: Washington, DC. Available on line http://www.dau.mil/pubs/gdbks/sys_eng_fund.asp (last accessed on February 22, 2009).

Edwards, P. (2006). *Arthur Tange: Last of the Mandarins*. Allen & Unwin: Sydney, New South Wales, Australia.

Flood, R.L. and Jackson, M.C. (1991). *Creative Problem Solving: Total Systems Intervention*. Wiley: Chichester, U.K.

Gorod, A., Sauser, B., and Boardman, J. (2008). System-of-systems engineering management: A review of modern history and a path forward. *IEEE Systems Journal*, 2(4): 484–499.

Grisogono, A. and Ryan, A. (2003). Designing complex adaptive systems for defence, in: *SETE Conference*, Canberra, Australian Capital Territory, Australia.

Hamel, G. (2000). *Leading the Revolution*. Harvard Business School Press: Boston, MA.

Hart, S. (1992). An integrative framework for strategy-making processes. *Academy of Management Review*, 17(2): 327–351.

Hart, S. and Banbury, C. (1994). How strategy-making processes can make a difference. *Strategic Management Journal*, 15: 251–269.

Hitchins, D.K. (1992). *Putting Systems to Work*. Wiley: Chichester, U.K.

Hitchins, D.K. (2000). World class systems engineering—The 5-layer model [web page]. http://www.hitchins.co.uk/5layer.html (accessed on March 7, 2001).

Hitchins, D.K. (2003). *Advanced Systems Thinking, Engineering and Management*. Artech House Inc.: Norwood, MA.

Hodge, R.J. (1998). Defence capability development—Learning from the future, in: *Proceedings of SETE Conference Systems Engineering Pragmatic Solutions to Today's Real World Problems—SE98*, Canberra, Australian Capital Territory, Australia, pp. 43–56.

Hodge, R.J. (2010). A systems approach to strategy and execution in national security enterprises, PhD thesis. Adelaide, Australia: University of South Australia.

Hodge, R.J. and Walpole, G.R. (1999). A systems approach to capability planning, in *Proceedings of SETE 1999 Conference Conceiving, Producing and Guaranteeing Quality Systems*, Adelaide, South Australia, Australia, pp. 21–32.

Kline, S.J. (1995). *Conceptual Foundations for Multi-Disciplinary Thinking*. Stanford University Press: Stanford, CA.

Maturana, H. and Varela, F. (1980). *Autopoeisis and Cognition: The Realization of the Living*. D. Reidel: Dordrecht, the Netherlands.

McIntosh, M., Brabin-Smith, R. et al. (1997). *Future Directions for the Management of Australia's Defence— Report of the Defence Efficiency Review*. Department of Defence: Canberra, Australia.

Michael, D.N. (1973). *On Learning to Plan—And Planning to Learn*. Jossey-Bass: San Francisco, CA.

Midgley, G. (2000). *Systemic Intervention: Philosophy, Methodology, and Practice*. Kluwer Academic/Plenum Publishers: New York.

Mintzberg, H. (1994). *The Rise and Fall of Strategic Planning*. The Free Press: New York.

Mintzberg, H., Ahlstrand, B., and Lampel, J. (1998). *Strategy Safari—A Guided Tour through the Wilds of Strategic Management*. The Free Press: New York.

Rasmussen, J. (1998). Ecological interface design for complex systems: An example: SEAD-UAV systems (EOARD Contract No. F61708-97-W0211). Wright Patterson Airforce Base: Dayton, OH.

Stevens R. (2011). *Engineering Mega-Systems*. CRC Press: Boca Raton, FL, ISBN 978-1-4200-7666-0.

Ulrich, W. (1983). *Critical Heuristics of Social Planning*. Haupt: Bern, Switzerland.

van der Heijden, K. (1996). *Scenarios—The Art of Strategic Planning*. John Wiley & Sons Ltd.: Chichester, U.K.

von Clausewitz, C. (1832). in: M. Howard and P. Paret (eds.), *On War* (1984). Princeton University Press: Princeton, NJ.

Warfield, J.N. (1994). *Science of Generic Design: Managing Complexity through Systems Design*, 2nd edn. Iowa State Press: Ames, IA.

Warfield, J.N. (1999). Twenty laws of complexity: Science applicable in organizations. *Systems Research and Behavioral Science*, 16: 3–40.

White, H. (1999). Personal conversation of deputy secretary strategy with Richard Hodge as Scientific Adviser Strategic Policy and Planning.

18 United States Military Partner Capacity

System Dynamics to Quantify Strategic Investments

John V. Farr, James R. Enos, and Daniel J. McCarthy

CONTENTS

KEYWORDS

Systems dynamics, Partner capacity, Military operations, Complex systems, Stability and reconstruction

ABSTRACT

Since the 1970s, the greatest threats to global security have come from both emerging ambitious states and nations that are unable or unwilling to meet the basic needs and aspirations of their people. Subsequently, since the end of the Cold War, the United States and its allies have begun a new stability and reconstruction (S&R) operation every 18–24 months. During the twenty-first century, only during 1968 was soldier in the United Kingdom military not killed in S&R operations. Recent history has shown that the armed forces of the most global powers will encounter significant challenges in S&R efforts—efforts that seek to establish a safe and secure environment in the assisted country, ensure government stability with democratic practices, including fair elections, rule of law, and human rights, to promote the development of a robust economy, and help the assisted country in becoming a respected member of the international community. We have had some success at affecting the social, governance, and economic fabric of a country. However, as recently demonstrated in Afghanistan and Iraq, this has come with a significant price tag in terms of human life and investments. Few will deny that we have struggled to efficiently and effectively invest our resources during S&R operations. Defense agencies and other members of the whole government team in concert with nongovernmental organizations must learn to better invest their resources before the outbreak of hostilities as well as throughout the spectrum of conflict and post-conflict operations. More importantly, we must better understand where and when to invest in building host nation capacity; bottom-up small projects versus top-down large projects; and the payoff for investments in governance, economic, and security efforts. This chapter uses formal systems theory to address some of these issues by describing the problem and then using system dynamics modeling to understand how these investments can affect the long-term legitimacy and economic capacity of a nation of concern.

GLOSSARY

AOV	Acts of Violence
BPC	Building partner capacity
CIA	Central Intelligence Agency
CLD	Causal loop diagrams
COIN	Counterinsurgency
DoD	Department of Defense
GDP	Gross domestic products
MPTs	Methods, processes, and tools
NGO	Nongovernmental organization
S&R	Stability and reconstruction
SD	System dynamics
SECDEF	Secretary of Defense
SFA	Security force assistance
US	United States

BACKGROUND AND PURPOSE

As the wars in Afghanistan and Iraq end, little has changed from a global security perspective. "The greatest threats to our national security will continue to come from both emerging ambitious actors/states and from nations unable or unwilling to meet the basic needs and aspirations of their people" (Department of the Army, 2008). This is true for most developed countries in the world. Since the end of the Cold War, the United States and its allies have begun a new stability and reconstruction (S&R) operation every 18–24 months. Recent history has shown that the military enterprise as a member of the whole government team will encounter significant challenges in future S&R efforts—efforts that seek to ensure government stability, a robust economy, democracy and fair elections, and human rights—and is a respected member of the international community for the nation(s) or region of interest. The global community has had some success in affecting the social, governance, and the economic fabric of a country. However, as recently demonstrated in Afghanistan and Iraq, this has not come without a significant price tag in terms of human life and investments. Few will argue that our solutions are often stovepiped by agency and we do not understand how to efficiently and effectively invest their resources. The military enterprise and other government and nongovernment organizations (NGOs) must better utilize their resources in S&R efforts. We need to gain a better understanding of when to invest in building host nation capacity, bottom-up small projects, top-down large projects, and the payoff for investments in governance, economic, and security efforts before the outbreak of hostilities or the cessation of counter insurgency (COIN) operations. This wicked problem is neither well understood nor easily studied.

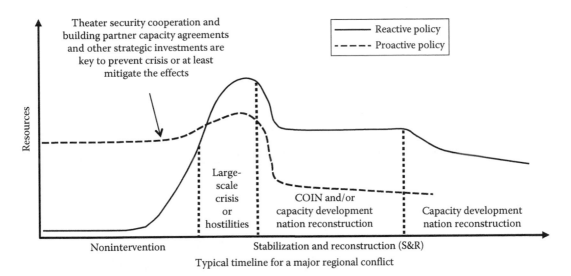

Typical timeline for a major regional conflict

FIGURE 18.1 Timeline as a function of resources for military intervention—the military interaction lifecycle.

A new philosophy is emerging where not only the United States but also other global powers use their industrial and military capacity to develop strategic partnerships designed for building relationships and host country capacity in lieu of simply selling military equipment. By proactively building capacity within partner countries we can hope to create and reinforce relationships, supporting mutual goals and interests, improve the quality of life in the partner nation, and perhaps prevent the emergence of nonstate actors who should seek to undermine lawful governance. As depicted in Figure 18.1, this type of proactive capacity building should occur early in the nonintervention stage.

Even without early intervention, we must be able to plan and resource S&R efforts throughout the conflict life cycle as shown in Figure 18.1. Each phase requires a different focus and priority of investment in security, governance, and economic development projects. The complex, adaptive environment our whole government team, partners, and NGOs face today requires an understanding of the conditions that exist in the partner nation in order to make smart investments in capacity development. The social, governance, and economic implications of these investments within the partner country and the resulting effects can often lead to unforeseen second-order effects. Quantitative means are needed for not only prioritizing investments but also understanding the short- and long-term effects of those investments. Formal systems engineering methods, processes, and tools (MPTs) provide us with the capability to not only capture the interactions but also quantify the investments and their effects. We hope that by applying these MPTs we can (1) better plan for these types of investments as part of the campaign-planning process, (2) better understand where and when to allocate resources in the governance, social, and economic domains, and (3) ultimately save lives by affecting the desired change.

Few quantitative techniques exist to allocate resources in support of building partner capacity (BPC) and a country's capacity to provide for its populace along with a fundamental understanding of how capacity is affected by economic, security, and governance investment. As shown in Figure 18.2, BPC from a US perspective can have a variety of diverse stakeholders, often with conflicting interests and perspectives, and a diverse set of programs that can produce a multitude of outcomes. Systems thinking tools and characterizations are key to help understand the interdependencies of the various agencies, actors, products, and constraints that can affect a country's capacity.

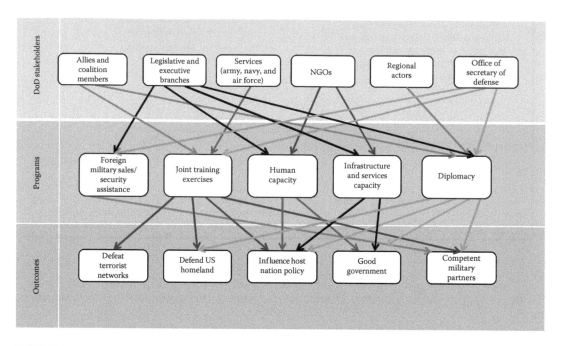

FIGURE 18.2   Stakeholders, programs, and capabilities swim lane chart for BPC.

Like all emerging wicked problems, competing definitions exist in the literature. For understanding this problem, we propose the following definitions:

- Nelson (2006) defines S&R as the process to achieve a locally led and sustainable peace in a dangerous environment. The military role in this process is halting residual violence and ensuring order and security, including those reconstruction efforts required to repair enough damage to enable restoration of the most essential services.
- BPC can best be defined as targeted efforts to improve the collective capabilities and performance of the military enterprise and its partners (Department of Defense, 2006). A major component of BPC is security force assistance (SFA). SFA is defined as supporting the development of the capacity and capability of foreign security forces and their supporting institutions (Department of Defense, 2010). BPC is accomplished mainly through training and equipping the partner nation's military and improving their quality of life through infrastructure improvements, education, and equipping the civilian workforce. Militaries are trained to fight and win their nation's wars. However, in modern conflicts to include S&R, nonmilitary capacity-building actions have become as important as any kinetic weapon system.
- Country capacity is the ability of a country and government to perform the functions of providing for the populace, solving problems, and achieving objectives in a sustainable manner. Figure 18.3 shows a systemigram of some of the essential elements of country capacity.

The approach presented in this case study shows how a complex system's problems can be structured and quantified using system dynamics (SD). This should lead to more defensible and transparent government policies and investments.

Figure 18.4 shows how governance and feedback control our complex system. In trying to substantiate our categorization of a nation state as a complex system, we first looked at complexity as presented in Table 18.1. We also studied the characteristics or key aspects of a complex system as shown in Table 18.2. The elements that comprise the nation state's efforts in each of these areas

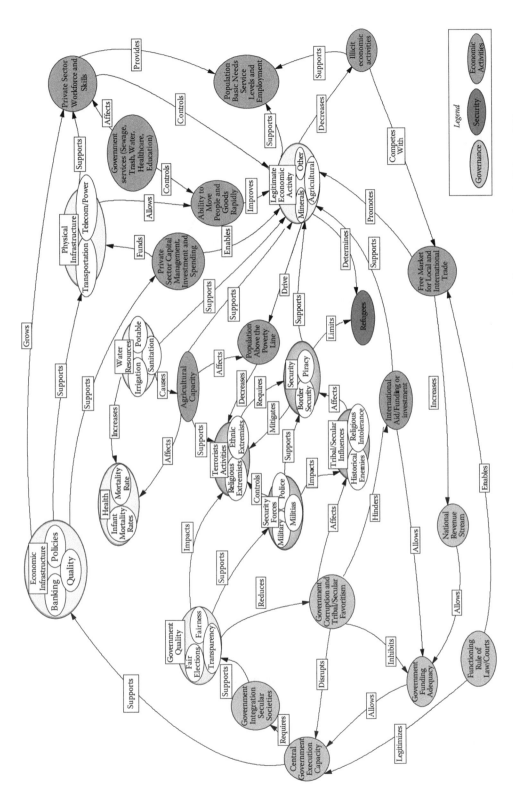

FIGURE 18.3 (See color insert.) Systemigram showing some of the essential elements of country capacity for a country in sub-Saharan Africa.

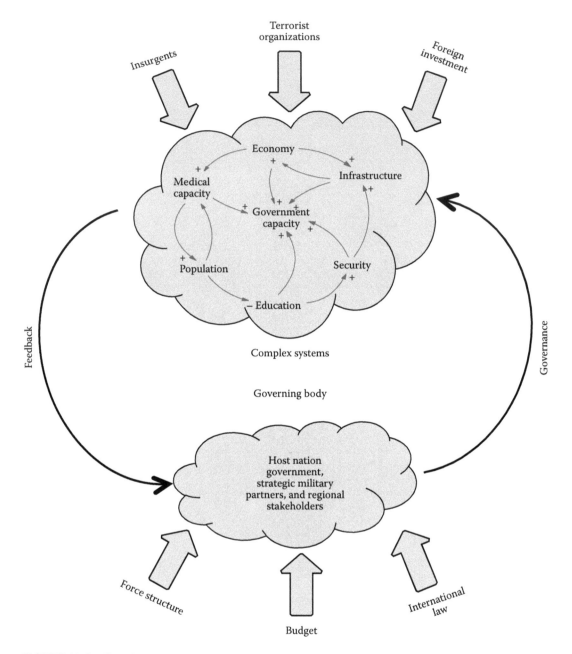

FIGURE 18.4 Complex system context.

can be considered as comprising some level of a system and so it is arguably more appropriate to consider these efforts as a *complex system* consisting of numerous enterprises for our modeling purposes. When one considers the many components involved as a nation state tries to achieve its objectives of providing security, stable governance, and economic opportunity to its citizens, it is easy to see that it qualifies as a complex system. Figure 18.3 was developed to capture some of the many external factors and interactions that control the behavior of our complex systems. The SD methodology presented is founded on the need for feedback for these types of systems.

Trying to categorize the complex interactions that comprise a nation, much less trying to mathematically model the dependencies of economic, security, and governance behavior on internal (religious and ethnic tensions, government policy and efficiency, etc.) and external stimuli (investments, regional

TABLE 18.1
Comparison of Various Views of the Nature of Complexity as Related to Complex Systems

Jackson (2003)	Ramalingam et al. (2008)	Mitleton-Kelly (2003)	Davies (2003)	Clemens (2002)	Sheard and Mostashari (2008)	Nation State Complex System Behavior
Based upon a particular view						Our view of ground truth is often based upon certain a priori assumptions such as values, loyalties, and rewards.
Inclusion of autonomous and independent systems					Autonomous interacting parts (agents) Energy in and out	The elements are heterogeneous, with building blocks (agencies, processes, external stimuli, etc.) that are individual agents of the systems. Nation states exist and are function-based upon intricate and multifarious interrelationships.
Identification of stakeholders						Often external intervention is required for governments to meet the needs of its people. Internal stakeholders often do not understand what must be done to improve government. The stakeholders are fluid with very different agendas.
Identification of boundaries						Boundaries at all levels are ill defined, especially for emerging nation states embroiled in COIN activities.
	Interconnected and interdependent elements and dimensions	Connectivity and interdependence			Nonhierarchical and central authority	No single group understands all of the activities or rational behind its mission, yet typically government is controlled by a central authority. Identifying and building common ground is necessary for progress. Elements are usually interconnected by common values and beliefs that may or may not translate to the next level.
	Feedback processes	Feedback				Feedback exists in any complex system and none more so than a nation state. Countries evolve over time based upon economic and social feedback.

(continued)

TABLE 18.1 (continued)
Comparison of Various Views of the Nature of Complexity as Related to Complex Systems

Jackson (2003)	Ramalingam et al. (2008)	Mitleton-Kelly (2003)	Davies (2003)	Clemens (2002)	Sheard and Mostashari (2008)	Nation State Complex System Behavior
	Emergence	Emergence		Emergence	Display emergent macro level behavior	The stability of being a valued member of the international community of a nation state is often driven mainly by leadership/governance. Security, governance, and development create synergies that are often difficult to capture.
	Nonlinear		Nonlinearity		Nonlinearity	Investments in security, governance, and economic development often do not translate directly to improved living conditions or a country's capacity to develop basic services.
	Sensitivity to initial conditions	Path dependence	Sensitive dependence on initial conditions			The initial state of a nation, especially with regard to security and counterinsurgency, can lead to massive differences in the effects of investments. Establishing a secure and safe environment is a prerequisite to effective reconstruction activities.
	Phase space					Nation states evolve over time as a function of internal and external stimuli (forces, resources, emotions, external and internal forces, etc.)
	Chaos and edge of chaos	Far from equilibrium	Edge of chaos	Punctuated equilibrium	Fuzzy Boundaries	Interactions vary dynamically with boundaries (responsibilities and products) that are often ill defined.
	Adaptive agents				Adapt to surroundings (environment)	Governments continue to adapt to leadership, external stimuli, and a host of other events either locally or at the global level. Internal and external relationships are the enablers of change.

					Description
Self-organization	Self-organization	Self-organization	Self-organization	Self-organizing	Nation states are continuously realigned based upon the affinity of the elements (leaders, organization, elected officials, etc.).
Coevolution			Coevolution	Coevolution — Become more complex with time; increasingly specialized	Ideally, a nation state evolves with the social-technical aspects of the global market and security environment. Elements of the complex system as a whole can evolve.
	Space of possibilities			Elements change in response to pressures from neighboring elements	Processes, products, people, etc., continue to evolve. Most nation sates have no status quo; otherwise, they cannot serve the basic needs of their people.
				Various scales	Processes and activities exist at all levels. Small changes at most levels can lead to significant changes in a nation state.
		Attractors			A nation state under the external influences of reconstruction often torn by internal strife is often under continuous tension. This dynamic system has multiple attractors (heritage, religion, economics, power, etc.) driving conflict and shaping society.
				Fitness Landscapes	Recent world events have shown that predator/prey may change overnight in any country. The continual power struggles can change quickly depending upon internal and external events.

TABLE 18.2

Relevant Aspects of a Complex System as Related to Stability and Reconstruction for a Nation State

Key Aspects of Complex Systems	As Related to a Nation State
Recognition of architecture	The architecture of the project played a key role: We started with the vision of taking action to improve cultural and business relationships rather than responding when a crisis occurs. We let the problem dominate how we structured the solution. We did not follow the steps in the architecture of the process shown.
Arrangement of these systems in a networked fashion rather than in a hierarchy	Nation states generally evolve into a hybrid complex system. Overlapping and interdependent functions of government often lead to network functionality even though they are typically organized hierarchically.
Recognition of external environment	See Figure 18.3 for some of the external environmental factors and internal context and their connections.
Management of complexity	Entities typically act in their own interest with regard to interventions. Few nation states are formally self-organized. However, informal relationships are key to their operation. People, organizations, and processes of a government seldom act independently and are instead driven by complex rules and hierarchies. Nation states can be best described as having a complicated hierarchical process with an obvious set of values (i.e., laws). However, in reality, subtle laws often govern the behavior of the system.
Modularity	The elements of a nation state in general are not tightly coupled. In most cases, organizations act independently. However, because they are typically hybrid in structure, some agencies, processes, etc., are required to be integrated. Nation states are composed of independent, domain-specific processes.
Views and viewpoints	The views that can be described using a systems architecture approach. See Farr et al. (2013), where a modified Zachman framework was used to describe architectural views for describing how a nation state operates and responds to internal and external stimuli such as S&R efforts.
Recognition of unknown unknowns	Unknown unknowns are more pronounced, in terms of number and effects, in nation states when compared with product-based systems. While mainly driven by complexity, the human element is more tightly coupled with the evolution of the system.
Definition of completion of the right action and creation of a control structure must at least have the range of the thing being controlled	No other complex system has the complexity of unknown unknowns as does a nation state. Perceived right actions can often lead to unforeseen and disastrous consequences. However, the control structure (i.e., government and the international community) has accepted norms governing behavior.
Leadership	Leadership styles vary among nation states. Too often, decisions are not driven by outcomes but instead by perception and power sharing. Chinese leaders often follow traditional leadership (analysis and response) whereas the more democratic the government, the more complex leadership styles emerge. At lower levels of government, leadership mirrors that of any project team for a product-based system.
Governance	Most nation states have a top-down means of governance (i.e., laws). However, laws do allow for provisional solutions and accountability, yet hinder flexibility and a problem-solving focus. Governance at the lower levels of government is often ruled by regulations and accountability like any product-based complex systems (i.e., ISO, CMMI, etc.).
Incremental commitment	Government by nature is incremental and follows any systems engineering process. Risk analysis is typically conducted at any phase throughout the external stimulus that would occur from an investment or event.

- Flexibility
- Adaptability
- Autonomy
- Independence
- Ability to evolve
- Ability to organize
- Resiliency
- Robustness

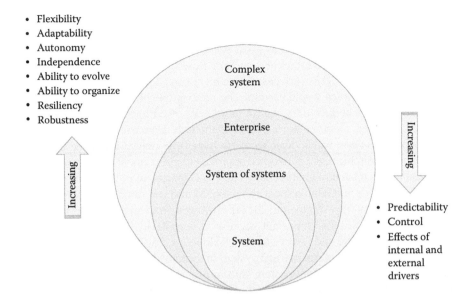

- Predictability
- Control
- Effects of internal and external drivers

FIGURE 18.5 Venn relationship among system types showing attributes as a function of hierarchy. (Modified from White, B.E., On principles of complex systems engineering-complex systems made simple tutorial, *INCOSE Symposium*, Denver, CO, June 20–23, 2011a; White, B.E. et al., On the importance and value of case studies—Introduction to the special session on case studies in system of systems, enterprises, and complex systems engineering, *2013 IEEE International Systems Conference (SysCon) 2013*, Orlando, FL, April 15–18, 2013. Copyright 2013 IEEE International Systems Conference. IEEE grants permission to reuse portions of this cited paper in this book.)

events, military actions, etc.), presents a whole set of unique challenges. In essence, there are three steps in which we can utilize formal systems theory to understand and model these interactions: (1) describe the problem to help understand its characteristics, (2) use diagramming tools to understand the interactions and dependencies, and (3) model the system with SD to understand its behavior over time.

In many respects, a nation state is the ultimate complex system. As shown in Figure 18.5, most countries exhibit all the attributes and behaviors of a complex system. They evolve, can self-organize, are resilient, flexible, etc. Yet they are not predictable, nor can they be controlled. Describing a complex system, understanding its interdependencies, and modeling a nation state are tremendous challenges.

In Chapter 2, Ireland and White use eight areas organized into four contexts to assess the complexity of a system. In every aspect, with the exception of acquisition environment, which is not relevant to our case, conducting S&R for a nation state is a messy affair. These seven areas include mission environment (mission is very fluid, ad hoc), scope of effort (extended enterprise), scale of effort (many different users), stakeholder involvement (multiple equities), stakeholder relationship (resistance to changing relationships), desired outcomes (build fundamentally new capability), and system behavior (system behavior will evolve). Note that this description could apply to almost any military exercise.

The role of architecture and requirements for a nation state present an interesting paradox. Most developing countries are governed by a robust architecture (i.e., agencies, laws, processes, etc.). One could argue that requirements cause the complex system (i.e., nation state) to adapt; however, it must evolve within the architecture (i.e., government). However, the recent Arab spring has shown us that requirements can cause major changes in a complex system even in the face of a very rigid architecture. In addition, the Arab spring is a greater example of how the neglected unknown unknowns can cause major changes in the complex system.

One could argue that the more developed countries are bound by architecture, whereas the less developed countries are more requirement-driven as they must adapt to meet the needs of the people. As described in Chapter 2 the critical details in the architect's design will be, first, the system's connections and interfaces (laws, governance, and agencies) and, second, the system's components (units that perform specific functions); this combination produces unique system level functions. An interesting research project would be to develop and quantify architecture for a nation state and see how strongly correlated variables such as gross domestic product (GDP), health, and education relate to the existence of a formal and robust architecture. Research is also needed into whether nation states that are being affected by S&R work best with a hierarchical government or a more networked form of government. North Korea would certainly be an example of a hierarchically ruled nation, whereas Afghanistan would be a country with a more networked form of government. Like most complex systems, developed countries are a hybrid of the two. Countries that seem to fail to meet the needs of their people and in which S&R seems to have little impact are those that are strongly dominated by either a hierarchical or a networked approach to government. Finally, the issue of scale is important especially as related to architecture. The US Department of Defense (DoD) budget for 2012 was about $700B, whereas Gambia spent about $4.5M and was ranked last out of 153 countries in the world in defense spending (Wikipedia, 2012a). However, from an architecture perspective, all countries still have a DoD. Scale is important because of complexity. With over 750,000 civilians and roughly three million active and reserve forces (Wikipedia, 2012b), the US DoD is many times more complex with many more enterprises and system of system elements than Gambia. Yet from a high-level architecture perspective, they contain similar processes, command and control structure, mission, etc. Table 18.2 summarizes these and other key aspects of a complex system.

Like any systems development problem, the *right* leadership style is critical to success. The emergence of Singapore, Rwanda, and China as economic powers can be traced to leaders with the right vision. The complex systems grew and adapted, often overcoming significant inertia to change, to meet the requirements of the people. One interesting aspect of a complex system and the role of leadership are views. Using Iran as an example, there are strong internal views about its role in regional politics, whereas the West has a very different view of Iran and its leadership as a complex system. Leadership is often able to foster national pride that can sustain and implement growth in the system.

Government, like project governance, must take into account the governance policies of many, primarily autonomous, organizations. There are checks and balances, overlapping turf, and power struggles between most government agencies. In theory, this drives the adoption of more democratic governing processes. When viewed from an architectural perspective (connections, interfaces, and lower-level systems), the way we describe a complex system that is a product versus a nation state is nearly identical.

Referring to Figures 18.3 through 18.5, SD is ideal to model the external and internal factors and the feedback that drive the behavior of a nation state. In the details of our SD model, we will present the internal and external factors we used in our SD representation. We will introduce the concept of government capacity, which is driven by external and internal factors, as the primary metric for whether a country is meeting the needs of its people. An SD representation is designed to capture feedback. Finally, by using stock and flows we were able to capture resource flows and constraints.

CHALLENGES

Understanding, quantifying, and building predictive tools for BPC are both complex and not well studied nor understood. Much research that has been conducted on components of the behavior of a nation state has been previously discussed (e.g., GDP models). The interdependencies of the layers, data collection, correct portrayal of the causal relationship and quantification of those relationships, and, more importantly, validating a comprehensive nation state model are tremendous challenges.

MODEL DEVELOPMENT

Systems diagramming tools such as systemigrams have been used to capture the complexities of governments and external influences. In our approach, we propose the development of a systemigram to represent the complex nature of a nation state. Based on the conceptual relationships captured by the systemigram, we used causal loop diagrams (CLDs) as a first step in building SD models. SD methodology is then used to model the behavior of the complex system over time and quantify the impacts of funding allocations. The resulting output can aid decision makers in determining policies on how the funding should be allocated between different capacity-building activities. We found the process of using systemigrams and CLDs to drive the development of an SD model to be highly iterative. Initially, we thought that the systemigram would be a valuable tool for developing the SD model. We had hoped that the systemigram would support the development of the CLD, which would ultimately make the development of the SD easier and more transparent. However, we found the systemigram to be a good tool for developing interdependencies but it did not translate well to developing an SD model.

As previously discussed, the purpose of this research is to demonstrate the utility of SD as an analytical tool for this class of complex systems problems. SD has been used extensively for military problems. Robbins (2005) developed an SD model focused mainly on combat. Much literature exists involving certain components of a government. For example, Sterman (2000) presents an SD model of GDP. Another paper deals with the role of water in the Manas Basin in Africa (Shanshan et al., 2009). Crane (2009) mainly used Likert scale ratings for the Democratic Republic of the Congo along with some open source data to develop a conceptual SD model. He (Crane, 2009) also used a set of what he termed Nation Building Elements that consisted of security, humanitarian relief, economic stabilization, and governance. Though a very useful paper in terms of grouping indicators, it does not contain any model results. We could not find any research that mentioned using government capacity/legitimacy as a primary indicator of a country being stable and meeting the needs of its people, nor could we identify any literature that used multiple layers in an attempt to build a government capacity model.

In this case study, we plan to demonstrate the utility of this approach using a hypothetical country in Africa. Trade-offs in allocating resources to security intervention, partner nation training, infrastructure and capacity development, weapons and technology sales, etc., will be analyzed to understand the effects on security/stability, governance, economic development, and societal issues (e.g., human trafficking and drugs).

The military is typically focused on those activities that are consistent with its mission—mainly security cooperation. The SD model presented is generic and represents those activities needed for a country to meet the needs of its populace. In some situations, such as the presence of significant counterinsurgency operations, the military would be involved in many of the activities shown in Figure 18.3. However, in general and as shown in Figure 18.2, the whole government team and members of the international community must conduct nonsecurity activities.

SD is a methodology to understand the dynamic behavior of complex systems through modeling and simulations. It explains the behavior of systems over time as a direct result of the system structure and aims to adjust individuals' mental models of the system to implement policies to improve system performance. Forrester (1961) described the potential for SD as an approach that should help in important high-level management problems (1961). He noted that solutions to small problems will only yield small results and that people get mediocre results by setting improvement goals too low. He suggested that the change must be at the enterprise level to achieve major improvement and that the goal should be to determine policies that lead to greater success (Forrester, 1961).

A useful technique in characterizing a complex system and understanding its requirements is to develop a description of the complex system *as is* and then develop a description of the system *to be*. These descriptions can include a characterization of the complex system and its component systems, the types of systems and their level of maturity, the objectives, requirements, and capabilities of the complex system and component systems, as well as the relationships between the component systems and their relationships with the environmental factors that impact them. In many cases, a high-level diagram can be developed in support of the description of the complex system. In the case of existing systems, it is often useful to develop graphs of performance indicators to create a sense of how the current *as is* system is operating and to establish performance objectives for the *to be* system. SD is uniquely able to capture the *as is* system and interject external stimuli (processes, events, resources, etc.) to predict the *to be* system.

This methodology provides an excellent tool for analyzing the complex interdependencies and feedback evident in nation reconstruction and capacity development that creates the dynamic behavior of the system. Dynamics are the behavior of a system over time, which are generally complex and nonlinear in nature (Forrester, 1961). This complexity comes from feedback within the system, time delays between decisions and effects, and the learning process of the system (Sterman, 2000). CLDs are a key element of the SD approach that are signed diagrams that represent the reinforcing or balancing feedback within a system. Causal loops are different from discrete, event-oriented perspectives of individual causes and effects in that they acknowledge that in a closed system, any cause is an effect and any effect is a cause (Richardson, 1991). One is able to describe the behavior of the system by talking through the loop to tell the story of the interactions within the system (Meadows et al., 2004).

SD is useful in describing and modeling the relationships within the complex system perspective of reconstruction efforts as well. The overall objective of BPC efforts is to increase the capacity of a country to a point at which it can sustain itself. This prevents the security situation and government legitimacy from degrading to a point where the country is thrown into a state of chaos. Figure 18.6 presents a CLD of government capacity. We propose that a government's capacity is based on the country's education level, medical capacity, security situation, infrastructure, and economic capabilities. The CLD represents the feedback between these individual elements that creates the dynamic behavior observed. Additionally, each of these individual system views of a government's capacity contains their own internal feedback structures that generate specific behaviors. By combining the behavior at the individual system level, the SD model is able to capture the behavior of the complex system over time.

We chose government capacity as our primary measure of a combination of stability, legitimacy, and people satisfaction (see Figure 18.6). Research needs to be conducted to understand quantitatively what a numeric value of country capacity means. Our research is focused on changes in country capacity but only from a qualitative and not an absolute perspective. As we think about the capacity of a nation state and its ability to provide a safe and secure environment, stable government, essential services, and a healthy economy, it is useful to take a systems thinking approach in order to better understand the various elements involved and the complex relationships involved.

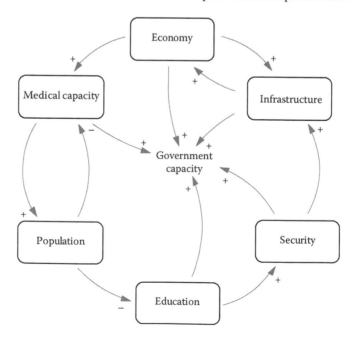

FIGURE 18.6 Complex systems of BPC using a CLD.

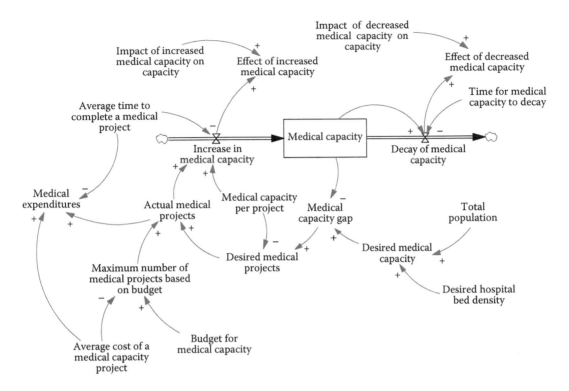

FIGURE 18.7 Medical capacity system structure.

Although the model includes a structure for each of these systems, an examination of one of the system level models demonstrates the structure seen in several of the systems. Figure 18.7 presents the system view of the medical capacity portion of the SD model. This view shows how the population system interacts with the medical capacity as the total population of the country has a positive relationship with the desired medical capacity. The model utilizes an average medical bed density of 10 developed countries* as the *Desired Hospital Bed Density* for the goal that the system must achieve (Central Intelligence Agency, 2013). Based on this desired medical capacity, the model calculates a gap and determines the number of projects to be completed. However, the reconstruction budget that is allocated toward medical capacity constrains the total number of medical projects that can be completed, thus limiting the increase in medical capacity. Through the use of auxiliary variables, any increase or decrease in the medical capacity has a similar effect on the overall capacity of the government.

This chapter does not present the education and infrastructure models because of space limitations. The model utilizes a similar structure for these systems with a desired capacity for each based on an average value from several developed countries; however, infrastructure is further subdivided into electrical grid, improved water, and improved sanitation.

Figure 18.8 presents the economic system structure that includes the nation's expenditures on infrastructure, medical, security, and education. Sterman (2000) provides a basic model of GDP, the proxy measurement for the nation's economy, which demonstrates how a nation's GDP will adjust to the aggregate demand for its goods and services. For this application, his model is manipulated to endogenize or internalize the variable of government expenditures by linking other system level models into the economy model.

A major component of government capacity is the security component and the ability for security forces to maintain the peace within a country. Choucri et al. (2006) built an initial model of state security

---

* Ten developed nations include the United States, Japan, Germany, United Kingdom, Canada, France, Russia, Brazil, Australia, and Spain.

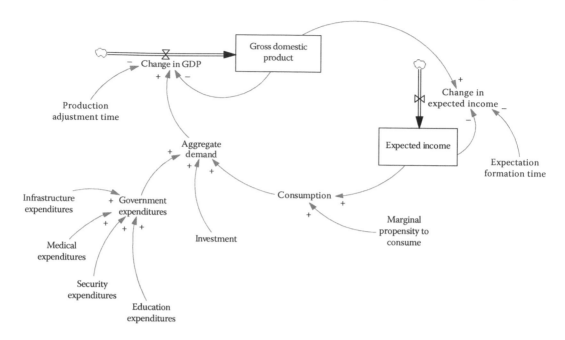

FIGURE 18.8 Economic system structure. (From Sterman, J., *Business Dynamics: Systems Thinking and Modeling for a Complex World*, Irwin McGraw-Hill, Boston, MA, 2000.)

that included the movement from the progovernment population to dissidents to insurgents that had the potential to be removed by security force. Their model provided the basis for the structure of the security view of this system of system model of a nation's capacity. This portion of the model links to government capacity and the budget for security forces and uses portions of the population model as well.

Figure 18.9 presents the stock and flow diagram of the security system that shows how the aged military population can become dissident based upon opposition recruitment. Additionally, some fraction of the dissident population will become insurgents. The model accounts for both the host nation security forces and external security forces. As the number of insurgents increases in the model, the acts of violence subsequently increase to a point that it triggers intervention from an external nation to help with security and reconstruction. Additionally, the model incorporates the decision to deploy additional forces as the acts of violence continue to rise. This decision rule is consistent with deployments to Iraq and Afghanistan, where an initial number of soldiers were deployed and then the United States increased the number of soldiers deployed as the violence in these nations increased. The model also incorporates the blowback from insurgent removal activities as it increases the effect of the opposition's recruitment message and builds increased support for the opposition.

The final system view to be discussed for our model is shown in Figure 18.10 and includes the population of the country divided into three stocks of individuals: children, adult population (15 years), and elderly (65 years). The structure is a simple aging chain between the different stocks with additional outflows for migration and deaths. The Central Intelligence Agency (CIA) World Factbook (2011) provides data for each of these populations, fractional birth rates, infant mortality rates, life expectancy, death rates, and migration rates, so the model can be easily manipulated to match any country's initial conditions. Again, to endogenize these variables, the country's medical capacity will have a direct impact on several of these variables and can be quantified in the mathematical relations within the model.

The power of SD models comes into force in their ability to model a system's behavior over time and to compare different policy recommendations. This model utilizes data from the CIA World Factbook for a typical sub-Saharan country as a base case to determine whether the model will generate observed historic behavior as government capacity decreases and sends the country into a state of chaos. As a country's security level decreases to a certain point, the model triggers intervention

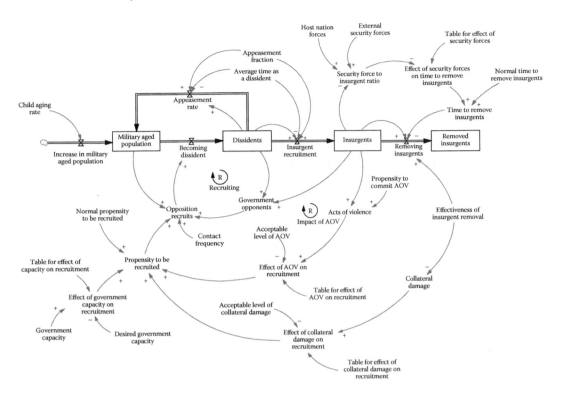

FIGURE 18.9 Security system structure. (Adapted from Crane, W.E., System dynamics framework for assessing nation-building in the Democratic Republic of the Congo, United States Army War College, Carlisle, PA, 2009.)

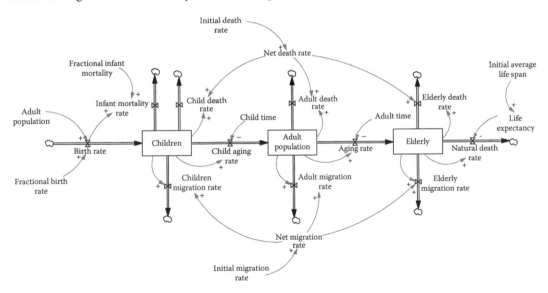

FIGURE 18.10 Population system.

on the part of a security force nation in the form of soldiers and money, which is the traditional method of intervention. The model will output the amount of capital invested by the country over time and determine a total amount of capital invested by the security force nation. Additionally, the model can also simulate the behavior over time for a nation's capacity and security if a preemptive intervention is executed in the form of BPC. A comparison of these two policies shows the advantages, both economic and social, of adopting a preemptive partner-building policy.

MODEL RESULTS

There is no measure for government capacity; however, several proxy measurements provide insight into the stability of a nation based on its capacity to support the people. The model utilizes a combination of the economic, infrastructure, medical, and security measurements to describe the government capacity. Metrics for these individual variables exist; for example, the GDP measures the country's economic performance. The model uses these individual, measurable metrics and an assigned weight to calculate the government's capacity. As government capacity is increased, the country is better prepared to support its population and the nation is more stable. Any policy, whether it is preemptive partner-capacity building or a reaction to a failed state, should aim to increase this capacity. Figure 18.11 presents the model output for government capacity. Both policies result in an initial decrease in government capacity. Although the reactive policy increases above that of the proactive policy, the latter will eventually overtake the former as the country begins to sustain the improvements made to its capacity.

From the security view of the model, the number of insurgents and the number of external security forces provide valuable insights into the overall capacity of the government. The number of insurgents indicates the nation's ability to conduct counterinsurgency operations and maintain a stable environment to develop their country.

Figures 18.12 and 18.13 present the output of the model, which includes both the proactive policy of becoming partners with the host nation and a reactive policy of deploying security forces after a crisis arises in the country. As shown, the level of insurgents in a proactive policy is drastically lower than the reactive policy. This is due to the improved training of host nation security forces when partner forces are deployed to a country to assist in building capacity. Figures 18.14 and 18.15 show the level of security forces that are deployed to the country in both policies. In a reactive policy, the security forces are deployed as the result of a crisis in the country and additional forces are deployed to combat the high level of insurgents in the nation. This is similar to the *surge* observed in Operation Iraqi Freedom. However, in the proactive policy, a constant force of 5000 soldiers from the partner nation is able to train host nation security forces with a much smaller number of soldiers.

The output of the budget portion of the model presents the monthly investment by security forces in the host nation and determines the total of all investments. Figure 18.14 presents the monthly

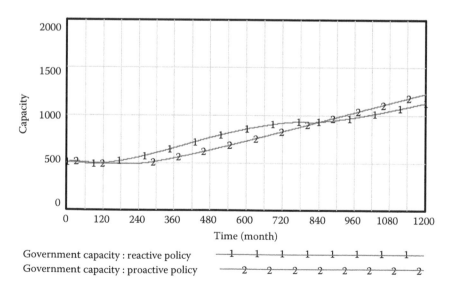

FIGURE 18.11 Model output—government capacity.

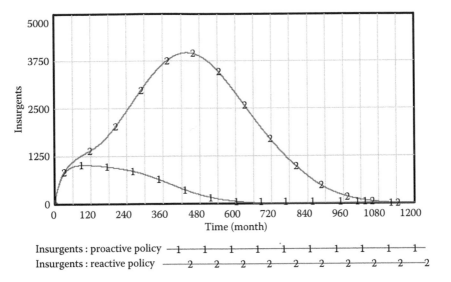

FIGURE 18.12 Model output—insurgents.

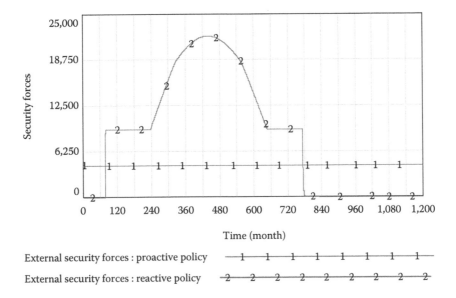

FIGURE 18.13 Model output—external security forces.

investment by security forces in the host nation. Although the output of the model does not exactly match the theoretical resources over time presented earlier, the behavior is generally the same. In a reactive policy, the partner nation rapidly increases investment in the country in reaction to a crisis and over time is able to slowly decrease its investment as the host nation's capacity is increased. For the proactive policy, a constant investment over the simulation demonstrates the requirements for a partnership to maintain the host nation's capacity. Figure 18.15 presents the total investment by security forces in the host nation. As shown, the proactive policy demonstrates significant savings over the reactive policy.

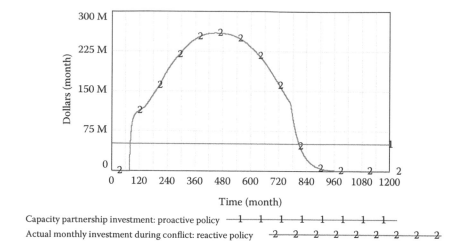

FIGURE 18.14   Model output—investment by security forces.

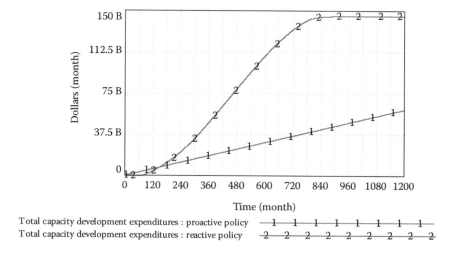

FIGURE 18.15   Model output—total investment by security forces.

ANALYSIS

Although the model functions without errors and provides an output, it has not been calibrated to historical data for use as a predictive model for future policies. The output appears to align with theoretical projections of investment of resources and security force levels; however, more work can be done to refine the model. Additionally, the model contains several assumptions that would need to be validated in order to provide a more accurate prediction of the behavior over time. However, the model demonstrates how SD can be utilized to model a complex system's behavior over time.

One of the challenges of this case study is to model a complex system's behavior over time against a measurement that does not exist. Although several proxy measurements of a government's capacity exist for some nations, such as government approval ratings, these do not measure the actual capacity of the nation. Without this measurement over time, it is difficult to calibrate an SD model to observed data and each individual submodel requires calibration to available data. We believe strongly that this is the right measure of the return on investment for S&R activities. However, more research is needed to quantify a value for country capacity. We have demonstrated lower-level relationships, for example, by quantitatively showing how increased security decreases terrorist activities. However, in order to truly be a meaningful analysis tool, we need to better understand what the acceptable levels of country/government capacity are.

SUMMARY

History has shown that as part of the whole government team, militaries will be involved in S&R actions for the foreseeable future. For example, in Afghanistan, the Internal Security Assistance Force includes 49 countries, with 28 nations providing combat troops. We need to better understand how military and other development actions contribute to the ability of a country to perform its most basic function—to provide for the needs of its people.

We used systems diagramming tools as a means to develop the dependencies as a first step in developing a quantitative model. As shown in Table 18.1, a nation state meets the criteria for a complex system. In general, SD is the best tool for trying to model a complex system. In reality, using diagramming tools to develop the dependencies for modeling using SD was highly iterative with the availability of data being the main driver for changing the model. In our SD model, we were able to replicate the behavior of a nation state. Moreover, our methodology, which utilizes country capacity as a measure of whether a country is meeting the basic needs of its populace, is novel.

CONCLUSIONS

Our model is a general representation of the methodology and seems to replicate the behavior of S&R activities for a generic nation state. The model needs to be modified for a specific scenario/country and then used as a baseline against historical data. We believe that this is the correct methodology to understanding resource allocation, secondary effects, priorities for the development of emerging countries, etc. However, building a detailed model is the only way to validate the interdependencies. Given that many of these S&R activities involve billions of dollars, a quantitative tool is needed to better understand how strategic investments can be best utilized. This research shows promise in that we can optimize/prioritize strategic investment alternatives that maximize the ability of a country to take care of its people.

SUGGESTED FUTURE WORK

The structure of the model facilitates the extension of the model to any developing country after some minor calibration to the model. The model uses inputs mainly from the CIA (2013), which are readily available for any country in the world. With these inputs alone, the model can replicate observed patterns of behavior in any country. Moreover, the use of additional data for other countries may provide an opportunity to better calibrate the model. Ideally, the model presented in this chapter can simulate the behavior of a government's capacity with minor modification to the initial variables. However, it is unlikely that the exact conditions exist in multiple countries of interest, so the model may not apply to all countries. The opportunity for true replication exists in the ability to apply the techniques and methodology presented in this chapter. Although the model of a country may differ slightly from that of the model presented, the idea of combining several measurable metrics, such as security, medical capacity, economy, and infrastructure, into a measure of a government's capacity is widely applicable. However, several basic research questions warrant further discussion:

1. How would the models differ for a developed versus a nondeveloped country?
2. Are the elements of country capacity presented as the right descriptors of behavior?
3. What is the right level of resolution for a nation state model?
4. How do we develop an architecture for a nation state? Can the same MPTs used for describing a traditional hardware/software system be used to describe a country?
5. Tools such as SD, agent-based modeling, and traditional event-driven simulation have matured to the point of possibly being usable for an insight into the behavior of complex systems such as a nation state. Can they be used to develop input for an SD model?
6. Using systems architecture, modeling, and other analytical techniques has been criticized by many in the social science community because of the complexity of nation building. Like any complex systems, are our MPTs mature enough to tackle these types of problems?

Research into the challenges of verification and validation, resolution, data, accurately capturing independences and synergies, second- and third-order relationships, etc., is starting to emerge. Unfortunately, complexity is evolving as fast as or faster than our ability to study and model. Meaningful analysis into the wicked problems of society will only be accomplished if we use the tools at our disposal and tackle these problems. As we attack these types of problems, gaps in our ability to model and analyze will emerge, paving the way for relevant research.

REFERENCES

Central Intelligence Agency. World factbook. https://www.cia.gov/library/publications/the-world-factbook/ (accessed January 3, 2013).

Choucri, N. et al. Understanding & modeling state stability: Exploiting system dynamics. MIT Sloan Working Paper CISL# 2006-02. Cambridge, MA: Massachusetts Institute of Technology, January 2006.

Clemens, W.C. Complexity theory as a tool for understanding and coping with ethnic conflict and development issues in post-Soviet Eurasia. *International Journal of Peace Studies, Autumn/Winter*, 7:2, 2002.

Crane, W.E. System dynamics framework for assessing nation-building in the Democratic Republic of the Congo. Carlisle, PA: United States Army War College, 2009.

Davies, L. Conflict and chaos: War and education, A1. London, U.K.: Routledge Falmer, 2003.

Department of the Army. FM3-07: Stability operations. 2008, http://usacac.army.mil/cac2/repository/FM307/FM3-07.pdf (accessed December 6, 2010).

Department of Defense. Quadrennial defense review—Building partner capacity. 2006, http://www.ndu.edu/itea/storage/790/BPC%20Roadmap.pdf (accessed January 13, 2012).

Department of Defense. Instruction 5000.68. Washington, DC: US Department of Defense, October 27, 2010.

Farr, J.V., Saltysiak, T., and Cloutier, R. An architectural framework for nation states in support of peace building operations. Draft White Paper, January 2013.

Forrester, J.W. *Industrial Dynamics*. Waltham, MA: Pegasus Communications, Inc., 1961.

Jackson, M.C. *Systems Thinking—Creative Holism for Managers*. Chichester, U.K.: John Wiley, 2003.

Meadows, D., Randers, J., and Meadows, D. *Limits to Growth: The 30-Year Update*. White River Junction, VT: Chelsea Green Publishing Company, 2004.

Mitleton-Kelly, E. *Ten Complex Systems and Evolutionary Perspectives on Organisations in Complex Systems and Evolutionary Perspectives on Organisations: The Application of Complexity Theory to Organisations*. Oxford, U.K.: Elsevier, 2003.

Nelson, C.R. How should NATO handle stabilisation operations and reconstruction efforts? Policy Paper, the Atlantic Council. Washington, DC: Atlantic Council of the United States, September 2006, http://www.acus.org/docs/061021-How_Should_NATO_Handle_SR_Operations.pdf (accessed January 20, 2012).

Ramalingam, B., Jones, H., Reba, T., and Young, J. Exploring the science of complexity ideas and implications for development and humanitarian efforts, Working Paper 285. London, U.K.: ODI, 2008.

Richardson, G.P. *Feedback Thought in Social Sciences and Systems Theory*. Philadelphia, PA: University of Pennsylvania Press, 1991.

Robbins, M.J.D. Investigating the complexities of nation building: A sub-nation regional perspective. Thesis, Air Force Institute of Technology, Wright-Patterson Air Force Base, Ohio, March 2005.

Shanshan, D., Lanhai, L., and Honggang, X. The system dynamic study of regional development of Manas Basin under the constraints of water resources. Sun Yat-sen University, Guangzhou, Guangdong, China, pp. 1–17, 2009.

Sheard, S.A. and Mostashari, A. Principles of complex systems for systems engineering. *Systems Engineering Journal*, 12(4):295–311, 2008.

Sterman, J. *Business Dynamics: Systems Thinking and Modeling for a Complex World*. Boston, MA: Irwin McGraw-Hill, 2000.

Wikipedia, http://en.wikipedia.org/wiki/List_of_countries_by_military_expenditures (accessed August 15, 2012a).

Wikipedia, http://en.wikipedia.org/wiki/United_States_Armed_Forces (accessed August 15, 2012b).

19 US Air Force Network Infrastructure System-of-Systems Engineering

Jeffrey Higginson, Tim Rudolph, and Jon Salwen

CONTENTS

CASE STUDY ELEMENTS

- *Fundamental essence*: This case study addresses the emergence of a large system-of-systems (SoS) information technology (IT) infrastructure and the methods and processes developed over time to acquire, manage, and operate this infrastructure for and by the stakeholder community.
- *Topical relevance*: As IT costs and consolidation become primary drivers in the infrastructure enterprise, there is increased pressure on infrastructure organization to adopt an SoS point of view and apply system-of-systems engineering (SoSE) methods and practices to improve infrastructure efficiency, lower cost, and improve the flexibility and speed of capability delivery.
- *Domain*: This case study examines a government (United States Air Force [USAF]) SoS; however, the lessons and observations are relevant to any large IT enterprise.
- *Country of focus*: United States.
- *Interested constituency*: Government and large institutional IT organizations.
- *Primary insights*:
  1. IT infrastructure can be operated, managed, and directed effectively as an acknowledged SoS.
  2. Centralized management and governance functions, shared across the stakeholder community, are effective mechanisms to establish requirements and deliver capability.
  3. Tacit acknowledgment and strong stakeholder support of the SoS is necessary to establish and execute strong SoSE principles.

KEYWORDS

Systems engineering, System of systems, Infrastructure, Enterprise engineering, Information technology

ABSTRACT

This case study focuses on a growing segment of the enterprise and system of systems (SoS) community—common infrastructure and specifically IT infrastructure. Common infrastructure represents a unique set of challenges for the engineering community in that IT infrastructure supports applications and services, but is not designed, acquired, or deployed with the full range of applications or services in mind. Instead, common infrastructure is designed to accommodate a generalized set of applications, services, users, and missions/business needs. It is also typically based on what is installed and/or used, instead of a *clean sheet* design to support an enterprise of possible activities. This intentionally *agnostic* nature of the infrastructure (not constrained by, but instead constraining mission/business functions) often leads to an autonomously architected and managed infrastructure capability and unexpected integration challenges when new applications or services attempt to utilize the infrastructure.

GLOSSARY

AFNET	Air Force Network
APC	Area Processing Center
DECC	Defense Enterprise Computing Center
DISA	Defense Information Systems Agency
DISN	Defense Information Systems Network
DoD	Department of Defense
ERP	Enterprise resource planning
IAP	Internet access point
IMS	Integrated Master Schedule

INOSC   Integrated Network Operations and Security Center
IT   Information technology
ITS   Information Transport System
MAIS   Major Acquisition Information System
MAJCOM   Major Command
MAN   Metropolitan area network
MOA   Memorandum of agreement
MOU   Memorandum of understanding
NIPRNET   Nonclassified Internet Protocol Router Network
NCC   Network Control Center
SA   Situational awareness
SE   Systems engineering
SoS   System of systems
SoSA   System-of-systems authority
SoSE   System-of-systems engineering
USAF   United States Air Force
VPN   Virtual private network

BACKGROUND

Initiated in the late 1990s, the USAF network infrastructure was established as a ground infrastructure–modernization effort to support the day-to-day information transport needs at fixed AF installations worldwide. At that time, infrastructure was managed at the base level. Off-base connectivity and network boundary protection was left largely to the individual base communications and networking organization. This led to numerous unique solutions, unpredictable configuration control, and decentralized management of the evolving IT enterprise. In light of these challenges, the AF began to consolidate infrastructure acquisition and management under the established Major Commands (MAJCOMs), thereby attempting to reduce and manage the growing diversity of implementations across the USAF. However, even this consolidation proved inadequate to fully control acquisition cost, manage and defend the infrastructure, and ensure mission success across MAJCOM boundaries. As an additional step, the USAF further consolidated the MAJCOM infrastructures into a single, USAF-wide IT enterprise—today's Air Force Network (AFNET). Figure 19.1 summarizes this evolution from 100+ base-level networks into a singly managed and defended infrastructure, providing transport, computing, security, and core services (such as email) to the AF community.

As future cost control and acquisition management initiatives continue across the United States Department of Defense (DoD), it is reasonable to expect further consolidation in DoD infrastructure, with a possible next phase (as shown in Figure 19.1) leading to the elimination of USAF IT infrastructures in lieu of a single, DoD-wide IT infrastructure. However, that end state is beyond the scope of this case study, though the lessons learned from the AFNET consolidation could offer insights into future challenges.

In this context, the AFNET emerged as an acknowledged SoS with recognized mission objectives and a compelling need for interoperability among the constituent MAJCOM systems (though each MAJCOM retained some ownership and control of their portions).

Over time, the AFNET has evolved into a directed SoS that is managed and funded centrally, with greater levels of interoperability and commonality (as evidenced by the common base boundary and gateway designs). As the AFNET has grown, greater expectations have been placed on the performance and capability of the infrastructure, which is expected to provide resources and services to all members of the AF community and support operations in normal and stressed environments. Figure 19.2 shows the general services and capabilities provided by the AFNET today. These diverse services illustrate the complexity and breadth of the AFNET, and hint at the challenges associated with the acquisition and management of a large infrastructure.

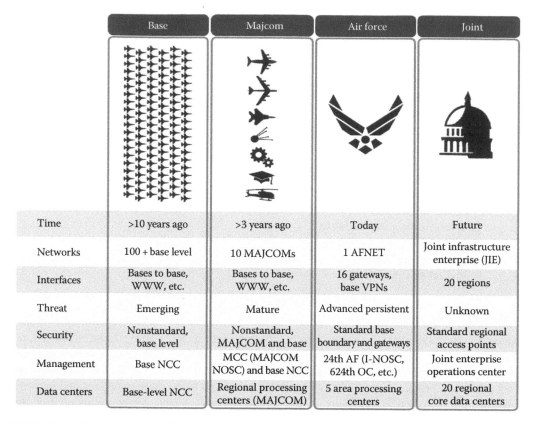

	Base	Majcom	Air force	Joint
Time	>10 years ago	>3 years ago	Today	Future
Networks	100 + base level	10 MAJCOMs	1 AFNET	Joint infrastructure enterprise (JIE)
Interfaces	Bases to base, WWW, etc.	Bases to base, WWW, etc.	16 gateways, base VPNs	20 regions
Threat	Emerging	Mature	Advanced persistent	Unknown
Security	Nonstandard, base level	Nonstandard, MAJCOM and base	Standard base boundary and gateways	Standard regional access points
Management	Base NCC	MCC (MAJCOM NOSC) and base NCC	24th AF (I-NOSC, 624th OC, etc.)	Joint enterprise operations center
Data centers	Base-level NCC	Regional processing centers (MAJCOM)	5 area processing centers	20 regional core data centers

FIGURE 19.1   Evolution of the AFNET.

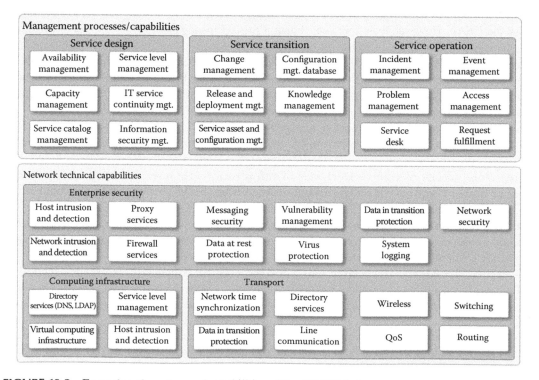

FIGURE 19.2   Examples of services and capabilities of the AFNET SoS.

PURPOSE

The systems that compose the AFNET portfolio provide an end-to-end capability to create, store, transport, manipulate, archive, protect, and defend information for the USAF. The AFNET portfolio is composed of multiple systems, including two large component programs:

1. The Information Transport System (ITS), which provides the basic wired and wireless communications infrastructure at each AF and Air National Guard base.
2. The AF Intranet, which provides the needed base-to-base connectivity and management, and network defense tools to protect the AF infrastructure.

In addition, there are several smaller systems and sustainment activities procured separately that are a part of the AFNET. All of these systems are integrated with non-AF infrastructure (provided by the broader DoD community) and aligned with DoD-level guidance and architecture requirements.

Figure 19.3 provides a high-level overview of the AFNET, illustrating the significant components of the information enterprise, which include the following:

- USAF bases, which may include data-processing centers
- AFNET *gateways* that provide control over data entering or leaving the AFNET
- AFNET network operations and security centers (located in several locations)

As suggested in Figure 19.3, the AFNET transports traffic via a secure virtual private network (VPN) mesh that rides over the nonclassified Defense Information Systems Network (DISN), often referred to as the nonclassified Internet Protocol Router Network (NIPRNET). Any traffic entering or leaving this USAF VPN mesh is routed through a security gateway for inspection and monitoring. In addition, the AFNET includes non-USAF systems, including the Defense Information Systems Agency (DISA)-provided Defense Enterprise Computing Centers (DECCs) and access to non-DoD networks via monitored and protected Internet access points.

The overview in Figure 19.1 makes it evident that the AFNET is really a large SoS that spans multiple levels of the USAF enterprise, from individual user services at the base to network-wide

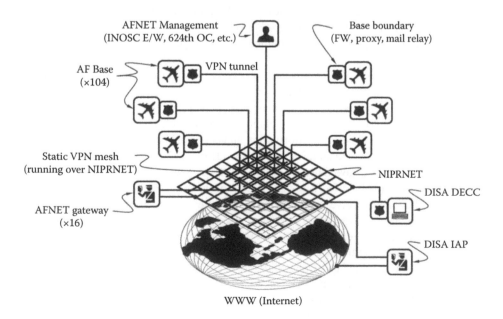

FIGURE 19.3 High-level overview of the AFNET.

management and defense functions, providing core network connectivity and services to more than 1.4 million users within the United States and across all theaters of operations. Management and defense of this infrastructure are provided at several levels. The components of this management capability are described in detail later in this case study.

CONSTITUENTS

In its most successful employment, an SoS implementation maximizes value across the stakeholder spectrum—value to acquirers, customers, users, etc. Each member of the stakeholder community, to be fully engaged and participating in the successful outcome of the SoS activity (including engineering), needs to be identified and brought into the process via effective and clear communication. Not every stakeholder requires the same level of communication (or communication vehicles) or interaction, but each requires early and enduring insight and input into the SoS activities.

Stakeholder analysis at every level creates the basis for the essential understanding to set outcomes and influence the development of an SoS. From our perspective, a stakeholder is a person, group, organization, or entity that affects or can be affected by the development and deployment of an SoS from an acquisition, operation, or support and maintenance perspective. Stakeholder analysis includes the identification of the key contributors and the continuing management of this relationship through an active communications program.

A comprehensive stakeholder analysis includes the following key benefits:

- It enables conscious and rational relationship management between key stakeholders by identifying stakeholder types and needs by enabling the SoSE governance process to determine the proper roles and responsibilities for each stakeholder group.
- It enables tailoring strategies for key stakeholders to take greater advantage of opportunities, and avoid or mitigate unwanted risks when they become apparent. This leads to the enablement of the SoSE management process and determination of the proper risk strategy at an SoS level.
- It supports changes in the stakeholder community without disruption to the SoS.
- It enables SoS evolution (change management) to be aligned with changes in the needs and outcomes of the stakeholder. This enables the SoSE management process to determine the best change management strategy at an SoS level.
- It enables continual improvement and adaptation to change.

Figure 19.4, sometimes referred to as a *power grid*, provides an example of the core AFNET, categorizing stakeholders in a four-quadrant chart that bins them by influence and interest. For example, stakeholders with low influence and low interest (quadrant 1) are not likely to be significant partners in the effort. Thus, they can be monitored for changing interest. Generally, they will receive minimal attention (beyond access to general information and occasional courtesy contacts), corresponding to their relatively low influence (and interest) in the effort.

Stakeholders in quadrant 2 have a higher interest, but, like the first group, a low influence. This group may wish to participate actively, but is unlikely to have a significant impact in the effort's success. As such, they may receive more attention than the first group, but still do not receive a significant investment. Here, most efforts are aimed at keeping this group informed and supportive.

The third group of stakeholders (quadrant 3) is the opposite of the second group. They are influential, but have little interest in the effort. This group may represent an opportunity to build support if they can be moved into the final quadrant (high interest, high influence) through targeted communication and education. At a minimum, this group must be kept satisfied via carefully crafted communications (possibly via the use of governance agreements such as Memorandums of Agreement [MOAs] and Memorandums of Understanding [MOUs]); otherwise, they can become significant impediments to the successful implementation of the effort.

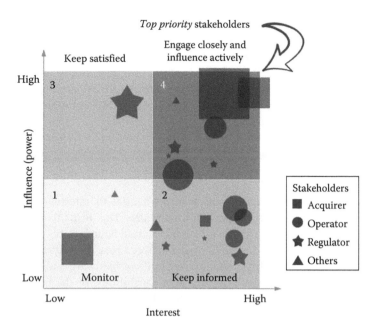

FIGURE 19.4   Stakeholder *power grid*.

The final group (quadrant 4) is most critical to early success. This group has high interest in the effort. They exert a great amount of influence over the outcome (positive or negative). These stakeholders require close and continuous engagement and carefully crafted communications products that allow for strong advocacy and support continuous feedback and collaboration. This group should command most of the attention and effort; stakeholders in this group are likely to persist throughout the effort.

The final piece of information deals with the size of the shapes plotted on the grid. In this presentation, the size of the shape is proportional to the time invested in the stakeholder. Logically, one would imagine that most of the time (hence, larger shapes) would be associated with stakeholders in the top right quadrant. However, as seen in this case, significant amounts of time may still be spent on lower-impact stakeholders, requiring careful monitoring to ensure that resources are best distributed among core stakeholders. It is also worth noting that some stakeholders can migrate from one quadrant to another, sometimes very quickly (e.g., *regulator* stakeholders who may become acutely interested in an effort during certain program milestones).

SYSTEM-OF-SYSTEM DESCRIPTION

Figure 19.3 provides a high-level overview of the AFNET and helps illustrate that the AFNET is an SoS that provides information transport, processing, storage, and infrastructure management and defense capabilities to more than one million users worldwide. Far from being a stand-alone infrastructure enterprise, the AFNET is closely integrated into the larger DISN and associated computing and storage components, such as the DISA DECCs. In addition, the AFNET provides connections to other DoD and government components as well as nongovernmental entities (e.g., commercial contractors involved in DoD business).

Though initially a set of independent base-level networks, today's modern network is managed as a single SoS with an operational structure that allows for the management and defense of the infrastructure in a coordinated fashion. Access into and out of the AFNET is provided via several security gateways, each supporting a set of primary bases and able to route traffic from other bases in the event of a gateway failure. These gateways represent the principal boundary into the AFNET through which all traffic entering or exiting the network is routed. Additional boundaries at each base further inspect and route network traffic. This *defense-in-depth* strategy is intended to assure mission success, while ensuring that bases can continue to perform missions under isolated conditions.

The general components of the AFNET include the base-level network infrastructure and network control center (NCC), gateways, area processing centers (APCs), and the integrated network operations and security centers (INOSCs). Each is described briefly and represents a unique set of capabilities supported by multiple systems.

Bases (often referred to as *enclaves*) provide a metropolitan area network (MAN) supporting USAF and joint users (Army, Navy, and Marine Corps). Bases significantly vary in size, from a few thousands to several tens of thousands of users executing various IT tasks across business, human resources, and mission domains.

At their boundaries, bases have security mechanisms similar to gateways, including firewalls, proxies, antivirus scanners, and traffic inspection capabilities. This helps to achieve a defense-in-depth security architecture where the gateway acts as an external boundary between the AFNET and other networks, and the base boundary acts as a border between base users/systems and the AFNET.

Each base has an NCC that provides a standardized suite of hardware and application packages that assure network security and network management capabilities for the base internal network. The NCC also includes a perimeter (internal and external filtering routers and a firewall) and other traffic inspection points.

The USAF gateway is the security boundary between the *world* and USAF Intranet. The USAF gateway implements a defense-in-depth architecture with multiple levels of security. These gateways contain the equipment to monitor, filter, and collect traffic for threat detection, forensic analysis, and performance management. The AF gateway is the secure connection point for the AF Intranet to the DISA NIPRNET and Secure Internet Protocol Router Network, and provides a centrally managed and standard connection to other government agencies and the Internet.

An APC is a data center that houses enterprise services shared across the USAF user base (i.e., email, shared websites/services, etc.). An APC is highly dependent on connectivity, bandwidth, and hardware, and may contain as many as 1000 servers per location.

The INOSC manages the base boundaries, including firewalls, bandwidth, infrastructure routers and switches, and end user devices (i.e., desktop configurations and wireless devices). In addition, the INOSC conducts incident investigation and response for the base boundaries and coordinates with base NCCs and APCs.

Figure 19.5 illustrates some of the factors at work in the AFNET system of systems, including external forces that may have a significant and enduring impact on the final configuration and operation of the deployed systems.

    A: *Complex system*—The autonomous and independent systems include the common infrastructure components of the AFNET SoS designed for a range of applications and not specifically for each base or operation. The systems that are installed and available provide security, transport (network), computing, storage, situational awareness, and command and control functions designed to accommodate the flexible and efficient application of infrastructure services to a wide range of mission needs. The components include non-USAF systems and potentially non-DoD networks and Internet access.

    B: *External factors*—External factors are represented by the context of individual operations, which include geography, legal and cultural systems recognized by international treaties, budget and funding instabilities, and often rapid changes in political and social situations.

    C: *Feedback*—This includes the normal USAF chain of command mechanism as well as internal government direction and occasionally media/nongovernmental responses.

    D: *Governing body*—The governing body is generally the USAF hierarchy of control as well as the military hierarchy to which the USAF is responsible. Within the USAF, there are

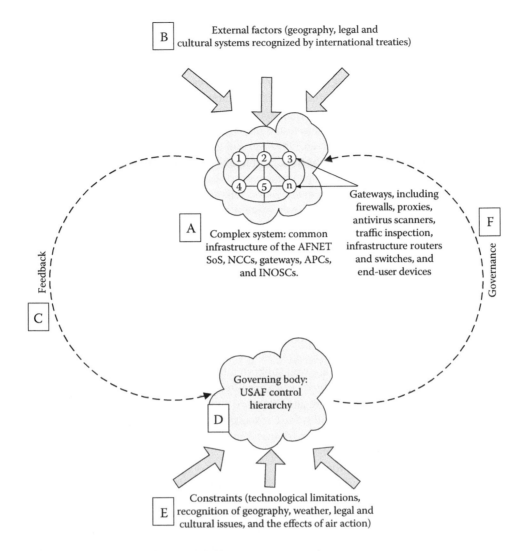

FIGURE 19.5 Generic depiction of AF infrastructure system of systems.

various bodies responsible for specific components of the SoS, operations, support, funding, and program planning. This diverse group of governing bodies illustrates one of the fundamental challenges of managing an SoS within a large organization like the AF, and highlights the need and importance of a string SoSE governance process.

E: *Constraints*—These may include many factors, including geography (especially in large/diverse theaters of operation), local operational agreements, available funding vehicles, situational awareness limitations, and technological challenges (though these are almost always the least imposing of the challenges listed). In addition, integration of USAF assets with other service assets and coalition partners can also result in interoperability constraints driven by both technical and operational diversity.

F: *Governance*—Governing issues are dominated by the need to migrate from base/MAJCOM infrastructure priorities toward a single USAF view of a common, shared infrastructure (SoS) or even to a single DoD vision. As noted in item D, a number of base-level, MAJCOM, and AF level governing bodies are involved in managing the AF IT infrastructure, making SoS effective governance a continuing challenge, and among the most promising areas for rapid improvement through structure SoSE.

CHALLENGES

As it was not originally conceived as an SoS, but rather as a group of independent and interoperable base-level systems, the AFNET has had to adopt an SoS approach during deployment and modernization. This has resulted in a number of challenges and stresses on the traditional DoD acquisition processes. For example, early in the infrastructure development, funds were provided to the base and MAJCOM levels to develop, deploy, and operate their portions of the infrastructure. As further consolidation occurred, for improved security and management and cost efficiencies, reassignment of funds was necessary (away from a MAJCOM level and toward a single, cognizant infrastructure organization, which today resides in AF Space Command). Reassignment of funds can be a difficult and time-consuming process, as DoD budgets and expenditures are often planned years in advance, and contracts already in place may span many years. This inertia in the acquisition and budgeting processes tends to lengthen the transition from system to SoS operations and complicate future planning and resourcing. Further challenges resulting from this migration toward an SoS include the following:

- *Common situational awareness (SA)*
  With a variety of management systems in use (composing the SoS) and limited access into system performance/operation, it is difficult to collect and integrate system-level performance data, making full enterprise situational awareness a challenge. This could be addressed through the adoption of a common tool set or standardized reporting requirements shared across the operational community to support the development of a common operations picture. Both options are well supported by an SoS approach.
- *Enterprise management of compute/storage resources*
  In light of consistent budget pressure and increased attention of cost savings, virtualization and consolidation of computing and storage resources have become an important initiative. A challenge for the AFNET is to provision and manage these resources on an enterprise scale across a diverse hardware profile.
- *Governance and baseline management*
  As with many merged networks, the AFNET operates a *patchwork* construct for managing the technical baseline from end to end. While the *operational baseline* is managed by the network operators, changes to the technical baseline occur rapidly and with little time for coordination. This is further complicated by the complex IT governance structure across the DoD and the legacy of allowing local control of IT assets. An efficient and streamlined change management process supported by consistent and effective governance remains a significant challenge.

Finally, the unique nature of any IT enterprise requires an agile approach to acquisition and support that can adapt quickly to threats and opportunities (e.g., technology improvements). As shown in Figure 19.6, a typical large acquisition program such as a Major Acquisition Information System (MAIS)

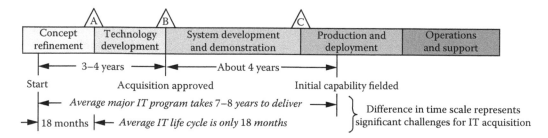

FIGURE 19.6   Comparison of typical system delivery to the IT life cycle.

may take 7–8 years to deliver, despite the average IT life cycles of 12–18 months. Often, the result is the delivery of systems that contain end-of-life components the day they are delivered. Then, the challenge for SoSE is its ability to maintain, or even accelerate, the speed of capability delivery to the customer, rather than introduce constraints or overhead requirements that add further delay.

SoSE DEVELOPMENT

As illustrated in Figure 19.7, systems engineering (SE) spans the full range of scales (and associated perspectives) from the simplest component to the most complex enterprise. At the extremes (component versus enterprise engineering), the distinctions are easily discernible, though at any one boundary, the distinctions can become less obvious. For example, an SoS may encompass smaller SoSs or overlap a portion of another SoS.

The characterizations in Figure 19.7 approximate boundaries for each scale and help illustrate the differing perspectives of engineers working on associated tasks across this spectrum. Note that Figure 19.7 also includes a representation of enterprise engineering, a scale that is even larger than the SoS scale discussed. At the enterprise level, the integration of multiple SoSs may be considered, along with strategic investment decisions and broader capability areas beyond the scope of this discussion.

For example, a traditional system-engineering problem emphasizes stable requirements, fixed designs, and decomposition of the solution into its simplest components, which can then be *engineered* and assembled into a *system*. This approach sharply contrasts the SoS perspective, which focuses not on requirements and decomposition to components, but on capability and aggregation of systems to a larger scale. These contrasts are helpful in understanding the different engineering tactics that must be embraced. In addition, these contrasts are helpful in cementing the relationship and importance of applying multiple scales of engineering to the solution of difficult (and interdependent) problems.

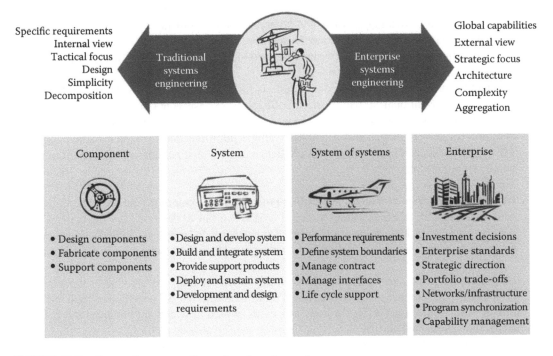

FIGURE 19.7 Contrasting perspectives of engineering scales.

Some of the more important contrasts in perspective include the following:

- *Requirements versus capabilities*
  In the SoSE perspective, focus is not on a specific set of rigid system requirements (i.e., traditional SE), but on the capabilities needed by the end user to execute the mission or achieve objectives. This capability focus is applied across the set or portfolio of systems to evolve solutions that can meet the changing capability needs and support new or evolving mission threads.
- *Control versus collaboration*
  A traditional system or component engineering stresses the importance of controlling the variables and managing sources of *risk* or dependencies in dimensions such as cost, schedule, and performance. Often the developing or acquiring agency seeks to control these variables by gathering the constituent components of the system and creating an internally managed (i.e., stovepiped) development. However, as the scope of the effort extends beyond a system into the SoS realm, attempts to control the variables quickly become intractable. In the SoSE domain, total control is not feasible. Instead, a network of collaborators (stakeholders) must be formed (via an integrated product team, for example) to jointly manage the aspects of engineering and acquisition across system boundaries.
- *Internal versus external*
  The SoSE view looks outside the boundaries of traditional systems and their requirements at the full spectrum of systems (and nonmaterial solutions) that could support the delivery of a capability. In contrast, traditional SE looks toward specific solutions bounded by a set of specific (internal) requirements or needs, and often sacrifices future SoS reuse for system-specific optimization.
- *Design versus architecture*
  In traditional SE, the end point is often a design (solution) that complies with a specific set of requirements. In the SoSE context, such a final design is often not practical due to the dynamic and interrelated characteristics of an SoS solution. Instead, emphasis on architecture is critical. This architecture defines the existing and *to be* capabilities of the SoS and supports the development of roadmaps, which can be applied to constituent systems, to achieve the desired end states (mission capabilities). In addition, the architecture establishes and defines key integration points between the systems needed to support the final SoS instantiation.
- *Decomposition versus aggregation*
  Reducing a problem into its component parts (which can then be easily engineered) is a foundation of traditional SE. However, in SoSE, the emphasis is not on the decomposition of the problem, but on the aggregation of many similar problems, systems, and needs into a common framework, which is aimed at delivering capabilities that span the mission needs. This aggregation and recognition of its challenge are key distinctions between traditional and SoS engineering.
- *Simplicity versus complexity*
  As noted, aggregation of systems and similar problems into a common (architectural) framework is a hallmark of SoSE. However, this approach can fail to describe fully the extremely complex problems with many interactions and dependencies. In large-scale SoSE, this aggregation must be augmented with complex SE methods applied to SoS and enterprise-level problems.

These foundational principles helped form the SoSE construct that was applied to the growing AFNET. Our first priority was to leverage as much of the existing *core* SE processes as possible, and extend them with several *new* SoSE capstone processes, which are needed to address the broader SoS perspectives described earlier.

Our model for SoSE implementation, shown in Figure 19.8, provides a continuous execution chain from a high-level strategy to improve SoSE and associated capability delivery to system-level

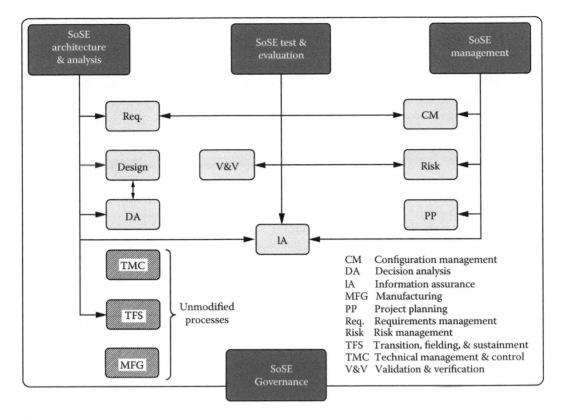

FIGURE 19.8   SoSE implementation model.

implementation. The focus in this model is directed at the interface between the proposed SoSE capstone processes (governance, architecture and analysis, test and evaluation, and management) and the existing SE processes already widely used at the system level. In this model, the SoSE processes are mapped onto the SE process at the system level. Here, they may modify the existing SE processes and their implementation at the system level. These SoSE processes are based on the SoSE strategy and objectives, and provide a means to engineer SoSs to support mission capabilities that transcend the boundaries of a single system. To achieve this cross-system scale of engineering, it is necessary to articulate the SoSE processes and understand (define) how they impact (modify) current SE processes and their implementation.

For example, the architecture and analysis SoSE process drives changes to the requirements and design SE processes at the implementation level (and how they are applied to the system). Similarly, the SoSE management process indicates modifications to a number of SE processes, including configuration management, risk management, and program planning.

Regardless of the specific *mapping* between SoSE processes and existing or new SE processes, modifications to the SE processes support the objectives of the SoSE effort (improved system-to-system interoperability, integratability, affordability, and mission effectiveness) by providing specific process direction, supported by policy and guidance, at the system level. Initially, policy and guidance occur locally and provide technical perspectives and engineering doctrines (e.g., SoSE *building codes*), which guide acquisition-program system engineers to execute their traditional SE role, but with an SoSE perspective that supports SoS mission objectives. Policy and guidance occur at higher levels in the form of policy instructions. Regardless of the level that the policy or direction originates, it should be consistent with the SoSE processes and conveyed via the existing SE processes to the systems engineers who effect the SoSE processes and their implementation. Supporting the engineering efforts, the acquisition planning, financial, contracting, and program management communities have similar policies and processes that represent SoS efforts within their domains/communities.

The basic SoSE approach coordinates the system-level SE activities of constituent systems. A small number of SoS-level activities focus on identifying stakeholders, developing the SoS architecture, and evaluating the SoS. This approach addresses the entire life cycle of system development and the SoSE-related actions that can occur as early as system concept development and as late as system sustainment and disposal.

An SoS exists as a collection of integrated/interoperable systems that support a given capability. It evolves as the capability evolves and as the constituent systems evolve. SoSE efforts can start at any point in the life cycle of any individual constituent system and continue for the lifetime of the capability. They feed into the life cycles of the constituent systems in a continuous, iterative manner. The constituent systems in any given SoS have independent life cycles. They are at different stages of maturity and can evolve at different rates. Some are in the early stages of concept development, while others are in sustainment (or even disposed).

The described SoSE approach covers a range of engineering concerns, including the following:

- Identifying and tracking the effects of change across an SoS's constituent systems
- Tracking risks across an SoS's constituent systems
- Allocating requirements across an SoS's constituent systems
- Ensuring stakeholder coordination with regard to engineering across an SoS's constituent systems

The overall approach begins with the SoSE governance process and includes the identification of an SoS, an appropriate SoS authority (SoSA) when necessary (not all SoSs require an SoSA), and the need for the associated SoSE. This results from the identification of a capability need and the realization that the satisfaction of that capability may require more than one system. Occasionally, SoSs (and the associated need for SoSE) are driven by a desire to realize certain nonfunctional attributes, such as affordability or efficiency, associated with a capability. For example, the drive to leverage common infrastructure and services can result in the formation of SoSs, as in the case of the AFNET.

SoSE Governance Process

Once started, the SoSE governance process manages the execution of the other three SoSE processes and the system-level project planning of constituent systems. The primary inputs to the SoSE governance process are relevant policy/guidelines, products from the other SoSE processes (most notably, the SoS Integrated Master Schedule [IMS] from the SoSE management process and the SoS architecture from the SoSE architecture and analysis process), and process status from the other SoSE processes. The primary outputs show the direction to the other SoSE processes and agreements to various stakeholders. Some of these stakeholders may be external to the execution of the SoSE processes.

SoS Architecture

The SoS architecture, derived from the architectures of the constituent systems (among other sources), models the existing and *to be* capabilities of the SoS and defines the technical basis for the evolution of the SoS. The SoS architecture highlights essential SoS qualities. Feedback from the SoSE architecture and analysis process supports the development of roadmaps, which can be used to synchronize related system acquisition efforts, to identify required changes to constituent architectures, and to gauge progress toward capability objectives.

In the SoSE architecture and analysis process, SoS engineers develop the SoS architecture* and analyze it to identify issues, risks, and opportunities. The primary inputs to the SoSE architecture

---

* The SoS architecture is an abstract representation of the supported capability and the constituents of the SoS, as they are arranged to form the SoS.

and analysis process include direction from the SoSE governance process, description of the associated capability, technical baselines of the constituent systems (to include their architectures), interests of the relevant stakeholders (e.g., relevant concept of operations and higher-level architectures), and various SoS artifacts (e.g., SoS-level requirements, risks, and test results). The primary outputs of this process are the SoS architecture,* which feeds the SoSE management and SoSE test and evaluation processes; the SoS analysis results, which support investment and engineering decision-making; and various courses of action for evolving the SoS. The SoSE architecture and analysis process interacts with the system-level design, verification and validation, and decision analysis processes.

SoSE Management Process

In the SoSE management process, engineers manage various SoS-level artifacts, including requirements, risks, test results, the SoS architecture, analysis findings, and courses of action. The SoSE management process works primarily through interactions with the system-level requirements management, configuration management, risk management, and technical management and control processes. The primary inputs to the SoSE management process include outputs from those system-level processes, direction from the SoSE governance process, and outputs from the SoSE architecture and analysis, and SoSE test and evaluation processes. The primary outputs of this process are the SoS IMS, which feeds the SoSE governance process, and the various managed SoS-level artifacts that feed subordinate processes.

SoS Test and Evaluation

The objectives of the SoS test and evaluation are similar to those for an individual system; however, they are aggregated at the SoS level. These include evaluation of design, assessment of risk and deficiencies (allowing early mitigation and resolution), and assuring mission capabilities have been achieved.

The SoSE test and evaluation process achieves these objectives using the following strategies:

- Planning and coordination of SoS test resources. As the capabilities provided by the SoS are distinct from the capabilities provided by the constituent systems, testing the SoS requires resources (people test infrastructure) that were not foreseen when planning the constituent systems. Without proper planning and coordination of SoS test resources, impact on cost, schedules, and performance are virtually assured.
- Problem and deficiency reporting and awareness of SoS issues. While a variety of problem-tracking tools exist among the constituent systems, it is difficult to identify which problem reports from the constituent systems impact the SoS and to what extent. The test and evaluation process provides a vehicle to federate problem-tracking processes and tools used for the constituent systems and ensures situation awareness of SoS issues that surface through the evaluation process.
- Linkage to risk management. Problem awareness enables better risk assessment across the constituents of the SoS and supports the identification and assessment of mitigation plans at the SoS level.

RESULTS

Problem Awareness

By applying the SoSE test and evaluation process, an end-to-end test strategy can be invoked across the constituent systems, resulting in SoS deficiency-problem identification and awareness (prior to deployment). This stems from using problem-tracking and monitoring tools throughout

---

* Note that the SoS architecture can (and should) be supplemented with more detailed models, as needed, to represent detailed process flows, information flows, constraints, etc.

the life cycle and by providing opportunities to report deficiencies to the SoSE architecture and analysis process, allowing modification of the SoS architecture and supporting system requirements.

Currently, the USAF operates a *production* laboratory that closely duplicates the fielded AFNET. Maintaining this capability allows high-fidelity assessment of constituent systems and the impact on SoS performance of changes (additions and deletions) to the technical baseline. It also permits the exploration of technical and operations alternatives to extending mission capability or addressing urgent needs.

Dependency Discovery

By using the SoSE architecture and analysis process, the dependencies of the constituent systems within the SoS are discovered early in the life cycle, allowing for proactive SE and early problem identification (refer to "Problem Awareness" section). This results in lower integration and maintenance costs at a system level (where much of the SoS cost is ultimately borne) and SoS level.

For the AFNET, current and *to be* SoS architectures help identify dependencies and interoperability issues. In addition, these resources can help project potential gaps in capabilities and identify needs for future system development.

Common Framework

By providing a common framework for an SoSE approach and strategy, and for a new infrastructure capability for the AF, the outcome is more affordable, interoperable, integratable, and robust. By understanding boundaries and information flows within and external to the SoS, information assurance can be addressed better and risks mitigated.

Continuous Monitoring

A major result of the SoS practices applied to the use case was recognizing that a snapshot in time is insufficient to ensure successful SoS execution. The initial investment to understand and achieve significant results requires continued monitoring throughout the SoS life cycle. While this may appear simple, it requires a full understanding of the SoS architecture and a long-term commitment to collect, process, understand, and act on measures of performance at the SoS level (measures collected only at the system level do not reliably predict SoS performance).

Improved Stakeholder Communication

The ability to initiate and maintain productive dialog with the core stakeholders is a necessary component of any successful SoS effort. Ensuring that core stakeholders are actively engaged in the development, execution, and continuous improvement of the SoSE activities and are fully vested in its success becomes a dominant task early in the SoSE effort. Our communication, built on the stakeholder identification described previously, was based on a simple, four-step approach:

1. Establishing the key messages
2. Determining communication objectives for each stakeholder group
3. Determining appropriate communication mechanisms to achieve objectives
4. Establishing and monitoring feedback

For each step, the SoSE carefully examined the options available, given the target stakeholder, desired outcomes, and available resources, to arrive at the optimum communication vehicles for each audience.

ANALYSIS AND OUTCOMES

Adopting an SoS approach for the IT infrastructure provided a new perspective on the acquisition and management of the many components that comprise the infrastructure portfolio, including hardware and software items (licenses), management functions, support contracts, and training and enterprise services (e.g., help desks, backup, and recovery capabilities). In each area, the SoS perspective supports a more complete view of the trade space and offers an opportunity for improved efficiencies and expanded capabilities.

For example, at the system level, server capacity and storage are purchased, maintained, refreshed, and licensed by individual systems via system acquisition programs. Multiple programs independently procuring and supporting these components do not achieve the economy of scale of a consolidated approach. In addition, it could potentially introduce incompatibilities, extra licensing costs, and additional training and support requirements. Initial analysis conducted on the IT portfolio demonstrated that an SoS approach provides considerable efficiencies. For example, storage utilization was examined across the IT enterprise and found to be below 20% (compared to industry targets of 60%–80%). Therefore, by segmenting the IT infrastructure into unique systems, significant unused capacity had developed within the enterprise, but was not shared across system boundaries. Adopting an SoS perspective allowed the acquisition community to quickly recognize this excess latent capacity and move to utilize it fully before purchasing more enterprise storage.

Enterprise licensing for management tools provides another example where significant cost reductions were identified by fully using licenses already available within the enterprise and converging (initially for cost, later for improved interoperability) on a common set of network management tools. A similar argument can also be made for backup and restoral services. When providing common services at the SoS (or enterprise) level, individual systems are freed from the requirement to independently provide these capabilities and the associated training, support, and licensing costs, which can be provided more efficiently at the SoS scale.

Initial analysis suggested capital cost reductions up to 20% over 5 years by migrating toward an SoS approach for the acquisition and management of IT infrastructure components and capabilities. Additional operation and sustainment economies are expected, but are difficult to quantify, as this cost covers a larger stakeholder base and is subject to the operation's tempo, which historically is difficult to predict.

In addition, abstraction of IT capabilities to an SoS level supports the synchronization of technical refresh cycles across the enterprise and the gradual reduction of conflicting technologies (and multiple generations of technology) within the infrastructure. This simplifies support requirements and improves baseline control.

Finally, embracing an SoS view also allows the emergence of *latent capabilities*—those composable from existing system elements with minimal engineering and integration costs—that support great agility and significantly help reduce the acquisition timelines that are critical to a flexible, responsive SoS. As with any large IT infrastructure, when the AFNET was first conceived, many of today's capabilities were not envisioned. Today, leveraging an SoS perspective, we can feed management tools and systems, and implement new capabilities with data from existing sensors, or we can leverage spare processing capacity to host new applications. This *composable* nature is a strong motivator for the SoSE approach, allowing for the continuous reengineering and composition of capability on demand to support rapidly evolving user needs in the cyber domain.

EPILOGUE

The effort to embrace, leverage, and benefit from SoSE requires a significant change in an organization's culture, which historically has been mapped to programs and resources. Programs are often characterized as the *system*, even if the system boundaries do not meet traditional system definitions.

In addition, the historic tendency to *control* systems and *eliminate* risk by completely controlling all elements of the SoS system makes success a challenge, due to the technical complexity of large, modern IT systems and the policy and management constructs built over the decades to enforce and protect the domain of the traditional program manager.

Cultural and organizational behavior changes required should not be underestimated. Behavior typically follows and is rewarded with resources.

By overcoming these challenges, the SoSE approach for the use case has highlighted effective, specific practices and outcomes, including the following:

- An architecturally driven system engineering that supports SoSE understanding and communication to the full range of stakeholders.
- Recognition that SoS analysis is a learned process that takes continued reinforcement to be established.
- SoS analysis supports the prioritization of capabilities under sponsor management and the assignment of requirements to component systems.
- SoSE management practices are different, but enable existing governance/engineering processes to work and support alignment within portfolios.
- Despite the inferred scope of SoSE activities, the processes can (and should be) agile and supportive of rapid decisions.*
- SoSE practices can be effectively expanded with enterprise architecture and enterprise engineering; SoSE must be part of a holistic approach, not a competing set of processes.

CONCLUSIONS

SoSE showed significant value in multiple outcomes:

- SoSE practices helped meet multiple (sometime overlapping, sometimes competing) requirement sets, while minimizing excess cost and duplicative capabilities across the USAF IT network enterprise.
- SoSE supported improved interoperability of capabilities and allowed for the delivery of *latent* capabilities—those available through the composition of existing system functions without significant engineering or integration costs.
- SoSE supports effective resource alignment.
- SoSE supports informed decisions for smart trade-offs by senior leaders. This dynamic leads to greater alignment of requirements and apportioning of limited resources.

SUGGESTED FUTURE WORK

Depending on the scope of the SoS, SoSE can be a focused process and art form. As the scope increases, the focused process must become a larger portion of the SoSE effort to ensure integration and interoperability between the constituent systems. The balance between teamwork of SoSE practitioners and subject matter experts (and system-level engineers) is very important and needs to be assessed continually; it also may be continually challenged. Too much emphasis on SoS perspectives can introduce excessive burdens at the system level and possibly introduce cost and schedule delays, and performance compromises. Too much emphasis on the system level can lead to an ineffective SoS. Discovering and maintaining this balance is critical to a successful SoS effort.

---

* The broad scope of AFNET still allowed SoSE principles to support 90-day rapid development sprints.

FURTHER QUESTIONS FOR DISCUSSION

As we explored the opportunities to migrate toward a common, managed infrastructure, we discovered that many of the most significant challenges were nontechnical in nature, and instead the product of *tradition* perceived threats to long-held positions or methods of operation, and a failure to properly appreciate opportunities and risks. In addition, without objective measures, it can be difficult to assess the progress of the SoSE effort, and collect the feedback necessary to identify and implement effective course adjustments—and overcome *tradition.*

For enterprise IT infrastructures (and likely any SoS endeavor), especially those that have evolved in a large, hierarchical organizational structure (with segmented responsibilities and occasionally constrained communications), it is worth asking several general questions, including the following:

1. What must organizations or individuals change in their operations, responsibilities, or organization to support the SoS outcomes?
2. What priorities and tasks of the existing stakeholders will be impacted, and how can the value of adopting an SoSE approach be articulated?
3. What objective measures can be adopted to monitor the SoS development and strategy, manage its implementation, and provide useful feedback?
4. What organizational/cultural reactions can be anticipated and how will these be supported or addressed?
5. What kind of people (especially early adopters) need to be engaged for a successful SoSE implementation, and how are they to be identified, developed, and encouraged?

Experience suggests that these fundamental questions are often overlooked in the hurry to implement SoSE, leading to an effort rich in process and structure, but lacking in advocacy and support, and, absent the outcomes and measures needed to demonstrate progress, failing to define success and motivate participation.

ADDITIONAL RESEARCH

Other interesting SoS case studies might include large enterprise resource planning (ERP) programs, which comprise significant internal and external dependencies on the IT infrastructure and associated data consumers and producers. Extending the SoS scope to include multiple ERPs or other applications being developed and deployed across the infrastructure could lead to powerful interoperability and performance observations. It may be worth noting that less structured attempts have been made to support ERP or other application integration in the IT infrastructure through the use of an infrastructure design document that provides an overview of the infrastructure, its performance, and the basic architecture—all intended to support the development of infrastructure-dependent applications.

Finally, despite success at the service level (USAF), the goal of the DoD is to migrate all services (Army, Navy, Air Force, and Marines) to a single, DoD enterprise IT infrastructure with common data centers and shared enterprise services. However, appealing from a business case perspective, the SoSE challenges are daunting. In particular, migrating toward a true DoD enterprise IT infrastructure requires merging multiple, large SoSs into a single *enterprise* SoS. While the perspectives described in this case study should remain valid, the process constructs need to be extended to support this larger enterprise scale. In addition, while certain common threads enabled a successful SoS construct (e.g., consistent SE process baseline, common funding mechanisms, and centralized SoS management), these would not exist to the same degree across the larger DoD domain. It is likely that these challenges will be posed, and these SoS constructs may provide the best starting point for the larger efforts.

DEFINITION RECOMMENDATIONS

(For reference when developing common reference definitions)

Affordability—The ability to deliver the requested capability or system within the available resources, including funding, manpower, etc.

Common interface—An interface that services multiple consumers or needs.

Effective—Meets or supports mission outcomes; provides the needed capabilities.

Efficient—Receiving the greatest mission capability for the invested resources, or not spending more than necessary to deliver the minimum required capability.

Integratability—A method of creating a larger and more complex entity by accumulation of individual components or systems.

Interoperability—The ability to exchange and/or share information among component systems.

System—An interacting mix of elements forming a whole that is greater than the sum of its parts.

Systems engineering—An interdisciplinary approach and process encompassing the entire technical effort to evolve, verify, and sustain an integrated and total life cycle balanced set of system, people, and process solutions that satisfy customer needs.

Systems of systems—Collection of systems that functions to achieve a purpose not generally achievable by the individual systems acting independently.

System of systems engineering—Planning, analyzing, organizing, and integrating the capabilities of several existing and new systems into an SoS capability that is greater than the sum of the capabilities of the constituent parts.

BIBLIOGRAPHY

1. Dahmann, J.S., Rebovich, G. Jr., and Lane J.A. Systems engineering for capabilities. *CrossTalk, The Journal of Defense Software Engineering*, 21(11), 4–9, November 2008.
2. Jamshidi, M., Editor. *System of Systems Engineering, Principles and Applications*. Taylor & Francis, Boca Raton, FL, 2008.
3. Daniels, M., Higginson, J., and Salwen, J. Proposed SoSE strategy for the USAF. *IEEE SoSE 2011 Conference*, Albuquerque, July 2011.
4. Defense Acquisition University. *Defense Acquisition Guidebook*. Department of Defense, Washington, DC, July 2011.
5. Baldwin, K., Dahmann, J., Lane, J.A., Lowry, R., and Rebovich, G. Jr. An implementers' view of systems engineering for systems of systems. *IEEE Aerospace and Electronic Systems Magazine*, 25(5), 11–16, May 2012.
6. Office of the Under Secretary of Defense for Acquisition, Technology and Logistics. Systems engineering guide for systems of systems, Version 1.0. ODUSD(A&T)SSE, Washington, DC, August 2008.

20 US Navy Submarine Sonar Systems

Layered Lateral Enterprise–Technical System Architectures for Defense Acquisition

John Q. Dickmann

CONTENTS

CASE STUDY ELEMENTS

FUNDAMENTAL ESSENCE

This case study examines the architectural transformation in the acquisition of US Navy submarine sonar systems. The technical transformation created a layered (sometimes described as modular) hardware/software system from a tightly integrated one.* The organizational (or enterprise) transformation created a more *open* and collaborative development and engineering enterprise, which would be able to field hardware and software upgrades to Navy submarines at substantial cost savings and on timescales that keep pace with the rate of change of information processing technology. This has been described as spiral or evolutionary development.

TOPICAL RELEVANCE

US defense acquisition programs have been notorious for their frequent cost and schedule overruns and for problems meeting technical performance requirements. Sixty-plus years' worth of effort to *reform* this system has failed; acquisition pathologies have proven remarkably resistant to change. A major theme of most acquisition reform efforts and reviews is process improvement, better and more stable definition of requirements, creation of appropriate incentives, and identification and application of best practices. This is a case study of successful acquisition reform; this case shows that mechanisms of success are rooted in areas that are not widely considered by most practitioners and scholars of defense systems development and acquisition. It highlights critical features of reform: (1) initiation and sustainment of change and (2) creation of a flexible development and engineering process and enterprise.

DOMAIN

US government and defense industry

COUNTRY OF FOCUS

United States

STAKEHOLDERS IN THE CASE STUDY

> US Navy's submarine fleet
> System prime contractors and subcontractors (Lockheed Martin)
> Small defense businesses (Digital Signal Resources [DSR], In-Depth Engineering, etc.)
> Government laboratory (Naval Undersea Warfare Center)
> Academic laboratories (JHU/APL; MIT/LL)
> Federally Funded Research and Development Centers (FFRDCs) (MITRE)
> Navy Systems Acquisition Headquarters (NAVSEA PMS-401)
> Submarine community programming and budgeting staff (OPNAV N77 [87])

INTERESTED CONSTITUENCY

This case study is of interest to program managers and other senior leaders responsible for managing complex system acquisition programs and enterprises. It applies to military and civilian engineers and managers, from the midlevel to senior executive, across government, academic, and

---

* Layered is modular but modular is not necessarily layered. Layered is preferred because this architectural characteristic more clearly separates similar, tightly coupled, functionality within a layer from other layers through loose coupling. This permits advances within the layers without affecting the interfaces, other layers, or the entire system; this is not always true with modular architectures.

commercial sectors. The lessons may be applicable to any organization responsible for sociotechnical system design and development.

PRIMARY INSIGHTS

The significant findings of this case study are that change in the acquisition of a complex, high dollar value system becomes a battle of organizational elites at the senior level and midlevel of the stakeholder organizations. The change must be catalyzed across the government–industry boundary, and there must be strong alignment at the senior levels within the government (especially the uniformed officers). In crafting a flexible system development, the testing and fielding process requires establishing lateral connections at multiple levels of organizational hierarchy: the working engineer level, the midlevel program managers, and the senior executive/flag officer level. These lateral connections in the form of working groups (WGs), integrated product teams (IPTs), and peer review groups help establish trust among participants, which removes barriers to problem solving while also speeding up information flow within the enterprise.

KEYWORDS

Acquisition Reform, APB, A-RCI, Architecture, COTS, Enterprise, Enterprise Architecture, Flexibility, Hierarchy, Innovation, Laterality, Layering, Organizational Design, Organizational Structure, Sociotechnical System, Process, Transformation

ABSTRACT

The US defense acquisition system is notoriously resistant to fundamental reform and improvement. This case study chronicles the genesis and evolution of fundamental change in the technical architecture of the Navy's submarine sonar system and its acquisition enterprise. Many studies and recommendations for acquisition reform focus on acquisition process and incentive structures; few focus on interdependencies among technical system architecture and enterprise architecture. Rather than focus on process, this case study focuses on the inter- and intraorganizational relationships, which were crafted by the key actors in the system, the larger-scale social and bureaucratic dynamics that enabled these actors to succeed, and on interactions with the technical system architecture. We find that the genesis of the change was similar to findings from many innovation and change case studies of organizational crises. We find that the technical architecture changed from tightly integrated hardware/software architecture to a layered architecture where hardware and software were decoupled by middleware. This enabled hardware upgrades and software upgrades to occur separately and supported a spiral development process, which was able to match system upgrades to commercial technology rates of change. We also find that lateral enterprise architecture, where different organizations were connected at different hierarchical levels, was critical to success. This enabled fleet operators to provide unfiltered feedback to development engineers, midlevel managers to coordinate priorities and key decisions, and senior leadership to provide consistent strategic guidance to the system.

GLOSSARY

AN/BSY-1 and BSY-2	Integrated Submarine Combat Systems deployed after the cancellation of the SUBACS system
AN/BQQ-10	Current version of Navy submarine sonar
APB	Advanced Processor Build
A-RCI	Acoustic-Rapid COTS Insertion
ASW	Antisubmarine Warfare
CNC	Computer Numerically Controlled

CNO	Chief of Naval Operations
COTS	Commercial Off-the-Shelf
DAPA	Defense Acquisition Performance Assessment
FFRDC	Federally Funded Research and Development Center
JTRS	Joint Tactical Radio System
NAVSEA	Naval Sea Systems Command
NR	Director of Naval Reactors; NAVSEA Code 08
NUSC	Naval Undersea Systems Center
NUWC	Naval Undersea Warfare Center
OPNAV	Chief of Naval Operations Staff
SSMC	Submarine Superiority Management Council
SSTP	Submarine Superiority Technology Panel
STRG	Submarine Tactical Requirements Group
TI	Technology Insertion
WSARA	Weapon Systems Acquisition Reform Act

BACKGROUND

Context

Loss of Acoustic Superiority in the Early 1990s

Since the early days of the Cold War, the US Navy had been locked in a struggle with the Soviet Navy for the control of the seas. A major component of this struggle was for acoustic dominance in the undersea domain—the ability to detect and track the opposition's submarines. This undersea competition consisted of a constant evolution of both operations and technologies to detect acoustic emissions and to minimize acoustic signatures—to gain *acoustic advantage*. Owen Cote chronicles this evolution in *The Third Battle: Innovation in the U.S. Navy's Silent Cold War Struggle with Soviet Submarines* (Cote, 2003). Through the first four decades of this competition, the United States generally had the upper hand, but by the late 1980s and early 1990s, that advantage was significantly eroded.

Undersea, submarine versus submarine competition is generally driven by acoustic advantage, the ability to hear the other guy before he hears you. When acoustic advantage is lost, submarine battles become duels, with outcomes generally depending upon chance. Through a long-standing effort, which consisted of a combination of espionage, research and development, and acquisition of embargoed technology, Soviet submarines gradually became quieter over the 1970s and 1980s. The combination of the Walker Spy Ring and the acquisition of computer numerically controlled (CNC) milling machines from Toshiba Corporation of Japan and Kongsberg in Norway enabled the Soviets to close the acoustic gap. The Walkers provided the Soviets knowledge of their acoustic vulnerabilities. The CNC milling machines allowed them to reduce the signature produced by submarine propellers, making it much harder for US submarines to find and track them (Cote, 2003; Friedman, n.d.; Parrish, 2004). This problem became obvious and acute in the early 1990s after two notably public submarine collisions (Associated Press Wire Service Report, 1993; Gushman, 1992).

Throughout the 1980s, as part of the undersea competition, the Navy had invested billions of dollars in acoustic research and development and advanced sonar system development, concentrating effort on custom-designed digital signal processing systems, with military-unique components and proprietary, highly (tightly) integrated hardware and software systems. Because of their complexity, these systems took a decade or more from conception to fielding, requiring billions of dollars with unit costs ranging in hundreds of millions. In 1986, the Submarine Advanced Combat System (SUBACS) was canceled and its developmental technologies restructured into two separate system development programs AN/BSY-1 and AN/BSY-2. The AN/BSY-1 was to be fielded on the latest submarines of the 688I class and the AN/BSY-2 was slated for installation on the new Seawolf submarine class. These systems represented the classic defense procurement scheme for submarine tactical systems at the time. Development specifications were done at the Navy Laboratory in New England (Naval Undersea Systems Center [NUSC], New London, and later in Naval Undersea Warfare Center [NUWC], Newport). The hardware and software detailed design and production were conducted at one of two prime contractors, either IBM Federal Systems (later Loral Space Systems, then Lockheed Martin-Manassas) or General Electric-Syracuse (later Lockheed Martin Syracuse).

The loss of acoustic superiority created great pressure to quickly improve signal processing performance of US submarine sonar systems to regain the acoustic (and hence, operational) advantage. Unfortunately, this pressure coincided with the end of the Cold War, which resulted in steep budget cuts, particularly to the antisubmarine warfare research and development program. With the constrained resources of the post–Cold War defense budget, the increased quieting of Russian submarines, worldwide proliferation of even quieter diesel submarines, and even questioning inside the Navy of the relevance of submarines in the *new world order*, the Submarine Force was in a difficult position: the operational problem was acute—the National Command Authority (President) was not allowing them to conduct their primary mission, some parts of the Navy were questioning their operational relevance, there was no money to fix the problem, and, even if funding existed, the development timeline was too long. This represented an organizational crisis, which ended up spurring what is arguably the most successful acquisition reform effort in US defense industry history.

Response: Gather Data and Evidence

The response to this technical and fiscal challenge by the submarine force was to examine the problem from a fact-based perspective. In early 1995, the senior leadership of the submarine force commissioned a four-member panel of acoustics experts to examine the technical performance of the submarine sonar system. Named the Submarine Superiority Technology Panel (SSTP), it generated a long list of recommendations, but there were two major conclusions: (1) the legacy sonar system architecture was ill-suited to keep up with changing technologies and threats: there was no way to leverage the power of Moore's Law to bring more computational power to bear on the problem; and (2) the signal processing architecture perspective inhibited experimentation with new algorithms. The panel observed that there was no viable means to test and evaluate and then integrate the plethora of advanced algorithms that had been developed by academic researchers using Navy acoustic research and development funding. Their overarching recommendation was to move toward a commercial off-the-shelf (COTS) processing architecture and toward a more *open* development process.

The recommendations of the SSTP generated 13 WGs, collectively called the Submarine Superiority Management Council (SSMC). Each WG was focused on a separate recommendation of the SSTP. The SSMC was established in the fall of 1995 and worked through the spring of 1996, by which time, the core problems of boosting processing power and developing a means to inject new algorithmic ideas into the sonar system were the main ones left requiring significant ongoing action. It is these problems that are the subject of this case study; these issues are (or were) at the core of the need to modify the sonar system technical architecture and to change the process by which the acquisition enterprise architecture designed, engineered, and fielded submarine sonar systems.

THEORETICAL VERSUS PRACTICAL INNOVATION

Case Study Framework

Our study consists of a narrative analysis based on interviews with key stakeholders involved with virtually all aspects of the technical and organizational evolution. We present a chronological narrative from both the technical and organizational perspectives. Our interviews form the basis of our analysis of both the architectural insights and the evaluation of the mechanisms for success in the transformation (Eisenhardt, 1998; Yin, 2003).

The case study emphasizes the structural/architectural relationships among the significant parts of both systems—technical and organizational. A key observation that arises out of the analysis is that interactions and battles among the elites in the organization, at multiple levels of organizational hierarchy, emerge as a key factor in the outcome—in both the dynamics and ultimate success of the change.

Existing Practices

The US Defense Acquisition System has become a *poster child* for arcane, bureaucratic, and often senseless system development processes. Despite several decades of reform effort, from the Packard Commission of 1986 to the Weinberger–Carlucci reforms that followed, to the recent Defense Acquisition Performance Assessment (DAPA) report and the Weapon System Acquisition Reform Act of 2009 (WSARA), the fundamental problems of defense acquisition have proven highly resistant to correction.* Large, complex, defense systems such as the Joint Strike Fighter, satellite programs, and Joint Tactical Radio System (JTRS) routinely fail to meet cost, schedule, and performance goals (Fox et al., 2011).

From this perspective, the changes to submarine sonar system acquisition are a significant anomaly. Because of the changed development process, resulting layered (modular) system architecture, and lateral enterprise architecture, development time for new software updates to the sonar system dropped by a factor of 5; development, production, and operating and support (O&S) costs dropped by a factor of 6–8; and ship set costs (the cost of the hardware installed on a single ship) dropped by a factor of 10. Initially, processing power increased by a factor of 13–80.[†] A system that used to cost hundreds of millions of dollars per unit and take upward of a decade to develop was now upgraded on a biannual basis for 1/10th the cost. How did this happen? We have already discussed the fundamental driving forces that catalyzed the change effort, but what was it that actually made the change happen? What made it work and why is it so hard to replicate?

Guiding Principles

The leaders, managers, engineers, and operators who initiated, oversaw, guided, and implemented the changes discussed in this case were not at all cognizant of the issues surrounding *Complex Systems Engineering* or *Complexity Theory*. They were, quite simply, involved in solving a problem that threatened the institution and were, to some extent, correcting a problem that they found professionally embarrassing. As discussed with the principal actors over several interviews, their guiding principles were trust, information transparency, and system-level incentives.

That said, they utilized several tools that have been identified as aspects of complex systems. First (and probably foremost), there were uncooperative stakeholders. They leveraged feedback processes to gain better information on existing problems and on the efficacy of potential solutions through the development process. They leveraged some aspects of self-organization, though our analysis does not support the nonhierarchical, bottom-up form of self-organization discussed by many complexity theorists. Potential solutions, the WG organization, the development process, as well as the objective for the process (the system requirements) all coevolved over time. The process was inherently feedback oriented and *spiral* in nature; and information flowing in, through, and around the process and enterprise impacted all aspects of it, causing it to change and evolve, essentially continuously. This was and continues to be a major source of friction between organizations involved in the Advanced Processor Build (APB)/Acoustic-Rapid COTS Insertion (A-RCI) process and those that are charged with oversight and funding of it. System solutions (new algorithms, applications) were adjudicated based on how well they solved a given problem—how *fit* they were. Sometimes, a solution was not good enough and had to go back for further development. The fitness landscape for the system overall was continuously in flux, based on changing operational needs and threats. Solutions had to meet a test of fitness in order to be passed through to system integration.

---

* Reasons for this resistance are complex and abundant, and a detailed discussion is beyond the scope of this study.
[†] Depending on which legacy sonar system is used for comparison.

CHARACTERIZATIONS

Type of System: Enterprise-Technical System

We classify the subject of this case study as an *enterprise-technical system.* The submarine sonar acquisition program consists of a single technical system with multiple subsystems (different sonar arrays, processing and display equipment, signal processing software, and applications and interfaces with other tactical systems). These subsystems are designed, built, delivered, supported, repaired, funded, and operated by a plethora of organizations across the industry and the government. The customers are submarine commanding officers and their crews. Funding is provided by the submarine program sponsor on the staff of the Chief of Naval Operations in the Pentagon (OPNAV). The Naval Sea Systems Command (NAVSEA) is the system program manager, the engineering and acquisition focal point for system development and fielding. Large commercial vendors (prime contractors), Navy laboratories, university research laboratories, and small businesses from across the country participate in the system design, development, and fielding.

System Maturity

The sonar system architecture is mature; the basic block diagram of a sonar system has not changed since the early 1960s. In this case, the *legacy* architecture of the highly integrated hardware and software processing system was inflexible—too expensive in terms of time, money, and engineering effort to upgrade within the bounds of available time and fiscal resources. The fundamental challenge was how to manage the evolution of both the technical system and the acquisition program to deliver both higher system performance and lower system cost on a timescale that could pace both the rate of change of technology and the rate of change of the operational threat with the urgency desired by the submarine force senior flag officer community.

Environment

The enterprise environment spans the range of contexts shown in Figure 20.1. Some aspects of the system change were very straightforward—the technical problem was well bounded, performance was measurable, the technical and operational environment, while changing, was not so unstable or dynamic that it was beyond the capability of the development process. Some aspects of the problem

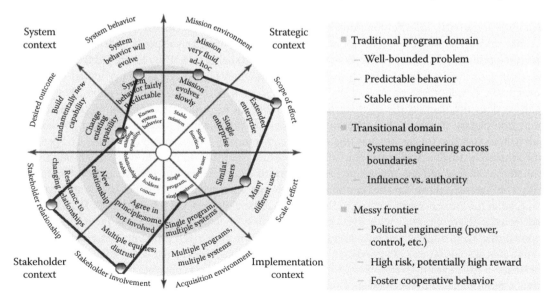

FIGURE 20.1 Enterprise systems engineering (ESE) profiler. (Courtesy of MITRE Corporation Bedford, Massachusetts.)

were *transitional*—the changed technical architecture and changed development process created the need to conduct systems engineering and integration across traditional organizational boundaries. The debate over the development process and the reallocation of fiscal resources away from the *legacy* prime contractor and government laboratory generated a battle over decision control in the system—who would make technical and resource allocation decisions? Some aspects of the process development and *rearchitecting* of the system acquisition architecture created problems at the *messy frontier.* There was considerable effort expended by some organizations in battling for power over the process. The program managers had to battle for this control and had to work continuously to build cooperative teams with people from diverse organizations across both large and small organizations, and across industry, government, and academia. A mapping of this case to the Enterprise Systems Engineering Profiler (Stevens, 2008) is shown in Figure 20.1.

Systems Engineering Activities

Systems engineering activities that were used in the evolution of the new sonar system architecture were (and remain) fairly standard. Technical rigor was mandated, performance was monitored, and constraints were met. The main point of interest was that the *classic* or *traditional* systems engineering *V* was modified to operate on a much shorter timescale to accommodate a *good enough* requirements definition, and the reprioritization of the delivery schedule over delivery of a complete list of system requirements.

This is not to say that system deliveries were substandard or not operationally viable. As the architecture of the sonar system moved toward a modular form, capabilities (e.g., processing algorithms, display alert functions) that were not ready on the schedule mandated by the development process were placed back in the development queue for integration during a later development cycle. The ability to bring in new development ideas and to recycle developments that needed more work is the hallmark of this SE process. The ability of the process to manage uncertainity in systems engineering activities while retaining technical rigor rather than vice versa, and the ability of the defense acquisition system to accommodate this type of uncertainty is unique.

As Is–To Be System Description

The evolution of the A-RCI/APB system, which came to be designated the AN/BQQ-10 sonar system, began with the *legacy* AN/BQQ-5 series sonar system. The BQQ-5 system was the Navy's first digital sonar system, designed in the late 1970s, and iterated through several upgrades over the decade of the 1980s. Its fundamental system architecture can be described as *fully integrated.* The hardware was custom-designed by the government laboratory and largely built to specifications by a prime vendor. Software and hardware were tightly integrated—it was difficult to upgrade system hardware (computer signal processing) without completely rewriting much of the software code. This architecture is reminiscent of the classic mainframe computer architectures of the late 1950s and early 1960s, before the IBM System/360, with its modular architecture, was introduced. A high-level diagram of this architecture is shown in Figure 20.2. In addition to the integrated hardware/software architecture, the sensors and displays were relatively tightly integrated as well. There was little technical and operational flexibility in the system. That is, only certain displays could present information from certain sensors, and to some extent, each sensor had unique processing associated with it.

As we noted earlier, the system change evolved over time, as success of the development and acquisition model was proved with at-sea performance. There was no explicit *to be* system architecture, but rather a thoughtful vision for iterative change that had, as a significant architectural component, the goal of separation of hardware and software interdependency via the use of middleware. Middleware is a set of software components aimed at providing a stable interface between hardware and software, so that one can be changed without excessive effort (and cost) in order to change the other. The BQQ-5 system was iteratively evolved over the course of several development cycles to encompass all of the sonar sensors on the ship (towed array, hull array, spherical

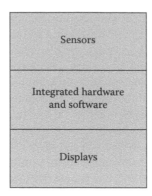

FIGURE 20.2   Legacy sonar system architecture.

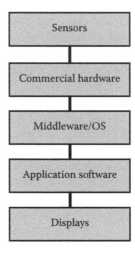

FIGURE 20.3   Layered sonar system architecture.

array, high-frequency array, and finally self-noise array) using commercial processors and displays, interfacing new signal processing algorithms with the commercial and legacy hardware via middleware. A representative diagram is shown in Figure 20.3. To guide the evolution, the Program Management Office (PMS-425) detailed upcoming system changes in its Acoustic Program Plan, updated semiannually. It is important to note that the sequencing of sonar array migration to the COTS architecture was guided by the direction given by the submarine fleet's senior leadership.

PURPOSE

SITUATION AND SOLUTION

The genesis of the change to the technical and enterprise architecture for submarine sonar systems has deep roots. The catalysts for action we discussed in the first section provided the impetus for change, but the ultimate dynamic was driven by the interplay of personal relationships, organizational culture, political dynamics, and resource constraints. The entanglement in these areas, the strands or threads that make up the fabric of the case demonstrate that focusing on a single, isolated factor as explanation is fraught with peril. Our discussion will describe, based on first-person interviews and primary source documents (where available), the evolution to the new process. Then we will abstract to glean what we assess are the high-level insights relevant to senior-level program managers with a responsibility for similarly complex acquisition programs.

The mechanisms for change in the submarine sonar system have deep roots. The sonar system program management community on both the government and commercial sides is a close-knit community, focused on delivering highly complex, operationally relevant systems to the US Navy's fleet operators. In the case of the uniformed officers, there were three program managers with cognizance over submarine sonar systems. The director of the Advanced Systems Technology Office (ASTO) was responsible for advanced development of algorithms and prototype systems. His job was to take advanced ideas from the academic community and bring them to a level of maturity where they could be tested at sea and transitioned to one of the main program offices. The sonar systems program office (PMS-425) at the time of the crisis was responsible for all fielded submarine sonar systems and for programs that were aimed at upgrading their capabilities. Then there was the Virginia Class Submarine Program Office, which was responsible for the newest class of submarine and all of the systems that were to be installed on it; one of those systems was the sonar system.

As the operational crisis developed, the uniformed program managers in all three of these offices were close colleagues. In fact, they had worked together in an advanced development program office several years before. While in this office, not only did they learn how to work together as individuals but they also learned how to connect advanced development scientists and engineers to operators—how to speed up the fielding of advanced systems. In addition, one of the key civilian engineering managers, the deputy program manager in the sonar systems program office, had been encouraged to investigate the use of COTS equipment in submarine sonar systems by his mentor, another Navy civilian engineer. This engineer, who was also a former program manager, had implemented COTS in another sonar system several years earlier. This program, in a prelude to the circumstances of the submarine sonar system, had undergone severe budget cuts in the late 1980s, as the contraction of the defense budget began toward the end of the Cold War.

It is also important to understand that the sonar program office had been trying to correct the performance shortfall before the SSTP was commissioned and announced its diagnosis and prescription for the problem. There were several candidate solutions under consideration to improve the processing performance of the sonar system, but none were ready for integration into existing sonar systems; either they were still at the prototype stage or integration costs were too expensive. These solutions all relied, to some extent, on using COTS processors to boost performance. Another factor that ultimately helped improve sonar system performance quickly was the large base of fundamental acoustic research that existed *on the shelf*. This research, commissioned during the robustly funded defense budget environment of the 1980s, consisted of advanced algorithms and processing ideas, which were demonstrated in laboratory experiments or theoretical calculations, but not implemented because the sonar systems lacked the requisite processing power.

Another factor supporting the change—another strand that came together—was the Small Business Innovative Research (SBIR) program. The SBIR program was established by Congress in the early 1980s, aimed at encouraging the United States Department of Defense (DoD) to invest in innovative ideas from small businesses. In this case, it was an SBIR effort by a small business named Digital System Resources (DSR) that provided a less expensive solution to upgrade the sonar system. Called the multipurpose processor (MPP), it was first conceived by DSR's founder and sponsored by the Virginia Program Office in 1992. The MPP SBIR effort was aimed at hosting a Navy sonar algorithm on a commercial processor. Successful demonstration of this capability by DSR allowed it to eventually be considered as a solution to the sonar performance problem.

All these threads were in existence and provided the context and options for the choices and actions of the Navy program managers and senior flag leadership as they moved forward with the SSTP recommendation in the SSMC WGs and beyond.

KNOWN PROBLEM(S)

The end of the Cold War saw dramatic reductions in the US defense budget. Coupled with the seeming loss of the primary purpose and ongoing search for a new military and Navy strategy, the performance problems of the submarine sonar systems became a significant threat to bureaucratic viability. The newest submarine class, Seawolf, had been truncated at three ships, as research and development funding had been slashed by at least an order of magnitude. Coupled with that, development timelines were long, measured in years for upgrades to existing systems and decades for new development systems. Though there were plans and technical solutions to upgrade existing submarine sonar system performance through engineering changes, they were too expensive and would take too long to install. A different approach had to be taken.

MISSION AND DESIRED OR EXPECTED CAPABILITIES

The problem was very straightforward, as noted earlier: to improve signal processing performance so that safe submarine operations could be conducted. As the studies recommended by the SSTP were conducted by the SSMC WGs, Sonar System program managers (who were also participants in the SSMC groups) began to work out an acquisition solution. They devised a process that would connect leading-edge academic and government laboratory acousticians with development engineers and fleet sonar operators. Their goal was to upgrade signal processing performance quickly, by cutting out layers of bureaucracy. This would solve the timeliness challenge. The technical solution came from the SBIR program. The developing prototype commercial signal processing systems would be used to display data from existing ship-towed array sonars. The goal was to have an at-sea demonstration within 18 months.

PROCESS FOR ACHIEVING THE OBJECTIVES NEEDED AND WHY

The concept of connecting algorithm developers, engineers, and operators required a disciplined, technically rigorous process. This process came to be called the APB process. Because complexity and technical difficulty were combined with urgency, the development process had to balance speed and technical excellence. This was achieved by creating multiple WGs, each concerned with a distinct aspect of the sonar performance problem. For signal processing specifically, the development process involved four steps (eventually named the *four-step process*). First, new ideas for signal processing were presented and vetted by experts in the WG. Ideas deemed worthwhile (technically viable and relevant to the problem) were funded for initial development of the algorithm and tested with at-sea data. Once the initial performance of the algorithm was judged sufficiently robust, it was migrated to the sonar system operating environment and tested with more challenging at-sea data—data that the developers had not seen before. This was the means chosen to ensure that algorithms were robust to variations in sonar and environmental signals.

COMPLEX ENTERPRISE TECHNICAL SYSTEM

ENVIRONMENT

The classic approach to changing complex military systems is the *as is/to be* paradigm. In this approach, the system as it exists is defined (as is). Then, its shortcomings are identified with corrections that fix them. The result is the *to-be* system architecture. Then a plan is developed to move the system from one state to the other. Unfortunately, these efforts rarely succeed. The reasons are numerous, from rapid change in underlying technology to insertion (or modification) of stakeholder interests over the development time, which leads to changes to the *to-be* system. Also contributing is the overall *top-down* approach to change, which is unable to identify, address, or control the innumerable microscale decisions that influence how the overall system develops in the face of uncertainty.

In practice, systems and organizations rarely move from an *as-is* to a *to-be* state. They change iteratively over time—they evolve. Even when a new program is begun, it normally starts with an existing system or design—an existing architecture. Even when beginning with a prototype design, a new program will change that design in a myriad of ways that are unpredictable.

Various stakeholders will act to influence design decisions; nontechnical factors such as budget turbulence, changing service priorities and preferences, and even the individual preferences of a program manager all drive technical choices.

This case study highlights to some extent the influence of key stakeholders, the tussles between stakeholders for access to and control over key decision authorities. It will also highlight architectural attributes that mitigated those influences to bring more technical objectivity to decisions and simultaneously shorten feedback loops from system operators (the fleet) to the development scientists and engineers.

SCOPE: THE TRANSFORMATION CHALLENGE

The case study involved a single system, with multiple subsystems, developed by a broad constituency of government, commercial, and academic organizations. Engineers, researchers, managers, and system operators were all involved. The problem was well known and understood, its fundamental nature and the general form of the technical solution were accepted by virtually all stakeholders. The end state was also agreed upon and, in large part, measurable with real data. The fundamental challenge for the technical and program management experts was creating a process that would deliver the required technical performance while giving the Navy customer (the submarine community's flag officers) an increased degree of visibility and control over technical and resource allocation decisions.

The scope of the technical system was initially limited to the submarine sonar system. Within this system, the initial change was limited to the towed sonar array, because this is the array that provided the best near-term prospect of regaining acoustic advantage. Over time, the technical scope was expanded, eventually encompassing processing for all sonar arrays on the submarine, the submarine tactical control system, and (today) the imaging and electronic warfare systems. The system was gradually re-architected as the success of the COTS model was proven by better performance of the system at sea.

On the enterprise side, the scope was initially (and intentionally) very broad. The recommendations of the SSTP were to bring in and test new ideas for signal processing. The method chosen by the program managers of ASTO and the Acoustic Systems Program Office* was specifically aimed at increasing the speed with which system developers got feedback from at-sea operations as well as bringing in new ideas. So the organizational change crossed government, industry, and academia boundaries. Following the evolution of change in the technical system, WGs were established and disestablished as dictated by the operational demand on technical decision making.

STRUCTURE

As discussed briefly earlier, the enterprise structure was transformed from a top-down hierarchy, where most interactions were considered *arm's length*, to what we term a *lateral hierarchy*, where interactions across organizational boundaries were the norm, rather than the exception (see "C" in Figure 20.4). This is not to say that in the legacy enterprise architecture, people from different organizations never interacted with each other. On the contrary, there were government representatives at the contractor's site and vice versa. Small business and academia contributed to the sonar development process as well. But in the legacy model, each organization was charged with a specific phase of system development, and the government laboratory and the prime contractor were the major decision makers when it came to technical design and engineering.

The enterprise architecture that evolved out of the sonar development working groups (SDWG) was one where more information flowed across organizational boundaries than in the legacy enterprise.

---

* At the time of the transformation, the sonar systems program office was charged with all backfit sonar systems. Over the initial years of the program, for reasons unrelated to the sonar system, the office's name was changed and at times it had cognizance over other systems.

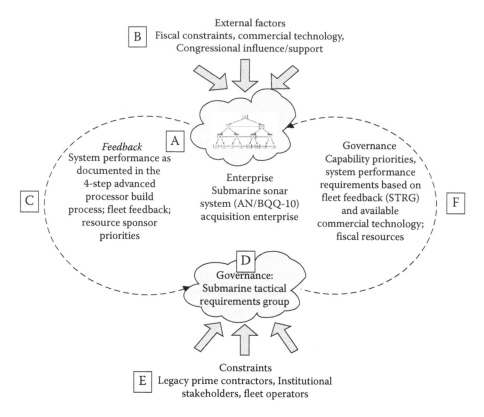

FIGURE 20.4   Context of the A-RCI/APB (A) enterprise.

In addition, technical decisions were largely made in these WGs and were reviewed by the chairs of the SDWG as well as independent peer review groups. So decisions were made at lower hierarchical levels and were informed more directly by information from multiple technical disciplines as well as fleet operators (Figure 20.4).

The cochairs of the SDWG acted as one of the *top-down* governing bodies for the WGs as shown in Figure 20.4. In addition to the SDWG, there remained the successor to the SSTP, now called the Technical Advisory Group (TAG), as well as the Submarine Tactical Requirements Group (STRG). These groups provided oversight and direction for both technologies under consideration for inclusion in future upgrades and for the operational problems that the A-RCI/APB process should be solving. So, though there was a decentralization of decision authority to lower hierarchical levels, top-down oversight and direction of the critical aspects of the process were retained ("D–E–C" in Figure 20.4). In short, there were strong feedback mechanisms in place to ensure that the best possible technical information was used to improve the systems and to keep the technologists focused on fleet-relevant technical solutions.

It should be noted that the process was not perfect. In our data gathering, some interviewees complained about favoritism in the process (especially in the later years, as the early participants moved out of the process) and that some of the technical implementations in the system were not addressing real fleet needs. But overall, based on the longevity of the process and the positive reception of the system by the fleet, we can say that the structure that evolved out of the IPT/WG process was effective.

BOUNDARIES

As discussed earlier, organizational boundaries became less important as the WG/IPT structure took hold and the cochairs of the SDWG held them accountable for developing solid technical solutions. Information flowed more freely across boundaries, and trust among participants grew. On the technical side, boundaries were much more important, and remain important to this day.* In the early days of the A-RCI/APB (A) process, the technical separation of the COTS portion of the system and the legacy portion was important. The whole system had to remain operationally useful, not just the newly modified portion. So technical boundaries were carefully observed, and decisions about which arrays to move inside the COTS boundary were based on operational need and the continued success of the operational performance of COTS-based systems. It may be observed, from a complex systems perspective, that this is a form of evolutionary change, where modifications are made, performance assessed, and the change kept or rejected. Though in this case, because of the integrity of the APB process, where changes were rigorously evaluated, it was rare that a change to the system was subsequently reversed.

At the higher level of the Defense Programming-Acquisition process, boundaries were sometimes not as porous as they were inside the development acquisition organizations. Some participants in the Pentagon sponsor for the sonar system, the office responsible for ensuring that the acquisition system conformed to budget and to the formal requirement process of DoD 5000, complained that the program managers were not addressing the formal requirements. This was largely because the new A-RCI/APB (A) process was responding to fleet input directly, as the APB process progressed. The formal requirements process operated at the speed of the Congressional Budgeting process and could not keep up. So there was (and likely remains) some frustration that the APB process is *out of control* and is not following the rules. We attribute this to the lack of visibility by the Pentagon into the inner workings of the APB process, and that the opaqueness is due to organizational boundaries.

INTERNAL RELATIONSHIPS

Internal relationships among organizations and WGs have been discussed in previous sections. In this section, we will address the issue of internal relationships among specific individuals who participated in the process. In the very early days of the development of the APB process, personal relationships among some of the government engineers and program managers were important for setting the conditions and engaging the dialogue that resulted in the process. These individuals formed a cohesive core of technical and managerial expertise and support as the change process evolved.

We also need to note the cohesiveness of the submarine force flag officers and senior Captains. This group was united in its goal of opening up both the sonar system to commercial processors and the development process to new, innovative ideas. This was a crucial factor, as individuals from the *legacy* organizations mounted a defense of their traditional areas of technical and managerial oversight. Specifically, the government laboratory and the prime contractor for the sonar system were removed from direct authority over many technical decisions as a result of the WG structure that was put in place. They mounted focused efforts to reverse the momentum of the change agents in the program offices. But the internal cohesion of the submarine force leadership behind its program managers, coupled with the early and continuing technical and operational success of the COTS-based system, was able to resist.

EXTERNAL FACTORS

One of the most significant characteristics of the enterprise in this case study is that it developed an architecture that is able to respond to external factors on a timescale at which they are relevant. The direct connection of fleet operators to developers and engineers provides unfiltered input regarding what is effective and important to them. The STRG letters, written semiannually, provide top-down

---

* Some would say that technical boundaries retain too much importance, a topic that will be discussed later in the case.

guidance on priorities, and these are fleet-driven.* At one level, these can be considered as external factors to the APB process, because they were directive in nature and served to constrain and to focus the effort of participants and of the process. At another level, they can be considered as internal factors—they represented internal guidance of the Submarine Force as to what to prioritize in the system that was to be delivered ("C" and "E" in Figure 20.4).

CONSTRAINTS

Constraints loom large in this case study ("G" in Figure 20.4). Without constraints, none of the innovations and transformations would have happened. It is arguable (and argued by some of the participants) that commercial processing would have been integrated into submarine systems eventually. While this is likely true, the shift in enterprise architecture and the inclusion of a wide array of participants and the more innovative technology solutions that were developed may not have occurred without the constraints.

The major constraints that drove the system change were fiscal and operational. These combined to create the urgency both to find a solution fast and to find a solution that would not cost millions or billions of dollars to implement. After the initial change, the need to keep up with the increasing computational power provided by Moore's Law created an embedded constraint of producing a hardware and a software upgrade every 2 years (for software, the upgrades were annual for the first several cycles).

CHALLENGES

ANTICIPATED

From the point of view of the leadership of the transformation effort, the challenges they anticipated were largely technical. They knew the technical solution; the challenge was going to be implementing it in a timely fashion. So, until the first at-sea demonstration, ensuring that the testing and integration was proceeding on schedule was the most important challenge. This was at the engineering and program management level. At the level of funding and senior-level support, the most important challenge was to maintain a steady funding stream for the A-RCI/APB (A) process. This challenge was made somewhat easier by the operational urgency of the problem. But it was also enabled by a serendipitous assignment of a particular flag officer, who ensured that the funding profile for the process was not made the object of perturbation. This admiral, Edmund Giambastiani, became the resource sponsor for the process right after it was established. Through three job assignments, he was in a position to ensure that his successors gave the A-RCI/APB (A) process the requisite funding to keep it from having to respond to fiscal uncertainty. This created a protected environment where the participants in the process were secure that their participation would not suddenly become subject to the vagaries of budget cuts.

The actual, largest, challenge was not *truly* anticipated by the change managers.

ACTUAL

The biggest and longest-lasting challenge turned out to be organizational. The degree of resistance by the legacy acquisition experts—the government laboratory and the prime contractor—was, according to the accounts that were given during our case study, huge. The major small business developer of the COTS solution saw relentless effort by the prime contractor to reverse programmatic decisions that arose as a result of the success of the COTS architecture. The government

---

* The STRG is chaired by the Commander of Submarine Development Squadron 12 and has membership from the senior post-command Captains in the submarine force both from Atlantic and Pacific Fleets and from the Acquisition Program Offices and the Pentagon.

program managers saw a relentless onslaught of attacks from the government laboratory. The work was so contentious that some of the engineers at the program office considered resigning; one of the deputy program managers said the change was so contentious that, "in my heart of hearts, I really thought we were only going to get one iteration of the development cycle."

DEVELOPMENT

PROJECT/PROGRAM MANAGEMENT

The managers of the A-RCI/APB (A) process, as it evolved, were establishing a management paradigm whereby planning was incremental on a 2-year cycle. Classical management tools were used—planning timelines, schedule updates, technical reviews, etc.—but the objectives and priorities of the process differed from a classical program management approach.

Planning

Because of the iterative nature of the software development process and the incremental upgrades necessary on the hardware procurement side, planning was conducted based on an annual upgrade tempo for software and biannual for hardware. Time was allotted for each step of the four-step development process, timed to coincide with established at-sea testing cycles. Hardware procurement was timed to provide enough time to establish the appropriate technical specifications for the next upgrade, vet alternative processor, memory, network switch selections, and then procure and conduct the minor alterations necessary for installation on board a submarine. The goal of the hardware selection and procurement process was to buy *state-of-the-practice* processing hardware, not the latest state of the art. The market would be used to work out bugs in leading-edge processing and bring down prices—the Navy would buy processors for commodity pricing, while they were in the mainstream of the mass market and before they became obsolete. Because of the routine upgrade to hardware across the fleet, this process of procuring computational power allowed the Navy to avoid excessive logistical and maintenance costs for processors that were no longer in the mainstream of the commercial market. It also resulted in reduced procurement costs; in the time between when hardware specifications were set and budget resources allocated and the processors actually purchased, the market costs dropped (recall the 18-month cycle of Moore's Law).

Budget, resource allocation, and roles (both of individuals and organizations) are the areas where the biggest changes were observed and where the most contentious arguments were to be found. Recall that the SSTP advised the submarine community to *open up* the system and the process to innovative ideas and to experiment with them, in addition to moving toward a COTS processing architecture. Along with this advice, the flag officers in charge of system procurement felt that they needed more visibility into the process, specifically in the area of selection of system integrators.

The four-step development process and the goal of bringing in more ideas and evaluating them inside an IPT/WG structure necessitated an increased role for working-level engineers and developers from small businesses, academic laboratories, and government laboratories. The modified, decentralized, technical decision-making structure also radically changed the roles of certain management positions, particularly those at the prime contractor and the government laboratory. These two organizations and the people who ran them were used to having a great deal of authority in technical and resource allocation for the development process. The instantiation of the APB process changed those roles, from both a bureaucratic power perspective and a resource allocation perspective.

In the new enterprise architecture, the larger players, the prime contractor, and the government laboratory saw not only reduced budget allocation power but also reductions in budget allocated to them. This was necessary because of the need to increase the role of small businesses and academia inside a fixed (and smaller) budget pie. Participants in the APB process were funded to bring ideas to the process, evaluated honestly and openly, and, if rejected, given the opportunity to improve them and resubmit them in a later cycle.

Schedule

Aside from the changed decision structure and budget allocation, the priority of the development schedule as the key driver in the process is probably the largest change from a more traditional systems engineering/program management paradigm. Meeting the deadlines to get the next iteration of APB and hardware upgrade installed and tested in the fleet became the paramount concern. If an algorithm was not up to the standards necessary for at-sea testing, it was rejected until the next development cycle, rather than delaying the cycle for further improvement. Because there were many aspects to an APB upgrade, it did not mean that if one algorithm or operator enhancement fell away, the upgrade was not an improvement. This prioritization of schedule over perfection is a key factor that characterized this type of spiral development process.

User/Operator Involvement

The other unique aspect of the A-RCI/APB (A) process is that it connects submarine fleet operators—sonar technicians—to the integration engineers and the algorithm developers. This form of direct feedback in both directions was a great morale booster. For the academic researchers, they were able to see the direct fruits of their labor. Their scientific work became more than just an abstract mathematical problem on a blackboard. They saw that it had tangible importance and value to the ultimate customer. For the sonar technicians, they were able to influence the tools that were delivered to them and saw tangible benefit in the improved operating capability of the system. In short, without layers of management review and information filtering between the fleet and the engineers/developers, technical quality and operational capability improved in ways not possible under the classic acquisition model.

Organization: Sonar Development Working Group(s)

The evolution of the organizational architecture for the A-RCI/APB program occurred over the course of about a year, as the efforts of the SSMC progressed and the various WGs were disbanded. The participation of the deputy program managers and program managers of ASTO and PMS-425 in these WGs, combined with their formal duties of sonar system development and production programs, placed them in a good position to migrate the SSMC effort into the formal acquisition process. Specifically, the computer processing working group (CPWG) and the research and development working group (RDWG) became the original WGs of the A-RCI/APB process.

The revised process for advanced development and production of submarine sonar systems evolved in conjunction with the work of the SSMC WGs over the course of late 1995 and early 1996. As the SSMC worked on the SSTP recommendations, sonar program managers incorporated the relevant WGs into its evolving process for vetting new signal processing algorithms and transitioning them to fielded systems.

As actions were taken to develop a new process, there was predictable resistance to the emerging changes. As one might expect, the organizations that held the major power in the old, legacy, development and production process wanted to retain this power. The fundamental choice was to move the locus of decision making on technical matters to the program office or to keep it with the government laboratory and prime contractor. In the face of this resistance, the flag officers in the submarine community remained aligned behind the change and the change agents who were leading the new process.

Following the SSMC WG format, the program managers implemented the four-step process using a similar structure. They created multiple WGs and peer review groups (for technical subject matters). The breakdown of these groups is shown in Figure 20.5. The operating rules for each of these groups were that they were to collaborate on solving their aspect of the sonar performance problem. As can be seen in the diagram, the peer review groups were connected to the execution groups.

Peer review is cited as the critical aspect of the WG-based process. Multiple participants in the process as well as the program management cited this factor as the key to success. The impact of peer review was to increase group objectivity in the assessment of technical solutions. Participants

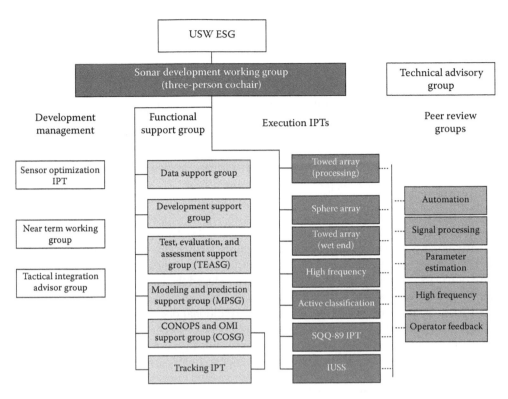

FIGURE 20.5 A-RCI/APB working groups. (From Navy Department, Sonar development working group annual report, 1999–2000, Naval Sea Systems Command, Unpublished, Washington, DC, October 31, 2000.)

were among *peers*, and their professionalism and professional reputation became the primary rationale for individual decision making. This increased the objectivity of the overall group decisions, because organizational interests of the participants were less important than their own professional reputations.

A second impact that the WG structure had was to connect the multiple participating organizations at the working level, where expert knowledge could be exchanged. This enabled increased information flow, unfiltered by the influence of hierarchical levels of management. The increased information flow was also a factor that made feedback in the process faster and more responsive. The IPTs and WGs acted to connect university and government laboratory scientists with development engineers and fleet operators. This connection had two impacts. First, it resulted in higher fidelity information exchanges and faster feedback between operators and the researchers/developers. Second, it served as a motivator for the academic researchers—their product (usually a signal processing or automation algorithm) was no longer some abstract software construct, but a product with value to a customer.* Also, leadership of the SDWG (a description that is used for the whole organization as well as the oversight function) was shared across three organizations, with rotating responsibility for management (cochairmanships). This connected program management across government stakeholders resulted in increased responsiveness and better alignment of interests.

Systems Engineering

The A-RCI/APB (A) process is not a *classic* systems engineering process. It is iterative, incremental, and spiral in nature. This creates disconnects with budgeting, programming, and requirement generation processes that are designed to support and mate with traditional systems engineering methods.

---

* Several participants and Working Group leaders emphasized this point during interviews.

Despite this deviation from what many would feel is good practice, technical rigor is not sacrificed. What is sacrificed is certainty in delivered capability.

Architecture

Beyond the insertion of middleware as a means of creating the ability to independently upgrade hardware and software, the architecture of the submarine sonar system developed by this process is unchanged from traditional sonar architectures. In fact, many aspects of the software (specifically the signal processing software) architecture remained unchanged.

The major benefits of the A-RCI/APB (A) approach result in challenges at the interfaces of other areas of the traditional systems engineering, testing, fielding, and sustainment methods and processes. From a logistics support perspective, the use of COTS equipment and an upgrade cycle of every 2 years enabled the submarine force to forego the stockpiling of 20 years' worth of spare parts. The upgrade and procurement cycle was such that, as equipment aged, new processors were installed rather than old replacements. As the process has evolved, it has become possible to remove early processors and use them (reinstall them) in submarines needing repair, as ships are upgraded. The Navy has calculated that up to $25 million in cost avoidance savings were generated by the use of COTS processing and other methods that reduced the logistics burden.

One of the most contentious (and ongoing) areas of systems engineering activity created by the A-RCI/APB (A) process is in the area of testing. Traditional operational test and evaluation of Navy systems is governed by comprehensive rules and policies to ensure that systems placed in a warfighter's hands will operate as advertised and expected. In the traditional acquisition model, where an entire system is specified and then developed, and engineering tested and then fielded, operational testing involves running a system (and the ship and crew) through its paces in virtually all conceivable warfighting situations. With an incremental development process, the entire system is not being replaced, but merely upgraded, so many of the operational testing rules may not apply.

This created a conundrum. The fleet was generally happy with the product delivered by the APB process. But by the letter of the law, a full operational test was required. With a fleet that is stressed operationally, there are often not enough assets (ships, targets, range time) to conduct a full test of the system for each APB. So often, the tests are truncated short of completion of all scenarios. This is an ongoing problem, where some in the testing community do not think the program managers are testing their systems adequately. Meanwhile, the force continues to operate and work through any technical problems.*

Four-Step Advanced Processor Build (APB) Process

As we have described earlier, the APB process originated as the result of a complex set of interactions between individuals and organizations. The overall goal was to deliver rapid performance improvements within the tight fiscal constraints of the post–Cold War defense budget (Navy Department, 1999). The development process consisted of two stages. The first was a technology insertion step, where the next update to hardware was chosen. The second stage was the APB. This process was an incremental build–test–build iterative process. The goal was to be able to bring in new ideas from academia, small business, and government laboratories and test them under increasingly strenuous performance conditions. If an algorithm failed, it could be resubmitted during a later APB cycle after further improvements. This process occurred in four steps and has come to be called the *four-step process*, depicted in Figure 20.6.

As shown, the process starts with review and selection based on a technical white paper. Selected white papers are then developed in the laboratory with initial testing done on selected data sets from at-sea operations. Transition to Step 3 testing is done based on performance with these data.

---

* *Note*: A full discussion of the detailed requirements and the differences of opinion regarding operational test and evaluation are beyond the scope of this case study. It is worth noting, however, that, in general, as systems become more information-intensive and are more likely to be managed in ways similar to the sonar system in this case, this problem will become more acute and more pervasive across the DoD.

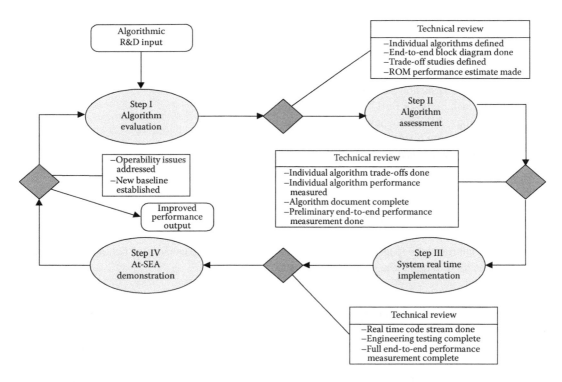

FIGURE 20.6 Four-step APB development process. (From Navy Department, Sonar development working group annual report, 1999–2000, Naval Sea Systems Command, Unpublished, Washington, DC, October 31, 2000.)

In Step 3 testing, the candidate algorithm is integrated to a full processor string, from sensor to display. Again, this step is performed with at-sea acoustic data. In Step 4, the algorithm is tested at sea in a fully integrated sonar system.

Technical System

As we described earlier, the technical system architecture was changed from a tightly integrated hardware–software system to a layered system, where the hardware and software were separated by middleware. Formally called multipurpose transportable middleware (MTM), it was originally used in a test to demonstrate the ability to host one of the Navy's sonar algorithms on commercial processing. This test proved successful and provided the foundation for the ongoing evolution of the system to a COTS processing model.

The development plan was to iteratively move the system boundary between the new COTS architecture and the legacy system one sonar sensor at a time. This strategy minimized risk by limiting the number of changes in each hardware and software update. The net result was a *spiral* development at two levels: there were a spiral at the level of the system transformation and a spiral at the level of improved capability for the system overall.

MANAGEMENT OF UNCERTAINTY

Philosophy

The overarching philosophy of the program managers toward the A-RCI/APB (A) process was that schedule had ultimate priority—the key was to deliver the best product on the schedule promised to the fleet. If something was not good enough, it would be worked into the next upgrade. The ability to achieve this goal was predicated on organizational speed, which required robust information flow (transparency) and trust among participants.

Policy

The program managers developed a set of axioms to guide the participants in and managers of the program:

1. Rapid COTS Insertion Means Just That.
2. Deliver Each Sensor's Full Theoretical Gain to the Operator.
3. All Bearings, All Frequencies, All the Time.
4. Avoid Modifying Successful Commercial Products.
5. Use the Lessons Learned.
6. Use State of the Practice, not State of the Art; Tactical Sonar Systems are not a Beta Test Site.
7. Configuration Management, vice Configuration Control.
8. Software Reuse Is Key to Affordability!
9. No One Organization Has the Full Story.

These axioms captured the technical, organizational, programmatic essence of behaviors and decision criteria needed to make the program successful.

Politics

Politics, both Congressional and organizational, played a large role in the success and near failure of the A-RCI/APB (A) program. Congressional support of small businesses quite literally provided many new innovative ideas that became the foundation of the program. Once the program was in place, the ability and willingness of the submarine force flag officers to defend it to Congress was essential in maintaining fiscal support. The support of the flag officers for the program managers who were under fire by the legacy organizations was also a crucial factor in maintaining momentum as the program grew and expanded across the multiple sonar arrays in the submarine.

RESULTS

Objectives Accomplished and Not Accomplished

By all accounts, the A-RCI/APB (A) process has successfully achieved its goal. Documented sonar system performance at sea has improved dramatically, as evidenced in both exercises and deployed missions. In a formal study of the cost of the ARCI program, ASSETT, Inc. documented the cost savings (discussed later). Perhaps the ultimate testimony to success is that, in 2001, the submarine Program Executive Officer expanded the COTS architecture and the APB process model to include the submarine tactical control system. Since 2001, there have been continuous upgrades to both the sonar and tactical system and, in recent years, the submarine imaging systems and the electronic warfare systems have been migrated to the model and incorporated into the overall COTS system architecture, now called Submarine Warfare Federated Tactical Systems (SWFTS).

Final System Description

The AN/BQQ-10 sonar system continued to evolve following the period covered in this case study. An overall diagram of how the evolution progressed is shown in Figure 20.7.

As noted in several briefings given by the program offices involved in A-RCI/APB (A) during that time, processing power increased, and procurement cost decreased. Specific numbers are given in Table 20.1.

Key Features and Benefits of the New Process and Architecture

The revised acquisition process and the altered technical architecture created several benefits. First, the decentralized decision making on technical issues, with approval by the program manager, was able to speed up the development process. The new structure also provided more direct feedback

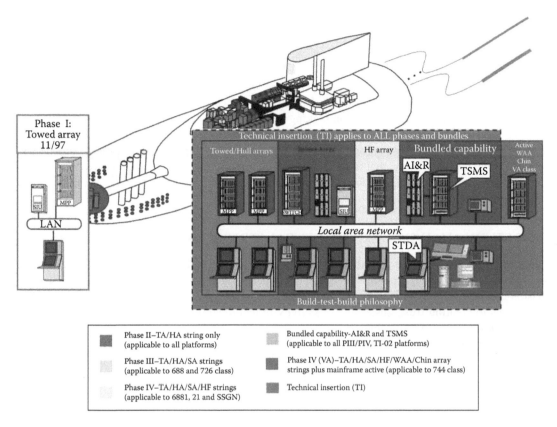

FIGURE 20.7 A-RCI system depiction. (From Kerr, G., Submarine acoustic systems and sensors, Briefing to the Deputy Chief of Naval Operations for Warfighting Requirements and Programs (N6/N7), January 31, 2005.)

TABLE 20.1
Submarine Sonar Cost–Performance Comparison

	SSN-688 (AN/BQQ-5E)	SSN-688I (AN/BSY-1)	SSN-21 (AN/BSY-2)	ARCI	NSSN
Development cost ($M FY98)	300	850	1250	100	150
Shipset cost ($M FY98)	85	100	190	5	35
Signal processing capability (MFLOPS)	498	1422	2920	39,100	76,800
Normalized values					
Development cost ($M FY98)	3.0	8.5	12.5	1.0	1.5
Shipset cost ($M FY98)	17.0	20.0	38.0	1.0	7.0
Signal processing capability (MFLOPS)	0.01274	0.03637	0.07468	1.0	2.0

from fleet operations, so that the system development was more data-driven than the legacy process, where much of system development was done via simulated threat data.

According to many of the participants in the process, more direct connection between algorithm developers in academia and government laboratories created greater motivation for them to ensure their work was the best it could be. The connection to fleet operators made the work much more tangible and direct. It was no longer a hard problem to solve or an academic exercise—it was a direct application to a real problem.

On the technical hardware side of the process, the focus was on leveraging the commercial market for processing power. The criterion of the program managers was *state of the practice, not state of the art*. This rule on procuring the processors enabled the system to be economically upgraded every 2 years (with individual ships upgraded every 4 years). The hardware was procured at an optimum time in the Moore's Law lifecycle, after the leading-edge users and before obsolescence by the next generation of processing.

The more *open business process* of the APB development process generated higher participation by small businesses, bringing in more ideas and, in general, spurring innovation. The WG/IPT structure, with its decentralized decision making and peer review enhanced technical rigor by prioritizing technical criteria over organizational interests. Over time, the WGs and IPTs became true teams, where each participant supported the others. It is important to note that the program managers emphasized to all participants that their ideas would get a fair and technically objective hearing. If any idea was deemed immature, it was welcome to be resubmitted for consideration in a later development cycle and be reevaluated.

CASE STUDY ANALYSIS—ARCHITECTURAL PERSPECTIVE

When discussing the change with practitioners—the program managers who invented and implemented the A-RCI/APB (A) process, WG and peer group leaders, and other participants—the discussion centers on issues of leadership, information transparency, and trust. They see these as key ingredients to creating trust, sustaining objectivity in technical decision-making situations where there are multiple methods of achieving success, and building and maintaining a process that is flexible enough to respond to changing requirements (threats) and technological developments.

KEEP PACE WITH MOORE'S LAW

The members of the SSTP and the program managers at the time said that it was critical to implement a COTS solution so that the system could keep pace with Moore's Law. It was also critical to be able to experiment with new signal processing algorithms in a *build–test–build* environment. Up to the point when the SSTP conducted their study, the *process* was closed to effective innovation. It was also conducted at what has been termed *arm's length*—highly formal interactions without effective interorganizational dialogue on critical issues such as technical and operational requirements, solution alternatives, etc. As an example, the various sonar systems in the submarine as well as across the wider Navy were developed in isolation. The Integrated Undersea Surveillance System, the suite of platforms, sonar systems, and shore processing stations that gathered and processed strategic and operational-level ASW data, were developed in isolation from submarine sonars and surface ship sonars. On the submarine, certain sonars were developed by university research laboratories while other sonar arrays were developed by the government laboratory. The prime contractor was responsible for integrating all of these disparate arrays, designing the hardware and software, and integrating both into a complete system that would operate at sea. This structure, where the Navy contracted mainly with the government laboratory and a handful of university research centers and a single prime contractor, was a system where many innovative ideas could not get an effective hearing. In the opinion of many participants in the old and the new process, the main reason for this was that organizational interests generally prevailed over technical rigor when the overall system was meeting specified performance.

All of these are valid and useful ways to view the effort and are effective guides to action. But if the payoffs for using commercial technology and for increased responsiveness to operational demands are so compelling, why has the process been so difficult to replicate?

VENTILATING THE DEVELOPMENT PROCESS

The objective of the program managers was to *open up* or *ventilate* the process so that the level of innovation could be elevated. As one of the program managers at the time stated: "We thought the system needed a little 'ventilation' … the system/process could use some shaking up."

There was an abundance of government laboratory–Federally Funded Research and Development Center (FFRDC) participation—but only a small industry contribution to the algorithmic (signal processing) problem.

The way the classic acquisition process works is that basic scientific research is funded by the Office of Naval Research. Once an idea or development seems promising, it usually migrates to a government laboratory, where much of the same work is repeated, but with a focus on practical application. This work is usually funded by the program office responsible for advanced development and often results in a working prototype. Once the prototype demonstrates promise, it transitions to the system acquisition program office, in this case PMS 425. This office again reworks the technology and science to some extent before fielding. In this process, the software code gets rewritten many times, with each person writing the code a little bit differently each time, at each stage. This creates systemic inefficiencies, especially when attempting to validate and test the software.

CONNECT SCIENTISTS TO FIELD OPERATORS

The situation with the new process was that the scientists were hands on in defining algorithms in real time as they were getting built for the target system. This creates interplay between science and engineering, so that what looks like risk becomes an opportunity when the program walls break down. If these three (J/P/S) did not have alignment (previous common experience in a development program), it would have been difficult to get the job done. They had to push this down into the organization—they did better on the process side than on the technical side.

TECHNICAL ARCHITECTURE

The *legacy* technical architecture, as we stated earlier, was a highly integrated design, controlled by two major entities in the acquisition process. These organizations were the government laboratory (NUWC) and the prime contractor (at the time of the crisis, this was IBM Federal Systems, based in Manassas, VA). The software code was written specifically to run on the hardware processors designed at NUWC and built by IBM Federal Systems. This architecture made it very difficult to either upgrade processing or modify software code to add a better algorithm. When the budget crunch came in 1992, this made upgrades of the existing legacy systems unaffordable.*

By the early 1990s, many participants, in and out of the government and the *legacy* acquisition system, knew that a way had to be found to use commercial processors to improve system performance. One small contractor, DSR, had been working on using COTS processors in sonar systems since the late 1980s.† Through a government program aimed at injecting small business innovation to Navy programs, DSR had been working since about 1993 on a way to host a towed array signal processing algorithm on a commercial processor. By the time of the SSTP and SSMC efforts, this SBIR had reached maturity and was ready for testing in a real system.

The key development (invention) of DSR in achieving the goal of hosting the government algorithm on a commercial processor was the *multifunction transportable middleware*—a software code that acted as a stable interface (translator) from the government software to the commercial processing. This technology is what enabled the architectural transformation in the system so that, as the COTS processors were successfully added to successive sonar arrays in the system, the processing could be upgraded while keeping the software program constant. It also allowed the software program to be upgraded without needing to upgrade the processing.‡

It is important to note that it is the hardware–software architecture of the sonar system which has changed, not the overall architecture of the sonar system itself—the fundamental block diagram of the sonar system is much the same as it was in the early 1960s, when advanced sonar systems were first fielded.

---

* Interviews with former engineers at NUWC.
† Interview with founder of DSR.
‡ In practice, the software programs and the hardware configurations are loosely coupled. In today's program, hardware and software updates are staggered, with a single hardware configuration matched to two successive software builds.

ACQUISITION ARCHITECTURE (ENTERPRISE ARCHITECTURE)

An architectural perspective on the organizational side of the sonar system provides us with a deeper insight to the relationship between organizational structure and technical architecture. Practitioners, both participants and managers, of the A-RCI/APB (A) process focus on critical issues such as decentralized decision making, trust, clear goals, protection of the innovators, and staying power. But an examination of the hierarchical relationships, especially at the middle and lower levels of organization, reveals an interesting pattern.

What we see is that within each level of hierarchy there were robust lateral interactions among peers (hence the term *peer review*). Members of the hardware selection team (the Technical Insertion IPT) and the main software/algorithm development team (the Signal Processing IPT) were drawn from several different organizations. Each of these members shared technical responsibility for the content and technical excellence of the product of the IPT.

A key factor in the decision to couple the IPT structure with giving IPT responsibility for technical decisions was the goal of refocusing participant priorities. As multiple participants noted, the technical challenge and an individual's technical reputation among peers became primary driver of decisions, not a particular organization's parochial interest. The *product* was developed by the group, not by a company that won a competition. A key feature, in our assessment of the organization, was that multiple organizations were brought together in teams so that decisions were made at lower levels in the work-breakdown structure of the systems engineering process. Architecturally, what this did was to create information flow paths at lower levels in the organizational hierarchy in addition to the more cooperative decision-making structure at the midlevel (program management) of the hierarchy. When coupled with the strong alignment at the flag officer level, this created both a cohesive structure with respect to the focus on the ultimate objective (improving signal processing performance) and a more flexible (responsive) organization overall due to the increased information flow among the participants.

In practice, these WGs met once a month to share information and to meet milestones. These frequent meetings and empowered decision structure were key factors in building up trust and transparency among participating organizations.

INITIATING AND SUSTAINING CHANGE: KEYS TO SUCCESS

As we observe the overall pattern of development and execution of this transformational change, there are two main phases: first, the initiation of change, and second, sustaining and evolving the enterprise toward a new process. It is important to note that there is really nothing new to learn regarding the impetus to change—the organization was in crisis. By multiple accounts of both observers and participants, it was an existential crisis: the submarine force had lost acoustic advantage, there were no funds to upgrade the system using the legacy system architecture and acquisition process, and there were serious questions about the need to have a substantial submarine force at all. In effect, the senior leadership (flag officers) was up against a wall, they had to follow the recommendation of the SSTP on how to improve the performance of the sonar system—move toward commercial processors.

Once the change was decided upon, it fell to the midlevel managers to implement the decision—these were the uniformed officers and government civilians charged with the program management and resource allocation decisions for the sonar system. As this implementation effort was evolved, there were multiple organizations jockeying for positions of decision authority in the process.

During this evolution from the SSMC toward the APB process, it was the program office management who desired to have a much more direct role in the system design and acquisition decisions. The *legacy* organizations—the government laboratory and prime contractor (IBM Federal Systems/Lockheed Martin)—were keenly interested in retaining the decision control that they had in the past. In addition, there were other government program offices that saw opportunity and were trying to gain control as well.

This power struggle created a lot of pressure from these legacy organizations on the midlevel management of the process. By many accounts, there was a sustained and often brutal attack on the performance and integrity of the change managers in the program offices. A key to successfully weathering this storm (some call it an *internal Cold War*) was the backing of senior flag officers and more senior program managers. One of these managers stated in an interview that the admiral always trumped DoD 5000.* When one of the chief engineers managing the COTS development process wanted to quit because of the relentless outside pressure, his program manager told him, "Don't worry, we'll protect you."

It is also important to understand that this protection had to be maintained across a sustained period of time. The first installation of a COTS processor on a submarine began in November 1997. The process was still being expanded through other sonar sensors and other tactical systems on the submarine in the 2003 time frame. As we discussed earlier, Admiral Giambastiani was a key factor in this protection. He became the main *protector* of the A-RCI system and APB process and was also its chief advocate in Congress and within the larger Navy.†

Overall Lessons

Many managers and researchers of complex engineering systems tend to view change from a process perspective. This is critical to success, but our case study here has shown that an architectural perspective is important as well and can provide some new insights for managers. What we have seen is that architecture can be viewed from multiple perspectives: technical and organizational.

On the technical side, we saw that breaking the system at the hardware–software boundary created an increased degree of flexibility. Hardware decisions were decoupled from software decisions. The use of middleware enabled developers to have to worry only about general performance dimensions of the hardware, such as processing power, and memory. The interface was *standardized*.

On the organizational side, mapping the new breakdown of the technical system to WGs and IPTs interacting with peer review groups created a new pattern of interactions among participants. The new process, the APB process, in effect, was mapped onto a new organizational structure. This structure connected experts from different organizations with different domain expertise to solve a common problem. Many WGs included operators from the fleet, scientists developing algorithms in the laboratory, and systems engineers responsible for integrating the algorithms into an operational system. By breaking down the barriers across organizations, the overall process became responsive to fleet operator needs and algorithm developers saw direct payoff of their work. This needed generating an increased level of trust across boundaries as well as job satisfaction and motivation.

Community Culture and Stamina

It is also important to consider the impact of the strong community culture in the submarine force. The alignment among senior and midlevel officers and their strong technical expertise were critical factors in both initiating and sustaining this change. As recounted by many of the mid- and senior-level people interviewed, the officer corps in the submarine force understood each other's roles and responsibilities, all were trusting in the technical competence of their counterparts, and all were committed to a common outcome, namely, the restoration of acoustic superiority at sea.

Extended Tour Lengths

The core A-RCI and APB program managers were involved in the process for nearly a decade. The earliest signs of a COTS-driven system can be seen in the program office's Acoustic Program Plan as early as mid-1994, before the SSTP was empanelled. The deputy program manager in that office retired from government service in 2007, while still involved in overseeing the vitality and health of the process. Another government manager/engineer was involved at the ASTO as a junior engineer in the 1994 time frame and left the government service in 2005 as the program manager of the APB process.

---

* DoD 5000 is the overarching policy document governing defense acquisition.
† When he became head of all Navy programs, one of his goals was to expand the A-RCI/APB model to aviation and surface ship systems.

Change Is a Battle among Elites for Control

As we noted earlier, this change in the acquisition process and in the technical architecture of the sonar system generated an *internal Cold War*. We are not aware of other case studies in defense acquisition that have highlighted this facet of change in as stark a manner as this one. The desire of the established *legacy* contractor and government technical authority to maintain control over both technical decisions and resource allocation was strong and their resistance forceful. Ultimately, the determinant of survivability of the process rested on technical and operational success buttressed by the strong backing of senior leadership in the submarine force and the stamina of the change agents.

PRACTICAL INSIGHTS

The insights discussed in the previous section can be considered strategic ones. At a more practical level, there are other insights that can be gleaned from this study. First, the overall systems engineering process is structurally different than traditional systems engineering and program management. The four-step APB process implements a truly *spiral* development process. Ideas can be deleted from the current round of development but may *go back to the drawing board* and be resubmitted in a follow-on APB. The annual nature of the APB process also enables it to be more responsive to fleet demand, which is inherently dynamic. As the operators absorb new capabilities and as the external threat environment changes, successive APB cycles can change selection criteria for system improvement.

Managing in this kind of environment requires a degree of comfort with change and uncertainty, which is difficult for many classically trained engineers and program managers.

From a tactical perspective with respect to managing through change, the managers we interviewed maintained that it was critical to have stamina in fights against the resistance to change. While this is not an unexpected insight, what was notable was the forcefulness of the resistance, the criticality of top-level protection, and the importance of leveraging bureaucratic opportunities to achieve tactical-level success that maintained momentum.

It was also critical to create and maintain contract incentives for vendor collaboration. One of these incentives was created by making contract award fees contingent on the successful performance of the whole system, not just individual vendor contributions to the system. In this way, an algorithm developer was motivated to ensure that their code was successfully integrated by Lockheed Martin and the supplier of the militarized COTS processors was motivated to ensure that they were able to be installed efficiently on board.

Key to Speed: Lateral Connections

In our view, a critical factor in the ability to speed up the process was the lateral connections across multiple organizations. Connecting academic researchers and production and design engineers with fleet operators and empowering them to make critical design and technical decisions was an indispensible factor in success. These changes, coupled with contract incentives, resulted, effectively, in disconnecting technical decisions from organizational interests. This generated trust in the process and in IPT/WG leaders and built team spirit around system-level performance.

SUMMARY—CONCLUSIONS

Organizational Elites: Control

We take two major insights from this case study. First, change in an enterprise-technical system is difficult, and can become a battle among organizational elites for control over the technical system decisions. These decisions range from technical to resource allocation areas of concern. We think this is a higher-order mechanism that can serve to explain many of the existing insights regarding the difficulty of change and innovation in organizations.

Layered Architectures: Flexibility

Our general thesis, drawn from the practice and experience in computer science, that layered architectures can enable flexibility is also supported by this case study. The critical technical decision and critical technology in this study comprise the use of middleware to separate the hardware and software development processes.*

The architecture of the acquisition process itself was also different. By breaking the development process up into smaller parts, the program office was provided more flexibility to manage contracts. Instead of awarding a monolithic contract to a single prime contractor, who then contracted with and managed the resources, the program office awarded smaller contracts to a wider array of small business—gaining more control over participation. Contracts were also written on shorter timescales, further increasing flexibility to add or delete participation as needed.

By placing multiple organizations on different IPTs and giving the IPTs authority to craft and make technical decisions, the process created an environment where organizational loyalties among participants were outweighed by technical objectivity. By subjecting the group's decisions to a separate peer review group, priority was maintained on getting the best technical answer, not necessarily one that was *good enough*, while also serving an organization's (or organizations') business or parochial interests. In this environment, over time connections among diverse organizations within the confines of the IPTs built trust relationships among members. Our observation is that this trust enabled richer information flows through the enterprise as a whole, shortening decision cycles and helping to make the overall process more responsive and efficient to changing demands and inevitable uncertainty.

Overarching Management Principle: Trust

In extensive discussions with the program managers, their overarching management principle was *leadership, transparency, and trust*. The other *best practices* cited by both leadership and participants were as follows:

1. Information transparency—technical objectivity
2. Protection of innovators (top cover)
3. Team-oriented performance incentives
4. Feedback, iterative improvement, use of field data
5. Prioritize schedule over specified performance

We find this list helpful, but our review, interviews, and analysis show us that these are insufficient to explain the root causes of A-RCI/APB (A) success. An architectural perspective shows us that lateral information flows among peers across the science, engineering, and operator domains (and, therefore, across organizational boundaries), is what allowed new insights and ideas to gain a hearing, to be tested and evaluated in a more objective manner, and, ultimately to improve the system's performance. But we also observed a cause–effect duality here. While the IPT structure and the APB operating process created the lateral information flows, trust seems to be both a cause and an effect of the structure and the flow. While the program managers created the structure, it strengthened and grew as trust grew among participants. But the initial structure would not have performed well had some initial degree of trust not been present.

---

* Middleware was seen as so critical to success that Admiral Giambastiani decided against making it widely available on the Internet. It was available to anyone who displayed an interest in contributing to the APB process on request. It is also important to understand that hardware and software are not completely independent—applications must be built so that they operate within the constraints of the processors.

FUTURE WORK

This study examined the genesis and evolution of the COTS insertion and advanced software development process (APB) for submarine sonar systems. Shortly after the successful completion of the second build cycle, the head of submarine acquisition directed that the process be expanded to the tactical and weapon control systems. This is part of what sparked off the *internal Cold War* discussed earlier. In the extension of the process, many observers have offered the opinion that either (1) it has not been implemented as effectively or (2) the APB process is not ideal for the tactical control system. Beyond the immediate extension, some participants and observers claim that the process as it exists in today's world is not as innovative or flexible as it was when it started—that organizational interests are now a primary concern of the participants and that little innovative development is occurring. Therefore, extending the initial phase of this study to include a more detailed examination of transition to another system (the tactical control system) would be a useful exercise.

Also, a more detailed examination of the architecture of the sonar system (the software and hardware architectures in relation to the enterprise) would be interesting. There exists conflicting evidence of the mapping of organization structure to technical structure and little to no studies that examine how multiple organizations (enterprises) fit this model.

Recall from the very beginning of the study that the driving force was to improve the sonar systems' performance—that was the singular technical deficiency that drove the operational crisis. Once that problem was solved, the senior leadership turned its focus to other priorities. Moreover, the secondary effect of implementing COTS in the APB process—shorter timescales for upgrades—increased innovation and cost savings and created a desire to apply the method to other systems to capture the same benefits. How the process maps or does not map well to other systems is a necessary future study.

The interaction between operational watch teams and the system is a scarcely studied topic. We examined the technical and acquisition architecture. But the structure of the technical system—especially as it relates to the flexibility of the operator displays an important area for future work. We can view the submarine's tactical systems as a layered architecture and the watch team as another layer above it. How does the systems engineering process (or how do layered architectures) help or hinder the ability of the submarine watch team to adapt to changing missions or change its overall structure as it searches for new ways to increase its effectiveness?

QUESTIONS FOR DISCUSSION

1. Do you think the method used by the managers in this case study is scalable to multiple systems in a larger system or, perhaps, to an entire warfighting platform (ship, submarine, aircraft, and ground vehicle)?
2. What are the issues that might have to be addressed as the technical and enterprise architecture and associated process are scaled to include multiple interdependent ship systems?
3. Discuss the significant bureaucratic and organizational barriers encountered by the mid-level managers in this case. What differentiates this success from the plethora of acquisition failures?
4. How did issues of community and cultural norms influence success in this case?
5. What are the practical challenges in crafting a lateral-hierarchical enterprise architecture? In other words, what are the social, organizational, bureaucratic, and policy hurdles that might stand in the way of building an enterprise architecture similar to the one described in this case?

RECOMMENDATIONS FOR ADDITIONAL RESEARCH

The APB process has been applied outside the submarine community. But many observers do not think that it has resulted in the same (or any) level of increased system flexibility, effectiveness, or cost savings. Case studies of these systems are necessary for the Navy and the defense acquisition community to understand more deeply the relevance and limitations of spiral development processes like this, to confirm or modify the insights regarding acquisition architecture and to develop generalizable insights.

It is also worth noting that, despite nearly 15 years of execution, there are no metrics for the RCI/APB process. The management has no indication of the health of the process: Are innovations being inserted into the system at an acceptable rate? Are small businesses still active participants? Is there a turnover in IPTs or are the same people involved now as were involved then? We must appreciate that the RCI/APB process is something completely different from the standard defense acquisition system; it operates on a fundamentally different systems engineering model and set of acquisition priorities. More in-depth understanding, with quantitative metrics, is necessary if its lessons are to be applied to other defense acquisition programs.

REVIEW OF RELEVANCE OF COMPLEXITY THEORY TO SYSTEM OF SYSTEMS (TABLE 20.2)

TABLE 20.2
Applying Complexity Theory to System of Systems

	Relevant Aspect	Issue and Support	Relevant Aspect of RCI/ APB Enterprise?
3.1	Recognition of wicked problems or messes including different worldviews (Churchman, 1967)	Supported by many including Checkland (1981); a key characteristic for the Western military operating in Afghanistan	No
3.2	Interconnected and interdependent elements and dimensions	Keating (2009:172) and Norman and Kuras (2006)	Yes
3.3	Inclusion of autonomous and independent systems	Sauser et al. (2009) and Norman and Kuras (2006)	Yes
3.4	Feedback processes	Keating (2003)	Yes
3.5	Unclear or uncooperative stakeholders	Often the case in projects	Yes
3.6	Unclear boundaries	Keating (2009:171), Keating (2003), and Mason and Mitroff (1981)	No
3.7	Emergence	Sauser et al. (2009), Keating (2009), and White (2007)	No
3.8	Nonlinearity	Keating (2003) and Norman and Kuras (2006)	No
3.9	Sensitivity to initial conditions and path dependence	Supported by Keating's Contextual Dominance (Keating, 2009:171)	Yes
3.10	Phase space history and representation	Previous history of states the SoS was in can be relevant	Yes
3.11	Strange attractors, edge of chaos and far from equilibrium	*Issue*: can an SoS be readily attracted to another state? *Possible initiator*: a change in requirements or another system being introduced	Yes
3.12	Adaptive agents	SoSs exhibit all characteristics a–r	Yes
3.13	Self-organization	Not providing top-down direction allows patterns to develop (DeRosa et al., 2008)	Yes
3.14	Complexity leadership	Uhl-Bien	Yes
3.15	Coevolution	Relevant	Yes
3.16	Fitness	There are satisfactory states, some better than others Norman and Kuras (2006)	Yes
3.17	Fitness landscapes	Norman and Kuras (2006)	Yes

REFERENCES

Associated Press Wire Service Report, U.S., Russian subs collide; damage slight. March 24, 1993.

Cote, O.W. The third battle: Innovation in the U.S. Navy's silent Cold War struggle with Soviet submarines. Naval War College Newport Paper No. 16. Newport, RI: Naval War College, Center for Naval Warfare Studies, 2003.

Department of Defense Directive. *The Defense Acquisition System*, DoDD 5000.1. Washington, DC, May 12, 2003.

Fox, J.R. et al. *Defense Acquisition Reform, 1960–2009: An Elusive Goal*. Washington, DC: U.S. Army Center of Military History, CMH Pub 51-3-1, 2011.

Friedman, N. *Technology and the New Attack Submarine*. Critical Issues Paper. Arlington, VA: Center for Security Strategies and Operations, Techmatics, Inc., 1996.

Gushman, J.H. Two subs collide off Russian Port. *The New York Times*, February 19, 1992.

Kerr, G. Submarine Acoustic Systems and Sensors. Briefing to the Deputy Chief of Naval Operations for Warfighting Requirements and Programs (N6/N7), January 31, 2005.

Navy Department. Sonar development working group (SDWG) 1998–99 year in review. Naval Sea Systems Command. Unpublished, Washington, DC, September 15, 1999.

Navy Department, Sonar development working group annual report, 1999–2000. Naval Sea Systems Command. Unpublished, Washington, DC, October 31, 2000.

Parrish, T. *The Submarine: A History*. New York: Viking, 2004.

Reuters News Service, US, Russian nuclear submarines collide in Arctic. Washington, D.C., 22 March 1993.

Stevens, R. Profiling complex systems. *IEEE Systems Conference*, Montreal, Quebec, Canada, April 7–10, 2008.

21 United States Air Power Command and Control

Enterprise Architecture for Flexibility and Decentralized Control

Michael W. Kometer and John Q. Dickmann

CONTENTS

CASE STUDY ELEMENTS

Fundamental essence: Intentionally designing command and control (C2) that turns policy goals into military action, which achieves these goals.

Topical relevance: Current doctrinal arguments among the services about C2 of airpower essentially talk past each other, focusing on certain aspects of the argument that are germane to the service's history and service culture. There is a great need to identify the factors that make particular C2 architectures more effective than others in particular situations.

Domain(s) (choose one)

Military

Country of focus: United States

Stakeholders

Passive/active

Primary/secondary

Primary insights: Commanders need to dynamically change the architecture of the C2 organization to handle changes in the context of war, which causes changes in the feedback loops present in the system.

KEYWORDS

Air Operations Center, Airpower, Architecture, Command and control, Feedback, Operation Allied Force, Operation Desert Storm, Operation Enduring Freedom, Operation Iraqi Freedom, Operation Odyssey Dawn, Operation Unified Protector, Organizational structure, Precision-guided munitions, Remotely piloted vehicles, Unmanned aerial vehicles

ABSTRACT

Since the end of World War II, command of military air power in the United States has been doctrinally invested in a single commander, usually an airman. This case study consists of four separate examinations of the use of combat air power between 1991 and 2003. In each case, the command and control architecture, while set up for centralized command and control of air power, was inevitably modified based on differing political and military contexts of each conflict as well as the need to adjust to the inherent uncertainties of conflict. It is these uncertainties that required responsive feedback from the Combat Air Operations Systems (CAOS), which was often difficult to achieve in the centralized command architecture. Our study observed responsive feedback loops at multiple levels of organization hierarchy, which are manifested by lateral connections across organizational boundaries. Our insight from the study of these conflicts is that senior commanders of complex organizations must devote time to organizational design; specifically they should consider establishing lateral connections across organizational boundaries and explicitly design these interactions and train their people to make decisions in a collaborative environment with multilevel feedback loops in mind. Creating lateral hierarchical organizations can enable military commanders to balance both efficiency and effectiveness in the application of military force to achieve political objectives. We think that these insights are relevant to senior leaders in the system design and acquisition world as well as in other areas of commercial endeavor.

GLOSSARY

ABCCC	Airborne Battlefield Command and Control Center
ACCE	Air Component Coordination Element
AF	Air Force
AFB	Air Force Base
AOC	Air Operations Center
AR	Army
ASOC	Air Support Operations Center
ATO	Air tasking order
AWACS	Airborne Warning and Control System
C2	Command and control
CALCM	Conventional Air-Launched Cruise Missiles
CAOC	Coalition (or Combined) Air Operations Center
CAOS	Combat Air Operations Systems
CAS	Close air support
CC	Component Commander
CENTCOM	CENTral COMmand
CFACC	Coalition Forces Air Component Commander
CIA	Central Intelligence Agency
CINC	Commander IN Chief
CJCS	Chairman of the Joint Chiefs of Staff
DASC	Direct Air Support Center
FAC-As	Forward Air Controllers-Attack

FECC	Fires Effects Coordination Cell
FM Field	Manual
FSE	Fire Support Element
G–N	Goldwater–Nichols
GPS	Global positioning system
HARM	High-speed Antiradiation Missiles
IADS	Integrated air defense system
ISR	Intelligence, surveillance, and reconnaissance
JAOC	Joint Air Operations Center
JCOA	Joint Center for Operational Analysis
JDAM	Joint Direct Attack Munition
JFACC	Joint Force Air Component Commander
JFLCC	Joint Force Land Component Commander
JSTARS	Joint Surveillance Target Attack Radar System
JTAC	Joint Terminal Air Controller
JTF	Joint Task Force
NAC	North Atlantic Council
NATO	North Atlantic Treaty Organization
OAF	Operation Allied Force
OEF	Operation Enduring Freedom
OIF	Operation Iraqi Freedom
OOD	Operation Odyssey Dawn
RPVs	Remotely piloted vehicles
SACEUR	Supreme Allied Commander, Europe
SCAR	Strike Coordination and Reconnaissance
SECDEF	Secretary of Defense
SLAM	Standoff Land Attack Missiles
TACC	Tactical Air Control Center
TACP	Tactical Air Control Party
TACS	Tactical Air Control System
TCP/IP	Transfer Control Protocol and Internet Protocol
TLAM	Tomahawk Land Attack Missiles
UAVs	Unmanned aerial vehicles
USAF	US Air Force

CONTEXT OF THIS CASE STUDY

A: The Combat Air Operations System (CAOS) is a system of systems that produces military air activity aimed at achieving policy goals. But we have expanded the boundaries of the system to include the components that apply not only to air activity—the attack and support aircraft and their support mechanisms—but also to subsystems that plan those activities, control them, and adjust them. The subsystems are named the strategic, planning, adjustment, and force application subsystems. As the study will show, these broad boundaries make analysis difficult, because they include factors like international politics that have inherent uncertainty and are often considered external to the system. However, the authors chose these boundaries because they allow us to examine command and control (C2) in a broad sense as part of the system, rather than the governance of that system (Figure 21.1).

B: There are still factors that are external to the system. Although we include the international politics involved, we treat economic and budgetary constraints as external to the system. In addition, domestic political pressures on the policy-makers are external to the system, yet their impact is significant. Many military analysts may disagree with the fact that the

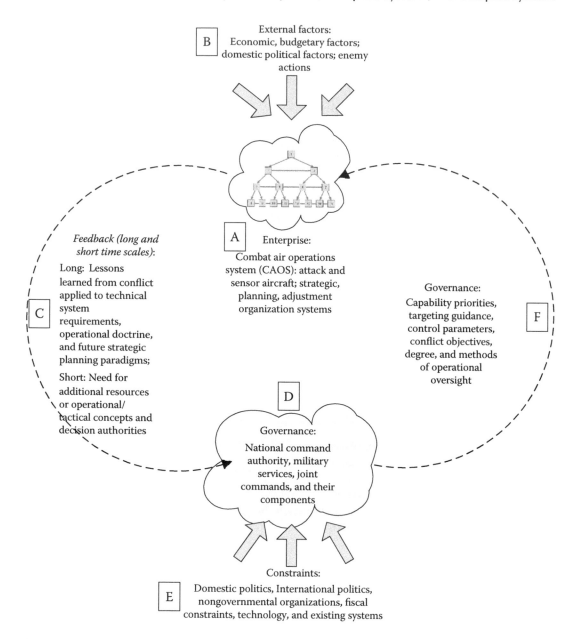

FIGURE 21.1   Context of complex system/system of systems/enterprise.

enemy is treated as an external factor in this analysis. However, the study analyzes the system that applies military airpower, and the enemy is not a part of that system—despite the fact that it is an important input to the system.

C: The CAOS provides feedback to the governing body mainly through the lessons learned from conflict and the requirements to fill gaps in organizations' capabilities to accomplish their missions. During a conflict, the governing body can affect the system by supplying more resources—aircraft, personnel, weapons—and, during longer conflicts, to develop new operating concepts or capabilities. But usually the feedback loop takes longer. The military organizations study the lessons learned and adjust their doctrine, training, and capabilities through formal systems that take years to complete.

D: This governing body is actually made of the military services, both separately and as part of the joint community. The military services are the ones responsible for organizing, training, and equipping the forces that deploy for the conflicts. They develop the doctrine and train the people who C2 the forces as well as those forces that apply the force on the battlefield. Since 1986, after the Goldwater–Nichols Defense Reorganization Act, the joint community has been a strong part of this governing body as well, with its own requirements, training, and doctrine.

E: So the governing body is actually not a unitary actor but a complex system of its own. The result, as the reader will see, is a lack of coherency in governance that often leaves the actors in different parts of the system responding to different interpretations. Thus, the Air Force's quest for centralized control was opposed even after the joint doctrine stipulated that there would be a Joint Force Air Component Commander (JFACC). The governing body is outside the warfighting chain of command, and therefore often unable to compel compliance with its doctrine and the standards to which it has trained its people.

INTRODUCTION

In early 2011, when the Arab Spring movement hit Libya by way of uprisings in Benghazi, several conditions in the international environment coalesced to make the situation ripe for international action to aid the rebel movement in Libya. Colonel Muammar Qadhafi's aggressive threats and military action against the uprisings prompted calls for international action from the League of Arab States, the African Union, and, eventually, the United Nations. In two military operations, coalitions led by first the United States and then the North Atlantic Treaty Organization (NATO) responded with air and maritime forces. The air operations suppressed Libya's air defenses and attacked Qadhafi's ground forces, giving the rebels time and space to organize and eventually overthrow the regime. But in doing so, the air effort languished far longer than most thought necessary. A preliminary assessment by the Joint Center for Operational Analysis (JCOA) found several challenges: (1) translating shifting and uncertain strategic guidance into coherent operational plans, (2) forming coalitions around underresourced commands, (3) protecting civilians without an adequate intelligence, surveillance, and reconnaissance (ISR) or ground perspective, and (4) adapting C2 processes and sharing information in what seemed to be an excessively ad hoc coalition (JCOA 2011).

This need to adapt to ad hoc situations with scarce resources is present in virtually every complex sociotechnical endeavor, especially military operations. The distinguished military commander and strategic thinker General Sir Rupert Smith has observed that every operation in his 40 years in the British Army required "chang[ing] our method and reorganize[ing] in order to succeed." He claimed that "[u]ntil this was done we could not use our force effectively," and even considered this "a normal—a necessary part of every operation" (Smith 2007).

Despite Smith's cogent observations and the actual experience of the US Air Force (USAF) over the past two decades, some interpretations of the USAF's tenet of *centralized control and decentralized execution* imply that there is a single right way to C2 airpower. With its roots in the battle history of World War II, airmen believe that control should be vested in an air commander, the Joint Force Air Component Commander (JFACC), through a centralized command node, the (Joint or Coalition) Air Operations Center (JAOC or CAOC) (Hinote 2009). While the *centralized–decentralized* tenet may still be a good conceptual guideline, the sociotechnical system that employs combat airpower (the Combat Air Operations System, CAOS) is so complex that no single method of controlling airpower will suffice for every situation.

This case study will show that in each of four instances from 1991 to 2003, employment of airpower required significant workarounds to reach the balance of efficiency and effectiveness necessary to achieve political goals. While the existing hierarchical C2 structure was set up to maintain coherency of tactical action with strategic political goals, workarounds were necessary to ensure effective combat operations. These workarounds created lateral connections that enabled lower organizational levels to coordinate effective responses to evolving situations. These lateral connections, in turn, threatened to do violence to the effort to maintain strategic coherency.

Our findings suggest that military commanders should devote time to deliberate organizational design, specifically to architectures that are more flexible than current organizational approaches. Flexible enterprise designs make use of vertical and lateral connections to solve complex problems by distributing control functions to a level appropriate to the situation and leveraging distributed expertise at that level to coordinate application of air power. In addition to the time devoted to this often-neglected organizational design activity, establishing lateral interactions and rules for appropriately decentralized control requires more investment in training and education. The people who operate the air operations system must have a higher level of knowledge and understanding as well as greater trust among each other so that they can perform these control functions from whichever location in the system at which they are posted.

DESCRIPTION OF THE CAOS

The primary purpose of what the military calls C2 is to produce unity of effort among all military actions, so that all effort leads to the achievement of common goals. However, even beyond that, military actions should be harmonized with those of other government agencies and coalition partners to produce unified action with them as well. If possible, unity of effort is achieved through unity of command (JP 3-0 2011). However, airmen claim that, in attempting to direct application or use of airpower, military commanders and, sometimes, policy makers have often misused it, nullifying the inherent advantages that air power brings to the battle.

The classic case to which airmen point when discussing this problem is the subordination of air units to ground commanders in North Africa early in World War II. During the planning for this campaign, airmen claimed that dividing the air units into separate chains of command decreased their ability to defend against the Luftwaffe (Syrett 1994). Ground commanders, however, wanted to ensure that each ground unit had air cover before taking any offensive. In the midst of intense discussions over the issue, Field Marshall Erwin Rommel attacked allied positions and nearly defeated the entire Allied Force in Tunisia in the Battle of Kasserine Pass. Hampered by weather, allied aircraft played only a minor role in this battle, but the fiasco ignited high-level discussions, and the Army Air Corps officially made the centralized control of aircraft under a theater-level airman a doctrinal tenet in FM 100-20, *Command and Employment of Air Power* (Syrett 1990).

Of course, there is an interservice component to this discussion about the control of airpower. USAF airmen have traditionally focused on the use of air power to win or at least create decisive effects in war at the theater, not the tactical, level. Marine and Navy aviators, on the other hand, have focused on the use of air power to support operations in their ground and sea environments, respectively. Marine and Navy commanders are therefore understandably reluctant to release the command of their aircraft to a single air commander, typically an Air Force officer. The challenge of coordination and control of air power has been a feature of every conflict and campaign since Kasserine Pass. The Air Force fought and won battles to create a single air manager in both Korea and Vietnam, yet the concept was implemented only at the end of those two wars and gained neither doctrinal standing nor interservice acceptance.*

---

* In Korea, the Commander in Chief Far East originally delegated *coordination control* to the Fifth Air Force when flying jointly with naval assets; however, the USAF and Navy interpreted this vague term differently. Essentially, the two services flew separate wars. Coordination improved, but an initiative to create a truly joint targeting board came a few days before the final armistice (Crane 2000). In Vietnam, the services again flew essentially separate wars, dividing the territory up into route packages. Then, in 1968, the effort to support Marines at Khe Sanh and the Army's First Cavalry Division overloaded the command and control capabilities in that region, prompting General Westmoreland to call for a single air manager. Again, the call came too late, and the Marines appealed all the way to the Secretary of Defense, who approved it only as a temporary expedient (Horwood 2006).

The Air Force finally achieved doctrinal standing for a single air commander in Operation Desert Storm, the campaign to remove Saddam Hussein's Iraqi Army from Kuwait. On the heels of several failed (or nearly failed) operations in the 1970s, the Goldwater–Nichols (G–N) Act of 1986 sought to create a more joint military, one which could capture cross-service synergies in battle. G–N put the Chairman of the Joint Chiefs of Staff (CJCS) in the operational chain of command and invested him with a powerful advisory role to the Secretary of Defense and President. It also gave geographic combatant commanders increased and substantial power (Kitfield 1995). General Norman Schwarzkopf, the commander of the US Central Command during Desert Storm, used this power to create the first ever JFACC, Lieutenant General Charles Horner of the USAF. Afterward, the Air Force understandably touted the success of the *centralized control, decentralized execution* construct, while the Army, Navy, and especially Marines pointed out that the JFACC had not actually possessed command of their air assets. Nevertheless, the Air Force went on to design its presentation of forces to Combatant Commanders around the JFACC concept and his major C2 node, the Air Operations Center (AOC; a *Combined* AOC [CAOC] in most cases).

In this endeavor, the Air Force was reassured by the political and technological environment of the 1990s. Desert Storm made use of precision weaponry and comprehensive battlefield sensors to an extent not seen before in warfare. Through the 1990s, this technology continued to mature, and the Air Force and other military services brought it to the field. Enabled by this newly fielded capability, the ability to precisely attack targets with a lower risk of collateral damage, political leaders used airpower in several instances where political goals concerned less than vital US interests. During the 1990s, narrow political objectives and, hence, narrowly defined combat goals and a global media that provided almost instantaneous information on military action combined with the technical capabilities of air power to promise, as Eliot Cohen put it famously, *gratification without commitment* (Cohen 1994). Especially after the disastrous political outcome of the operations in Somalia, airpower gave policy makers a way to do *something* without getting embroiled in a costly war or to risk casualties in cases where vital interests were not at risk, but at the same time, where it was inappropriate not to do anything. In fact, with precision weaponry, even the risk to the civilians in the area was minimized (O'Mara 2002). With the development of video sensors that could fly on unmanned aerial vehicles (UAVs, later called remotely piloted vehicles, RPVs), USAF commanders could seek targets of opportunity to attack. The core information transfer protocols of the Internet (Transfer Control Protocol and Internet Protocol, TCP/IP) allowed UAV video to be distributed just about anywhere. However, the AOC was the only place that was really capable of processing it all in a form and on a time scale that could support the tough, complex, trade-offs required of decision makers in these situations. All of this provided increased motivation to the USAF's push to make the AOC *the* place from which the JFACC would exercise centralized control of air power (Kometer 2007).

The tensions between civilian policy makers and military commanders, interservice tensions, the doctrinal predispositions of the services, continuous technological improvement, and the different situations that arise in the battlespace all affected the way C2 was applied through the period of our study. Because of this complexity, this study draws the boundaries of the CAOS much more broadly than most other representations of air power C2. For example, ground commanders need to have the capabilities that air power brings to battle, and their needs for employment of air power often conflict with the needs of the air component, Special Forces, and even ground commanders in other geographic locations in the theater. At the same time, the rapid diffusion of information technology across the battlespace has enabled other service component commanders as well as lower-echelon air, maritime, and ground commanders to process more and more ISR information, information previously only available to the CAOC. And, as we noted earlier, global media coverage of war has increased the desire (and the ability) of civilian policy makers to have access to battlefield information. It has also made them want to exercise finer (or, perhaps, a different style of) control over the use of air power.

To gain insight into the dynamics that result from these interactions among stakeholders, it is helpful to view both the feedback loops that comprise these tensions and the enterprise architecture that governs the actions of those in the system. Feedback loops show the interdependencies among people and concepts that we may or may not realize are interacting. Architecture is the overall scheme that governs the set of processes, communication channels, and business rules—both formal and informal—that people use to accomplish the mission as well as being governed by them.* We define enterprise architecture as the structural representation of the relationships among people, systems, and organizations that make up a system. We view hierarchical and lateral relationships as the aspects of enterprise architecture that are most important for determining the balance between strategic alignment of goals and flexibility to adjust to emergent tactical situations and opportunities.

A feedback representation of the CAOS is a multilevel model that incorporates the tensions discussed earlier. Figure 21.2 shows such a representation in multiple subsystems. We will not map out the actual feedback loops for each subsystem. However, the *Strategic Subsystem* includes the factors that drive the civil–military tensions in a typical war, like international and domestic politics, the media, and fiscal constraints. Policy makers are subject to pressures that make military action risky for them and often force them to retain ambiguity in their agendas and put restraints on military action. The *Plans Subsystem* would show that military commanders want to be given a clear goal and be left alone to achieve it as they deem appropriate. They would develop plans and command relationships to create unity of action among all coalition partners and services to accomplish these goals, and then assess their progress toward achieving them. Of course, nothing as complex as war ever goes as planned, so the *Adjustment Subsystem* includes the ways that military commanders have to adjust their plans and force employment schemes to react to emerging situations in war. Finally, the *Force Application Subsystem* would show the interaction of forces trying to execute orders in the battlespace (Kometer 2007, pp. 63–79). When analyzed as a whole, this representation shows that policy makers and military commanders have a tendency to focus on achieving direct results, often neglecting the need to account for unintended consequences or unforeseen circumstances. This requires the development of a learning organization—something that is generally achieved through empowerment and depth in command relationships. It is learning organizations that can adapt to and overcome the chaos and change inherent in war (Kometer 2007, pp. 272–275).

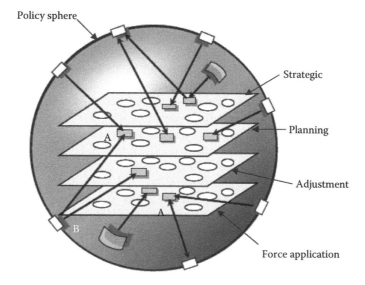

FIGURE 21.2   Multilevel feedback model of CAOS.

---

* More precisely, architecture is the *overall scheme* by which the functional elements of a system (technical, people, organizational) are *hierarchically partitioned* to subsystems/teams and how they are *arranged* with respect to each other. Architecture *sets the rules* that govern interactions in and among systems, both in *current operations* and as teams, organizations, systems, and relationships between and among them evolve *over time* in response to changing technology, operational demands, and external constraints.

Architecturally, we represent the relationships among the organizations that act to carry out the actions in each of these feedback representations by diagramming their interactions and interrelationships using hierarchical diagrams. Many organizational theorists typologize organizations as bimodal: hierarchical or networked, differentiating or integrating, top-down or bottom-up (Powell 1990; Romanelli 1991). But in military C2 situations, we are often confronted with complex and changing externalities, which require deviation from idealized organizational forms and modes of operation. This, of course, is addressed by organizational contingency theory and the information-processing view of organizations (Lawrence and Lorsch 1967; Galbraith 1969). For example, Galbraith's view of flexible lateral organizations is particularly applicable in the case of the CAOS (Galbraith 1993). Galbraith's approach is that to process uncertain and changing information in a responsive manner, an organization must be set up to have lateral interactions at the level where the information is relevant to the decisions. Galbraith's focus is on technical product development organizations; he does not address larger organizational problem solving such as coordination among multiple projects, coordination between policies, or resource allocation among different projects. In our conceptualization of organizations, we add a multilevel view to Galbraith's flexible lateral organization construct. In this model, interactions at multiple levels of hierarchy enable processing information among parts of the organization that deal with information at similar levels of the overall problem (i.e., strategy, plans, adjustment, and force application). Figures 21.3 and 21.4 are simple representations of the various types of organizational architecture we consider as we examine the conflicts in this case study.

Over the course of the major combat operations in the period from 1991 to 2003, the CAOS evolved and adapted to the political environment, interservice tensions, technological development, and changing missions. A quick overview of four of these operations will show how feedback loops and lateral interactions in the enterprise architecture affected the system's ability to respond to these contingencies.

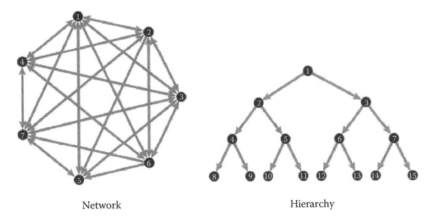

Network   Hierarchy

FIGURE 21.3   Representation of simple network and a four-level hierarchy.

FIGURE 21.4   Architecture with lateral connections (4–5; 5–6; 4–6; 4–7) and cross-layer connections (2–7; 6–10).

CAOS IN DESERT STORM (1991–1992)

For many in the US military, Operation Desert Storm is the one the United States got right. It excised the ghosts of Vietnam, gave the United States a clear victory, and put the world on notice that, in the post–Cold War environment, the United States was the lone superpower. Airmen still cite Desert Storm procedures when claiming that another war is not being prosecuted correctly. However, there were distinct factors that made the chosen architecture right for the situation of Desert Storm.

The context of the war after Iraq's August 1990 invasion of Kuwait lent itself to a major conventional war approach. The fall of the Berlin Wall had freed the United States from some of the constraints of the Cold War, and Iraq's invasion of another sovereign country was relatively easy for politicians to sell as a clear act of aggression.* However, at Camp David on August 2, 1990, General Schwarzkopf briefed President Bush that it could take 8–12 months to assemble a force to kick the Iraqis out. Nevertheless, by the weekend, the president declared the goal was indeed to oust the Iraqis, warning that "[t]his will not stand, this aggression against Kuwait" (Woodward 1991).

The President had articulated a condensed timeline for action. Because of this, General Schwarzkopf was forced to develop rapid options, options that became airpower-centric. Leaving the JFACC, Lieutenant General Horner, in Saudi Arabia as commander in theater, he called the Chief of Staff of the USAF asking for response options and received a briefing from Colonel John Warden and his Checkmate organization.† Warden's initial plan, Instant Thunder, was meant to defeat Iraq through airpower alone by shutting down the Iraqi *system*. Through a series of briefings and bureaucratic clashes, this plan transformed into the first, second, and third phases of an overall campaign plan to kick Iraq out of Kuwait. Thus, the stage was set for sequential employment of military force: air first, then air attacking ground, then air aiding a ground invasion (Olsen 2007).

The air planners, therefore, did not initially have to collaborate much with other component commanders. The Tactical Air Control Center (TACC), a precursor to today's AOC, was located in Riyadh alongside Schwarzkopf's headquarters, so Horner and his lead planners could get Schwarzkopf's guidance in person. Using paper charts and sticky notes, they assembled a Master Attack Plan that matched his (Schwarzkopf's) desired results (effects) with targets and the weapons and aircraft that would attack them. Then they produced an air tasking order (ATO)—the minute-by-minute plan for the operation. This plan had to be delivered through cumbersome means—including flying a hard copy out to aircraft carriers via helicopter.‡,§ For preplanned targets, the information in the ATO (and subsequent changes made in the TACC) was enough to accomplish the mission.

Of course, this caused problems among the service components: they often felt that their requirements for air power support were not being adequately (or fairly) addressed. As the air campaign unfolded and the ground portion of the campaign drew closer, ground commanders at the corps level were concerned that the air component was not attacking targets they requested—the forces directly across the border of Kuwait (Deptula 1991). Although this was largely a misperception,

---

* Of course, there was a tremendous diplomatic effort to shore up the coalition and secure Saudi Arabia's invitation to help. However, it was obvious to all that this was an act that should not be allowed to stand (Freedman and Karsh 1993).

† Checkmate is the organization on the Pentagon's Air Staff responsible for the development and analysis of strategic options for air campaigns. It has taken on various specific functions, but usually responds to the needs of the Air Force Chief of Staff. Colonel Warden had recently completed work at the National Defense University on a strategic concept of air power, and the current Chief, General Mike Dugan had hired him to bring that thinking to the Air Staff.

‡ The Air Tasking Order (ATO) is the overall directive for employment of air power in a campaign. It contains detailed information on aircraft allocations, ordnance requirements, targeting information and timing, air routes, and special instructions. The ATO covers a set period of time—usually 24 h—based on the planning cycle for the campaign.

§ Deptula (1998), presented to Major Michael Kometer by Major General Deptula, Hickam AFB, HI, April 22, 2004. Author's personal collection.

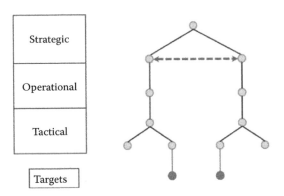

FIGURE 21.5 Desert Storm architecture.

it was important to ensure that these requirements were handled in a way that enhanced teamwork.* This situation was resolved by creating a formal coordinating board. Schwarzkopf put his deputy, General Calvin Waller, in charge of a formal target nomination process. This process gave the air component a prioritized list of targets, including those from the ground commanders (Putney 2004). Architecturally, this target nomination process represented a lateral connection at the strategic-operational level between the air and land components. It provided a means of coordinating and addressing different priorities between air and ground commanders on an appropriate time scale. A notional representation of this architecture is shown in Figure 21.5. In addition to target planning, there are numerous doctrinal and procedural interactions at multiple levels for the purposes of coordinating interdiction and CAS missions as they are executed. Unlike the lateral connection provided by Waller's board, these standardized interactions do not represent the type that facilitates strong coordination between the two organizations so that the enterprise as a whole can effectively respond to the dynamics of the battle. We will discuss this in more detail later.

As noted earlier, this targeting coordination process worked for preplanned targets, but the air component was responsible for reducing the capability of the Iraqi Army. Since twentieth-century armies are inherently mobile, the Iraqi Army presented many targets for which the ultimate location was unknown during planning.† This presented several problems for target coordination and the targeting process that butted up against the limits of technology of the time.

The first problem was that the air component did not have the sensing capability of finding these targets. The main method for finding them was through the eyes of either the ground troops, including Tactical Air Control Parties (TACPs), or the aircrews flying over the target area. Air operations in support of ground forces are classified into two categories: those close to the friendly ground troops are CAS while operations further away are classified as interdiction. For interdiction missions in Desert Storm, aircraft were directed to areas where enemy forces had been seen called *kill boxes*. The aircrew then had to search for the enemy and attack on their own. During the war, experienced pilots started serving as *Killer Scouts* who would loiter in a kill box to find the targets and direct other aircrews to them (Welsh 1993).

There were a few exceptions where sensors aided the aircrew in finding targets. Maverick missiles on A-10s and F-16s had infrared sensors and F-111s, F-15Es, and A-6Es had infrared sensors that helped them slew laser-guided bombs to a target. In fact, the latter were used to find and destroy

---

* The misperception was due to the fact that General Schwarzkopf was acting as his own Land Component Commander (JFLCC). Schwarzkopf wanted the sorties directed to attack the Republican Guard, an elite organization within the Iraqi Army, and his orders to Lieutenant General Horner overruled the Corps Commanders' requests. Horner thought he was still fulfilling these requests with the interdiction sorties that are discussed later in this section, so he did not worry about it.

† Targets on a battlefield change locations often, and their relative priority shifts as they become more or less threatening. This means aircraft sent to attack targets in a certain area will not know which targets they should attack until they arrive there.

tanks, which cooled at a different rate than the surrounding desert (Jamieson 2001). However, neither of these methods was useful for surveillance of a large area. For large-area surveillance, the experimental Joint Surveillance Target Attack Radar System (JSTARS) had this capability and performed this function. As the campaign evolved, JSTARS was able to detect moving traffic in the kill boxes and direct fighters to the area of interest. In fact, the JSTARS was the first to detect Iraqi forces moving into position for the famous Battle of Khafji (Watts and Murray 1993).

Detecting targets was the first part of the problem. Even when a sensor—human or machine—found a target, the other problem was using this information to attack it. While precision weapons simplify the process of transferring information to the weapon, most precision weapons at the time (laser-guided bombs, antiradiation missiles, and Maverick infrared-guided missiles) still required the aircrew to find the target on the battlefield. Besides this limitation, these weapons were in limited supply; precision weapons were only used on 9,117 of the 41,309 strikes in Desert Storm.* For the most part, information on target location was passed via the low-bandwidth, error-prone method of voice over the radio. Therefore, aircrew flying interdiction missions usually only received direction from an aircraft within line of sight—the Airborne Warning and Control System (AWACS), the Airborne Battlefield Command and Control Center (ABCCC), the JSTARS, or another attack aircraft (including Killer Scouts). Aircraft-flying CAS received directions from the Air Support Operations Center (ASOC), Direct Air Support Center (DASC), or TACPs—control nodes collocated with ground forces, which used targeting information generated by those forces. While this made for a relatively clean, hierarchical communications scheme, it limited the channels available to get information on emerging targets to aircrews with the weapons required to engage those targets.

The limitations of these information channels were evidenced in the inability of US forces to successfully engage Scud missiles in the Western Desert of Iraq.† All the information necessary to respond to a launch was available somewhere, but the process of getting information to the aircrew involved a telephone call from the US Space Command to the TACC, a radio call to AWACS, another radio call from AWACS to the nearest aircraft, and then a visual search by the aircrew for the target. This cumbersome process was unable to keep up with the Scud launchers' ability to *shoot and scoot* (Kometer 2007, pp. 215–216). The development of Killer Scouts demonstrates that the process was difficult in the interdiction domain as well.‡

In Figure 21.5, we showed that strategic and operational level coordination between air and land components was facilitated by General Waller's target coordination board. We also discussed that the Iraqi Army was mobile, compounding the problem of target location discussed earlier. An additional problem that this mobility created was that target priorities of the ground commanders changed as the Iraqi Army moved. It was difficult for the ATO process to adjust to these changing priorities. The eventual (and informal) solution to this problem was that the land component put ground liaison officers onboard AWACS and ABCCC in order to inform the airborne commander (an airman) which targets were most important to the ground force as they were detected or reported. This effectively created a second lateral connection at the operational level of organization, as shown in Figure 21.6. In this diagram, we can see that the ability to update targeting priorities in the airborne command node achieves flexibility to strike targets with either ground or air forces. The challenge becomes ensuring that the altered targeting decisions made at the middle level remain consistent with the strategic direction of the commander at the top.

---

* Precision weapons in Desert Storm included laser-guided bombs, Maverick missiles, electro-optically guided bombs, Skipper missiles, high-speed antiradiation missiles (HARM), Shrike, Walleye, standoff land attack missiles (SLAM), Tomahawk land attack missiles (TLAM), and conventional air-launched cruise missiles (CALCM) (Hill et al. 1993).

† Scud missiles represented a high-priority political target, because they could be aimed at Israel, and policy makers strongly desired to keep Israel neutral in the conflict for fear of broadening participants to the wider Arab world. By showing that the United States was forcefully addressing the issue, policy makers hoped to avoid Israeli retaliation against Iraq for Scud missile attacks on Israelis.

‡ The Killer Scout concept was originally developed by aircrews in the Vietnam War and was resurrected to solve the moving target problem in Desert Storm.

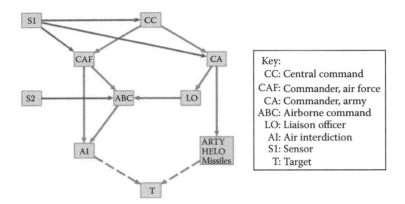

FIGURE 21.6 Operational-level lateral connection.

In Desert Storm, airmen considered many things about the employment of airpower revolutionary, including the use of stealth, precision, and a concept of effects-based thinking that linked their use to strategic goals (Deptula 2001). Even the use of a JFACC to lend unity of effort to airpower actions was new. However, we have observed that the use of lateral connections to collaborate among diverse parts of the overall organization created flexibility to respond to changing battlefield conditions and shifting priorities, especially for the ground forces. Strategic goals were accomplished because of a conventional, hierarchical method of planning and executing military action was applied to clear political goals and an enemy for which this hierarchical method (and the conventional warfighting style) was appropriate.

CAOS IN ALLIED FORCE (1999)

In Serbia in 1999, NATO found itself fighting a very different war against Slobodan Milosevic and his Serbian Army. Here the political goals were not nearly as clear-cut, and there were tighter restrictions on the use of military force. Given this situation, the conventional warfighting style was not nearly as effective. By 1999, communications and information technology had advanced to the point where many lateral connections were possible; however, political constraints drove a mainly hierarchical approach to the employment of airpower.

Political goals and constraints in the conflict were shaped by the fact that this was an operation executed under an alliance for less-than-vital interests. The conflict with Serbia was mainly humanitarian in nature, and it was clear the United States was not willing to step forward and lead the effort. Yet the humanitarian crisis was stark enough, and the precedent of conflict in the Balkans was ominous enough, that someone had to do something. Only NATO could marshal the forces to intervene, forming a coalition of the willing. But the NATO allies were deeply divided on whether there was a mandate for intervention. It was clear that both China and Russia would block a United Nations mandate, so alternative justification would have to suffice. But this route brought with it the necessity to constantly nurture the consensus of the alliance partners (Daalder and O'Hanlon 2000). To sustain the legitimacy afforded by NATO, member countries—even those who provided a majority of the capability, like the United States—would have to abide by the decisions of the North Atlantic Council (NAC).

In this environment, sustaining NATO solidarity was more important than achieving tactical effectiveness. NATO's political objectives were to gain acceptance of a political settlement negotiated at Rambouillet the previous two months, withdrawal of the Serb Army and Special Police Forces from Kosovo, end excessive and disproportionate violence in Kosovo, and prevent the spread of instability in the region (Henriksen 2007). But because the allies were not all on the same page regarding the legitimacy of force, the Supreme Allied Commander Europe (SACEUR), General

Wesley Clark, had to justify his target selections to NATO and often to individual governments. In addition, he had to be extremely concerned about the potential for friendly or civilian casualties that would destroy the already fragile NATO solidarity (Clark 2001).

This political reality severely restrained the air component's employment of airpower. The Coalition Forces Air Component Commander (CFACC), Lieutenant General Michael Short, wanted to use the same combination of effects-based operations, parallel warfare, and leadership attack that had been successful in Desert Storm (Tirpak 1999). He and his staff were unable to do this because they were only allowed to hit the targets that were approved and fed to them daily via video teleconference with Clark and his staff (Kometer 2007, p. 127). Short and Clark disagreed about the right way to employ airpower; the political leverage available to Clark dictated the slow, gradual approach that NATO employed. When a couple days of bombing did not force Milosevic to capitulate as many had expected, NATO turned to targeting Serb forces in Kosovo, then gradually brought the bombing closer to Milosevic and his cronies in Belgrade.

This centralization of control through restraint impacted the entire targeting chain. With authority for air strategy and targeting taken out of his hands, Short established rules of engagement that restricted the altitude at which his airmen could fly, the targets they could attack, and other details of the engagements (Grant 2002). Aircrew became extremely cautious, often calling back to the CAOC for permission when it may not have been necessary. According to some of the aircrew flying over Kosovo, this kept them from effectively engaging Serb forces.*

There was, however, evidence that lateral connection capability—if not the actual implementation—was growing. Without ground troops, the air component turned more extensively to sensors. The Predator UAV had seen its first combat use in Bosnia and was used extensively over Kosovo. The JSTARS was also widely used, and in fact the JSTARS crew often cooperated with ABCCC and airborne forward air controllers (FAC-As) to find and attack targets.† Aircrew also had a new way to transfer information to the weapon: the joint direct attack munition (JDAM). This GPS-guided weapon needed only coordinates instead of a visual identification of the target. However, as was the case in Desert Storm, getting all this information to the aircrew still required the CAOC's involvement. Short set up a highly classified fusion cell to get and disseminate information on Serbian integrated air defense system (IADS) targets and another on the combat operations floor to do the same with fielded forces in Kosovo (Kometer 2007, pp. 128–129). These teams had to take information from Predator sensors, validate them as targets, and relay the information through AWACS or ABCCC to the attack aircraft. They also validated many of the FAC-A targets, relayed by voice through AWACS and ABCCC in a cumbersome two-way trip that significantly delayed attacks on emerging targets.‡ Information traveled vertically to the CAOC and then down another vertical channel, rather than laterally among these sensors and actors.

By many assessments, this use of airpower was ineffective in accomplishing what should have been the primary objectives: stopping the ethnic cleansing and forcing Milosevic back to the bargaining table.§ In fact, the ethnic cleansing increased after the initial, anemic, bombing that policy makers had hoped would compel Milosevic. The attacks on the Serb Army were largely ineffective, as the Serbs were masters of deception. The restraints on air power and lack of ground troops made it almost impossible to stop enemy actions in an urban environment. Milosevic did not capitulate

---

* Phil Haun, Lt. Col., USAF (A-10 FAC-A in Allied Force), emails to author, March 11, 2004, and April 11, 2005, author's personal collection. Also Haave and Haun (2003).

† Mustafa Koprucu, Lt. Col., USAF (JSTARS senior director during Allied Force), telephone interview with author, April 6, 2004. Also Haave and Haun (2003, p. 138).

‡ Gary Crowder, Col., USAF (Deputy C-3 during OAF and senior offensive duty officer during OEF), email to author, September 22, 2004.

§ For example, Lambeth (2000) laments that to use airpower to project the appearance of "doing something" without a sound strategy is to "misemploy the air weapon inexcusably." Lieutenant General Short was vocal after the war in that NATO achieved its objectives in Kosovo "to some degree by happenstance rather than by design." He said it wasn't clear yet if NATO had won, because "the desired end state has never been articulated," Michael Short, Lt. Gen., USAF, quoted in Tirpac.

until a combination of increased intensity bombing (with the threat of more to come), Russian diplomacy, and a distant threat of ground invasion all increased the pressure on him (Hosmer 2001).

Although increased use of lateral connections would have been more appropriate for the tactical fight in this battlefield, the gradual approach may have been the only politically acceptable solution. The slow start was probably the limit of what was possible given the limited political leverage available from a divided NATO. Milosevic's reaction, to increase ethnic cleansing, and NATO's response, to focusing on hitting Serb forces, however ineffectively, gave NATO the moral high ground and helped cement its solidarity. The restraints ensured low friendly and civilian casualties, which also helped to maintain this solidarity. Solidarity ensured NATO did not crumble over the extended 78-day campaign, probably contributing to the Russians' losing patience with Milosevic and convincing him to come to the negotiating table.

So the method for commanding and organizing air power was a response to many of the feedback loops active in the political arena. However, had there been some way of ensuring this level of control at higher levels while still activating lateral connections at the lower levels, the campaign may have been more effective tactically and still politically acceptable. Destroying more Serb military forces would have been acceptable to NATO and would probably have put more pressure on Milosevic.

CAOS IN OPERATION ENDURING FREEDOM (2001)

A little more than two years later, the 9/11 terrorist attacks on the United States ignited a very different type of war. Having been attacked, the US public and its policy makers had none of the hesitancy to get involved that had characterized the lead up to Operation Allied Force. President Bush made it clear that this would not just be a reactive cruise missile attack but an all-out war that would put Americans on the ground—with all the risks that were implied (Woodward 2002).

But that did not mean the strategy was clear. In fact, due to the combination of speed with which the president wanted to respond, the difficulty in getting forces into place, and the unfamiliar character of the war, the strategy and plans for Operation Enduring Freedom (OEF) were immature at the outset. The National Security Council was unsure exactly how military action would create a political change that would end the operation—or what political change would actually be desirable.* All were convinced that this was a guerilla-type campaign against the Taliban, for which Special Forces and the Central Intelligence Agency (CIA) would be essential. But Secretary of Defense Donald Rumsfeld wanted to push the military to figure out a transformational approach—and transform itself in the process.[†] Commander of US Central Command General Tommy Franks saw the campaign in phases, with the initial phases setting conditions for an ultimate *decisive combat operations* phase once US troops were finally in place. However, this plan was still in draft form when the action began on October 7, 2011.[‡] In addition to these uncertainties, there were also significant political concerns regarding cultural issues, which led to tight controls on certain aspects of targeting, similar to the case of Operation Allied Force.

As a result of this ambiguity, the air component was again not empowered to develop its own air strategy aimed at accomplishing clear objectives. Franks' staff in Tampa, Florida, held most of the authority for selecting targets, leaving the officers at the CAOC frustrated that they were not trusted to perform this function.[§] However, preplanned bombing was not the decisive factor in this war.

---

* The initial thoughts were not necessarily to topple the Taliban but to create some kind of split that would result in a more moderate government (see Conetta, 2002).

† See Woodward (2002, pp. 43–44, 62–63, 99). On p. 135, Woodward quotes Rumsfeld saying, "if you're fighting a different kind of war, the war transforms the military."

‡ See Franks (2004). Kometer (2007, pp. 130–131) claims that the plan vastly underestimated the effects the initial air and special operations phase would create, and the toppling of the Taliban occurred before forces were in place for a phase 3.

§ David Deptula, Maj. Gen., USAF (CAOC Director in OEF), interview with Major Michael Kometer, Hickam AFB, HI, April 22, 2004 and David Hathaway, Lt. Col., USAF (Deputy Chief of Strategy Division in OEF and OIF), interview with Major Michael Kometer, March 26, 2004, Shaw AFB, SC. Author's personal collection.

The initial night's strikes were aimed at only 31 targets, using 50 Tomahawk missiles, 15 heavy bombers, and 25 attack fighters plus support aircraft (Leibstone 2001). Over the next two weeks, the campaign sputtered and stalled until the Special Forces and CIA teams were inserted, connected with the Northern Alliance, and then figured out how to bring airpower to bear on the Taliban.

When this happened, the interaction of Special Forces, indigenous forces, and air power showed the true value of lateral connections in this war. Special Forces teams that included trained terminal air controllers, either from Special Forces or conventional units, quickly learned to call airpower to support the Northern Alliance, earning their trust and confidence. These controllers had the capability, with laser range finders, to produce target coordinates that could be input directly into JDAMs, now the weapon of choice (Call 2007). Planning became an exercise in getting aircraft out to loitering stations, from which they could respond to calls from these controllers as the situation demanded.

Sensors had also been further developed by this time. Predator UAVs were still in demand, this time operated by both the USAF and the CIA. In addition, some Predators now had Hellfire missiles integrated into them, which allowed their crews to fire at a target they detected, eliminating one transfer of information between sensor and shooter. This was not an increase in laterality, however. Most of the time Predators were used to attack high-level targets such as important Taliban leaders. For these cases, authority to strike was held in Tampa—first by General Franks and later by one of his key staff officers, a USAF officer named Major General Gene Renuart.* At other times, when the targets were not as strategically or politically important, the CAOC served as the conduit to get the information to an attack aircraft and control the engagement. There were, however, cases where Predator video went directly to the aircrew of an attack aircraft. The AC-130s had incorporated a technical modification called Remotely Operated Video-Enhanced Receiver (ROVER) that allowed the crew to see Predator video and use it to find targets (Kometer 2007, pp. 198–201).

But in all of this, authority was either held very high or delegated very low. The middle layer languished, making the vertical planning and execution channels ineffective. The standard military strategy-to-task and assessment planning functions decayed. The portion of the CAOC that coordinated with the ground component atrophied as well.† But, most dangerously, the air component was operating without an ASOC, the part of the air power organization that is the formal coordination point with the CAOC for close air support of ground forces. An ASOC was not established because of political desire to minimize the number of personnel deployed to foreign soil for OEF.‡ Additionally, Special Forces units were not familiar with how to operate with a CAOC. Because conventional US forces were not yet in the fight, there was no ASOC in Afghanistan. Fortunately, there were only a small number of aircraft over Afghanistan during these first months, so the lack of an ASOC was not overly detrimental to operational effectiveness (Kometer 2007, pp. 136–137; Neuenswander 2003). Problems did not arise until the complexity of operations in Afghanistan increased during the run-up to Operation Anaconda in late January and early February of 2002.

The objective of Operation Anaconda was to round up the remaining Taliban and al Qaeda insurgents in southeastern Afghanistan's Shah-i-Kot Valley before they could escape across the border to Pakistan. This operation was conducted by elements of the 10th Mountain Division, various Special Operations Forces units, and indigenous Afghan fighters under the command of the Army's Major General Frank Hagenbeck. Unlike other operations in Afghanistan, where air support was loitering and coordinating directly with geographically dispersed Special Forces on the ground, Operation Anaconda developed into a major operation in a small geographic area. Because of the multiple service component organizations involved and the lack of standard organizational structures, lines of

---

* The fact that Renuart was a USAF officer did not make it easier for those at the CAOC to accept that the authority remained in Tampa. An officer on a joint staff is a member of that staff, regardless of service. The argument was that the joint staff should have released the authority to the component staff.

† Brett Knaub talking about Mazar-i-Sharif.

‡ The CAOC for Central Command was initially based in Saudi Arabia; in April 2003, it was moved to al Udeid air base in Doha, Qatar.

authority were confused, hampering coordination. This lack of appropriate coordination eventually created confusion on the battlefield.

For many reasons that are beyond the scope of this case study, the understaffed CAOC was unaware of the operation until only a few days before it was scheduled to commence. Consequently, officers at the CAOC did not have time to set up the necessary coordination measures that should have accompanied such an ambitious operation in a small area. During the operation, the CAOC supplied plenty of aircraft, but many ended up stacked up in the airspace waiting for clearance into the battle. Key airmen working with Special Forces at the time set up an ad hoc *ASOC-like* node to coordinate aircraft movement into and out of the battle area. These last-minute heroics paid off: although the operation did not go as planned, air support came to the rescue in several key instances. However, the chaos led to several near accidents and an obvious lack of combat effectiveness overall.

In summary, OEF represented a conflict where the appropriate structure should either have been an explicit hybrid of hierarchical and networked form at the start or have been changed as the character of the war evolved over time with changing priorities and strategic and political objectives. An unclear strategy and several key constraints at the strategic level early in the conflict led to a centralization of air power command normally delegated to the air component. At the same time, somewhat contradictorily, the need to work closely with Special Forces in unforeseen ways led to a delegation of a great deal of authority to lower (tactical) levels. This authority at lower levels enabled rapid reaction and adaptation to the uncertainties on the battlefield, while the centralization of authority at CENTral COMmand (CENTCOM) was possible because of the geographic dispersion and small proportion of US participation in the war. But as a consequence, the middle layers of the organization were neglected, so when the character of the war changed to a more densely populated conventional one during Operation Anaconda, the organization was not agile enough to exercise hierarchical channels necessary to make this shift or create the mid-level lateral connections necessary for coordination and deconfliction of the many details that arise as forces are concentrated for action.

CAOS IN OPERATION IRAQI FREEDOM (2003)

As a result of the problems encountered in Operation Anaconda, the C2 structure for air power evolved between OEF and the next major conflict, Operation Iraqi Freedom (OIF). The near-chaotic episode in Operation Anaconda prompted some changes, but two years of planning and preparing for what would obviously be a more high-intensity, conventional war had a lot to do with the change as well.

After Operation Anaconda, Air Force and Army officials engaged in some finger pointing. To the Army, the Air Force had not provided responsive close air support. In response, the Air Force replied that the Army units involved had not brought the air component in on the planning for the operations until a couple days before its execution. In addition, it claimed that close air support had been responsive; episodes of longer responses were actually dynamic targeting of specific moving targets (Kometer 2007, pp. 141–142).

Despite the interservice squabbling, the Air Force responded with an organizational adaptation to improve its integration with the land component. It developed a formal lateral connection between the JFACC and other components. Named the Air Component Coordination Element (ACCE), this liaison element was intended to keep the air component abreast of all operations in the affected component and facilitate better synchronization and support from the air component. For OIF, there were multiple ACCEs: at the Joint Force Commander's headquarters, at the Land Component headquarters, at the Special Operations Component Headquarters, and at the Maritime Component Headquarters. These were meant to establish lateral connections at a fairly high level for better collaboration among the component commanders.

But there were also operational feedback loops at work to improve this level of collaboration. The US strategy in OIF was much clearer than it had been in the initial portion of OEF. From the beginning, it was clear to General Franks that this was to be a war to remove Saddam Hussein from power using large conventional forces (although not as large as many would have liked) (Franks 2004, p. 350). At the operational level, there were many unknowns about how to achieve this, and Secretary Rumsfeld constantly pressed Franks with questions about these details. While some perceived this as top-down hierarchical control, it served the purpose of establishing a clear understanding of the strategic level concerns, and in the end the military plan belonged to General Franks in both form and substance.* President Bush made it clear that this war was important enough that the United States would risk the loss of life to obtain it, and that the military was empowered to accomplish the mission in the best way it knew.

The military began preparing for this war well in advance. At multiple planning conferences and rehearsals, the service components began to work together in what some of them described as an extremely *joint* manner, creating plans that emphasized teamwork over single-service solutions.† This seemed to create an atmosphere where the component commanders developed a higher level of trust and were more willing to support each other than in past conflicts. When General Franks changed the original plans for a preliminary air bombardment at the last minute so that he could get his ground forces in early to seize oil refineries before they could be damaged, the CFACC, Lieutenant General Moseley responded quickly and enthusiastically that the air component could do whatever was needed (Franks 2004, p. 439).

As with Desert Storm, in OIF there was still a great deal of hierarchical structure to the enterprise architecture, but in this case, the components were fully empowered. CENTCOM still controlled targeting for the air component, but this time it included a formal targeting board (reminiscent of the one established in Desert Storm) where all components got input—with CENTCOM's desired targets having equal status with the other components' targets. The process became more operations-oriented and less driven by Franks' intelligence staff. In fact, now air component planners actually had some leeway to pick the targets they thought would create the right effects, then get them approved by the target management board.‡ Unlike in Afghanistan, OIF saw a full use of the Tactical Air Control System (TACS). Aircraft that were sent to perform CAS were managed by either the ASOC or the Marine Direct Air Support Center (DASC) and eventually directed by a Joint Terminal Air Controller (JTAC—a new name for one of the members of a TACP).§ This kept the middle level of the hierarchy energized.

However, in this environment, many lateral connections were exercised at the lower levels as well. Kill box interdiction was again a big part of the plan to prepare the way for the ground troops' advance. The former Killer Scouts' mission to fly in open kill boxes and find targets was now called Strike Coordination and Reconnaissance (SCAR). However, now SCAR pilots had improved ways to find targets. Some had listening pods to aid them—using the pods, the crews could get even better coordinates to share with other attack aircraft. But the CAOC could also get involved, relaying information from other sensors like JSTARS or UAVs. Because aircrew dropping JDAMs needed only coordinates to engage a target, the synthetic aperture radar of Global Hawk and JSTARS could feed this information at night or during a sandstorm, allowing the crews to attack when they otherwise could not have. Even the ASOC could now direct aircraft remotely based on these same sensors. And the same pairing of aircraft and Special Forces that had been so potent in Afghanistan

---

* One of these strategic concerns was that forces remain small and light, driving the transformation that Rumsfeld sought. However, the picture remains one of an appropriate relationship between policy and military strategy. See Woodward (2002, p. 98) and Thomas and Brant (2003). Franks (2004, pp. 350–356) describes a meeting with the NSC where Franks and policy makers probed each other about details to ensure a common understanding, yet each also understood his lane.

† David Hathaway, Lt. Col., USAF (Deputy Chief of Strategy Division for CENTAF CAOC in OEF and OIF), email to author, March 21, 2005.

‡ Brett Knaub, Lt. Col., USAF (CENTCOM J33 during OEF and OIF), emails to author, September 22 and November 5, 2004, and Hathaway email March 21, 2005.

§ There were differences in procedures between the Army's area of operations (using an ASOC) and the Marines' (with a DASC). The Marines opened kill boxes much closer to their own troops than the Army did—probably because the Army was trying to integrate aircraft with its long-range artillery much farther out than the Marines, who do not have such long-range weapons. See Kometer (2007, p. 143).

operated again in northern Iraq, out of reach of conventional coalition ground forces for much of the war (Kometer 2007, pp. 235–237).

In addition to the lateral connections established by the ACCE, OIF saw more robust lateral connections at the operational and tactical levels as well. Because of the last-minute reduction in the size of the planned invasion force, the V Corps Commander, Lieutenant General William Wallace, had to find a way to leverage air power to compensate for the smaller firepower of his reduced-size force. As the invasion plan was evolving, the V Corps ASOC unit developed a new organizational structure and targeting approach. Because of their increased information access, the V Corps' ASOC and its Corps counterpart, the Fires Effects Coordination Cell (FECC) had better situational awareness of the battle space. The ASOC proposed a more integrated structure with the FECC (tighter lateral interactions between the two organizations) and a more collaborative targeting process. The lateral connections and targeting process the ASOC developed were the solution to Wallace's problem of a smaller fighting force (Kirkpatrick 2004). The lateral interconnections between the air and land components in OIF are represented in Figure 21.7. The ACCE connections are shown between the air component (AF) and the land component and the V Corps (AR and V). The V Corps' Fire Support Element (FSE) and the V Corps' ASOC have a much stronger lateral connection than the standard doctrinal interaction of past conflicts.*

Of course, this did not all happen seamlessly. There were times when this mix of hierarchy and lateral connections caused confusion, raising the risk of accidents. However, air operations in OIF demonstrated a better combination of the use of both vertical and lateral connections than in OEF.

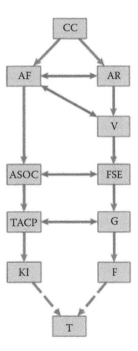

FIGURE 21.7 Iraqi freedom architecture (here G is the ground commander, F are the ground forces, KI are the kill box interdiction aircraft, and T is the target).

---

* The nature of the decision and information flow structure in the FSE/ASOC organization was such that information that came in via USAF information sources could quickly and easily be transferred to the FSE for targeting by V Corps assets or vice versa—targets could just as easily be passed to a TACP for engagement via air assets. This represents a dramatic difference from conventional ASOC operations where decision making was minimal and deconfliction and allocation of air to ground forces was the main goal, not the evaluation and assignment of targets to ground or air power.

CONCLUSIONS AND IMPLICATIONS

After the initial 21 days of OIF, the United States entered a period of combat operations in both Afghanistan and Iraq that looked *different* than those the aircrews had seen before. Within days of the taking of Baghdad, aircrew stopped executing classic CAS from the overcrowded airways over the city. Instead, they moved toward helping ground troops find targets, including enemy troops lying in wait for them (Kometer 2007, pp. 241–242). Over time, this type of *nontraditional ISR*, ISR from UAVs (eventually called remotely piloted vehicles—RPVs), and airlift became more important to ground troops in most cases than bombing. In this type of war, preplanned targeting was almost unheard of, and aircrews were not allowed under any circumstances to find and drop bombs on targets that had not been cleared by either higher headquarters or the ground commander.* In a sense, the robust information and sensing technology that enabled increased laterality and flexibility in early OIF operations became the means to constrict this flexibility as was done in early OIF operations and in Kosovo.†

Then came the 2011 Libya air operations, which were a shift in another direction. Suddenly, the air component was called upon to dismantle an IADS, set up a no-fly zone, and attack conventional ground troops—including the SCAR mission.‡ But in NATO, a lack of planning expertise hampered the air component's takeover of air operations. In both the US-led Odyssey Dawn and the NATO-led Unified Protector operations in Libya, aircrew had difficulty at first accepting the authority they received to execute SCAR. It took a while to transition from the lower level of responsibility to which they had been accustomed.

Meanwhile, the United States has been busy preparing for other scenarios that are more likely in the future. The well-publicized shift of focus to the Pacific and several other likely scenarios for force employment in the near term require the US military to fight in contested environments. Not only would likely opponents have more robust conventional military forces, they would be better able to make use of the electromagnetic spectrum to hamper US forces. US C2, which for over two decades has enjoyed almost unfettered freedom to obtain and pass information, would be subject to various methods of attacking that information. Under these circumstances, it is very likely that the C2 apparatus would have to degrade from its full manifestation to modes with less information sharing (or at least information sharing over mediums with lower bandwidth). Commanders would have to adapt to delegating some responsibilities, as they may be unable to exert the level of control they now enjoy. They might, in essence, be required to rely on coordination across organizational boundaries at middle and lower levels of command, just as they did in the early days of OEF and OIF, but without the robust sensing and information transport capability on which they have come to rely.

If these degraded modes are not developed and fleshed out now, before a crisis happens, they will most likely be developed ad hoc in the heat of battle. While not a predetermination of failure, it is certainly not an ideal way to develop a C2 architecture. Attention to an intentional design of technical and enterprise (organizational) architectures will, at the very least, prepare airmen to define these degraded control modes. At best, it could provide well-defined and therefore more capable modes of operation and degradation that are tailor-made for use in these contested environments.

SUGGESTED FUTURE WORK

This chapter has proposed a very different way to look at the overall challenge of achieving policy goals through military air power. The combined evaluation of feedback and architecture seems to be very powerful in its explanatory and prescriptive potential. Here are some ways to take the research further.

---

* Need to cite this somehow.
† It is important to note that these examples demonstrate that the impact of technology can be to both inhibit and enable lateral interactions and/or vertical control. We think this indicates that senior leaders must pay closer attention to their organization/enterprise architecture than in the past.
‡ SCAR: Strike Coordination and Reconnaissance.

TABLE 21.1

Review of Relevance of Complexity Theory to System of Systems

	Key Aspects	Relevance to CAOS	Notes
3.1	Recognition of wicked problems or messes including different worldviews (Churchman 1967)	Yes	Often this recognition is only implicit for the major stakeholders.
3.2	Interconnected and interdependent elements and dimensions	Yes	
3.3	Inclusion of autonomous and independent systems	Yes	
3.4	Feedback processes	Yes	
3.5	Unclear or uncooperative stakeholders	Yes	Media, other military services.
3.6	Unclear boundaries	Yes	Sometimes boundaries are disputed from a command–control perspective.
3.7	Emergence	Yes	Warfare is an emergent phenomenon at multiple scales.
3.8	Nonlinearity	Yes	War is inherently nonlinear.
3.9	Sensitivity to initial conditions and path dependence	Yes	War is inherently a chaotic system.
3.10	Phase space history and representation	No	This concept has been applied to CAOS, but only at the level of metaphor and analogy.
3.11	Strange attractors, edge of chaos and far from equilibrium	Yes	Same as 3.10.
3.12	Adaptive agents	Yes	
3.13	Self-organization	No	
3.14	Complexity leadership	No	
3.15	Coevolution	Yes	This concept has been applied to CAOS, but only at the level of metaphor and analogy.
3.16	Fitness	Yes	This concept has been applied to CAOS, but only at the level of metaphor and analogy.
3.17	Fitness landscapes	Yes	This concept has been applied to CAOS, but only at the level of metaphor and analogy.
4.1	Recognition of architecture	Yes	
4.2	Arrangement of these systems in a networked fashion rather than in a hierarchy	Yes	We actually view the system as a hybrid at all levels. There is a hierarchy, but at each level of the hierarchy there is also a network of some sort—partly formal, partly informal. Organizational theory speaks of hierarchy versus network. But there's also the element of *laterality*. Jay Galbraith talks about *flexible lateral organizations*, but only at one level. What I see (in effective C2 organizations and elsewhere) is laterality at multiple levels, which can also be described as networks at each level. Then when you get to multiple organizations (or even when a single organization gets very large), you have to consciously create networks at multiple levels of abstraction. In some contexts, this might be called *communities of practice* but I don't think it's a really good description of what we have here, especially in a combat organization. Again, this is because it is partly formal and partly informal. The way the participants activate the informal parts and integrate them with the formal is the critical point to consider.

(continued)

TABLE 21.1 (continued)
Review of Relevance of Complexity Theory to System of Systems

	Key Aspects	Relevance to CAOS	Notes
4.3	Recognition of external environment and project context	Yes	
4.4	Management of complexity	Yes	
4.5	Modularity	No	
4.6	Views and viewpoints	Yes	
4.7	Recognition that there are unknown unknowns	Yes	
4.8	Complexity requires the definition of completion of the right action and creation of a control structure must at least have the range of the thing being controlled	Yes	
4.9	Recognize different organizational structures due to scale and complexity	Yes	
4.10	Recognition of the importance of a different leadership style	Yes	
4.11	Governance	Yes	
4.12	Incremental commitment initiating the project normally	Yes	
4.13	Systemigram	Yes	
4.14	What is the most appropriate definition of a complex system: Type A Boardman and Sauser (2008) describe a system of systems (SoS) as having the following characteristics: • Autonomy and independence • Belonging • Connectivity • Diversity • Emergence These concepts require qualification: for example, in a traditional SoS, in which existing defense assets have been included, these systems would have been initially independent, but are now locked into the SoS by engaging a particular version that is not independent with regard to change. Belonging also needs clarification: does an adult person belong to their parent's family? The answer is both yes and no, thus requiring qualification. Type B		We both think these definitions are, to some extent, describing a different part of the elephant. However, type B resonates best with us. In fact, military planners are using the language of *wicked problems* to underpin the movement toward *design* of military operations. In these cases, where there is such ambiguity about what is to be done, defining the problem is the problem. However, there is also an issue with multiple and often conflicting purposes and uncertainty about course of action effectiveness in the face of dynamic (changing) external constraints and internal resource limitations. The multiple organizations involved often charge ahead regardless of whether there is uncertainty; the problem is they do this in conflicting ways.

TABLE 21.1 (continued)
Review of Relevance of Complexity Theory to System of Systems

Key Aspects	Relevance to CAOS	Notes
An alternative approach is provided by Jackson (2003, p. 138), who describes complex systems as those that are "interconnected and complicated further by lack of clarity about purposes, conflict, and uncertainty about the environment and social constraints. In tackling wicked problems, problem structuring assumes greater importance and problem solving using conventional techniques. If problem formulation is ignored or badly handled, managers may end up solving, very thoroughly and precisely, the wrong problem." Mason and Mitoff (1981) added that such an ill-structured problem situation is made up of highly interdependent problems.		
Type C		
A further description is provided by Anne-Marie Grisogono (2006), of DSTO Australia, who describes complexity in the following terms:		
1. Causality is complex and networked: that is, simple cause-and-effect relationships don't apply—there are many contributing causes and influences to one outcome; and conversely, one action may lead to a multiplicity of consequences and effects.		
2. The number of plausible options is vast: so it is not plausible to optimize.		
3. System behavior is coherent: there are coping patterns and trends.		
4. The system is not fixed: the patterns and trends vary, for example, the *rules* seem to be changing—something that worked yesterday may not do so tomorrow.		
5. Predictability is reduced: for a given action option, it is not possible to accurately predict all its consequences, or for a desired set of outcomes, it is not possible to determine precisely which actions will produce it.		

QUESTIONS FOR DISCUSSION

1. Where do the authors draw the boundaries for the system that commands and controls combat airpower, and is this the classical way to think of this system?
2. What relative contribution do you think the use of feedback and architecture makes to the understanding and potential redesign of the system?
3. In what positive and negative ways does the *governance* constrain and restrain the system?
4. How might the *governance* change if the scope of the system is narrowed (e.g., might C2 itself be considered governance in some cases)?
5. What type of redesign do the authors propose for the system, and how do you think that might be realized?

Additional research on both the feedback and architectural analysis needs to be pursued further. If the analysis is to be truly useful, it needs to highlight the specific factors and situations that should affect the design of the situation and then give recommendations for the commanders' responses to those situations. For example, different mission threads would most likely require different architectures. Different phases of a given fight may change the architecture for a given mission thread, and different types of wars may necessitate different approaches in all phases. The feedback analysis should be done for each factor that may affect the architecture. Finally, we presented only one type of architectural view in this chapter. The architecture should be analyzed using multiple views that get into relationships, systems, and overall flow of information.

Readers may benefit from perusing Table 21.1 showing the authors' assessment of how well the relevant aspects of complex systems from the complexity theory (refer to Chapter 2) applied to this case study.

REFERENCES

Call, S. *Danger Close: Tactical Air Controllers in Afghanistan and Iraq.* College Station, TX: Texas A&M Press, 2007, pp. 19–20.

Clark, W. *Waging Modern War: Bosnia, Kosovo, and the Future of Combat.* New York: Public Affairs, 2001, pp. 200–201.

Cohen, E. The mystique of American Airpower. *Foreign Affairs* 73(1), 109, 1994.

Conetta, C. *Strange Victory: A Critical Appraisal of Operation Enduring Freedom and the Afghanistan War.* Project on Defense Alternatives, Research monograph no. 6. Cambridge, MA: Commonwealth Institute, January 30, 2002, http://www.comw.org/pda/0201strangevic.pdf. Accessed July 20, 2012.

Crane, C. *American Airpower Strategy in Korea, 1950–1953.* Lawrence, KS: University Press of Kansas, 2000, pp. 28, 162–163.

Daalder, I. and M. O'Hanlon. *Winning Ugly: NATO's War to Save Kosovo.* Washington, DC: Brookings Institute, 2000, pp. 43–45.

Deptula, D. SAF/OSX and Chief of Iraq/MAP cell in campaign plans during desert storm, Memorandum for record, subject: Feedback from SECDEF/CJCS meeting with CINC and component commanders, CHP-5A in desert story collection. Maxwell AFB, AL: AFHRA, February 9, 1991.

Deptula, D. Reflections on desert storm: The air campaign planning process, Briefing, Version 8, October 20, 1998.

Deptula, D. *Effects Based Operations: Change in the Nature of Warfare.* Arlington, VA: Air Force Association's, Aerospace Education Foundation, 2001.

Franks, T. *American Soldier.* New York: Regan Books, 2004, pp. 249–251.

Freedman, L. and E. Karsh. *The Gulf Conflict, 1990–1991: Diplomacy and War in the New World Order.* Princeton, NJ: Princeton University Press, 1993, pp. xxix, 69–75.

Galbraith, J.R. Organization design: An information processing view. Sloan working papers. Cambridge, MA: MIT, 1969.

Galbraith, J.R. *Competing with Flexible Lateral Organizations.* Reading, MA: Addison-Wesley, 1993.

Grant, R. Reach-forward. *Air Force Magazine* 85(10), 43–47, October 2002.

Haave, C. and P. Haun. *A-10s over Kosovo: The Victory of Airpower over a Fielded Army as Told by the Airmen Who Fought in Operation Allied Force.* Maxwell AFB, AL: Air University Press, 2003, pp. 147–148.

Henriksen, D. *NATO's Gamble: Combining Diplomacy and Airpower in the Kosovo Crisis 1993–1999*. Annapolis, MD: Naval Institute Press, 2007, pp. 7–8.

Hill, L., D. Cook, and A. Pinker. *A Statistical Compendium*, Vol. 5, Part 1 in E. Cohen and T. Keaney, Gulf War Air Power Survey. Washington, DC: Government Printing Office, 1993, pp. 513–515.

Hinote, C. *Centralized Control and Decentralized Execution: A Catchphrase in Crisis*? Research paper 2009-1. Maxwell AFB, AL: Air Force Research Institute, 2009, pp. 13–20.

Horwood, I. *Interservice Rivalry and Airpower in the Vietnam War*. Ft. Leavenworth, KS: Combat Studies Institute Press, 2006, pp. 146–165.

Hosmer, S. *Why Milosevic Decided to Settle When He Did*. Santa Monica, CA: RAND, 2001, pp. xiii–xxi.

Jamieson, P. *Lucrative Targets: The U.S. Air Force in the Kuwaiti Theater of Operations*. Washington, DC: Air Force History and Museums Program, 2001, p. 81.

Jansen, J., N. Dienna, T. Bufkin II, D. Oclander, T.D. Tomasso, and J. Sisler. JCAS in Afghanistan: Fixing the Tower of Babel. *Field Artillery* 2, 24–29, March/April 2003 (Call 60–61).

Joint Center for Operational Analysis (JCOA). Libya: Operation Odyssey Dawn (OOD) executive summary. Suffolk, VA: JCOA, September 21, 2011, p. 12.

JP 3-0. Doctrine for Joint Operations, August 11, 2011.

Kirkpatrick, C.E. Joint fires as they were meant to be: V Corps and the Fourth Air Support Operations Group during operation Iraqi freedom. Land Warfare Papers. Arlington, VA: Association of the U.S. Army, Institute of Land, 2004.

Kitfield, J. *Prodigal Soldiers: How the Generation of Officers Born of Vietnam Revolutionized the American Style of War*. Washington, DC: Potomac Books, Inc., 1995, pp. 296–298.

Kometer, M. *Command in Air War: Centralized versus Decentralized Control of Combat Airpower*. Maxwell AFB, AL: Air University Press, 2007, pp. 108–109.

Lambeth, B. *The Transformation of American Airpower*. Ithaca, NY: Cornell University Press, 2000, p. 232.

Lawrence, P.R. and J.W. Lorsch, Differentiation and integration in complex organizations. *Administrative Science Quarterly* 12(1), 1–47, 1967.

Leibstone, M. War against terrorism and the art of restraint. *Military Technology* 25(11), 18–21, November 2001.

Neuenswander, P. JCAS in operation Anaconda: It's not all bad news, Letters to the editor. *Field Artillery* 3, 2, May/June 2003.

O'Mara, R. Stealth, precision, and the making of American foreign policy, Master's thesis. Maxwell AFB, AL: School of Advanced Airpower Studies, Air University Press, June 2002.

Olsen, J.A. *John Warden and the Renaissance of American Air Power*. Washington, DC: Potomac Books, 2007, pp. 145–192.

Powell, W.W. Neither market nor hierarchy: Network forms of organization. In *Research in Organizational Behavior*, L.L. Cummings and B.M. Staw (eds.). Greenwich, CT: JAI Press, 1990.

Putney, D. *Airpower Advantage: Planning the Gulf War Air Campaign 1989–1991*. Washington, DC: Air Force History and Museums Program, 2004, p. 100.

Romanelli, E. The evolution of new organizational forms. *Annual Review of Sociology* 17(1), 87–103, 1991.

Smith, R. *The Utility of Force: The Art of War in the Modern World*. New York: Vintage, 2007, p. x.

Syrett, D. Tunisian campaign. In *Case Studies in the Development of Close Air Support*, B.F. Cooling (ed.). Washington, DC: Air Force History and Museums Program, 1990, pp. 169–172, 184–185.

Syrett, D. Northwest Africa, 1942–1943. In *Case Studies in the Achievement of Air Superiority*, B.F. Cooling (ed.). Washington, DC: Air Force History and Museums Program, 1994, pp. 227–331.

Thomas, E. and M. Brant, The education of Tommy Franks. *Newsweek* 141(20), 24–29, May 19, 2003.

Tirpak, J. Short's view of the air campaign, Washington watch. *Air Force Magazine* 87(8), 43–45, September 1999.

Watts, B. and W. Murray, *Operations*, Vol. 2, Part 1 to E. Cohen and T. Keaney, Gulf War Air Power Survey. Washington, DC: Government Printing Office, 1993, pp. 260–261.

Welsh, M.A. Day of the Killer Scouts. *Air Force Magazine* 76(4), 67–68, April 1993.

Woodward, B. *The Commanders*. New York: Simon and Schuster, 1991, pp. 260–261.

Woodward, B. *Bush at War*. New York: Simon and Schuster, 2002, pp. 44, 51–52.

Section IX

Transportation

22 NextGen
Enterprise Transformation of the United States Air Transport Network

Hamid R. Darabi and Mo Mansouri

CONTENTS

CASE STUDY ELEMENTS

Fundamental Essence

- In the last decades, the role, structure, technology, and complexity of Air Transportation Network (ATN) changed significantly. The modern ATN is a complex, large-scale critical infrastructure.
- According to Rebovich and White, an enterprise system is "an entity comprised of inter-dependent resources … that interact with each other … and their environment to achieve goals" [1]. Because in ATN many players interact to deliver transportation service and share the infrastructure, it is a loosely connected complex enterprise.
- Notwithstanding the Federal Aviation Administration (FAA) regulatory position, often, there is not a central decision-making authority in enterprise systems like ATN [2].

- FAA started modernization of the US ATN through a revolutionary project entitled "The Next Generation Air Transportation System (NextGen)." However, because of neglecting the enterprise-wide complexity of the project and lack of attention to the unique characteristics of governance of this project, it faces many challenges. The aim of this case study is to highlight the need for a new enterprise transformation governance framework.

Topical Relevance

- Air transportation is a significant part of the US economy. In 2009, the airlines carried 793 million passengers, and air freight transportation generated more than 53 billion revenue ton-miles (RTMs) [3]. The aviation industry sustained 10.2 million jobs, contributed $1.3 trillion in economic activity, and generated 5.2% of the gross domestic product (GDP) in the same year [3].
- Because of this huge, massive scale, understanding the whole system is impossible for a group of people without access to modern models, which capture the complexity of the system.

Domain(s)

- The distributed governance structure of ATN and its transformation are the focus of this case study.

Country of Focus

- The ATN is studied in the United States, which has the most complex air transport network on the globe.

Stakeholders

- The ATN includes commercial airlines, the US Air Force, the national airspace management system, numerous aircrafts (military, commercial, and private), and a network of airports and air bases.
- The FAA is the main institution (representing both civil and military interests, in contrast to other nations) that regulates the US ATN. Airport authorities are able to set and enforce the rules pertaining to their regional resources. The International Civil Aviation Organization (ICAO) sets the standards and best practice guidelines for managing the international interactions with the US ATN.

Primary Insights

- The current conditions of a complex and interconnected system like the US ATN result from adaptive interactions among multiple players in different structural layers.
- The growth in this system did not happen randomly or by fiat but through an unpredictable, nonspecifiable evolutionary path.
- Therefore, in changing any structural part of this system like rules and regulations, network structure and infrastructure, and the players, it is essential to observe and try to interact with the impacts of change in the overall conditions of this complex adaptive system. These outcomes include (unexpected) emergent properties, especially disruptive events that might be unexplainable even after they occur.

KEYWORDS

Air transport network; Air transport management; Regulating network industry; Complex socio-technical system; FAA NextGen; Enterprise transformation

ABSTRACT

ATN in the United States is an important enabler for the US economy. This document highlights the complexity in the US ATN, dynamics of interactions in the network, regulatory and commercial players' challenges and interests, and technical constraints of this broad network. Moreover, the role of FAA NextGen as a complex enterprise, which aims to revolutionize the US air transportation network, is studied. This study highlights the requirement for a comprehensive governance framework for enterprise transformation and proposes a three-layered framework that includes authority layer, organizational layer, and physical layer to facilitate a theoretical understanding of this complex enterprise.

GLOSSARY

ADS-B	Automatic Dependent Surveillance Broadcast
ALPA	Airlines Pilots Association
ATC	Air traffic control
ATN	Air Transportation Network
ATO	Air Traffic Organization
CAA	Civil Aviation Authority
CAS	Complex adaptive system
Data Comm	Data Communication Program
ERAM	En Route Automation and Modernization
FAA	Federal Aviation Administration
FAA NextGen	Next Generation Air Transportation System
GAO	General Accountability Office
GDP	Gross domestic product
GPS	Global positioning system
IATA	International Air Transport Association
ICAO	International Civil Aviation Organization
INS	Inertial navigation system
JPDO	Joint Office of Planning and Development
NAS	National Airspace System
OEP	Operational evolution partnership
SoS	System of systems
TCAS	Threat Alert and Collision Avoidance
UAS	Unmanned aircraft systems

BACKGROUND

CONTEXT

The demand for air transportation shows exponential growth during the last decades. In 1980, the airlines carried 281 million passengers [4]. Within 10 years, the revenue passenger miles in the US market increased from 540.7 million in 1995 to 779.0 million in 2005 [5]. This demand is expected to show an 82.4% increase from 2010 to 2030 [6]. This increment in traffic will cause congestion and delay, and in the current network would cause up to a $6.5 and $19.6 billion economic loss by 2015 and 2025, respectively [7].

Moreover, the catastrophic events of 9/11 had an enormous impact on the US ATN. One impact of these events was increasing security at all levels. Another impact was a sharp decline in the demand of air transportation for almost 2 years until its partial recovery by the end of 2003 [3]. Moreover, this change in the business environment caused a change in the shape of the ATN. As Hua et al. found, there is a jump in the number of ATN nodes (active commercial airports) after 2001 [8]. They attributed this change to the effort of airlines to find new profitable city-pairs.

These major changes motivated President Bush to sign the Vision 100-Century of Aviation Reauthorization Act to revolutionize the US ATN in 2003 [9]. The integration plan of Next Generation Air Transportation System (FAA NextGen) was released by the Department of Transportation on December 15, 2004. The FAA NextGen project was estimated to end in 2025 [10].

NextGen is the transformation of the US ATN led by the FAA. The objectives of NextGen are to (1) retain the US leadership in global aviation, (2) expand the capacity, (3) ensure safety, (4) protect the environment, (5) ensure the national defense, and (6) secure the nation [11]. This complex project is designed to meet the new-century needs of air transportation.

PERTAINING THEORIES

The ATN includes different entities like airlines, airports, and air traffic control (ATC) service, which are working together. In the abstract form, this network is a *complex adaptive sociotechnical system* (CASS). The idea of CASS is derived from complex adaptive systems, which according to Holland are "systems that have a large numbers of components, often called agents, that interact and adapt or learn" [12]. We define CASS as "large-scale systems in which humans and technical constituents are interacting, adapting, learning, and coevolving. In these systems technical constraints and social and behavioral complexity are of essential essence." In air transportation, the airlines and airports adapt to the changes in economic, social, and technical environment. Therefore, the ATN exemplifies the characteristics of a *complex network*. The *evolutionary economic theory* is the main theory that explains the interactions within an economic complex adaptive network [13]. The characteristics of this complex adaptive network are presented in the following sections.

EXISTING PRACTICES

Because of the importance of ATN, there are various approaches to studying the impact of a regulatory institution on this infrastructure. Analyzing the economics of the network through statistical analysis is one method. For example, the impact of deregulating ATN on the welfare of the United States is analyzed by this method [14]. However, this method is incapable of capturing networks effects and analyzing the effects of particular policies on the whole network. Moreover, this method requires long-term and accurate data about the system.

Comparative studies and benchmarking is another method [15]. In this perspective, effective regulation of ATN requires understanding and applying best practices of similar systems. Winston and de Rus collected numerous case studies to serve this purpose. However, the possible ignorance of the subtleties of CASS and the different nature of the complexity in different cases limit the applicability of this approach.

A noble and more recent approach to understanding ATN is through the application of game theoretic models. Aguirregabiria and Ho used this method to study the impact of network shape on entry deterrence in the market and entertaining different policies [16,17].

The last method is more similar to our approach of studying ATN as a CASS. This approach tries to fulfill a notable gap in the scholarly work about the air transportation system. Applying powerful methods of complexity science in understanding US ATN is still in its inception period, although there are noteworthy works (see Refs. 18–20).

GUIDING PRINCIPLES

In a useful categorization of the flexibility of organization in a system of systems (SoS), Gorod et al. have defined four classes of SoS: (1) rigid organization, (2) planned organization, (3) flexible organization, and (4) chaotic organization [21]. According to their categorization, because there are many stakeholders in the ATN and the outcome is dependent on their collective decisions, the ATN is a flexible SoS [21]. This flexibility demands a systemic enterprise framework to govern this SoS [22]. Fortunately, the guiding principles that players use to make critical decisions are limited. Some of these strategic decisions are changing the structure of the network, the network capacity, and the segmentation of the market by diversifying services. Therefore, some of the system dynamics can be anticipated based on these guidelines. For example, the general strategy of air passenger carriers is to develop hub-and-spoke or point-to-point networks, and the low-cost airlines use the strategy of serving lucrative parts of the network. Using these guiding principles enables researchers to study the network.

CHARACTERIZATIONS

In comparison to other transportation networks, there is one major similarity and one major difference between the ATN and other networks. The similarity is that decision-making is fragmented in this widespread network. The difference is that safety and environmental impacts are a major concern in the ATN.

Type of System

According to Mansouri et al., there are four types of systems to govern: (1) assembly systems, (2) traditional systems, (3) flexible SoSs, and (4) chaotic systems [22]. For the management of assembly systems, the hierarchical command and control management style is appropriate. Management of traditional systems can be done by applying five classical principles of management, which are planning, organizing, staffing, directing, and monitoring [23]. The flexible SoSs must be managed by an enterprise system governance structure, while enforcing and influencing are the main methods for directing chaotic systems [22]. The ATN exhibits the characteristics of a flexible SoS, and as a result, an enterprise system governance structure is appropriate to coordinate it.

System Maturity

The Capability Maturity Model Integration (CMMI) standard defines five maturity levels for a system: (1) initial, (2) managed, (3) defined, (4) quantitatively managed, and (5) optimizing [24]. A very simplistic analysis of the air transportation industry in the United States can identify its maturity within the framework of these stages. We consider the period from the invention of the airplane (1903) to the formation of the Civil Aeronautics Authority (1938) as the initial stage. In this stage, the US Postal Service provided the airmail service (1908). In the second stage (from 1938), the central management bureau and the first passenger aircraft developed (1949). From 1965, when the FAA took charge, the third stage of maturity started. In this stage, safety, security, and navigation standards were defined, and the period of regulating air transportation began. Moreover, the navigation organization, workforce unions, and airline associations were established in these years. Therefore, we refer to this stage as the defining stage. The fourth stage began from the Air Transportation Deregulation Act (1978), in which the decentralized airplanes provided more efficient services and a more competitive environment. The last and final stage started from the establishment of the FAA NextGen project to optimize the air transportation infrastructure and to maximize the utilization of its resources.

Environment

The US ATN is bonded within different suprasystems. For example, it is part of the global air network, it serves the US economy, and it is part of the transportation network in the United States.

ICAO recommends regulations for the global ATN. Therefore, the US air industry is obliged to follow ICAO's rules and regulations. The flow between the US ATN and the global network is enormous; US and foreign carriers transported 160.1 million passengers and 9.73 million freight tons in 2010 between the United States and foreign destinations [25].

The role of the ATN in the US economy and the general transportation network in the United States has already been discussed. Airports are an important part of intermodals, which are nodes to change the mode of transport like rail and bus, and air transportation is the fastest way to move perishable and precious goods from one part of the United States to another. Therefore, this interconnectivity enhances the capacity and reliability of the US supply chain.

Systems Engineering Activities

Although different parts of the US ATN are analyzed using systems engineering methods, there is no comprehensive review about the current network in general. A proposed architecture of the *to be* processes of the infrastructure is documented in "Concept of Operations for the Next Generation Air Transportation System" [11].

As Is System Description

The US ATN consists of 547 commercial airports (from 19,782 airports in total) as of March 2012 [26]. Out of 76 active airlines in the US market, 19 air carriers are considered *major* airlines with more than $1 billion annual revenue, 35 are *national* airlines with annual revenues between $100 million and $1 billion, and 22 are *regional* airlines [27]. The D1B1 database [28], which includes 10% ticket information issued in the domestic US market, shows that about 75% of the total domestic air traffic is carried by 15 major passenger carriers among the 26 busiest cities—as depicted in Figure 22.1.

To Be System Description

FAA NextGen will result in major changes in the ATN. The structural changes in the ATN after NextGen will be (1) from ground-based navigation to satellite-based navigation, (2) from the current

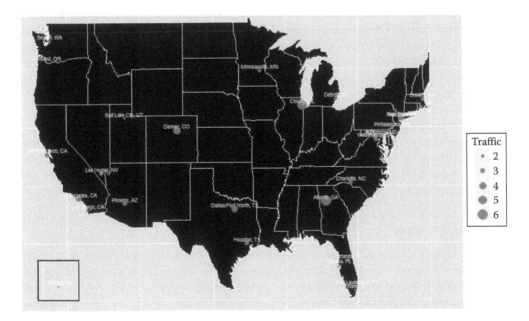

FIGURE 22.1   **(See color insert.)** Major hubs in the United States and the percentage of total network traffic.

disconnected system to a net-centric connected architecture, (3) from a fragmented weather fore-cast system to a single authoritative one, from voice communication to digital communication, and improvement in formerly visibility-limited operations [29].

The purpose of these changes is to increase the utilization of the current infrastructure. However, there are some factors that might cause emergent behavior. Airlines would have the ability to expand their hub-and-spoke network, which would cause disproportionate changes in the demand. Moreover, a change in the aircraft technology and increase in the demand for smaller aircraft [6] will have unpredictable effects on the shape of the network in the future. These dynamics require further research and attention.

PURPOSE

In classifying SoS, the Department of Defense introduces four types of SoS according to their purpose: (1) virtual SoS, which does not have a centrally agreed purpose; (2) collaborative SoS, in which constituent component systems voluntarily fulfill an agreed-upon purpose; (3) acknowl-edged SoS, in which there is a recognized purpose; and (4) directed SoS, which is built to fulfill a predetermined purpose [30]. From this perspective, the US ATN is a collaborative SoS, in which different constituents deliver services, manufacture products, and share knowledge to provide transportation service. In other words, the US ATN is a collaborative effort of its constituents to deliver a safe and secure air transportation service. The mission of the constituents is discussed in the "Mission" section.

HISTORY

As mentioned earlier, US Air Transportation came into being since the birth of the first airplane in 1903. In the early twentieth century, the Air Mail Act facilitated the development of the airline industry followed by the Air Commerce Act, which gave the Secretary of Commerce the power to establish airways, certify aircrafts, and license pilots along with issuing and enforcing air traffic regulations. The first commercial airlines in the United States were Pan American, Western Air Express, and Ford transport service [31].

The mid-twentieth century saw the Civil Aeronautics Act being passed which established the Civil Aeronautics Board. The Board performed several functions including determining airline routes of travel and regulating prices for passenger fares. It operated to generate competition among the airlines by striving to offer the best-quality service. Later, this organization evolved into the FAA, which was created in 1958, in response to a mid-air collision over the Grand Canyon, to man-age safety operations [31].

The late twentieth century was the postderegulation era in which new carriers rushed into the market and new routes directly connected cities that were previously accessible only via a string of layovers. The number of customers and competition increased due to a drop in airfares, which gave rise to a new problem of ATC. The air traffic controller strike in 1981 brought a temporary setback to the growth of the ATN. All of these setbacks resulted in a subsequent recession in the early 1990s. The surviving airlines rode out of this recession to record reasonable profits by the late 1990s [32].

THEN CURRENT SITUATION

The trend of traffic in the US ATN shows that although there are around 560 commercial airports, the traffic is consolidated on a small subset of these airports. The analysis shows that during the period 2000–2008, the total traffic in the 35 most congested airports was 63.1%–64.5% of the total traffic [6]. Moreover, 91.6%–92.1% of the traffic was carried by the top 100 congested airports [6]. The aforementioned prediction of increment in the traffic highlights the fact that the problem of congestion in particular airports will worsen if future market demands materialize.

KNOWN PROBLEM(S)

As discussed, there are two major problems in the ATN: (1) the increase in traffic congestion and delays and (2) the demand for safety and security. The terrorist attacks of 9/11 highlighted the requirement for new security measures. The Vision 100-Century of Aviation Reauthorization Act reiterates this requirement [9]. The traffic congestion problem is discussed in previous sections.

Furthermore, there are other issues like environmental and energy preservation considerations. The CO_2 emission from aircraft is a subject of discussion, and there are attempts to substitute fossil fuels with green fuels and to develop engines suited for green energies. The urban pollution and noise is a major concern in increasing the capacity of airports and is highly regulated by various acts and rules. The role of the NextGen project in answering these issues is discussed in the following sections.

MISSION AND DESIRED OR EXPECTED CAPABILITIES

Mission

As already stated, the US ATN is a collaborative SoS in which there are different constituents with various missions. For example, the mission of the FAA is defined "to provide the safest, most efficient aerospace system in the world" [33], while the purpose of the Air Transport Association of America "is to foster a business and regulatory environment that ensures safe and secure air transportation and enables US airlines to flourish stimulating economic growth locally, nationally and internationally" [34].

Based on the current status of the US air transportation system, adapting our present air transport paradigm will prove insufficient to meet the challenges discussed earlier. In order to account for these challenges, the mission of NextGen is to ensure a healthy, environment-friendly, nationally interoperable air transportation system by 2025. It would possess the potential to ensure safe, efficient, and reliable movement of large numbers of people and goods throughout the air transportation system in a way that is consistent with the national security objectives. Moreover, the FAA NextGen Vision includes an underlying set of principles and several key desired capabilities, which support a much wider range of operations.

Desired Capabilities

The objective of the NextGen project is to use satellite-based navigation, which will enable a significant increase in US airspace capacity. In NextGen, many airplanes, especially those that fly at higher altitudes, will have the global positioning system (GPS) as well as an inertial navigation system (INS). Therefore, pilots will know their positions more precisely without having to rely so much on ground controllers. This will permit airplanes to engage in flights without direct ground control throughout the vast airspace, instead of being forced to fly in-trail under the direction of ground controllers. This change in the direction will cause major fuel efficiency. Moreover, in congested areas, planes will fly closer together safely because of the Automatic Dependent Surveillance Broadcast (ADS-B) system, which provides excellent situational awareness to the pilots regarding the whereabouts of nearby airplanes. This revolution in navigation technology is estimated to reduce the delays by 30% and to cause 1.4 billion gallons cumulative reduction in fuel use by 2020 [35].

PROCESS FOR ACHIEVING THE OBJECTIVES NEEDED AND WHY

Because of the aforementioned factors, the current air transportation infrastructure cannot be depended upon to meet the growing demand for air travel and airfreight transportation. Moreover, the increase in smaller general aviation aircraft in future years will reinforce the congestion problem [6]. This and the pervasive use of antiquated ground system technology (much of it is from the 1970s or earlier) are the primary reasons why the overall US air transportation system needs modernization. FAA NextGen promises a comprehensive overhaul of the National Airspace System (NAS) to ensure as many safe, secure, and hassle-free flights as possible.

COMPLEX ENTERPRISE

In this section of the case study, we analyze the ATN through the lenses of complex enterprise. Mansouri and Mostashri define an extended enterprise system as a "complex network of distinctive yet distributed and interdependent organizational systems that are connected in an autonomic way to achieve objectives beyond reaching capacities of each" [36]. Incorporating this definition, the main constituent elements within the ATN are FAA, airlines, airport authorities, and international organizations like ICAO and International Air Transport Association (IATA). Furthermore, the project is dependent on the collaboration of other systems like local communities and labor unions. The roles and interplays among these organizational systems are discussed further in the "Stakeholder" section.

Similar to any complex system, the US ATN is not working in vacuum but in interaction with a complex environment. There are numerous external factors that might affect the long-term prospective of the industry. The airplanes manufacturing technology is moving in two opposite directions: bigger airplanes, with new generation of the design of wings, and smaller affordable aircrafts. This shift in the air manufacturing technology will change the market demand. Tighter environmental regulations and the movement for support of green fuels will also have an impact on the conditions of the industry. A more detail discussion is presented in the "Environment" section.

To deal with these changes, the governing body uses its leverages to influence the behavior of the systems. The observations about the behavior of the systems, through feedback, will result in changes in the policies (Figure 22.2).

In Figure 22.2, the main systems of this complex enterprise are airlines and airport authorities. The whole system is under the influence of external factors like change in technology, international regulations, and treaties, change in demand, and change in the competition. The governing body, which is mainly the FAA, gets the feedback from the system by observations, market studies, airline reports, regular technical inspections, and presidential and congressional oversights.

Here is a brief explanation of factors A–F in Figure 22.2.

A: *Complex system* is the ATN, which consists of all US airlines, commercial services of airports and their physical infrastructure, air traffic management systems, the general aviation community (including the military), and the people most closely associated with them.

B: *External factors*—Antiquated and gradually more modernized technology, international regulations, air traffic demands of the flying public, air cargo, and international flights. Equipment manufacturers are considered external factors. Moreover, competition outside of the airline industry, the general economic situation, weather conditions, and the environmental concerns and regulations are also a part of external factors.

C: *Feedback* consists of almost everything that happens in the ATN, including air traffic management and rerouting successes, and many negative impacts such as airport (primarily runway) congestion, limitations to *free flight* caused by traditional in-trail ATC methods, flight delays induced by weather and busy tower control areas, clogged air–ground communication channels due to limited spectrum and antiquated multiple airborne radios, etc.

D: *Governing body*—Includes the FAA, ICAO, IATA, Air Line Pilot Association, Air Traffic Controllers Union, local airport authorities, etc. All these organizations had considerable impact on the success of air traffic management improvements, particularly those associated with the ATN.

E: *Constraints*—There are many; to name a few, safety (the primary concern), security (especially to assure the flying public against terrorist threat), environmental impact (e.g., of airports in populated areas), infrastructure (e.g., inadequate number and length of runways), the legacy enterprise (e.g., antiquated automation equipment and old voice radios), and budget (despite collected landing traffic fees, Congress diverts the funds for other uses).

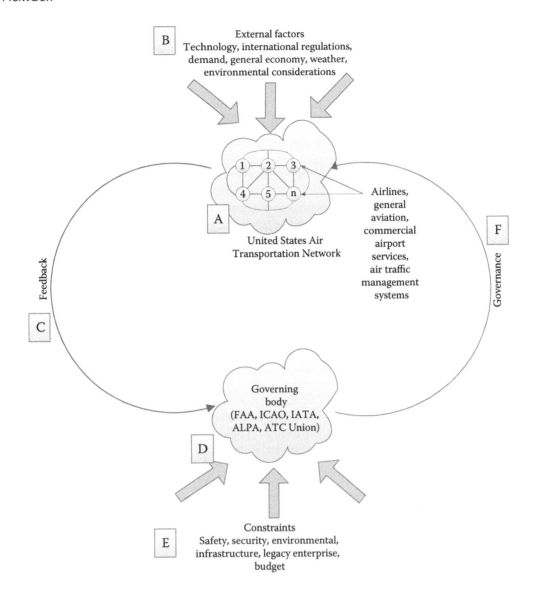

FIGURE 22.2 Framework to understand the governing body and US ATN interactions.

F: *Governance*—This consists mostly of the bureaucratic and slowly evolving introduction of streamlined regulations (such as smaller vertical separation requirements over the Atlantic Ocean, newer technology, and improved system concepts such as more modern ATC methods like ADS-B.

Emergent behavior, evolutionary path, and adaptability are among some of the differences between systems and complex systems. In ATN, there are many sources of emergence. For example, the terrorist attacks of 9/11 caused tightening of security policies, reduction in tourist travel, and bad economic performance in general. Another example is the volcanic ashes of 2010, which caused a cascade of schedule problems in air transportation. Because of the ever-increasing complexity of global air transportation network, these emergent behaviors will increase in the future.

The evolutionary path of ATN and the adaptability of players are interconnected. For example, during the last decades, the change in oil prices caused the demand for new fuel-efficient airplanes, which itself caused a new cost structure and change in business strategies. The evolutionary economics is very suitable to explain these changes, while it is hard to understand the change by merely studying the market.

ENVIRONMENT

Nowadays, the airlines are working in a very hostile economic environment. Two major airline passenger segments are business travelers and tourists; the demands of both groups are dependent on the condition of the general economy. In addition, the freight market is dependent on the growth of other industries. For example, during the financial crisis, from December 2007 to its end in June 2009, the airline revenue faced a major decline even more than the general economy.

Another force of change in the airline industry is the improvement in the technology. Rapid improvement in the technology in airline industry has been common from the birth of the industry. The fuel efficiency of a modern jet has improved by 70% in comparison to the first generation of jet engines in the 1930s [37]. The planning, scheduling, network design, revenue management, airport design, and other managerial and engineering aspects of the airline business have significantly improved by the development of new mathematical models. Moreover, navigation technology has provided safe flight conditions in adverse environmental and weather conditions.

Safety issues also have a significant impact on the airline business. For example, the volcanic ash in Iceland affected 462,000 passengers and caused $957 million economic loss in America. This occurred because the airplane routes, which are enforced by the safety regulations, were unusable and the domino effect of rescheduling flights caused huge economic loss.

Moreover, there are other economic concerns. Fuel cost is the biggest part of the airline operating costs [38]. Therefore, a trivial change in the price of jet fuel can affect the profit margin of the industry, although this risk in the change of fuel price is often mitigated by sophisticated financial strategies. As a labor-intensive industry, the airlines should negotiate with powerful labor unions. Labor strikes in the industry are not uncommon and cause significant economic losses.

SCOPE

Various numbers that can represent the enormous scope of the US ATN have already been discussed. A complex network of 600 commercial airports and 76 airlines carries about 800 million passenger miles and 53 billion ton-miles of freight. Thirty-nine million annual air traffic landing and takeoff operations in this huge network are managed by 35,000 ATC organization personnel [26].

STRUCTURE

Assuming that airports are nodes and airways are edges of a complex network, we can use graph theory and network analysis to understand the structure of the air transportation system. Similar to many real-life networks, the ATN is a small-world network. Small-world networks are highly clustered graphs in which the distance between any pair of randomly selected nodes is very low [39].

Hua et al. used data about the itinerary of domestic US passengers in 2007 to analyze this network. They studied some critical measures of this network and arrived at the following conclusions [8]:

- The density of the network is 0.015. This number is between 0 and 1 and shows the number of edges to the total possible number of edges. The low density shows that the network is far from a fully connected network.
- The average shortest path in the network is 3.23 air flights, and most of the airports can be connected within two flights. This should be compared to the 1372 cities in the network, which highlights the high connectivity of this small-world network.

- Any two airports can be connected by fewer than eight flights. The largest number (eight) of connecting flights must be taken between Ophir Airport in Alaska and Jackson Carroll Airport in Kentucky.
- Each airport is connected, on average, to eight other airports, and the maximum number is in Atlanta International Airport with 371 connections.
- The 9/11 terrorist attacks caused a structural change in the network. From 2001 to the end of 2002, the number of nodes in the network rose from 691 to 1283, which shows a shift in the business to serve new destinations.

The airlines use different strategies to develop and maintain their own network within this complex structure. Generally, there are two network design strategies that airlines use [40]: (1) point-to-point strategy, and (2) hub-and-spoke strategy. In a point-to-point network, the traffic between each node of the network is directly handled by flights, while in a hub-and-spoke network, the traffic is accumulated to central nodes (hubs) and is directed from there. The hub-and-spoke network structure provides economy of scale [41] while it creates disutility because of delays and connecting flights. A critical strategy of airlines is to expand their network by collaborations and code-sharing alliances with other airlines. This strategy is discussed in the "Internal Relationships" section. Figure 22.3 shows the hub-and-spoke network of the Sun Country Airline in winter 2012 of the routes, with more than 5000 passengers. The point-to-point network visualizes the Southwest Airlines network in the same season, with the routes of more than 20,000 passengers. Both networks are derived from the 10% ticket information issued in the domestic US market available at the DB1B database [28].

BOUNDARIES

The NAS can be considered as the boundary of the system. The NAS "is the network of US airspace: air navigation facilities, equipment, services, airports or landing/takeoff areas, aeronautical charts, information/services, rules, regulations, procedures, technical information, manpower, and material" [42]. The NAS in the United States is a shared space between the commercial airlines and military services. This shared space creates a boundary for the airspace that commercial airlines can use. After the 9/11 attacks, new regulations were enforced to protect the military-specific airspace.

INTERNAL RELATIONSHIPS

To facilitate the use of air transport networks, especially between domestic and foreign networks, airlines create different types of code-sharing alliances and collaborations [40]. The main idea of

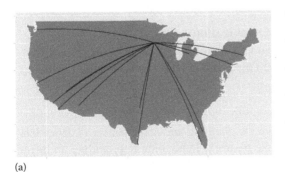 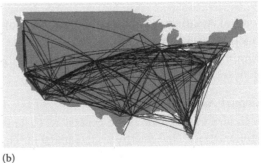

(a) (b)

FIGURE 22.3 Sample hub-and-spoke versus a point-to-point network. (a) Sun Country Airline hub-and-spoke network and (b) Southwest Airlines point-to-point network.

the deregulatory reform in the industry was to facilitate the competition because it would increase economic efficiency and improve the general society welfare [43]. Since then, airlines have developed different business models to compete in new segments of the market [44]. However, some scholars mentioned competition in this industry as *destructive competition* [45] that hinders utilizing the network capacity.

To improve the load factor, which is the ratio of total number of passengers to the total number of available seats, and for the purpose of serving new areas, airlines started developing single and multi-alliances. Different forms of alliances provide the ability to use the networks of other airlines. There are also critiques about these alliances. For example, Lin highlighted the entry barrier of these alliances [46], and Sundman pinpointed the increase in monopolistic behavior [47]. This dichotomy between competition and collaboration is the subject of different studies [e.g., 48].

External Factors

Although market price and airway regulations decreased after the Airline Deregulation Act of 1978, the total number of regulations increased dramatically. Noise and safety regulations constitute the main portions of these new regulations [45]. Aircraft noise is subject to many regulations, and takeoffs and landings in airports are ruled to reduce the impact on the local communities. Safety regulations are of even more importance. Although there is an economic debate that the risk in air transportation industry is overly mitigated in comparison to other risks like fatal car crashes [45], this emphasis on safety has made air transportation the safest mode of travel.

Constraints

Due to the limited capacity of airports, the IATA proposed a three-level categorization mechanism for airport management. In highly congested airports, Level III, the airport capacity is divided into time slots for flight arrival or departure. In Level II airports, there is a potential congestion in specific periods, and "[a] Level 1 airport is one where the capacity of the airport infrastructure is generally adequate to meet the demands of airport users at all times" [49]. Historically in the United States, airports like Kennedy, LaGuardia, Newark, Chicago O'Hare, and Ronal Reagan National Airport in Washington were under traffic restrictions [50], but currently, LaGuardia and Ronald Reagan Washington National are the only first-come, first-served slot reservation airports [51].

The environmental impact of the airports is another major concern, particularly for airport decision makers. Although, currently, the airports are under fewer market regulations, their level of environmental and safety regulations are increased significantly in current decades [52].

Governing Body

Federal Aviation Administration and Air Traffic Organization

The role of the FAA in regulating the air transportation system in the United States is discussed. Within the FAA, 35,000 personnel of the Air Traffic Organization (ATO) govern daily operations of air traffic to achieve safety and efficiency.

Airport Authorities

Each airport is a node in the general ATN. Therefore, its planning and management is dependent on the national and international demand of the whole network. In addition to providing an infrastructure for arrival and departure of airplanes, airports are ATN hubs that accumulate passenger and freight traffic [52]. The role of airports as intermodals and hubs for change of the mode of transportation is already discussed.

International Civil Aviation Organization

The ICAO is the specialized agency under the United Nations that recommends the principles, technical standards, and regulations for international air transportation. A hundred and ninety-one countries are members of this organization [53].

Airlines and Air Carriers

The first airline in the United States was established in 1925 by the legislation of the Air Mail Act. Since then, 76 major, national, and local airlines have been operating in the United States till date [27]. While some of the major airlines are specialized in transportation of cargo, others are active in moving passengers and products around the United States.

International Air Transport Association

IATA is the trade association of global airlines. Currently, 240 airlines are members of this organization, which carry 84% of the world air traffic. The mission of IATA is "to be the force for value creation and innovation driving a safe, secure and sustainable air transport industry that connects and enriches our world" [54]. IATA represents and leads the collective interests of its member airlines.

Labor Unions

Similar to other service industries, the airline industry is a labor-intensive industry. Labor cost is between 25% and 30% of the total cost of the airlines [32], and airline productivity is directly related to its workforce quality [55]. The unions provide 70% of the airline industry workforce in the United States and the airline industry is one of the most highly unionized industries in this country [56].

Aircraft Manufacturers

The aircraft-manufacturing industry is an important part of the US economy, which generates about 1.4% of the total GDP. More than 500,000 jobs are sustained by this industry, which signifies its importance. Boeing and European company Airbus primarily produce large commercial jets. The market for regional air jets is dominated by Brazilian Embraer and Bombardier from Canada, but these companies are highly dependent on US-manufactured parts [57]. Because of the oligopoly in this market and its dependence on the US ATN, aircraft manufacturers are important stakeholders in the system.

Air Passengers

Air passengers are broadly divided into two market segmentations. The first segment comprises the business travelers, who are very sensitive to the time and duration of the flight and whose employers pay their ticket price. The second market segment includes the leisure travelers, who are able to book their flights weeks before and are price-sensitive [58]. These two segments guide the revenue management activities of the airlines and scheduling of flights.

Logistic Service Companies

As already stated, because of the fast speed of air transportation, it is a valuable part of the global supply chain. Between 1980 and 2000, air cargo increased by five times [59]. In the United States, the main business of three of the major airlines—the airlines with more than $1 billion annual revenue—is transporting cargo: Astar Air Cargo, Federal Express, and UPS. This growth in the air cargo transportation emphasizes its importance for businesses.

CHALLENGES

One inherent cause of complexity in ATN is the presence of many stakeholders. This one factor is probably the most significant challenge in revolutionizing air transportation through FAA NextGen. The Joint Planning and Development Office determined the following most important risks to achieve FAA NextGen goals in 2025 [60]:

- Achieving interagency collaboration: The viability of FAA NextGen is dependent on the success of partner agencies to deliver their promises. It also requires integrating the work of multiple organizations to reach its purpose. Therefore, the Senior Policy Committee (SPC) is providing the required leadership to ensure delivery of the project.
- Capacity, delay, and environment: NextGen sets very high standards to reduce the delays, to increase the capacity of the US ATN, and to reduce undesirable environmental impacts. To achieve these objectives, many technologies should be integrated into current aircrafts, airports, facilities, and organizations. As a result, determining objectives that are both viable and favorable is important. This cannot be realized without reaching and maintaining consensus between important stakeholders.
- Airport and airspace security: To guarantee the security of national air space, many governmental agencies act together to detect suspicious activities and to protect the airspace. The design and planning of the required activities is an important interorganizational task. The Department of Homeland Security (DHS) leads the partner agencies in handling the challenges of this area.
- Information sharing: FAA NextGen is not viable without its net-centric capability of information sharing between important stakeholders. Lack of consensus about who is responsible for development and maintenance of this capability is an important challenge in implementation. Moreover, a change in the current legacy organizations' perspective toward sharing information is necessary.
- Research and development, verification and validation: The current practice of verification and validation is tailored toward testing and accepting a single large-scale complicated system. However, this practice falls short in integrating and testing a complex widespread SoS. FAA NextGen is a complex multistakeholder SoS and demands new methods to test not just the elements, but the relationships between them for the purpose of increased safety and security.
- Balance of human versus automation: Although the increase in ATC automation increases the operating capacity of the US ATN, it raises policy challenges. "Workforce adjustments, consolidation and alignment of facilities, assuring an equivalent level of safety to performance by a human operator, coordination of investments in complementary ground and aircraft automation, and shifts in liability" are the most important results of the improvements in technology.
- Local community support: This is also important for the success of FAA NextGen because airports are an important element of local communities and changes in airports require their support and collaboration.
- Communication, navigation, and surveillance backups: Having a backup to satellite navigation is mandatory to ensure the required level of reliability in the system. But the current ground infrastructure of ATC does not provide digital communication capabilities of NextGen (ADS-B).
- Integrating unmanned aircraft systems (UAS) into NAS: The demand for UASs is expected to increase for commercial and security purposes. This will raise a challenge to develop processes that are required to integrate UASs into FAA NextGen. The stakeholders are developing a research, development, and demonstration (RD&D) plan to address this issue.
- Achieving a critical mass of equipped aircraft: Realizing the full potential of any innovative technology requires a critical mass of technology adopters, and FAA NextGen is not an exception. This challenge has significant importance because it can reduce the total

realized profits of the project. The Joint Planning and Development Office (JPDO) has provided various options such as providing financial incentives, paying manufacturers' nonrecurring costs, providing operational incentives, and mandating equipage.

DEVELOPMENT

Involvement of multiple stakeholders in FAA NextGen is the key unconventional aspect about this program. Unlike many other systems, in the air transportation system, the stakeholders are not just the users of the system, as operating, maintaining, and improving the system requires collaboration among various players. Therefore, from the initial steps, the FAA accommodated the enterprise perspective in implementing NextGen [61]. This enterprise approach replaced the operational evolution partnership (OEP) term to highlight the role and responsibility of key agencies [61].

PROGRAM MANAGEMENT

The NextGen program includes a portfolio of implementation projects. The FAA categorized its implementation portfolios into (1) improved surface operations, (2) improved approaches and low-visibility operations, (3) improved multiple-runway operations, (4) performance-based navigation, (5) time-based flow management, (6) collaborative air traffic management, (7) separation management, (8) on-demand NAS information, and two supportive portfolios, which are: (9) environment and energy, and (10) system safety management [35].

While the detailed scheduling of these portfolios is beyond the scope of this book (see Ref. [35] for more information), the interdependency between the services and operations in FAA NextGen is noteworthy. In a complex program like NextGen, the act of dividing tasks into cohesive yet independent portfolios is a challenging one. For example, because of technical and budget difficulties, the use of the En Route Automation and Modernization (ERAM) program was postponed. But ERAM is itself dependent on the Data Communication Program (Data Comm) and enables ADS-B [35]. As a result of these interdependencies, program management in NextGen is a daunting task.

Contingencies

Change in the fiscal budget priorities is a major source of contingencies in FAA NextGen, similar to any other multiyear and large-scale government-funded project. As Forman puts it, "The process of systems architecting requires two things above all others: value judgments by the client and technical choices by the architect. The political process is the way that the general public, when it is the end client, expresses its value judgments. High-tech, high-budget, high-visibility, publicly supported programs are therefore far more than engineering challenges; they are political challenges of the first magnitude" [62].

FAA NextGen is not an exception to the rule. The US Government Accountability Office (GAO) has published several reports on the status of FAA NextGen on the performance of the project. The September 2012 testimony to the House of Representatives summarizes the main challenges of project implementation as "(1) delivering and demonstrating NextGen's near-term benefits; (2) developing a cost-effective mechanism to encourage operators to equip with NextGen technologies; (3) maintaining timely delivery of acquisitions; (4) clearly defining NextGen leadership roles and responsibilities; and (5) balancing the needs of the current radar-based systems and NextGen systems through the transition" [63]. This and other reports of the GAO highlight the uncertain benefits and costs of NextGen and the huge impact of implementation and delays on the final benefits of the program.

Information Management

Realizing the capabilities of NextGen is founded on a net-centric information-sharing infrastructure. The aim of this infrastructure is to provide the relevant information to authorized people both reliably and quickly. This change requires modifications in the underlying processes of the enterprise that provides these data [11]. Therefore, the JPDO designed a three-layered community model for the FAA NextGen enterprise. The first tier of this model depicts key concepts, actors, roles, and their relationships. The second tier provides information exchange between these stakeholders and enterprise segments. The third layer details the segments and the services, operations, and actors that provide those services or operations [64].

The NextGen enterprise architecture uses both the federal enterprise architecture (FEA) and the Department of Defense Architecture Framework (DoDAF). Different layers of the enterprise architecture (EA) are summarized in the strategic, operational, and system and program layers similar to DoDAF recommendations. The FEA is used to describe five architecture models: performance reference model (PRM), business reference model (BRM), service component reference model (SRM), technical reference model (TRM), and data reference model (DRM). The JPDO provides details of each model [64]. To ensure the security of information, the information services are divided into two separate secure and nonsecure areas.

Strategy

In addition to the traditional challenges of program management, one key challenge in implementing NextGen is the involvement of the airlines and international players. The reason is that lack of their participation in the project can delay achieving the full potential of its benefits. In any other innovative technology, a critical mass of early adopters is required for growth. In NextGen, the airlines and air carriers need to be equipped with costly systems to use the new satellite navigation system. The need for better communication about the project, its risks, and creating a consensus are recommended by RTCA, and the uncertainty in the benefits of this investment is highlighted by the GAO [63].

The FAA took action by initiating a proposal for the aviation community engagement strategy to implement NextGen [65], and developing a business case for implementing NextGen. The aim of these initiatives is to increase the key players' engagement in the process of adopting NextGen, and to make transparent the key decision factors for them. However, this challenge in FAA NextGen is unprecedented and the evolution of this part of the strategy can provide a good case study in the future.

Resources

FAA NextGen is a collaborative effort of thousands of experts in various organizations. In the FAA, the vice president of NextGen and operations planning in ATO were responsible for the project, while currently the assistant administrator for NextGen is the highest ranked individual reporting to the FAA deputy administrator. In addition, many subcontractors are working on portfolio programs to ensure on-time and on-budget delivery of the project.

While the initial estimate of the required budget for the program was $40 billion, there are critical remarks that in some scenarios it might be expanded to over $160 billion [66]. Until February 2012, the project cost overrun was 60% (from $11 billion estimated to $17.7 billion actual) [67]. To manage the uncertainty in cost, the FAA management shifted its perspective in achieving a mid-term milestone in 2020. The total expected cost of the midterm project by 2030 is forecasted as $37 billion, and the benefits are assumed to be worth $106 billion [68].

Schedule

As mentioned earlier, the project was initiated by Vision 100-Century of Aviation Reauthorization Act in 2003 and the integration plan started by 2004. The project was initially estimated to last until 2025 [10]. The FAA published a document Destination 2025 in which the vision for operation capabilities of NextGen are described [69].

SYSTEMS ENGINEERING

FAA NextGen is not just the transformation in air transport infrastructure, but also a change in the enterprise that provides those capabilities. The significant enterprise transformation is a challenge to systems engineers and requires a comprehensive framework. Although the EA document provides an overview of the desired capabilities [64], it does not provide guidance to enterprise transformation and to integrating new technologies in the legacy system.

Architecture

The FAA used both the DoDAF and the FEA in developing the FAA NextGen enterprise architecture. For educational purposes, we focus on the DoDAF framework and explain the EA using this framework. The DoDAF prescribes three different views for understanding an enterprise: operational view (OV), systems view (SV), and technical standard view (TV). Moreover, the all view (AV) provides a comprehensive perspective toward the enterprise.

From different DoDAF views, FAA NextGen EA includes the following elements: (1) overview and summary information (AV-1), integrated dictionary (AV-2), community model (OV-1), operational node connectivity description (OV-2), operational information matrix (OV-3), activity model (OV-5), operational event/trace description (OV-6c), system functionality description (SV-4), and operational activity to system functionality traceability matrix (SV-5) [64].

For example, Figure 22.4 shows a graphical representation of the community model. As briefly explained in the role of information in the program, a community model with three tiers is developed to highlight the position of each actor in this enterprise. Figure 22.4 shows the involved stakeholders and the high-level concept of operations that creates the core of the FAA NextGen.

Although the DoDAF is a suitable framework for capturing a snapshot of a working enterprise, it is not enough to understand the complexity of *enterprise transformation*. The change from the current as-is system to the future to-be system is a daunting task in complex systems and requires much attention to the evolutionary path of the system. Moreover, the DoDAF and similar frameworks are not tailored to address the challenges of governance of extended enterprise. In addition, process of integrating new technologies in an SoS, including their test and verification, is neglected in the DoDAF. Because of the lack of this framework, the project is faced with fundamental challenges (see "Challenges").

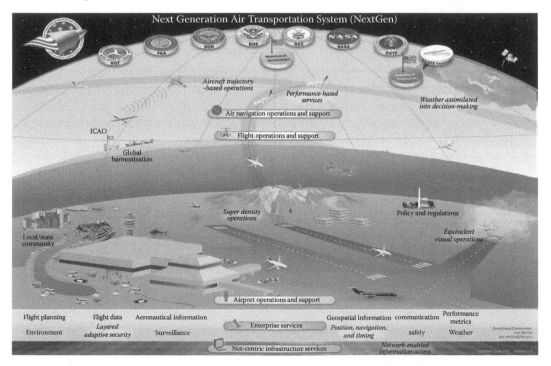

FIGURE 22.4 Graphical representation of the first tier of the community model. (From JPDO, Enterprise architecture V2.0 for the next generation air transportation system, Joint Planning and Development Office, 2007.)

Program Acquisition

In system engineering, a big challenge in making FAA NextGen a reality is the acquisition of programs from various subcontractors. As explained in the sections on challenges and budgeting, lack of coherence in acquisition can lead to cost overruns and increased project complexity because of interdependencies. As a first step to handle this issue, the FAA developed an acquisition workforce plan to enhance its human resource to deal with this issue. The role of 1490 acquisition professionals is to ensure the acquisition of new technologies with respect to safety and efficiency [70].

RESULTS

The rapid improvement of the technology and aforesaid environmental factors provided an opportunity for change in the air transportation infrastructure. The objective of FAA NextGen is to make this radical improvement happen.

OBJECTIVES ACCOMPLISHED

Currently, the architecture of the NAS is a disparate system. In the current structure of NAS, different entities navigate airplanes within separated parts of airspace. The NextGen four-dimensional trajectory (4DT) system is based on the satellite-based navigation in which the routing of airplanes is done through the digital communication among all the relevant entities on a net-centric system. The transformation that is required for applying new architecture is summarized [11]:

- Network-enabled information access: This transformation is central to the FAA NextGen mission in making information available, secure, and usable in real time.
- Performance-based services: This will enable a clear and precise definition of service tiers that allow the government to switch from equipment-based regulations to performance-based regulations.
- Weather assimilated into decision making: Leveraging the benefits of network-enabled information access; this capability will help in providing a common *weather* picture in order to support decision making.
- Layered, adaptive security: The current air transportation system employs security as an *add-on* dimension. The layered, adaptive security system will integrate security functions into the FAA NextGen system while moving large numbers of people and goods with the requirement of proportionally fewer resources to achieve it. It will exist in layers of defense designed in a way to detect threats relatively early.
- Aircraft trajectory-based operations: This capability would not only improve system efficiency but also meet the system goals of safety, security, and environmental compatibility.
- Super density operations: Through this capability, arrival and departure spacing would be reduced, and simultaneously capacity would be increased with closely spaced and converging approaches at distances closer than currently allowed and through simultaneous operations on a single runway.
- Equivalent visual operations: This capability would enable the controllers to delegate responsibility to aircraft to maintain separation when the aircraft is in the airport area. It would permit more airports to reliably serve their community or region, whether for commercial service, business aviation, air taxi services, air cargo, or general aviation [11].

FINAL SYSTEM DESCRIPTION

The system description of the FAA NextGen transformation process can be illustrated using its systemigram. Systemigram is a powerful tool to capture the main entities and relationships within a system. The FAA NextGen project systemigram is presented in Figure 22.5.

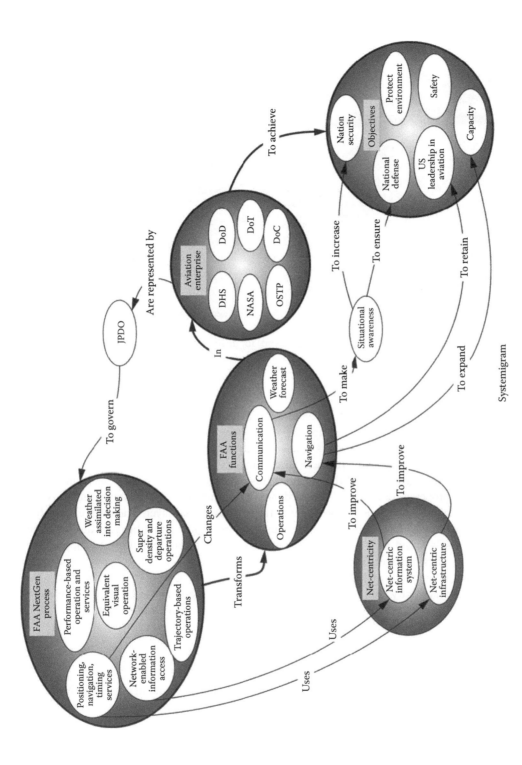

FIGURE 22.5 FAA NextGen process systemigram. (From Darabi, H.R. et al., Federal Aviation Administration NextGen: Applying systems thinking in practice, in *Proceedings of IEEE System of Systems Engineering Conference (SoSE)*, Albuquerque, NM, 2011.)

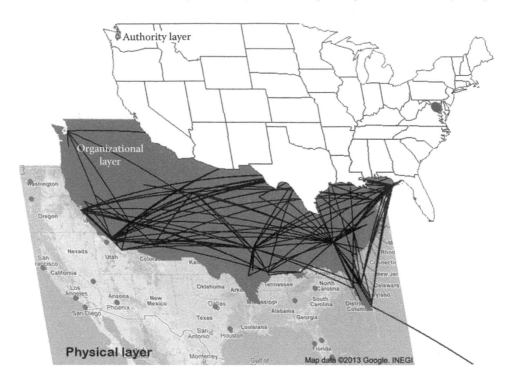

FIGURE 22.6   (See color insert.) Three-layered framework for analyzing network industries.

ANALYSIS

To understand the complexity of the US ATN, a three-layered framework, which is presented by Darabi and Mansouri [71], can be used (Figure 22.6). The lowest layer is the physical layer, in which the flow of the people, materials, and goods creates the dynamics. For example, the airports, aircrafts, and physical facilities create the physical layer of the ATN. The middle layer is the organizational layer, which might carry most of the intangible resources such as knowledge and funds. The interorganizational processes create the complex dynamics in this layer. For example, in the ATN, the airways and airspace management system are in the organizational layer. The interrelationship between the organizational and physical layers in this network is interesting because although the presence of airways is based on conventional contracts, the physical technology to support the protocols are in the physical layer. The top authority layer includes different regulatory authorities, which regulate the interactions and flows in the other layers. In the US ATN, this comprises the different state, county, and city regulatory organizations, which set restrictions on airports, airways, or other aspects of the network. Together, these players create the fluent and reliable ATN in the United States.

ANALYTICAL FINDINGS

Analyzing the three-layered framework, it is apparent that most of the challenges of FAA NextGen are in the mid-layer. Four different problems in organizational layers are notable: (1) creating the consensus among organizational players, (2) dealing with net-centricity and information-sharing capabilities required, (3) harmonizing the actions of different contractors and organizations involved, and (4) lack of a coherent map for transforming the legacy enterprise to NextGen enterprise. Therefore, although many technologies are involved in applying FAA NextGen, the main challenges lie within the enterprise and intangible strings of interdependencies in this layer.

LESSONS LEARNED

According to the previous discussion, the biggest challenge in FAA NextGen arose from legacy enterprise and its inflexibility to meet the demand of change. An important strategic shift in implementing NextGen is the decision to focus on a midterm intermediate system. In uncertain environments with highly complex tasks, it is a general recommendation to focus on developing intermediate systems [72]. There are many benefits to developing such systems, ranging from examining the assumptions and concepts to creating a prototype for communicating and creating consensus. It would be an interesting future research topic to analyze the success of this key decision.

However, attempting to manage FAA NextGen through the limited perspective of technological change was not a successful endeavor. The budget overruns and schedule delays highlight the requirement of synergy between players and the impact of politics on a large-scale multibillion-dollar infrastructure project.

BEST PRACTICES

In the first step of conceptual design of the FAA NextGen project, the JPDO was established. The purpose of this organization is to develop and maintain the integrated plan for the project. JPDO is responsible for studying the requirements and capabilities of all of the major stakeholders to ensure the project success. Establishment of the JPDO and its work as the integrator of the requirements of stakeholders is an important practice in developing NextGen. The experience of JPDO and its challenges in governing the relationship between different stakeholders and solving technical challenges can be an important body of knowledge for further governance of any type of multistakeholder complex project.

REPLICATION PROSPECTS

FAA NextGen is unique from different aspects: (1) this project underlies an unprecedented transformation in a large-scale infrastructure; (2) there is a combination of legacy systems with cutting-edge technology systems in the project, and this hybrid nature of the system raises many more concerns, especially about safety; and (3) the implementation of the project requires involving and leading various established organizations. The best practices of running this project can be applied in a wide range of large-scale multistakeholder projects, especially in infrastructure systems.

SUMMARY

The US ATN is a complex adaptive network in which airlines, airport authorities, airspace navigation and management, and other stakeholders work together to provide a safe and secure commercial transportation service. To meet the future demands of this network and to enable the national economy, it is essential to expand the capacities of this infrastructure.

The FAA NextGen project is an attempt, led by the FAA, to enhance air transportation capabilities by applying modern technologies and techniques. Changes in navigation from ground-based to satellite-based and digital information sharing in a net-centric infrastructure are some of the technological and technical components of this project. Many governmental and commercial organizations are involved to ensure its success.

An important requirement in running this project is the need to transform the extended enterprise of the ATN. Because the required processes and organizational changes should accompany the technological change, mapping the transformation framework for this evolution is essential. And because humans are part of this complex system, and the need for human ingenuity and integrity is so important to success, the established knowledge, skills, and techniques of many other *soft* disciplines, such as philosophy, psychology, sociology, and organizational change management, must also be applied.

CONCLUSIONS

FAA NextGen is an ambitious, large-scale, and complex transformation of the US ATN. Implementing this project is beneficiary and mandatory for the long-term economic leadership of the United States in air space. However, this project might face major drawbacks during implementation. First, utilizing the real benefits of NextGen might be delayed because of a lack of interest and motivation in key stakeholders like airlines to bear the upfront investments. Second, implementing NextGen requires the net-centric information-sharing capability and the desire and ability of the involved organizations to make information accessible. The current culture and process of doing business limit this capability. Third, various programs in the system are highly interdependent and any change in the schedule or requirement results in cost overruns and time delays. The fourth and final big challenge is integrating legacy enterprise into this cutting-edge technological advancement.

We have provided a three-layered framework to simplify understanding of the US ATN. The authority layer, the organizational layer, and the physical layer are three distinct layers in the system. This framework shows that most of the challenges in implementing FAA NextGen do not occur in the physical layer and technology, but in the organizational layer. The recommendation, then, is to acquire the enterprise transformation perspective into adapting FAA NextGen. Developing a coherent enterprise transformation map enables active and passive stakeholders to envision their future position in the system. It also facilitates their adaptation to necessary changes.

FAA NextGen is a unique important case study because it provides an unprecedented opportunity to study the implementation of a complex, widely dispersed, highly technological project with multistakeholders. Moreover, studying the role of legacy enterprise and the way to handle the challenge of its transformation is valuable, because in upcoming decades we will see more challenges of this kind.

SUGGESTED FUTURE WORK

The US ATN is a complex adaptive system that includes various organizations. To understand the dynamics of this complex sociotechnical system and to be able to govern it, there is an increasing need for novel models. Developing abstract models and simulations of this network has two main benefits: (1) it would increase the ability of authorities to govern the system and to predict its behavior under different regulations, and (2) it would provide a new communication method to facilitate the shared understanding between different stakeholders of the system.

As discussed earlier, a comprehensive framework for the transformation of extended enterprises like the US ATN is another gap in the current practice. Although there are some frameworks for managing the transformation in individual organizations, these models should be further adjusted to provide a method for transforming interconnected network organizations. Having this framework will enhance the practitioners' ability of governing large-scale complex projects like FAA NextGen.

1. What are the similarities and differences between ATN and the Internet? Try to categorize your answers regarding (a) network structure, (b) stakeholders, (c) technology, (d) rules and regulations, and (e) economic impact.
2. What are some of the important technologies within and outside of ATN that are able to change the structure of ATN during the next 30 years? How will the shape of airports and aircrafts be with these changes in the technology? What about airlines and ticketing processes?
3. How will the change in global trade and international economics affect ATN in the next 30 years? What emergent behaviors might occur by an increase/decrease in global economy?
4. What parts of the ATN will not change significantly during the next 30 years? After you come up with a list, assume you are living in 1980 and do the same analysis. Compare your answers with reality. Why were you right about the changes and why did you fail to predict the future?

5. What are the similarities and differences between the Single European Sky project and FAA NextGen. How does the presence of various regulatory institutions separate the European experience from that of the United States? What is the difference in the structure of the physical network and does that cause any difference in decisions?

ACKNOWLEDGMENTS

The authors of this section would like to thank Mayur Chikhale and Nazanin Andalibi for their prior contributions. They would also like to thank the book editors for their kind corrections and insightful comments.

REFERENCES

1. G. Rebovich Jr., Systems thinking for the enterprise, in *Enterprise Systems Engineering: Advances in Theory and Practice*, G. Rebovich Jr. and B.E. White, eds. Boca Raton, FL: CRC Press, 2011, p. 51.
2. H.R. Darabi, M. Mansouri, N. Andalibli, E. Parra, A framework for decision making in extended enterprises: The FAA NextGen case, in *International Congress on Ultra Modern Telecommunication and Control Systems (ICUMT)*, Moscow, Russia, 2010.
3. FAA, The economic impact of civil aviation on the U.S. economy, Federal Aviation Administration, Washington, DC, 2011.
4. JPDO, NextGen overview. Available at: http://www.jpdo.gov (accessed 6/1/2013), 2013.
5. US Census, Table 1073. Available at: http://www.census.gov/compendia/statab/2012/tables/12s1073.pdf (accessed 6/1/2013), 2013.
6. FAA, FAA aerospace forecast, fiscal years 2010–2030, Federal Aviation Administration, Aviation Policy and Plans, Washington, DC, 2011.
7. E. Wingrove, S. Hasan, D. Ballard, R. Golaszewski, J. Cavalowsky, L.A. Oslong, Extent and impact of future nas capacity shortfalls in the united states: A socio-economic demand study, in *Proceedings of Sixth US/Europe Air Traffic Management Research and Development Seminar*, Baltimore, MD, 2005, pp.1–10.
8. G. Hua, Y. Sun, D. Haughton, Network analysis of US air transportation network, *Annals of Information Systems*, 12, 75–89, 2010.
9. Vision 100-Century of Aviation Reauthorization Act, 2003.
10. JPDO, Next generation air transportation system integration plan, Joint Planning and Development Office, Washington, DC, 2004.
11. JPDO, Concept of operations for the next generation air transportation system, Joint Planning Development Office, Washington, DC, 2007.
12. J.H. Holland, Complex adaptive systems, *Daedalus*, 121, 17–30, 1992.
13. R.R. Nelson, S.G. Winter, *An Evolutionary Theory of Economic Change*. Cambridge, MA: Belknap Press of Harvard University Press, 1982.
14. C. Winston, U.S. industry adjustment to economic deregulation, *The Journal of Economic Perspectives*, 12, 89–110, 1998.
15. C. Winston, G. de Rus, *Aviation Infrastructure Performance: A Study in Comparative Political Economy*. Washington, DC: Brookings Institution Press, 2008.
16. V. Aguirregabiria, C.-Y. Ho, A dynamic game of airline network competition: Hub-and-spoke networks and entry deterrence, *International Journal of Industrial Organization*, 28, 377–382, 2010.
17. V. Aguirregabiria, C.-Y. Ho, A dynamic oligopoly game of the US airline industry: Estimation and policy experiments, *Journal of Econometrics*, 168, 156–173, 2012.
18. G.L. Donohue, Air transportation is a complex adaptive system: Not an aircraft design, presented at *the American Institute of Aeronautics and Astronautics International Air and Space Symposium and Exposition: The Next 100 Years*, Dayton, OH, 2003.
19. B.E. White, Complex adaptive systems engineering (CASE), in *2009 Third Annual IEEE Systems Conference*, Vancouver, British Columbia, Canada, 2009, pp. 70–75.
20. B.J. Holmes, Transformation in air transportation systems for the 21st century, presented at the *24th Congress of the International Council on the Aeronautical Sciences*, Yokohama, Japan, 2004.
21. A. Gorod, S.J. Gandhi, B. Sauser, J. Boardman, Flexibility of system of systems, *Global Journal of Flexible Systems Management*, 9, 21–31, 2008.

22. M. Mansouri, A. Gorod, B. Sauser, A typology-based approach to adopting effective management styles for enterprise systems, in *Fourth Annual IEEE Systems Conference*. San Diego, CA, 2010, pp. 300–305.
23. H. Koontz, *Principles of Management: An Analysis of Managerial Functions*, 4th edn. New York: McGraw-Hill Book Company, 1972.
24. R. Kneuper, *CMMI: Capability Maturity Model Integration: A Process Improvement Approach*. Santa Barbara, CA: Rocky Nook Inc., 2008.
25. DoT, U.S. international air passenger and freight statistics, DoT—Office of the Assistant Secretary for Aviation and International Affairs, Washington, DC, November 2011–2012.
26. FAA, Administrator's fact book, Federal Aviation Administration, Washington, DC, 2012.
27. RITA, Air carrier groupings, 2012. Available at: http://www.bts.gov/programs/airline_information/accounting_and_reporting_directives/number_299.html (accessed 6/1/2012).
28. RITA, Origin and destination survey: DB1B ticket, 2013. Available at: http://www.transtats.bts.gov/Fields.asp?Table_ID=272 (accessed 6/1/2013).
29. H.R. Darabi, N. Andalibi, M. Chikhale, M. Mansouri, Federal Aviation Administration NextGen: Applying systems thinking in practice, in *Proceedings of IEEE System of Systems Engineering Conference (SoSE)*, Alburquerque, NM, 2011.
30. DoD, Systems engineering guide for systems of systems, Deputy Under Secretary of Defence (Acquisition and Technology), Washington, DC, 2008.
31. FAA, *Federal Aviation Administration: A Historical Perspective 1903–2008*. Washington, DC: U.S. Department of Transportation, Federal Aviation Administration, 2011.
32. R. Doganis, *The Airline Business*, 2nd edn. London, U.K.: Routledge, 2006.
33. FAA, Federal Aviation Administration, Mission, Available at http://www.faa.gov/about/mission/ (accessed 2/1/2014)
34. ATA, Air Transport Association, A4A, 2013. Available at: http://www.airlines.org/Pages/About%20A4A.aspx (accessed 6/1/2013).
35. FAA, FAA's NextGen implementation plan. Federal Aviation Administration, 2012.
36. M. Mansouri, A. Mostashari, A systemic approach to governance in extended enterprise systems, in *Fourth Annual IEEE Systems Conference*, San Diego, CA, 2010, pp. 311–316.
37. P.M. Peeters, J. Middel, A. Hoolhorst, Fuel efficiency of commercial aircraft: An overview of historical and future trends. Amsterdam, the Netherlands: National Aerospace Laboratory NLR, 2005.
38. B. Vasigh, K. Fleming, T. Tacker, *Introduction to Air Transport Economics: From Theory to Applications*. Burlington, VT: Ashgate Publishing Limited, 2008.
39. A.L. Barbasi, *Linked: The New Science of Networks*. Cambridge, MA: Perseus Publishing, 2002.
40. A. Cento, *The Airline Industry: Challenges in the 21st Century*. Segrate, Italy: Physica-Verlag Heidelberg, 2009.
41. M.E. O'Kelly, D.L. Bryan, Hub location with flow economies of scale, *Transportation Research Part B*, 32, 605–616, 1998.
42. FAA, National airspace system, in *Instrument Flying Handbook*. Available at: http://www.faa.gov/regulations_policies/handbooks_manuals/aviation/media/FAA-H-8083-15B.pdf, 2012.
43. P. Hanlon, *Global Airlines: Competition in a Transnational Industry*, 3rd edn. Oxford, U.K.: Elsevier, 2007.
44. S. Holloway, *Straight and Level: Practical Airline Economics*, 3rd edn. Burlington, VT: Ashgate Publishing Company, 2008.
45. E. Ben-Yosef, *The Evolution of the US Airline Industry: Theory, Strategy and Policy*, vol. 25. New York: Springer, 2005.
46. M.H. Lin, Airline alliances and entry deterrence, *Transportation Research Part E: Logistics and Transportation Review*, 44, 637–652, 2008.
47. T. Sundman, Airline code-share alliances with antitrust immunity and their competitive effects on international passenger output: An application to monopolistic and oligopolistic network structures on the trans-Atlantic market, Master thesis, Department of Economics, Hanken School of Economics, Helsinki, Finland, 2009.
48. H.R. Darabi, A. Mostashari, M. Mansouri, Modeling competition and collaboration in the airline industry using agent-based simulation, *International Journal of Industrial and Systems Engineering*, 16(1), 30–50, 2014.
49. IATA, *Worldwide Slot Guidelines*. Geneva, Switzerland: International Air Transport Association, 2012.
50. H. Fukui, An empirical analysis of airport slot trading in the United States, *Transportation Research Part B: Methodological*, 44, 330–357, 2010.
51. ARO, Airport Reservation Office. Available at: http://www.fly.faa.gov/Products/Information/ARO/aro.html (accessed 6/1/2013), 2013.

52. N. Ashford, P.H. Wright, *Airport Engineering*, 3rd edn. New York: John Wiley & Sons, Inc., 2011.
53. ICAO, 2013. Available at: http://www.icao.int/about-icao/Pages/default.aspx (accessed 2/1/2014).
54. IATA, IATA—About us, 2013. Available at: http://www.iata.org/about/Pages/index.aspx (accessed 6/1/2013).
55. J.H. Gittell, *The Southwest Airlines Way: Using the Power of Relationships to Achieve High Performance*. New York: McGraw-Hill, 2003.
56. P. Belobaba et al., *The Global Airline Industry*. Chichester, U.K.: John Wiley & Sons, Inc., 2009.
57. M.D. Platzer, U.S. Aerospace Manufacturing: Industry overview and prospects, Congressional Research Service, Washington, DC, 2009.
58. M.W. Tretheway, T.H. Oum, *Airline Economics: Foundations for Strategy and Policy*. Vancouver, British Columbia, Canada: Center for Transportation Studies, University of British Columbia, 1992.
59. ICAO, Annual civil aviation report, *ICAO Journal*, 56, 1–13, 2001.
60. JPDO, Targeted NextGen capabilities for 2025, Joint Planning and Development Office, Washington, DC, 2011.
61. FAA, FAA's NextGen Implementation plan, Federal Aviation Administration, Washington, DC, 2008.
62. B. Forman, The political process and systems architecting, in *The Art of Systems Architecting*, M.W. Maier and E. Rechtin, eds., 2nd edn. Boca Raton, FL: CRC Press LLC, 2000.
63. G.L. Dillingham, Next generation air transportation system, FAA faces implementation challenges, United States Government Accountability Office, Washington, DC, September 12, 2012.
64. JPDO, Enterprise architecture V2.0 for the next generation air transportation system, Joint Planning and Development Office, Washington, DC, 2007.
65. FAA, FAA-aviation community engagement strategy for implementing NextGen, Federal Aviation Administration, Washington, DC, 2010.
66. GAO, Integration of current implementation efforts with long-term planning for the next generation air transportation system, United States Government Accountability Office, Washington, DC, 2010.
67. GAO, Air traffic control modernization: Management challenges associated with program costs and schedules could hinder NextGen implementation, U.S. Government Accountability Office, Washington, DC, 2012.
68. FAA, The business case for the NextGeneration air transportation system, Federal Aviation Administration, Washington, DC, 2012.
69. FAA, Destination 2025, Federal Aviation Administration, Washington, DC, 2012.
70. FAA, Acquisition workforce plan, Federal Aviation Administration, Washington, DC, 2012.
71. H.R. Darabi, M. Mansouri, Studying regulatory institutions competition and collaboration dynamics in network industries: Using agent-based modeling and simulation, in *IEEE International Conference on Systems Engineering (SysCon '13)*, Orlando, FL, 2013, pp. 131–136.
72 M.W. Maier, E. Rechtin, *The Art of Systems Architecting*. Boca Raton, FL: CRC Press, 2000.

23 Airbus A380 and Boeing 787

Contrast of Competing Architectures for Air Transportation

Michael J. Vinarcik

CONTENTS

CASE STUDY ELEMENTS

Fundamental essence

This case explores two diametrically opposed architectures developed to serve customers in the same industry: commercial air travel. The Airbus A380 is the largest commercial airliner currently in production; the Boeing 787 is a smaller airliner designed to maximize efficiency and improve passenger comfort.

Topical relevance

This case explores how two major aircraft manufacturers pursued different architectures and the pitfalls each has encountered while fielding its system. It discusses emergent behavior associated with the introduction of these aircraft.

Domain
Industry

Countries of focus
United States of America/France/Germany/United Kingdom/Spain

Stakeholders
Airlines, governments, passengers

Primary insights
Systems architectures are driven, in part, by corporate strategy. The Airbus A380 and Boeing 787 are targeted at widely differing market segments and illustrate extreme expressions of differing corporate approaches. Both aircraft cost billions to develop, both experienced development and production delays, and both had significant engineering issues that needed to be resolved after production began. These aircraft are elements within the complex system of systems (SoS) that is the global airline network, and both impacted and were impacted by the emergent behaviors of that SoS.

KEYWORDS

Commercial aviation, Manufacturing issues, Program management, Supplier management, Systems architecture, Technology readiness

ABSTRACT

This case study explores the radically different approaches taken by Airbus and Boeing as they developed the latest generation of their passenger aircraft. The giant Airbus A380 and the efficient Boeing 787 reflect two fundamentally different approaches to air travel and highlight the different cultural climates, corporate structures, and assumptions made by both organizations as they bet billions of dollars and, in some ways, their futures.

The comparison and contrast of these SoS approaches, constraints (e.g., the A380 is profoundly impacted by airport infrastructure), triumphs, and pitfalls of the two aircraft will illustrate fundamental systems architectural and systems engineering principles.

GLOSSARY

ADS-B	Automatic dependent surveillance—broadcast
APU	Auxiliary power unit
ATC	Air traffic control
ATM	Air traffic management
DELAG	*Deutsche Luftschiffahrts-Aktiengesellschaft* (the German airship travel corporation)
FAA	Federal Aviation Administration
GPS	Global positioning system
ICAO	International Civil Aviation Organization
ILS	Instrument landing system
INS	Inertial navigation system
MW	Megawatt
OEM	Original equipment manufacturer
SAS	*Société par actions simplifiée* (simplified joint-stock company/corporation)
SIA	Singapore Airlines Limited
SoS	System of systems
TCAS	Traffic collision avoidance system
VAC	Volts alternating current

BACKGROUND AND PURPOSE

The development of modern complex systems is a costly and complicated process. Original equipment manufacturers (OEMs) must align their development efforts to meet the needs of their customers. The modern product marketplace is becoming increasingly polarized. At one extreme, the lifecycles of consumer goods (cell phones, computers, televisions, etc.) continue to shorten. At the other, owners and operators of heavy equipment continually seek ways to extend the life of existing systems (often past the initial design life). For example, the service life of the Boeing B-52 Stratofortress may exceed 90 years. Delta Air Lines recently purchased a number of decade-old MD-90 airframes, is adding them to its fleet, and will likely operate them for an extended period [1].

Commercial airlines often operate aircraft for decades. Therefore, acquisitions are carefully planned to ensure that each airline's fleet meets its current and forecast airframe needs. Supporting analyses must take into account each airline's routes, expected passenger and freight volumes, passenger expectations (impacting seating density in a given airframe), applicable government regulations, and the infrastructure available for landing and servicing aircraft.

Southwest Airlines has standardized on the Boeing 737 to minimize training, maintenance, and logistic complexity. It shed Boeing 717 aircraft associated with its acquisition of AirTran and allowed Delta Air Lines to assume the leases; it provided $137 million in retrofits to prepare them for Delta's use [1]. Southwest's focus on relatively short, point-to-point domestic flights to carefully selected destinations allows it to adopt this strategy. Other airlines, serving broader markets with a variety of capacity and infrastructure needs, typically require a suite of airframes to economically provide their services. Therefore, aircraft manufacturers must develop a family of products that meet the current and future needs of these customers (much as automakers offer a wide range of vehicles to suit a broad range of consumer needs and tastes).

When an OEM commits to the development of a new aircraft, it is making a substantial commitment of time, resources, and effort; a given airframe will often be in production for decades and may spawn an entire family of civilian and military derivatives (e.g., cargo, extended range, short-range with increased passenger capacity, in-flight refueling tanker, anti-submarine platform, or radar surveillance). It is important for the architecture of the system to be flexible enough to support this follow-on development. For example, the latest Boeing 737 derivative, the 737 MAX, will be delivered in 2017, nearly 50 years after the first 737 took to the skies. It will boast new engines and promises a 10%–12% reduction in fuel consumption; Southwest Airlines was the first customer to place a firm order (the largest firm order in Boeing history, totaling $19 billion and 150 aircraft) [2].

Two of the major aircraft OEMs, Airbus SAS and the Boeing Company, recently developed radically different aircraft that showcase diametrically opposed systems architectures. The Airbus A380 is a double-deck widebody airliner, the largest commercial airliner in the world, created to carry large numbers of passengers on long routes. The Boeing 787 Dreamliner is a midsize, long-range airliner developed with fuel efficiency as a key performance parameter. As Maier and Rechtin state, architecture can be seen as the physical (or technical) embodiment of strategy [3, p. xxiii]; in the case of these airliners, radically different corporate strategies are driving their development.

AIRLINE ARCHITECTURE HISTORY

The commercial airline industry was born in 1909 with the founding of the *Deutsche Luftschiffahrts-Aktiengesellschaft* (DELAG, the German Airship Travel Corporation). By the outbreak of World War I, DELAG had carried over 30,000 passengers on 1,500 flights. Subsequent airlines shifted away from lighter-than-air craft but continued to serve passenger and freight markets. The industry benefited from the wars of the twentieth century, which produced thousands of experienced aviators and caused the development of new technologies that were directly applicable to commercial aviation.

Improvements such as aluminum bodies, pressurized cabins, improved engine technology (including the now-ubiquitous turboprop and turbofan engines), and improvements in weather forecasting allowed commercial airlines to thrive. Throughout the twentieth century, numerous companies leveraged these advances and developed and marketed aircraft to airlines. Notable producers included:

- de Havilland (maker of the Comet, the first jet airliner, whose initial triumph was later marred by infamous metal fatigue issues)
- Vickers (maker of the Viscount, the first turboprop airliner)
- Glenn L. Martin Company (producer of the Clipper flying boats)
- Douglas Aircraft (creator of the DC-3, the first widely successful airliner)
- Lockheed (developer of the iconic Constellation and TriStar)
- Boeing (manufacturer of the 314 Clipper, Stratocruiser, and the 7X7 series of jetliners)
- Airbus (manufacturer of the A3XX series of airliners)

Today, after mergers, acquisitions, and market exits, only Boeing and Airbus remain as Western manufacturers of large airliners. Because of the large development costs and long development cycles associated with the commercial airliner market, neither of these companies can afford an unsuccessful product. For that reason, most airliner development in the past 50 years has been more evolutionary than revolutionary, with the general layout and architectural features well established (just as the fundamental architecture of the automobile has been relatively constant for much of the twentieth century). However, each OEM interprets market research and stakeholder needs differently; differing views on how to best serve customers led to the contrasting developments of the Airbus A380 and the Boeing 787.

For most of human history, travel was a laborious endeavor. Generations of humans lived and died within a few miles of their birth; warfare, trade, and occasional exploration were typically the only reasons that people traveled long distances. Economic and military necessity drove improvements in transportation, logistics, and associated infrastructure, leading to paved roads, canals, and continuously improved ships. By the nineteenth century, railroads and steamships provided regular passenger service and accommodated the needs of the growing number of individuals affluent enough to purchase tickets and brave enough to endure the rigors of longer journeys.

At the dawn of the twentieth century, airships and aircraft with sufficient range and payload capacity became available to begin regular passenger service. Airfields began to appear; initially simply flat patches of ground, they evolved into the complex facilities we know today, capable of handling large volumes of passenger and cargo traffic. Approach and landing lights, radar, instrument landing systems, passenger terminals, fuel depots, and a host of other innovations allowed continuous improvements in efficiency and safety. The midcentury development of the hub-and-spoke system revolutionized air travel and the routing of passengers from origin to destination.

However, it is the handling and queuing of aircraft on the ground that is the key constraint on the air traffic system. Airliners are fundamentally large vehicles, with associated hangar, terminal, and runway needs. The capacity of an airport is dictated by its terminals and runways. The mandated separation between planes during takeoff and landing limits the throughput of aircraft, and the length of the runway dictates the size of aircraft that may land. Airports are expensive to construct (a contributing factor to the use of flying boats that took off and landed on water to service developing countries during the early years of commercial aviation) and require a sizeable commitment of valuable and desirable land area (typically near highways and major metropolitan areas). This makes airport capacity the bottleneck in this SoS.

Airbus was founded in 1970 as an initiative between the French, German, and British governments. This collaborative approach was taken to compete effectively with the US aircraft industry, which dominated commercial aviation at the time. This effort bore fruit in 2003, when, for the first time, Airbus shipped more planes than Boeing. Airbus currently directly competes with Boeing with its A3XX family of airliners (currently including the A318, A319, A320, A321, A330, A350, and A380).

Boeing was founded in 1917 and served both the commercial and military markets with its aircraft. After several decades of innovation, Boeing introduced the 707 in 1958; it was the first American jetliner. Boeing merged with McDonnell Douglas in 1997 and has continued to expand its commercial jetliner offerings. It currently produces 7X7 airliners seating 100–500 passengers in addition to its cargo and military aircraft. As of this writing, the 737, 747, 767, 777, and 787 are still in production.

Both Airbus and Boeing serve a worldwide client base and operate within a global patchwork of government regulations and competing client needs. The airline industry has shifted to a deregulated model, carriers have consolidated, and fuel costs continue to spiral upward. In addition, the terrorist attacks of September 11, 2001, had a significant impact on US air travel and constricted the cash flow and liquidity of airlines. These events stimulated renewed efforts to better integrate military and commercial aviation in the United States, in part to be able to coordinate bringing aircraft together as well as keeping them apart, respectively, for example, fighters from Otis Air Force Base were late in intercepting the two airliners headed for New York City's World Trade Center. Throughout their history, US airlines have regularly been in bankruptcy, driven by a lack of income to offset the high operating and capital expenses needed to operate successfully [4, p. 185].

MISSION AND DESIRED CAPABILITIES

Air travel (and the OEMs and operators associated with the industry) serves a variety of purposes in the modern world: express freight, auxiliary military airlift capability, and premium passenger services are all aspects of this industry. Because of the commercial importance of maintaining a healthy air travel industry, governments worldwide regulate safety, infrastructure, and operating conditions. These regulations often drive airport capacity limits (e.g., the separation distance for A380s was increased due to its large turbulent wake, reducing the throughput of the airports handling the behemoth) and personnel needs (e.g., the number of mechanics required to service a fleet and meet mandated maintenance schedules).

Passengers, whose expectations continue to evolve, expect amenities and services unheard of a few decades ago. Paradoxically, they will accept reduced legroom, baggage charges, and the discontinuation of hot food service while simultaneously demanding in-flight wireless Internet access and personalized entertainment. To complicate matters, the payload constraints of an airliner challenge airframe manufacturers to find weight-efficient ways to provide in-flight entertainment, wider seats, and improved carry-on cargo capacity. Every pound added to provide these capabilities (e.g., seatback displays or larger overhead stowage solutions) must be offset with weight reductions elsewhere in the system or performance, range, or operating efficiency will be degraded.

Airlines, who often lease their airframes, require an efficient way to move passengers and meet these comfort and convenience expectations. The focus on reducing costs per passenger-mile has led to a variety of industry practices, such as Southwest's ruthless focus on maximizing flight hours by reducing the time needed to load passengers and prepare each plane for its next flight. Hardware solutions also abound, from retrofitting aircraft with fuel-efficient engines to the adoption of winglets to reduce drag and improve fuel efficiency. Delta Air Lines has vertically integrated by buying its own refinery (to stabilize fuel prices) [5] and a maintenance subsidiary (allowing it to maintain its aging fleet economically) and purchasing rather than leasing its planes (it owns 75% of its fleet) [1]. It is profiting using this contrarian strategy.

The push for efficiency by airlines has impacted their employees. Pilots benefit from cockpit commonality (typically within an OEM's family of aircraft), and maintenance workers are reaping the benefits of ongoing design-for-maintainability efforts and improved service practices.

To manage these highly complex and highly constrained systems, aircraft OEMs have been forced to develop and adopt traditional or conventional systems engineering best practices. In many ways, systems engineering has grown up alongside the aerospace industry. Many of the systems engineering tools and processes in widespread use today were developed to manage the increasing complexity of aircraft. Consider the contrast between the aircraft (developed 40 years apart) of Table 23.1.

TABLE 23.1
Ford Tri-Motor and Boeing 737 Comparison

	Ford Tri-Motor	Boeing 737
Year introduced	1925	1968
Crew	3	2–3
Passengers	10	85–215
Cabin	Unpressurized	Pressurized
Top speed (mph)	150	544
Range (miles)	550	1,500–5,500
Maximum altitude (ft)	18,500	35,000–41,000
Propulsion	Three Pratt & Whitney nine-cylinder wasp radial engines	Various (twin turbofan)

In a mere 40 years, maximum altitude doubled (driving pressurized cabins and associated complexity), range increased three to tenfold, passenger capacity increased 20-fold, and a host of safety and operating regulations emerged. Avionics, communications, radar, and a host of other systems had to be integrated into the later aircraft. Systems engineering practices, such as requirements traceability and interface management, were developed and adopted by aircraft OEMs. They also drove the creation of advanced engineering techniques such as finite element analysis, computer-aided design, and computer-aided manufacturing.

However, the field of *complex* systems engineering, where people (the principal stakeholders, in particular) are intentionally and purposefully considered part of the system to be engineered, is emerging. In these complex environments, one can only intervene with actions for potential improvements, observe carefully and objectively what happens, and be ready to make another course correction at some point. One difficulty derives from not knowing exactly when to make the next decision because of the systemic time delays in the evolution of the complex system. One is never in full control since people cannot be sufficiently modeled. Thus, it is impossible to prespecify a solution or accurately predict what will happen next. When considered as a whole, the worldwide air travel network is a complex SoS of which aircraft are merely one component. It meets Nicolis and Prigogine's definition of a complex system [6], in that it is made up of multiple elements that are dependent upon each other and their environment.

COMPLEX SYSTEM

Figure 23.1 provides a tailored overview of some of the complexities detailed earlier that are associated with commercial aircraft design, development, manufacture, and operations. With reference to Figure 23.1, consider the following factors:

A—*Complex SoS*: Many interacting components, both internal and external to an aircraft, must be coordinated and orchestrated, including the airframe, fuel-efficient engines, various technologies for efficient operation (e.g., winglets), flight control systems (auto pilot, INS, GPS, ADS-B, ILS, TCAS, etc.), cabin air conditioning, pilot-controller communications systems (for ATC, ATM, flow control, etc.), passenger seating, and in-flight service capabilities (Internet, entertainment systems, food and beverages, baggage storage, etc.). These are traditional subsystems on the aircraft and defined as complicated. However, other systems are complex and have both autonomous and independent aspects. For example, logistics support (e.g., refineries, maintenance hangers, training facilities) must satisfy the needs of other aircraft, which in turn may be impacted by a variety of other factors, both natural (e.g., the weather) and manmade (introducing a new engine or aircraft). Some of these systems are mandated by regulations (out of the control of the aircraft), others are controlled by external entities (such as the airport and fueling infrastructure), and still others are *messy* and difficult to forecast, such as emerging

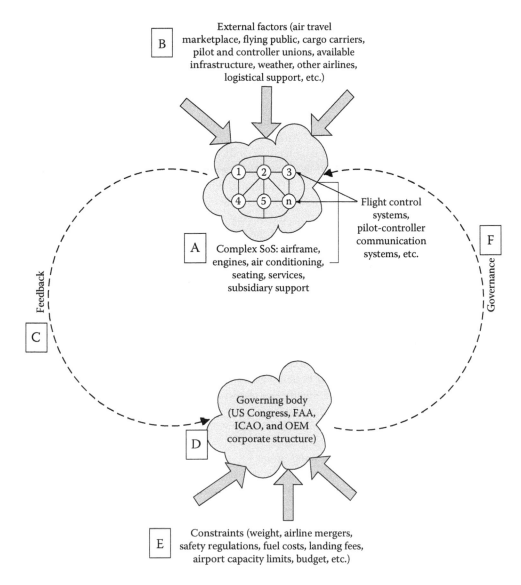

FIGURE 23.1   Context of the Airbus A380 and Boeing 787 complex system of systems.

entertainment technologies and evolving customer expectations. All must be considered, managed, and addressed by both the aircraft manufacturer and end user.

B—*External factors*: These include the air travel marketplace; the flying public; shippers of cargo, pilot, controller, and other worker union demands; available infrastructure (airports, runways, gates, etc.); bad weather; other airlines; fuel prices; and suitable logistical support.

C—*Feedback*: This includes perceived requirements; development and equipment cost, performance, and schedule; aircraft flight and passenger compartment architectures; and make-buy options.

D—*Governing body*: A collection of agencies or organizations consisting of the US and world-wide governments, the FAA, ICAO, and each OEM's corporate structure.

E—*Constraints*: Aircraft weight, results of airline mergers, government airline safety regulations, fuel costs, landing fees, airport capacity limits (e.g., landing spacing due to large aircraft–induced turbulence), budget, etc.

F—*Governance*: Air routes, speed, and waypoint air traffic management, aircraft separation and spacing, vertical aircraft separations, collision avoidance and conflict resolution advisories and regulations, OEM architectural guidance, etc.

In this complex environment, at the dawn of the twenty-first century, both Boeing and Airbus embarked on new aircraft development. Boeing had considered offering longer variants of its successful 747 (the 747–500 and 746–600), but by 1997, it had discontinued development [4, p. 180]. This strategic shift left an opening in the market and Airbus subsequently announced its A380, a double-deck behemoth that would provide supersized long-haul capability (see Figure 23.2). Boeing insisted there was no market for the A380 and, by 2001, had unveiled a *Sonic Cruiser* proposal. The Sonic Cruiser would speed transoceanic flights, cruising at 750 mph and with a 9000 mile range [4, p. 184]. It was presented as "a sure thing, the biggest innovation in air travel since the invention of the jet engine." This radical, tail-first jet was canceled by 2002, and Boeing's focus shifted to a smaller, highly efficient concept initially dubbed the 7E7 (later rechristened the 787; see Figure 23.3: Boeing 787).

It should be noted that in 1984, 62% of transatlantic flights were made by Boeing 747s; by 2004, 747s only accounted for 16% of that market [4, p. 180]. That drop in large-body market share, plus the continued fragmentation of air routes, had convinced Boeing that the bulk of opportunities for growth lay with smaller, more fuel-efficient planes. Airbus adopted its contrarian strategy to fill a perceived gap in its portfolio and allow a full spectrum of offerings (much as General Motors once spanned the automobile market with brands ranging from Chevrolet to Cadillac).

Both the A380 and the 787 are similar in overall architecture; the fundamental differences are in capacity and range. Both are low-wing monoplanes with turbofan engines slung below the wings (four on the A380 and two on the 787). If both are analyzed with techniques proposed by Firesmith et al. [7], both are shaped by related architecturally significant requirements such as

- Operate from existing airports (located around the world)
  - Fit within gate/hangar size constraints
  - Do not exceed runway load-bearing limits
  - Avoid incidental damage to/from signage and facilities
  - Use existing fueling and replenishment infrastructure
- Interoperate with existing air traffic control systems

FIGURE 23.2   Airbus A380. (Courtesy of David Varga/Shutterstock.com.)

FIGURE 23.3 Boeing 787. (Courtesy of Sascha Burkard/Shutterstock.com.)

- Meet global noise, emissions, and safety standards
- Interface with major third-party subsystems
  - Avionics and communications
  - Engines
- Meet reliability and cost requirements

Passenger aircraft such as these operate in a demanding duty cycle, with significant pressure and temperature changes between airport and cruising altitude and the possibility of significant shock loading to the airframe upon landing. Given the extended service life modern aircraft experience, fatigue and cumulative damage are factors that must be considered when designing an aircraft and its associated maintenance and inspection schedule. For example, in 2011, a Boeing 737 operated by Southwest Airlines made an emergency landing after a 5 ft fracture in a fuselage lap joint caused the cabin to depressurize. The root cause was fatigue cracking in an area not part of the normal inspection regimen. The aircraft was 15 years old and had accumulated nearly 40,000 takeoff/landing cycles; inspections of 136 other 737s in the fleet discovered five other planes with similar cracks (with 40,000–45,000 cycles). Inspection routines were subsequently modified to verify aircraft were not experiencing cracking in this area [8]. It should be noted that because of Southwest's efficiency in *turning around* aircraft (deplaning passengers, resupplying/refueling, and boarding passengers can be done in 20 versus 90 min for most competitors [9]), its planes accrue cycles faster than most other operators.

The end users of aircraft, viz., the airlines, often provide feedback to the OEMs about specific requirements that new airliners must meet, such as range, capacity, noise, or internal configurability. In addition, a web of international operational and safety regulations must be met. It should be noted that some efforts have been made to harmonize the regulatory environment. For example, in 1999, the International Civil Aviation Organization (ICAO), in cooperation with manufacturers, airports, and member agencies, expanded the 65 m wingspan limit originally established to accommodate the 747. The new *80 meter box* was created so airport gates could allow planes with wingspans and lengths up to 80 m to be accommodated [10]. This constraint would impact aircraft almost as surely as the size of the Panama Canal's locks limited ship sizes for decades. There are other, less obvious requirements that must be considered, such as whether the runway length and other facilities at Bangor, Maine, can accommodate the aircraft. This unlikely constraint is due to the airport's role in handling numerous diverted flights, particularly those that develop emergency situations on transatlantic flights. The airport recently had to purchase an upgraded heavy-duty tug to be able to move A380s, if necessary [11].

DEVELOPMENT CHALLENGES

Airbus officially approved the A380 in December 2000 and promised first delivery in early 2006 (after a year of verification and validation testing). The development effort was expected to cost $11 billion, a significant sum that would impact the company's viability if it failed.

The project and program management was complicated by the international makeup of the parent company itself. Airbus's facilities are spread across four countries: Germany, France, Spain, and the United Kingdom. The diffusion of engineering activities across these facilities leads to coordination and integration issues, particularly when cultural and language differences are considered. On occasion, Airbus management has increased tensions between these disparate groups of engineers. For example, Charles Champion, the head of the A380 program for much of its development, decided to put engineers in Spain in competition with those in Germany to design the aircraft's wing. The Madrid-based team came up with a concept that wrapped composite ribs in a metal frame. This was ultimately adopted for the larger 24 of the 61 wing ribs (the balance were made from aluminum). However, the concept did not fully consider the impact of manufacturing. Aluminum ribs were machined out of a single piece of aluminum, but composites are made with floppy, multilayer sandwiches of fiber strips that are embedded in a polymer matrix. Workers found it impossible to build up the desired geometry without ruining floppy areas of the ribs; the ultimate solution was to assemble and harden each side of the rib separately and then glue them together. However, workers at the assembly plant (located in Wales) damaged the first delivery of ribs by walking on them (a common practice that was harmless to traditional metal ribs but damaging to the composites) [12].

The Boeing 787 was initiated after the Sonic Cruiser was scrapped; by 2003, Boeing had identified a potential shift in airline routing practice from the hub-and-spoke system toward more direct, point-to-point flights. It should be noted that Southwest Airlines, the only major carrier to be profitable for 33 consecutive years, follows this strategy. Its standardization on one airframe (the 737) simplifies maintenance complexity (a savings of millions of dollars in training and spare parts inventory), allows it to shift planes effortlessly within its network, and ensures its pilots and crews are experts at flying and servicing all the planes in its fleet. Instead of collecting passengers at crowded hubs (with their associated flight delay stackups) and having planes waiting at the terminal to load stragglers, its planes carry passengers directly to their destinations. These point-to-point operations allow its planes to spend more than an extra hour in the day airborne vs. its competitors; coupled with fast turnarounds, this enables each plane to fly an extra revenue-producing flight every day [9].

According to Mary Bentrott, Boeing vice-president of sales, marketing, and in-service support for the 787, between 1990 and 2006, the number of city pairs more than 3000 nautical miles apart connected by air routes, the number of flights, and the quantity of available seat kilometers (seating capacity multiplied by miles flown) have all doubled. There has been a simultaneous slight decline in average aircraft size. This demonstrates that market forces are pushing airlines to offer more direct, point-to-point flights, flown more frequently, on smaller planes. Bentrott stated, "Our strategy has been to design and build an airplane that will take passengers where they want to go, when they want to go, without intermediate stops; do it efficiently while providing the utmost comfort to passengers; and make it simple and cost-effective for airlines to operate" [13].

To achieve this vision, Boeing focused on providing improved fuel economy by relying on technological advances in engine and airframe technology (primarily fuel-efficient engines and lighter composite materials for the plane's structure); to improve the passenger experience, it conducted a decade's worth of focus groups to understand the design parameters that influence the parameters (both tangible and intangible) that define perceived passenger comfort [13].

To develop the 787, Boeing chose to outsource much of its detailed design and engineering work to its supplier network, focusing instead on the integration aspects of product development.

Organizations that manufacture products are typically viewed favorably for outsourcing design, engineering, and manufacturing tasks. This is due to the perception that costs will be lowered (primarily due to lower labor and materials costs) and that the company has identified its core competencies to keep in-house and outsourced the remainder of its operations.

This approach was tried in the automotive industry; *full-service suppliers* were given increased design responsibilities for the components they produced, and the automotive OEMs focused their efforts on integration and testing. This approach did not yield the cost savings expected and resulted in a brain drain of critical skills from the OEMs. By the time of this writing, the design of key components (particularly safety-related) is being insourced at least one OEM.

The 787 ultimately was more than 3 years late and billions of dollars over budget. The reason, in large part, was this decision to outsource development. The following quotes illustrate what transpired:

> Acquired on the job and over time, tribal knowledge is…the shared method of performing countless daily tasks efficiently and in coordination with colleagues. In short, **tribal knowledge is the grease that cuts friction throughout the design and assembly process**. [14]

> **As we outsource part of this work, we're removing opportunities for learning** this trade, for learning these skills…**As we reduce these opportunities to learn how to do these jobs, the Boeing Company becomes less capable** to do the job. [14]

> [Before outsourcing, individuals] had easier access to the program engineers who worked in the same building and could quickly address problems as they arose. [14]

> Boeing's goal, it seems, was to convert its storied aircraft factory near Seattle to a mere assembly plant, bolting together modules designed and produced elsewhere as though from kits. The drawbacks of this approach emerged early. Some of the pieces manufactured by far-flung suppliers didn't fit together. [15]

> Boeing executives now admit that the company's aggressive outsourcing put it in partnership with suppliers that weren't up to the job…we spent a lot more money in trying to recover than we ever would have spent if we tried to keep many of the key technologies closer to Boeing. [15]

> Boeing assumed that its suppliers would share its commitment to quality and meeting ambitious delivery deadlines. But this did not happen and the 787 missed at least seven deadlines and went way over budget. [16]

The root cause of these issues can be traced to two fundamental concepts drawn from materials science and military strategy. First, every interface in nature has a *surface energy*. Creating a new surface (e.g., by cutting a block of steel into two pieces) consumes energy that is then bound up in that surface (or interface). Interfaces in human systems (or organizations), a critical aspect of complex systems such as these, also have costs in the effort to create and maintain them. Second, friction reduces performance. Carl von Clausewitz, the noted military strategist, defined *friction* as the disparity between the ideal performance of units, organizations, or systems, and their actual performance in real-world scenarios. One of the primary causes of friction is ambiguous or unclear information.

By creating additional interfaces among individuals responsible for designing components, Boeing impeded the free flow of information, expertise, and situational awareness that is critical for success. It Balkanized the engineering and integration activities and generated the same types of issues that plagued Airbus during the development of the A380.

SYSTEMS ENGINEERING

A380 SYSTEMS ENGINEERING

The A380 was severely constrained by a number of factors and it required robust systems engineering to manage the associated trade-offs. For example, the *80 meter box* restricting the overall size of the plane dictated that the wing length-to-fuselage diameter ratio deviates from optimal (the wings are stubbier than desirable for a plane of this size). This, in turn, constrained the lift that could be generated, requiring more engine power (and, therefore, more fuel, which reduced payload capacity).

The aircraft faced numerous weight challenges throughout development, including an early demand to reduce seat weight by 15%. To help alleviate the weight issues, 60% of the plane's wiring is made from aluminum instead of copper [17]. This resulted in a 20% weight savings but added complications related to corrosion and arcing that were solved by plating the wiring with nickel and copper.

Airbus also elected to use a higher-pressure hydraulic system (5000 psi instead of the traditional 3000 psi) to save weight. This reduced the weight of the system by 1 ton (with additional indirect weight savings) and the reduced size of the plumbing simplified routing of the piping in congested areas of the airframe [18].

In addition to the relatively straightforward challenges of designing an airworthy craft within the imposed size and weight constraints, the Airbus team had to accommodate a variety of disparate customer wants and needs. For example, Emirates Airlines wanted more cargo space, Quantas wanted more range, and Singapore Airlines wanted the plane to be quieter. To meet this noise requirement, engine diameter was increased by 5 in. with an associated weight increase of 1 ton. Customer demands for interior flexibility also drove complexity increases; for example, changing seat spacing impacts wiring harness design. The net result is an airframe suboptimal by objective engineering standards, but that gives customers what they want, according to Champion [12].

Difficulties in design and production led to the abandonment of a planned freighter version, ceding the large cargo plane market to variants of Boeing's 747.

As the A380 has entered service, airlines have capitalized on its size not to maximize capacity by cramming passengers into the available space but instead to increase luxury amenities. Some internal configurations feature waterfalls, luxury suites, and spacious seating arrangements. Korean Air has installed a duty-free showcase, Emirates Airlines provides showers in first class, Quantas installed a lounge, and SIA (Singapore Airlines Limited) included double beds in first class. Not since the demise of the large passenger airships (such as the *Graf Zeppelin* and *Hindenburg*) have air passengers had access to luxuries such as these [19].

Airbus faced many challenges while executing the A380; in addition to the never-ending pressure to reduce weight, engineers were challenged to design components larger than those they had typically created. This resulted in a variety of design challenges, since the components could not simply be scaled up from smaller designs and fundamental parameters, such as a component's natural frequency, are different. "Because of the size, you cannot find one piece which has been done before," said Robert Lafontan, Senior Engineer on the A380 [12].

One common problem experience by the Airbus and Boeing teams was the need to coordinate the dispersed engineering teams. As time pressure increased, Airbus's Spanish engineers and British manufacturing staff began to quarrel. Charles Champion referred to the spats as typical and said "I'm sure I don't know 5% of the arguments among staff" [12]. By 2003, as the first wing was lifted into position, Champion said, "No one would have believed it – not even me" [12].

787 Systems Engineering

The success of the 787 would depend heavily on its fuel efficiency; fuel costs make up more than 40% of an airlines operating costs [9]. To address this issue, Boeing made some interesting architectural choices in its up-front design; fuel economy was the key architecturally significant requirement. Its original goal was to reduce fuel costs by 15%–20%. To achieve that, it developed a multitiered strategy.

Two to three percent of the savings would be achieved by reducing the aircraft's weight, primarily through the use of composites in the wings. It would also use new, fuel-efficient engines from Rolls-Royce (the Trent 1000) and General Electric (the GEnx-1B). These engines achieved a 15% reduction in fuel consumption vs. the 767's engines by incorporating several technological advances:

- Higher propulsive efficiency through increased bypass ratio
- Higher engine thermal efficiency through increased overall pressure ratio and improved component efficiencies

- Improved thrust-to-weight ratio through the application of advanced materials
- Introduction of a novel dual-use electrical power generation system that doubled as the engine start system [20, p. 5]

Two other noteworthy requirements were added for the engine developers to address: First, the interface between the plane and the engines would be common for both engine vendors. This simplified testing, integration, and engine management and would allow airlines to reengine their planes at will without the need for complicated retrofitting. As straightforward as this appears, it is the first time in modern jet airliner history that the engine interface has been standardized. Second, it eliminated bleed air systems from the engines. This impacted the following systems:

- Engine start
- Auxiliary power unit (APU) start
- Wing ice protection
- Cabin pressurization
- Hydraulic pumps [21, p. 20]

Bleed air is normally pulled from the engine core at 1000°F, is cooled to 400°F so that it reaches the fuselage at 300°F, and is then further air-conditioned to pressurize and heat the cabin. Replacing bleed air with an electrical compressor enabled Boeing to remove the ducting and associated bleed air system for a net weight savings and an overall 3% fuel efficiency increase.

However, this conversion to more electric aircraft had longer-term consequences. First, more generators were needed (six total: two per engine, two in the tail-mounted APU). Second, total power generation increased to 1.5 MW, distributed by a 235 VAC system instead of the traditional 115 VAC system. The need to provide additional high-power, high-voltage battery backups led to the inclusion of the lithium-ion battery pack that caused the 2013 grounding of the 787 fleet.

Boeing elected to use a high-energy cobalt-based lithium-ion battery chemistry susceptible to thermal runaway events (fires). The 63 lb battery packs were subcontracted to GS Yuasa, a Japanese manufacturer [22]. After two battery fires grounded the 787 fleet, the National Transportation Safety Board investigated the incidents. At the time of this writing, the root cause has not been determined, and Boeing has retrofitted the planes with improved battery enclosures and venting to prevent a recurrence of the incident. It should be noted that the original battery pack was in an unsophisticated enclosure with no active cooling or fire mitigation provisions. Other design features (such as smaller cells, the use of heat-absorbing materials, or active cooling) most likely would have prevented this costly and embarrassing series of incidents. Boeing's lack of experience with lithium-ion batteries prevented it from identifying the need for more stringent safety and containment requirements.

The 787 also faced challenges related to some of its smallest parts: the fasteners that hold each plane together. By 2007, the 787 was the fastest-selling jetliner ever developed (567 orders). Suppliers had consolidated, merged, and reduced production following the September 11–related slump and lacked the capacity to support both Airbus and Boeing as they ramped up production. Boeing also finalized specifications for its fasteners late in the design process, robbing the suppliers of time to tool up and certify the fasteners (some of which required nonstandard coatings) [23]. The composite construction used in some areas of the 787 reduced the total number of fasteners needed, but tens of thousands are still needed for each plane.

In addition, 8000 fasteners had to be reworked on the first twelve 787s built. Bolts used to fasten titanium structures to composites have bevels between the head and the shank; the mating hole must be chamfered to provide clearance. The fastener installation specification referred mechanics to another specification if composite plastic was being drilled; that specification referred the mechanic back to the first specification if the fastener head was on the titanium side, as it was in this case. However, an additional subspecification appeared to supersede the second specification and provided a contradictory table containing different and inaccurate information. Yet another separate document provided even more conflicting information. This confusing mix of instructions led the mechanics to countersink the holes too shallowly; this caused the bolt bevels to bind on the chamfers and prevented the bolt heads from properly carrying load [24].

RESULTS

The first A380 was not delivered to Singapore Airlines until October 2007, one year later than initially expected. Issues with the composite wings surfaced. Cracks were detected in the wing brackets and a costly ($629 million) replacement program is under way to correct this issue [25].

Airbus has received 262 orders for the A380 (90 from Emirates) to date. Its marketing research anticipated a market of more than 1200 A380s in a 20-year period (up to 1700 if airport congestion drives the need to increase per-flight capacity). It has overspent its original $12 billion budget by several billion dollars and is delivering the aircraft 25% slower than originally forecast. The original break-even point for the program was 270 units, increased to 420 units by 2007, and increased further to an unspecified number by 2013. Some experts feel the program will never break even [25].

The first Boeing 787 was not delivered to All Nippon Airways until September 2011, four years later than initially anticipated. As of this writing, Boeing has received 930 net orders (there were also numerous cancellations). Each 787 currently costs $200 million to manufacture (with a selling price estimated to be $116 million); Boeing expects to break even on each plane by 2015 and has set its break-even point at 1100 units [26].

LESSONS LEARNED

Mason and Mitroff define *wicked problems* as ill-structured problem situations made up of highly interdependent problems [27, p. 137]. The choice of what sort of airliner to develop is certainly a wicked problem and as Jackson writes, "Most organizations fail to deal properly with wicked problems because they find it difficult to challenge accepted ways of doing things. Policy options that diverge from current practice are not given serious consideration" [27]. The Boeing Sonic Cruiser was a radical departure from conventional aircraft design; it may have succeeded if it had been developed (perhaps creating new niches just as the A380 has allowed the creation of super premium luxury flight experiences).

Both the 787 and A380 are successfully serving airlines as both manufacturers resolve *teething pains* related to failure modes (wing cracks, battery fires) not detected and resolved via modeling or testing. Time will tell which model is ultimately more successful and profitable; if either plane is replaced by a similar, upgraded airframe, it will demonstrate that the market vision that drove its introduction was correct.

Both OEMs suffered from excessive interface management issues within their development teams. A greater focus on partitioning based on the connections and interactions inherent in the systems' architectures and a data-driven insourcing/outsourcing plan would have helped both Airbus and Boeing avoid lengthy, costly delays. In addition, both vendors did not adequately manage the introduction of newer technologies (composites and lithium-ion batteries) and incurred expensive retrofits and negative public perception.

CONCLUSIONS

Both Airbus and Boeing completed development of new aircraft that have found customers in the worldwide airline industry. However, both the A380 and 787 suffered from cost overruns, delays, and highly visible engineering issues (cracking and fires). These issues highlight the difficulties in launching extremely complicated products and show the impact of organizational and strategic decisions on the development process. Whether either aircraft is ultimately profitable will depend on how well the market visions that shaped their architectures stand the test of time. The introduction of each was a significant milestone for their parent companies; only history will show what emergent behaviors each will spark within the airline network.

FUTURE WORK

The author suggests that the organization of product development teams, particularly when considering collocation and insourcing/outsourcing decisions, should be data driven. The systems architecture should be analyzed with suitable tools, such as partitioning via design structure matrices, and the results used to drive strategy. Appropriate analysis of successful and failed product development efforts would substantiate this hypothesis.

QUESTIONS:

1. Firesmith states that, in general, the complexity of a system increases with the
   * Size of the system
   * Number of different missions supported or performed by the system
   * Number of requirements implemented by the system
   * Number and complexity of its individual subsystems
   * Number and complexity of the relationships, interfaces, and interactions between the system and
     - Its subsystems (internal interfaces)
     - External systems (external system interfaces)
     - Human users and operators (human interfaces)
   * Heterogeneity of the system's subsystems (e.g., in terms of application domain and technologies incorporated)
   * Complexity of the system's technologies [28, p. 16]
       Discuss the airliners from the pre–World War I, 1920–1930s, and modern eras in terms of their relative complexity using these metrics (e.g., LZ-13 *Hansa*, Ford Tri-Motor, DC-3, Lockheed Constellation, and Boeing 707/737/747).
2. Consider a SoS with which you interact (e.g., the Internet, telephone system, food delivery). Discuss architecturally significant requirements that apply to any system added to this SoS.
3. There are indications that Boeing is still quietly working on the Sonic Cruiser [29]. Discuss the ramifications and potential emergent behavior if this aircraft were introduced into the global airline network.
4. Discuss Boeing's decision to outsource engineering tasks related to 787 development. What experiences have you had with organizations that have made similar choices?
5. Boeing's choice to standardize the 787 engine interfaces seems logical in hindsight; discuss what factors internal or external to aircraft OEMs might have delayed this significant interface requirement. Do not neglect nontechnical factors (such as politics) or issues external to the aircraft OEMs.

REFERENCES

1. S. Carey, Delta flies new route to profits: Older jets, *Wall Street Journal*, November 12, 2012. Available at http://online.wsj.com/news/articles/SB10001424052970203406404578072960852910072. Accessed March 9, 2014.
2. Boeing, Boeing 737 MAX logs first firm order from launch customer southwest airlines, December 13, 2011. [Online]. Available at: http://boeing.mediaroom.com/index.php?s=43&item=2072. Accessed July 9, 2013.
3. M.W. Maier and E. Rechtin, *The Art of Systems Architecting*, 3rd edn. Boca Raton, FL: CRC Press, 2009.
4. K. Kemp, *Flight of the Titans*: *Boeing, Airbus and the Battle for the Future of Air Travel*. London, U.K.: Virgin Books, Ltd., 2006.
5. S. Carey and A. Gonzalez, Delta to buy refinery in effort to lower jet-fuel costs, *Wall Street Journal*, April 30, 2012. Available at http://online.wsj.com/news/articles/SB100014240527023040503045773763542 88927594. Accessed March 9, 2014.

6. G. Nicolis and I. Prigogine, *Exploring Complexity*. New York: W.H. Freeman, 1989.
7. D.G. Firesmith, P. Capell, D. Falkenthal, C.B. Hammons, D.T. Latimer IV, and T. Merendino, *The Method Framework for Engineering Systems Architectures*. Boca Raton, FL: Taylor & Francis Group, LLC, 2009.
8. National Transportation Safety Board, NTSB continues investigation of Southwest Airlines Flight 812, April 25, 2011. [Online]. Available at: http://www.ntsb.gov/news/2011/110425.html. Accessed July 08, 2013.
9. J. Brancatelli, Southwest Airlines' seven secrets for success, July 8, 2008. [Online]. Available at: http://www.wired.com/cars/futuretransport/news/2008/07/portfolio_0708. Accessed July 8, 2013.
10. M. Milstein, Superduperjumbo, *Air & Space Smithsonian*, 21(2), 22, May 7, 2006. [Online]. http://www.airspacemag.com/flight-today/Superduperjumbo.html?c=y&page=2. Accessed July 8, 2013.
11. S. McCartney, In case of emergency, fly to one tiny airport in Maine, *Wall Street Journal*, June 20, 2012. Available at http://online.wsj.com/news/articles/SB10001424052702304441404577478603178669654. Accessed March 9, 2014.
12. D. Michaels, For Airbus, making huge jet requires new juggling acts, *Wall Street Journal*, May 27, 2004. Available at http://people.wku.edu/indudeep.chhachhi/519files/519hout/Boeing%200504.pdf. Accessed March 9, 2014.
13. M.E. Babej and T. Pollak, Boeing versus Airbus, *Forbes*, May 24, 2006. [Online]. Available at: http://www.forbes.com/2006/05/23/unsolicited-advice-advertising-cx_meb_0524boeing.html. Accessed July 8, 2013.
14. K. Peterson, *A Wing and a Prayer: Outsourcing at Boeing*. Everett, WA: Reuters, 2011.
15. M. Hiltzik, 787 dreamliner teaches Boeing costly lesson on outsourcing, *Los Angeles Times*, February 15, 2011. Available at http://articles.latimes.com/2011/feb/15/business/la-fi-hiltzik-20110215. Accessed March 9, 2014.
16. P. Cohan, Why 787 Dreamliner battery woes are the tip of Boeing's iceberg, January 21, 2013. [Online]. Available at: http://www.forbes.com/sites/petercohan/2013/01/21/why-787-dreamliner-battery-woes-are-the-tip-of-boeings-iceberg/. Accessed July 9, 2013.
17. Aviation Today, Coming soon: The innovative Airbus A380, *Aviation Today*, April 1, 2006. [Online]. Available at: http://www.aviationtoday.com/regions/usa/Coming-Soon-The-Innovative-Airbus-A380_206.html#.UbIyiJx_2y8. Accessed July 8, 2013.
18. Eaton Corporation, Eaton's aerospace case history Airbus A380 aircraft, September 2010. [Online]. Available at: http://www.eaton.com/ecm/idcplg?IdcService=GET_FILE&dID=406955. Accessed July 9, 2013.
19. Business Traveller, Airbus A380: How the airlines compare, March 23, 2013. [Online]. Available at: http://www.businesstraveller.com/news/airbus-a380-the-layouts. Accessed July 8, 2013.
20. S.F. Clark, 787 propulsion system, *Aero Magazine*, pp. 4–13, QTR3, 2012.
21. J. Hale, Boeing 787 from the ground up, *Aero Magazine*, pp. 15–23, QTR4, 2006.
22. J. Paur, How a battery grounded Boeing's revolutionary dreamliner, January 17, 2013. [Online]. Available at: http://www.wired.com/autopia/2013/01/boeing-787-electric-fire-grounding/. Accessed July 8, 2013.
23. J. Wallace, Fastener shortage may hurt production schedule for 787, May 23, 2007. [Online]. Available at: http://www.seattlepi.com/business/article/Fastener-shortage-may-hurt-production-schedule-1238349.php. Accessed July 8, 2013.
24. D. Gates, Boeing 787 fastener problems caused by Boeing engineers, *The Seattle Times*, November 20, 2008. [Online]. Available at: http://seattletimes.com/html/businesstechnology/2008413389_dreamliner20.html. Accessed July 8, 2013.
25. D. Michaels, Airbus wants A380 cost cuts, *Wall Street Journal*, July 13, 2012. Available at http://stream.wsj.com/story/farnborough-international-airshow/SS-2-20493/; #FIA12#FF @DanMichaelsWSJ @PlaneDame @sivag @ghimlay @AirlineReporter @BrandiLAX @BernieBaldwin @AvWeekGuy @MaxABEd @FG_STrim @APEXmary; http://translate.google.com/translate?hl=en&sl=zh-TW&u=http://aviationhk.365d.info/t739-topic&prev=/search?q%3DD.%2BMichaels,%2BAirbus%2Bwants%2BA380%2Bcost%2Bcuts,%2BWall%2BStreet%2BJournal,%2BJuly%2B13,%2B2012%26biw%3D1366%26bih%3D643. Accessed March 9, 2014.
26. A. Fontevecchia, Boeing bleeding cash as 787 dreamliners cost $200M but sell for $116M, but productivity is improving, May 21, 2013. [Online]. Available at: http://www.forbes.com/sites/afontevecchia/2013/05/21/boeing-bleeding-cash-as-787-dreamliners-cost-200m-but-sell-for-116m-but-productivity-is-improving/. Accessed July 8, 2013.
27. M. Jackson, *Systems Thinking—Creative Holism for Managers*. Chichester, U.K.: John Wiley, 2003.
28. D.G. Firesmith, *Profiling Systems Using the Defining Characteristics of Systems of Systems (SoS)*. Pittsburgh, PA: Software Engineering Institute, 2010.
29. L. Xavier, Boeing's Sonic Cruiser is back, May 2, 2012. [Online]. Available at: http://www.mobilemag.com/2012/05/02/boeings-sonic-cruiser-is-back/. Accessed July 8, 2013.

24 US Air Traffic System
Autonomous Agent-Based Flight for Removing Central Control

Sergio Torres

CONTENTS

CASE STUDY ELEMENTS

FUNDAMENTAL ESSENCE

This case examines the Air Transportation System of Systems (ATSoS), the subsystems and stakeholders involved, the complex interactions among them, and the main data flows. Particular attention will be placed on the complexities of the ATSoS, the inefficiencies of the current approach to manage air traffic, and the solutions that have been proposed to address inefficiency problems. A swarm theory–inspired alternative approach to resolve the inefficiencies of the ATSoS will be explored in depth.

The stakeholders of the ATSoS span the commerce and government domains. Although the main focus of this case study is the United States, the study can be applied to other regions because the current air traffic management paradigm is fairly common throughout the major airspaces contributing to air traffic.

TOPICAL RELEVANCE

The transportation of goods and passengers is an essential activity for the world economy. In the United States, civil aviation creates 10 million jobs and contributes 1.3 trillion dollars to the US economy. Despite reduced demand for air travel due to global recession, the Federal Aviation Administration (FAA) forecast projects that by 2032 commercial air carriers will fly 1.9 trillion available seat miles (ASM) and transport 1.23 billion passengers for a total of 1.57 trillion passenger miles per year (FAA 2012a). Clearly air transportation is vital for the US economy.

Recognizing the degree of obsolescence of the current air traffic automation infrastructure and the projected increase in demand, the US government in partnership with other stakeholders is embarking in an unprecedented overhaul of the air transportation infrastructure. The plan, called NextGen and enacted by the president and Congress in 2003 under the VISION 100 Century of Aviation Reauthorization Act, demands the largest investment ever made in civil aviation: between US$20 and US$27 billion for equipment, software, and training by 2025 (Huerta 2011). The European civil aviation authorities are also involved in a major modernization effort of their air traffic infrastructure, the Single European Sky ATM Research (SESAR). NextGen and SESAR provide an opportunity to critically examine the current approach to manage air traffic in order to reevaluate the fundamental principles of air traffic operations, understand the sources of deficiencies, and identify solutions.

STAKEHOLDERS

An examination of the user needs and the services required for air transportation helps expose the difficulties inherent in planning and executing flight operations. The functions executed by different actors and the relationships among stakeholders are covered in the Background section; the stake-holders involved in the ATSoS are as follows:

1. *The users*: Flying public, postal/cargo/carrier service providers, and the military. On any given day, more than 85,000 flights are in the skies in the United States. One-third of those operations are commercial carriers. The remaining two-thirds include general aviation, with private planes and business jets, air taxi flights, military flights, and air cargo flights (http://www.faa.gov/nextgen/snapshots/nas/). It is the demand for air travel that drives the need for air traffic services.
2. *Airlines and operators*: Commercial airlines, air cargo operators, and private jet operators have business objectives to satisfy while complying with regulations.
3. *Flight crew*: The pilots in control of the flight have ultimate responsibility for the safety of the passengers and the fulfillment of the operation's goals.
4. *Air traffic controllers*: They strive to provide a safe, orderly, and expeditious flow of traffic on a first-come, first-served basis.
5. *Traffic flow managers*: They take measures to balance demand against capacity.
6. *Government regulators*: They provide the legal and regulatory framework required to per-form air travel in a way that protects the safety and the interest of the public, the economic interests of the country, and the environment.

Primary Insights

This case study highlights the conclusion that the inefficiencies seen in the ATSoS and subsequent escalating costs of air travel stem from the lack of integration between the systems that compose the System of Systems (SoS). The study also shows that a systemic approach that deals with the inher-ent complexity of the ATSoS provides a viable solution to the inefficiency problems. The key points motivating a case study of the ATSoS are as follows:

- The commercial *airline industry* operates in a highly competitive environment, and with increasing demand it is *facing perplexing problems*: Increased operating costs, inefficiency and uncertainty in air traffic services, tighter environmental demands, very narrow profit margins, labor issues, and coupling to the global cycles of the economy. In 2007, there were 4.3 million hours of flight delay in the United States, which raised airlines' operating costs by US$19 billion and consumed about 740 million additional gallons of jet fuel (JEC 2008). In the current operational environment, these flight delays are due to air traffic control delays (28% of delays) or to causes that are under control of the airlines (26% of delays) (RITA-BTS 2013).
- *The air transportation system is fragmented and not well integrated*: Actors in the system have independent (at times contradicting) goals, operate on different strategic domains, and make decisions based on information that is not shared among the relevant stakeholders.
- *Airspace and landing capacity is limited* and the current approach to balance capacity against demand is based on central control and relies on demand predictions, but there is a limit to predictability and overcontrol renders the system suboptimal.
- *Swarm intelligence can leverage system complexity*: These issues are symptoms of the lack of a systemic approach that considers the complexity of the system. At any given time, there are around 7000 active flights over the United States whose flight paths and times over control waypoints are being centrally planned and controlled. With this approach, traffic saturation at bottlenecks generates excess fuel consumption, emissions, and flight delays. In contrast, an agent-based approach where operators can independently choose the routing that is more compatible with their business objectives promotes optimality of the system as a whole and can be implemented in a self-organizing fashion, a key behavior trait of complex systems.

KEYWORDS

Airline operations, Air traffic complexity, Air traffic control, Air traffic management, Air traffic optimization, Swarm theory applied to air transportation, Traffic flow management

ABSTRACT

This case study examines the current approach to the ATSoS, identifies its intrinsic limitations, and applies complexity science to reconceptualize the architecture of the ATSoS enabling significant performance improvements of the system as a whole. The current approach is based on a traditional systems engineering partition of the larger (airspace operations) system into functionally separated services lacking robust feedback mechanisms between them. The subsystems are structured as extensions of disjoint legacy concepts that base strategic planning on demand predictions and central control. The lack of feedback mechanisms, the inherent uncertainty in traffic predictions, and the disjoint strategic domains of the subsystems are the principal sources of inefficiency and interoperability issues.

Air traffic is inherently a complex system, but the traditional systems engineering approach does not manage the complexity dimension—complexity is detrimental. It is shown that by removing central control and implementing an autonomous agent-based flight concept, complexity can be reoriented to extract benefit. By providing real-time, accurate, and complete airspace availability information, operators acting autonomously—under the guide of their own business interests—can realize *swarm* emergent behavior: flexibility, robustness, resilience, self-organization, and natural tendency toward optimality of the system as a whole.

GLOSSARY

4D	Four-dimensional (latitude, longitude, altitude, and time)
ANSP	Air navigation service provider
AOC	Airline Operations Center
ASAS	Airborne separation assistance system
ASM	Available seat miles
ATC	Air traffic control
ATM	Air traffic management
ATSoS	Air transportation system of systems
C-ATM	Collaborative air traffic management
CDM	Collaborative decision making
CDTI	Cockpit display of traffic information
Data Comm	Data communications
DoDAF	Department of Defense Architecture Framework
ERAM	En Route Automation Modernization
FAA	Federal Aviation Administration
FIM	Flight deck interval management
FMS	Flight management system
FOC	Flight Operations Center (in this study FOC will be used to refer to the combined functions performed by the Flight Operations Center, the Airline Operations Center, and the Wings Operations Center)
FP	Flight plan
GDP	Ground delay program
ICAO	International Civil Aviation Organization

ISEF	Integrated systems engineering framework
JPDO	NextGen Joint Planning and Development Office
MAT	Managed air traffic
MIT	Miles-in-trail
NAS	National Airspace System
NextGen	FAA's next generation air transportation system
nm	Nautical mile (1 nm = 1.852 km)
NNEW	NextGen Network Enabled Weather
PBN	Performance-based navigation
PSO	Particle swarm optimization
RNAV	Area navigation
RNP	Required navigational performance
SAA	Special Activity Airspace
Self-MAT	Self-managed air traffic
SESAR	Single European Sky ATM Research
SI	Swarm intelligence
SoS	System of systems
SWIM	System-wide information management
TBO	Trajectory-based operations
TCAS	Traffic alert and collision avoidance system
TFM	Traffic flow management
TMI	Traffic management initiative
VFR	Visual flight rules

BACKGROUND

CONTEXT

In the 1970s, air traffic control (ATC) automation consisted of the use of computers and radar screens to help controllers focus on the critical task of safely maintaining aircraft separations. In the later part of that decade, the Airline Deregulation Act was passed, which caused the air traffic load to significantly increase, resulting in the emergence of new challenges. To address this, the National Airspace System (NAS) Plan of 1982 introduced enhanced automation in the areas of navigation, surveillance, and communications and started an ambitious revamping of the entire air traffic system. The renewal effort—the Advanced Automation System (AAS)—cost billions of dollars and included more than 200 projects. The excessive growth of system requirements and escalating costs drove the project to a politically unsustainable situation and the project was rescoped.

By the early 1990s, the aging ATC automation was showing signs of saturation. With seven million flights per year and a positive trend in traffic growth, it was clear that the legacy automation was not going to be able to support the demand for air traffic services. To address this technology gap, the En Route Automation Modernization (ERAM) system was planned and developed to replace the current 1960s technology automation system. ERAM is (2014) being deployed and will serve as the automation platform to integrate future concepts envisioned as part of the FAA's Next Generation Air Transportation System (NextGen). NextGen is an unprecedented overhaul of the air transportation infrastructure that will introduce advanced communication, navigation, and surveillance technologies to support air traffic demand well into the year 2025.

NextGen provides an opportunity to critically examine the current approach to manage the ATSoS in order to identify potential deficiencies and to reevaluate the fundamental principles of air traffic operations. In this case study, the traditional hierarchic architecture approach to NextGen will be contrasted with a complex system approach—Self Managed Air Traffic (Self-MAT)—that allows operators more freedom to choose air routes according to their business objectives while complying with safety and capacity constraints.

HIGH-LEVEL DESCRIPTION OF THE ATSoS

We begin with an introductory description of the current approach to manage air transportation operations, which is embodied in NextGen. Subsequent sections expand the main concepts and provide context and analysis.

The ATSoS involves a large number of stakeholders, each exerting influence guided by their own disparate interests and acting upon different operational domains. The onion-like layered structure of the ATSoS organized according to the operational domain (from tactical to strategic to long-term planning) is shown in Figure 24.1. The diagram is used as a map to guide the reader through the complex maze of functions, actors, interactions, and data flows that comprise the ATSoS. Later on in the study, other views of the ATSoS would be necessary to show the control loops (Figure 24.4) and SoS context (Figure 24.3).

The diagram shows the relationships—that arise at each layer of the structure—among users and service providers and the functions that each side performs. On the users' side, one finds the passengers, the dispatcher at the Flight Operations Centers (FOC), and the air crew. The corresponding actors on the service providers' side are government officials, traffic flow managers, and air traffic controllers. The functions executed by the service providers are rigidly divided into two realms: ATC and Air Traffic Management (ATM). ATM is the aggregate of the traffic flow management, air traffic services, and airspace management functions.

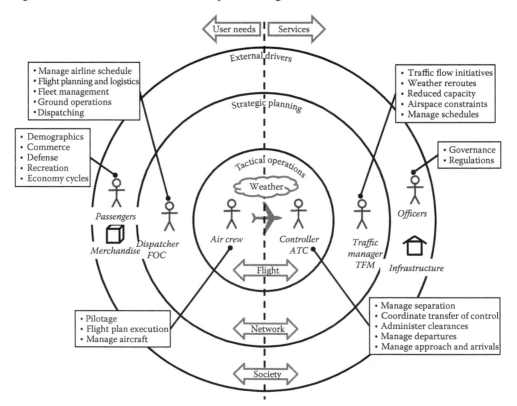

FIGURE 24.1 High-level structure of the air transportation SoS.

ATC functions operate in the tactical domain. Air traffic controllers continually monitor the progress of flights and communicate with pilots providing clearances (routing changes, holdings, and interim altitudes) in order to keep the aircraft safely separated. To facilitate the task of the controller, the airspace is split into terminal, en route, and oceanic airspaces. The terminal airspace comprises the region around major airports, 9–74 km from the airport and up to 10,000 ft altitude. The en route airspace is the main airspace above land and is further divided into sectors that are assigned to separate controllers (a sector defines the area of jurisdiction). Another task of the controllers is to coordinate and transfer control when an aircraft crosses sector boundaries.

TFM operates in the strategic domain. Traffic flow managers analyze the traffic load expected at sectors and airports hours in advance and implement traffic management initiative (TMI) aimed at reducing the flow of aircraft flying into congested airspace due to weather, special events, or traffic volume.

On the users' side of the SoS (left side of the diagram), the functions follow a hierarchical fashion that mirrors the ATC/TFM functions. The dispatcher at the FOC handles the logistics related to preparing the flight and managing the airline's resources. The air crew manage the aircraft while executing the flight plan generated by the FOC and striving for passport safety and comfort.

The normal communication process is also split according to the tactical/strategic domains. Controllers interact with pilots, dispatchers with traffic flow managers, but there is no channel of communication between the FOC and ATC. Traffic management personnel relay traffic advisories to air traffic controllers and interact with the FOC via the collaborative decision making (CDM) process, which is a group effort to share information and analyze situations that may affect air traffic.

RELEVANT DEFINITIONS

The high-level description of the ATSoS in the previous section introduced the main actors and functions. To complete the description, it is important to go over a few operational concepts and other frequently used terms (Table 24.1).

THEORETICAL BACKGROUND AND PRIOR RESEARCH

The high-level description of the ATSoS started revealing the cracks in the system and the high cost of interoperability issues resulting from loosely coordinated actions taken by traffic flow managers, on the one hand, and air traffic controllers, on the other hand. The inefficiency problems of the ATSoS motivated a look at autonomous flight concepts such as *free-flight* (which will be covered later) and to develop the concepts adopted by the NextGen plan. In this section, we will take a closer look at the current approach to manage air traffic operations and will examine autonomous flight concepts and the insight provided by swarm theory into ways to implement agent-based traffic concepts such as Self-MAT.

The structure of the ATSoS presented thus far (Figure 24.1) reflects the fragmented, loosely integrated approach currently used to manage air transportation operations. Important aspects to keep in mind for further analysis are as follows:

- In the tactical domain, *the user has minimal input into the decision process*. ATC does not know the business priorities of the operator.
- *TFM and ATC are not fully integrated*; thus, plans designed by traffic flow managers to ensure efficient utilization of the airspace can be undone by air traffic controllers when conflicts arise (which is a very likely occurrence in congested airspace).
- This situation suggests that if the information currently handled by ATC and TFM could be made available in real time to the pilot, then the pilot could take decisions that satisfy not only separation and capacity constraints but also business objectives. Is that possible? The answer provided by complexity science is explored later.

TABLE 24.1
Definition of Terms

Term	Definition	Source
4D trajectory	A description of the movement of an aircraft, both in the air and on the ground, including position, time, and, at least via calculation, speed and acceleration.	ICAO (2005)
ATC	A service operated by appropriate authority to promote the safe, orderly, and expeditious flow of air traffic.	FAA (2012)
ATM	The aggregation of the airborne functions and ground-based functions (air traffic services, airspace management, and air traffic flow management) required to ensure the safe and efficient movement of aircraft during all phases of operations.	ICAO (2010)
Aircraft intent	Information on planned future aircraft behavior, which can be obtained from the aircraft systems (avionics). It is associated with the commanded trajectory and will enhance airborne functions. The aircraft intent data correspond either to aircraft trajectory data that directly relate to the future aircraft trajectory as programmed inside the avionics, or the aircraft control parameters as managed by the automatic flight control system (AFCS). These aircraft control parameters could either be entered by the flight crew or automatically derived by the flight management system (FMS).	ICAO doc 9854
Area navigation (RNAV)	A method of navigation that permits aircraft operation on any desired flight path within the coverage of ground- or space-based navigation aids or within the limits of the capability of self-contained aids, or a combination of these.	FAA (2012c)
CDM	It is a group effort involving the system stakeholders in determining the best approach to a given traffic management situation.	FAA (2009)
Data communications (Data comm)	Controller pilot data exchange links (voice or digital).	
NAS	The common network of US airspace; air navigation facilities, equipment and services, airports, or landing areas; aeronautical charts, information, and services; rules, regulations, and procedures, technical information, and manpower and material. Included are system components shared jointly with the military.	JPDO-JPE
NextGen	FAA's Next Generation Air Transportation System. A substantial and long-term change in the management and operation of the national air transportation system.	FAA (2012b)
Swarm intelligence (SI)	SI is the emergent behavior that results from the interactions of a group of goal-seeking agents competing for resources.	Beni and Wang (1989)
TFM	Traffic management is the craft of managing the flow of air traffic in the NAS based on capacity and demand.	FAA (2009)
Trajectory-based operations (TBO)	The use of 4D trajectories as the basis for planning and executing all flight operations supported by the air navigation service provider. TBO is the underpinning concept in NextGen.	JPDO
User-preferred trajectory	The user-preferred trajectory is the set of consecutive segments linking waypoints and additional pseudo waypoints computed by the FMS to build the vertical profiles and lateral transitions.	EUROCONTROL (2010)

LEGACY SYSTEMS AND EXISTING PRACTICES

The current ATC/TFM can be modeled as a resource network of queuing servers. For final approach and landing, aircraft are guided by controllers to line up in a service queue where they *wait* (slow down, circle in holding patterns, or path stretch) for the clearance to land. Queueing theory analysis indicates that the wait time in the queue grows exponentially with inter-arrival time variance and with the inverse of available capacity (Torres and Delpome 2012). Traffic flow is adjusted based on demand predictions computed with 4D trajectories, but the accuracy of flight trajectory estimation is limited by multiple factors of a stochastic nature (Torres 2010).

To develop NextGen concepts and implementation plans, the FAA has established the National Airspace Systems Integrated Systems Engineering Framework (NAS ISEF) as a vehicle to describe the Enterprise Architecture and the NAS requirements. It provides a set of architectural views (executive, financial, operational, etc.) and requirement documents derived from the Department of Defense Architecture Framework (DoDAF).

TOWARD AUTONOMOUS FLIGHT

Emerging air traffic concepts, even before NextGen, are moving toward granting greater autonomy to operators. The idea of autonomous flight has been considered in the past based on evidence that there are significant benefits when operators can choose the routing that satisfies their business objectives (Allen et al. 1998). Motivated by the projected gains, the concept of *free-flight* aroused significant interest during the 1990s. The concept of free-flight gives operators the freedom to follow wind-optimized direct routes to their destination (RTCA 1994). Separation assurance needs to be delegated to the pilots in order for free-flight to work (Duong and Hoffman 1997), but the technology that allows self-separation and the maturity of the operational concept is far from real-world requirements and complexities (Ruigrok and Hoekstra 2007).

In limited situations, delegated separation is permissible as long as the technology supports it (Jackson et al. 2005). It is envisioned that information provided by the cockpit display of traffic information (CDTI) function can support separation applications such as alerting the flight crew of potential conflicts with other aircraft so that conflict avoidance maneuvers can be initiated (RTCA 2008). Airborne separation assistance systems (ASAS) (Ruigrok and Korn 2007) applications, such as the Airborne Spacing—Flight deck Interval Management (ASPA-FIM) concept being developed by the FAA (RTCA 2010), establishes a merging and sequencing procedure by aligning one aircraft behind another and maintaining spacing by means of position reports broadcast by the leading aircraft.

Implementations of autonomous flight concepts have been thought of as unstructured free flow similar to flight by visual rules (VFR) but extended to the entire airspace (Hoekstra et al. 2001). In these approaches, pilots constantly monitor for possible conflicts and react tactically with conflict avoidance adjustments to their flight path, similar to the traffic alert and collision avoidance system (TCAS) but with a larger look-ahead. One such approach is the Distributed Air/Ground Traffic Management (DAG-TM) concept developed by NASA Langley Research Center and the National Aerospace Laboratory of the Netherlands (NLR) (Ballin et al. 2002). This research demonstrated the feasibility of free-flight operations given the appropriate flight deck decision support automation to handle self-separation. DAG-TM experiments showed that by letting pilots take routing decisions and delegating separation responsibility to pilots, system capacity and safety are significantly enhanced. Furthermore, by distributing decision making, resources are added to the system as demand increases.

The type of unstructured, unplanned autonomous flight operation, discussed thus far, incurs complex technological demands and high risks in dense airspace where de-conflicting maneuvers result in phase transitions of the fluid that could trigger chaotic behavior (Sawhill et al. 2011). For example, what happens if one pilot reacts to a potential conflict following self-separation rules but he is surrounded by other aircraft (assuming high-density operations in this example), thus triggering in turn their reactions and so forth to subsequent neighboring aircraft? Not difficult to imagine the highly unstable, and potentially unsafe, perturbation wave that results in the flow. For autonomous flight to be a practical solution, there needs to be a mechanism to naturally coordinate self-agency. For that, we will turn to swarm theory.

Swarm Theory

The insight gained by queueing theory considerations provides a clear indication that the high cost and inefficiencies inherent in the current approach to air traffic is rooted in great measure in the lack of an explicit way to manage complexity. A key point with respect to the inefficiency of the system is that the overall efficiency of the air traffic system depends on the collective behavior of all players. There is a strong coupling feedback between the 4D trajectories that are members of the ensemble forming air traffic flows. The factors that influence the execution of a flight from gate to gate depend on multiple interactions with other flights. Because the coordination of arrivals and departures at an airport are operations that involve the entire collection of aircraft in the inflows and outflows, these factors motivate an in-depth examination of the air traffic system in light of complexity science principles.

Several researchers have demonstrated the benefits of a systemic approach to the ATM problem (Azzopardi and Whidborne 2011; Sawhill et al. 2011; Nedelescu 2012). The dependence on collective behavior and the highly complex and tightly linked network of agents that characterizes the air traffic flow optimization problem led Teodorovic (2003) to examine the principles of natural SI (Beni and Wang 1989) to solve complex problems in air traffic.

In nature, it is observed that SI exhibits remarkable efficiency in solving optimization problems (Bonabeau et al. 1999). The air traffic system exhibits traits similar to natural systems made of simple interacting social agents. Recognizing that air traffic in many respects is a multiswarm system, it has been proposed to examine the principles of SI theory to improve air traffic operations (Torres 2012). The principles of self-organized systems and intelligent-agent technologies are particularly suited to the multiagent, multiobjective structure of the ATSoS (Pechoucek and Sislak 2009; Azzopardi and Whidborne 2011).

SI has inspired the development of global optimization algorithms suitable for complex multidimensional systems. One such algorithm, particle swarm optimization (PSO), emulates swarm behavior to probe the solution space of an objective function and efficiently find the location where the optimal solution resides (Kennedy and Eberhart 1995; Parsopoulos and Vrahatis 2010). PSO and related algorithms have been applied to various complex air traffic problems: aircraft departure sequencing (Fu et al. 2008), conflict prevention and resolution (Burdun and Padentyev 1999), route optimization for Unmanned Air Systems (Jevtic et al. 2010), optimization of arrival sequences (Bing and Wen 2010), optimization of airway network design (Cai-long and Wen-bo 2011), and optimization of airline cargo operations (Bonabeau and Meyer 2001).

Self-MAT proposes to remove the line that separates the ATSoS into the two hemispheres of *users* and *service providers* (Figure 24.1) and also to remove the line dividing the tactical and strategic domains and making it a continuum. The integrated Self-MAT structure of the ATSoS is represented in Figure 24.2. Note that the air traffic controllers retain an important function, their role will shift from *management by tactical intervention* to a more strategic *management by planning and intervention by exception*, and similarly the function of the air traffic manager will shift to more interesting strategic functions shared with the operators. How can this concept be implemented is the subject of the Development section, for now it is important to highlight the fact that developing simple interaction rules between members of the swarm (aircraft) and providing full and accurate information to pilots so that they have situational awareness the system becomes and instantiation of SI. The 4D airspace grid shown in the diagram serves as the information sharing mechanism needed for this approach to work (details will be provided). The recognition that air traffic

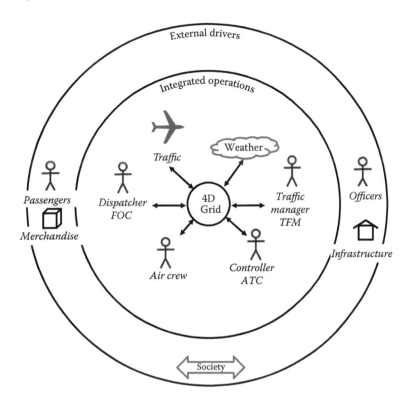

FIGURE 24.2   Self-MAT efficient integration of the ATSoS.

is the result of decisions independently made by goal-seeking agents suggests that the tendency to optimality found in natural swarms should also be realizable in the air transportation system, if the system is allowed to operate as an intelligent swarm (implying agent-based autonomous flight). This case study further develops this thread and shows that desirable swarm behaviors (distributed functioning, tendency to optimality, resilience, efficiency, etc.) can be realized in the air traffic system by removing central control.

GUIDING PRINCIPLES

Whereas the NextGen approach views the air transportation system as a hierarchical architecture of separate functions (i.e., separation management, capacity management, etc.), Self-MAT considers the system as a whole, guided by the theory of SI, for which the following principles apply:

- *Nonlinearity*: Interactions among trajectories create complexity that does not scale linearly with trajectory aggregation; global optimization is not possible in the presence of networked causality and independent agency.
- *System of systems*: The air traffic system is extensive, highly dynamic, composed of physically distributed, preexisting, heterogeneous, autonomous, and independently governed systems. As such, it is expected to exhibit unpredictable emergent behavior that is compounded by human behaviors affecting the system (i.e., unruly passengers, distracted pilot, technician error, etc.).
- Because *emergence* is an outcome of complex systems, overcontrolling or top–down approaches are detrimental and could potentially inhibit positive emergent behaviors.

TABLE 24.2
Characterizations of NextGen and Self-MAT

Characterization	NextGen	Self-MAT
System maturity	Modernization plan currently being implemented by the FAA and envisioned to cover the period up to 2025	Self-MAT is at the moment a research effort at the concept stage. It has not been incorporated officially into any future plan. Self-MAT could gradually be phased in within the performance-based navigation (PBN) scheme (initially the service is provided into special corridors to aircraft that are properly equipped)
System engineering approach	Traditional systems engineering hierarchic architecture, top–down approach that does not take into account the complexity dimension	Integrated, holistic, and systemic
Structure	NextGen modernizes and extends the legacy functions but does not change the legacy paradigm to air traffic, which relies on predictions and central control (Figure 24.1)	Self-MAT is an agent-based implementation that removes the boundaries between strategic and tactical domains (Figure 24.2)

- *Self-organization* can be oriented toward optimality by exploiting existing market forces. Optimality, however, cannot be achieved or at least any optimum would be short lived because of the dynamics of most situations. Resilience and efficiency of the air traffic system result as emergent behaviors of a swarm of autonomous agents navigating in a maze of conflicting constraints and under a high-dimensional multiobjective function.

CHARACTERIZATIONS

To finalize the exposition of the NextGen and Self-MAT approaches, Table 24.2 contrasts characterizations related with engineering approaches, maturity, and structure.

PURPOSE

HISTORY

The current ATSoS emerged from a process of gradual automation of decades old manual procedures. The aim of computer automation was restricted to provide tools to the controller to facilitate repetitive tasks and to help in decision making. The lack, therefore, of a systemic approach resulted in an aggregate of *ad hoc* and disjoint components designed to handle specific functions but poorly interoperable and unable to handle the complexity arising from the stochastic nature of operations and the nonlinear interactions between the components.

Manual procedures were developed to accommodate the limitations of early technology (such as radio beacons), but as new technologies get introduced, the same old procedures continue to be applied. Instead of taking the most efficient path, which can be computed with the computer power available to operators, airplanes are forced to fly along rigidly defined airways.

With NextGen, the concept of TBO seeks to improve the predictability of the system by providing a more accurate and shared view of the flight in the form of the predicted four-dimensional (4D) trajectory that includes the horizontal path, the vertical profile, and the time or arrival at key points along the path.

The Current Situation

The FAA, following NextGen plans, is implementing a series of operational improvements with the goal of providing enhanced air traffic services able to support the increasing demand while maintaining safety (FAA 2012b). As mentioned in the introductory section, the ERAM automation platform is currently (2014) being deployed and enhanced to integrate the technologies called for by NextGen: the Automatic Dependent Surveillance—Broadcast (ADS-B), Data Comm, and the System-Wide Information Management (SWIM) system.

The ADS-B technology provides accurate aircraft position to controllers based on data obtained from the Global Positioning System (GPS). The gained accuracy provided by GPS-based position information will allow controllers to reduce the separation standards and hence increase capacity.

Data Comm will replace voice communications with electronic transmissions that enable granting more complicated clearances that are required for more efficient routing. SWIM is an information platform that will allow information to be disseminated among systems with an open, flexible, modular, and secure data architecture.

Known Problems

The problems of loose ATC-TFM integration and the limited participation of the FOC in the ATM/ATC decision-making process were mentioned in the Background section. These and related problems are analyzed further in the following paragraphs.

Interoperability Issues

A fundamental problem of the current system is that the ATC and TFM functions operate within orthogonal domains (strategic vs. tactical), respond to different goals, and are not tightly coordinated among themselves nor with the operators. This situation results in suboptimal performance, increasing cost, and detrimental environmental impacts.

Limits to Predictability

The benefits of TBO are based on the assumption that the availability of 4D trajectories increases the predictability of the system. In practice, this assumption has severe limitations. Because of the multiplicity of factors that affect the flight (weather, human behavior, fidelity of mathematical models, unknown aircraft intent, tactical intervention, deviations in aircraft performance, etc.) and their stochastic nature, there is a limit to the accuracy of a 4D trajectory (Torres 2010). A flight cannot be planned and executed as an individual and isolated flight, but rather as a network of interrelated events. However, while developing TBO, most of the efforts to increase the accuracy of predictions have focused on individual trajectories, but not on the ensemble of trajectories. In addition to the intrinsic limitations of trajectory predictions, an even more severe limitation is the fact that the variance of traffic flows is an ensemble attribute that has to do with the behavior of the system as a whole, not its individual components. If the decisions regarding how a flight is executed depend on predictions of its future path, those predictions should take into account the possible interactions of the flight with the other flights and with the resources that other flights are also competing for.

Integration Issues

The NextGen plan implements a range of technologies that encompass the entire gamut of air transportation services: weather, communications, surveillance, and navigation. An area of concern, however, is that these technologies are being developed as scaled extensions of legacy components without full consideration of integration issues and systemic behaviors resulting from the intrinsic complexity of the system. The same fragmented architecture is being updated (scaled) with additional technology, but the underlying approach remains basically unchanged. This is a fundamental problem because complexity is not explicitly addressed; complexity and uncertainty do not scale linearly.

Inability to Incorporate User Preferences

A report by the FOC Study Team finds that the current NextGen approach is focused on ATC modernization rather than what is best for the NAS and TBO; the key operating paradigm of NextGen lacks full consideration of FOC decision processes (JPDO 2012).

NextGen concepts take into account user preferences but do not offer a mechanism to let operators have direct choice. Operators can provide, for example, a set of desired trajectories for a given flight ranked in order of preference, but in the event of a conflict, air traffic controllers have the last say about which trajectory (if any) receives a clearance for flight. The issue of limited effect of user preferences is further complicated by the fact that airline business parameters (which are key inputs to optimization drivers) are not within the purview of controllers and traffic managers. It is unlikely that operators disclose proprietary information (such as cost index, scheduling priorities, and aircraft load factors) that is required for cost-efficient decision making. For efficient operations, it is imperative that decision-makers base their decisions using shared information, but this is not possible in a fragmented system. Airlines do not know about specific ATC decisions that may affect their flights, and in turn, ATC cannot accurately estimate the impact of their choices to airline business goals. In Self-MAT, on the other hand, operators know their business objectives and have the freedom to make routing choices based on these priorities.

Unachievable Optimality

Optimization of the NAS is not explicitly addressed by any NextGen function or operational improvement. Optimization is severely limited by the central planning approach to TFM. The fact that TFM and ATC are not tightly integrated means that strategic planning to implement traffic flow initiatives can be undone by ATC when safety requirements so demand. TFM initiatives aimed at metering the inflow of aircraft to an arrival point cannot control the inter-arrival variance, which is one of the factors determining throughput.

Limited Use of Onboard Avionics Capabilities

Modern FMS systems are capable of computing trajectories that minimize cost, distance, fuel burn, and other business factors important for the operators. It is envisioned that with future Data Comm mechanisms some of the information generated by the FMS could be utilized by ground automation systems. Even with such mechanisms in place, however, at the moment of a routing change due to a TMI or a tactical intervention, it is not possible to use those powerful capabilities.

MISSION, DESIRED OR EXPECTED CAPABILITIES, AND PROCESS FOR ACHIEVING THE OBJECTIVES

In general terms, the NextGen objectives are to increase the efficiency of air traffic services while maintaining safety. The ATM network envisioned in NextGen will make use of 4D trajectories to share information and manage operations, will implement aircraft self-separation procedures when the aircraft equipage allows it, will increase CDM among stakeholders, will make use of advanced capabilities on board the aircraft, and will developed more efficient design of airspaces and routing choices (such as RNAV/RNP approaches) that exploit advances in navigation.

To achieve NextGen goals, the FAA has developed a series of operational improvements organized in groups of strategic activities (or *service roadmaps*) as follows (FAA NASEA 2013):

- Trajectory-based operations (TBO)
- Collaborative air traffic management (C-ATM)
- High-density operations
- Terminal environment efficiency

- Reduced weather impact
- Safety, security, and environmental performance
- Facilities

TBO is a key enabler of NextGen. With TBO, the predicted 4D trajectory of the flight is used as the basis for planning and executing the flight. The rationale behind TBO is the belief that sharing the trajectory among users and air service providers allows for better coordination, informed decisions, robust strategic planning, and more efficient use of resources. The TBO concept supports an important number of NextGen operational improvements, such as

- ADS-B separation (2010)
- Oceanic in-trail climb and descent (2010)
- Increase capacity and efficiency using RNAV/RNP (2010)
- Delegated responsibility for in-trail separation (2013)
- Flexible entry times for oceanic tracks (2013)
- Automation support for separation management (2014)
- Point-in-space metering (2014)
- Initial conflict resolution advisories (2015)
- Flexible airspace management (2017)
- Provide interactive flight planning from anywhere (2015)
- Automation support for trajectory negotiation (2018)
- Automated support for conflict resolution (2018)
- Reduce separation standards (3 nm from 5 nm en route) (2019)
- Reduce oceanic separation and enhanced procedures (2019)
- Dynamic airspace performance designation (2019)
- Flexible routing (2019)
- Flow corridors—Level 1 static (2020)
- Self-separation airspace-oceanic (2022)
- Self-separation airspace operations (2022)
- Automated negotiation/separation management (2023)
- Trajectory-based management gate-to-gate (2025)
- Automation-assisted trajectory negotiation and conflict resolution
- Flow corridors—Level 2 dynamic (2025)

C-ATM provides mechanisms to incorporate user preferences in air traffic operations. Key operational improvements included in the C-ATM solution set are

- Continuous flight day evaluation (2012)—real-time and postevent traffic monitoring for strategic planning
- Flight plan constraint evaluation and feedback (2013)
- On-demand NAS information (2013)
- TMI with flight specific trajectories (2014)
- Improved management of Special Activity Airspace (SAA) (2017)
- Full CDM (2017)

COMPLEX SYSTEM OF SYSTEMS

The scope of the ATSoS includes the airborne functions and all the service infrastructure and systems required to make air transportation possible. The structure of the ATSoS and the subsystems comprising the ATSoS (ATC, TFM, FOC, aircraft/crew) were introduced in the high-level description provided in the Background section (Figure 24.1). The context diagram of

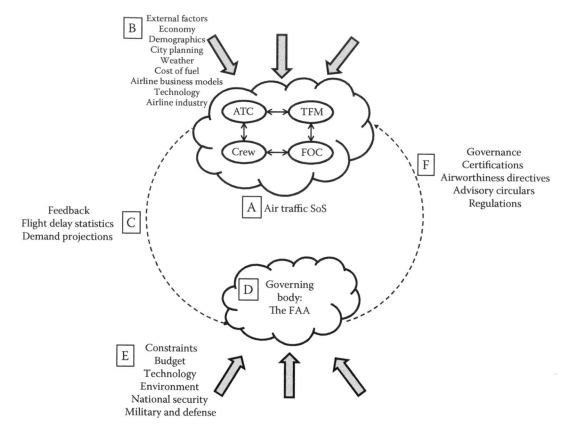

FIGURE 24.3   Air traffic SoS context diagram showing the ATSoS (A) with its environment and constraints (B, E, F), governance (D) and feedback loop (C).

the ATSoS (Figure 24.3) presents the subsystems in relation to the operational environment, external factors, and feedback loops.

ENVIRONMENT AND CONSTRAINTS

Air traffic is driven by socioeconomic demands and the expectations of the end users (the flying public, business, and cargo carriers) and is affected by multiple factors (B, E, and F). The relevance of these factors (Table 24.3) is manifested at different planning horizons that span a large range, going from minutes (convective weather) to years (demographics and city planning).

FEEDBACK AND GOVERNANCE

The control loops and relationships between the ATSoS subsystems are described in a separate section. Here we are concerned with interactions of the ATSoS with the socioeconomic ecosystem that surrounds it.

The US Department of Transportation (DOT) Bureau of Transportation Statistics (http://www.transtats.bts.gov/) keeps a record of air traffic operations and tracks the on-time performance of domestic flights operated by large air carriers. These statistics inform the flying public about the quality of services provided by the airlines and can be used by the airlines to make adjustments to their operations.

In the United States, the Federal Aviation Administration (Department of Transportation) is the main governing body for air transportation. The FAA regulations (http://www.faa.gov/

TABLE 24.3

External Factors and Constraints Affecting the ATSoS

Factor	Description
Demographics	Airlines open or discontinue routes based on the population patterns that dictate demand for air travel.
World and local economy	Fuel prices, business activities, recreation, and events are all tied to the cycles of the economy.
International regulation	Airlines must implement procedures to comply with international regulations such as handling of hazardous materials, handling international passengers, etc.
Government and environmental regulations	Safety regulations, limits on emissions, national security, etc.
City ordinances	Noise abatement regulations, limitations imposed by the city to build new runways or expand airports and/or number of operations into an airport will affect air traffic.
Labor demands	Labor unions (pilots, air traffic controllers, etc.).
Technological advances	Availability of navigation, communications, and surveillance technologies such as ASD-B enable increased capacity.
Airline business objectives	Type of operation (hub and spoke or city-pair), airline scheduling influence traffic patterns.
Aircraft capabilities	The PNB concept foresees that certain procedures designed for more efficient operations can only be granted to aircraft that are properly equipped. Airlines need to balance the cost of getting their aircraft equipped with the benefits received by participating in PBN services.
Weather	Visibility around airport areas affects the Airport Acceptance Rate (AAR), reduced capacity could trigger ground stop (GS) or ground delay program (GDP) initiatives; presence of convective weather cells could trigger integrated collaborative rerouting (ICR) measures.
Airspace usage	Special use airspace (SUA) are reserved airspaces (typically military) that limit the traffic of civilian aircraft. SUAs are not permanently active (some have an on/off schedule).
Civil infrastructure	Maximum capacity at an airport (AAR) is determined by the number of runways.

regulations_policies/faa_regulations/) include advisory circulars, airworthiness directives, aircraft certification, hazardous materials regulations, operating requirements, flight rules, ATC procedures, and separation standards.

The US Congress created the Joint Planning and Development Office (JPDO: http://www.jpdo.gov) to manage the partnerships designed to bring NextGen online. These partnerships include private-sector organizations, academia, and government agencies:

- Department of Transportation (DOT)
- Department of Commerce (DOC)
- Department of Defense (DOD)
- Department of Homeland Security (DHS)
- Federal Aviation Administration (FAA)
- National Aeronautics and Space Administration (NASA)
- White House Office of Science and Technology Policy (OSTP)
- Office of the Director of National Intelligence (ODNI)

BOUNDARIES

For the purposes of the analysis performed in this case study, the boundaries of the ATSoS are defined by the *strategic planning* circle in the high-level diagram in Figure 24.1. From an operational point of view, this boundary spans time horizons that go from the long-term planning functions performed by the FOC—days or weeks before the flight departs—to the tactical interventions by ATC that takes place during flight.

The *strategic planning* boundary also encompasses the internal boundaries defined by the *tactical operations* circle and the *services/users* line dividing the ATSoS structure into two distinct hemispheres. Those boundaries reflect the current structure of the ATSoS.

The Self-MAT swarm theory approach removes the *user/services* line (it is a self-service approach) and also removes the separation between strategic and tactical domains. The strategic plan developed by operators according to their own business criteria becomes the plan that gets executed—not without continuous dynamical adjustments in response to changing capacity and safety constraints.

INTERNAL RELATIONSHIPS AND DECISION PROCESS

The communication channels and control loops in the ATSoS mirrors the fragmented arrangement of the subsystems comprising the ATSoS. Figure 24.4 shows the interactions that take place as the flight progresses.

Before departure, the Airline Operations Center (AOC) checks with the traffic flow managers for potential traffic flow constraints that could result in GS or GDP activities. If GS or GDP initiatives apply, the CDM process is invoked to give operators the chance to provide their preferences as to which flights get affected. Dispatchers go through the flight preparation process and pilots get the predeparture briefing.

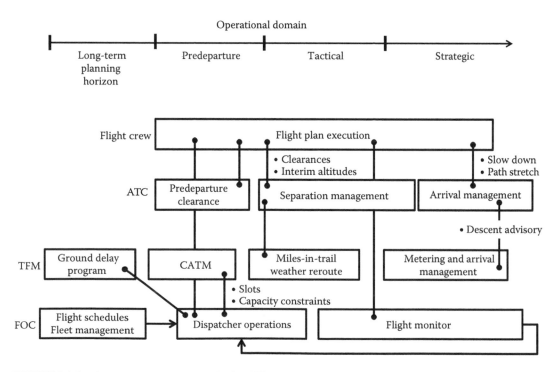

FIGURE 24.4 Actors and relationships in the ATSoS.

After departure clearance is granted, tower controllers take care of managing the departure queue and controlling ground movements until the aircraft takes off. Departure operations could be limited by a Departure Sequence Program (DSP), which is a TMI designed to achieve a constant flow of traffic over a common point. A DSP involves coordination between the controllers and traffic management personnel. Once the flight departs, most of the interaction takes place between the pilot and controllers, the FOC does not talk to ATC.

Once the flight takes off, control is transferred to the terminal area controller, which takes care of controlling to a point at the boundary of the terminal area and up to (typically) 10,000 ft. At this point, the aircraft enters en route airspace, where en route controllers monitor the flight while under radar contact. Controllers communicate with pilots (currently using voice channels and, in the future, using electronic communications when the Data Comm infrastructure gets deployed) providing instructions to change altitude levels, momentarily level off (if the aircraft was undergoing an altitude transition), change speeds, or deviate from the planned route of flight. These actions could be the response to conflict avoidance, convective weather, avoidance of flow constrained areas, or arrival sequencing directives generated by traffic flow managers. During flight, pilots can also request altitude changes or *direct-to* shortcuts to the route. In current operations in the United States, the flight plan submitted by the operators to the FAA only contains a requested cruise altitude (ICAO flight plans accommodate multiple cruise altitudes), but as the aircraft burns fuel, it becomes more advantageous to flight at a higher altitude. In that case, pilots need to request an altitude change (which they may or may not get, depending on traffic load).

During cruise, as the aircraft crosses sector boundaries, control is transferred to different controllers. While the flight is in en route control, traffic flow managers can initiate TMI that affect the flight, such as miles-in-trail (MIT), minutes-in-trail (MINIT), En Route Sequencing Program (ESP), Airspace Flow Programs (AFP), and ICR. These TMIs are coordinated with air traffic controllers.

When the aircraft is within 200–250 nm from the arrival point (or *meter fix*—a designated waypoint in the route where traffic flow is sequenced for delivery to the terminal area controller), the Traffic Management Advisor (TMA)—which is a tool to sequence and schedule aircraft to the meter fix—assigns an arrival slot to the flight. The Scheduled Time of Arrival (STA) to the meter fix is presented to en route and terminal area controllers to ensure that the aircraft meet the assigned time. TMA is part of TFM, and it will be replaced by a more powerful system, the Time-Based Flow Management (TBFM), currently being deployed.

The TMA planning and sequencing program is an instantiation of a centrally controlled queue. TMA plans who gets into the queue when so that service can be provided. This approach works without any delays only under perfect *just-in-time* conditions (when the client gets to the queue at the exact assigned time). Just-in-time conditions are not realistic in aircraft traffic flow. It was noted in the Background section that the expected wait time of the queue grows exponentially with the variance of the arrival times to the meter fix.

In Self-MAT (as will be described in detail in the Development section), there is no arrival queue variance, and traffic flow is self-regulated and adjusts dynamically to accommodate to the available capacity. Whereas in TMA/TBFM, flow is centrally planned, in Self-MAT, each user takes care of its own plan.

CHALLENGES

Reconceptualizing the legacy approach to the ATSoS that has maintained air transportation functioning in the United States since the first days of the Central Flow Control Facility in 1970 presents significant challenges. The analysis performed here has exposed the limitations of the legacy approach and proposed the concept of autonomous flight and self-management in order to make the system more efficient. The biggest challenges facing the Self-MAT approach are maintaining safety, developing viable implementation strategies, preventing airline gaming, guaranteeing equity, maintaining system stability, and gaining stakeholders acceptance.

- *Safety*: Safety is the number one concern, and it must be guaranteed by any future development of the ATSoS. The Self-MAT implementation plan, or any new concept, must include a safety net such that any deviation that impacts the safety can be mitigated. In Self-MAT, flights choose their routing but those routes are still subject to clearances by air traffic authorities, and once airborne, the flights are still under monitoring by air traffic managers that can take the flight off the autonomous flight clearances and back into manual control in the event of a deviation from the clearance.
- *Implementation*: Implementing a completely new operational approach cannot be done overnight, it has to be phased in gradually. The Self-MAT autonomous flight concept relies on flight management technologies and, as is usually the case, these new technologies will be adopted by operators slowly as the benefits become apparent. NextGen concepts face similar challenges. For that reason, PBN is seen as the door to introduce new operational concepts in a nondisruptive fashion. Self-MAT could be initially deployed as a service made available on a PBN basis.
- *Gaming*: Mechanisms must be established to prevent operators from gaming the system to obtain benefits that unfairly limit access to the airspace to other users. This problem can be mitigated by introducing a cost to 4D clearance change requests (i.e., penalties to operators that submit bogus requests for clearances). Commoditization of airspace has the advantage of introducing market forces that help drive the swarm toward optimality. The concept also helps to distribute capacity in a more efficient manner: the cost of a landing slot at peak time should not be the same as a slot at an off-peak time. Operators can engage in airspace trading.
- *Equity*: How can rationing of resources be performed equitably in a self-managed environment? The current CDM process, as well as the evolution of CDM approaches envisioned in C-ATM, can be applied to Self-MAT via a package of rules that incorporate the 4D-Grid.
- *Stability*: Since in Self-MAT operators are agents that react independently, a situation could arise by which a change in the system could trigger unstable behavior.
- *Acceptance*: Implementing a completely new operational approach changes the way users and service providers work. It will not be surprising if air traffic controllers, traffic flow managers, engineers, and managers perceive the new system as a threat to their jobs. In light of the long tradition of central control, the idea of autonomous flight could be seen by the engineers that developed and now maintain the legacy system as a farfetched, far from real, and nonpractical proposition. The developers of Self-MAT have a tremendous challenge to demonstrate the concept in compelling and crisp engineering terms.

DEVELOPMENT

Self-MAT recognizes that operators are goal-seeking agents that can be considered members of a swarm, that operators need access to the airspace, that operators interact among themselves following relatively simple conflict avoidance rules, and that they can take decisions that best fit their goals if sufficient information and situational awareness is provided to them. Therefore, when operators are set free, SI traits emerge. The focus of this section is to describe how Self-MAT works and how it can be implemented with existing or planned NextGen technologies.

SELF-MAT SOLUTION TO IMPROVE THE ATSoS

At a fundamental level, the air traffic system can be described as an ensemble of independent agents that have a well-defined goal: to transport passengers and goods from point A to point B safely and in the most economical way possible and subject to capacity constraints and safety considerations. As such, it exhibits the traits of a complex system:

- Networked causality
- Independent agency
- Nonlinear feedback processes
- Unforeseen (and not manageable by any single actor) interactions between components
- Nonobvious interdependencies (e.g., intricate couplings between arrival and departure flows)
- Disparate goals and operating domains of independent actors
- Effects of multiple environmental factors of a stochastic nature, such as weather and human behavior

With the air transportation system, however, the desired emergent behaviors of SI systems (autonomy, distributed functioning, efficiency, and tendency to optimality) are inhibited due to central control. The natural solution to the inefficiency problems is to *acknowledge that air traffic is a swarm system* and to let it act as such by *removing central control*. This objective can be achieved by implementing an autonomous flight concept where pilots/operators are provided with a real-time, continuous, accurate, and complete view of airspace availability, including capacity constraints and use of airspace by other aircraft (shared situational awareness) so that they can plan their operations as autonomous agents that respond to their business goals while complying with constraints. Another key technology that enables agent-based air traffic is 4D-NAV, namely, the capability of modern FMS to use a 4D trajectory (4DT) for guidance to within known tolerances. Agent-based air traffic is possible by designing the 4D trajectories used for guidance by the FMS based on shared situational awareness (that includes airspace used by other aircraft).

This approach allows managing complexity directly and could be accomplished by a rearrangement of NextGen concepts and technologies as shown in Figure 24.5.

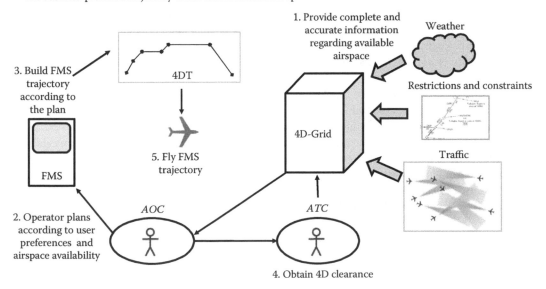

Self-MAT
The business-preferred trajectory closes the information loop

FIGURE 24.5 Self-MAT operational concept, (1) information sharing, (2) mission-driven plan, (3) high-fidelity trajectory built according to user preferences, (4) controller acceptance and (5) flight execution based on coordinated trajectory.

The Self-MAT diagram (Figure 24.5) shows the operational concept, interactions, and actions that take place under a self-organized air traffic paradigm (a more detailed exposition of the Self-MAT concept can be found in Torres (2012), and Torres and Delpome (2012)):

1. *Provide complete and accurate information regarding available airspace*: The flow of information between members of the swarm is necessary to drive the emergent behavior toward nonchaotic, coherent flows. This is achieved by sharing constraints and trajectories via a 4D-Grid. The 4D-Grid is used to store and share airspace utilization (including weather) and constraints. Since space is four dimensional, time is also parceled into slots. The size of the cells is determined by separation rules and by buffers to accommodate the required navigation performance (RNP) of the aircraft. The concept of a 4D-Grid has been used in trajectory-based methods to integrate gridded weather avoidance fields (Avjian et al. 2009) and to develop efficient conflict detection algorithms (by storing a 4D trajectory in the grid, a doubly occupied cell indicates a conflict) (Jardin 2005; Avjian and Dehn 2012). A 4D-Grid concept to store weather information, the *4D-Cube*, is also planned for the NextGen Network Enabled Weather (NNEW).

2. *Operator plans according to user preferences and airspace availability*: Cells in the 4D-Grid are marked as occupied or available. For example, the cells corresponding to a SAA will be occupied during the times that apply. Moving convective weather fronts are reflected in cells by changing state as the front moves. Metering to a terminal area appears in the 4D-Grid by making the cell corresponding to the meter fix as available only for 1 min at a time, for instance, if the flow capacity is one aircraft per minute. The 4D-Grid is also populated with the trajectories from other flights that have received clearances. Cells reserved by other flights acquire an occupied state during the time interval the aircraft is planned to spend flying that cell. The 4D-Grid provides common situational awareness across facilities, time domains, and actors.

3. *Build FMS trajectory based on information in the 4D-Grid* so that the 4D RNAV path goes through nonoccupied cells. This means that by construction, the 4D trajectory does not encounter conflicts with other aircraft or restricted airspaces and it arrives at the meter fix when there is service available avoiding queue wait time. Since the FMS trajectory is used for closed-loop guidance by way of the AFCS, the 4D path stays within a known tolerance. Aircraft flying with VNAV/LNAV engaged can stay laterally within the RNP, longitudinally can meet required time of arrival (RTA) to within a few seconds (Klooster et al. 2009), and in the vertical dimension, the vertical RNP standards are maturing.

4. *Obtain 4D-clearance*: Once the AOC has generated a flyable 4D trajectory that complies with ATC and TFM constraints, the AOC can submit it for clearance. Note that once a 4D clearance is granted, it becomes a contract: ground control strives to facilitate the execution of the plan according to the clearance and the operator commits to stay within established tolerances of the cleared 4D path. Dynamic and continuous replanning is necessary to accommodate changing conditions.

5. *Fly FMS trajectory*: Upon clearance granted, the AOC uplinks the trajectory to the aircraft and the flight proceeds accordingly. The feasibility and limitations of aircraft guidance controlled by a 4D trajectory has been shown by Vaddi et al. (2012).

6. *Monitor the evolution of the system* and make adjustments to trajectories when opportunities for further savings arise or if unforeseen conditions develop that block parts of the airspace originally allocated to flights. The cycle of monitoring and adjustment is a highly dynamic process. Stability of the system becomes a concern.

With Self-MAT, the control loops of the legacy system (Figure 24.4) are replaced by a *reserve/commit/fly* control loop mechanism (Figure 24.6) that is entirely driven by users.

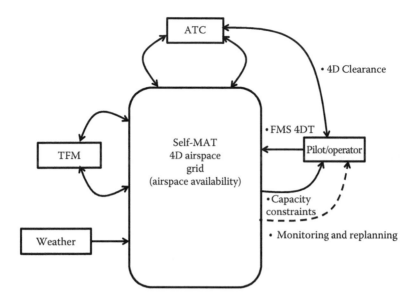

FIGURE 24.6 Control loops in Self-MAT.

The *reserve/commit/fly* construct is a compact representation of the Self-MAT concept: operator analyses the available airspace (via access to the 4D-Grid), designs the best possible routing given the available 4D airspace, submits the routing choice for clearance, and commits to execute the cleared 4D path and flies.

RESULTS

Having described the mechanisms by which autonomous flight can take place, we proceed to analyze how Self-MAT solves the problems with the legacy approach to the ATSoS. To summarize, the main problems of NextGen that have been described earlier are

- Interoperability issues
- Integration issues
- Reliance on predictions (predictability is limited and exponentially expensive to improve)
- Unable to fully incorporate user preferences
- Unachievable optimality

INTEROPERABILITY ISSUES

An essential function to enable TBO is the ability to coordinate and grant 4D clearances. The 4DTRAD service (RTCA 2012) relies on 4D clearances; however, under the legacy paradigm, 4D clearances are not guaranteed to be flown as initially granted because they are affected by trajectory prediction uncertainty. In Self-MAT, the 4D FMS trajectory is imprinted in the 4D-Grid and thus can be used to *probe*, visualize, and understand the interaction of the flight's entire path in relation to the other flights and constraints. For the internal operation of the ground ATC systems, the FMS trajectory that has been cleared can be synchronized as described in Torres et al. (2011) and Klooster et al. (2010).

INTEGRATION ISSUES

The TFM and ATC functions are naturally integrated in Self-MAT. Capacity constraints and weather events are incorporated in the 4D-Grid. All the other traffic cleared for flight by ATC is also incorporated in the 4D-Grid by marking as *reserved* the cells traversed by the respective 4D trajectories. The operator uses that information to design a strategy that *goes around* those constraints; in other words, the operator plans the flight to follow a conflict-free trajectory that uses available airspace. If the aircraft accurately tracks the cleared 4D trajectory, then the flight is executed as planned, free of tactical interventions by controllers that are solving separation or capacity issues. In Self-MAT, the operator/pilot is converted into an agent that maintains separation and meets meter time; both goals are achieved by following the cleared 4D FMS trajectory.

RELIANCE ON PREDICTIONS

In Self-MAT, a trajectory is not used as a prediction but rather as an expression of aircraft intent. Instead of relying on predicted demand in order to manage traffic flow, in Self-MAT, capacity is automatically allocated based on the 4D path aircraft use for guidance. Thanks to the ability to track a 4D trajectory, aircraft can follow the cleared 4D trajectory to within a known RNP. Remaining uncertainties (wind, for instance) are *absorbed* by the 4D guidance approach (corrections are continuously being applied by the aircraft guidance and control system). In summary, with Self-MAT, uncertainty is managed: at any given time, the state of the 4D-Grid reflects current and projected airspace utilization in a deterministic way.

UNABLE TO FULLY INCORPORATE USER PREFERENCES

In Self-MAT, the base for the 4D clearance is the FMS trajectory (business reference trajectory) built by the user according to business objectives. Since the trajectory is generated to traverse only available airspace, the likelihood that the flight will proceed unperturbed increases significantly, uncertainty is reduced, and business objectives are met. By knowing constraints up front, the airline can find a solution that is more cost-effective and takes into account their business objectives. The resulting 4D plan may not be the absolute optimal path for a given individual flight—but it is conflict free, so it reduces the uncertainty and deficits associated with delays and reroutes. The alternative presented by TFM launching a flight with an unknown delay time to be incurred in order to meet time at the metering fix generates uncertainty that limits the strategic options available to the users.

UNACHIEVABLE OPTIMALITY

If all aircraft participating in Self-MAT are compliant with their cleared 4D trajectory, there will be no conflicts by construction and no need for tactical intervention. Furthermore, the ability of aircraft to follow their 4D paths within the RNP allowed by their equipage on board can be exploited to reduce separation buffers and accommodate more aircraft in a given volume of airspace.

When all aircraft are following the committed FMS 4D trajectories to within the known RNP tolerances, no conflicts are expected and traffic merging and flow to arrival fixes will be metered because the trajectories were built with those constraints. The end result is a tight and harmonized integration of ATC, TFM, and user preferences, in other words, an instantiation of TBO. Removing central control and providing complete situational awareness to operators endows the air traffic system with SI attributes that, driven by market forces, generate pressure on the system to adjust continuously toward an optimal use of resources. Opportunity costs (i.e., savings attained by being proactive and the high cost of not participating in Self-MAT) provide the market forces.

The *reserve/commit/fly* process (Figure 24.6) is a dynamic loop that allows adjustments as new opportunities to optimize business objectives emerge or new constraints, such as weather events, materialize. In Self-MAT, each operator is searching for local minima of their objective function based on their business objectives. With these mechanisms in place, the emergent behavior is the tendency to optimize the use of resources. The search for optimality is distributed to the operators.

Computational load and resources needed to derive optimal solutions are distributed; resources are instantly created when the demand increases (each FOC brings its own analysis and decision-making infrastructure). The first fundamental theorem of welfare economics indicates that such a system tends to a Pareto efficient allocation of resources.

ANALYSIS

The potential benefits of Self-MAT need to be evaluated and the factors that limit its effectiveness need to be understood. To achieve these goals, a simulation environment that incorporates realistic conditions is being developed. The simulation platform is capable of running air traffic scenarios (based on recorded live traffic) in pseudo-real time. To establish a baseline, the same scenarios are also run in legacy platforms. Performance metrics are computed in both runs and compared. Relevant metrics are

- Fuel consumption
- Delay time (relative to the airline schedule)
- Number of participating flights that deviate from the committed 4D trajectory

The agent-based air traffic simulation platform (Figure 24.7) includes the following components:

- *Operator*: A module that manages operator functions: Generates flight plans, provides fleet prioritization, and defines cost function parameters.
- *Aircraft Performance Module (APM)*: Computes aircraft dynamics, thrust, and fuel burn required to execute a flight plan according to a selected 4D trajectory.
- *Trajectory Synthesizer (TS)*: Generates flyable 4D trajectories that comply with airspace availability obtained from the 4D-Grid.
- *Optimizer*: Evaluates the cost function based on candidate 4D trajectories and finds the trajectory that minimizes cost.
- *Monitor*: Verifies that the flight has not deviated from the cleared 4D trajectory.
- *ATC*: Takes flights that deviate from the cleared 4D path out of participation in Self-MAT operations.
- *Weather*: Injects disturbances to the simulation.
- *Metrics*: Collects performance metrics.

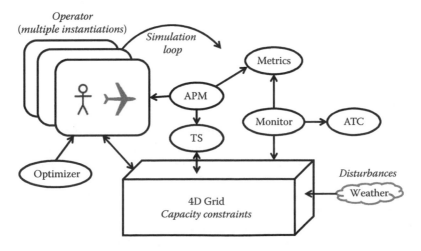

FIGURE 24.7 Self-MAT simulation platform.

The simulation includes disturbances (weather) that trigger operator reevaluation of flight plans. The software runs in a pseudo-real-time loop that continuously monitors flight conformance and allows operator adjustments to extract additional savings. Weather effects are not only convective weather events that pop up but also errors in wind forecast data. In order for a flight to track to the committed 4D trajectory from time to time, guidance corrections need to be applied (thrust and drag settings that impact fuel consumption) due to wind forecast errors. This is one of the factors that limits the efficiency gains in Self-MAT. In Self-MAT, uncertainties are removed, but the price is paid by the necessary thrust adjustments so that the flight follows the committed 4D path. The simulation allows testing the sensitivity of this cost to the magnitude of disturbances.

The operator side of the simulation consists of several instantiations of flight plan generators. The operator module generates flight plans and consults the 4D-Grid to find out the airspace availability. The optimizer processes the flight plan against the options allowed by the airspace availability (per the 4D-Grid) and evaluates the airline cost function in order to find the optimal solution. The Self-MAT framework is particularly suitable for operator side optimization algorithms that search for cost-effective trajectory solutions that go through hazard fields and restrictions (Dougui et al. 2010; Vela 2010; Taylor and Wanke 2011). Furthermore, when the operator has the capability to manage the use of available airspace, operator-initiated slot swapping is possible and can deliver additional savings.

SUMMARY

We have examined the limitations of centralized control of traffic flow into congested airspaces and highlighted the fact that due to intrinsic limitations stemming from the complexity of interactions between the factors affecting flights and to the stochastic nature of the system, variances in the demand would render ineffective any traffic management plan. The proposed agent-based alternative is not subject to the effects of variance because traffic flow is self-managed: users simply execute the flight plan that best fits their goal. That plan is coordinated via shared information so that it accommodates to airspace availability.

A cursory look at the radically different approach to air traffic operations proposed in the Self-MAT concept could result in quickly dismissing the idea as nonpractical. But a closer examination of its technological requirements reveals that most of the technologies being developed as part of NextGen do indeed provide what is required for autonomous flight to take place, for example, modern airborne technologies, ADS-B, SWIM, NNEW, and Data Comm. Similarly, several concepts being advanced for NextGen are also aligned with Self-MAT. For example, C-ATM and the move toward delegated separation. Self-MAT could be construed as a concept that negates NextGen, but that is not correct. Just as delegated separation is being considered when certain circumstances are present and within designated PBN corridors, Self-MAT could be thought as a PBN service embedded in a pool of technologies and operational concepts that get used when the circumstances allow. In summary, Self-MAT is a viable solution that because of the great potential benefits that it offers should be given serious consideration.

CONCLUSIONS

We have developed a new air traffic concept that removes central control and applies principles of complexity science. Central control and reliance on predicted aircraft trajectories and airspace load is the source of most of the observed inefficiencies in the current system. An agent-based system where the pilot self-separates from other aircraft and complies with capacity constraints, while following its own business objectives, is not affected by the inherent uncertainty of predictions. The combined behavior of all agents generates desirable emergent behavior that drives the system as a whole toward Pareto optimality. Agent-based air traffic is enabled by the capability of modern FMS to guide the aircraft based on a 4D trajectory, which in turn is built based on shared situational awareness that includes flow constraints, weather, and air traffic restrictions.

SUGGESTED FUTURE WORK

QUESTIONS FOR DISCUSSION

What are the most difficult challenges likely to be encountered by an autonomous flight approach such as the proposed Self-MAT concept?

Can an agent-based autonomous flight concept (such as Self-MAT) be implemented in a mixed equipage environment where not all operators can participate?

Within an autonomous flight approach, can operators game the system to extract additional use of resources at the expense of reduced service to other users? What mitigation strategies could be implemented?

What safety issues could arise with autonomous flight and what approaches could be implemented to mitigate potential safety risks?

If autonomous flight is allowed, what would be the role of air traffic controllers and traffic flow managers?

What kind of issues could arise with autonomous flight in high-density airspaces? And how can those issues be mitigated?

In Self-MAT, by design, flights that are compliant to the committed 4D trajectory are conflict free. To what extent would it be beneficial to allow delegated separation strategies when deviations from compliance arise?

ADDITIONAL RESEARCH

There are a number of open research questions related to the implementation of an agent-based self-MAT concept, to mention a few:

- What is the optimal 4D-Grid design (cell size, etc.)?
- What cell reservation mechanisms could be implanted to manage the 4D-Grid?
- What is the cost incurred in 4D-guidance versus cost of current inefficiencies?

ACKNOWLEDGMENTS

The author would like to thank Steve Hansen, Sid Rudolph, Jon Dehn, and Kevin Hightower of Lockheed Martin for technical review, useful discussions, and feedback.

REFERENCES

Allen, D.L., Haraldsdottir, A., Lawler, R.W., Pirotte, K., and Schwab, R., 1998, The economic evaluation of CNS/ATM transition, April, CNS/ATM Projects, Boeing Commercial Airplane Group, Seattle, WA.

Avjian, R.M. and Dehn, J., 2012, Trajectory-based integration of aircraft conflict detection and weather avoidance fields, *89th Annual Meeting of the American Meteorological Society: Technology in Research and Operations, How We Got Here and Where We're Going*, January 22–26, 2012, New Orleans, LA.

Avjian, R.M., Dehn, J., and Stobie, J., 2009, NextGen trajectory-based integration of grid-based weather avoidance fields, *ATCA 54th Annual Conference and Exposition*, October 2009, Washington, DC.

Azzopardi, M.A. and Whidborne, J.F., 2011, Computational air traffic management, *30th Digital Avionics Systems Conference*, October 16–20, 2011, Seattle, WA.

Ballin, M.G., Hoekstra, J.M., Wing, D.J., and Lohr, G.W., 2002, NASA langley and NLR research of distributed air/ground traffic management, *Proceedings of the AIAA Aircraft Technology, Integration, and Operations Conference*, AIAA-2002-5826, Los Angeles, CA.

Beni, G. and Wang, J., 1989, Swarm intelligence, *Proceedings of the 7th Annual Meeting of the Robotics Society of Japan*, Tokyo, Japan, RSJ Press, pp. 425–428.

Bing, D. and Wen, D., 2010, Scheduling arrival aircrafts on multi-runway based on an improved artificial fish swarm algorithm, *IEEE International Conference on Computational and Information Sciences*, Chengdu, Southwest China.

Bonabeau, E., Dorigo, M., and Theraulaz, G., 1999, *Swarm Intelligence: From Natural to Artificial Systems*, Oxford University Press, New York.

Bonabeau, E. and Meyer, C., 2001, *Swarm Intelligence: A Whole New Way to Think about Business*, May, Harvard Business Review, New York. pp. 106–114.

Burdun, I.Y. and Padentyev, O.M., 1999, AI knowledge model for self-organizing conflict prevention resolution in close free-flight air space, *IEEE 1999 Aerospace Conference*, Snowmass at Aspen, CO.

Cai-long, C. and Wen-bo, D., 2011, A multi-objective crossing waypoints location optimization in air route network, *IEEE 2011 3rd International Workshop on Intelligent Systems and Applications*, May 28–29, 2011, Wuhan, People's Republic of China.

Dougui, N., Delahaye, D., and Mongeau, M., 2010, A new method for generating optimal conflict free 4D trajectory, *4th International Conference on Research in Air Transportation*, June 1–4, 2010, Budapest, Hungary.

Duong, V.N. and Hoffman, E.G., 1997, Conflict resolution advisory service in autonomous aircraft operations, *Digital Avionics Systems Conference, 16th DASC, AIAA/IEEE*, Vol. 2, Irvine, CA, pp. 9.3–10-17.

EUROCONTROL,2010, SESAR Trajectory Management Document, SESAR Joint Undertaking, Project ID B.04.2, Edition 00.02.90, September 17, 2010

FAA, 2009, Traffic Flow Management in the National Airspace System: A presentation explaining terms, techniques, and programs associated with traffic flow management in the National Airspace System, Federal Aviation Administration (FAA) Air Traffic Organization, Washington D.C., 2009-AJN-251, http://www.fly.faa.gov/Products/Training/Traffic_Management_for_Pilots/TFM_in_the_NAS_Booklet_ca10.pdf. Accessed October 2009.

FAA, 2012a, FAA aerospace forecast—Fiscal years 2012–2023, U.S. Department of Transportation, FAA, Aviation Policy and Plans. http://www.faa.gov/about/office_org/headquarters_offices/apl/aviation_forecasts/aerospace_forecasts/2012–2032/. Accessed February 26, 2014.

FAA, 2012b, NextGen implementation plan, March 2012. http://www.faa.gov/nextgen/. Accessed February 26, 2014.

FAA, 2012c, Aeronautical information manual. http://www.faa.gov/air_traffic/publications/atpubs/pcg/index.htm. Accessed February 26, 2014.

FAA NASEA, 2013, Portal 8.0, federal aviation administration, national airspace system enterprise architecture (NAS EA) products. https://nasea.faa.gov/. Accessed February 26, 2014.

Fu, A., Lei, X., and Xiao, X., 2008, The aircraft departure scheduling based on particle swarm optimization combined with simulated annealing algorithm, *IEEE Congress on Evolutionary Computation*, Hong Kong, People's Republic of China.

Hoekstra, J.M., Ruigrok, R.C.J., and van Gent, R.N.H.W., 2001, Free flight in a crowded airspace? *Air Transportation Systems: USA/Europe Air Traffic Management Research and Development*, AIAA Inc., Napoli, Italy.

Huerta, M, 2011, Statement of Michael Huerta, Deputy Administrator Before the Committee on Transportation and Infrastructure, Subcommittee on Aviation on the Benefits of the Next Generation Air Transportation System, October 5, 2011, http://www.faa.gov/news/testimony/news_story.cfm?newsId=13134. Accessed February 26, 2014.

International Civil Aviation Organization (ICAO), 2005, *Global Air Traffic Management Operational Concept*, 1st ed., ICAO, Montreal, Quebec, Canada.

International Civil Aviation Organization (2010), Eighteenth Meeting of the AFI Satellite Network Management Committee, SNMC/18– WP/12, 1–4 June 2010.

Jackson, M.R.C., Sharma, V., Haissig, C.M., and Elgersma, M., 2005, Airborne technology for distributed air traffic management, *44th IEEE Conference on Decision and Control, and the European Control Conference*, December 12–15, 2005, Seville, Spain.

Joint Planning and Development Office—Joint Planning Environment, http://jpe.jpdo.gov/ee/request/folder?id=21967. Accessed July 2013.

Jardin, M.R., 2005, Grid-based air traffic control strategic conflict detection, *AIAA Guidance, Navigation and Control Conference and Exhibit*, San Francisco, CA.

Jevtic, A., Andina, D., Jaimes, A., Gomez, J., and Jamshidi, M., 2010, Unmanned aerial vehicle route optimization using ant system algorithm, *5th International Conference on System of Systems Engineering*, Loughborough, England.

Joint Economic Committee (JEC) Majority Staff, 2008, Flight delays cost passengers, airlines and the U.S. economy billions, May. http://www.jec.senate.gov/public/?a=Files.Serve&File_id=47e8d8a7-661d-4e6b-ae72-0f1831dd1207. Accessed February 26, 2014.

JPDO, 2012, Flight operations centers: Transforming NextGen air traffic management—FOC study team report, Report Number: A382865, July.

JPDO-JPE   Joint Planning and Development Office – Joint Planning Environment. http://jpe.jpdo.gov/ee/request/folder?id=21967.Accessed July 2013

Kennedy, J. and Eberhart R.C., 1995, Particle swarm optimization, *Proceedings of IEEE International Conference on Neural Networks*, Vol. 4, IEEE Service Center, Piscataway, NJ, pp. 1942–1948.

Klooster, J., Torres S., Earman, D., Castsillo-Effen, M., Subbu, R., Kammer, L., Chan, D., and Tomlinson, T., 2010, Trajectory synchronization and negotiation in trajectory based operations, *29th Digital Avionics Systems Conference*, October 3–7, 2010, Salt Lake City, UT.

Klooster, J.K., Del Amo, A., and Manzi, P., 2009, Controlled time-of-arrival flight trials, *8th USA/Europe Air Traffic Management Research and Development Seminar (ATM2009)*, June 29–July 2, 2009, Napa, CA.

Nedelescu, L., 2012, Accelerating the air transportation system's transformation: A systems engineering, systems thinking and complexity science based rational framework to unify heretofore dissociated lines of inquiry, *12th AIAA Aviation Technology, Integration, and Operations (ATIO) Conference*, September 17, 2012, Indianapolis, IN.

Parsopoulos, K. and Vrahatis, M., 2010, *Particle Swarm Optimization and Intelligence: Advances and Applications*, IGI Global, New York.

Pechoucek, M. and Sislak, D., 2009, Agent-based approach to free-flight planning, control, and simulation, *IEEE Intelligent Systems*, 24(1), 14–17.

RITA-BTS, 2013, Airline on-time statistics and delay causes. http://www.transtats.bts.gov/OT_Delay/ot_delay-cause1.asp?type=21&pn=1. Accessed February 26, 2014.

RTCA, 1994, Final report of RTCA task force 3, free flight implementation, RTCA Inc., Washington, DC.

RTCA Inc., 2008, Minimum operational performance (MOPS) for aircraft surveillance applications system (ASAS), RTCA, Inc, Washington, DC.

RTCA Inc., 2010, Safety, performance and interoperability requirements document for airborne spacing flight deck interval management (ASPA-FIM), Interim Draft V 0.6, RTCA, Inc, Washington, DC, July 16.

RTCA, SC-214 and WG-78, 2012, Data communications safety and performance requirements, Version I, RTCA SC-214/EUROCAE WG-78, RTCA Inc, Washington, DC, February 1.

Ruigrok, R. and Korn, B., 2007, Combining 4D and ASAS for efficient TMA operations, *7th AIAA Aviation Technology, Integration and Operations (ATIO) Conference*, Belfast, Northern Ireland.

Ruigrok, R.C.J. and Hoekstra, J.M., 2007, Human factors evaluations of Free Flight: Issues solved and issues remaining, *Applied Ergonomics*, 38(4), 437–455.

Sawhill, B.K., Herriot, J., and Holmes, B.J., 2011, Complexity science tools for interacting 4D trajectories and airspace phase transitions, *11th AIAA Aviation Technology, Integration, and Operations (ATIO) Conference*, September 20–22, 2011, Virginia Beach, VA.

Taylor, C. and Wanke, C., 2011, Dynamically generating operationally acceptable route alternatives using simulated annealing, *Ninth USA/Europe Air Traffic Management Research and Development Seminar*, *ATM2011*, Berlin, Germany.

Teodorovic, D., 2003, Transport modeling by multi-agent systems: A swarm intelligence approach, *Transportation Planning and Technology*, 26(4), 289–312.

Torres, S., 2010, Determination and ranking of trajectory accuracy factors, *29-th Digital Avionics Systems Conference*, October 3–7, 2010, Salt Lake City, UT.

Torres, S., 2012, Swarm Theory Applied to Air Traffic Flow Management, Procedia Computer Sciences, Elsevier (Ed. C.H. Dagli), Vol. 12, pp. 463–470, 2012, Complex Adaptive Systems Conference, November 14–16, Dulles, Virginia, http://www.sciencedirect.com/science/article/pii/S1877050912006965.

Torres, S. and Delpome, K.L., 2012, An integrated approach to air traffic management to achieve trajectory based operations, *31st Digital Avionics Systems Conference*, October 14–18, 2012, Williamsburg, VA.

Torres, S., Klooster, J.K., Ren, L., and Castillo-Effen, M., 2011, Trajectory synchronization between air and ground trajectory predictors, *30th Digital Avionics Systems Conference*, October 16–20, 2011, Seattle, WA.

Vaddi, S., Sweriduk, G.D., and Tandale, M., 2012, Design and evaluation of guidance algorithms for 4D-trajectory-based terminal airspace operations, *12th AIAA Aviation Technology, Integration, and Operations (ATIO) Conference*, September 17, 2012, Indianapolis, IN.

Vela, P., 2010, Topologically based decision support tools for aircraft routing, *29th Digital Avionics Systems Conference*, October 3–7, 2010, Salt Lake City, UT.

25 Heathrow Terminal 5
Cost Management for a Mega Construction Project

Claire Cizaire and Ricardo Valerdi

CONTENTS

CASE STUDY ELEMENTS

FUNDAMENTAL ESSENCE

One does not easily get the approval for the expansion of an airport, however much congested the airport may be. British Airport Authority Limited (BAA), very much aware of this fact, decided that if it were to build a new terminal for London Heathrow (LHR), it would have to be large enough to accommodate the demand for the next few decades to come. Terminal 5 (T5) was thus designed to increase the capacity of London Heathrow by a sharp 50%. This expansion project, a system of systems embedded in an airport in operations, was very ambitious, and its budgeted cost was

£4.3 billion. BAA had never undertaken a project of this scope before. Against all odds, the project was delivered not only on time but also within budget.

Topical Relevance

London Heathrow, the world's third busiest airport, is an important element of the United Kingdom's economy. However, it was becoming more difficult for LHR to accommodate the growing demand for air transportation. The airport had been badly congested for years, and other rising European airports with great ambitions were positioning themselves to capture the demand and ultimately overthrow Heathrow. The expansion program approved by the UK Secretary of State in 2001 was thus very much needed. However, the scope and complexity of the project put its sponsor BAA at risk. The budgeted cost of Terminal 5 represented two-thirds of BAA's equity value at the time. In addition, the evolution of the project was closely monitored by the media, and BAA was keen on protecting its image. To ensure the timely and successful completion of the extension work, BAA took a novel approach to project management, which can be considered a turning point for the construction industry.

Domains

Construction industry, transportation industry, project management

Country of Focus

United Kingdom

Stakeholders

Primary stakeholders

(a) *Project sponsor*: The airport owner and operator, BAA, now called Heathrow Airport Holdings. The BAA management team, which acted as the project leader, was very active throughout the program.
(b) *Project contractors*: 20,000 suppliers worked on the T5 program. The first-tier contractors, including Laing O'Rourke, Turn and Townsend Group, Mott Mac Donald, and EC Harris Group, were active throughout the project.
(c) *Main user of the new terminal*: British Airways (BA), the carrier chosen to have the exclusive use of the new terminal. BA was involved during the design phase and at the end of the program, for the handover.
(d) *Local and central government*: The authorities were actively involved during the inquiry and some agencies remained implicated during the project.

Secondary stakeholders

(a) *Local communities*: Residents of the airport's surrounding areas took part in the public inquiry.
(b) *End users*: The passengers' requirements were taken into account during the design phase.

Primary Insights

Well aware of the negative track record of the construction industry, the British airport operator BAA took a new approach to the Terminal 5 project from the beginning. The BAA management team developed a unique contract with its suppliers called the *T5 Agreement*. The contract's mechanisms laid the foundation for a collaborative environment throughout the project. This agreement was instrumental, but it alone could not have accounted for the success of a 5-year program. BAA took additional measures. A dedicated team was put together to take the leadership of the project. This team imposed processes and combined traditional cost management methodologies with new techniques to ensure the successful completion of an ambitious and risky project.

KEYWORDS

Air transportation, British Airport Authority, Contracting, Earned value management, Integrated Baseline Review, London Heathrow, Megaproject, Project management, Terminal 5, United Kingdom

ABSTRACT

Cost overruns and delays are so common in the construction industry that they are almost expected. The success of LHR Terminal 5 (T5) thus came as a pleasant surprise. The main objective of this case study is to explore the factors that enabled this remarkable success.

The T5 expansion project was needed to increase passenger capacity and was of strategic importance for Heathrow Airport to remain competitive with other European hub airports. Its colossal budgeted cost of £4.3 billion reflected its scope and complexity but also foreshadowed a myriad of challenges and risks of delays and cascading additional costs. Nevertheless, the new terminal was delivered not only on time in March 2008 but also, against all odds, within budget.

Well aware of the negative track record of the construction industry and the financial risks of the project, the sponsor, British airport operator BAA, adopted a novel approach. BAA developed the *T5 Agreement*, a unique contract that provided suppliers with strong incentives to outperform the industry's standards, both individually and collectively. The contract also gave BAA the authority to operate as the sole project leader throughout the program. BAA used this legitimacy to institute efficient processes and well-aligned incentives early on. The collaborative mindset setup by the *T5 Agreement* influenced how traditional cost management methods, such as earned value management (EVM), were used to incentivize the right behavior. The company also used modern techniques to reduce the potential for wasted time and cost by streamlining processes. The combination of these factors contributed to the surprising but welcomed success of the T5 project.

GLOSSARY

3D	Three-dimensional graphics
ATC	Air traffic control
BA	British Airways
BAA	British Airport Authority Limited, now known as Heathrow Airport Holdings
EVM	Earned value management
IATA	International Air Transport Association
IBR	Integrated baseline review
LHR	London Heathrow Airport
T5	Terminal 5 at London Heathrow Airport
UK	United Kingdom

BACKGROUND

London Heathrow Airport (IATA designation: LHR), owned and operated by BAA, was built in the 1940s, 22 km (14 miles) west of Central London. This airport is an extremely important element of the UK economy as an enabler for transportation, tourism, and commerce. With an annual traffic of over 60 million passengers since 1998, it has been one of the world's four busiest airports for many

TABLE 25.1

Terminal Capacity of the Four Largest European Airports in 2007

| | 2007 | | |
Airport	Number of Terminals	Estimated Annual Passenger Capacity (Millions)	Passenger Traffic (Millions)
Heathrow	4	55	68
Charles de Gaulle	3	62	60
Frankfurt	2	54	54
Schiphol	1	60	58

Sources: Aeroports de Paris, Document de reference, 2007; Schiphol Group, Annual report, 2007.

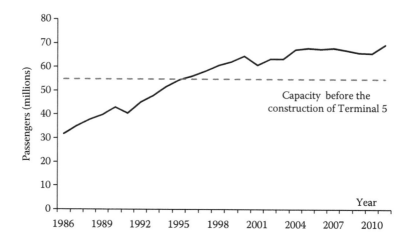

FIGURE 25.1   Number of passengers handled at Heathrow.

years. LHR is also the world's largest international hub, as it welcomes more international passengers than any other airport in the world.

LHR has also been one of the most congested airports, both in terms of airfield capacity and passenger throughput capacity. Historically, LHR only has two runways. In comparison, Frankfurt Main Airport has three runways, Paris Charles de Gaulle four, Amsterdam Schiphol five, and most American airports at least three. In addition to operating very close to its airfield capacity, Heathrow was operating well above its designed passenger throughput capacity, contrary to its European competitors, as shown in Table 25.1. The LHR's facilities have been accommodating over 70 million passengers annually when it was only designed for about 55 million passengers, as shown in Figure 25.1. Delays and customer complaints kept increasing.

BAA was well aware of the threat these two separate capacity issues posed to Heathrow. However, the attempts the company made to expand the airport inevitably set off a passionate debate in the country. Opponents and supporters were numerous. They had the government as a referee and the media as a relay.

PURPOSE

With an ever-growing demand for air transportation, congestion problems were starting to undermine London Heathrow's supremacy [7]. Nearby airports were positioning themselves to become the next European gateway to the world. It was strategically important to increase

London Heathrow's capacity and secure the airport's regional domination for the years to come, not only for BAA but also for the local economy.

The construction of a third runway had been a recurring debate for decades, but the neighboring communities' stand against it had so far prevailed. To build another runway, the airport's boundaries would have to be pushed either north or south. Not only would hundreds of houses have to be torn down but the noise envelope would have also have inexorably grown larger.

On the other hand, building a new terminal to accommodate more passengers comfortably and thus improve the service level at LHR was in the realm of possibilities. There was enough space at LHR for a fifth terminal, which could increase passenger throughput without significantly impacting the noise envelope.

The congestion of Heathrow airport had been predicted many years before. In fact, the development of a fifth terminal at LHR was encouraged by the UK Department for Transport as early as in 1985, in the Airports Policy White Paper [7]. After this white paper was published, BAA started planning for a new terminal and presented its plan to local, regional, and national stakeholders in 1992. The application for Terminal 5 was submitted in February 1993. The Secretary of State for Transport appointed an inquiry inspector a year later, and after a few preinquiry meetings, the planning inquiry finally started in May 1995 to end in 1999. It took no less than 525 days. This became the longest inquiry in the UK planning history [8,9]. Nevertheless, BAA's wait was not yet over, as the Secretary of State still had to approve the report. He eventually granted his consent in November 2001, almost 9 years after BAA's official application.

SYSTEM/SYSTEM OF SYSTEMS/ENTERPRISE

The expansion program approved by the Secretary of State encompassed much more than the construction of an additional building at London Heathrow. Terminal 5 was not designed to be a single-hall, low-cost building for the occasional surplus of traffic. BAA, the owner and operator of LHR, had coveted this expansion for years and it intended Terminal 5 to be an autonomous terminal the size of the fourth largest European airport.

The construction program was divided into 16 projects, as shown in Figure 25.2. Each project addressed problems of different natures and involved different parties. The chosen construction site, shown in red in Figure 25.3, was the size of London's Hyde Park (350 acres). The Twin Rivers running through the site first had to be diverted. The main building of Terminal 5 and its two satellites were then to be built [15], connected to the existing transportation network and equipped with the

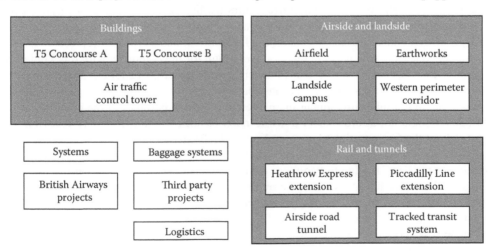

FIGURE 25.2 Sixteen projects of Terminal 5. (From Doherty, S., *Heathrow's Terminal 5: History in the Making*, John Wiley & Sons, Chichester, England, 2008.)

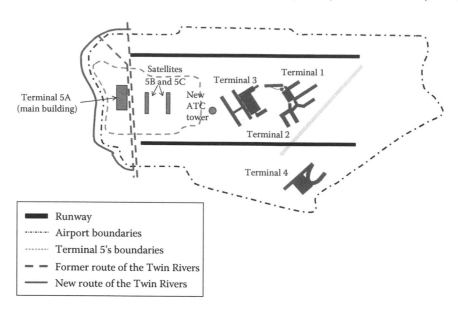

FIGURE 25.3 Overview of the construction site.

modern systems found common to most major airports of the world, for example, the baggage handling system. Airside—part of an airport directly involved in the arrival and departure of aircraft—work was required to allow aircraft to taxi from the runways to the T5 gates. The new terminal profoundly changed the layout of the airport and a new air traffic control (ATC) tower had to be erected. As the future occupant of the terminal, BA was involved during the design phase as well as the end of the project to coordinate the handoff. Overall, 20,000 suppliers were selected through a competitive bidding process to carry on this five-year construction project.

As shown in Figure 25.4, operating an airport such as London Heathrow is a highly complex business, with a large number of stakeholders, clients that are very different in nature, heavy investments, and a highly regulated environment [2]. In addition, due to the size of the airport and its role in the British economy, the management of LHR is under strong public scrutiny, both nationally and internationally [3].

BAA is, however, a private company and as such answers to its shareholders; one of the management team's primary objectives is to maximize its profitability. BAA is a capital-intensive business, with significant investments being made every year to build, maintain, and continuously improve terminals, runways, hangars, and other facilities. These investments typically represent 30%–50% of the annual revenues. The stakes are therefore very high for the company: with this kind of relative amounts, mistakes can have disastrous financial consequences. In addition, these investments often concern facilities in operations and must be conducted with minimal impact on the daily operations. The high level of investment in an operating facility partakes in the complexity of the business. BAA could not afford to shut down London Heathrow and the airport kept running as usual throughout the five years it took to complete the T5 Program.

BAA had strong financial and strategic motives for the expansion of Heathrow. The company has a large number of clients leading to two types of revenue streams: the aeronautical income and the retail income. The aeronautical fees are paid by the airport's main clients, operators in the aviation and logistics business. They pay departing passenger charges as well as landing and parking fees. The retail income is a secondary source of revenues from the various retailers housed within the terminals, car parking fees, advertising spaces, car rental companies, duty free, bureaux de change, catering, etc. Both revenue streams are strongly linked to the number of passengers going through Heathrow. BAA thus has a strong incentive to both increase passenger traffic, which in the case of Heathrow was close to saturation when T5 was built, and improve its attractiveness to its end users, the passengers.

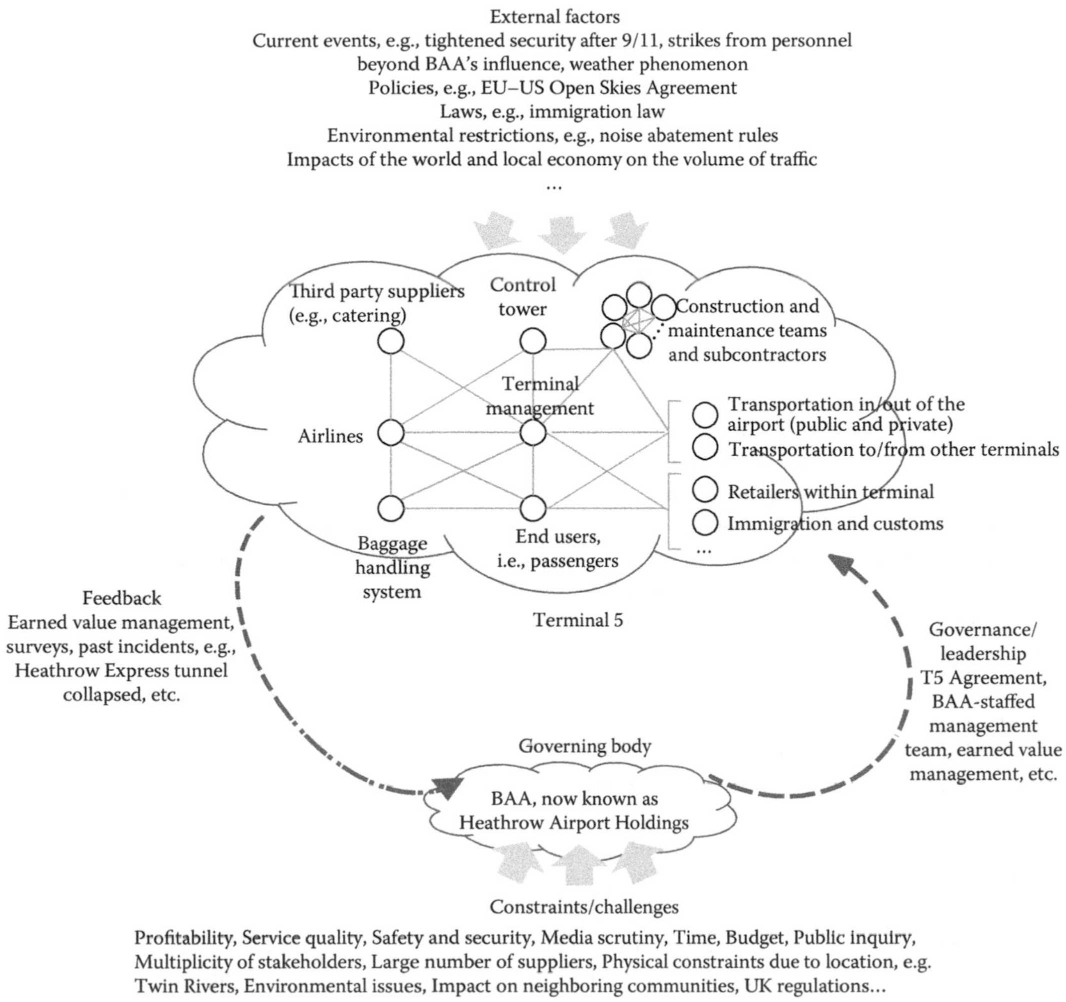

External factors
Current events, e.g., tightened security after 9/11, strikes from personnel
beyond BAA's influence, weather phenomenon
Policies, e.g., EU–US Open Skies Agreement
Laws, e.g., immigration law
Environmental restrictions, e.g., noise abatement rules
Impacts of the world and local economy on the volume of traffic
...

Third party suppliers (e.g., catering)

Control tower

Construction and maintenance teams and subcontractors

Terminal management

Airlines

Transportation in/out of the airport (public and private)
Transportation to/from other terminals

Retailers within terminal
Immigration and customs
...

Baggage handling system

End users, i.e., passengers

Terminal 5

Feedback
Earned value management, surveys, past incidents, e.g., Heathrow Express tunnel collapsed, etc.

Governance/ leadership
T5 Agreement, BAA-staffed management team, earned value management, etc.

Governing body

BAA, now known as Heathrow Airport Holdings

Constraints/challenges
Profitability, Service quality, Safety and security, Media scrutiny, Time, Budget, Public inquiry,
Multiplicity of stakeholders, Large number of suppliers, Physical constraints due to location, e.g.
Twin Rivers, Environmental issues, Impact on neighboring communities, UK regulations...

FIGURE 25.4 Complex system of systems.

While an airport's primary function is to provide an interface between airlines and passengers, Heathrow airport quickly overgrew this definition under BAA's supervision to become a system of systems offering sophisticated services to its main users, the air carriers and the passengers. BAA turned LHR into a large intermodal platform allowing passengers to efficiently switch from one means of transportation to another: taxis, buses, express trains, subway, private cars, and of course airplanes.

A strong emphasis was therefore put, right from the inception phase, on the quality of the integration of Terminal 5 into the existing transportation network. Since the Heathrow Express Operating Authority is a subsidiary of BAA, it was relatively easy to connect Terminal 5 to the Heathrow Express rail link, allowing passengers to travel from Paddington station in Central London to T5 in only 21 min. In addition, the management team strived to offer a direct and efficient connection between the new terminal and the public London Underground as well as the London Orbital motorway (M25). The Piccadilly Line was extended and a new station was opened within Terminal 5. The alterations of the M25 motorway requested by BAA to service the new terminal were, however, very controversial and led to an extensive press coverage [20,23]. The plan was to build a connecting road to Terminal 5 and widen a portion of an already large M25, making it the widest road in Britain. The Highways Agency was in favor of the extension but environmental groups and local politicians opposed the development scheme. The Roads Minister stepped in and eventually gave his approval. All these subprojects were critical to Terminal 5's success but required endless negotiations with privately owned companies, local public authorities, and various lobbying groups.

As shown in Figure 25.4, operating an airport the size of Heathrow requires the involvement of a number of autonomous entities. Safety and security are ensured by the governmental agencies, which include the immigration and customs authorities or the control tower granting pilots the right to move an aircraft away from a gate. To keep up with other European airports, BAA also needs to continuously enhance the passenger's experience within the airport with additional retailers and catering companies, an improved layout, better baggage handling systems, etc. BAA also has to work hand in hand with its own clients, the airlines, to improve the overall service quality to the end users. BA was designated to have the exclusive use of the new terminal. BAA believed that Britain's largest and best-known airline was in an ideal position to realize the full potential of the new terminal. Indeed, the United Kingdom's flagship carrier's vision was to turn T5 into its flagship terminal. All the entities mentioned earlier are independent, with their own emergent behaviors and business objectives, yet they collaborate on a daily basis to enhance the passengers' experience.

The competiveness of the different airlines operating from LHR and the combination of routes they offer is also an important driver of passenger traffic and Terminal 5 can be affected by local or international policies. For example, the Open Skies Agreement between the United States and Europe liberalized the aviation services and thus leading to long-term changes in airline schedules, prices, and traffic, which directly affected Heathrow.

BAA and Terminal 5 are also subject to many other external and unpredictable factors. Strikes from personnel beyond BAA's influence such as air traffic controllers or pilots, severe weather conditions, prolonged icing or snow storms, and even volcanic ash clouds, can disrupt the daily operations and construction work for a prolonged period of time. Catastrophic events can also be very detrimental to the operations and thus the profitability of the company. For example, the N1H1 outbreak or the 9/11 attacks in New York City set a series of unprecedented challenges at Heathrow.

Over time, BAA managed to turn LHR into a small city. Over 80,000 jobs are directly associated with the operations of the airport. While the airport provides a large pool of employment for the local area, its expansion is closely monitored by the neighboring communities. BAA has learnt to work with local authorities to minimize the negative impact of the airport on its environment. The diversion of the Twin Rivers required by the T5 Program is a good example of this partnership. The Twin Rivers are artificial waterways that were originally created 400 years ago to increase the water supply of royal residential estates. As shown in Figure 25.3, the rivers went right through the middle of the chosen site for the new terminal. In order to build T5, the rivers first had to be diverted. This subproject was highly environmental sensitive. Local communities and environmental groups expressed their concerns regarding the water quality and the biodiversity. To overcome these objections, BAA collaborated with the Environment Agency, the Royal Parks Agency, and the London Borough of Hillingdon throughout the T5 Program. Together, they developed a tailored solution to this very specific engineering problem. A special emphasis was put on the preservation and enhancement of the local habitat. This subproject was awarded a Civil Engineering Environmental Award for the high environmental standards sustained throughout the construction, and BAA now has representatives on the Twin River Management Committee.

Heathrow-related employment accounts for approximately 3.4% of total jobs in the Greater London area, and 2.6% of its gross-value-added [17]. Due to its particular role in the British economy, LHR is subject to the close scrutiny of the both the authorities and the general public. Any attempt to develop the airport sets off a heated debate relayed by the media and involving the British government, local authorities, environmental groups, and local communities, as well as a myriad of companies. The neighboring communities and some lobby groups believe that BAA still has plenty of room to improve the efficiency at LHR. However, a number of observers are concerned that if the supremacy of Heathrow were to be questioned, not only would BAA be negatively affected but so would the country.

TABLE 25.2

Review of Relevance of Complexity Theory to Systems of Systems (See Chapter 2)

	Key Aspect	Relevant	Notes
3.1	Recognition of wicked problems or messes including different world views	Yes	Cost-based contracts are standard practice in the construction industry. However, they encourage a short-term vision that can compromise the quality of the output. BAA stepped back and challenged this approach. They drafted a new contract based on white papers and best practices from other industries.
3.2	Interconnected and interdependent elements and dimensions	Yes	By essence.
3.3	Inclusion of autonomous and independent systems	Yes	Some of the building blocks are independent of others (e.g., baggage handling system).
3.4	Feedback processes	Yes	The transmission of information was crucial for the success of the project: the management team expected suppliers to tell them about any arising issue, in order to find an appropriate solution. The new directions were then sent back to the suppliers. The two-way communication ultimately created a virtuous circle.
3.5	Unclear or uncooperative stakeholders	Yes	The potential impact of uncooperative stakeholders was acknowledged and mitigated with the T5 Agreement.
3.6	Unclear boundaries	Yes	Also acknowledged early on.
3.7	Emergence	Yes	
3.8	Nonlinearity	Yes	
3.9	Sensitivity to initial conditions and path-dependence	Yes	The beginning of the collaboration and the first months of BAA as the leader were key.
3.10	Phase space history and representation	Yes	
3.11	Strange attractors, edge of chaos, and far from equilibrium	Yes	Issues jeopardizing the schedule, quality, or budget.
3.12	Adaptive agents	Yes	
3.13	Self-organization	No	
3.14	Complexity leadership	Yes	
3.15	Co-evolution	Yes	Suppliers and contractors influenced each other.
3.16	Fitness	Yes	
3.17	Fitness landscapes	No	

The Terminal 5 project was not a standard construction program. Most aviation professionals believed that a fifth terminal at LHR was long needed. However, the topic was so important, so controversial, and involved so many stakeholders that the T5 public inquiry was the longest one in the United Kingdom's history. To get the approval of the Department of Transport and win the support of the public, BAA had to go well beyond its comfort zone. Every detail, from the widening of M25 to the diversion of the Twin Rivers and the erection of a control tower amidst flying aircraft, required a tailored approach and a particular attention to all constituents. For all these reasons, the T5 program was a unique and highly complex construction program (Tables 25.2 and 25.3) that required success on the execution front as well as the management of autonomous entities in order to be completed on time and within budget.

CHALLENGES

The T5 program was truly interdisciplinary with extraordinary proportions. It was to span over five years. During the end of the planning phase in 2001, the chosen completion date was set for March 14, 2008. The project had a budget of £4.3 billion, which gave it the status of a megaproject [5].

TABLE 25.3

Key Aspects of Complex Systems from an SoS Perspective

	Complex Systems Key Aspect	Relevance	Notes on T5
4.1	Recognition of architecture	Yes	BAA, the client, spent a lot of time and efforts on the design of a project strategy. A framework was developed to address unknown unknowns as they materialized and a set of processes was put in place to measure progress throughout the project.
4.2	Arrangement of these systems in a networked fashion rather than in a hierarchy	No	The T5 project was hybrid, but with an emphasis on a hierarchical structure. Collaboration was fostered, but the management team and top-tier suppliers were clearly identified.
4.3	Recognition of external environment and project context	Yes	The environment was a major constraint, as T5 had to be built without interrupting any operations at Heathrow. Furthermore, the collaboration environment fostered during the project required a dedicated management team.
4.4	Management of complexity	Yes	Special processes were put in place to address complexity.
4.6	Modularity	Yes	
4.7	View and view points	Yes	
4.8	Recognition that there are unknown unknowns	Yes	The T5 Agreement was designed to best tackle the unknown unknowns that would inevitably occur during the project.
4.9	Recognize different organization structure due to scale and complexity		
4.10	Recognition of the importance of a different leadership	Yes	The BAA team brought on board experienced project leaders from other industries.
4.11	Governance		
4.12	Incremental commitment initiating the project normally	Yes	Getting the approval for the projects took several years.
4.13	Systemigram	Yes	

BAA had to face different types of challenges, as shown in Figure 25.4. The first one was to get the approval of the Secretary of State. It was a long uphill battle [18]. The airport owner and operator had to negotiate with all kinds of stakeholders, some of which remained involved even after the inquiry. For example, the Environment Agency, the Royal Parks Agency, and the London Borough of Hillingdon were implicated in the Twin Rivers diversion. BAA had to manage the expectations of these stakeholders throughout the project.

Another difficulty arose from the very large number of suppliers involved, as shown in Figure 25.7. The workforce consisted of no less than 50,000 persons [7] working for about 20,000 suppliers. First-tier suppliers included Laing O'Rourke, Turn and Townsend Group, Mott Mac Donald, and EC Harris Group. These suppliers were expected to hold all their subcontractors to the standards set by BAA. BAA had never undertaken a project of such scale before and very few projects in the UK construction industry could compare to Terminal 5. The contractors were all hired to work on the same project but this did not mean that they would all work hand-in-hand. With about 8000 workers on site during peak times, the situation needed careful orchestration. The army of suppliers needed a clear leadership and strong incentives to perform well throughout the program. BAA identified these two points as a critical success factors very early on.

Safety was also an important concern throughout the project. The scale of the construction project and the larger number of independent contractors put at risk the lives of the individuals involved in the project.

To add to the intrinsic complexity of this colossal project, Terminal 5 had to be constructed without disrupting the daily operations of LHR. For example, the construction of the new air traffic

FIGURE 25.5 New air traffic control tower. (Image courtesy of Tamás Kolos-Lakatos.)

control tower, shown in Figure 25.5, between Terminals 5 and 3 was not to interfere with a single scheduled flight, otherwise airlines and passengers would complain.

These challenges compounded the universal budget and delivery time constraints. Cost overrun and delays are so frequent in the construction industry that they are almost considered to be unavoidable as illustrated in Figure 25.6 [6,12]. The T5 initial budget estimate was so high that any small overrun from the targeted cost could quickly reach a nine-digit figure. With 16 projects and 147 subprojects [24,25], the risks of quickly running behind schedule were significant. Studies from past projects showed that Terminal 5 was likely to be delivered two years late with a 40% cost overrun [19].

Not to mention, the media was waiting to pounce on the project. The airport expansion had many detractors [16] and the project had already set a record in the UK planning history. Any delay or additional cost would immediately make international headlines and tarnish BAA's image.

DEVELOPMENT

BAA financed the construction program. In most construction projects, the sponsor is passive. However, this unique project called for innovative measures. There was too much at stake. BAA was striving to reduce the risk of cost overrun and/or delays.

The management team thus looked for novel approaches that would mitigate the risks without compromising the quality or the safety of the project. Many experts, aware of the poor track record of the construction industry, had already suggested a few solutions. *Constructing the Team* [14] and *Rethinking Construction* [11] were two papers that inspired the T5 management team. The core principles outlined in these two key documents were (1) the sponsor, not the supplier, should bear the risk of the project; and (2) partners are more valuable than traditional suppliers.

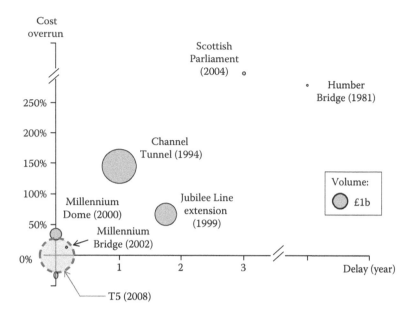

FIGURE 25.6 Cost overruns and delays of previous construction projects in the United Kingdom. (From Flyvbjerg, B. et al., *Megaprojects and Risk: An Anatomy of Ambition*, Cambridge University Press, Cambridge, U.K., 2003.)

The company tried to put these two principles into action. This decision was reinforced by the observation that even though BAA would certainly be able to transfer the liability to a contractor if a risk were to materialize, BAA would still be the one suffering from its consequences. Any delay would have a cascading impact on other parallel projects and any unplanned costs would inexorably result in bad publicity for the London airport operator.

This lesson was painfully learned from a past event that left a black eye on BAA when the Heathrow Express (an airport rail link from LHR to London Paddington Station in central London) tunnel collapsed in 1994 [13]. BAA's first reaction was to try to identify the suppliers responsible for the incident. However, while the different parties were arguing, the media and ultimately the public were waiting for the problem to be solved. In the end, BAA was the company under the spotlight and was blamed for focusing on liability instead of having the tunnel repaired. When the management realized this, they quickly changed strategies and first tried to get all the suppliers together to fix the problem. BAA learned the lesson and took it one step further with the Terminal 5 program.

BAA thus decided to change the reward mechanisms for their suppliers and remain intimately involved with all aspects of the project as the leader. A BAA-staffed management team was designated to take the lead of the expansion program. This long and thorough inquiry had already allowed BAA to develop a relationship with all the stakeholders and become familiar with all the details of the project, including the financial aspects. The management team was therefore well qualified to make quick and efficient decisions whenever a risk arose and negotiate the changes with other stakeholders. Furthermore, the large number of suppliers required a clear and uncontested leadership from start to finish.

T5 AGREEMENT

The management team set out to work on new contract mechanisms that would maximize the chances of success. With a traditional contract, BAA would have parceled out the risk and transferred pieces to each contractor. Instead of this, BAA looked for a way to relieve its suppliers of most of the liability and leave them with only one concern: ensure the eventual success of the T5 program. BAA intended to align its contractor's objective with its own. This would lay solid

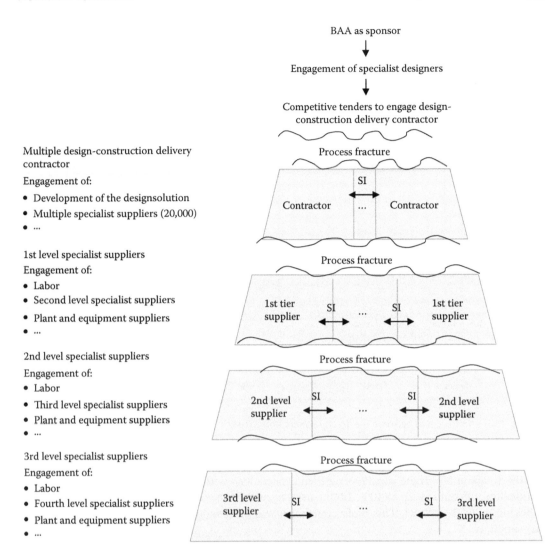

Multiple design-construction delivery contractor

Engagement of:

- Development of the designsolution
- Multiple specialist suppliers (20,000)
- ...

1st level specialist suppliers

Engagement of:

- Labor
- Second level specialist suppliers
- Plant and equipment suppliers
- ...

2nd level specialist suppliers

Engagement of:

- Labor
- Third level specialist suppliers
- Plant and equipment suppliers
- ...

3rd level specialist suppliers

Engagement of:

- Labor
- Fourth level specialist suppliers
- Plant and equipment suppliers
- ...

SI stands for sponsor integration

FIGURE 25.7 Multiple design-construction contractor tender process.

foundations for a collaborative environment but would not guarantee it. To ensure that contractors would indeed work together throughout the project, an additional measure was necessary.

BAA developed the *T5 Agreement*, a unique commercial contract for its first-tier suppliers, who were carefully selected based on their track records. The *T5 Agreement* was a cost-reimbursable contract with two components designed to foster the performance of suppliers as individual entities as well as their performance as a group. The financial compensation proposed in the *T5 Agreement* consisted of three layers, as exhibited in Figure 25.8.

According to the first layer, suppliers were paid their actual costs plus a fixed ring-fenced profit of 5%. The actual costs had to register within a targeted cost range set at the beginning of the project. This predetermined 5% profit was guaranteed. A cost-verification team was in charge of validating all incurred costs. This first layer transferred the risk away from suppliers to BAA.

The second layer was an additional variable profit margin, ranging from 0% to 10% [7]. This component was an incentive for suppliers to outperform themselves in the long run. If the final

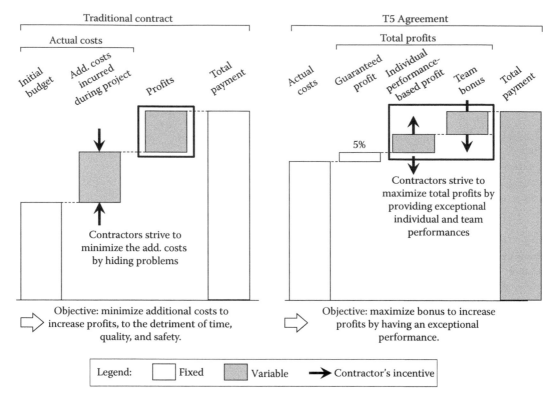

FIGURE 25.8 Differences between the *T5 Agreement* and a traditional contract.

performance of a supplier was close to the industry's best practice, the additional profit would increase from 0% to about 5%. If the supplier exceeded expectations with an outstanding performance, the additional profit could go up to 10%. Both parties agreed on the terms and performance metrics at the beginning of the project. This profit component was independent of the overall performance of other suppliers.

A third and final layer ensured that contractors would look after each other. For the project to be successful, suppliers had to account for the potential impacts of their actions on the other parts of the program. The *T5 Agreement* instituted, a bonus, based on group performance. This bonus was a financial incentive for the group of suppliers to deliver the best product possible on time and on budget. The bonus was a predetermined share of a common incentive fund, which was set up by BAA at the beginning of the program. If a risk, an unexpected expenditure, materialized during the project, the incentive fund was used by BAA to cover the unplanned cost. The money remaining at the end of the project was distributed to the different actors, according to a predetermined share. This last incentive mechanism ensured that suppliers would keep their costs in check, in an attempt to preserve the fund, and focus instead on improving their margins by delivering an exceptional performance.

With traditional contracts, suppliers maximize their profits by minimizing the additional costs incurred during the project, as shown in Figure 25.7. This trade-off can drive suppliers to hide a problem from other parties for as long as they can, in the hope that the issue be may be fixed or go unnoticed. This type of behavior affects the chances of coming up with a timely solution and seriously compromises the success and the quality of a project. The *T5 Agreement* neutralized the adversarial tendencies that contractors may have. Their objectives were aligned with those of BAA, as shown in Figure 25.9. Another consequence of this arrangement was that it gave BAA an active role in the project.

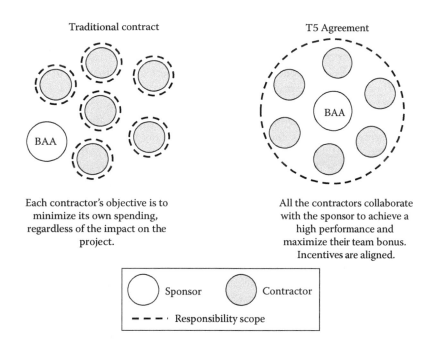

Traditional contract

T5 Agreement

BAA

BAA

Each contractor's objective is to minimize its own spending, regardless of the impact on the project.

All the contractors collaborate with the sponsor to achieve a high performance and maximize their team bonus. Incentives are aligned.

Sponsor   Contractor

– – – · Responsibility scope

FIGURE 25.9 Comparison of incentives between a traditional contract and the *T5 Agreement*.

LEADERSHIP

By transferring all the risk back to BAA, the *T5 Agreement* turned BAA into the legitimate leader for the project. BAA made the most of this legitimacy and this responsibility. A special task force was put together. The prospects of escalating costs drove the BAA management team to seek the appropriate expertise from other fields. The automotive industry is renowned for its attention to detail when it comes to cost savings. Similarly, the oil industry is experienced in managing very large projects. Experts from both fields were invited to join the BAA team and contribute lessons learned. They brought their vision and experience to the client's team. This T5 management team was of a significant size: about 100 people worked on the project full time.

The team kept track of everything during the five-year project and acted as the conductor. Frequent meetings were instituted to track progress and identify potential problems. The team had a long-term vision for the program. Whenever a risk would materialize, they would look for sound solutions that, in the end, would reduce delays, save money, and guarantee the quality of the project. This is not always the case. With traditional contracts, suppliers are the ones dealing with the encountered issues and they usually focus on short-term solutions.

IMPACT ON MANAGEMENT TECHNIQUES

The uncontested authority of the T5 management team allowed them to institute processes right from the beginning of the project. The team put a lot of emphasis on the EVM and Integrated Baseline Review (IBR) methods [7,19]. These techniques are two well-known complementary methods used to plan and measure the progress of a project. They require breaking down the project into small packages before the beginning of the contract work. This was not an easy task as the T5 program consisted of 16 projects, ranging from the construction of a control tower amid flying airplanes to the diversion of two rivers. However, these projects were eventually broken down into 1500 work packages [25]. Allocating the appropriate resources to each work package and then assembling the pieces of the puzzle to define a reasonable sequence of event was time-consuming. However, the BAA management team was keen on planning the project very meticulously. They bought an off-the-shelf software to assist them in their task and help them integrate all the dimensions of the projects.

The efficiency of EVM or IBR varies. One of the major reasons why these methods can fail is not the absence of data but the lack of accurate data. Some suppliers can be tempted to report *optimistic* data and temporarily cover up some problems. However, in the context of the *T5 Agreement*, it was in the interest of each supplier to proactively report problems as early as possible. The earlier a risk was identified the more time there was to find a solution and reach the final target of exceptional performance. The *T5 Agreement* lifted some of the reasons one may have had to dissimulate the facts, but this was not foolproof and the BAA management team put together a dedicated cost-verification team to ensure the accuracy of the cost data.

It enabled the management to be aware of and therefore deal with the different risks and issues much earlier than before. The cost and time performance measures were reviewed on a regular basis. When a risk was identified, a solution was sought and the specific performance indicators were tracked on a weekly basis. The main contractors moved their teams into the same offices, thus facilitating the communication process.

In most projects, an IBR is only run once, shortly after the start of the project. The T5 management team instituted periodic IBRs. To conduct these reviews, BAA and the main contractors first had to agree on a feasible sequence of work packages that would lead to the successful completion of the project and then validate the allocated resources (budget and time). This process was very cumbersome, but BAA used an off-the-shelf software package to help integrate. IBRs were conducted every six months. This allowed the sponsor and the teams to review the progress and detect potential problems early on.

Like any other project, Terminal 5 had milestones. In addition to the final deadline of March 2008, 70 intermediate milestones were first set to keep the project on schedule [7]. Meeting individual deadlines is critical for a project when there are 147 subprojects with interrelated schedules. The management observed that no team of suppliers wanted to be publicly known and therefore blamed for a delay. An official ceremony was thus organized for each milestone. The ceremony consisted of passing over to the next team a stone symbolizing responsibility for the next phase of T5. The official character of the handover put some pressure on the teams and forced them to meet the deadlines or they would suffer from negative publicity. The different suppliers were also presented as visible actors of the projects during these ceremonies.

The management also put in place a variety of small measures that likely contributed to the overall success. 3D design and off-site prefabrication reduced the amount of construction effort

FIGURE 25.10 London Heathrow Terminal 5. (Image courtesy of Tamás Kolos-Lakatos.)

spent while ensuring the safety of the workforce. For example, the teams in charge of the main's terminal roof assembly were first sent to Yorkshire, a town a few hours north of London, to stage the system. The roof, visible in Figure 25.10, was the largest single span roof in Europe and assembling it was a delicate task. The process was first tested off-site where workers were able to rehearse the sequencing of the different tasks several times without compromising the other ongoing T5 projects at LHR. This off-site practice training had a cost, but it enabled the teams to put the roof together safely and quickly.

As a result, the new terminal, shown in Figure 25.10, and its satellites were delivered not only on time on March 2008, but also within budget. Operations at the much awaited Heathrow Terminal 5 started as planned, on March 30, after an inauguration ceremony led by Queen Elizabeth II.

LESSONS LEARNED

The T5 program was very well planned. Arguably, the 525 day-long inquiry gave BAA plenty of time to devise a sound strategy.

First of all, BAA was able to align all parties' objectives from the very beginning of the program through the *T5 Agreement*. Its two-layer bonus mechanism provided suppliers with a valuable reason to focus on delivering a high-quality output, both individually and as a team.

Second, BAA took the leadership of the project. A dedicated team was put together, with expertise from many different fields. BAA was able to institute processes, such as recurring IBRs. They were willing to invest the time and energy necessary to ensure the efficiency of traditional project management methods.

The BAA management team enforced the contract's essence on a daily basis. BAA had a long-term interest in the project. The management team ensured the early detection of problems by tracking everything. Their uncontested leadership and the fact that they had long since developed a relationship with the stakeholders and suppliers helped them negotiate changes whenever a problem occurred.

First-tier suppliers shared offices, thus enabling better communication in practice and a series of innovative cost-saving measures were put in place.

SUMMARY

Heathrow Terminal 5 was delivered on time and within budget. Queen Elizabeth II inaugurated the terminal on March 14, 2008. However, the operational opening of Terminal 5 at LHR was a disaster [22]. Baggage handling problems led to the cancellation of over 30 flights. Some passengers were not reunited with their luggage until a week after their arrival. BA, Terminal 5's occupant, later admitted that the lack of training of its own employees was the origin of the chaos. This catastrophic opening day has, unfortunately, overshadowed the remarkable performance of BAA and its contractors in the execution of the Terminal 5 construction project that mark a definitive point in the construction industry. Time will tell whether these lessons will carry over to other megaprojects in the United Kingdom.

CONCLUSIONS

The T5 public inquiry was the longest in the United Kingdom's planning history and the T5 program became an outlier in the UK construction history of megaprojects. BAA and its contractors proved that a construction project of great scale is not doomed to long delays or budget mishaps: the £4.3 billion project was delivered on time, without cost overruns. This success is the result of great teamwork and great leadership. The seeds for an efficient collaboration were planted in the *T5 Agreement*, and BAA cultivated the benefits of this agreement from the onset by actively working with all its contractors to keep the project on track. The teamwork increased the efficiency and effectiveness of the management techniques traditionally used for construction projects.

SUGGESTED FUTURE WORK

This case study barely scratched the surface of the myriad of complexities that existed in the T5 project. The civil engineering challenges could have taken an entire chapter themselves. Not to mention the human dynamics involved with the process of negotiating the T5 Agreement between stakeholders and the intricacies in navigating sometimes competing forces of airlines, passengers, environmental regulators, residents of London, and connecting transportation networks. Additional areas to explore in this case study also include the various *ilities* involved in the design of T5 such as interoperability, scalability, reliability, sustainability, and changeability. These are important characteristics of systems of systems and are of growing importance as complexity increases.

Furthermore, the consideration of the Law of Requisite Variety can be further explored in the context of T5. The law states that "for appropriate regulation the variety in the regulator must be equal to or greater than the variety in the system being regulated." In other words, the more complex a system is, the more complex its control mechanisms must be in order to avoid failure or destruction. The variety of autonomous systems in the T5 project leads one to conclude that the control mechanism needs to be hugely complex. But at what point does complexity reach a level where it can no longer be managed? How can this be measured and resolved? Such open questions will help advance our understanding of systems of systems and the effort necessary to bring them to fruition.

QUESTIONS FOR DISCUSSION

- How should the boundary of a complex project be determined?
- In which cases is risk sharing between sponsor and contractors most beneficial?
- How does stakeholder power and legitimacy enable and at the same time impede progress?
- What insights does Heathrow T5 provide about managing interfaces between autonomous systems?
- How can airports like Heathrow balance the tension between interrupting the residents in its immediate surroundings and responding to increasing needs for transportation into and out of a world-class airport?

ADDITIONAL RESEARCH

This case study surfaces many interesting issues in the area of mega-project coordination, incentives for performance, risk management, stakeholder salience, labor agreements, and global competition. Future areas of research include the feasibility of governance models—such as the one prescribed in the T5 Agreement—to other mega projects (e.g., high-speed rail in California, Panama Canal expansion, Masdar City in the United Arab Emirates, etc.). It would also be interesting to do retrospective studies to determine whether or not such governance models would have led to better cost and schedule outcomes for already completed but disastrous project (e.g., Boston's Big Dig, Sydney Opera House, Channel Tunnel, etc.).

Another possible area of research involves the management of uncertainty. There are numerous difficulties involved with adequately estimating the amount of time it would take to negotiate with all stakeholders, obtain approval from regulatory agencies, and complete the construction. Since many of these organizations are autonomous and sometimes have conflicting objectives, the process of quantifying uncertainty and managing it becomes inherently complex. Beyond traditional construction engineering skills, the success of projects like these depend on interpersonal skills that are often not taught to project leaders. Identifying such soft skills is another area of research that would greatly improve the likelihood of success of such mega projects.

REFERENCES

1. Aeroports de Paris. 2007. Document de reference.
2. CAA. 2002. Heathrow, Gatwick and Stansted airports' price caps, 2003–2008: CAA recommendations to the Competition Commission.
3. Caves, R.E. and Humphreys, I. 2002. After terminal 5: Where next for UK airport policy? *Journal of Air Transport Management* 8(4): 199–200.
4. Churchman, C. W. (1967). Guest editorial: Wicked problems. Management Science 14(4): 141–142.
5. Davies, A., Gann, D., and Douglas, T. 2009. Innovation in megaprojects: Systems integration at London Heathrow Terminal 5. *California Management Review* 51(2): 101–126.
6. Dempsey, P.S. 1998. Denver's new airport finally takes wing. *Forum for Applied Research and Public Policy* 13(1): 19–25.
7. Department for Transport. 1985. Airports policy white paper. http://www.theguardian.com/business/2006/oct/26/theairlineindustry.lifeandhealth. Accessed 9 March 2014.
8. Department for Transport. 2005. Heathrow Terminal Five article. *Planning Inspectorate Journal.*
9. Department for Transport. 2005. Heathrow Terminal Five planning process—Response to FOI request.
10. Doherty, S. 2008. *Heathrow's Terminal 5: History in the Making.* Chichester, England: John Wiley & Sons.
11. Egan, J. 1998. *Rethinking Construction.* London, U.K.: Her Majesty's Stationery Office (HMSO).
12. Flyvbjerg, B., Bruzelius, N., and Rothengatter, W. 2003. *Megaprojects and Risk: An Anatomy of Ambition.* Cambridge, U.K.: Cambridge University Press.
13. Kletz, T. 2012. *Learning from Accidents.* London, U.K.: Routledge.
14. Latham, M. 1994. *Constructing the Team: Final Report.* London, U.K.: Her Majesty's Stationery Office (HMSO).
15. McKechnie, S., Dervilla, M., Frankland, W., and Drake, M. 2008. Heathrow Terminal 5: Terminals T5A and T5B. *Proceedings of the Institution of Civil Engineers—Civil Engineering* 161(5): 45–53.
16. Milmo, D. 2006. Ryanair—The world's least favourite airline. *The Guardian*, October 26, 2006. Available at http://www.theguardian.com/business/2006/oct/26/theairlineindustry.lifeandhealth. Accessed 09 March, 2014.
17. Optimal Economics. 2011. Heathrow related employment.
18. Pellman, R. 2008. Heathrow Terminal 5: Gaining permission. *Proceedings of the Institution of Civil Engineers—Civil Engineering* 161(5): 4–9.
19. Potts, K. 2008. *Construction Cost Management: Learning from Case Studies.* Oxon, U.K.: Taylor & Francis.
20. Ramesh, R. 1997. Heathrow plan hit by block on M25 widening. *The Independent* (London, England) July 23, 1997.
21. Schiphol Group. 2007. Annual report.
22. Up, up and away. 2004. *The Economist* 372: 62.
23. Wolmar, C. 1997. Minister gives green light to widen M25. *The Independent* (London) March 21, 1997.
24. Wolstenholme, A. 2008. Introduction. *Proceedings of the Institution of Civil Engineers—Civil Engineering* 161(5): 3.
25. Wolstenholme, A, Fugeman, I., and Hammond, F. 2008. Heathrow Terminal 5: Delivery strategy. *Proceedings of the Institution of Civil Engineers—Civil Engineering* 161(5): 10–15.

26 US Airways Flight 1549
Root Cause Analysis for the Miracle on the Hudson

Karunya Rajagopalan and Danny Kopec

CONTENTS

CASE STUDY ELEMENTS

- *Fundamental essence*: Analysis of the US Airways Flight 1549 incident of January 15, 2009, including root cause and lessons learned.
- *Topical relevance*: Dealing with a complex system that includes interactions among technical systems, humans, and coordination of various agencies and nature.
- *Domain(s)*: Government.
- *Country of focus*: United States.
- Stakeholders
  - *Passive*: FAA.
  - *Active*: US Airways, crew members, passengers.
  - *Primary*: US Airways, crew members, passengers.
  - *Secondary*: FAA, NTSB, NYPD.

ABSTRACT

On January 15, 2009, a cold afternoon at LaGuardia Airport in New York City, US Airways Flight 1549 was cleared for takeoff. The aircraft was an Airbus A320 manufactured in France that was a US domestic flight flying from New York City to Charlotte, North Carolina UNRC (2011). The flight took off at 3:25 PM EST. As it reached an altitude of 3200 ft, a flock of migratory Canadian geese struck the aircraft, damaging the engines and resulting in a loss of the aircraft's thrust. All 150 passengers and 5 crew members, 2 pilots, and 3 flight attendants evacuated the aircraft and survived this incident.

According to a report published by the *National Transportation Safety Board* (NTSB) (NTSB 2010), the probable cause for this accident was the ingestion of birds by the engines. However, there were many contributing factors to the incident. In this chapter, we investigate root causes, consider what system elements may have helped to avoid a catastrophe, and document lessons learned from the case.

GLOSSARY

AOA	Angle-of-Attack
APU	Auxiliary Power Unit
ASOS	Automated Surface Observing Systems
ATC	Air Traffic Control
ATP	Airplane Transport Pilot
CFR	Code of Federal Regulations
DGAC	Direction Générale de L'Aviation Civile
ECAM	Electronic Centralized Aircraft Monitoring
EOW	Extended Over Water
EST	Eastern Standard Time
FAA	Federal Aviation Administration
FBW	Fly-By-Wire
HPC	High-pressure compressor
HPT	High-pressure turbine
JAA	Joint Aviation Authorities
LGA	LaGuardia Airport
LPC	Low-pressure compressor
LPT	Low-pressure turbine
NTSB	National Transportation Safety Board
NYC	New York City
QRH	Quick Reference Handbook
RPM	Revolutions per minute

BACKGROUND

Bird Strikes

The first recorded aircraft incident that was caused by bird strikes occurred in 1912 (Sodhi 2002). Bird aircraft collisions are becoming more common today due to the increased number of aircraft and an increase in the prevalence of certain kinds of birds. As systems designers, we must consider bird strikes on aircraft systems as they can cause human and economic losses. In the United States, the Federal Aviation Administration (FAA 2013) maintains a database of wildlife strikes. The FAA also maintains a program to address wildlife strikes. However, strike reporting is voluntary, and the FAA wildlife strike database has information that has been received from airlines.

Bird strikes are a serious threat to aviation safety and have indeed caused a number of accidents, including many deaths. Bird strikes typically happen during takeoff or landing. Flocks of birds striking an aircraft can damage engines, and sometimes, no recovery is possible, eventually leading to a plane crash. There are many theories that attempt to find the causes for such collisions that may lead to accidents (Sodhi 2002). Airfields are usually sites for bird habitats. When an airfield or the surrounding area has not been used for an extended period of time, birds tend to get less vigilant about aircraft and hence collisions will occur more frequently. Modern aircraft design is also cited as a cause since today's aircraft body is wider as opposed to that of earlier narrow-bodied aircraft. Canada geese are the most massive birds that can strike an aircraft. These geese range in length from 30 to 43 in. and 50 to 73 in. in wingspan with a weight in the range of 6.6–19.8 lb (Dolbeer and Seubert 2013). These birds typically move in large flocks, resulting in higher impact on civilian and military aircraft than any other bird strikes.

Before 1960, bird strikes were not considered serious threats. However, after an October 1960 incident with an Eastern Airlines plane, bird strike data started to be collected. It was a Lockheed Electra L-188 aircraft taking off from Boston's Logan International Airport and en route to Philadelphia, which crashed into the sea when a flock of about 20,000 starlings flew onto the aircraft's path. Tragically, hundreds of birds were ingested into the engines, causing severe damage (Kalafatas 2010). The aircraft crashed into the water, killing 62 people on board, with only 10 survivors. Michael N. Kalafatas, in his book *Bird Strike: The Crash of the Boston Electra* (2010), mentions that the incident in the 1960s was a catastrophe and he references the US Airways Flight 1549, known as *the Miracle on the Hudson* in 2009, stating "...such an event is not likely to happen again with so positive an outcome." He touches on some important aspects of systems such as the methods used to control birds and the conflict that occurs between humans and machines when they share airspace. *The Miracle on the Hudson* case is usually viewed as a success story, although there were a number of lessons that could have been learned from the 1960s case such as engine certification tests and simulation training.

In April 2008, soon after its departure, a Challenger 600 (a Canadian business jet series) suffered from multiple bird strikes. Both engines ingested birds, and one engine lost power. Since the other engine was functional, the pilot was able to land safely, but repair costs amounted to millions of dollars. Effective bird control measure may have saved economic losses to this aircraft.

OTHER POSSIBLE CAUSES

Another more recent case that occurred in Indonesia in April 2013 had the pilot land the aircraft onto water. Preliminary investigations suggest that landing in this way was necessary because the pilot had missed the runway. Investigators also point out that extreme weather conditions could also have contributed to the incident. The aircraft was a Boeing 737-800 flight operated by Lion Airlines in Thailand. Nonetheless, all 101 passengers and 7 crew members survived. It is interesting that this operator had a number of previous nonfatal accidents that did not affect the airline's ability to stay in operation.

This leads us to think critically about the various systems involved, as the management of every airline system must face similar situations. Human error, weather, simulation of environmental conditions for training, wildlife, and safety are some of the concerns that systems designers must take into consideration.

SUBJECT INCIDENT

As we reflect on the emergency landing of the US Airways Flight 1549 on the Hudson River, we realize that a plane crash has many possible expected or unexpected factors that may have caused the accident. But, in the case of this incident, there were many factors that helped avoid a major catastrophe. As the flight left LaGuardia Airport (LGA), it was hit by a flock of migratory geese that caused damage to both engines, leading to losses of power. We first look into the construction and functions of the aircraft so that we develop some background and understanding for the incident.

The A320 was manufactured in 1999 by Airbus, which is headquartered in France. According to the weight and balance manifest provided by US Airways, the plane departed LGA with a weight of 151,510 lb (NTSB 2010). The maximum limitation on takeoff weight is 151,600 lb. The airbus A-320 uses a fly-by-wire (FBW) control system, which is an electronic interface, equipped with a sidestick to fly the aircraft. The FBW system converts flight movements into electronic signals, and the control computers determine how to provide response and stabilize the aircraft without pilot intervention. There are three modes of operation for the FBW flight's control system, namely, normal law, alternate law, and direct law. Normal law is the mode in which the computer control system has control while the aircraft is flying. Normal law mode, thus, is a protection for the aircraft against stalling, overspeed, and protection against parameters such as pitch attitude, high angle-of-attack, etc.

The captain Chesley B. "Sully" Sullenberger, aged 57, held single- and multiengine dual Airplane Transport Pilot (ATP) certification. He had accumulated 19,663 total flight hours, including 8,930 h as pilot-in-command, 4,765 h of which were in A320 air planes (NTSB 2010). The captain also stated that he was in good health. On the day in question, the captain had arrived from Charlotte Douglas International Airport to LGA around 1400 h and had a quick turnaround that day. The first officer also had a lot of flying hours and was with the captain for a 4-day trip leading to the final leg of the trip scheduled to depart from LaGuardia direct to Charlotte. The three flight attendants were well trained.

The aircraft had two engines. These engines were two CFM* International dual-rotor turbofan engines. The engines were certified by a joint bilateral certificate agreement between the United States and France in accordance with Joint Aviation Authorities (JAA) engine regulation requirements.

Figure 26.1 shows us a cross-sectional view of the engine. As air enters the engine inlet area, it passes through the fan and then moves through two distinct parts. Most of the ingested air bypasses the core of the engine through the bypass duct and provides 70% of the thrust (NTSB 2010). The remaining air enters the engine's core. Here the air is compressed and combusted (mixed with fuel). This energy provides the rotational power to the fan.

A bird encountered by the airplane may strike any part of the engine area. If a bird enters near the outer radius of the engine, it will not cause the engine to stop working as the birds

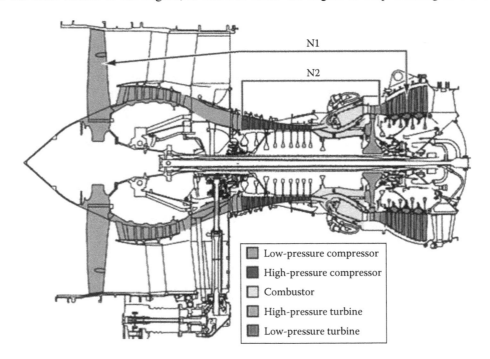

FIGURE 26.1   Cutaway of a turbofan engine showing the LPC, HPC, combustor, HPT, and LPT.

---

* CFM is the name of an international corporation that manufactures jet engines for commercial airplanes, (http://www.cfmaeroengines.com). Accessed on January 10, 2013.

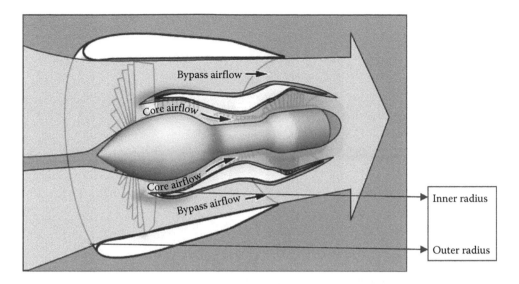

FIGURE 26.2   Showing the two paths of ingested air.

would continue to move toward the bypass flow path and get expelled through the rear of the engine. However, thrust production may be affected, but the engine will still be able to operate. If the bird strikes near the inner radius of the engine inlet, a portion of the bird may get ingested into the engine's core, which could lead to damage of internal components. As we can see from Figure 26.2, this is the path for core airflow. Depending on the severity of the damage, the engine might burn out.

The aircraft engine certification process consists of many tests that demonstrate compliance with certification requirements. The engines of this aircraft were certified in 1996. The certification requirements at that time included (NTSB 2010)

1. Ingestion of a 4 lb [large-sized] bird…may not cause the engine to
   a. Catch fire
   b. Burst (release hazardous fragments through the engine case)
   c. Generate loads greater than those specified in Section 33.23(a) of the Code of Federal Regulations (CFR)
   d. Lose the capability of being shut down
2. Ingestion of 3 ounce birds or 1½ lb [medium-sized] birds…may not
   a. Cause more than a sustained 25% power or thrust loss
   b. Require the engine to be shut down within 5 min from time of ingestion
   c. Result in a potentially hazardous condition (NTSB 2010)

The medium bird-ingestion criterion for the engine certification required volleying seven 1½ birds into the engine's core. Similarly, the large bird-ingestion certification criterion for these engines volleyed a four lb bird into the fan of the engine. In 1992, the test report for the engine stated that the aircraft met requirements for medium-bird tests. The FAA and French Direction Générale de L'Aviation Civile (DGAC) jointly approved the report. Again, the engine was certified for large bird ingestion by the FAA and DGAC. In early 1993, the CFM added a special condition for certification that changed the medium-bird weight to two-and-a-half pounds and required that the engine should be able to operate with no more than 25% loss of thrust for 20 min after the bird ingestion. The engines were subjected to these tests, and in March 1993, a test report stated that the engines operated as expected with the special condition test.

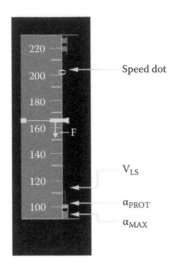

FIGURE 26.3   A320 airspeed scale showing protection speeds.

In 1996, some changes were made to the core design of the engine, and the engine had to be recertified for these hardware changes. Again, the engine was granted certification after it was subjected to retesting. Since 1996, there have been changes made to the testing criteria. The size of the birds used was increased.

Weather is observed at LGA by Automated Surface Observing Systems (ASOS). These systems record information continuously on wind speed, cloud formation, temperature, precipitation, and visibility. A report for these conditions is transmitted every hour. The closest ASOS for LGA was located 1.6 miles away in Central Park, Manhattan. The weather condition on that day was as follows: winds 190° at 8 kt, visibility 10 miles, cloud ceiling broken at 3700 ft, and temperature 6°C.

Figure 26.3 shows us the airspeed scale of the aircraft. The following is the aircraft speeds and their significance (NTSB 2010):

Speed dot: The speed that provides best lift over drag ratio.
F: Target speed of the aircraft on approach.
V_{LS}: Lowest selectable airspeed providing an appropriate margin to the stall speed and is based on aerodynamic data.
α_{prot}: Speed at which α-protection mode becomes active. This is based on plane weight and configuration.
α_{max}: The max angle-of-attack (AOA) that may be reached in pitch normal law. This is based on plane weight and configuration.

For safe operation, the speed of the aircraft should be between the speed dot and V_{LS}.

PURPOSE

Various investigation reports cited that the flight and crew members were properly certified. Examination of the aircraft also revealed no existing system or engine issues. The airplane was found compliant with existing federal laws.

The NTSB report classifies this incident as "ditching." The term ditching usually means a planned water event in which the flight crew, with the aircraft under control, knowingly attempts to land in water. In contrast to an inadvertent water impact, in which there is no time for passenger or crew preparation, ditching allows some time for preparation.

NTSB (2010)

This event allowed some time for the pilots to alert the attendants and perform quick checks. It is important to note at this point that the airplane was equipped for Extended Over Water (EOW) operation, which was purely coincidental for an aircraft flying from NYC to Charlotte. The incident happened in daylight, which means nearby ferries could view the incident and rush to help. The crew and flying officers did not have any specific training for incidents such as this one but instead had many years of experience that helped them by allowing use of their judgment when needed.

SYSTEM/SYSTEM OF SYSTEMS/ENTERPRISE

A generic but tailored specific instantiation of the complex/complicated transportation system being discussed is depicted in Figure 26.4.

"A system is an assemblage if elements or components that are interconnected so as to harmoniously perform one or more functions or serve one or more purposes" (DeLaurentis 2007).

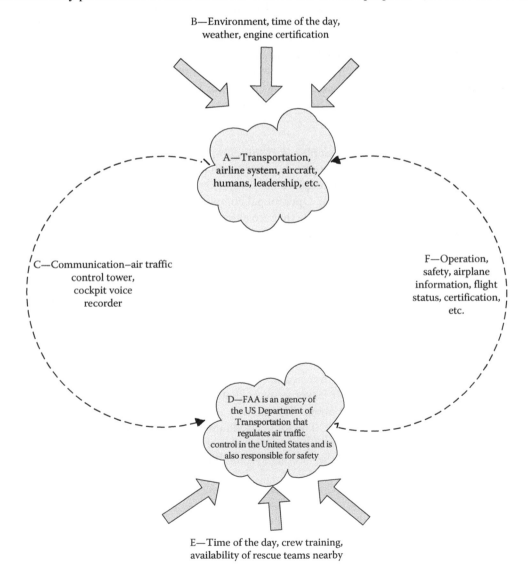

FIGURE 26.4 Context of a transportation system.

We, as human beings, are embedded in systems around us (DeLaurentis 2007). Understanding these systems is critical to improvement and sustenance of the human race. "A system-of-systems consist of multiple, heterogeneous, distributed occasionally independently operating systems embedded in networks at multiple levels that evolve over time" (DeLaurentis 2007).

A: This *Complex System* involves an airline system. Some of the subsystems that make up this complex system are the aircraft and its components, humans in the aircraft, the weather system, etc.

B: Some *External Factors* that affect this complex system are the environment, time of the day, weather, and engine certification.

C: *Feedback*—There was constant communication between the aircraft and the air traffic control tower that enabled exchange of feedback.

D: *Governing Body* FAA—It is an agency of the US Department of Transportation that regulates ATC in United States and is also responsible for safety. In this case study, the FAA is the governing body for civil aviation in the United States and is primarily responsible for implementing policies and regulations, issuance of pilot certifications, operating the ATC, and promoting safety.

E: *Constraints*—Although the FAA is the primary governing body for civil aviation, there are other factors such as crew training, time of the day, availability of rescue teams nearby, etc., that were also part of this case but were constraints to FAA.

F: The *Governance* provided by the FAA includes flight operation, safety, airplane information, flight status, certification, etc.

System of Systems (SoS) is always complex, and this complexity may be measured across various dimensions such as operational complexity, system complexity, algorithmic complexity, and managerial complexity. Continuing to explore our aircraft system we can elaborate on some of these measures in our enterprise system. Operational complexity involves the skill to operate the aircraft and control the various subsystems that are essential for the successful operation of the aircraft. The aircraft system has many subsystems such as the electrical systems, communication systems, and computer control systems. The properties and interactions of these systems can lead to a different property that is exhibited at the enterprise level, which is the interactions of all these system in the efficient performance of the aircraft. System complexity relates to many subsystems in our enterprise, and each subsystem needs its own management and have their own properties but yet contribute together toward the operation of our SoS. People play a central role in all such systems (Eisner 2012). Humans may take on many roles in a system, including that of a customer, a builder, an acquirer, a user, an operator, a maintainer, and so on (DeLaurentis 2007).

As we have seen, this incident is not just an incident about an aircraft. There were many underlying systems that were part of larger systems that again make up the whole SoS, which is the enterprise that we are talking about. The humans in this SoS had many roles. The flying officers were primarily flight operators. The flight attendants were there to support the operation of the flight. The other humans on board (passengers) were the customers. The other systems that were part of this enterprise included the weather system, the ATC, and communication systems, among others.

This case study falls into the category of *complex systems* because it embodies a number of systems that in themselves must perform competently. Hence, the case is also complicated in that it depends on subsystems that are not usual elements of the day-to-day operations of an airline. These include the engine malfunction, the ingestion of birds (a highly unusual event), and the consequent necessity of an emergency landing on water. Finally, when the passengers were evacuated on the plane in the water, it was important to ensure their safety as well.

These issues explain why the case and its potential catastrophic outcome, was complicated. The complex aspect of the case involved the necessity of efficiently dealing with a number of other systems, including the flock of birds, the engine failure and subsequent necessity of landing safely on water, communicating with the emergency water services such as ferries, ambulances, police, etc., and evacuation procedures.

CHALLENGES AND DEVELOPMENT

As this flight left NYC's airport on January 15, the aircraft met with a flock of migratory Canadian geese that stuck the engines and damaged them. The captain realized that there was a loss of thrust and decided that the only option was to take control of the aircraft. He also activated the auxiliary power unit (APU) and reported the situation to the LaGuardia ATC. At this point, the ATC instructed the captain to land at a nearby airport and gave him the option of landing at LGA or the nearby Teterboro Airport. However, the captain quickly analyzed the options and realized that there was no time to make a landing. The aircraft was continuing to lose thrust. The first officer had started to run through the Quick Reference Handbook (QRH) dual-engine failure checklist. During the next 2 min, the flying officers were in constant communication with the ATC at LGA for landing clearance. The speed of the aircraft dropped and was now close to V_{LS}. This meant that that flight would go into α_{prot} mode, which is the mode that is automatically activated by the computer control system to prevent flight stalling. The captain instructed the crew to prepare for a water landing. Thus, 3 min after the bird strike had passed and the aircraft was skidding over water. The crew helped the passengers get outside and onto the wings of the aircraft while they awaited help. Ferries arrived within a few minutes and passengers were rescued. The captain was the last to leave the accident area, making sure that every passenger has exited the aircraft and received help. A few passengers were hospitalized for hypothermia, but all the passengers and crew members survived (Langewiesche 2009).

ANALYSIS

According to the investigation report published by the NTSB, the cause of this incident was the ingestion of birds into the engines. This ingestion caused a loss of thrust, and the flight was force-landed on the Hudson. We have already seen details about the aircraft and the systems involved. Further analysis into this incident provides us with interesting insights. The pilot guided the aircraft into the river and stayed clear of populated areas. He was able to land smoothly on the water, which again is a skill that the pilot displayed. Passengers could use the wings of the aircraft while they waited for help. The crew of the flight displayed extraordinary commitment and courage during the whole incident. It is to be noted that no specific emergency training was given to any of the crew members. However, the captain, as part of his regular job, had routinely exercised an important role of training other pilots to handle many types of in-flight emergencies, including loss of engine power. Clearly, he was his own best teacher and displayed remarkable poise, courage, and confidence in handling the situation. For their parts, the crew also used their good judgment and many years of flying experience to act properly in the moment. The captain was very prompt in activating the APU and was also instrumental in sharing his knowledge with other pilots, especially during his offline and routine training of other pilots in handling emergencies. Without this critical step, the flight would have descended with a different speed and without any control. This step is not mentioned in the QRH. Incidentally, the first step in the QRH was to attempt to restart the engines. However, the checklist assumed that at least one engine would be functional, which was not so in this case. Checklists are often beneficial to ensure completeness, but how do we measure their effectiveness especially in times of an emergency or in a situation where the underlying assumption for using the checklist does not exist? In this case, both the engines of the aircraft were damaged and no document or training was given for handling such a situation. Also, the aircraft was one that was equipped for EOW operations. Thus, it had life vests, emergency slide rafts, etc. Passengers used these rafts for standing as they waited for help. Without these rafts, passengers could have died from the effects of the cold water. After the accident, the NTSB recommended emergency and engine failure training to all flight crew staff.

Despite so many contributing factors to this incident, this is a case of a disaster that was averted from killing many passengers. There were seemingly unrelated factors in this case that supported the *miracle* outcome:

1. Exceptional management by the crew during the accident.
2. Excellent decision making by the captain.
3. The accident occurred in daylight, allowing for better visibility.
4. Fortunate use of an aircraft equipped for an extended overwater flight, although not required for this route.
5. Crew members had many years of experience, allowing them to act on the situation and perform evacuation.
6. Rescue teams were in close proximity and could get there within a few minutes without which the passengers and crew members would have suffered, and possibly died, because of the cold waters.

CONCLUSIONS

Complex systems such as an aircraft may have many subsystems that may be the potential cause of an accident or disaster. Likewise, in a complex system such as a hospital, software errors may be one of a number of system errors (such as payroll, admissions, staff, medications, and treatment), which could be one or more of the critical underlying factors of a disastrous situation (Leveson 1995). Perhaps surprisingly, when an *executive* decision is needed, the experience and know-how that is accrued by an expert human operator (possibly based on instincts) can result in the decision to take a critical course of action, thereby generating the most reliable and, possibly, most suitable solution. The flight, in this case, was equipped with a computer-controlled navigation system. The captain, however, realized the need for human intervention and took control. In unique situation such as this, there is nothing that will effectively replace the higher level judgment and decision-making process that is developed from know-how and experience. That is, nothing based on either human effort (determination) or computer calculation could have replaced the sheer know-how and experience of Captain Sully.

Many years of flying experience and the excellent communication skills demonstrated by the crew also helped prevent the incident from becoming a disaster. However, this does not replace the need for a contingency plan at all times.

Large projects often grow out of scope. We need plans to track progress and manage scope. These plans also need to address events that may involve low-probability contingency events and their risks. Nearly all projects require a certain amount of risk analysis under operating conditions.

Today, we live in a world dominated by complex systems software, where there needs to be at least one person or a small team, responsible for the safety of the system that will be part of the larger system under development. In addition, testing at all levels should be integral to any software development. Testing should not only map the requirements to the features under development but also account for the use of the software in the enterprise. Training should be an integral part of the software development lifecycle. As software development moves through various phases in its lifecycle, its functionalities and features may often be interpreted differently. This makes it important for sufficient training by the potential stakeholders on the features and functionalities of the product.

QUESTIONS FOR DISCUSSION

- What were the diverse complex systems involved in this potentially catastrophic case?
- How would a complicated system, operating in a complex environment become complex?
- How safe are our software systems and applications in a complex environment?
- How do we measure risk factors and validate our risk analysis?
- What is the importance of operator training?
- What were the leadership qualities that the captain showcased?

EPILOGUE

by Vernon Ireland

The Sully Sullenberger incident is an interesting example of the captain of a commercial jetliner taking charge of the flight, when the on-board computer system indicates the plane cannot land at an airport. In Sullenberger's case, he had pressure to assess the situation very quickly and make a decision before the plane lost too much altitude. The on-board computer system could only react in terms of the inputs and data it had been programmed to address. By comparison, the pilot had multiple sensors and a first-rate contextual awareness through experience. If Sullenberger had taken longer to investigate the options, discussed them with his copilot and the control towers at the three New York airports, and formed a committee as often happens in business and government, he would not have had altitude or speed in this plane without effective engines, to turn and glide for a successful landing on the Hudson. This is an example of the plane's computer system not being in a position to take in the whole context of the complex problem and recommend action. It is an interesting case given the importance of context and the environment for decision making in complex systems. The response of the computer was purely in terms of the areas it had been programmed to respond to whereas the pilot had all his senses operating and was guided by his extensive experience.

Another potentially fatal incident occurred in which the Captain Richard de Crespigny had to overrule the on-board computing system. Qantas Flight QF32, an Airbus A380, with 479 passengers and crew on board, was flying from Singapore to Sydney on November 4, 2010, when one of the engines exploded and severely damaged the plane.

It was an unusual flight in that de Crespigny, a very experienced Qantas Captain who had been flying for over 35 years and had over 15,000 h of flying experience plus at least 1,000 h in simulators, including considerable experience captaining an A380, was having his regular in-flight review by three other captains. This led to some complex leadership issues.

A few minutes after takeoff, there was a massive explosion in Engine 2. Hundreds of pieces of the turbine blade cut through the wings and fuselage damaging or destroying a large number of the flight control systems on the 570 tons aircraft, including 104 tons of jet fuel (de Crespigny 2012, p. 139).

An oil feed-pump on Engine 2, which was a Rolls-Royce Trent 900 engine designed for the A380, with over 34,000 parts and weighing 16.4 tons, and costing US$18.5 million, exploded. It supplied cooling and lubricating oil to a bearing support in the turbine section of the engine, had been assumed to fracture by de Crespigny, 3 min after takeoff and 1 min before the engine failed. When it fractured, oil under pressure ignited, and a flame of approximately 2600°C, surrounded by about half of the turbine disc, overheated and weakened it.

The turbine producing 63,000 horse power, and traveling at 11,200 RPM, sheared off its shaft and exploded faster than 1.5 times the speed of sound (de Crespigny 2012, p. 321). One large piece of the turbine pierced straight through the top of the wing, ripping out the flight controls. Another large piece cut through the plane's belly, cutting at least 400 wires and numerous services (de Crespigny 2012, p. 322). The third piece traveled back, splintering into five pieces that hold the forward wing spar, creating devastating shock waves and carnage within the fuel tanks (de Crespigny 2012, p. 322).

Over 600 wires were cut, there were *hundreds of holes in the wing*, there were 100 impacts on the leading edge of the wing flaps and tailplane and at least 14 holes on the multiple fuel tanks (de Crespigny 2012, p. 320).

These comments were started by comparing this flight with that of Sullenberger's ability to evaluate the flight options in a manner that the plane's computers could not. The QF32 Captain de Crespigny, sitting only 4 m from the exploded engine, was immediately bombarded with warnings from the A380's automatic warning system Electronic Centralized Aircraft Monitoring (ECAM). Engine 2 indicated fire although this was obvious. The ECAM list of warnings also indicated Engines 1, 3, and 4 were all degraded in different ways (de Crespigny 2012, p. 162).

The engine had exploded and shrapnel had torn through everything in its path, with multiple wires being cut. The brakes were reduced to 28% on the left and right wings (de Crespigny 2012, p. 186). Of the 22 systems on the A380, 21 were damaged (de Crespigny 2012, p. 187).

There are 250,000 sensors on the A380, and the piloting team were being confused by the various alerts. They had already lost 50% of their hydraulic pumps, and now, the computer was recommending that they shut down a further 25% of the pumps (de Crespigny 2012, p. 201). These pumps control the ailerons and slats and severely affect the ability of the aircraft to roll (de Crespigny 2012, p. 201). However, de Crespigny did not want to be left with only two pumps out of eight as the ECAM was recommending, so he chose not to follow the recommendation.

Although de Crespigny had experienced stressful situations (de Crespigny 2012, pp. 167–168), this crisis put him and his copilot under a lot of extra stress. During this stage of the flight, the cabin crew were carefully managing passengers who were getting very nervous. Many had seen flames emerging from the engine, and some could see the damaged fuselage and wing through their windows.

While the emergency return to Singapore is also a long story, there are a number of key incidents to be highlighted.

Readers will remember that unusually there were three additional captains on board, their purpose being to conduct a regular operational check on Richard de Crespigny. Once the plane was initially stabilized, at 7,400 ft, Richard's reaction was to climb to 10,000 ft and, if necessary, conduct an Armstrong spiral, named after the astronaut, and essentially attempt to glide into Singapore. But when he announced he was going to do this, the other captains all said, "No!" Leadership of complex organizations needs to recognize that a command and control style is usually inappropriate. So de Crespigny, although he had formal control of the plane as the operating captain, recognized their views and did not attempt to gain altitude. He called back to the Singapore control tower to cancel the command that he would seek to attain 10,000 ft.

The next issue was handling the ECAMS warning system, which was saying the plane could not land, among a three-page list of double-column warnings. The plane was 42 tons heavier than its maximum landing weight (de Crespigny 2012, p. 232). The issue was that the 4000 m airstrip in Changi was too short to stop the badly damaged A380 with its severely degraded braking system.

Added to the stress of managing the plane, de Crespigny was being queried by Qantas emergency center in Sydney asking what had happened as they had heard on CNN that a Qantas plane had dropped engine parts over Indonesia.

As the flight progressed, a number of adjustments were made, including losing a lot of fuel due to leakages and movements of fuel around the plane. On landing the plane was still overloaded with fuel, but the crew did not take the chance of circulating over the ocean for an hour or more to dump fuel as they were not sure how much fuel they really had.

The automatic computer system then calculated that the plane had 100 m to spare in the 4000 m landing strip. This required very careful piloting with airspeed needing to be within 1 kt of the ideal; otherwise, either the airstrip would be too short or the plane would stall during landing. In fact, the automatic warning systems called STALL, STALL, STALL during the landing (de Crespigny 2012, p. 260). This also had to be neglected.

History indicates that QF32 did land and the passengers all clapped. But passengers did not yet realize that jet fuel was leaking past white hot brakes, and passengers needed to stay on board for a further 30 min while the brakes cooled (Crespigny 2012, p. 269). By the way, Engine 1 could not be shut down even though it was flooded with water.

This is a fascinating example of the ability of people to sift through multiple warnings by the 250,000 sensors, to decide which to accept and respond to, and which ones to disregard. The skill of de Crespigny and his crew may be able to be built into a computer sensing system, but with 250,000 sensors, this had not happened on the A380 at this point in time. It is doubtful that a computer program will be able to assess the various opinions of people in the flight deck, to recognize stress levels, and to respond appropriately. This is likely why we still have pilots flying planes rather than the autopilot doing the whole job.

Material for these comments comes from two sources. The first is the excellent book by Richard Champion de Crespigny titled *QF32*, published by Pan Macmillan Australia in 2012. The second source came from a chance encounter by Professor Vernon Ireland (one of the five coeditors of this book) when Richard de Crespigny was flying back to Sydney directly after the QF32 flight. The author was seated next to Richard, who provided some of the context of the previous QF32 flight. Remarkably he also reported a minor incident of an engine exploding on the 747.

ACKNOWLEDGMENTS

As an instructor the greatest joy one can have is to see that your students are motivated, involved, and do something productive related to the course subject that goes beyond the classroom. As I have said, most student projects *die in my office*.

I take great satisfaction in the fact that in September 2012 these five masters students—Karunya Rajagopalan, Bustamante Brathwaite, Eranga Gamage, Shawn Hall, and Mariusz Tybinski—accepted my offer to make a contribution to this book. We were just studying complex systems failures, and the students were preparing their individual case studies for my course in Software Methodology. Furthermore, a year earlier Shawn Hall had completed a master's thesis related to requirements for hospital systems. As it turned out, the next year, Shawn found a job whereby the ideas and skills he developed in his thesis are directly employed. Shawn also happens to write very well.

Mariusz Tybinski had already published in the area of medical errors and was pursuing a thesis in this area when this project came about. So it was a natural fit for him, and it helped him progress on his thesis. Bustamante Brathwaite is a superb student who has assisted me in many ways, and Eranga Gamage took on many unpleasant tasks to help the medical errors team finish its work.

And then there is Karunya Rajagopalan. Karunya not only led the development of the Miracle on the Hudson Case Study, given in this chapter, but also made a full contribution to the Medical Errors Case Study (see Chapter 12). This also led to her writing and completing a thesis in the area of mobile medical systems.

There were many meetings our team had during the course of the 2012–2013 academic year. I enjoyed working with all of them and feel they can be proud of what was produced.

I have been publishing well (and poorly) for nearly 40 years. Never in the course of those years have I been through so many editing cycles. The publishers can rest assured that the editors of this book made every effort to achieve *Total Quality Management* in the delivery of our material. I would particularly like to thank Brian White for being kind, approachable, supportive, and responsive throughout this project. I learned we have a number of interests in common. I would also like to thank Professor Vernon Ireland for his thoughts and contributions, as well as Professor Jimmy Gandhi for his suggestions.

REFERENCES

de Crespigny, R. 2012. *QF32: The Captain's Extraordinary Account of How One of the World's Worst Air Disasters Was Averted*. http://www.alibris.com/booksearch?browse=0&keyword=QF32&mtype=B&hs.x=33&hs.y=20&hs=Submit. Accessed on March 9.

DeLaurentis, D. 2007. The role of humans in complexity of systems of systems. In Duffy, V.G. (ed.), *Digital Human Modeling. Human Computer Interface Interaction International*, LNCS 4561. Springer-Verlag, Berlin, Germany, pp. 363–371.

Dolbeer, R. and J. Seubert. 2013. Canada goose populations and strikes with civil aircraft: Positive trends for aviation industry. www.birdstrike.org. Accessed on May 4.

Eisner, H. 2012. *Topics in Systems*. Mercury Press.

Federal Aviation Administration (FAA). 2013. FAA wildlife strike database. http://wildlife-mitigation.tc.faa.gov/wildlife/default.aspx. Accessed on April 25.

Kalafatas, M. 2010. *Bird Strike: The Crash of the Boston Electra*. Brandeis University Press, Waltham, MA.

Langewiesche, W. 2009. *Fly by Wire: The Geese, the Glide, the Miracle on the Hudson*. Farrar, Straus and Giroux, New York.

Leveson, N. 1995. *Safeware*. Reading, MA: Addison-Wesley.

National Transportation Safety Board (NTSB). 2010. Aircraft accident report. NTSB/AAR-10/03, PB2010-910403. National Transportation Safety Board, Washington, DC.

Sodhi, N. August 2002. Competition in the air: Birds versus aircraft. *Auk*, 119, 587–595.

United States Nuclear Regulatory Commission (USNRC). 2011. US Airways Flight 1549: Forced Landing on Hudson River Case Study 2 August 2011. Safety culture communicator.

Index

Milton Keynes UK
Ingram Content Group UK Ltd.
UKHW020828141024
449569UK00008B/589